THE OXFORD HANDBOOK OF

NUCLEAR SECURITY

THE OXFORD HANDBOOK OF

NUCLEAR SECURITY

Edited by

CHRISTOPHER HOBBS,

SARAH TZINIERIS,

and

SUKESH K. AGHARA

OXFORD
UNIVERSITY PRESS

OXFORD
UNIVERSITY PRESS

Great Clarendon Street, Oxford, OX2 6DP,
United Kingdom

Oxford University Press is a department of the University of Oxford.
It furthers the University's objective of excellence in research, scholarship,
and education by publishing worldwide. Oxford is a registered trade mark of
Oxford University Press in the UK and in certain other countries

Published in the United States of America by Oxford University Press
198 Madison Avenue, New York, NY 10016, United States of America

British Library Cataloguing in Publication Data
Data available

Library of Congress Control Number: 2023948183

ISBN 978-0-19-284793-5

DOI: 10.1093/oxfordhb/9780192847935.001.0001

Printed and bound by
CPI Group (UK) Ltd, Croydon, CR0 4YY

MIX
Paper | Supporting responsible forestry
FSC® C013604

Contents

PART V NATIONAL APPROACHES TO NUCLEAR SECURITY

PART VI COMBATTING NUCLEAR SMUGGLING

PART VII PERMANENT THREAT REDUCTION

PART VIII NUCLEAR SECURITY DURING TIMES OF CRISIS

Contributors

Irman Abdul Rahman is an Associate Professor in the Department of Applied Physics at the National University of Malaysia (UKM).

Gary A. Ackerman is an Associate Professor and Associate Dean in the College of Emergency Preparedness, Homeland Security and Cybersecurity at the University at Albany, SUNY.

Hamisu A. Adamu is an instrument engineer and Lecturer in Electronics and Telecommunication at Ahmadu Bello University in Zaria.

Sukesh K. Aghara is Professor and Director of Nuclear Engineering, Director of the Integrated Nuclear Security and Safeguards Laboratory, and Associate Dean of Research for the Francis College of Engineering at the University of Massachusetts Lowell.

Yasemin Balci is Co-Programme Director for National Implementation at the Verification Research, Training and Information Centre (VERTIC) in London.

Wyn Q. Bowen is Professor of Non-Proliferation and International Security and Co-Director of the Freeman Air and Space Institute at King's College London, and a William Penney Fellow with the UK Atomic Weapons Establishment (AWE).

Matthew Bunn is the James R. Schlesinger Professor of the Practice of Energy, National Security, and Foreign Policy in the Harvard Kennedy School at Harvard University.

Geoffrey Chapman is a Visiting Research Fellow in the Centre for Science and Security Studies (CSSS) in the Department of War Studies at King's College London.

Jinho Chung is a Senior Researcher at the Korea Institute of Nuclear Nonproliferation and Control (KINAC).

Matthew Cottee is a British civil servant. He completed his PhD in the Department of War Studies at King's College London.

Ferenc Dalnoki-Veress is the Scientist-in-Residence and an Adjunct Professor in the James Martin Center for Nonproliferation Studies (CNS) at Middlebury Institute of International Studies at Monterey.

Karl Dewey is a Research Associate at the International Institute for Strategic Studies (IISS) in London.

Michael Duitsman is a Research Associate in the James Martin Center for Nonproliferation Studies (CNS) at Middlebury Institute of International Studies at Monterey.

Bryan R. Early is Associate Dean for Research, Associate Professor, and Vice Chair of the Political Science Department in the Rockefeller College of Public Affairs and Policy at the University at Albany, SUNY.

Chris Englefield is a radioactive substances regulation specialist and a former regulator at the United Kingdom's Environment Agency.

Vitaly Fedchenko is a Senior Researcher in the Weapons of Mass Destruction Programme at the Stockholm International Peace Research Institute (SIPRI).

Trevor Findlay is a Principal Fellow in the School of Social and Political Sciences at the University of Melbourne.

Friederike Frieß is a Senior Scientist at the Institute of Safety/Security and Risk Sciences at the University of Natural Resources and Life Sciences (BOKU) in Vienna.

Andrew Futter is Professor of International Politics at the University of Leicester.

Amandeep Gill is Project Director and CEO of the International Digital Health and AI Research Collaborative (I-DAIR) and a Professor at the Geneva Graduate Institute, and previously served as India's Ambassador to the Conference on Disarmament in Geneva.

Martin Grigg is a physical security consultant, project lead, and technical advisor for large-scale and high-security projects, as well a technical author and training provider for the security industry.

Xiaobing Guo is a Research Professor and Director of the Center of Arms Control Studies at the China Institutes of Contemporary International Relations.

Jason T. Harris is Associate Professor of Radiological Health Sciences in the School of Health Sciences and Director of the Center for Radiological and Nuclear Security (CRANS) at Purdue University.

Christopher Hobbs is Professor in Science and International Security in the Department of War Studies and Director of the King's Institute for Applied Security Studies (KIASS) at King's College London.

James R. Holmes is J. C. Wylie Chair of Maritime Strategy at the Naval War College of the United States and a Nonresident Fellow at the University of Georgia.

Zenobia Homan is a Research Fellow and Project Coordinator in the Centre for Science and Security Studies (CSSS) in the Department of War Studies at King's College London.

Styrkaar Hustveit is a Senior Adviser at the Norwegian Radiation and Nuclear Safety Authority.

Gregory B. Jaczko is a Lecturer in the School of Public and International Affairs at Princeton University and former Chairman of the United States Nuclear Regulatory Commission.

Sunday A. Jonah is Professor of Physics and the Director of the Centre for Energy Research and Training (CERT) at Ahmadu Bello University in Zaria.

Margarita Kalinina-Pohl is Director of the CBRN Security Program in the James Martin Center for Nonproliferation Studies (CNS) at Middlebury Institute of International Studies at Monterey.

Igor Khripunov is a Nonresident Fellow at the Stimson Center in Washington, DC.

Benjamin Kienzle is Reader in Security Cooperation in the Department of Defence Studies at King's College London.

Dmitry Kovchegin is an independent researcher and consultant, studying and supporting implementation of international nuclear security cooperation.

Moritz Kütt is a Senior Researcher in the Institute for Peace Research and Security Policy at the University of Hamburg.

Na Young Lee is Director General of the Office of Nuclear Nonproliferation at the Korea Institute of Nuclear Nonproliferation and Control (KINAC).

Angela E. Leek is Director of Radiological Solutions and Regulatory Affairs at Summit Exercises and Training.

Craig Marianno is Assistant Professor of Nuclear Engineering and Deputy Director of the Center for Nuclear Security Science and Policy Initiatives at Texas A&M University.

Marco Marzo is Secretary-General of the Brazilian-Argentine Agency for Accounting and Control of Nuclear Materials (ABACC).

Faizal Mohamed is Professor of Nuclear Science at the National University of Malaysia (UKM).

William M. Moon is an independent consultant, retired from the Defense Threat Reduction Agency, and former Cooperative Threat Reduction (CTR) Program Manager.

John Mueller is Professor Emeritus of Political Science at Ohio State University and a Senior Fellow at the Cato Institute.

Egle E. Murauskaite is a Senior Researcher and Simulation Designer for the ICONS Project at the University of Maryland.

Bob Officer is Head of Security at Nuclear Transport Solutions in the United Kingdom.

Sindre Øvergaard is Head of Department at the Institute for Energy Technology and formerly Head of Section at the Norwegian Radiation and Nuclear Safety Authority.

Tom Parkhouse is a former Superintending Inspector and Professional Lead for Civil Nuclear Cyber Security at the United Kingdom's Office for Nuclear Regulation.

Julia M. Pearce is Reader in Social Psychology and Security Studies in the Department of War Studies and Associate Dean (Impact and Innovation) in the Faculty of Social Science and Public Policy at King's College London.

Ross Peel is a Research and Knowledge Transfer Manager in the Centre for Science and Security Studies (CSSS) in the Department of War Studies at King's College London.

Rajeswari Pillai Rajagopalan is Director of the Centre for Security, Strategy and Technology (CSST) at the Observer Research Foundation in New Delhi.

Miles A. Pomper is a Senior Fellow in the James Martin Center for Nonproliferation Studies (CNS) at Middlebury Institute of International Studies at Monterey.

Shraddha Rane is a Systems Research Analyst in the Department of Global Security and Simulation Studies at Sandia National Laboratories.

Scott Roecker is Vice President of the Nuclear Materials Security Program at the Nuclear Threat Initiative (NTI) in Washington, DC.

Nickolas Roth is Senior Director of the Nuclear Materials Security Program at the Nuclear Threat Initiative (NTI) in Washington, DC.

Daniel Salisbury is a Visiting Senior Research Fellow in the Centre for Science and Security Studies (CSSS) in the Department of War Studies at King's College London and was formerly a Leverhulme Trust Early Career Fellow.

Jeffrey D. Semancik is Director of the Radiation Division at the Connecticut Department of Energy and Environmental Protection.

Tone-Mette Sjømoen is a Director at the Norwegian Radiation and Nuclear Safety Authority.

David K. Smith is a Visiting Senior Research Fellow in the Centre for Science and Security Studies (CSSS) in the Department of War Studies at King's College London.

Håvar A. Sollund is a Senior Advisor at the Norwegian Radiation and Nuclear Safety Authority.

Christopher Spirito is a Nuclear Security Consultant at Idaho National Laboratory.

Johannes H. Sterba is Head of the Center for Labelling and Isotope Production at the TRIGA Center Atominstitut, TU Wien.

Amelie Stoetzel is a Researcher in the Centre for Science and Security Studies (CSSS) in the Department of War Studies at King's College London.

Sarah Tzinieris is a Research Fellow in the Centre for Science and Security Studies (CSSS) in the Department of War Studies at King's College London.

Şebnem Udum is an Associate Professor in the Department of International Relations at Hacettepe University in Ankara.

Ben Whittard is Managing Director, Solutions at Nuclear Transport Solutions in the United Kingdom.

Abbreviations

AI	artificial intelligence
CBRN	chemical, biological, radiological, nuclear
Covid-19	Coronavirus disease 2019
CPPNM	Convention on the Physical Protection of Nuclear Material
A/CPPNM	Amendment to the Convention on the Physical Protection of Nuclear Material
CTR	Cooperative Threat Reduction (programme that derived from the 1991 Soviet Threat Reduction Act)
DBT	design basis threat
DPRK	Democratic People's Republic of Korea; North Korea
EU	European Union
EURATOM	European Atomic Energy Community
FSU	Former Soviet Union
G7/G8	Group of Seven/Group of Eight
GICNT	Global Initiative to Combat Nuclear Terrorism
Global Partnership	G7/G8 Global Partnership against the Spread of Weapons and Materials of Mass Destruction
HEU	highly enriched uranium
IAEA	International Atomic Energy Agency
ICONS	International Conference on Nuclear Security
ICSANT	International Convention for the Suppression of Acts of Nuclear Terrorism
IND	improvised nuclear device
INFCIRC	International Circular (of the IAEA)
INSSP	Integrated Nuclear Security Support Plan
INTERPOL	International Criminal Police Organization
IPPAS	International Physical Protection Advisory Service
ITDB	Incident and Trafficking Database
LEU	low enriched uranium
LWR	light water reactor

MOX	mixed oxide (fuel)
NMAC	nuclear material accounting and control
NPT	Treaty on the Non-Proliferation of Nuclear Weapons
NSF	Nuclear Security Fund
NSG	Nuclear Suppliers Group
NSP	Nuclear Security Plan
NSS	Nuclear Security Series
PF	proliferation financing
PSI	Proliferation Security Initiative
Pu-239	Plutonium-239 (a fissile isotope of plutonium)
RDD	radiological dispersal device
RED	radiological emission device
RID	radiological incendiary device
ROK	Republic of Korea; South Korea
UK	United Kingdom
UN	United Nations
UNODA	United Nations Office on Drugs and Crime
UNSC	United Nations Security Council
UNSCR	United Nations Security Council Resolution
US	United States
U-235	Uranium-235 (a fissile isotope of uranium)
WMD	weapons of mass destruction
9/11	Terrorist attacks of 11 September 2001

PART I

INTRODUCTION

CHAPTER 1

NUCLEAR SECURITY

CHRISTOPHER HOBBS AND SARAH TZINIERIS

Introduction

The purpose of *The Oxford Handbook of Nuclear Security* is to impart a broad understanding of how security has been developed around what are among a nation's most sensitive assets. Depending on the country, these might include nuclear weapons, nuclear materials, radiological sources, and civil nuclear facilities. Nuclear security as explored in this volume encompasses the mitigation of a wide range of threats and malicious acts, including theft, sabotage, unauthorized access, and illegal transfer.[1] It includes preventative measures that can be taken in advance of an incident, including permanent threat reduction and steps to build national and organizational resilience. In addition, it covers the development of systems to detect and respond to ongoing events and enable longer-term post-incident recovery.

The envisioned readership of *The Oxford Handbook of Nuclear Security* includes both nuclear security specialists working in the industry itself and a broad audience of policy makers, government officials, non-governmental organization (NGO) staff, academics, and students, among others. It should be noted that, in order to provide an in-depth treatment of nuclear security, the Handbook purposely excludes strategic issues involving nuclear weapons, including arms control, deterrence, disarmament, and non-proliferation, detailed studies on which can be found elsewhere.[2]

The academic and policy literature on nuclear security has burgeoned since the 1990s, triggered initially by the collapse of the Soviet Union, which led Russia and other former Soviet Union (FSU) countries struggling to protect vast stockpiles of nuclear weapons and materials. Several studies have explored how new security measures and programmes could be enacted to mitigate the risk of nuclear leakage.[3] A decade later, the 11 September 2001 ('9/11') attacks on the United States (US) served to dramatically heighten both political concerns and independent analysis regarding the possibility and pathways to nuclear terrorism.[4] However, despite this increase in scholarly attention, the body of literature on nuclear security remains disparate and the subject matter under-explored. This is in part a result of the different disciplinary approaches that can be applied to the study of nuclear security, given its broad legal, political, and technical foundations. Academic articles on nuclear security can be found in a wide array of journals, from political science, to law, to engineering. In addition, a long-standing culture of secrecy around nuclear security within various national policy-making communities has served to stymie independent research in this area, although this has started to change in recent years.

The Oxford Handbook of Nuclear Security seeks to provide a comprehensive and state-of-the-art examination of nuclear security, taking a multidisciplinary approach to exploring its different dimensions—bringing together academics from a range of different disciplines, including political science, international relations, law, physics, and engineering. Contributions include not only those from Western scholars (who have tended to dominate nuclear security debates to date) but also from academics from other parts of the world, both from developed and developing nuclear countries. Here, the editors have engaged not just established scholars but a new generation of emerging experts. A small number of chapters have also been authored or co-authored by practitioners, former government and regulatory officials, and nuclear security professionals, to gather important practical insights into nuclear security at the 'coal face'.

In developing their chapters, authors have taken different approaches to shed light on the key aspects relevant to the subject matter. For those where there exists already a significant body of literature, efforts have focused on summarizing debates, elucidating key concepts, and examining disputed aspects. This includes discussions over the relative importance of nuclear security and how its different elements should be prioritized and approached. For those writing on subjects typified by an absence of information, the chapters present new empirical findings, such as in those that explore different national approaches to developing nuclear security systems.[5] In some cases, authors have also taken the opportunity to advance new theoretical frameworks and typologies that offer potential new ways to understand nuclear security-related phenomena, including on topics as diverse as the application of nuclear forensics and how and why states may choose to sponsor nuclear terrorism.[6]

A Brief History of Nuclear Security

Nuclear security is not a static concept, but rather one that has evolved over time in response to changes in threat perceptions, major incidents, and new political priorities. Consequently, a brief history of nuclear security is provided here to set the scene for the Handbook. This captures key events, international initiatives, and other developments in nuclear security, many of which are subsequently covered in greater detail in other chapters of the volume.[7]

Early Focus on Non-Proliferation and Nuclear Safety

The need to secure nuclear materials has been recognized since the very dawn of the nuclear age, signalled to the world with the atomic bombings of Hiroshima and Nagasaki in 1945. At this time, the nature of the international security climate was dominated by great power competition between the US and the Soviet Union, which served to shape government thinking around nuclear issues, including security. As a consequence, discussions related to nuclear security were predominantly focused on protecting national nuclear weapon arsenals from state actors, encompassing scenarios that included 'sabotage, theft, espionage or loss of control'.[8] There was some exploration of other security threats although these would only become prominent many decades later. For example, a closed US Senate hearing in 1946 debated the possibility of nuclear materials being smuggled across national borders, as a means by which Soviet operatives might covertly deliver an atomic bomb into New York City.[9] The hearing led to the infamous 'Screwdriver Report' (which remains classified to this day), a study by the US Atomic Energy Commission that investigated how radiation emissions could be used to detect nuclear materials hidden within a container while in transit.[10] However, it was not until the collapse of the Soviet Union that the illicit trafficking of nuclear materials would become a significant international security issue, with hundreds of incidents reported in the early 1990s.[11] To combat this, radiation detection systems for border monitoring, as initially envisioned in the 1940s, were deployed in Eastern Europe, and later expanded worldwide.[12]

During the early part of the Cold War, nuclear issues were dominated by efforts to limit the vertical and horizontal proliferation of nuclear weapons, with nuclear security essentially an add-on to international discussions rather than a distinct area of focus. The close relationship of nuclear security to nuclear safety also became apparent, as fears over the possibility of accidents involving nuclear weapons led to the inclusion of safety features that served the additional function of making unauthorized access more difficult.[13] In the 1950s, the civil application of nuclear technology started to become a reality with the US and the Soviet Union providing support for countries interested in establishing research and development programmes, as outlined in President Dwight D. Eisenhower's 'Atoms for Peace' speech before the United Nations (UN) in 1953.[14]

By the late 1950s, several countries had launched commercial nuclear power programmes. However, despite these developments, the international focus remained on limiting the spread of nuclear weapons and avoiding nuclear accidents. To this end, nuclear safeguards—systems designed to control and account for sensitive materials—were developed and deployed, although on a largely bilateral basis. At the same time, nuclear safety was promoted as a concept fundamental to nuclear operations. This was driven by the occurrence of a number of serious nuclear incidents, most notably the 1957 Windscale fire that occurred when two reactors at a nuclear facility in northwestern England caught alight, resulting in a significant atmospheric release of radioactive materials.[15]

The prevalence of concerns about protecting nuclear materials at the time was reflected in the Statute of the International Atomic Energy Agency (IAEA), an intergovernmental organisation established in 1957 to support international cooperation in the nuclear field. The Statute made direct reference to the importance of collaborative efforts to develop and apply nuclear safety standards. It also provided the IAEA with a remit to, upon request, apply safeguards to nuclear materials in civil use to help ensure that these were limited to peaceful purposes. This legislative basis was later codified in the Non-Proliferation Treaty (NPT), which entered into force in 1970. Article III of the NPT required safeguards administered by the IAEA to be applied 'on all source or special fissionable material in all peaceful nuclear activities' in non-nuclear weapon states.[16] In contrast, nuclear security was largely absent from discussions at major international forums and the text of treaties on nuclear issues.

Nuclear Security Synonymous with Physical Protection

The 1960s and 1970s saw a significant rise in serious international security incidents, including hijacking and terrorism. The deterioration of the security landscape, coupled with a rapidly expanding nuclear industry, created impetus on the need to establish international guidelines to protect nuclear materials and facilities from non-state actors. This task was taken on by the IAEA, which in 1975 published INFCIRC/225, 'Recommendations for the Physical Protection of Nuclear Material'.[17] While not legally binding, as the first international guidance in this area INFCIRC/225 was to become the cornerstone of an emerging international nuclear security regime.

Initially international governance was limited in scope to the international trade and transportation of nuclear materials, an area that grew significantly during the 1970s. This precipitated the development of both informal initiatives and formal instruments on nuclear security, provisions of which were mostly linked to the guidance outlined in INFCIRC/225. For example, the Nuclear Suppliers Group (NSG), created in 1974 in response to India's nuclear test, constituted an effort to prevent nuclear proliferation through controlling the supply of civil nuclear technology. While focused on non-proliferation, members of the NSG were also required to ensure the effective implementation of physical protection in order to prevent the theft or illicit transfer of nuclear material. These became conditions of supply, with the specific measures enacted drawing on the recommendations contained within INFCIRC/225.[18]

The 1970s also saw the development of the first legally binding instrument directly focused on nuclear security, the Convention on the Physical Protection of Nuclear Material (CPPNM), which opened for signature in 1980 and entered into force in 1987. While achieving international agreement over a legislative instrument was a significant step in the development of international nuclear security governance, the Convention was limited in scope to civil nuclear materials in international transport. This particular shortcoming would later be addressed in a 2005 Amendment to the Convention

(A/CPPNM), which extended it to domestic use, storage, and transport (although the A/CPPNM would not come into force until 2016). The Convention also lacked an oversight body or verification mechanism to ensure that states were meeting their obligations, with emphasis instead placed on state responsibility. Indeed, this remains the case today with the development and assessment of nuclear security systems performed at the national level.

Nuclear Leakage Triggers a Broadening of Nuclear Security

The early 1990s saw a significant increase in political attention on nuclear security, following the break-up of the Soviet Union, which led to a degradation in the security of nuclear weapons and other sensitive materials, creating concerns that the so-called loose nukes might fall into the hands of rogue states, terrorist groups, or criminal gangs.[19] This situation resulted in unprecedented multilateral support to Russia and other FSU countries—mainly led by Washington—with hundreds of millions of dollars spent on a Cooperative Threat Reduction (CTR) programme, among other similar projects. Most visibly, these included support for upgrading physical protection systems. As before, the intervention remained packaged as part of broader arms control efforts, which included, for example, the dismantlement of stockpiles of chemical, biological, and nuclear weapons.

What was particularly significant about these initiatives was a broadening out from the traditional focus on physical protection to also encompass new nuclear security areas, such as border monitoring and nuclear forensics in an effort to combat nuclear smuggling, as well as permanent threat reduction through the elimination of high-risk materials. The latter activity included an agreement between the US and Russia known as the 'Megatons to Megawatts Program'.[20] Under this arrangement, excess Russian weapons-grade highly enriched uranium (HEU) from Soviet-era nuclear warheads was down-blended to low enriched uranium (LEU) and shipped to the US for use in nuclear power plants.

The expansion of nuclear security as a concept was also reflected in discussions at the IAEA, with, for instance, its member states expressing in the 1994 General Conference Resolution a wish for the Agency to take measures against the illicit trafficking of nuclear material. This resulted in the IAEA in 1995 establishing its Incident and Trafficking Database (ITDB).[21] Today the ITDB remains a highly valuable resource which, drawing on information supplied by its member states, records incidents of nuclear and radiological materials lost from regulatory control, with the aim of identifying common threats and trends. The 1990s also saw a significant increase in research by academics and analysts on nuclear security issues, which served to shape and influence policy making. Most notably, a series of studies in the early 1990s by epistemic communities in the US and the FSU 'drove the thinking behind the concept of the CTR Program'.[22]

Mass Casualty Terrorism Becomes a Reality

By the late 1990s, growing concerns about international terrorism resulted in an initiative to develop a new treaty specifically focused on nuclear terrorism. This was led by Russia, at the time facing a significant threat from a Chechen separatist movement, which had demonstrated in 1995 its ability to use radioactive materials in terror attacks.[23] Negotiations started in 1998 on the International Convention for the Suppression of Acts of Nuclear Terrorism (ICSANT), although progress on this was initially slow, with some states expressing concern about the definitions, scope, and overlap with existing treaties such as the CPPNM. However, this all changed with the 9/11 attacks on the US, which sharply brought into focus how mass casualty terrorism was now a reality—heightening concerns that terrorist groups might seek other types of weapons of mass destruction (WMD), including nuclear capabilities.

Over the next decade, the events of 9/11 spurred the development of a range of new international instruments, informal processes, and academic and policy research. For example, adopted just over two weeks after 9/11, UN Security Council Resolution 1373 required states to adopt key 'international conventions and protocols to combat terrorism', while also acknowledging the threat posed by international terrorism and the trafficking of nuclear materials. In June 2002 at the Kananaskis Summit, the G8 launched the Global Partnership Against the Spread of Weapons and Materials of Mass Destruction, funded with a commitment of US$20 billion over 10 years, to prevent terrorist groups or states that support them from acquiring WMD. The nuclear security role of the IAEA was also strengthened with the establishment of a dedicated Office of Nuclear Security in 2002, together with a rolling multi-year Nuclear Security Plan and a dedicated Nuclear Security Fund.[24] Another milestone for nuclear security in the 9/11 aftermath was the adoption in April 2004 of UN Security Council Resolution 1540, which sought to prevent the proliferation of WMD to non-state actors.

The urgency of the ICSANT negotiations also increased following the events of 9/11, with the Convention finally entering into force in July 2007. ICSANT criminalized acts related to nuclear terrorism and encouraged international cooperation to prevent, investigate, and prosecute involved parties. Significantly, it was broader in scope than the CPPNM, encompassing not only civil nuclear materials and facilities, but radioactive material and nuclear materials in military use. It also provided great specificity on the nuclear terrorist threat, making a distinction between a 'nuclear explosive device' and 'radioactive material dispersion'.[25]

During this period, potential nuclear terrorism scenarios were further fleshed out within the academic and policy communities, with Ferguson and Potter's 2004 seminal work on this topic which outlined the four broad 'faces' of nuclear terrorism.[26] This was also a time of intense debate regarding the likely probability of a major act of nuclear terrorism, with certain scholars viewing this as a near certainty, while for others it remained a virtual impossibility.[27] Research into the implementation of nuclear security measures as a means of mitigating the risk of nuclear terrorism also increased significantly during this period, aided by the foundation of organizations focused on this issue such as the Nuclear Threat Initiative (NTI) and the establishment of international non-governmental networks such as the Fissile Material Working Group (FMWG).[28]

Nuclear Security Assumes Political Primacy

The continued rise throughout the 2000s and 2010s of transnational terrorist groups with a propensity for mass casualty attacks—for whom the acquisition and use of WMD could further, as opposed to hinder, their goals—kept the threat of nuclear terrorism towards the forefront of the international security agenda.[29] This was reflected in US President Barack Obama's first major foreign policy speech in Prague in 2009, when he described nuclear terrorism as the 'most immediate and extreme threat to global security'.[30] The US subsequently initiated the Nuclear Security Summit process—comprising four summits between 2010 and 2016—aimed at galvanizing international cooperation, with the stated goal of 'secur[ing] all vulnerable nuclear materials around the world'.[31]

What was unique about the Nuclear Security Summits in contrast to the leader-led summits that had gone before, such as Kananaskis in 2002, was that they were focused on nuclear security rather than being framed around broader arms control and non-proliferation issues. The summit process also expanded the geographical spread of past initiatives—which had previously been focused on a particular region, namely the FSU—to the entire world. The summits were also notable in their involvement of NGOs and the nuclear industry, for which a Nuclear Industry Summit was held in conjunction with the 2016 Nuclear Security Summit. The summits were also unusual in their emphasis on national pledges, termed 'house-gifts', and multinational commitments, termed 'gift baskets', which targeted key areas of nuclear security. This aspect served to expand the breadth and depth of discussions into operational security areas that included insider threats, security culture, cyber security, among many others.[32]

Following the cessation of the summit process, the international nuclear security agenda was taken on by various other initiatives and international organizations. Action Plans were developed at the 2016 Summit for the IAEA, the UN, the International Criminal Policy Organization (Interpol), the Global Initiative to Combat Nuclear Terrorism (GICNT), and the Global Partnership. Since then, these five organizations and other nuclear security initiatives have helped to maintain state-level focus on nuclear security. For example, every three or four years the IAEA has convened International Conferences on Nuclear Security (ICONS), whose 2022 edition featured both ministerial and parallel technical sessions on scientific, legal, and regulatory issues. The political salience of nuclear security has also been elevated following Russia's invasion of Ukraine in 2022, and the ensuing fighting that has occurred around the Zaporizhzhia nuclear power plant which has sustained significant damage.[33] This development has served to vividly illustrate how state actors can undermine security at nuclear facilities, as well as reaffirmed the close connection between nuclear security and nuclear safety when it comes to mitigating the risk of a radioactive release.

Structure of the Handbook

The Oxford Handbook of Nuclear Security is divided into eight thematic sections. The opening section of the Handbook (Part I: Introduction) provides a broad overview of the key concepts of nuclear security, including a basic grounding in atomic physics and the unique physical properties of nuclear and radioactive materials. There is information about the nuclear fuel cycle, including details about the chemical operations involved in enrichment and reprocessing, and about how the potential use of nuclear and other radioactive materials by malign actors can represent a serious human health and environmental risk. Part I also makes the distinction between nuclear materials involved in civil nuclear processes and in weapons programmes, as well as the related concepts of nuclear non-proliferation and safeguards. In addition, it introduces the concept of nuclear safety, acknowledging the overlapping principles that it shares with nuclear security in the protection of nuclear and radiological materials, such as design basis, emergency preparedness, and organizational culture. The aim of this part of the book is to familiarize readers with the technical background knowledge underpinning many of the nuclear security concepts discussed in more detail in other chapters.

The second section of the Handbook (Part II: Evolution of Nuclear Security Threats) explores the threat posed by nuclear terrorism and other threats relating to the unauthorized access of nuclear and radiological materials and associated facilities. Although much of the Handbook is focused on *response*—i.e., the implementation of nuclear security—to understand the necessity and appropriateness of such measures, it is also vital to conceptualize the *basis* of threats to nuclear assets and the attendant risks and vulnerabilities. Within Part II, the major scenarios widely perceived to embody the concept of nuclear terrorism are discussed, as well as their potential to cause mass casualties and their severe disruptive and economic effects. There is also an examination of how threat perceptions have played a key role in driving and shaping nuclear security efforts—and how such perceptions have changed over time—tracing their evolution and considering possible future risks. Included here is a discussion of the considerable disconnect that exists within the academic and policy communities regarding the likelihood of a nuclear terrorist incident, with predictions ranging from a near certainty to an almost impossibility. This section also explores other risks to nuclear assets, including cyber-attacks and the insider threat, which are widely considered within industry as the preeminent threats.

In Part III: International Nuclear Security Regime, attention turns to global responses to nuclear security threats. In contrast to the global response to proliferation risks, nuclear security has not been dominated by a single unifying legal instrument, such as the NPT. Rather, over the years an array of actors, events, and threat perceptions have shaped national and international approaches. Part III examines the legal, political, and multilateral aspects of nuclear security, and their interplay on the international stage. The international nuclear security architecture comprises a diverse array of binding and non-binding policy and legal instruments, IAEA guidance, and informal initiatives; it is the implementation of these by individual governments, regulators, and operators that determines the ultimate success of nuclear security. Chapters explore both the international regime as whole and key initiatives such as UN Security Council Resolution 1540 and the Nuclear Security Summit process. Part III also explores prospects for future reform of the international nuclear security regime, capturing debates in developing countries where nuclear security can be viewed as a pretext for restricting access to nuclear technology.

The spotlight of the Handbook then turns to the operational aspects of nuclear security in Part IV: Securing Nuclear Facilities and Assets. This section covers in detail the implementation of nuclear security at the facility level with a focus on key technical tools and systems, such as physical protection technologies, cyber security, and nuclear material accounting and control. It also explores the concept of security culture, in terms of how the human factor is increasingly recognized as a critical component of nuclear security. In addition, it discusses how security is implemented in different environments and for different types of assets, as this can vary considerably; from protecting critical systems at a nuclear power station from cyber-attacks, to mitigating the risk of theft of highly sensitive nuclear material while in transport, to preventing the dispersal of radioactive sources at a medical facility. This section of the Handbook explores nuclear security in a wide variety of settings, building on concepts previously discussed.

The fifth section of the Handbook (Part V: National Approaches to Nuclear Security) provides a detailed examination of how nuclear security has developed at the national level. This section recounts the experiences of different countries in terms of their legislative and regulatory approaches to nuclear security, together with background on the varying historical paths that shaped the development of their respective nuclear sectors. States have adopted radically different national approaches to one another in the area of nuclear security, and the benefits and challenges of these are explored through comparing the experiences of a select group of countries. For example, the UK has recently transitioned to a system of goal-setting nuclear security regulation, taking a highly differentiated approach to the prescriptive regulatory systems of most states with established nuclear sectors. Other countries selected for in-depth study include Russia, China, India, and the Republic of Korea, as well as Turkey and the United Arab Emirates—the latter newcomers to the civil nuclear industry that are in the process of establishing legislative and regulatory systems for the first time.

Part VI: Combatting Nuclear Smuggling examines global efforts to combat the trafficking of nuclear materials within and across national borders. The scenario of nuclear materials going out of regulatory control through theft is an enduring global concern, having evolved from opportunity-driven amateurs to more organized efforts involving transnational criminal networks. This section also explores the role of intelligence and law enforcement in countering the threat of nuclear smuggling and in locating lost or stolen materials. The use of radiation detection systems for border monitoring is also examined. These systems are in operation at thousands of locations around the world; however, there exists serious concerns over their ability to detect smuggled nuclear materials. This chapter is complemented with an exploration of the emerging discipline of nuclear forensics and its key technical and legal aspects. Here, case studies of real-life forensic investigations are utilized to highlight how the provenance of recovered nuclear material can be traced.

Part VII: Permanent Threat Reduction then turns to more radical efforts to mitigate the risk of nuclear terrorism, through 'permanent threat reduction'. This concept refers to ongoing global efforts to eliminate—or at least minimize to the extent possible—the existence of high-risk materials in civil nuclear activities. This section explores international efforts, mostly led by the US, to minimize the use of HEU in civil nuclear programmes, examining the key activities involved such as down-blending, converting research reactors to use LEU, and repatriating high-risk materials. The case study of Nigeria provides a real-life example of what a successful permanent threat reduction programme can achieve. This section also looks at the challenges to permanent threat reduction, focusing on the use of 'alternative technologies' to replace radioactive sealed sources. Again, a case study is employed to illustrate the actual use of alternative technologies, in this case the removal of gamma blood irradiators in Norway.

Finally, Part VIII: Nuclear Security during Times of Crisis explores the delivery of nuclear security during extreme events. This topic has been brought into sharp focus by the global Covid-19 pandemic, with facilities having to be either shut down or secured while meeting the challenges of a reduced workforce and staff transitioning to remote working. This section introduces the concept of emergency preparedness and response as a means of building resilience in advance of a nuclear event. It also examines the lesser-explored topic of responding to incidents involving radioactive materials, such as the detonation of a crude radiological dispersal device. Two case studies help elucidate these concepts and draw lessons for nuclear security from past experiences: the 2011 earthquake and tsunami in Japan, and the Covid-19 pandemic. While nuclear security crises, inevitably, are characterized by widely varying scenarios, there exist key factors, such as the development of an effective organizational culture, that determine the success—or failure—of recovery.

A Nuclear Security Agenda

Nuclear security is a vast and complicated topic to study and understand given its legal, political, and technical components, the wide range of materials and facilities requiring protection, and the different international, national, and organizational approaches available towards this end. *The Oxford Handbook of Nuclear Security* seeks to be as exhaustive as possible in its treatment of nuclear security, covering all major aspects and enlisting a wide range of contributors. However, there exists plenty of scope for further research.

Nuclear security threats are continually evolving and warrant close attention in different contexts. For example, in the US, a recent rise in far-right extremism may supplant the threat posed by foreign terrorist groups—a focus over the past two decades that has largely driven and shaped US-led nuclear security initiatives.[34] New research can help policy makers in understanding this threat and other emerging issues and assessing their potential relevance for nuclear security. Moving beyond the threats posed by non-state actors, Russia's invasion of Ukraine and the damage incurred at its nuclear facilities has profound implications for not just international nuclear security cooperation but also for the international governance regime as whole, and for how nuclear security and safety can be ensured under wartime conditions. This is an area of discussion that until recently has been largely absent from international forums.

There is also much to be done when it comes to exploring national approaches to nuclear security. The Handbook seeks to further discussions in this area through examining the approach taken by several countries at very different stages of their nuclear development. However, this remains a furtive area, with much more research needed to understand how and why states develop national security systems, and their likely effectiveness. At the operational level, key aspects of nuclear security—cyber security, the insider threat, and security culture—have risen to the fore in recent years and are now widely recognized as essential. However, there is currently little research on how effective programmes in this area can be developed and tailored to different environments, the likely challenges encountered, and how these can be overcome. It is the hope of the editors that the Handbook will serve to stimulate more research into these and other nascent areas of this vital, but too often neglected, subject.

Notes

1. Nuclear security as explored in this Handbook has close parallels with the internationally accepted definition of nuclear security provided by the International Atomic Energy Agency (IAEA): 'The prevention and detection of, and response to, theft, sabotage, unauthorized access, illegal transfer or other malicious acts involving nuclear material, other radioactive substances or their associated facilities'. See IAEA, *IAEA Safety Glossary: Terminology Used in Nuclear Safety and Radiation Protection* (Vienna: IAEA, 2016). https://www-ns.iaea.org/downloads/standards/glossary/iaea-safety-glossary-draft-2016.pdf
2. For an exploration of strategic issues involving nuclear weapons and nuclear technology, see for example, Lawrence Freedman and Jeffrey Michaels, *The Evolution of Nuclear Strategy* (New York: Palgrave MacMillan, 2019); Scott D. Sagan and Kenneth N. Waltz, *The Spread of Nuclear Weapons: An Enduring Debate*, 3rd ed. (New York: W. W. Norton, 2012); Andrew Futter, *The Politics of Nuclear Weapons*, 2nd ed. (New York: Palgrave MacMillan, 2020).
3. For example, see John M. Shields and William C. Potter, eds. *Dismantling the Cold War: US and NIS perspectives on the Nunn-Lugar Cooperative Threat Reduction Program* (Cambridge, MA: MIT Press, 1997).

4. For example, see Michael Levi, *On Nuclear Terrorism* (Cambridge, MA: Harvard University Press, 2007).
5. For example, the chapters The United Kingdom's Transition from Prescriptive to Outcome- Focused Nuclear Security Regulation and India's Approach to Nuclear Security extensively apply findings from research interviews held with industry practitioners to generate new insights on the subject matter.
6. For example, Vitaly Fedchenko's chapter 'The Evolution of the Legal and Technical Aspects of Nuclear Forensics' employs social construction of technology (SCOT) theory as an explanatory model, while in their chapter 'State Sponsorship of Nuclear Terrorism', Gary A. Ackerman and Bryan R. Early develop a typology on state-sponsored nuclear terrorism containing four different scenarios in which this phenomenon might occur.
7. For example, in 'The Evolution of Global Nuclear Security Governance', Wyn Q. Bowen, Matthew Cottee, and Sarah Tzinieris explore nuclear security treaties, initiatives, and international guidance, while in 'The Four Nuclear Security Summits', Amandeep Gill discusses in detail the origins and impact of the US-led Nuclear Security Summit process.
8. Amandeep Gill, *Nuclear Security Summits: A History* (New York: Palgrave MacMillan, 2019), 20.
9. Kai Bird, 'The First Line against Terrorism', *Washington Post* [online], 12 December 2001. www.washingtonpost.com/archive/opinions/2001/12/12/the-first-line-against-terrorism/4a53ee41-f528-4a4f-9f9b-64d38353a03c/?utm_term=.ae16fd1a0f95
10. Jerome I. Friedman and William A. Little, 'Robert Hofstadter', *Biographical Memoirs*, 79, 175. www.nap.edu/read/10169/chapter/11#175
11. P. Williams and P. N. Woessner, 'Nuclear Smuggling: Adaptability, Organized Crime and Undercover Operations', International Conference on Measures to Prevent, Intercept and Respond to Illicit Uses of Nuclear Material and Radioactive Sources, Sweden, 7–11 May 2001, 309–35. https://inis.iaea.org/collection/NCLCollectionStore/_Public/33/050/33050277.pdf?r=1
12. Tyson Gustafson, 'Radiological and Nuclear Detection Devices', Nuclear Threat Initiative Report [online], 18 April 2007. https://www.nti.org/analysis/articles/radiological-nuclear-detection-devices/
13. Fred Charles Ikle in collaboration with Gerald J. Aronson and Albert Madansky, 15 October 1958, 'On the Risk of an Accidental or Unauthorized Nuclear Detonation', RAND RM-2251, 32–28.
14. Dwight D. Eisenhower, 'Atoms for Peace Speech', 8 December 1953 (IAEA, Vienna). https://www.iaea.org/about/history/atoms-for-peace-speech
15. Lorna Arnold, *Windscale 1957* (New York: Palgrave Macmillan, 1995), 154; Richard Wakeford, 'The Windscale Reactor Accident–50 Years On', *Journal of Radiological Protection*, 27 (2007), 211–15.
16. Treaty on the Non-Proliferation of Nuclear Weapons (NPT), 22 April 1970 (New York: United Nations). https://www.iaea.org/publications/documents/treaties/npt
17. Christopher Price, 'Development of the IAEA Nuclear Security Recommendations on Physical Protection of Nuclear Materials and Nuclear Facilities (INFCIRC/225/Rev.5)', *Journal of Nuclear Materials Management*, 39/3 (2011), 11.
18. Communications Received from Certain Member States Regarding Guidelines for the Export of Nuclear Material, Equipment or Technology, IAEA INFCIRC/254, February 1978. https://www.iaea.org/sites/default/files/infcirc254.pdf

19. George Bunn, 'Raising International Standards for Protecting Nuclear Materials from Theft and Sabotage', *The Nonproliferation Review*, 7/2 (2000), 146.
20. Charles B. Yulish, 'Assessing the Megatonnes to Megawatts Program', *Science*, 295/5564 (2002), 2368–70.
21. Incident and Trafficking Database (ITDB), IAEA, Vienna. https://www.iaea.org/resources/databases/itdb. Please note that the ITDB was originally named the Illicit Trafficking Database, but its name was changed in 2012 to better reflect its role of capturing of all incidents (including non-malicious) in which nuclear and radioactive materials have fallen outside of regulatory control.
22. Sara Zahra Kutchesfahani, *Politics & The Bomb: Exploring the Role of Epistemic Communities in Nuclear Non-Proliferation Outcomes*, PhD dissertation (London: University College London, 2010). https://discovery.ucl.ac.uk/id/eprint/767199/
23. R. M. Frost, 'Dirty Bombs: Radiological Dispersal and Emission Devices', *Adelphi Papers*, 45/378 (2005), 76–7.
24. Piet De Klerk, 'How Nuclear Security Arrived at the IAEA', in Joseph F. Pilat, ed. *The International Atomic Energy Agency* (New York: Taylor & Francis, 2021), 189–201.
25. International Convention for the Suppression of Acts of Nuclear Terrorism, 13 April 2005, (New York, United Nations), 2. https://treaties.un.org/doc/Treaties/2005/04/20050413%2004-02%20PM/Ch_XVIII_15p.pdf
26. Charles D. Ferguson and William C. Potter, *The Four Faces of Nuclear Terrorism* (New York: Routledge, 2005).
27. For an alarming look at the threat posed by nuclear terrorism, see for example, Graham T. Allison, *Nuclear Terrorism: The Ultimate Preventable Catastrophe* (New York: Times Books/Henry Holt, 2004); for a less alarming look, see, for example, John Mueller, *Atomic Obsession: Nuclear Alarmism from Hiroshima to Al-Qaeda* (Oxford: Oxford University Press, 2012).
28. Nuclear Threat Initiative. https://www.nti.org/; International Nuclear Security Forum, https://www.stimson.org/project/insf/ that builds on the working of the Fissile Material Working Group. https://armscontrolcenter.org/fmwg/about-us/
29. Sammy Salama and Lydia Hansell, 'Does Intent Equal Capability? Al-Qaeda and Weapons of Mass Destruction', *The Nonproliferation Review*, 12/3 (2005), 615–53.
30. Barack Obama, 'Remarks by President Barack Obama, Hradcany Square, Prague, Czech Republic, 5 April' (Washington, DC: The White House, Office of the Press Secretary, 2009). https://obamaadministration.archives.performance.gov/content/continue-make-progress-toward-securing-most-vulnerable-nuclear-materials-worldwide-0.html
31. Jesse Lee, 'Expanding the Effort to Achieve Nuclear Security, 12 April', (Washington, DC: The White House, Office of the Press Secretary, 2010). https://obamawhitehouse.archives.gov/blog/2010/04/12/effort-achieve-nuclear-security-0
32. Kelsey Davenport, June 2018, Nuclear Security Summits at a Glance, Arms Control Association. https://www.armscontrol.org/factsheets/NuclearSecuritySummit
33. For detailed information on the status of Zaporizhzhia Nuclear Power Plant and nuclear security and safety in Ukraine, see IAEA, 'Nuclear Safety and Security in Ukraine', 2023. https://www.iaea.org/nuclear-safety-and-security-in-ukraine
34. Rebecca L. Earnhardt, Brendan Hyatt, and Nickolas Roth, 'A Threat to Confront: Far-right Extremists and Nuclear Terrorism', *The Bulletin* [online], 14 January 2021. https://thebulletin.org/2021/01/a-threat-to-confront-far-right-extremists-and-nuclear-terrorism/

Suggested Reading

Barrow Alex, Bowen Wyn, Cottee Matthew, Hobbs Christopher, Lentini Luca, Moran Matthew, and Tzinieris Sarah, (2022), *Nuclear Security Briefing Book, 2022 Edition* (London: King's College London).

Bowen Wyn, Cottee Matthew, and Hobbs Christopher, (2012), 'Multilateral Cooperation and the Prevention of Nuclear Terrorism: Pragmatism over Idealism', *International Affairs*, 88/2, 349–68.

Bunn Matthew, Roth Nickolas, and Tobey William H., (2019), *Revitalizing Nuclear Security in an Era of Uncertainty* (Cambridge, MA: Belfer Center for Science and International Affairs, Harvard Kennedy School).

Ferguson Charles D. and Potter William C., (2005), *The Four Faces of Nuclear Terrorism* (New York: Routledge).

Hobbs Christopher and Moran Matthew, (2022), 'Exploring the human dimension of nuclear security: the history, theory, and practice of security culture', *The Nonproliferation Review* [online]. https://doi.org/10.1080/10736700.2020.1811532

CHAPTER 2

OVERVIEW OF NUCLEAR AND RADIOACTIVE MATERIALS

FRIEDERIKE FRIEß AND MORITZ KÜTT

Introduction

Radioactive materials decay through the emission of particles. The resulting radiation is invisible to human senses but can cause significant harm to living organisms. Consequently, malicious actors could use radioactive materials for nefarious purposes. At the same time, radioactive materials have advantageous properties for many medical, industrial, and research applications. Today, commonly used radioactive materials for such purposes include cobalt-60 (Co-60), caesium-137 (Cs-137), iridium-192 (Ir-192), and strontium-90 (Sr-90).

Nuclear materials[1] are a subset of radioactive materials. These materials fuel civilian nuclear power reactors for electricity production and research reactors for teaching and scientific investigations. However, they are also used in nuclear weapons, which, if detonated, cause catastrophic consequences through heat radiation, pressure waves, immediate radiation, and delayed radioactive fallout.

This chapter first introduces key technical concepts for these materials of relevance to nuclear security and describes the historical discoveries that provide our basic understanding of atoms and radioactivity. The next part of the chapter focuses on nuclear materials in nuclear weapons, including a description of their production, most common civilian applications, and current material stockpiles. Subsequently, the chapter outlines the production and current use of radioactive materials. The chapter ends with a discussion of the effects of radioactivity, radioactive materials, and nuclear weapons use.

A Historical Introduction to Radioactive Materials

The characteristic of radioactivity is the disintegration (decay) of material after a certain time. Properties of the nucleus govern this disintegration. The nucleus is the inner part of an atom. It is positively charged and surrounded by negatively charged electrons. Nuclei consist of two types of more fundamental particles: positively charged protons (their count—atomic number—defines to which element a nucleus or atom belongs); and neutrons (neutral particles, of which atoms typically contain a number equal to or larger than the number of protons). Atoms from the same element that have varying numbers of neutrons are called isotopes. The total number of protons and neutrons (mass number) defines a specific isotope. Cobalt-60, for example, is the isotope of the element cobalt with 27 protons and 33 neutrons. Most of the more than 4,000 isotopes known today are radioactive, with only about 250 stable.[2]

At the turn of the twentieth century, physicists discovered additional structures within the atom. In 1911, Ernest Rutherford was the first to propose a model containing negative electrons surrounding a small, positive nucleus.[3] The nucleus contains almost the total mass of the atom. The negatively charged electrons travel in orbits around the nucleus, with the number of electrons in the outer orbits defining the element's chemical properties. Niels Bohr refined this model in 1913 based on recent developments in quantum mechanics. In the 1920s, researchers postulated the existence of neutral particles within the nucleus.[4] Later named neutron, this particle carries no charge and is slightly heavier than the proton. It decays into a proton and an electron outside of a nucleus. In 1932, James Chadwick proved the existence of neutrons in experiments.[5]

In parallel to the discovery of sub-atomic structures, researchers learned about radioactivity. During the disintegration processes, nuclei emit particles that carry away energy released in the decay. When the French scientist Henri Becquerel investigated the fluorescence of minerals—in 1896, Becquerel had exposed potassium uranyl sulphate to sunlight—he expected the uranium to absorb energy from the sun and emit it later as X-rays, which he wanted to measure with photographic plates. After developing plates without prior exposure to sunlight, Becquerel was surprised to discover an image from the uranium, nevertheless. The radiation originated from a process within the atom.[6] Marie Skłodowska Curie and her husband Pierre Curie discovered polonium and radium, other radioactive elements. Both are present in uranium ore, produced by the constant but slow decay of uranium itself. Marie and Pierre Curie were awarded the Nobel prize for physics in 1903 for their work, jointly with Becquerel.[7]

Further research revealed differences in the radiation emitted by decay. First, Rutherford classified the radiation according to the emitted particle's ability to ionize matter into alpha rays and beta rays.[8] In 1900, Paul Villard discovered a third category, later dubbed gamma rays by Rutherford.[9] The three labels are still used today. Besides describing emitted particles, one can use them to describe decays (for example, 'alpha decay').

If an atomic nucleus undergoes alpha decay, it ejects two protons and two neutrons in one package, a helium nucleus called an alpha particle. Alpha particles are emitted with discrete energies ranging from 1 to 10 MeV and have a double positive charge. They penetrate into matter only short distances and can be shielded by approximately 10 cm of air or a sheet of paper. During beta decay, the nucleus emits an electron. Electrons have a higher ability to penetrate matter: several metres in air and several millimetres in solid materials.

In 1935, Irène and Frédéric Joliot-Curie shared the Nobel prize for chemistry for their discovery of new radioactive elements and of the second kind of beta decay with a positron being emitted.[10] The positron is an 'antimatter' particle, having the same properties as the electron but a positive charge.[11] They are named beta minus (electron) and beta plus (positron) decay, to distinguish both types. Furthermore, high-energy photons (gamma rays) are emitted when the nucleus releases excess energy from an 'excited state'. Photon interactions with matter are weaker than alpha and beta interactions. They are, therefore, very difficult to shield. Excited nuclei result from both alpha decays and beta decays.

In 1903, Rutherford and his colleague Frederick Soddy had postulated that radioactivity leads to the transmutation of elements, that is, one element becomes a different element after radioactive decay.[12] Based on this idea, scientists later experimented with induced transmutation. They produced new, completely artificial, radioactive isotopes by irradiating material with alpha particles and neutrons.[13] In 1939, Otto Hahn and Fritz Straßmann discovered a new reaction, nuclear fission, induced by neutron bombardment. Lise Meitner explained their observations theoretically:[14] in the nuclear fission reaction, the nucleus of a (heavy) atom splits into two (or more) nuclei. Nuclear fission releases huge amounts of energy, mostly in the form of kinetic energy of the resulting nuclei, and gamma radiation.

Nuclear fission can occur spontaneously or can be induced by neutrons. One source of neutrons is another nuclear fission. The number of free neutrons per fission and the likelihood of fission in the surrounding material determines whether a 'chain reaction' is possible. In a chain reaction, neutrons from one fission induce, on average, at least one other fission. In a nuclear power plant, the chain reaction occurs in a controlled manner; the number of free neutrons stays constant over time. In a nuclear weapon, the numbers of neutrons and fissions increase exponentially, releasing large amounts of energy in a very short period of time, about one microsecond.

All isotopes that undergo spontaneous fission, alpha decay, beta decay, or gamma decay are called radioactive. Radioactive decay is a probabilistic process, as the prediction of exact decay times of an individual atom is impossible. However, it is possible to specify the average number of disintegrations in a given time. Common units to measure that activity are becquerel (Bq; one disintegration per second) or curie (Ci;

one Ci equates 37,000 Bq). An isotope's 'half-life' describes its radioactivity in a different way. Half-life is the time after which, on average, half of all initial atoms have decayed. The energy of the emitted particles is another important property. Nuclei emit alpha and gamma radiation at discrete, characteristic energies. These energies can be measured and are available in tabulated form. Beta rays are emitted with a continuous energy spectrum.

Nuclear Materials and Nuclear Weapons

At the outset of the Second World War, scientists worldwide quickly realized nuclear fission's potential for enormous energy release. Out of the fear of Germany developing a nuclear weapon, the United States (US) clandestinely started a massive industrial-scientific programme, the 'Manhattan Project', which led to the first operational reactors, the production of fissionable materials, the first test of a nuclear explosion on 16 July 1945, and eventually, the use of nuclear weapons on the Japanese cities of Hiroshima (6 August 1945) and Nagasaki (9 August 1945).[15] The US programme set the stage for a subsequent nuclear arms race with the Soviet Union. Over time other states for a variety of strategic, domestic, and normative reasons also sought to develop nuclear weapons. At the time of writing, nine countries own nuclear weapons, and several more possess the necessary materials to manufacture them.

The most weapon-relevant nuclear materials are fissile materials. Fissile isotopes are able to sustain a fission chain reaction. The isotopes uranium-235 (U-235) and plutonium-239 (Pu-239) are the fissile isotopes commonly used in nuclear weapons. Table 2.1 lists important properties of select fissile materials,[16] and includes a column for the critical mass, which describes the amount of material in which a chain reaction, once started, can be sustained. If less material is present, too many neutrons from fissions will escape the configuration without causing additional fissions. The values shown in the table have been calculated for bare spheres of the material. Specific nuclear weapon configurations (for example, neutron 'reflectors' surround a fissile sphere) can reduce the mass needed. The International Atomic Energy Agency (IAEA) defines so-called significant quantities (SQ) for nuclear materials; these quantities give 'the approximate amount of nuclear material for which the possibility of manufacturing a nuclear explosive device cannot be excluded'.[17]

Natural uranium contains 99.3 per cent of the isotope uranium-238, with the fissile uranium-235 only present at a level of 0.7 per cent. Higher fractions of uranium-235 are necessary for weapon applications. The process of increasing the uranium-235 fraction in uranium is called enrichment. The enrichment level describes the fraction of uranium-235, that is, 5 per cent enriched uranium contains 5 per cent of uranium-235. A small fraction of power reactors is fuelled with natural

Table 2.1. Important fissile material isotopes and weapon-related properties.[a] Data for half-life and main decay modes is from the IAEA Nuclear Data Service.[b] Critical mass, spontaneous fission neutron rates, and specific power are based on the authors' own calculations. Dose conversion coefficients are from the ICRP.[c]

Unit	Half-life [years]	Specific activity [Bq/kg]	Main decay mode -	Critical mass [kg]	Spontaneous fission neutron rate [neutrons/kg/s]	Specific power (from alpha decay) [W/kg]	Effective dose coefficient for ingestion [Sv/Bq]	Effective dose coefficient for inhalation [Sv/Bq]
U-233	159,190	3.57×10^{11}	alpha	15	0.38	0.27	5.1×10^{-8}	3.6×10^{-6}
U-235	0.70 billion	8.00×10^{7}	alpha	46	0.01	<0.01	4.7×10^{-8}	3.1×10^{-6}
U-238	4.47 billion	1.24×10^{7}	alpha	-	13.5	<0.01	4.5×10^{-8}	2.9×10^{-6}
Pu-238	87.7	6.34×10^{14}	alpha	8	2.6 million	557	2.3×10^{-7}	4.6×10^{-5}
Pu-239	24,110	2.30×10^{12}	alpha	10	15.4	1.89	2.5×10^{-7}	5.0×10^{-5}
Pu-240	6,561	8.40×10^{12}	alpha	37	1.0 million	6.94	2.5×10^{-7}	5.0×10^{-5}
Pu-241	14.329	3.83×10^{15}	beta minus	13	1.72	7.36×10^{-2}	4.8×10^{-9}	9.0×10^{-7}

a Table courtesy of the authors.

b IAEA, n.d. Nuclear Data Service. https://www-nds.iaea.org.

c K. Eckerman, J. Harrison, H-G. Menzel, and C. H. Clement. 2012. 'Compendium of Dose Coefficients Based on ICRP Publication 60'. International Commission on Radiological Protection. *ICRP Publication.* No. 119. Ann. ICRP 41 (Suppl.).

uranium. However, most use uranium with enrichment levels of 3 to 5 per cent. For use in nuclear weapons, enrichment levels of 90 per cent (weapons-grade uranium) or higher are typical.

Technologies that have been used to separate uranium isotopes include gaseous diffusion, electromagnetic separation, and gas centrifuges. All rely on the approximately 1 per cent weight difference of the two main uranium isotopes. Today, all commercial enrichment facilities use gas centrifuges. In nuclear weapons, the amount of material required to achieve a critical mass is lower for higher enrichment levels. The IAEA defines uranium with enrichment levels below 20 per cent as low enriched uranium (LEU), assuming that it is impractical to construct a weapon with such material. Uranium with higher levels is highly enriched uranium (HEU).[18] For HEU containing 5 kg or more of uranium-235, the IAEA recommends the highest levels of physical protection (Category I).[19] HEU containing 25 kg of uranium-235 is considered the above-defined SQ.[20]

To reach different enrichment levels, only slight modifications in enrichment facilities are necessary. Enriching uranium to 20 per cent already requires most of the separative work necessary to achieve weapon-usable HEU. Gas centrifuges, as currently installed in enrichment plants, are complex pieces of equipment. So far, only state actors have produced HEU for a nuclear weapon, while at the same time several enrichment facilities for commercial purposes are owned and operated by private companies.

It is simpler to build a nuclear weapon with HEU than plutonium. In 'gun-type' weapons, one mass of HEU is shot into another using conventional explosives to create a critical mass. The first nuclear weapon used in warfare, the Hiroshima bomb, was an HEU weapon. Developers were so convinced that the design would work that no prior testing of this weapon type was deemed necessary.[21]

The second common nuclear-weapon material is plutonium, for which only trace amounts exist in nature. Nuclear reactors produce plutonium-239 through neutron capture of uranium-238 followed by two radioactive decays of the resulting uranium-239 and then its decay product, neptunium-239 (Np-239). Non-fission neutron captures in plutonium-239 lead to the production of higher plutonium isotopes, starting with plutonium-240. Plutonium-238 is also produced in nuclear reactors from uranium-235. Production of these plutonium isotopes is inevitable in a nuclear reactor. The longer the irradiation period, the higher the fractions of the other plutonium isotopes. For plutonium quantities of 2 kg and more, the IAEA recommends the highest levels of physical protection (Category I).[22] A SQ of plutonium is 8 kg.[23]

Certain plutonium isotopes have physical properties that reduce their attractiveness for use in nuclear weapons (Table 2.1). High radioactivity, as in the case of plutonium-238, requires a more complex weapon design capable of dissipating radioactive heat.[24] High spontaneous fission rates, as in the case of plutonium-240, also increase the challenge of building a nuclear weapon, as neutrons produced in spontaneous fission events might start the chain reaction prematurely once a mass becomes critical, reducing the explosive yield. Such a pre-initiation might not be as important, however, for non-state actors who aim at spreading terror. Commercial nuclear power plants typically operate with irradiation periods in the range of one to two years. To produce plutonium with a high fraction of plutonium-239, and a low fraction of plutonium-240, the fuel elements must be removed from a reactor core after a couple of months. For many reactor designs, this can only be done when the reactor is shut down. In the US, plutonium with more than 93 per cent of plutonium-239 is called weapons-grade.[25] Plutonium with a typical composition resulting from commercial reactor use is called reactor-grade.[26]

Weapons containing plutonium use 'implosion' to create a supercritical mass. In modern designs, a hollow sphere of plutonium is compressed using conventional explosives. A critical mass is reached through the increase in density. Building an implosion-type weapon is more difficult than the simple gun-type design, which is unsuitable for plutonium-based nuclear weapons. The US tested its implosion design in 1945 before dropping a weapon of this design on Nagasaki. The production of a plutonium implosion-type nuclear weapon would present a far greater technical challenge to a non-state actor group than a HEU gun-type nuclear weapon.[27]

Nuclear Material Stockpiles and Uses

Civilian energy production is the largest consumer of fissile materials. Power reactors need regular refuelling. At the time of writing, there are approximately 450 nuclear power reactors currently in operation.[28] They consume almost 60,000 tons of natural uranium annually, mostly after enrichment to LEU. The LEU consumption for research reactor fuel is significantly less. Small amounts of LEU are also used as irradiation targets in accelerator and research reactor facilities for the production of radioactive isotopes.

Many research reactors were originally built to be fuelled with weapons-grade uranium. Since the 1970s, international initiatives have sought to reduce proliferation risks by supporting their conversion to LEU fuel.[29] According to the International Panel on Fissile Materials, there are currently 108 research reactors with HEU fuel, including two reactors that use HEU targets.[30]

Military reactors consume the largest fraction of HEU. It is estimated that there are currently 135 submarines and 13 military surface ships with a total of 185 HEU-fuelled reactors in four countries (India, Russia, the United Kingdom, and the US).[31] At the time of writing, only states with nuclear weapons use HEU for military naval propulsion, but the 2021 AUKUS pact may see the US and the UK transfer HEU-fuelled submarines and associated technology to Australia.[32]

Several countries have operated reprocessing plants to separate plutonium from spent nuclear fuel for civilian use. Initially, these efforts were driven by the motivation to use plutonium as start-up fuel for breeder reactors that could produce more plutonium than they consumed. Most states, though, abandoned plans for breeder reactors over time, as more cheap uranium was discovered and because of concerns about civilian plutonium being diverted to weapons. At the time of writing, Russia is operating two prototype breeder reactors, and Chinese and Indian prototypes should become operational soon. France and Japan use separated plutonium in mixed oxide (MOX) fuel (with uranium) for light water reactors. In France, this replaces approximately 10 per cent of uranium fuel. Despite using MOX fuel, France's and Japan's separated plutonium stockpiles grow. The UK, which owns overall the largest separated stockpile of about 120 tons of civilian plutonium, currently has no policy for how to use or dispose of this plutonium.[33] This raises various security issues ranging from unintended loss to theft and sabotage.[34]

Table 2.2 lists current stockpiles of highly enriched uranium and plutonium for countries with relevant stocks. The table does not separate between military and civilian

Table 2.2. Global stocks of highly enriched uranium and plutonium as of the beginning of 2022.[a] The stockpiles include both military and civilian material, as they are, in principle, all weapon-usable. For France, Japan, India, Russia, and the UK, civilian plutonium stockpiles are significantly larger than military stockpiles. The figures for Iran date to the beginning of 2022 as the country only started HEU production in 2021.[b]

Country	HEU stockpile (metric tons)	Currently producing HEU	Plutonium stockpile (metric tons)	Currently producing military plutonium
China	14 ± 3	No	2.9 ± 0.6	No
France	28.8 ± 6	No	90.9 ± 1	No
India	4.9 ± 2	Yes	9.6 ± 5.1	Yes
Iran	0.03	Yes	-	No
Israel	0.3	Unknown	0.84 ± 0.1	Yes
Japan	-	-	45.8	No
North Korea	Uncertain	Uncertain	0.04	Yes
Pakistan	4.9 ± 1.5	Yes	0.5 ± 0.17	Yes
Russia	680 ± 120	Yes	191.5 ± 8	No
United Kingdom	22.6	No	119.7	No
United States	487	No	87.8	No
Other states (combined)	4	No	-	No

a Table courtesy of the authors. See Kütt, Moritz, Mian, Zia, and Podvig, Pavel, 'Global Stocks and Production of Fissile Materials, 2022', in *SIPRI Yearbook: Armaments, Disarmament and International Security* (Oxford: Oxford University Press, 2023).

b IAEA, *Verification and Monitoring in the Islamic Republic of Iran in Light of United Nations Security Council Resolution 2231 (2015)*, Report by the Director General, 3 March. GOV/ 2022/ 4 (Vienna: IAEA, 2022). https://www.iaea.org/sites/default/files/22/03/gov2022-4.pdf.

use. It should be noted that certain countries, for example, Japan, only use fissile material for civilian applications. In states that own nuclear weapons, the distinction is often blurred, as the categorization is mostly of declaratory nature. In principle, nearly all material listed in the table is weapon-usable.

Production and Use of Radioactive Materials

Radiation emitted from radioactive decays of certain isotopes has medical, industrial, and research applications. Initially, scientists obtained naturally occurring radioactive materials through chemical separation. Later, they discovered ways for artificial production. The earliest medical use of such 'sources' dates back to the beginning of the twentieth century. The Curies first separated radium, which is present in uranium ores. Doctors started to discover that particles emitted from radium sources could kill cancerous tissue.[35] Radium and its direct decay product radon remained the most commonly used radioactive materials for practical applications during the first half of the twentieth century. Today, naturally occurring radioactive materials play only a minimal role.

Bombarding targets with particle beams produced in accelerators can produce artificial radioactive materials.[36] Similarly, nuclear fission reactions inevitably produce radioactive fission products, which can be chemically separated (for example, caesium-137 and strontium-90). Fission reactors also contain very strong neutron flows. Target materials placed in reactors absorb neutrons, turning stable isotopes into radioactive materials. This production pathway is used, for example, for the production of cobalt-60.

Shortly after the first nuclear reactors became operational, scientists realized the potential of artificially produced radioactive materials.[37] The first medical source of iridium-192 was used in 1950, of cobalt-60 in 1951, and of caesium-137 in 1955.[38] Early patents pointed to the possibility of radioactive sources in industrial settings for well logging and level gauges. For example, combinations of radioactive sources and radiation detectors inserted into oil wells reveal geological structures while drilling. Directing radiation through material (for example, a tank) provides information on fill levels.[39] Such radioactive gauges are particularly useful if no direct contact with the material is possible. They also allow for the measurement of densities and the thickness of material (for example, paper production),[40] and provide imaging information (for example, quality checks of welds). The strongest radiation sources can be used to sterilize food, seeds, and medical equipment.[41]

Today, it is estimated that more than 10,000 hospitals around the world use radioisotopes, both in sealed sources and as tracers.[42] Trace materials are short-lived isotopes that are injected into patients or processes. As their typical half-lives are of the order of hours, they are of limited concern to nuclear security—the dangerous material decays away before it might have negative effects. Sealed, longer-lived, radioactive sources are a greater concern, because they remain dangerous for periods of months and years, over which appropriate security measures must be ensured.[43]

Today, most radioactive isotopes are produced in a small number of research reactors.[44] Production in research reactors is preferable because they allow for more flexibility in operating cycles. The IAEA's Research Reactor Database lists 77 reactors in the category of isotope-production reactors.[45] A 2021 study by the US National Academy of Sciences listed 21 reactors used for the production of cobalt-60 in five countries (Argentina, Canada, China, India, and Russia), and only two reactors for the production of caesium-137 (Russia and India).[46]

After irradiation, special processing facilities handle highly radioactive materials using glove boxes or shielded hot cells.[47] After separating the radioactive sources from targets or fuel, they are placed into small containers and sealed. Sources lose their utility over time due to radioactive decay, but the remaining activity can still pose significant risks. Secure management, including disposal or recycling, for disused sources, is a necessity. Such services are often provided by the initial source distributors.

Particle accelerators also produce radioactive material. Because accelerator beams can consist of positive protons or other nuclei, they can produce materials that later undergo beta plus decay emitting positrons. Cyclotrons are the most common form of accelerator for radioactive material production. Since they are of limited size, it is possible to produce radioactive materials directly at locations of use, like hospitals. Production near usage location is particularly relevant for radioactive materials with very short half-lives. In 2015, approximately 1,200 cyclotrons around the world produced radioactive materials, at least occasionally.[48]

Several properties of radioactive sources are relevant to potential security concerns. The mobility of a source ranges from fixed installed irradiators to sources that are by design mobile, for example, sources used in well logging. The chemical form of the material influences whether it is easily distributable (for example, in powder form or water soluble) or harder to separate (for example, embedded in glass/ceramic). The source's activity is also key: the higher, the more dangerous. Lastly, the type of radiation (alpha/beta/gamma) influences whether sources have a wide range of effects when distributed in the environment or only internal effects after ingestion/inhalation.

After the terrorist attacks of 11 September 2001, the international community strengthened efforts to secure access to radioactive sources. The IAEA defined five source categories based on source activity and the health risks posed by specific isotopes.[49] The Agency recommends that states create national registers for sources of Category I and Category II. No such register exists on the international level. The authors estimate that there are 1–2 million radioactive sources worldwide, of which 0.25–0.5 million fall in Categories I and II.[50]

The IAEA operates three databases with regard to radioactive sources. The DIrectory of RAdiotherapy Centers (DIRAC) lists more than 7,000 centres where patients are treated with radiation and nearly 15,000 radiotherapy machines.[51] The IAEA's Incident and Trafficking Database (ITDB) lists incidents of unauthorized access to nuclear material or loss thereof, as reported by IAEA Member States.[52] The International Catalogue of Sealed Radioactive Sources and Devices (ICSRS) collects data on source types, radioactive devices, and suppliers to help in the identification of unknown sources.[53]

Radioactive materials cannot sustain a neutron chain reaction and thus cannot be used for an exponentially increasing energy release as in a nuclear weapon. To cause harm, two different forms of attack are possible. First is spreading the material, for example, using conventional explosives. Here, gamma emitters pose the greatest risk because of the relatively long range of the radiation. Second is using the material in an i^3 (inhalation, ingestion, immersion) attack. In such cases, the radioactive material is in direct contact with the human body.[54]

Effects of Radioactive Materials and Nuclear Weapons

Radiation above a certain threshold energy can separate electrons from atoms. Such radiation is called ionizing radiation, and the resulting charged atom is called an ion. The interaction of radiation with living tissue and, in particular, DNA can cause serious damage to the whole organism. The biological effects of radiation exposure depend on the dose. The 'absorbed' dose describes how much energy is absorbed per mass unit. It is measured in units of 'gray' (Gy) (joule per kg) and scales proportionally to the number of absorbed particles: twice as many particles of a given energy double the dose.

Different particles depositing the same energy deposition can do different damage to tissue. Alpha particles are densely ionizing and can therefore cause greater harm than the same dose of beta particles. This difference is taken into account by the quantity coined 'equivalent dose'. The unit of measuring the equivalent dose is the 'sievert' (Sv). Conversion from absorbed dose to equivalent dose is done by multiplying values with particle-specific weighting factors.

Neither absorbed dose nor equivalent dose cover the fact that organs and tissue have different susceptibility to harm caused by ionizing radiation. These differences are included in the 'effective dose'. Effective doses are also measured in sieverts. Dose coefficients describe the relationship between the physical quantity (number and energy of incoming particles) and the effects they have on human tissue. The International Commission on Radiological Protection (ICRP) regularly publishes dose coefficients and the above-mentioned weighting factors.[55]

Radiation effects can be separated into deterministic or early health effects and stochastic or delayed health effects. Early health effects are the result of high exposure in short times. They lead to immediate and often easily visible symptoms such as skin burns and loss of hair. The higher the dose, the more severe are the effects. After whole-body exposure to doses higher than 50 Gy, death usually occurs within days.[56]

Stochastic effects result from continuous or low-level exposure. The damage done to cells might or might not cause cancer or hereditary illness. The number of certain effects of low-level exposure can be estimated among a large group of people. It cannot be predicted for single individuals. For individuals, only probabilities for the occurrence of an effect can be specified. A typical example of a stochastic effect is radiation-induced cancer, which can occur decades after exposure. Higher exposure makes the effect more likely to occur but does not necessarily imply a more severe effect.

Much of the data regarding the health effects of radiation are derived from studies of Japanese atomic bomb survivors. Large-scale studies, also including the children of atomic-bomb survivors, are still ongoing.[57] But research is not limited to this group. Deterministic and stochastic effects of ionizing radiation can also be observed in nuclear workers and the victims of nuclear weapons testing.[58]

The basic principles to protect people from the hazardous consequences of ionizing radiation are the same for planned and emergency situations:[59] the duration of exposure should be minimized, the distance to the radiation source maximized, and as much shielding should be applied as possible. Naturally, those principles are most complicated to adhere to in emergency situations like accidents and attacks. For planned therapeutic exposures such as destroying cancerous tissue, doses should be localized to the target region.[60] Radiation exposure cannot be completely avoided. For health practitioners working with radioactive sources or for maintenance workers at nuclear power plants, for example, the ALARA principle applies. According to this principle, exposure and dose are kept 'as low as reasonably achievable'.[61]

Table 2.3. Important radioactive isotopes in medical, industrial, and research use.[a] Half-lives and main decay modes are from the IAEA Nuclear Data Service.[b] Dose conversion coefficients are from the ICRP.[c]

Nuclide Unit	Half-life [years]	Specific activity [Bq/kg]	Main decay mode	Effective dose coefficient for ingestion [Sv/Bq]	Effective dose coefficient for inhalation [Sv/Bq]
Cobalt-60	5.27	4.19×10^{16}	beta minus / gamma	3.4×10^{-9}	1.0×10^{-8}
Strontium-90	28.91	5.09×10^{15}	beta minus	2.8×10^{-8}	3.6×10^{-8}
Caesium-137	30.08	3.21×10^{15}	beta minus / gamma	1.3×10^{-8}	9.7×10^{-9}
Iridium-192	0.21	3.41×10^{17}	beta minus / gamma	1.4×10^{-9}	5.2×10^{-9}

a Table courtesy of the authors.

b IAEA, Nuclear Data Service.

c Eckerman et al., 'Compendium'.

Depending on their radiation characteristics, the commercially used radioactive sources are assigned to different categories by the IAEA. The highest intensity sources are considered Category I.[62] Typically, those are cobalt-60, strontium-90, and caesium-137.[63] Relevant characteristics of important isotopes in radiation protection are listed in Table 2.3.

The effects of nuclear weapons include the effects of radiation as well as blast and heat. One nuclear weapon might easily release energy equivalent to thousands of tons (kilotons) of conventional high explosives. The temperature of the nuclear material reaches millions of degrees. A fireball of superheated material is formed and grows rapidly. About 35 per cent of the energy is released as thermal radiation, which travels with the speed of light, preceding the subsequent shock wave.[64] Thermal radiation causes burning and can ignite flammable materials.

The shock wave or nuclear blast carries about 50 per cent of the energy.[65] It starts moving outwards from the fireball directly after the explosion, roughly with the speed of sound. The overpressure can destroy structures. The shock wave also produces strong winds (dynamic pressure). Injuries can result from impacts. The effects of the shock wave can also ignite fires indirectly by destroying electrical circuits or gas lines. The nuclear reactions in the weapon result in a blast of neutrons and gamma rays, and radioactive materials. For nuclear explosives that release more than the equivalent of about 1 kiloton of chemical explosive energy, the lethal effects of the thermal radiation and the shock wave are more significant than the initial nuclear radiation.

A longer-term effect of a nuclear explosion is residual nuclear radiation. Radioactive fallout (delayed nuclear radiation) originates from radioactive isotopes created in the fission process. For a surface or near-surface burst, radioactive isotopes mix with dirt and debris in the fireball, which quickly rises, due to its buoyancy, to altitudes of kilometres or tens of kilometres, depending on the yield. Lighter particles can travel long distances before they return to earth. Locally, that is, within a hundred kilometres or so, radioactive fallout can cause lethal doses within days or weeks. Further away, it can increase the likelihood of cancer.

Beyond direct effects, researchers also have warned of long-term climatic consequences of nuclear weapon use. First mentioned in the early 1980s, some of these effects are summarized by the label 'nuclear winter'.[66] According to the theory, nuclear weapon-induced firestorms in cities would produce large amounts of soot, which would rise into the stratosphere and be distributed throughout the northern hemisphere due to prevailing winds.[67] The soot particles would reduce the sunlight reaching the earth's surface, resulting in cooling. Initial studies focused on large-scale, global nuclear war.

More recent works claim that even regional nuclear wars could have significant global effects. Specifically, it is claimed that the use on cities of a total of 100 nuclear weapons in South Asia would lead to a reduction in global average temperatures of approximately 1.25 degrees for several years, with widespread effects on grain production.[68] Studies on the effects are ongoing using newer climate models. Nitrogen oxides created by nuclear explosions could also significantly reduce the stratospheric ozone layer that shields the earth's surface from solar UV radiation.[69]

Conclusion

Radioactive and nuclear materials are widely used around the globe. Nuclear materials are present throughout the fuel cycle of civilian nuclear energy production. Both categories pose risks if used for malicious purposes. The most significant impact would be the use of fissile material in nuclear weapons. HEU and plutonium are the necessary materials to develop these weapons, of which kilogram quantities are potentially sufficient for that purpose. Given that the stockpiles of certain countries amount to hundreds of tons of these materials, tight controls are necessary to prevent theft.

For radioactive material, mostly used in the form of radioactive sources, the global distribution is even more widespread. Hospitals use radioactive sources, as do industrial enterprises and research institutions. Malicious actors could steal these sources and create radioactive contamination by dispersing the material. While the impact would be significantly lower than with the use of nuclear weapons, there could still be major physical, disruptive, and economic effects. Consequently, significant focus must also be placed on securing radioactive materials, in particular Categories I and II, throughout their lifecycle.

Notes

1. According to the Statute and a Glossary of the IAEA, nuclear materials include the elements thorium and uranium, and the isotope plutonium-239, as well as materials containing those elements and isotopes listed before. IAEA, 28 December 1989, *Statute*, Vienna. https://www.iaea.org/sites/default/files/statute.pdf; IAEA, *IAEA Safeguards Glossary*, 2001 Edition, IAEA International Nuclear Verification Series, No. 3 (Vienna: IAEA, 2002). https://www.iaea.org/sites/default/files/iaea_safeguards_glossary.pdf
2. Zsolt Sóti, Joseph Magill, and Raymond Dreher, 'Karlsruhe Nuclide Chart: New 10th Edition 2018', *EPJ Nuclear Sciences & Technologies*, 5/6 (2019), 1–11. https://doi.org/10.1051/epjn/2019004
3. Ernest Rutherford, 'The Scattering of α and β Particles by Matter and the Structure of the Atom', *The London, Edinburgh, and Dublin Philosophical Magazine and Journal of Science*, 21 (1911), 669–88. https://doi.org/10.1080/14786440508637080
4. Niels Bohr, 'On the Constitution of Atoms and Molecules', *The London, Edinburgh, and Dublin Philosophical Magazine and Journal of Science* 26/153 (September 1913), 476–502.
5. The Nobel Prize Foundation, 'James Chadwick—Biographical', *Nobel Prize Outreach AB* 2022 [online]. https://www.nobelprize.org/prizes/physics/1935/chadwick/biographical; James Chadwick, 'Bakerian Lecture: The Neutron', *Proceedings of the Royal Society of London A*, 142 (1933), 1–25. https://doi.org/10.1098/rspa.1933.0152; James Chadwick, 'Possible Existence of a Neutron', *Nature*, 129 (1932), 312. https://doi.org/10.1038/129312a0
6. Henri Becquerel, 'Sur quelques propriétés nouvelles des radiations invisibles émises par divers corps phosphorescents', *Comptes rendus de l'Académie des sciences*, 122 (1896), 559–64.
7. The Nobel Foundation, 'The Nobel Prize in Physics 1903', *Nobel Prize Outreach AB* 2022 [online]. https://www.nobelprize.org/prizes/physics/1903/summary

8. Ernest Rutherford, 'Uranium Radiation and the Electrical Conduction Produced by It', *Philosophical Magazine*, 47 (1899), 109.
9. L. Gerward, 'Paul Villard and His Discovery of Gamma Rays', *Physics in Perspective*, 1/4 (1999), 367–83. https://doi.org/10.1007/s000160050028
10. The Nobel Foundation, 'The Nobel Prize in Chemistry 1935', *Nobel Prize Outreach AB* 2022 [online]. https://www.nobelprize.org/prizes/chemistry/1935/summary
11. Irène Curie, 'The Nobel Prize in Chemistry 1935—Nobel Lecture', *Nobel Prize Outreach AB* 1935 [online]. https://www.nobelprize.org/prizes/chemistry/1935/joliot-curie/lecture; Irène Curie and Frédéric Joliot, 'Un nouveau type de radioactivité', *Comptes rendus hebdomadaires des séances de L'Academie des Sciences*, 198/3 (1934), 254–6.
12. Ernest J. Rutherford, *Radio-activity* (Cambridge: Cambridge University Press, 1904).
13. Enrico Fermi, et al., 'Artificial Radioactivity Produced by Neutron Bombardment', *Proceedings of the Royal Society London A*, 146 (1934), 483–500. https://doi.org/10.1098/rspa.1934.0168
14. Lise Meitner and O. R. Frisch, 'Disintegration of Uranium by Neutrons: A New Type of Nuclear Reaction', *Nature*, 143 (1939), 239–40. https://doi.org/10.1038/143239a0
15. Richard Rhodes, *The Making of the Atomic Bomb* (New York: Simon and Schuster, 2012).
16. Uranium-233 is another fissile material, produced through thorium use in nuclear reactors. While thorium is considered a nuclear material, thorium or uranium-233 are not discussed further in the chapter, as there is no active use of these materials beyond research.
17. IAEA, *IAEA Safeguards Glossary*.
18. IAEA, *Nuclear Security Series Glossary*, Version 1.3 (Vienna: IAEA, 2015), 17.
19. IAEA, *Nuclear Security Recommendations on Physical Protection of Nuclear Material and Nuclear Facilities (INFCIRC/225/Revision 5)*, IAEA Nuclear Security Series, No. 13 (Vienna: IAEA, 2011).
20. IAEA, *IAEA Safeguards Glossary*.
21. Jessie Kratz, 'Little Boy: The First Atomic Bomb', *U.S. National Archives* [blog]. https://prologue.blogs.archives.gov/2020/08/06/little-boy-the-first-atomic-bomb
22. IAEA, *Nuclear Security Recommendations*.
23. IAEA, *IAEA Safeguards Glossary*.
24. Because of the high heat production, the IAEA excludes plutonium with a plutonium-238 content of more than 80 per cent from safeguards requirements and from its physical safety recommendations for plutonium.
25. The formal US definition is that anything with more than 19 per cent plutonium-240 is reactor grade, and material in between is considered fuel grade.
26. Office of Arms Control and Nonproliferation, *Nonproliferation and Arms Control Assessment of Weapons-Usable Fissile Material Storage and Excess Plutonium Disposition Alternatives*, Report no. DOE/NN-0007, Oak Ridge, TN: Office of Scientific and Technical Information, US Department of Energy, 1997). https://doi.org/10.2172/425259
27. Matthew Bunn and Anthony Wier, 'Terrorist Nuclear Weapon Construction: How Difficult?', *The Annals of the American Academy of Political and Social Science*, 607/1 (September 2006), 133–49. https://doi.org/10.1177/0002716206290260.
28. IAEA, *Nuclear Power Reactors in the World—2021 Edition*, Reference Data Series, No. 2 (Vienna: IAEA, 2021).
29. L. Gil, 'Countries Move towards Low Enriched Uranium to Fuel Their Research Reactors', *IAEA Bulletin*, 60-4/November (2019), 26–7.

30. International Panel on Fissile Materials, 'Facilities: Research Reactors', IPFM [online], 4 September 2021. https://fissilematerials.org/facilities/research_reactors.html
31. International Panel on Fissile Materials, 'Facilities: Naval Reactors', IPFM, [online] 4 September 2021. https://fissilematerials.org/facilities/naval_reactors.html; Russia also operates six civilian vessels with nine reactors. Two more countries, China and France, operate naval reactors but they use LEU.
32. Frank N. von Hippel, 'The Australia–UK–U.S. Submarine Deal: Mitigating Proliferation Concerns', *Arms Control Today* [online] November 2021.
33. Christopher Fichtlscherer, Friederike Frieß, and Moritz Kütt, 'Assessing the PRISM Reactor as a Disposition Option for the British Plutonium Stockpile', *Science & Global Security*, 27/2–3 (2019), 124–49. https://doi.org/10.1080/08929882.2019.1681736
34. Paul Brown, 'UK's Plutonium Stockpile is an Embarrassing Risk', *The Energy Mix* [online] 3 September 2020; Nuclear Decommissioning Authority (NDA), 2019, 'Progress on Plutonium Consolidation, Storage and Disposition, Progress Report'. https://www.theenergymix.com/2020/09/03/uks-plutonium-stockpile-is-an-embarrassing-risk/
35. Compare, for example, Robert H. Herbst, 'Cancer of the Prostate: A Combined Surgical and Radium Method of Treatment', *Journal of the American Medical Association*, 72/22 (1919), 1610. https://doi.org/10.1001/jama.1919.02610220028010; J. M. H. MacLeod, 'Further Observations on the Therapeutic Value of Radium and Thorium', *The British Medical Journal*, 1 (1904), 1366–9. https://doi.org/10.1136/bmj.1.2267.1366
36. Ernest O. Lawrence and N. E. Edlefsen, 'On the Production of High Speed Protons', *Science*, 72 (1930), 376.
37. J. S. Mitchell, 'Applications of Recent Advances in Nuclear Physics to Medicine', *The British Journal of Radiology*, 19/228 (1946), 481–7. https://doi.org/10.1259/0007-1285-19-228-481
38. H. F. Freundlich, J. L. Haybittle, and R. S. Quick, 'Radio-iridium Teletherapy', *Acta Radiologica*, 34/1–2 (1950), 115–34. https://doi.org/10.3109/00016925009176568; Roger F. Robison, 'The Race for Megavoltage X-Rays versus Telegamma', *Acta Oncologica*, 34/8 (1995), 1055–74. https://doi.org/10.3109/02841869509127233; Francisco Comas and Marshall Brucer, 'First Impressions of Therapy with Cesium 137', *Radiology*, 69/8 (1957), 231–5. https://doi.org/10.1148/69.2.231
39. J. A. Good, 'Radiation-Absorption Type Liquid Level Indicator', Patent Number 2662985, Application Number 50908 (Alexandria: US Patent Office, 1953).
40. Geir Anton Johansen and Peter Jackson, *Radioisotope Gauges for Industrial Process Measurements* (Chichester: Wiley, 2004). https://doi.org/10.1002/0470021098
41. Mario Finkiel, 'Sterilization by Gamma Irradiation', *Tuttnauer.com* [blog], 6 December 2016. https://tuttnauer.com/blog/sterilization-by-gamma-irradiation; J. L. Roy, 'Gamma-Ray Sterilization of Medical Utensils', *Nature*, 197 (1963) 134. https://doi.org/10.1038/197134b0
42. World Nuclear Association, April 2022, 'Radioisotopes in Medicine' [online]. https://world-nuclear.org/information-library/non-power-nuclear-applications/radioisotopes-research/radioisotopes-in-medicine.aspx
43. Unauthorized removal or loss of such sources can pose significant harm to individuals—or even larger groups. In 1987, a caesium-137 source was removed from an abandoned clinic in Goiânia, Brazil. Parts of the source and the source assembly were spread over the city. It took two weeks before authorities recognized the symptoms of radiation exposure. Four people died, and hundreds had to be decontaminated. See IAEA, 'Radiation Sources: Lessons from Goiania', *IAEA Bulletin*, 30/4 (1988).
44. IAEA, *Nuclear Security Systems and Measures for the Detection of Nuclear and Other Radioactive Material out of Regulatory Control*, IAEA Nuclear Security Series, No. 21

(Vienna: IAEA, 2013). https://www.armscontrol.org/act/2021-11/features/australia-uk-us-submarine-deal-mitigating-proliferation-concerns

45. IAEA, 2022, Research Reactor Database (RRDB) [online]. https://nucleus.iaea.org/rrdb/#/home
46. Committee on Radioactive Sources, *Radioactive Sources: Applications and Alternative Technologies* (Washington, DC: The National Academies Press, 2021). https://doi.org/10.17226/26121
47. IAEA, *Strengthening Control Over Radioactive Sources in Authorized Use and Regaining Control Over Orphan Sources*, IAEA-TECDOC-1388 (Vienna: IAEA, 2004).
48. Paul-Emmanuel Goethals and Richard Zimmermann, 2015, 'Cyclotrons used in Nuclear Medicine: World Market Report and Directory', Edition 2015, MEDraysintell. As cited in IAEA, *Radiation Safety of Accelerator Based Radioisotope Production Facilities*, Specific Safety Guide, No. SSG-59 (Vienna: IAEA, 2020).
49. IAEA, *Categorization of Radioactive Material*, IAEA Safety Standards Series, No. RS-G-1.9 (Vienna: IAEA, 2005).
50. Not only is there no international registry, but also many national registers consider the number of sources sensitive information. The estimate is based on a comprehensive email survey of radiation protection agencies around the world. Only 18 countries responded. The lower value is the total of these responses; the upper value was calculated based on respondents' total gross domestic product (GDP) compared to the global GDP.
51. IAEA, 2022, DIrectory of RAdiotherapy Centers (DIRAC) [online]. https://www.iaea.org/resources/databases/dirac
52. IAEA, 2022, Incident and Trafficking Database (ITDB) [online]. https://www.iaea.org/resources/databases/itdb
53. IAEA, 2022, International Catalogue of Sealed Radioactive Sources and Devices [online]. https://www.iaea.org/resources/databases/international-catalogue-of-sealed-radioactive-sources-and-devices
54. James M. Acton, M. Brooke Rogers, and Peter D. Zimmerman, 'Beyond the Dirty Bomb: Re-thinking Radiological Terror', *Survival*, 49 (2007), 151–68. https://doi.org/10.1080/00396330701564760
55. Eckerman et al., 'Compendium'.
56. United Nations Environment Programme, *Radiation Effects and Sources: What is Radiation? What Does Radiation Do to Us? Where Does Radiation Come From?* (Nairobi: UNEP, 2016),
57. K. Kodama, K. Ozasa, and T. Okubo, 'Radiation and Cancer Risk in Atomic-Bomb Survivors', *Journal of Radiological Protection*, 32/1 (2012), N51–4. https://doi.org/10.1088/0952-4746/32/1/N51; Radiation Effects Research Foundation (RERF), 'Overview of RERF Research Programs'. https://www.rerf.or.jp/en/programs/research_activities_e/outline_e
58. H. G. Claycamp, N. D. Okladnikova, T. V. Azizova, Z. D. Belyaeva, B. B. Boecker, V. S. Pesternikova, and B. R. Scott, 'Deterministic Effects from Occupational Radiation Exposure in a Cohort of Mayak Pa Workers: Data Base Description', *Health Physics*, 79/1 (2000), 48–54. https://doi.org/10.1097/00004032-200007000-00009; Steven L. Simon and André Bouville, 'Health Effects of Nuclear Weapons Testing', *The Lancet*, 386/9992 (2015), 407–9. https://doi.org/10.1016/S0140-6736(15)61037-6
59. IAEA, *Radiation Protection and Safety of Radiation Sources: International Basic Safety Standards*, IAEA Safety Standards Series, No. GSR Part 3 (Vienna: IAEA, 2016).
60. IAEA, *Occupational Radiation Protection*, IAEA Safety Standards Series, No. GSG-7 (Vienna: IAEA, 2018).

61. IAEA, *Radiation Protection*; A. Oudiz, J. Croft, A. Fleishman, J. Lochard, G. Lombard, and G. Webb, *What Is ALARA?* (Fontenay-Aux-Roses: CEPM, 1986). https://inis.iaea.org/collection/NCLCollectionStore/_Public/19/012/19012877.pdf
62. IAEA, *Categorization of Radioactive Material.*
63. IAEA, *Strengthening Control Over Radioactive Sources in Authorized Use and Regaining Control Over Orphan Sources.*
64. US Office of Technology Assessment, *The Effects of Nuclear War* (Washington, DC: United States Congress Office of Technology Assessment, 1979).
65. Samuel Glasstone and Philip J. Dolan, *The Effects of Nuclear Weapons* (Washington, DC: US Department of Defense and the Energy Research and Development Administration, 1977).
66. P. Crutzen and J. Birks, 'The Atmosphere after a Nuclear War: Twilight at Noon', *Ambio*, 11/2–3 (1982), 114–25.
67. Alan Robock, 'Nuclear Winter', *WIREs Climate Change*, 1/3 (2010), 418–27. https://doi.org/10.1002/wcc.45
68. International Committee of the Red Cross, 'Climate Effects of Nuclear War and Implications for Global Food Production', ICRC Information Note, No. 2 (2013); Jonas Jägermeyr, et al., 'A Regional Nuclear Conflict Would Compromise Global Food Security', *Proceedings of the National Academy of Sciences of the United States of America*, 117/13 (2020), 7071–81. https://doi.org/10.1073/pnas.1919049117; Lili Xia and Alan Robock, 'Impacts of a Nuclear War in South Asia on Rice Production in Mainland China', *Climatic Change*, 116/2 (2013), 357–72. https://doi.org/10.1007/s10584-012-0475-8; Alan Robock, Luke Oman, and Georgiy L. Stenchikov, 'Nuclear Winter Revisited with a Modern Climate Model and Current Nuclear Arsenals: Still Catastrophic Consequences', *Journal of Geophysical Research: Atmospheres* [online], 112/D13 (2007), 2006JD008235. https://doi.org/10.1029/2006JD008235
69. Charles G. Bardeen, et al., 'Extreme Ozone Loss Following Nuclear War Results in Enhanced Surface Ultraviolet Radiation', *Journal of Geophysical Research: Atmospheres* [online], 126/18 (2021), e2021JD035079. https://doi.org/10.1029/2021JD035079

Suggested Reading

Acton, James M., Rogers, M. Brooke, and Zimmerman, Peter D., (2007). 'Beyond the Dirty Bomb: Re-thinking Radiological Terror', *Survival*, 49, 151–68. https://doi.org/10.1080/00396330701564760

Committee on Radioactive Sources, (2021), *Radioactive Sources: Applications and Alternative Technologies* (Washington, DC: The National Academies Press). https://doi.org/10.17226/26121

Glasstone, Samuel and Dolan, Philip J., (1977), *The Effects of Nuclear Weapons* (Washington, DC: US Department of Defense and the Energy Research and Development Administration).

Kütt, Moritz, Mian, Zia, and Podvig, Pavel, (2023), 'Global Stocks and Production of Fissile Materials, 2022', in *SIPRI Yearbook: Armaments, Disarmament and International Security* (Oxford: Oxford University Press).

Jorgensen, Timothy J., (2017), *Strange Glow: The Story of Radiation* (Princeton: Princeton University Press).

United Nations Environment Programme, (2016), *Radiation Effects and Sources: What is Radiation? What Does Radiation Do to Us? Where Does Radiation Come From?* Nairobi: UNEP.

CHAPTER 3

THE INTERFACE BETWEEN NUCLEAR SAFETY AND NUCLEAR SECURITY

JASON T. HARRIS AND SHRADDHA RANE

INTRODUCTION

NUCLEAR safety and nuclear security are two closely related disciplines as both have the common goal of protecting human life and health and the environment from the harmful effects of ionizing radiation.[1] The International Atomic Energy Agency (IAEA) defines nuclear safety as 'the achievement of proper operating conditions, prevention of accidents or mitigation of accident consequences, resulting in protection of workers, the public and the environment from undue radiation hazards'.[2] Nuclear security is defined as 'the prevention and detection of and response to, theft, sabotage, unauthorized access, illegal transfer or other malicious acts involving nuclear material, other radioactive substances or their associated facilities'.[3] The key difference between the two is that in nuclear safety, the concern of radiological risk stems from human error, equipment malfunction, internal events (such as fires), or external events (such as flooding or earthquakes). Nuclear security, on the other hand, involves intentional, malicious action taken by an individual (adversary) or group of individuals. While the initiating events that can lead to these risks may be quite different, the outcomes may be the same.

The interface of nuclear safety and security has received considerable attention in recent years. This point comes from the broader examination of industrial safety and security. Going back to 1990, in his book *Human Error*, James Reason applied the concept of safety barriers.[4] Here, he described the isolation between the hazard and its consequences as slices of Swiss cheese. Each slice represents risk controls while the holes in each slice represent failures or weaknesses in each protective barrier. In this representation, coincidental failure of several barriers designed to prevent an event from escalating to a major outcome is shown as the holes in the barriers lining up to give clear line of sight to allow the escalation. The model shows no single failure causes a major accident; rather multiple failures are needed to create such catastrophic events.

Security adds a complicating element to the model Reason describes, as protective barriers could specifically be targeted to initiate, accelerate, or directly cause a catastrophic event. An example to illustrate this point involves the safety instrumented systems of a nuclear power plant. These systems are necessary to safely operate the plant to prevent or minimize an event that could lead to the release of radiation. In the case of newer, digital systems that are often connected to a network, a targeted cyber-attack could disable or affect the performance of the safety system, possibly causing or creating a catastrophic event. Integrating security using software and hardware solutions and staff protocols and requirements, with safety systems, is essential to lower the overall risk of such an event from occurring.

Nuclear safety and security share many commonalities and thus should be treated as interrelated subjects, mutually reinforcing, and fully integrated. Without integration, the potential for a catastrophic event leading to the release of radiation could occur. While this is a rather extreme case, more subtle examples exist. For example, the response of emergency personnel to a safety-related incident in a secured area could be delayed if their access to that area is restricted for security reasons. Transportation of radioactive materials must also reconcile the need for safety, by using appropriate signage and labelling to display the hazards while driving at 'slow' speeds, with the need for security, which stresses not using signage and labelling that would reveal the contents of the transport and render it a potential target while driving at 'fast' speeds to reduce the likelihood of successful attack. These are just two examples of where safety and security may come into conflict, but the situations that can be conceived are endless. Ultimately, nuclear safety and security are interdependent, which brings up an interesting question about the relationship between these two elements: without nuclear safety, do we have nuclear security? The same question could be asked another way: without nuclear security, do we have nuclear safety?

The Need for an Integrated Approach

Benefits of Nuclear Safety and Security Integration

Several studies have lauded the benefits of having an integrated approach for the implementation of nuclear safety and security.[5,6,7] Integration enhances the safe and secure operations of nuclear facilities and ensures effective delivery of radiation protection. Such efforts provide efficiency benefits for the safe, secure, and peaceful operation of existing and future nuclear fuel cycle systems. Many solutions have been discussed about how to best integrate nuclear safety and security. For example, multiple studies have emphasized the importance of establishing joint awareness raising, education, and training programmes for stakeholders as a means of overcoming integration challenges.[8,9] Other studies have focused on how to establish effective systems for facilitating cooperation and information sharing between key national nuclear security and safety stakeholders. There is also clear benefit to achieving harmonization and promoting internationally agreed standards for nuclear security and safety.[10]

In their research, Sonal Gandhi and Jungmin Kang showed that the interface between nuclear safety and security requires integration across many elements, including legal and regulatory frameworks, responsibility, design concepts and criteria, graded approach, operations, emergency preparedness and response, and training and education.[11] As such, this integration benefits when conflicts are minimized and synergies are maximized. An integrated approach towards nuclear safety and security needs to be adopted at every level, from the international legal framework down to facility operations.

Challenges and Conflicts in Nuclear Safety and Security

Although nuclear security and safety share a common purpose, they are often implemented differently at the site level. Measures for both must be designed and delivered in an integrated manner, so that nuclear security measures do not compromise nuclear safety and, conversely, nuclear safety measures do not compromise nuclear security.[12] While integration efforts have been made by both existing and emerging nuclear countries, these have not been without some challenges at the international, national, and operational levels.

The international nuclear safety and security regimes are made up of several different instruments, with the integration of these challenging in part due to the fact that they evolved separately. Inherently, safety and security are different, and thus, these sets of international instruments differ widely in scope; they also treat safety or security largely in isolation. Safety is focused on accidental events whereas, in the case of nuclear power plants, security is aimed principally at preventing acts of sabotage that could result in releases of radioactive materials. For safety, openness to information sharing by states has led to greater cooperation. In security, however, the need to keep information confidential has resulted in challenges in designing and implementing international programmes. Security has appropriately received increased attention, but not until the aftermath of the 11 September 2001 terrorist attacks in the United States (US).

The global safety regime dates back to the early development of nuclear power. Some of the international instruments that relate directly to nuclear safety include the Convention on Nuclear Safety (1994), the Joint Convention on the Safety of Spent Fuel Management and on the Safety of Radioactive Waste Management (1997), and the Code of Conduct on the Safety of Research Reactors (2004). The focus on operational safety occurred after the nuclear reactor accidents at Three Mile Island in the US and Chernobyl in the former Soviet Union.[13] The global nuclear security regime is not as mature as its safety counterpart, beginning roughly in the 1980s with the adoption of the Convention on the Physical Protection of Nuclear Material (CPPNM) (1980). Other important instruments related to nuclear security include the United Nations Security Council Resolutions 1373 (2001) and

1540 (2004), the Amendment to the CPPNM (2005), and the International Convention for the Suppression of Acts of Nuclear Terrorism (2005).

Although the international nuclear safety and security regimes are based on several legally binding instruments, participation is mainly voluntary in respect to detailed implementation.[14] Some efforts have been made with integration through the development of legal instruments that include both nuclear safety and security, including the Convention on Assistance in the Case of a Nuclear Accident or Radiological Emergency (1986) and the Code of Conduct on the Safety and Security of Radioactive Sources (2004).[15,16] However, neither of these documents contain the breadth of requirements needed to cover all aspects of both nuclear safety and security.

At the national level, states take different approaches in establishing nuclear security and safety systems, although there exists a number of common challenges to integration. For example, it can take years, if not decades, to align separate well-established legislative frameworks for nuclear security and safety.[17] Meanwhile separate regulatory regimes and bodies may exist for nuclear safety and security. In the case of the United Kingdom (UK), parallel approaches to nuclear safety and security regulation existed for decades, until being gradually integrated in the early 2000s—a process that took over a decade to achieve.[18] Additionally, implementation is burdensome for countries with smaller facilities or those without a large national nuclear infrastructure. Countries that only possess radioactive sources held in hospitals, universities, and research institutions may not see a need for integration.

On the part of operators or licensees, reluctance for integration tends to stem from the greater attention traditionally given to nuclear safety, negative perceptions about safety and security integration, and poor nuclear safety and security cultures. This reluctance is compounded by the fact that operators may not be ready or willing to invest in non-technological solutions like staff training, examinations, and assessments due to limited resources and expertise. Nuclear safety and security are typically separate and competing departments within an organization, with personnel from each often having different backgrounds, expertise, and technical languages.[19] Consequently, there may be no direct connections, responsibilities, or methods for readily communicating key information. When it comes to nuclear safety and security training, this may be done in silos, without consideration given to the other discipline.

Recommendations at a Holistic Level

Sustainable Traits for the Nuclear Safety and Security Interface

Interface management is based on the tenet that products or organizations should be considered holistically to achieve the best organizational outcomes, with comprehensive recommendations that cut across key regulatory, operational, and cultural elements. Organizations that operate complex, large-scale technological infrastructure can thrive when management understands how each organizational element contributes to and affects the whole. Being mindful of crucial interfaces between parts of the organization enables everyone to work towards common goals. Continually improving the integration of nuclear safety and security should remain a high priority, with the recognition that adopting a holistic approach to nuclear safety and security generally provides additional opportunities to identify and exploit potential synergies between the two disciplines.

For the nuclear security and safety interface to play its proper role, it must be sustained over time at three distinct levels: the strategic level, the operational level, and the cultural level. The different levels of the interface must work together in a consistent and complementary manner to promote functional excellence. At the strategic level, countries and their competent authorities must address the elements of the legislative and regulatory framework to strengthen the confluence between nuclear safety and security. It is the responsibility of leadership at both the national and facility levels to establish overarching policy and strategy to attain a unified management system for the two regimes.

Guidance on how this can be achieved can be found within the IAEA report 'The Interface between Safety and Security at Nuclear Power Plants',[20] and the best practice guide of the World Institute for Nuclear Security (WINS), 'An Integrated Approach to Nuclear Safety and Nuclear Security'.[21] These documents recommend establishing objectives and expectations, led by management, to address the strategic issues of policy, governance, and responsibilities towards integrating nuclear safety and security. To ensure the adequacy of emergency response system, the documents also emphasize the importance of maintaining a close relationship between safety and security regulators—where security regulators are strongly encouraged to be included in all safety-related legislative meetings and forums as well as inspections and other reporting arrangements. The development of integrated systems may also be beneficial within other areas of operations, such as emergency management, job performance review, and independent oversight and assurance.

The 2011 Fukushima Daiichi nuclear disaster in Japan, and the authorities' subsequent response to this incident, demonstrated the importance of a holistic approach to nuclear safety and security regulation. In this case, public trust in the ability to assess risk associated with nuclear power plummeted significantly. The Fukushima accident underscored the importance of regulatory independence from any economic interests, policy interests, and political interests, as well as the need for a more integrated approach to nuclear safety and security. The lack of security in the immediate aftermath of the Fukushima crisis is also an example that highlights the need for harmonizing accident and incident reporting and planning for such combined nuclear safety and security dangers. In response to the accident, Japan fundamentally restructured its nuclear regulatory framework, establishing the Nuclear Regulatory Authority (NRA) in 2012 in an effort to bring together safety and security regulation and to promote independence and transparency.

Because of the close correlation between the two disciplines of safety and security, many countries see advantages in having a single regulator responsible for both (or at least a more integrated regulatory approach). French regulatory officials, for instance, have expressed interest in having safety and security fall under the same nuclear regulatory body.[22] Currently, the independent French regulatory agency Nuclear Safety Authority (ASN) is responsible for nuclear safety and oversight of nuclear security (for example, security of radioactive materials, including transportation of such materials), but not for oversight of physical security of nuclear power plants. The regulation and control of monitoring physical protection is the responsibility of its technical support organization, the Institute for Radiological Protection and Nuclear Safety (IRSN). Integration is currently managed through biannual meetings between the Department of Nuclear Security and ASN to communicate any findings related to security observed during their inspections on safety issues and to improve upon other safety and security coordination issues.[23] Sweden has also taken steps in this direction, with the regulatory bodies forming a joint industry initiative coordination group, the KSKG (Kärnkraftss äkerhetskoordineringsgrupp), to coordinate critical nuclear safety and security issues, conduct stress tests, and work on other upcoming regulatory requirements.[24]

National efforts aimed at assessing nuclear risks for the purpose of security and safety planning should be integrated. Expanding and promoting a shared database of design basis threats (DBT) and design-based accidents (DBA) should also be encouraged.[25] For example, inter-departmental meetings between groups of nuclear safety and nuclear security managers—in an effort to share ideas, experiences, and corrective actions—can enhance management process integration. This might be further supported through joint operational-level guidance and training, which cuts across processes and procedures, and joint exercises and learning opportunities to create mutual understanding and to help build relationships between different groups.[26]

The IAEA technical report 'The Nuclear Safety and Nuclear Security Interface: Approaches and National Experiences' summarizes the views expressed by participating countries on the safety and security interface.[27] This report, consistent with best practice guides produced by WINS, encourages developing performance indicators for monitoring the effectiveness of the nuclear safety and security interface and conducting regular meetings between security, engineering services, and maintenance staff to monitor progress and address emerging issues. It is also recommended that safety operational issues like radiation protection, emergency planning, fire protection, environmental monitoring, and chemical safety are discussed jointly between safety and security staff. To ensure that balanced attention is given to both parties, the organization should find ways to cross-train the nuclear safety and security staff under the same management line. A centralized information system and command centre would benefit the operators' direct operations during a safety or security event.

All facility staff and other personnel who work on site should receive security training to educate them about the credible threats and associated protection measures. Devising a combined training and assistance programme with collaboration by both safety and security experts is a good practice that can support integration. This can be further enhanced by appropriate cross-sector knowledge and information transfer. For example, France's ASN requests opinions and recommendations from various advisory safety and security committee experts, in different areas of reactor, waste, nuclear equipment, etc. With these insights, the ASN prepares conclusions based on operational events and shares operating experience within the nuclear industry.[28] Similarly, the US Nuclear Regulatory Commission (NRC) through internal processes, safety and security advisory panels, and working groups ensures that its activities in one discipline (safety or security) do not adversely affect the other. The US NRC also requires licensees to consider safety and security implications of all aspects of nuclear power plant operation and encourages staff members from both disciplines to meet periodically to identify, discuss, and resolve potential interface issues.[29]

It is also essential to pursue integration at the cultural level of a nuclear organization, which includes the values and beliefs of employees, shared norms, behaviours, and practices.[30] Safety and security are based on the same principle of limiting risk and therefore it should be possible to develop organizational cultures that reinforce both in a mutually supporting manner. However, thanks to lessons learned from major safety events such as the accidents at Chernobyl and Three Mile Island, safety culture is a more embedded concept than security culture and as such might take primacy in situations when they come into conflict. Consequently, organizations may need to focus attention on how security and safety cultures interact through, for example, in-depth cultural assessments.

Promoting periodic nuclear safety and security culture self-assessments can enable all facility personnel to develop and maintain a basic understanding and awareness of important knowledge that may be outside their areas of specialization, such as: (a) security threats to the facility; (b) security plans, arrangements, and measures that address the threat; and (c) safety issues at the facility and how safety is ensured and maintained.[31] These activities may require a significant investment of resources for personnel to develop unified perceptions and views about their responsibilities and appropriate behaviours with respect to the interface of safety and security. Here, certain personnel who are appropriately qualified and experienced in both safety and security, such as health physicists, can be best positioned to involve and encourage staff in making use of the site corrective action programmes to report safety- and security-related deficiencies or threats. By using all three approaches, the disciplines of nuclear safety and nuclear security can integrate the various contributory elements of the interface, helping to meet the overall strategic objectives of the organization.

Safety and Security Integration over the Lifecycle of a Nuclear Facility

Incorporation of good practices to strengthen the nuclear security and safety interface in the early stages of nuclear design and installation provides an excellent opportunity to minimize vulnerabilities and develop an effective interface that cuts across both disciplines. Integration between security and safety should begin at the conceptual design phase and extend through the life of the facility. Examples of good practices in specific areas that support this integration are described here.[32,33]

Siting, Design, and Commissioning

The selection of a nuclear installation site needs to be based on both safety and security criteria. This should include screening that takes into account a facility's proximity to populated areas, topographic obstructions, public transport infrastructure, high crime areas, sites of political unrest, high hazard installations, and any other relevant issues. During all phases of construction, there should be careful oversight to ensure that safety and security systems are assembled without latent defects. Construction process safety is vital to preventing accidents that could result in a radioactive release event. Construction site security should be maintained to prevent security breaches that could result in the theft of materials or the malicious introduction of structural defects.

Special attention should also be paid to the interface between safety and security measures during the construction and commissioning period, as a large and diverse population of workers other than the facility staff may be working in the vicinity of the nuclear installation. Here, closely coordinated joint working between facility staff, contractors, and safety and security personnel will decrease the potential for incorrect sequencing, programme delays, and repeated testing—and help optimize test schedules for key safety-related and security-related readiness activities. This should be underpinned by a graded approach that seeks to provide the greatest protection to those targets with the highest potential consequences.

Operation, Maintenance, and Inspections

The operating period is the longest nuclear facility lifetime phase in which the safety and security interface needs to be managed and sustained. For a nuclear power plant, this might be between 20 and 40 years.[34] Due to the presence of fresh, irradiated reactor fuel, inventories of many radioactive and nuclear materials, and the numerous operational experimental hazards, the operational phase is also the phase during which safety and security risks are arguably the greatest. Maintenance operations, including surveillance activities and inspections, should be carried out on a regular basis to ensure compliance.

Compensatory measures should be initiated whenever safety or security resources are rendered unavailable, for example, when electrical maintenance requires shutting off power to an area, possibly compromising alarm or surveillance systems.[35] To address such issues, each operating facility may require both compensatory safety–security measures and an inventory of communication interfaces between departmental levels. It is also crucial at each shift change that all incoming personnel are briefed on any safety or security concerns that are likely to emerge during their shift due to recent changes to the operating activities or plans of the facility.

A 'one-size-fits-all' approach to manage or address safety–security interface is neither equitable nor practical. The approach should be risk based and tailored to the facility's needs, culture, capabilities, and resources. For the mutual benefit of safety and security, the facility should engage in frequent reviews and inspections where safety and security specialists are jointly consulted to ensure that interface issues are appropriately addressed; these activities can document best practices and enforce appropriate penalties or other responses for non-compliance. Each of these processes should include targeted inspection guidance, with inspections conducted according to an integrated inspection plan. Establishing joint assessments and inspections by mixed teams of safety and security staff encourages collaboration and greater breadth of knowledge.

Transport and Storage

Balancing transparency and confidentiality concerns in the delivery of safety and security measures during both normal and emergency situations, along with strengthening the coordination of safety and security elements between relevant authorities, are some of the challenges observed in relation to the transport and storage of nuclear and radioactive materials. Integrated assistance must also be provided to address a lack of coherence associated with the national transport nuclear security regulations and internationally harmonized dangerous goods regulations.[36]

These issues and how to combat them are outlined in an IAEA technical guidance 'Managing the Interface between Safety and Security for Normal Commercial Shipments of Radioactive Material', which focuses on encouraging dialogue and agreement between the competent authorities and the operators undertaking shipments.[37] The guidance recommends an adaptive process that can be tailored to fully address and bridge the need for harmonized transport safety and security regulations. To overcome the potential challenges posed by overlapping multiple governmental agency jurisdictions, a conceptual framework based on 'shared regulatory space' is suggested. The assignment of roles and responsibilities between the different agencies should be well coordinated to ensure a smooth safety and security framework.

Emergency Response

An integrated approach to contingency planning can help an organization respond effectively to an emergency, ensuring that all critical functions are performed in a timely manner. This approach supports strategies for rapidly determining the origin of events and deploying appropriate first responders (safety personnel, security forces, or a combination of both), thereby reducing any conflicts or delays. During a response to a security breach, inadequate training and awareness may lead security forces to be unnecessarily exposed to radiation hazards or cause unintentional damage to safety-related system controls.

An organizational crisis occurred when the Cerro Grande wildfire of May 2000 threatened nuclear security operations and nuclear storage facilities at Los Alamos National Laboratory (LANL) in the US. Emergency response and preparedness by LANL failed at many levels as the crisis came as a surprise and developed rapidly, requiring quick decision-making skills and solutions. The lack of detailed contingency plans and performance testing, overlapping lines of authority, and ineffective leadership were some of the challenges that the facility faced while responding to a fire of this magnitude.[38] It is good practice for security forces to receive periodic training, including facility walk-throughs on the safety aspects of the facility, and vice versa.

Decommissioning

The interface between safety and security should continue to be vigilantly managed by the facility even when it enters a decommissioning phase. Challenges that have occurred during construction and commissioning may re-emerge during the decommissioning stages of an installation. Indeed, both nuclear safety-related risks and nuclear security-related risks tend to increase when a facility undergoes major changes.[39] This may occur from dismantling activities and the removal of physical protection systems confining radioactive materials.

To minimize the risk of inadvertent criticality in the reactor core during the decommissioning phase, it is a common practice to store either a partial amount or the entire amount of fuel elements in the reactor pool or the spent fuel storage pool. These safety provisions may, however, increase the vulnerability of the facility from a security perspective, both because of limited staff in attendance and because the fuel may be more attractive to theft due to its accessibility and decreased radioactivity. To ensure adequate attention is given to both safety and security, such situations need to be addressed in coordination between safety and security specialists. Furthermore, emergency and contingency plans should be updated and made consistent with the conditions of the reactor shutdown. In addition to the aforementioned recommendations, regularly measuring and monitoring the status of the nuclear safety and security cultures using independent oversight, peer reviews, and staff surveys helps contribute to successful integration.

Implementation of a Hybrid Top-down and Bottom-up Approach

The key to successful implementation of best practice guidelines and lessons learned on integrating nuclear safety and security is to place emphasis on leadership and management systems. Effective planning, established at the strategic level and implemented at the direction of the relevant authorities, translates national-level priorities into a set of objectives and actions at the operational and cultural levels in the organization. In a top-down approach, the whole system is successively partitioned into a hierarchy of subsystems. At the highest level, global objectives are initially defined and then gradually communicated to the lower subsystems of the hierarchy to be developed and specified. By comparison, the bottom-up approach is where relatively narrow goals are initially set at the lower levels of the organizational hierarchy. The ideas and perceptions from employees at lower levels of the organization are then gradually integrated into the global goals and strategy framework of the higher levels.

Implementing a hybrid blend of top-down and bottom-up elements to manage the interface of nuclear safety and security has the benefit of fostering continuous bi-directional planning and minimizing the differences and disconnects that may occur between the leadership and the rest of the facility staff. For example, Scotland's and Ireland's healthcare regulators—Healthcare Improvement Scotland (HIS) and the Health Information and Quality Authority (HIQA), respectively—are configured to deliver quality healthcare improvement support through a hybrid of top-down deterrence and bottom-up persuasive approaches. Top-down approaches include deterring breaches of protocol and enacting penalties where they occur (e.g. monitoring and accountability); bottom-up approaches include capacity building, staff empowerment, and persuasive encouragement of adherence to guidance.[40]

To facilitate this hybrid framework, the leadership at regulatory bodies and facility managers should take prime responsibility for making safety and security work coherently. Both safety and security are built on a regulatory framework pertaining to a set of functional components of management systems: management responsibility, quality management, performance assessment management, and resource management. For each of these functional components, decisions should be made about how to coordinate safety and security. Experts in one of either the safety or the security fields may be expected to show a bias in favour of their own field when they contribute to decisions that involve unavoidable trade-offs between safety and security objectives. Safety professionals may tend to emphasize safety-related procedural management of reactor operations, fuel handling and core management, and maintenance of components important to safety and emergency preparedness. By contrast, security professionals may tend to emphasize personnel security, informational and cyber security, access controls, security event reporting, and security-related management issues. Bearing in mind that some actions undertaken to advance one purpose could adversely affect the other, any decisions about implementing any safety- or security-related changes at the facility should be supported by the equal consultation of experts in each discipline.

In addition to management using a systematic process for evaluating and implementing changes, it is also equally important for employees at different levels of the organization to participate and engage in an iterative process of continuous improvement. Employee feedback mechanisms give personnel an opportunity to provide upper management with insights from a more personal and targeted level. The hybrid management framework calls for the induction and training of personnel representing each level of the organizational hierarchy to participate in decision making with an eye to both safety and security, ensuring that an optimal balance of safety and security objectives is achieved.

Integrated Safety and Security Risk Models

Risk models play an increasingly important role in dealing with both unintentional/non-malicious and intentional/malicious threats. However, the delivery of integrated safety and security risk models by industry has been slow, predominantly because of the different approaches taken by safety and security practitioners. In recent years, researchers from the two disciplines of safety and security have been working in conjunction to identify risk initiating events, estimate their likelihoods and impacts, and then devise frameworks to prevent, mitigate, and assess safety- and security-related risk.[41,42] For example, there have been integrated safety and security risk assessment methods developed for automotive cyber physical systems, where a combined analysis method is applied to intelligent and cooperative vehicles to examine attack possibilities and failure scenarios.[43]

In the nuclear safety discipline, deterministic and probabilistic assessment methodologies have been traditionally used to estimate the frequency of accidents and failures, as well as to analyse the accident sequence with event and fault trees. Probabilistic safety assessment (PSA) has contributed significantly to the understanding of how best to ensure the safety of nuclear power plants.[44] The US NRC uses the science of probabilistic risk assessment (PRA) to examine the extent of a system's risk and identify problems that could have most impact on safety. However, safety-focused risk analysis and assessment approaches struggle to adequately include malicious and intentional acts against the nuclear power industry. Unlike the evolution of safety PRA methods, security assessments have yet to achieve the level of consensus that is required to accept a standard set of approaches.[45]

Furthermore, adversary modelling presents a substantial challenge to security risk assessments that utilize the canonical triplet definition of risk: the product of threat, vulnerability, and consequences.[46,47] Factors that increase the difficulty of applying the triplet definition of risk to nuclear security include the large uncertainty in the adversary attack probability and the dynamic nature of intelligent adversaries who can adapt to defensive measures. Unlike safety, security is a shared responsibility of facility, national and regional government, law enforcement, among others—increasing interdependencies and thus creating a complex non-linear decision-making system that may challenge classic PRA logic.

Despite the many challenges in security risk analysis, especially with the paucity of information about the occurrence or likelihood of an event, various types of possible security risk assessment approaches have been proposed. Garrick and colleagues advocate the use of PRA for assessing terrorism risk.[48] Elisabeth Pate-Cornell presents a model for setting priorities among threats and countermeasures based on PRA, decision analysis, and elements of game theory.[49] Meanwhile, the authors of this chapter Shraddha Rane and Jason T. Harris have developed a quantitative risk-based methodology that radiological facilities can employ to conduct self-assessments and gain a better understanding of the threats that they face.[50]

Because of the different development processes and maturity levels of nuclear safety and security risk assessment methodologies, an integrated risk framework is appropriate to build better risk models. The independent nuclear safety and security risk model, at best, may not take explicit advantage of measures that provide benefits against multiple risk domains, and, at worst, it may lead to applications that increase overall risk due to incongruities.[51] Productive efforts aimed to quantify the possible aspects of nuclear safety and nuclear security integration are currently underway. The vulnerability evaluation simulating plausible attacks (VESPA) approach leverages similarities across safety, security, and safeguards (3S) to develop a 'what if', scenarios-based comparative tool for the selection of plausible targets that, from the perspective of a terrorist, shows which vulnerabilities have a high potential to be exploited.[52]

Another framework called dynamic probabilistic risk assessment (DPRA) uses dynamic event trees that evaluate 3S risk by dynamically evaluating the interactions between uncertainties in real simulation time.[53] The UK's Office for Nuclear Regulation (ONR) uses the 'bow-tie method'—a structured approach, shaped like a bow-tie, creating a clear differentiation between proactive and reactive risk management[54]—as one of its methodologies to improve the visibility and understanding of safety cases, complementing the other more rigorous safety analysis techniques used in the nuclear industry.[55] The bow-tie method, for example, if extended to include security and safety synergistically, may help control the critical central (or top) event by putting in place safety and security primary barriers on the left side of the top event, while mitigating the escalation of consequences into a large safety- and security-related catastrophe on the right side of the top event if prevention measures are breached.

Research Needs and Future Considerations

The integration of nuclear safety and security is essential, and efforts to coordinate these two disciplines are beneficial to both. There is a growing need for both an international consensus on policy analysis and quantitative risk-based research on how to continually work and improve the nuclear safety and security infrastructure. With respect to the degree to which nuclear safety and nuclear security are to be integrated, special attention is needed in relation to the following areas.

Integrated Framework

The fundamental question of how to formally integrate the nuclear safety and security risk assessment techniques remains. Further research is needed to understand safety and security trade-offs objectively and to integrate several distinct analyses into a large risk assessment and decision-making framework. Sandia National Laboratories, for example, proposes to operationalize a unified analysis framework—potentially in terms of an extended probabilistic risk analysis (ePRA) to understand overall risks across the domains of safety, security, and safeguards.[56]

Interface Metric

A central well-defined, measurable, and actionable metric is needed for integrated assessments. Creation of metrics necessitates a value judgement between safety and security, while carefully considering implicit trade-offs between the two disciplines. For example, the Indian government has recently introduced 19 key performance indicators for security and safety to grade chemical and petrochemical industries based on their practices.[57] Likewise, a score tool for the interface of safety and security would help facility staff work in conjunction with external inspectors or regulators to identify all available options for integration and select those offering the greatest benefits for implementation.

Stakeholder Engagement and End-User Application

Users or customers like to be heard, particularly when their business is impacted by a decision where their input has not been taken into consideration. The best way to improve external stakeholder satisfaction and increase interoperability and usage among end users is to regularize dialogue and interaction among all stakeholders: nuclear operators, regulators, international organizations, and policymakers. Further research is needed on developing a stakeholder input forum that would meet periodically to advance nuclear governance and safe and secure plant operations. And more research is needed to consider both technical and social value judgements in the integrated risk assessment and management process to bridge the gap between the actual risk of harm and society's perception of the risk of harm. Possibly, misconceptions about the risk of harm would take hold in the sphere of public opinion due to incorrect or misinformed commentaries on published reports from integrated nuclear safety and security risk metrics; this problem could be addressed by institutional responses to educate the public about the actual levels of existing risk and risk mitigation measures.

Conclusion

Nuclear safety and security have developed as two distinct disciplines for many years. Despite their differences, nuclear safety and security share many commonalities. It is only practical and logical to reconcile and leverage their similarities first, and then move into integrating differences to develop an integrated nuclear safety and security framework. Deploying a unified platform for safety and security would develop the level of organizational resilience necessary to adequately address an environment with dynamic risk. The dynamic risk presents itself in the form of a rapid change in threat level emerging from an unpredictable event, where the ultimate harm is different than the initially expected harm.[58] Having radiological and nuclear organizations align their elements of safety and security will help them to both manage threats and hazards on multiple fronts and to optimize responses to an unfolding crisis.

The combination of increased harmonization of nuclear safety and security at the regulatory level, transparent and free exchange of information across all departmental lines, and regular monitoring of safety and security cultures may assist an organization to harness a centralized platform that coordinates consistent and swift emergency responses. Organizations must be prepared to face continually evolving risks. And to keep ahead, organizations must become more proactive in addressing the need for integration without compromising either nuclear safety or nuclear security.

Notes

1. IAEA, *The Nuclear Safety and Nuclear Security Interface: Approaches and National Experiences*', IAEA Technical Reports Series, No. 1000 (Vienna: IAEA, 2021).
2. IAEA, *Management of the Interface between Nuclear Safety and Security for Research Reactors*, IAEA TECDOC Series, No.1801 (Vienna: IAEA, 2016).
3. Ibid.
4. John Reason, *Human Error* (Cambridge: Cambridge University Press, 1990).
5. Kenneth Luongo, Sharon Squassoni, and Joel Wit, 'Integrating Nuclear Safety and Security: Policy Recommendations' (Washington, DC: Center for Strategic and International Studies, 2011), 1–6. https://www.csis.org/analysis/integrating-nuclear-safety-and-security-policy-recommendations. http://csis.org/files/publication/111214_Integrating_Nuclear_Safety_and_Security_Memo_2.pdf
6. Solymosi Máté, 'Assessing and Enhancing Nuclear Safety and Security Culture for Small Facilities that Handle Radioactive Material', *International Journal of Nuclear Security*, 3/1 (2018), article 12. https://doi.org/10.7290/ijns030112

7. WINS, 'An Integrated Approach to Nuclear Safety and Nuclear Security', WINS International Best Practice Guide' (Vienna: WINS, 2016). https://www.wins.org/document/4-2-managing-interfaces-between-nuclear-safety-and-security-at-nuclear-facilities/
8. Jason Harris and Edward Waller, 'Initiatives to Integrate Nuclear Security with Radiation Protection Education', in *ETRAP 2017 Conference Proceedings 30 May–2 June 2017, Valencia, Spain* (Brussels: European Nuclear Society, 2017), 45–52. https://www.euronuclear.org/download/etrap2017-proceedings/#
9. Edward J. Waller and Jim Van Maanen, 'The Role of the Health Physicist in Nuclear Security'. *Health Physics*, 108/4 (2015), 468–76. https://doi.org/10.1097/HP.0000000000000258
10. M. Suzuki, Y. Izumi, T. Kimoto, Y. Naoi, T. Inove, and B. Hoffheins, 2008, 'Investigating 3s Synergies to Support Infrastructure Development and Risk-Informed Methodologies for 3s by Design', IAEA-CN-184/64, IAEA Symposium on International Safeguards: Preparing for Future Verification Challenges (Vienna: IAEA, 2008), 1–8. https://inis.iaea.org/collection/NCLCollectionStore/_Public/42/081/42081506.pdf
11. Sonal Gandhi and Jungmin Kang, 'Nuclear Safety and Nuclear Security Synergy', *Annals of Nuclear Energy*, 60 (2013), 357–61.
12. IAEA, 'The Nuclear Safety and Nuclear Security Interface: Approaches and National Experiences'.
13. Pierre Tanguy, *Three Decades of Nuclear Safety*, IAEA Bulletin 2/1988: Special Report (Vienna: IAEA, 1988).
14. Paul Murphy and Ira Martina Drupady, 'International Conventions and Legal Frameworks on Nuclear Safety, Security, and Safeguards', in Victor Nian, ed., *Advanced Security and Safeguarding in the Nuclear Power Industry: State of the Art and Future Challenges* (London: Academic Press, 2020), 175–205. https://doi.org/10.1016/b978-0-12-818256-7.00007-6
15. IAEA, *Convention on Assistance in the Case of a Nuclear Accident or Radiological Emergency*, INFCIRC 336 (Vienna: IAEA, 1986).
16. IAEA, *Code of Conduct on the Safety and Security of Radioactive Sources* (Vienna: IAEA, 2004).
17. Gandhi and Kang, 'Nuclear Safety and Nuclear Security Synergy'.
18. M. Sims, 'ONR's Experiences from Adopting an Outcome Focused Approach to Civil Nuclear Security Regulation', Paper #303: International Conference on Nuclear Security (ICONS) (Vienna: IAEA, 2020).
19. Gandhi and Kang, 'Nuclear Safety and Nuclear Security Synergy'.
20. IAEA, *The Interface between Safety and Security at Nuclear Power Plants: A Report by the International Nuclear Safety Group*, INSAG Series, No. 24 (Vienna: IAEA, 2010).
21. WINS, 'An Integrated Approach to Nuclear Safety and Nuclear Security'.
22. US Government Accountability Office (GAO), *Countries' Regulatory Bodies Have Made Changes in Response to the Fukushima Daiichi Accident*, GAO-14-109 (Washington, DC: US Government Printing Office, 2014).
23. IAEA, Integrated Regulatory Review Service (IRRS)—Mission to The Republic of France, IAEA-NS-IRRS-2014/09 (Vienna: IAEA, 2014).
24. Swedish Radiation Safety Authority, 'Sweden's Eighth National Report under the Convention on Nuclear Safety', Swedish Government Official Reports (SOU) and the Ministry Publications Series Ds 2019:16 (Stockholm: Swedish Government Offices, 2019). https://www.regeringen.se/4adae6/contentassets/c8c431c94efb4c4abefb38ca36272b5a/swedens-eighth-national-report-under-the-convention-on-nuclear-safety-ds-201916.pdf
25. IAEA, 'The Interface between Safety and Security at Nuclear Power Plants'.

26. WINS, 'An Integrated Approach to Nuclear Safety and Nuclear Security'.
27. IAEA, 'The Nuclear Safety and Nuclear Security Interface: Approaches and National Experiences'.
28. IAEA, 'Integrated Regulatory Review Service (IRRS)—Mission to The Republic of France'.
29. IAEA, n.d., 'Interfaces with Nuclear Security', Integrated Regulatory Review Service Mission to the United States. Module 12 (Vienna: IAEA, 2014). https://www.nrc.gov/docs/ML1125/ML112510464.pdf
30. Edgar H. Schein, *Organizational Culture and Leadership*, 3rd edn (San Francisco: Jossey-Bass, 2004).
31. IAEA, 'Management of the Interface between Nuclear Safety and Security for Research Reactors'.
32. Ibid.
33. IAEA, 'Managing the Interface between Safety and Security for Normal Commercial Shipments of Radioactive Material'.
34. Stanislav Novak and Milan Podest, 'Nuclear Power Plant Ageing and Life Extension: Safety Aspects: An Overview of Issues and the IAEA's Symposium', *IAEA Bulletin*, 4 (1987), 31–3. https://www.iaea.org/sites/default/files/29402043133.pdf
35. IAEA, 'The Interface between Safety and Security at Nuclear Power Plants'.
36. IAEA, 'The Nuclear Safety and Nuclear Security Interface: Approaches and National Experiences'.
37. IAEA, 'Managing the Interface between Safety and Security for Normal Commercial Shipments of Radioactive Material'.
38. Geoffrey Chapman, Rebecca Earnhardt, Christopher Hobbs, Nickolas Roth, Daniel Salisbury, Amelie Stoetzel, and Sarah Tzinieris, *Nuclear Security in Times of Crisis* (London: Centre for Science & Security Studies, 2021).
39. IAEA, 'Management of the Interface between Nuclear Safety and Security for Research Reactors'.
40. Aoife M. McDermott, Lauren M. Hamel, David Steel, Patrick C. Flood, and Lorna McKee, 'Hybrid Health Care Governance for Improvement? Combining Top-Down and Bottom-Up Approaches to Public Sector Regulation', *Public Administration*, 93/2 (2015), 324–44.
41. Yean-Ru Chen, Sao-Jie Chen, Pao-Ann Hsiung, and I-Hsin Chou, 'Unified Security and Safety Risk Assessment: A Case Study on Nuclear Power Plant', in *TSA 2014, First International Conference on Trustworthy Systems and Their Applications*, 9–10 June, *Taichung, Taiwan* (New York: IEEE, 2014), 22–8.
42. Sabarathinam Chockalingam, Dina Hadžiosmanović, Wolter Pieters, André Teixeira, and Pieter van Gelder, 'Integrated Safety and Security Risk Assessment Methods: A Survey of Key Characteristics and Applications', in Grigore Havarneanu, Roberto Setola, Hypatia Nassopoulos, and Stephen Wolthusen, eds., *Critical Information Infrastructures Security* (London: Springer, 2017), 50–62. https://doi.org/10.1007/978-3-319-71368-7_5
43. Christoph Schmittner, Thomas Gruber, Peter Puschner, and Erwin Schoitsch, 'Security Application of Failure Mode and Effect Analysis (FMEA)', in Andrea Bondavalli and Felicita Di Giandomenico (eds.), *Computer Safety, Reliability, and Security, 33rd International Conference, SAFECOMP 2014, Florence, Italy, 10–12 September* (Cham: Springer, 2014), 310–25.
44. IAEA, *Probabilistic Safety Assessment*, INSAG Series, No. 6 (Vienna: IAEA, 1992).
45. Ibid.
46. Ibid.

47. John B. Garrick, James E. Hall, Max Kilger, John C. McDonald, Tara O'Toole, Peter S. Probst, Elizabeth Rindskopf Parker, et al., 'Confronting the Risks of Terrorism: Making the Right Decisions', *Reliability Engineering and System Safety*, 86/2 (2004), 129–76.
48. Ibid.
49. Elisabeth Paté-Cornell, 'Risk and Uncertainty Analysis in Government Safety Decisions', *Risk Analysis*, 22/3 (2002), 633–46.
50. Shraddha Rane and Jason Harris, 'Development of a Potential Risk Index for Radiological Security', *Risk Analysis*, 41/8 (2020), 1257–73. https://doi.org/10.1111/risa.13625
51. Robert Forrest, Jason C. Reinhardt, Timothy Wheeler, and Adam D. Williams, 'Summary of a Sandia National Laboratories Workshop on Extended Probabilistic Risk Assessment (ePRA)', Report No. SAN2018-6151C (Albuquerque, NM: Sandia National Lab, 2018). https://www.osti.gov/servlets/purl/1526342
52. Antonio Cipollaro and Guglielmo Lomonaco, 'Contributing to the Nuclear 3Ss via a Methodology Aiming at Enhancing the Synergies between Nuclear Security and Safety', *Progress in Nuclear Energy*, 86 (2016), 31–9. https://doi.org/10.1016/j.pnucene.2015.09.013
53. Elena Arkadievna Kalinina, Adam David Williams, and Douglas Osborn, 'Final Results from a System-Theoretic Framework for Mitigating Complex Risks in International Transport of Spent Nuclear Fuel', 14th Probabilistic Safety Assessment and Management (PSAM14), Los Angeles, 16–21 September (2018). https://www.osti.gov/servlets/purl/1592056
54. Alex de Ruijter and Frank Guldenmund, 'The Bowtie Method: A Review', *Safety Science*, 88 (2016), 211–18.
55. Marc Vannerem, 'Bow Tie Methodology: A Tool to Enhance the Visibility and Understanding of Nuclear Safety Cases', Report no. NEA/CSNI/R(2012)4. OECD/NEA WGFCS Workshop (Bootle: Office for Nuclear Regulation, 2012). https://inis.iaea.org/collection/NCLCollectionStore/_Public/46/114/46114462.pdf?r=1
56. Forrest et al., 'Summary of a Sandia National Laboratories Workshop on Extended Probabilistic Risk Assessment (ePRA)'.
57. Chemicals Legislation in the World, 'Current Chemicals Legislative Framework in India', 2013. https://chemycal.com/dap/files/MCM%20World/201405_MCM-world_india.pdf
58. Dustin Radtke, 'Managing Dynamic Risk: How Your Campus Can Achieve Organizational Resilience in 2022', *Campus Security & Life Safety* [online], 22 March 2022. https://campuslifesecurity.com/articles/2022/03/22/how-your-campus-can-achieve-organizational-resilience-in-2022.aspx

Suggested Reading

Cipollaro, Antonio and Lomonaco, Guglielmo, (2016), 'Contributing to the Nuclear 3Ss via a Methodology Aiming at Enhancing the Synergies between Nuclear Security and Safety', *Progress in Nuclear Energy*, 86, 31–9. https://doi.org/10.1016/j.pnucene.2015.09.013

Gandhi, Sonal and Kang, Jungmin, (2013), 'Nuclear Safety and Nuclear Security Synergy', *Annals of Nuclear Energy*, 60, 357–61.

IAEA, (2010), *The Interface between Safety and Security at Nuclear Power Plants: A Report by the International Nuclear Safety Group*, INSAG Series, No. 24 (Vienna: IAEA).

IAEA, (2016), *Management of the Interface between Nuclear Safety and Security for Research Reactors*, IAEA TECDOC Series, No.1801 (Vienna: IAEA).

IAEA, (2021), *The Nuclear Safety and Nuclear Security Interface: Approaches and National Experiences*, IAEA Technical Reports Series, No 1000 (Vienna: IAEA).

PART II

EVOLUTION OF NUCLEAR SECURITY THREATS

CHAPTER 4

THE RISK OF NUCLEAR TERRORISM

JOHN MUELLER

Introduction

Nuclear weapons, the most destructive implements of destruction ever invented, have inspired great anguish over the last three-quarters of a century. In 1979, the prominent political scientist Hans J. Morgenthau was heard to proclaim, 'The world is moving ineluctably towards a third world war—a strategic nuclear war. I do not believe that anything can be done to prevent it.'[1] However, it could be argued in contrast that, whatever their impact on rhetoric, strategic theorizing, defence budgets, and political posturing, nuclear weapons have had (at best) a quite limited effect on history, have been a substantial waste of money and effort, do not seem to have been terribly appealing to most states that do not have them, and are unlikely materially to shape much of our future—and it would be possible as well to point out that, although no one has been killed by a nuclear explosion since 1945, hundreds of thousands of people have perished from wars and economic sanctions designed to oppose nuclear proliferation.[2]

Nonetheless, alarm about the weapons persists, and much of it has focused on the possibility that small stateless groups could obtain nuclear weapons and then set them off. Accordingly, atomic bombmaker J. Robert Oppenheimer contended in 1946 that if three or four men could smuggle in units for an atomic bomb, they could 'blow up New York'[3]—although, while expanding fires and fallout might increase the effective destructive radius, a ground burst Hiroshima-size device would actually 'blow up' about 1 per cent of the city's area.[4] And in 1974, nuclear physicist Theodore Taylor proclaimed the problem to be 'immediate', and explained 'how comparatively easy it would be to steal nuclear material and step by step make it into a bomb'. At the time he thought it already too late to 'prevent the making of a few bombs, here and there, now and then', or 'in another ten or fifteen years, it will be too late'.[5]

Concerns about the atomic terrorist surged in the wake of the 11 September 2001 attacks ('9/11'), and it has been widely predicted for two decades now that, because al-Qaeda operatives used box cutters so effectively in the attacks, they would, despite being under siege, soon apply equal, if previously undisclosed, talents in science and engineering to fabricate nuclear weapons and then detonate them on enemy cities. Brian Jenkins has run an Internet search to discover how often variants of the term 'al-Qaeda' appeared within 10 words of 'nuclear'. There were only seven hits in 1999 and 11 in 2000, but this soared to 1,742 in 2001 and to 2,931 in 2002.[6]

In this spirit, Graham Allison in 2004 produced a thoughtful, influential, and well-argued book, *Nuclear Terrorism: The Ultimate Preventable Catastrophe*, relaying his 'considered judgment' that 'on the current path, a nuclear terrorist attack on America in the decade ahead is more likely than not'.[7] He has had company in his alarming conclusions. In 2007, physicist Richard Garwin put the likelihood of a nuclear explosion on an American or European city by terrorist or other means at 20 per cent per year, which would work out to 87 per cent over a 10-year period.[8] And US politicians and officials have preached to an anxious, appreciative, and numerous choir on the subject. For example, former US president Barack Obama proclaimed nuclear terrorism to be 'the most immediate and extreme threat to global security'[9] even as his defence secretary, Robert Gates, contended that every senior leader in the government was being kept awake at night by 'the thought of a terrorist ending up with a weapon of mass destruction, especially nuclear'.[10]

Impelled by such fears, the threat posed by international terrorism has been repeatedly rendered in the most extreme terms. In 2002, then-national security adviser Condoleezza Rice insisted that 'after 9/11, there is no longer any doubt that today America faces an existential threat to our security—a threat as great as any we faced during the Civil War, the so-called "Good War", or the Cold War'.[11] Following this line of thinking, it became fashionable in some circles to extravagantly denote the contest against Osama bin Laden and his sympathisers as (depending on how the Cold War is classified) World War III or World War IV. The *New York Times* editorial board declared on 23 April 2008, that 'the fight against al-Qaeda is the central battle for this generation', and Senator John McCain more expansively, and repeatedly, pointed to what he called 'radical Muslim extremism' as the 'transcendental challenge of the 21st century', and expressed the belief that, if it prevailed, it can affect 'our very existence'.[12] On that same wavelength in 2015, the CIA's Michael Morell published a book proclaiming the conflict with terrorism to be *The Great War of Our Time*.[13]

The mindset has had disastrous policy consequences. In particular, the 2003 war in Iraq, which has resulted in the deaths of more people than perished at Hiroshima and Nagasaki combined, was importantly impelled by a fear that Iraq's leader, Saddam Hussein, would develop nuclear weapons and then hand them off to groups like al-Qaeda to detonate.

However, 9/11 remains an extreme outlier: no terrorist act before or since, in war zones or outside them, has inflicted even one-tenth as much total destruction.[14] Moreover, international terrorist groups seem to have exhibited only limited desire and even less progress in going atomic. This may be because, after brief exploration of the possible routes for obtaining such a weapon, they, unlike generations of alarmists, have discovered that the tremendous effort required is scarcely likely to be successful.

Routes to a Terrorism Nuclear Bomb

There are three potential routes a terrorist group might take to obtain a nuclear weapon. One would be for it to be given or sold a bomb by a generous like-minded nuclear state for delivery abroad—the impelling fear about Iraq in 2003. This is highly improbable, however, because there would be too much risk, even for a country led by extremists, that the ultimate source of the weapon would be discovered. As prominent analyst Matthew Bunn puts it, 'A dictator or oligarch bent on maintaining power is highly unlikely to take the immense risk of transferring such a devastating capability to terrorists they cannot control, given the ever-present possibility that the material would be traced back to its origin.' Important in this last consideration are deterrent safeguards afforded by 'nuclear forensics', the rapidly developing science (and art) of connecting nuclear materials to their sources even after a bomb has been exploded.[15]

Moreover, there is a very considerable danger to the donor that the bomb (and its source) would be discovered even before delivery or that it would be exploded in a manner and on a target the donor would not approve—including on the donor itself. Another concern would be that the terrorist group might be infiltrated by foreign intelligence.[16]

A second route would be for the terrorist group to steal or illicitly purchase a bomb. In the wake of the Cold War, there was great worry about such 'loose nukes' in unstable post-Communist Russia. However, both Russian nuclear officials and experts on the Russian nuclear programme point out that those bombs are difficult to maintain and have a lifespan of one to three years, after which they become 'radioactive scrap metal'.[17] Even some of those most alarmed by the prospect of atomic terrorism have concluded that, 'It is probably true that there are no "loose nukes", transportable nuclear weapons missing from their proper storage locations and available for purchase in some way.'[18]

It might be added that Russia and other nuclear powers have an intense interest in controlling any weapons on their territory. Stephen Younger, former head of nuclear weapons research and development at Los Alamos National Laboratory, notes, 'Regardless of what is reported in the news, all nuclear nations take the security of their weapons very seriously.'[19]

Moreover, as technology has developed, finished bombs have been outfitted with devices that will trigger a non-nuclear explosion that will destroy the bomb if it is tampered with.[20] And there are other security techniques: bombs can be kept disassembled with the component parts stored in separate high-security vaults, and procedures can be organized so that two people and multiple codes are required not only to use the bomb, but also to store, to maintain, and to deploy it.[21]

Since terrorists are unlikely to be able to buy or steal a useable bomb, and since they are further unlikely to have one handed to them by an established nuclear state, the most plausible route for terrorists would be to manufacture the device themselves from purloined materials. This is the route identified by a majority of leading experts as the most likely to lead to nuclear terrorism.[22] Because of the dangers and difficulties of transporting and working with plutonium, a dedicated terrorist group, it is generally further agreed, would choose to try to use highly enriched uranium.[23] The idea would be to obtain as much of this stuff as necessary and then to fashion it into an explosive.

The likely product of this effort would not be a bomb that can be dropped or hurled, since this would massively complicate the delivery problem. Rather, the terrorists would seek to come up with an 'improvised nuclear device' (IND) of the simplest design—one that could be set off at the target by a suicidal detonation crew. This would be a 'gun' type of device in which masses of highly enriched uranium are hurled at each other within a tube. At best, such a device would be, as even the deeply concerned Allison acknowledges, 'large, cumbersome, unsafe, unreliable, unpredictable, and inefficient'.[24] The process is a daunting one, even in this minimal case. In particular, the task requires a considerable series of difficult hurdles to be conquered and in sequence.

At the present time and likely for the foreseeable future, stateless groups are simply incapable of manufacturing the required fissile material for a bomb because the process requires an effort on an industrial scale.[25] Moreover, they are unlikely to be supplied with the material by a state for the same reasons a state is unlikely to give them a workable bomb. Thus, they would need to steal or illicitly purchase this crucial material.

The terrorist thieves would also need to know exactly what they want and where it is, and this presumably means trusting bribed, but not necessarily dependable, insiders. And to even begin to pull off such a heist, the terrorists would need to develop a highly nuanced street sense in foreign areas often filled with people who are suspicious of strangers.[26] This approach requires the terrorists to pay off a host of greedy confederates, including brokers and money transmitters, any one of whom could turn on them or, either out of guile or incompetence, furnish them with stuff that is useless. Insiders might also come to ruminate over the fact that, once the heist had been accomplished, the terrorists would (as Jenkins puts it none too delicately) 'have every incentive to cover their trail, beginning with eliminating their confederates'.[27]

In addition, because of improving nuclear safeguards and accounting practices, it is decreasingly likely that the theft would remain undetected.[28] This is an important development because, once it is noticed that some uranium is missing, the authorities would investigate the few people who might have been able to assist the thieves, and one who seems suddenly to have become prosperous is likely to arrest their attention right from the start.

It is also relevant to note that over decades, known thefts of highly enriched uranium totalled fewer than 16 pounds or so.[29] This is far less than needed: for a crude bomb, over 100 pounds are required to produce a likely yield of one kiloton.[30] Moreover, none of the thieves had buyers lined up—nearly all were caught while trying to peddle their wares. Indeed, concludes analyst Robin Frost, 'There appears to be no true demand, except where the buyers were government agents running a sting.'[31]

After procuring the required fissile material, the terrorists would likely need to cross international borders. For a considerable fee, opium traders (for example) might provide expert, and possibly even reliable, assistance. However, the routes are not as chaotic as they appear and are often under the watch of suspicious and careful criminal regulators.[32] Moreover, if they became suspicious of the commodity being smuggled, some of these might find it in their interest to disrupt passage, perhaps to collect the bounteous reward money offered by alarmed governments if the uranium theft had been discovered.

Once outside the country with their precious booty, the terrorists would need to set up a large and well-equipped machine shop to manufacture a bomb and then to populate it with a very select team of highly skilled scientists, technicians, and machinists. The process would also require good managers and organizers. The group would have to be assembled and retained for the monumental task while simultaneously raising no consequential suspicions among friends, family, and police about curious and sudden absences from normal pursuits back home. Pakistan, for example, maintains a strict watch on many of its nuclear scientists even after retirement.[33] Members of this bomb-building team would also have to be utterly devoted to the cause. And in addition, they would have to be willing to put their lives, and certainly their careers, at high risk because after their bomb was discovered, or exploded, they would likely become the targets in an intense worldwide dragnet operation facilitated by the fact that their skills would not be common ones.

Some observers have insisted that it would be 'easy' for terrorists to assemble a crude bomb if they could get enough fissile material.[34] However, Christoph Wirz and Emmanuel Egger, two senior physicists in charge of nuclear issues at Switzerland's Spiez Laboratory in 2005, bluntly conclude that the task 'could hardly be accomplished by a subnational group'.[35] They point out that precise blueprints, not just sketches and general ideas, are required, and that even with a good blueprint the terrorist group 'would most certainly be forced to redesign'.[36] They also stress that the work, far from being 'easy', is difficult, dangerous, and extremely exacting, and that the technical requirements 'in several fields verge on the unfeasible'.[37]

Younger makes a similar argument, expressing his amazement at 'self-declared "nuclear weapons experts", many of whom have never seen a real nuclear weapon', who 'hold forth on how easy it is to make a functioning nuclear explosive'. Information is readily available for getting the general idea behind a rudimentary nuclear explosive, but none of this is detailed enough for 'the confident assembly of a real nuclear explosive'. Younger is quick to enumerate the difficulties the group would confront when attempting to fabricate one on their own. He stresses that uranium is 'exceptionally difficult to machine' while 'plutonium is one of the most complex metals ever discovered, a material whose basic properties are sensitive to exactly how it is processed', and both require special machining technology. Younger concludes, 'To think that a terrorist group, working in isolation with an unreliable supply of electricity and little access to tools and supplies' could fabricate a nuclear bomb or IND 'is far-fetched at best.'[38]

Under the best of circumstances, the process could take months, or even a year or more, and it would all, of course, have to be carried out in utter secret even while local and international security police are likely to be on the intense prowl. In addition, people, or criminal gangs, in the area may observe with increasing curiosity and puzzlement the constant coming and going of technicians unlikely to be locals. Unless they lived constantly in the shop (itself a noticeable event), the conspirators would eat, sleep, drink, and recreate elsewhere, constantly bumping into curious, and potentially inquiring, people.

The process of fabricating an IND requires, then, the effective recruitment of people who at once have great technical skills and will remain completely devoted to the cause. In addition, a host of corrupted co-conspirators—many of them foreign—must remain utterly reliable, international and local security services must be kept perpetually in the dark, and no curious outsider must get consequential wind of the project over the months (or even years) it takes to pull off.

The finished product could weigh a ton or more.[39] Encased in lead shielding to mask radioactive emissions, it would then have to be transported to, and smuggled into, the relevant target country. And, finally, at the target site, the crew, presumably suicidal, would have to set off its improvised and untested nuclear device. While doing so, they would have to hope that the machine shop work had been perfect, that there were no significant shakeups in the treacherous process of transportation, and that, after all this effort, the device did not prove to be a dud.

The financial costs of the extended operation in all its cumulating, or cascading, entirety could easily become monumental. There would be expensive equipment to buy, smuggle, and set up, and people to pay—or pay off. Moreover, any criminals competent and capable enough to be an effective ally in the project are likely as well to be both smart enough to see boundless opportunities for extortion and psychologically equipped by their profession to be willing to exploit them.

Even the very alarmed Bunn and Wier contend that the atomic terrorists' task 'would clearly be among the most difficult types of attack to carry out' or 'one of the most difficult missions a terrorist group could hope to try'.[40] And, as the Gilmore Commission, a special advisory panel to the US President and Congress, stresses, setting off a nuclear device capable of producing mass destruction presents not only 'Herculean challenges', but also it requires that a whole series of steps be accomplished. If each is not fully met, the result is not simply a less-powerful weapon, but one that cannot produce any significant nuclear yield at all, or cannot be delivered.[41]

There could easily be at least 20 steps or hurdles and *all* of them must be surmounted by the atomic aspirant. If it is assumed that the various barriers are independent, or effectively independent, from each other and if it is further generously assumed that the terrorists have a fighting chance of 50 per cent of overcoming each of the 20 obstacles, the chances that a concerted effort would be successful comes out to be less than one in a million.[42] Moreover, all this focuses on the effort to deliver a single bomb. If the requirement were to deliver several, the odds become, of course, even more prohibitive.

Capacity, Progress, and Interest

Any notion that the actual al-Qaeda enemy has ever had the capacity to come up with nuclear weapons (even if it wanted to) looks extremely questionable, as does the notion that they could effectively handle one that has been pilfered or given to them by a state. Moreover, the degree to which al-Qaeda and other terrorist groups have pursued, or even had much interest in, nuclear weapons may have been exaggerated. There is some occasional evidence to indicate that the group might have some interest in atomic weapons, but this is limited and often ambiguous. The same can be said about evidence that it has actively sought to achieve an atomic capacity. Khalid Sheikh Mohammed, the apparent mastermind behind the 9/11 attacks, reportedly said that al-Qaeda's atom bomb efforts never went beyond searching the Internet.[43]

As a key indication of al-Qaeda's desire to obtain atomic weapons, many have focused on a set of conversations in Afghanistan in August 2001 that two Pakistani nuclear scientists reportedly had with bin Laden and three other al-Qaeda officials. Pakistani intelligence officers characterize the discussions as 'academic', and they also maintain that term to be the descriptor the scientists 'insisted' on using. The officers do report, however, that the scientists 'described bin Laden as intensely interested in nuclear, chemical and biological weapons'. But his questions do not seem to be very sophisticated, and as the scientists themselves have reportedly insisted, the discussion was rather basic, and they provided no material or specific plans. Moreover, the scientists likely were incapable of providing truly helpful information because their expertise was not in bomb design, which might be useful to terrorists, but rather in the processing of fissile material, which is almost certainly beyond the capacities of a non-state group (as discussed previously).[44]

Rumours and reports that al-Qaeda managed to purchase an atomic bomb, or several, have been around now for decades, beginning around 1998. Louise Richardson catalogues a number of these stories, and concludes, 'There can be little doubt that most of these reports are as reliable as the reports that Saddam Hussein was developing and stockpiling weapons of mass destruction.'[45] And in his assessment of these 'vague rumors and reports', Jenkins points out that, 'Although the facts of these reported contacts, deals, and deliveries were shrouded in mystery, the "shocking revelations" were often shorn of uncertainty and offered as "empirical proof", presented as fact and accepted on faith.'[46]

Anne Stenersen of the Norwegian Defence Research Establishment concludes after an exhaustive study of available materials, that although 'it is likely that al-Qaeda central has considered the option of using non-conventional weapons', there 'is little evidence that such ideas ever developed into actual plans, or that they were given any kind of priority at the expense of more traditional types of terrorist attacks'. As she also notes, evidence from a recovered al-Qaeda computer indicates that the group had earmarked some US$2,000–US$4,000 for research on weapons of mass destruction (WMD) at the time, all of it apparently for (very crude) chemical work with some potentially for biological weapons.[47] Evidence seized during the bin Laden raid of 2011 indicates that, as of 2006, al-Qaeda central had about 176,000 euros to its name.[48]

When examined, then, the evidence of al-Qaeda's desire to go atomic, and about its progress in accomplishing this exceedingly difficult task, is remarkably skimpy, if not completely negligible. The scariest stuff—years of loose nuke rumour and chatter and hype—seems to have no substance whatever. Moreover, the terrorist use of unconventional weapons can be especially counterproductive, by turning off financial supporters, dividing the movement, and eliciting major crackdowns.[49]

In addition, the experience with unconventional weapons cannot be too encouraging to the would-be atomic terrorist. One group that tried, in the early 1990s, was the Japanese apocalyptic group Aum Shinrikyo. Unlike al-Qaeda, the group was not under siege, and it had money, expertise, and a remote and secluded haven in which to set up shop—even a private uranium mine. After making dozens of mistakes in judgement, planning, and execution in its nuclear quest, Aum Shinrikyo tried to produce biological weapons which, as it happened, did not work either. In the end, it turned to chemical ones, resulting eventually in a somewhat botched release of sarin gas in a Tokyo subway that managed to kill a total of 12 people.[50]

For the most part, terrorists seem to be heeding the advice found in a memo on an al-Qaeda laptop seized in Pakistan in 2004: 'Make use of that which is available ... rather than waste valuable time becoming despondent over that which is not within your reach.' In other words: keep it simple, stupid. And, in fact, it seems to be a general historical regularity that terrorists tend to prefer weapons that they know and understand—not new, exotic ones.[51] 'Say what you want of terrorists', observes pilot Patrick Smith, 'they cannot afford to waste time and resources on schemes with a high probability of failure.'[52] As Jenkins notes, 'None of the bloodiest attacks involved any weapons that had not already been in the terrorist arsenal for many years.'[53] Indeed, the truly notable innovation for terrorists over the last few decades has not been in qualitative improvements in ordnance at all, but rather in a more effective method for delivering it: the suicide bomber.[54]

There have also been concerns that another terrorist or insurgent group, ISIS or Islamic State, which emerged on the international scene in 2014, might obtain or fabricate atomic weapons. For a while Islamic State held considerable territory in Iraq and Syria, and it possessed quite a bit of looted wealth. However, it does not seem to have explored the opportunity, in part perhaps because the group was not able to do much of anything clandestine and because it was under relentless siege as it progressively generated more and more enemies.[55]

The apparent aversion of terrorists and terrorist groups to unconventional weapons has extended as well to such scenarios as sabotaging or stealing radioactive material from nuclear reactors. Reactors are rather secure facilities, although there have been a few instances of attempts at break-ins, in particular, one in South Africa. However, the incident barely got past some perimeter defences, seems not to have involved efforts to get at uranium supplies at all, and, regardless, never got close to them.[56] Nor has there been much terrorist interest in obtaining radioactive material to fashion into 'dirty bombs'—devices that could cause considerable disruption if relatively little loss of life.[57] In this case, technical and logistic hurdles seem to have been prohibitive.

Controls

Despite the challenges faced by atomic terrorists, it does not follow that policies designed to inconvenience them are necessarily unneeded or unwise. Although the atomic terrorists' task under present conditions is already monumental, and although their success is most improbable, efforts to further enhance this monumentality, if cost-effective and accompanied with tolerable side effects, are generally desirable. Establishing a reliable inventory of fissile material may be almost as important as securing it. If authorities are able to detect immediately when some goes missing, international policing efforts would be triggered. These are likely to considerably hamper the terrorists' prospects for success, not only by setting investigators into action, but also by alerting the public worldwide (including the criminal element) to what has happened and by establishing and publicizing generous rewards for productive tip-offs.

It would also be sensible to devote funds for further development of the science of nuclear forensics to identify the source of the fissile material in a bomb or device. Indeed, this seems to be a no-brainer. An inexpensive project, it may also have desirable scientific spin-offs. Nuclear sting operations are rather inexpensive, and it probably makes sense to continue these in order to deter sellers and to prevent a viable black market from emerging. However, sting efforts have thus far come up with little: even if all of the sellers rolled up by the police had been successful in peddling their illicit product to a single dedicated terrorist group, the group would not have been able to accumulate enough for even a single bomb.

An area that needs much more attention concerns education about the effects of radiation. This concern is relevant not only to an atomic explosion, particularly a ground burst one, but also to the effects of a somewhat more likely 'dirty bomb' explosion and of accidental radiation releases. There are no immediate dangers from even fairly substantial increases in radiation, but people are simply not aware of that fact, and radiation fears may induce overly evasive behaviour in the immediate and short term. This should very much be a primary concern. In fact, if radiation levels are raised only somewhat above normal background levels in a small area by a dirty bomb, accident, or nuclear fallout, a common recommendation from nuclear scientists and engineers is that those exposed should calmly walk away. Those in charge have failed to advance this information much—or even, perhaps, at all. Effectively, therefore, they encourage panic, and the danger is, as one nuclear engineer puts it, 'If you keep telling them you expect them to panic, they will oblige you. And that's what we're doing.'[58]

Conclusion

The risk of nuclear terrorism is low—perhaps vanishingly so. It is unlikely that 'loose nukes'—nuclear weapons missing from their proper storage locations and available for purchase in some way—exist, and it is equally unlikely a true black market in nuclear materials exists. And, because terrorists face a host of organizational and technical hurdles, fears that they will be able to build or acquire an atomic bomb or device are unjustified. In addition, the evidence of any desire of terrorists to go atomic and of any progress in accomplishing this exceedingly difficult task is remarkably skimpy, if not completely negligible.

Sensible, cost-effective policies designed to make the probability even lower may be justified, given the damage that can be inflicted by an atomic explosion. But obsessive alarmism about the likelihood and imminence of atomic terrorism has had policy consequences that have been costly and unnecessary—not only the 2003 war in Iraq as noted earlier, but the focus on WMD that seduced agencies away from due preparation for disasters that have actually happened, such as Hurricane Katrina in the US. Indeed, the existential bombast about a terrorist bomb that follows so naturally from decades of atomic obsession may actually encourage the most diabolical and murderous terrorists to investigate the possibility of obtaining one. Fortunately, however, would-be atomic terrorists are exceedingly unlikely to be successful in such a quest, however intense the inspiration and encouragement they may receive from the unintentional cheerleaders among their distant enemies.

Notes

1. Francis Anthony Boyle, *World Politics and International Law* (Durham, NC: Duke University Press, 1985), 23.
2. These propositions are developed at length in John Mueller, *Atomic Obsession: Nuclear Alarmism from Hiroshima to Al Qaeda* (Oxford: Oxford University Press, 2010). Also see John Mueller, 'Nuclear Weapons Don't Matter, but Nuclear Hysteria Does', *Foreign Affairs*, 97/6 (2018), 1–6.
3. Kai Bird and Martin Sherwin, *American Prometheus: The Triumph and Tragedy of J. Robert Oppenheimer* (New York: Knopf, 2005), 349.
4. A ground burst Hiroshima-sized bomb would destroy about three square miles. New York City occupies 300 square miles. Mueller, *Atomic Obsession*, 9.
5. John McPhee, *The Curve of Binding Energy* (New York: Farrar, Straus and Giroux, 1974) 7, 195–7.
6. Brian Michael Jenkins, *Will Terrorists Go Nuclear?* (Amherst, NY: Prometheus, 2008), 250–1.
7. Graham T. Allison, *Nuclear Terrorism: The Ultimate Preventable Catastrophe* (New York: Times Books, 2004), 15.
8. Graham T. Allison, 'The Ongoing Failure of Imagination', *Bulletin of the Atomic Scientists*, 62/5 (2006), 36–45.

9. Barack Obama, 'Remarks by President Barack Obama, Hradcany Square Prague, Czech Republic, 5 April' (Washington, DC: The White House, Office of the Press Secretary, 2009).
10. Jim Mannion, 'Gates Warns States Against Giving Terrorists Nuclear Bomb'. *AFP* [online] (11 June 2008).
11. Condoleezza Rice, 'Dr. Condoleezza Rice Discusses President's National Security Strategy, 1 October,' (Washington, DC: White House, Office of the Press Secretary, 2002).
12. Editorial, 'All the Time He Needs', *New York Times* (13 April 2008); Fareed Zakaria, 'True or False: We Need a Wartime President', *Washington Post* (6 July 2008).
13. Michael Morell, *The Great War of Our Time: The CIA's Fight against Terrorism from al Qa'ida to ISIS* (New York: Twelve, 2015).
14. For the argument that al-Qaeda central resembles Lee Harvey Oswald, the assassin of John Kennedy, in that both were fundamentally trivial entities that got horribly lucky once, see John Mueller and Mark G. Stewart, *Chasing Ghosts: The Policing of Terrorism* (New York: Oxford University Press, 2016).
15. Matthew Bunn, *Securing the Bomb* (Cambridge, MA: John F. Kennedy School of Government, Harvard University, 2007), vi. Also see William Langewiesche, *The Atomic Bazaar: The Rise of the Nuclear Poor* (New York: Farrar, Straus and Giroux, 2007), 20; Matthew Bunn, 'A Mathematical Model of the Risk of Nuclear Terrorism', *Annals of the American Academy of Political and Social Science*, 607/1 (2006), 103–20; Matthew Bunn and Anthony Wier, 'Terrorist Nuclear Weapon Construction: How Difficult?', *Annals of the American Academy of Political and Social Science*, 607/1 (2006), 133–49; Jenkins, *Will Terrorists Go Nuclear?*, 198; For an excellent discussion of nuclear forensics, see Michael A. Levi, *On Nuclear Terrorism* (Cambridge, MA: Harvard University Press, 2007), 127–33.
16. Robin M. Frost, *Nuclear Terrorism After 9/11* (Abingdon: Routledge, 2005), 64. Jenkins, *Will Terrorists Go Nuclear?*, 143; Also see Kier A. Lieber and Daryl G. Press, 'Why States Won't Give Nuclear Weapons to Terrorists', *International Security*, 38/1 (2013), 80–104.
17. Anna Badkhen, 'Al Qaeda Bluffing about Having Suitcase Nukes, Experts Say', *San Francisco Chronicle* (23 March 2004).
18. Anna M. Pluta and Peter D. Zimmerman, 'Nuclear Terrorism: A Disheartening Dissent', *Survival*, 48/2 (2006), 56; See also Stephen M. Younger, *The Bomb: A New History* (New York: Ecco, 2009), 152; Frost, *Nuclear Terrorism After 9/11*, 17–23.
19. Stephen M. Younger, *Endangered Species* (New York: Ecco, 2007), 93.
20. Jenkins, *Will Terrorists Go Nuclear?*, 141.
21. Joby Warrick, 'Pakistan Nuclear Security Questioned: Lack of Knowledge about Arsenal May Limit U.S. Options', *Washington Post* (11 November 2007).
22. Levi, *On Nuclear Terrorism*, 26; Also see Charles D. Ferguson and William C. Potter, *The Four Faces of Nuclear Terrorism: Threats and Responses* (New York: Routledge, 2005), ch. 3–4; Peter D. Zimmerman and Jeffrey G. Lewis, 'The Bomb in the Backyard', *Foreign Policy* [online] (16 October 2009).
23. Richard L. Garwin and Georges Charpak, *Megawatts and Megatons* (New York: Knopf, 2001), 314; Allison, *Nuclear Terrorism*, 96–7; Christoph Wirz and Emmanuel Egger, 'Use of Nuclear and Radiological Weapons by Terrorists?', *International Review of the Red Cross*, 87/859 (2005), 500; Frost, *Nuclear Terrorism After 9/11*, 27–8; Bunn and Wier, 'Terrorist Nuclear Weapon Construction', 135; Langewiesche, *Atomic Bazaar*, 21–3; Levi, *On Nuclear Terrorism*, 73–81; Younger, *The Bomb*, 142–3.
24. Allison, *Nuclear Terrorism*, 96–7; also see Garwin and Charpak, *Megawatts*, 350.

25. Wirz and Egger, 'Use of Nuclear and Radiological Weapons', 499–502; Bunn and Wier, 'Terrorist Nuclear Weapon Construction', 136–7; Langewiesche, *Atomic Bazaar*, 20; Levi, *On Nuclear Terrorism*, 15.
26. Langewiesche, *Atomic Bazaar*, 33–48.
27. Jenkins, *Will Terrorists Go Nuclear?*, 140, 152–3.
28. Levi, *On Nuclear Terrorism*, ch. 5.
29. Dafna Linzer, 'Nuclear Capabilities May Elude Terrorists, Experts Say', *Washington Post* (29 December 2004).
30. Garwin and Charpak, *Megawatts*, 314, 350.
31. Frost, *Nuclear Terrorism After 9/11*, 9.
32. Langewiesche, *Atomic Bazaar*, 54–65.
33. Daniel Pearl and Steve LeVine, 'Pakistan Has Ties to Group It Vowed to Curb – Military State's Elite Is Linked to Activities of Nuclear Scientist'. *Wall Street Journal* (24 December 2001).
34. Graham T. Allison, Owen R. Coté Jr., Richard A. Falkenrath, and Steven E. Miller, *Avoiding Nuclear Anarchy: Containing the Threat of Loose Russian Nuclear Weapons and Fissile Material* (Cambridge, MA: MIT Press, 1996), 12.
35. Wirz and Egger, 'Use of Nuclear and Radiological Weapons', 499–502.
36. Ibid.
37. Ibid.
38. Younger, *Endangered Species*, 86, 93, 88; Younger, *The Bomb*, 146.
39. Bunn and Wier, 'Terrorist Nuclear Weapon Construction', 142.
40. Ibid., 133–4, 147.
41. Gilmore Commission (Advisory Panel to Assess Domestic Response Capabilities for Terrorism Involving Weapons of Mass Destruction), First Annual Report: Assessing the Threat (15 December 1999), 31.
42. For a more detailed discussion about the steps and this calculation, see Mueller, *Atomic Obsession*, 184–9; in his 'Mathematical Model' study of 2005, Bunn has performed a somewhat similar exercise and assigns probabilities that are much more favourable to the terrorists. With parameters like that and with some additional considerations, he concluded that there was a 29 per cent chance of a terrorist atomic bomb being successfully detonated in the next decade—that is, by 2016.
43. Jenkins, *Will Terrorists Go Nuclear?*, 27.
44. Kamran Khan and Molly Moore, '2 Nuclear Experts Briefed Bin Laden, Pakistanis Say', *Washington Post* (11 December 2001).
45. Louise Richardson, *What Terrorists Want: Understanding the Enemy, Containing the Threat* (New York: Random House, 2006), 162.
46. Jenkins, *Will Terrorists Go Nuclear?*, 274.
47. Anne Stenersen, *Al-Qaida's Quest for Weapons of Mass Destruction: The History behind the Hype* (Saarbrücken: VDM Verlag Dr. Müller, 2008), 35–6, 39.
48. Nelly Lahoud, *The Bin Laden Papers* (New Haven, CT: Yale University Press, 2022), 92.
49. John Parachini, 'Putting WMD Terrorism into Perspective', *Washington Quarterly*, 26/4 (2003), 43–6.
50. Linzer, 'Nuclear Capabilities'; Frost, *Nuclear Terrorism After 9/11*, 38–40.
51. Craig Whitlock, 'Homemade, Cheap and Dangerous: Terror Cells Favor Simple Ingredients In Building Bombs', *Washington Post* (5 July 2007); see also Parachini, 'Putting WMD Terrorism into Perspective', 44–6; Stenersen, *Al-Qaida's Quest for Weapons of Mass*

Destruction; David C. Rapoport, 'Terrorists and Weapons of the Apocalypse', *National Security Studies Quarterly*, 5 (1999), 51; Gilmore Commission, 37.

52. Patrick Smith, 'The Airport Security Follies,' *New York Times* (28 December 2007).
53. Jenkins, *Will Terrorists Go Nuclear?*, 189.
54. Robert A. Pape, *Dying to Win: The Strategic Logic of Suicide Terrorism* (New York: Random House, 2005).
55. John Mueller and Mark G. Stewart, 'Misoverestimating ISIS: Comparisons with Al-Qaeda', *Perspectives on Terrorism*, 10/4 (2016), 32–41.
56. Scott Pelley, 'Assault on Pelindaba', *60 Minutes* (23 November 2008), CBS Television; Micah Zenko, 'A Nuclear Site Is Breached: South African Attack Should Sound Alarms', *Washington Post* (20 December 2007).
57. J. A. Geisecke, et al., 'Assessment of the Regional Economic Impacts of Catastrophic Events: CGE Analysis of Resource Loss and Behavioral Effects of an RDD Attack Scenario', *Risk Analysis*, 32/4 (2012), 583–600.
58. Theodore Rockwell, 'Radiation Chicken Little', *Washington Post* (16 September 2003); see also Rupert Wingfield-Hayes, 'Fukushima: Is Fear of Radiation the Real Killer?', *BBC News* [online], 11 March 2014.

Suggested Reading

Allison, Graham T., (2004), *Nuclear Terrorism: The Ultimate Preventable Catastrophe* (New York: Times Books).

Frost, Robin M., (2005), *Nuclear Terrorism After 9/11* (Abingdon: Routledge).

Jenkins, Brian Michael, (2008), *Will Terrorists Go Nuclear?* (Amherst, NY: Prometheus).

Langewiesche, William, (2007), *The Atomic Bazaar: The Rise of the Nuclear Poor* (New York: Farrar, Straus and Giroux).

Levi, Michael A., (2007), *On Nuclear Terrorism* (Cambridge, MA: Harvard University Press).

Lieber, Kier A. and Press, Daryl G., (2013), 'Why States Won't Give Nuclear Weapons to Terrorists', *International Security*, 38/1, 80–104.

Mueller, John, (2010), *Atomic Obsession: Nuclear Alarmism from Hiroshima to Al Qaeda* (New York: Oxford University Press).

CHAPTER 5

STATE SPONSORSHIP OF NUCLEAR TERRORISM

GARY A. ACKERMAN AND BRYAN R. EARLY

Introduction

ACADEMICS and policymakers have typically viewed the nuclear security threats posed by state actors as constituting a distinct set of threats from those posed by non-state actors. Some have argued, however, that those threats could converge in the form of state-sponsored nuclear terrorism.[1] Policymakers, for example, have expressed concern about the potential for insecure, aggressive, or unstable states like North Korea, Pakistan, or potentially Iran to provide terrorists with nuclear weapons or the materials needed to create them for use in terrorist attacks.[2] While scepticism exists about whether terrorists would actually use nuclear weapons and states' incentives to sponsor nuclear terrorism,[3] the unprecedented potential consequences of such an attack mean that policymakers must continue to consider this threat and analyse its likelihood.[4] Scholars have generally paid less attention to the risks of state-sponsored terrorist attacks against nuclear facilities—including in the cyber realm. This chapter takes stock of what the existing literature reveals about the various types of state-sponsored nuclear terrorism, and then provides a comparative analysis of the likelihood and characteristics of the four main scenarios of state-sponsored terrorism.

Defining the concept of 'state sponsorship' is critical to understanding how governments could contribute to the use of radiological and nuclear terrorism. While the basic understanding of there being some connection between a state and a terrorist group is uncontroversial, beyond this there is much ambiguity surrounding the term.[5] The primary axis of contention is the degree to which the support given by the state is active and deliberate. Active state sponsorship occurs when a state knowingly and intentionally provides support to a terrorist actor, including such activities as the provision of weapons, finances, training, ideological guidance, or safe harbour. Passive state sponsorship occurs when a state takes no direct action to support terrorists, but knowingly permits terrorist activities in the areas it controls.[6]

Within the context of nuclear terrorism, however, the primary concern is active state sponsorship. Given that nuclear weapons and materials are both politically salient and environmentally hazardous, it is almost axiomatic that states pay close regard to their disposition. It is one thing for a state's leaders to claim that the state did not realize that a terrorist organization was recruiting in their territory. It is another matter entirely to argue that the government did not realize that a terrorist group stole a nuclear weapon or fissile materials. What this boils down to is that, given the salience of nuclear materials and weapons, wilful neglect of nuclear security is tantamount to active sponsorship and would be regarded by most other countries as such.

As for states that lack the capacity to safeguard nuclear and radiological materials, that situation is better covered under the banner of nuclear and radiological security (or lack thereof) than state sponsorship of terrorism. Therefore, this chapter limits state sponsorship to the more active approach and defines it as: *the deliberate provision by a government institution of resources that offer concrete advantages or material support to a terrorist organization.*[7] Although terrorism-like violence (e.g. covert bombings) undertaken by state agents could be considered a type of state-sponsorship terrorism, this chapter argues such activities blur the lines between covert warfare and terrorism.[8] Instances where a state actor is directly involved in managing terrorist attacks or operations conducted by non-state proxies are thus excluded from the concept of state-sponsored terrorism discussed here.

An important insight to emerge from the theoretical literature on state sponsorship is that the relationship is inherently delegative in nature, wherein the state (acting as the principal) obtains the services of a terrorist group (acting as its agent) to help the state achieve its goals.[9] This creates a 'principal–agent' dynamic, which is explored in several important studies by borrowing from the broader literature on principal–agent systems.[10] Part of a state's calculations in sponsoring terrorism must be whether the terrorist group might act against the interests of the state (what is termed 'agency loss').[11] Agency loss can range from deviations from sponsor preferences (such as in the types of targets attacked by the terrorist group) to the terrorists actually turning their violence against their erstwhile sponsor.[12] There is also the possibility of shirking, wherein a terrorist group reaps the benefits of sponsorship but does not do much to further the goals of the sponsor.[13]

Analysing the threats posed by state-sponsored nuclear terrorism is inherently challenging and speculative due to the lack of past cases. Acknowledging this, the chapter reinforces existing arguments that states have minimal incentives to help terrorists engage in attacks using nuclear weapons, while also suggesting that state-sponsored terrorist attacks against facilities in both the physical and cyber nuclear realms are improbable but, in comparative terms, much more likely to occur. This suggests that governments should allocate more attention and resources to mitigate those risks. The chapter also broadens nuclear security debates over whether terrorists *would* use nuclear weapons if they *could* obtain them to include more plausible scenarios that could play to non-state actors' more typical strengths and capabilities.

The Typologies of Nuclear State-Sponsored Terrorism

This section analyses the most plausible scenarios by which states might seek to sponsor nuclear terrorism. Terrorists have various options in seeking to use nuclear weapons or target nuclear infrastructure, varying in terms of the challenges and consequences associated with their use. In their seminal study, Charles Ferguson and William Potter identified four archetypes of radiological and nuclear (RN) terrorism: 1) use of a military-grade nuclear weapon; 2) use of an improvised nuclear device (IND); 3) use of a radiological dispersion device (RDD) or 'dirty' bomb; and 4) attacks against nuclear-related infrastructure.[14] The use of nuclear weapons as part of terrorist attacks would involve significant destruction, including tremendous loss of life and property damage. In contrast, RDDs are generally considered to be 'weapons of mass disruption' rather than of mass destruction. Attacks on nuclear facilities may cause mass disruption or destruction, depending on their severity.

This section focuses solely on state-sponsored nuclear terrorism that involves the use of a military-grade nuclear weapon, an IND, or an attack against nuclear infrastructure. The latter case is further differentiated between physical attacks (or sabotage) and cyber-attacks. Terrorism involving RDDs is excluded in the analysis because such attacks could rely on any radioactive materials and available sources are much more widely available than fissile materials; they, therefore, require a separate analysis from the narrowly nuclear-focused analysis addressed here.

Threat Scenario 1: State provides viable nuclear weapon to a terrorist organization

The type of state sponsorship of nuclear terrorism that has received the most attention from scholars and commentators is where a state deliberately provides a terrorist group with one or more viable nuclear weapons. To fall within the scope of the analysis here, this must involve conscious authorization by state command authorities to transfer the weapon(s); a 'broken arrow' or 'empty quiver' incident[15] in which a rogue element of the state assists terrorists in acquiring a weapon, while important, is regarded here as a failure of nuclear security, rather than an instance of sponsorship. The likelihood of Threat Scenario 1 almost exclusively depends on the motivations of the various parties, with the assumption that the state possesses the capability to transfer the weapon(s) and the terrorist the capability to receive them.[16] The motivation is assessed through the lens of the aforementioned concepts of cost–benefit trade-offs and principal–agent dynamics.

Starting from the perspective of the terrorist actor, it should be recognized that only a minority of terrorists would seek a nuclear weapons capability in the first place.[17] Among the potential benefits that such a group might perceive are: the elevation of the group's status far above their peers to be on par with that of the most powerful states; the ability to strike their enemies regardless of any physical protection; and the opportunity to realize their most extreme catastrophic goals, often of an apocalyptic character.[18] For such groups, the easiest route to a nuclear weapons capability is to receive a ready-to-use weapon from a state. Moreover, the infamy and notoriety stemming from a state providing a terrorist group with the most powerful of weapons could make such an operation attractive to prestige-seeking terrorist leaders.

Several major disincentives exist, however, that prevent most terrorists from seeking nuclear weapons. In most cases, it can be assumed that terrorists seeking to acquire nuclear weapons intend to use them.[19] Using nuclear weapons in an attack would inflict indiscriminate devastation that would be incompatible with many terrorist groups' strategic goals. Such attacks would also entail overwhelming retribution against themselves or their purported constituencies. This limits the number of terrorist organizations for which pursuing nuclear weapons is an attractive option. Even for the subset of terrorist actors who would theoretically be interested in obtaining a nuclear capability, they must perceive themselves as organizationally capable of taking advantage of a bequeathed weapon. Transporting, storing, and deploying a nuclear weapon without being interdicted by authorities is a tall order and many organizations would be aware of their own operational deficiencies in this regard. Even if confident in their tradecraft generally, they would still need to worry that if counter-terrorism agencies got wind of their possession of a weapon, their group would probably be the target of the largest search and destroy mission in history.

In addition, for the even smaller subset of terrorist organizations that would actively pursue a nuclear weapon, there might be some reluctance about accepting it from a state. While this scenario represents the simplest way to acquire such a weapon, the more astute terrorist leaders would realize that such beneficence usually comes with strings attached, for example, limitations on how, when, and against whom the weapon could be used. For terrorists who might have specific, singular preferences in this regard (e.g. an apocalyptic cult), this likely divergence in interests with any state sponsor might convince them that they would be better off going it alone.

Overall, while there might be some terrorist groups who want a nuclear weapon and for whom the benefits of accepting a nuclear weapon from a state sponsor would outweigh the costs, this is not likely to be large subset of the world's terrorist groups. In the entire history of terrorism, only a handful of violent non-state actors have shown any serious interest in obtaining a nuclear weapon (including the Islamic State in the 2010s and Chechen rebels in the 1990s) and only two—Aum Shinrikyo and al-Qaeda—are known to have actively pursued them.[20]

Turning to the more complex considerations of the state actor, there are only a limited number of benefits to transferring a nuclear weapon. First, handing these weapons off to proxies to use, no matter how trusted or competent they are, entails states relinquishing operational control of the weapons and would presumably only be considered when the terrorist recipient possessed operational capabilities that the state itself could not muster.[21] This benefit would thus only materialize in the—even theoretically—extremely unusual case where a state possessed sufficient capacity and resourcefulness to succeed with its own nuclear weapons programme but lacked the wherewithal to deliver these weapons using its own conventional means or covert agents.

Another potential benefit for the state is pecuniary, where the state transfers weapons in exchange for money or other desired goods or services. This is marginally more feasible than the previous benefit, in that a poor state that has expended significant resources to develop nuclear weapons might find itself lacking resources for other needs and decides to fundraise by selling some of its nuclear weapons. North Korea is the threat most referenced in this regard. North Korea's moribund economy and extremely low GDP create a strong need for outside resources, which create greater incentives for the regime to sell its weapons of mass destruction.[22] Graham Allison highlights North Korea's long history of being willing to, in the words of former Secretary of Defense Robert Gates, 'sell anything it has to anybody who has the cash to buy it.'[23] Such sentiments are summed up by nuclear scholar Michael Levi, who as far back as 2008 maintained that 'No leader of any nuclear country *other than North Korea* has any meaningful incentive to deliberately transfer nuclear weapons or materials to a terrorist group' [emphasis added].[24]

The final possible benefit is more esoteric, and it concerns the remote possibility that, since a state cannot absolutely control a terrorist proxy and terrorists can be more volatile in their behaviour, the state could utilize this to its advantage as a bargaining chip in crisis negotiations (discussed in the non-nuclear context by Jeremy Berkowitz[25] and alluded to in Brittnee Carter et al.).[26] Under such a scenario, a state might perceive a benefit in advertising a 'latent' capacity to transfer nuclear weapons to its terrorist proxies in order to introduce a 'wild card' into their rivals' decision making or to establish a 'dead man's switch' as a form of deterrence.

Arrayed against these rather meagre benefits are several powerful disincentives. The first is rooted in deterrence theory and reflects the idea that providing nuclear weapons to terrorist organizations would bring upon the state the most severe forms of retribution from any state or states that are targeted (or even threatened) by the nuclear-armed terrorists.[27] Indeed, if one assumes that the sponsoring state was not powerful enough in the first instance to challenge the targeted state directly, this retribution might be expected to result in regime change or even the wholesale destruction of the sponsoring state. The potential costs for any sponsor are thus enormous.

Two key elements are required, however, before such a deterrent can operate: 1) the sponsoring state would need to believe that the targeted state could link the nuclear terrorism back to them; and 2) the sponsoring state must regard the implicit or explicit threat that the targeted state would engage in this retribution as being credible. With respect to linking the nuclear weapon back to the sponsoring state, there has been much policy discussion and government investment (at least in the United States) in the domain of nuclear forensics. This involves technical procedures that analyse the isotopic by-products of a nuclear detonation to establish where the fissile material within the weapon came from.[28] Assessments of the ability of nuclear forensics to reliably identify the source of weapons material vary, with optimists[29] pointing to the improvement in nuclear forensic capabilities and pessimists[30] stressing their limitations, which include the need for reference samples from all facilities that produce or hold fissile material. The latter is, naturally, unlikely to be forthcoming from many countries, least of all those who sponsor terrorism.

Yet, even beyond technical forensics, there are multiple other avenues by which to establish attribution, including a variety of intelligence assets. Indeed, Keir Lieber and Daryl Press argue that attribution for nuclear terrorism is highly likely.[31] First, the history of (non-nuclear) terrorism shows that the more people killed in a terrorist act the more likely attribution becomes. Second, there are a limited number of state sponsors of terror, and for each of these, their specific terrorist proxies are largely known. While there are debates regarding evidentiary standards in international relations for establishing state sponsorship of terrorism,[32] it is doubtful that a state which has fallen victim to nuclear terrorism would refrain from responding if it could not trace the attack back to a state sponsor 'beyond a reasonable doubt.'

If attribution is likely, the question then becomes whether a prospective nuclear state sponsor believes that any retribution from the targeted state would indeed result in costs that it could not bear. In the language of deterrence, how credible is the threat of punishment? Most experts appear to judge that the catastrophic consequences of a nuclear attack would create both a desire on the part of the leaders of the targeted state and a political imperative from their citizens to respond as quickly and forcefully as possible, with all means at their disposal.[33] It is possible that in certain theoretical cases the target of the nuclear terrorist attack may be incapacitated by it, lacks credibility in responding to attacks in general,[34] or that the potential costs of retribution may be perceived as being too high or too difficult. If present, these factors could undermine the deterrent threat of retribution. While probable retribution should be a powerful disincentive in most cases, it may thus not be sufficient to completely dissuade potential state sponsors in all circumstances.

There are other potential costs beyond retribution that can act to dissuade would-be nuclear sponsors. Aside from the danger of direct retribution, there are normative[35] and legal strictures against transferring a nuclear weapon to terrorists, as exemplified by United Nations Security Council Resolution 1540. If discovered, such actions would delegitimize the sponsoring state, almost certainly inviting international opprobrium and multilateral sanctions, as well as potentially alienating ideological constituencies and even inciting domestic opposition against the sponsoring government.

Perhaps most significantly, the principal–agent dynamic looms especially large. Agency loss could mean that a nuclear weapon is used against a state that the sponsor does not want to see harmed or, worse still, is deployed against the sponsor itself.[36] Moreover, the terrorist organization could either wittingly or unwittingly (if infiltrated by counter-terrorism agencies) reveal where the weapon came from.[37] Bestowing an autonomous nuclear capability on an unfit—either incompetent or untrustworthy—proxy thus carries acute risk for the sponsor and constitutes an enormous potential cost.

Most states are unlikely to have greater confidence in any terrorist proxy than in their own covert forces. State sponsors, if they were even willing to take this risk, would only share a nuclear weapon with their most trusted and capable proxies. Relying on trusted partners, however, would increase the likelihood of attribution, as discussed previously.[38] Therefore, in most circumstances, there would be few benefits for the state to arming a terrorist group with a nuclear weapon and a significant likelihood of enormous costs materializing. This makes it extremely unlikely that a state would intentionally give one or more nuclear weapons to a terrorist group.

It is not, however, infeasible (even Christopher McIntosh and Ian Storey, who are very sceptical, accede to this possibility)[39] and there are certain extreme conditions that might enhance the benefits and diminish the costs to the point where the balance tips in favour of going ahead with the transfer of a weapon to a terrorist. These are exceptionally remote possibilities, but there are at least three scenarios offered to indicate that the probability is not quite zero. These include strategic miscalculations by an insolvent North Korea, a state conducting a 'false flag' sponsorship operation to implicate another state, and a head of state simply ceasing to care about consequences.

Threat Scenario 2: State assists a terrorist organization to acquire an IND

The remaining types of nuclear terrorism sponsorship can be analysed by comparison to the most extreme type just described—provision of a viable weapon. The first of these comparisons examines the possibility of a state assisting a terrorist organization to acquire an IND, primarily by helping the terrorists access the required fissile materials.[40] This could entail the state simply handing over fissile material from its own stocks but could also include the state using its resources to facilitate the terrorists gaining access to fissile material in other states.

From the terrorist's perspective, while obtaining the requisite amount of highly enriched uranium (HEU) or plutonium is by far the primary obstacle to gaining a nuclear weapons capability, the other elements of production are non-trivial and require engineering and explosives expertise, specific equipment, a location to operate undisturbed, and so on. This would increase both the resources the group would need to muster, as well as the risks of failure or interdiction by authorities. The potential disincentives for the group are thus almost certain to be larger than in Threat Scenario 1. The only benefit to constructing its own weapon would be potentially providing the terrorists with a greater degree of operational control, although most state sponsors would likely try to have a say in how the resultant weapon was deployed. Overall, the motivation for the terrorist is likely to be a little lower than in Threat Scenario 1, or at most the same, and would apply to the same small subset of terrorist organizations.

Turning to the incentives and disincentives for the state, there are no real additional benefits to providing terrorists with fissile material over providing them with an intact weapon. With respect to the potential costs, attribution under Threat Scenario 2 might be marginally more difficult because material, rather than an intact weapon, is being transferred, which might be easier to conceal from foreign intelligence agencies. However, the nuclear forensics issue would persist and there is still no guarantee that the state's assistance would go unnoticed by external parties or that the terrorist group would not be infiltrated, and so forth.

The problem of agency loss, however, is even more acute with this type of nuclear sponsorship since there would be no real opportunity for the state to remotely control the disposition of the resulting weapon. If the terrorist group produced a weapon itself, even in the unlikely event that it installed its own version of command-and-control safeguards, there would be little chance that it would provide these to their state sponsor. The chances that the terrorist group would deviate from the state's intended preferences with respect to the use of the weapon are thus substantially increased.

On balance, there is likely to be a lower incentive compared with Threat Scenario 1, since the state would be gaining relatively little in terms of deniability but sacrificing a lot in terms of control. Combined with the same or less incentive for the terrorist group, the overall likelihood of Threat Scenario 2 is therefore logically lower than for Threat Scenario 1, at least for existing nuclear weapons states.

The only major difference here is that there are many more states that possess fissile materials than those with working nuclear weapons, which might broaden the scope of possible nuclear sponsors, or the opportunities for complex false flag-type operations. However, of the non-nuclear weapons states who possess stocks of fissile material, the only known state sponsor is Iran. It is difficult to envision countries such as Japan, Brazil, or the Netherlands knowingly handing fissile materials over to a terrorist group. Moreover, the same strictures that apply to nuclear weapons states in this regard would apply to Iran.

Threat Scenario 3: State assists the terrorist organization in attacking a nuclear facility

This type of sponsorship entails state assistance to the terrorist group in carrying out an attack on a nuclear facility in the targeted state. A few provisos are necessary for an attack to be included in this category. First, the support must consist of direct assistance in aid of the specific attack on the nuclear facility, rather than general assistance that the state is providing to the terrorist group over a longer period. Therefore, a state that generally provides funds to a terrorist group that, as part of its campaign against a target state, decides to attack a nuclear facility in that state without any direction or specific assistance from its sponsor is not considered an example of Threat Scenario 3.

Second, there must be nuclear material present at the facility and the intent of the attack must be some type of large-scale release of radioactive material (such as a 'meltdown' event).[41] Therefore, an attack on a facility that is still under construction or one where the objective is to disrupt the normal operation of the facility (both of which have occurred on numerous occasions)[42] would not be included in this category. The type of assistance could vary greatly, ranging from merely providing intelligence to the terrorist group on how and where to attack, to training, equipping, and providing logistical support to the actual attack, such as fake passports or secure staging areas. These types of support for individual attacks are not uncommon in the annals of state sponsorship, a prime example being the extensive Iranian logistical assistance provided in the 1990s to Hezbollah operatives who attacked Jewish and Israeli targets in Argentina.

For the terrorist group, the main benefit would simply be to increase the chances of such an attack succeeding, since any attack against a nuclear facility (which in almost all cases represent hardened targets) would be an operationally complex endeavour. The costs for the organization could include sacrificing some autonomy in tactics and targeting, and an increased operational risk from foreign intelligence agencies due to interactions with the state sponsor's representatives. Moreover, several of the more sophisticated terrorist organizations might very well have the operational capability to conduct an attack against a nuclear facility on their own, so the additional risk and diminished freedom of action might not be worth it. Therefore, it is likely that only terrorist groups who would otherwise lack the capability to launch an attack against a nuclear facility would welcome a sponsor's assistance.

From the state's perspective, an attack of this type could provide a good opportunity to disrupt and weaken a rival while maintaining at least some plausible deniability. It allows for considerable economic and psychological consequences to be inflicted on an enemy, while utilizing an attack that does not carry stakes as high as those of nuclear weapons. With respect to potential costs, the issues of attribution and agency loss are still present but do not act as strongly as in Threat Scenario 1 and Threat Scenario 2. For attribution, nuclear forensics do not come into play, so the sponsor need only worry about other aspects of attribution (such as intelligence), which does not differ from non-nuclear state sponsorship cases.

Lieber and Press's argument[43] that attribution rates for high-casualty terrorist attacks are extremely high suggests that the more successful a state-sponsored terrorist attack against a nuclear facility, the more likely the state sponsor's identity would be revealed. Given the lower stakes when compared to a nuclear detonation, such attacks would not expect the same existential level of punishment from the targeted state even if the sponsor's role gets discovered. Moreover, while the principal–agent concerns that are present in any sponsorship dynamic would remain, they also have lower potential downsides in this scenario because the terrorists could not easily turn a nuclear weapon back on its sponsors. These factors would all lower the potential costs of this type of sponsorship as opposed to expectations for Threat Scenarios 1 and 2 and make Threat Scenario 3 substantially more likely to occur than the preceding scenarios.

Threat Scenario 4: State assists a terrorist organization in launching a cyber-attack against nuclear infrastructure

Given the increasingly networked and digital nature of all aspects of society, there have been concerns that state or non-state actors could undermine the security of nuclear infrastructure utilizing cyber means. Although digital systems in the nuclear sector are meant to be kept separate from other systems, or 'air-gapped' from external networks, this has rarely been the case in practice. Experts have called into question the cyber security of both civilian and military nuclear infrastructure.[44] Indeed, the Nuclear Threat Initiative (NTI) reports that many nuclear facilities worldwide have little to no cyber security measures and existing efforts lag behind the evolution and urgency of the threat.[45]

This category of nuclear sponsorship entails a state providing measurable assistance to a terrorist actor to attack nuclear infrastructure using cyber means. Note that the actors cannot simply be mercenaries or criminals hired by the state or covert organs of the state, because this would not constitute sponsorship. The defining factor here would be that the group would potentially seek to attack nuclear infrastructure independent of the state's involvement.

The three main forms of attack considered here are, in order of increasing severity but decreasing probability: 1) using cyber means to disrupt operations at a nuclear facility (which would not, strictly speaking, be considered 'nuclear terrorism'); 2) bringing about the uncontrolled release of radiation by cyber means, such as disabling cooling and safety systems to cause a meltdown; and 3) compromising nuclear weapons command, control, and communications systems to 'spoof' countries into erroneously launching a nuclear attack or, in extremis, to actually gain operational control of weapons through digital means.[46] The rather antiquated legacy systems utilized by the US, some of which might still be analogue in nature, probably makes this last form of cyber-attack there less of a threat currently than in some other countries. This could change, however, as systems are modernized.

From the perspective of terrorist actors, assistance from a state (which need not itself possess any weapons or fissile material for this type of attack) would, as in Threat Scenario 3, likely increase the chance of success. This would be especially useful if the state provided intelligence, specialized cyber intrusion tools, or access to an insider that could circumvent cyber security systems at the facility.

As for prospective state sponsors, two features might make these types of attacks attractive. First, cyber-attacks are relatively cheap compared to sponsoring physical attacks and, second, all cyber systems have vulnerabilities that can be exploited if these are discovered before being patched. Indeed, even low-resource states likely have superior cyber-attack capabilities to almost any terrorist organization. Furthermore, assisting terrorists can create additional barriers obfuscating the state's involvement, especially if the state assists terrorists with whom it has not traditionally had a relationship. After all, attribution of cyber-attacks is a difficult endeavour even in the non-nuclear context. Assistance with cyber-attacks can conceivably even be provided anonymously by the state to the terrorist group, thus further obscuring attribution.

Unlike the scenarios involving physical weapons or materials, there are fairly easy work-arounds to the principal–agent issue in cyberspace since, for example, the state can only provide terrorists with intelligence on or establish back doors into the specific facilities and systems that it wants targeted, minimizing the possibility that the terrorists would be able to go off-script. For all these reasons, assisting a terrorist group with cyber-attacks against nuclear infrastructure is even more likely than Threat Scenario 3, and thus the most probable of the four Threat Scenarios. However, it should be reiterated that only in extreme cases would either the state or the terrorist seek to initiate a nuclear war, so the first two forms of nuclear cyber-attack listed previously are far more likely than the third.

Comparing the Profiles of State-Sponsored Nuclear Terrorism

Table 5.1 provides a comparative summary of the relative costs and benefits for terrorists and sponsors for each nuclear terrorism Threat Scenario and links that to the estimated probability of each type occurring. Given that an instance of state sponsorship will only occur if the benefits exceed the costs for *both* parties, the overall likelihood of each type of sponsorship is determined by the lower likelihood assigned to each party. The costs, benefits, and likelihood are ranked on a basic seven-point scale with the values: Extremely Low, Very Low, Low, Moderate, High, Very High, and Extreme.[47]

The assessment suggests that states are unlikely to sponsor terrorist attacks through the provision of a military-grade nuclear weapon or fissile materials. While the benefits of these types of assistance to the few terrorist groups who seek to use nuclear weapons would be high, the costs, especially retribution, associated with such attacks would similarly be high for both sponsors and terrorist organizations. Given the additional principal–agent challenges for a state helping a terrorist develop an IND, this type of sponsorship is the least likely. The analysis comes to similar conclusions as those of other nuclear security experts.[48] In addition, a key set of differences between physical and cyber-attacks on nuclear facilities is that there are even greater challenges for attribution with respect to the latter in terms of identifying both the sponsor and terrorist organization—making this type of sponsorship the most likely.

The next step is to consider how the likelihood of each type of state sponsorship affects the risk of the associated types of nuclear terrorism. This calculation is based on the standard conception of risk as a combination of consequences and likelihood, with the latter determined by motivation and capability of the adversary. The consequences element is represented by estimating the *Target Consequences* as the severity of the consequences for the target of a successful nuclear terrorist attack. The motivation element is represented by an estimate of the *Number of Terrorists with Intent* to engage in that particular category of nuclear-related terrorism. The *Attack Difficulty* dimension captures our assessment of how hard it would be for the terrorist actor to carry out a

Table 5.1. Comparison of the likelihood of state-sponsored nuclear terrorism.[a]

Threat Scenario (TS)	Terrorist Org[b] Expected Benefits	Terrorist Org Expected Costs	Sponsor Expected Benefits	Sponsor Expected Costs	Likelihood of Sponsorship[c]
TS1: Provision of Military-Grade NW	Very High	Low to Moderate	Low to Moderate	Very High[d]	Very Low
TS2: Provision of Fissile Material	Very High	Moderate	Low to Moderate	Extreme	Extremely Low
TS3: Assist with Physical Attack on Nuclear Facilities	Moderate to High	Moderate	Moderate to High	Moderate to High	Moderate
TS4: Assist with Cyber-attack on Nuclear Facilities	High	Low	High	Low	Highest

a Table courtesy of the authors.

b Assuming the terrorist organization desires this nuclear capability.

c Likelihood estimate based on the benefits less costs for each party, with the lowest likelihood across both parties constituting the overall estimate.

d Except for certain extreme, highly improbable scenarios discussed in the text, for example miscalculation, intentional attribution deceit, and regime 'swan song'.

specified type of nuclear terrorist attack and is assumed to be inversely related to the likelihood of the terrorist possessing the capability required to do so. In determining the estimate of the *Overall Risk*, each tactic is considered to be the rough aggregate of the *Target Consequences*, the *Number of Terrorists with Intent*, and the reciprocal value of the *Attack Difficulty* (i.e., higher values of *Attack Difficulty* reduce the overall risk). The same scale is utilized as previously.

Notably, two versions of the risk are considered: the risk without state sponsorship being involved and then the risk if state sponsorship occurs. The main difference between them is the level of *Attack Difficulty* for the terrorist in each case. However, in the state sponsorship case, the overall risk must also be moderated by the likelihood that the sponsorship will indeed occur. This enables an assessment of the marginal risk of state sponsorship, in other words, the difference between the overall risk with and without state sponsorship. It must be emphasized that this exercise is not intended as a formal, probabilistic risk assessment with precise estimates, but rather as a structured theoretical examination of the various dynamics involved. As such, it is primarily orientated towards providing a comparative look at relative likelihoods and risks across the categories of nuclear-related terrorism, without making any determinations regarding precise probabilities or risk magnitudes.

Table 5.2 compares the marginal risks of each threat scenario. For the use of military nuclear weapons, the overall risk remains very low in both cases, even if state sponsorship

Table 5.2. Comparison of marginal risk of categories of nuclear terrorism.

			No State Sponsorship		With State Sponsorship		
Category Threat Scenario (TS)	Target Consequences	Number of Terrorists with Intent	Attack Difficulty	Overall Risk	Likelihood of Sponsorship (from Table 5.1)	Attack Difficulty	Overall Risk
TS1: Use of Military-Grade NW	Extreme	Low	Extreme	Very Low	Very Low	Low	Very Low
TS2: Use of an IND	Extreme	Low	Extreme	Very Low	Extremely Low	Moderate	Extremely Low
TS3: Physical Attack on Nuclear Facilities	Moderate to High	Moderate	High	Low to Moderate	Moderate	Moderate	Moderate
TS4: Cyber-attack on Nuclear Facilities	Low to High	High	High	Low	High	Low	Moderate to High

Table courtesy of the authors.

could reduce the difficulty for terrorists in carrying out such attacks—given how unlikely that is to occur. In the case of an IND, the even lower likelihood of sponsorship occurring, and the somewhat lesser diminution of *Attack Difficulty* makes the overall risk a little lower than without state sponsorship. Including state sponsorship in the risk calculation for physical attacks on nuclear facilities actually raises the risk, albeit not by much. It is in the realm of cyber-attacks, however, where state sponsorship appreciably increases the overall risk. That risk is still, at most, moderate–high, given the more limited consequences of most cyber-attacks against nuclear facilities. In sum, the provided analysis suggests that the higher likelihood of state sponsorship and lower potential difficulty of such attacks, especially in countries with poor cyber safeguards, means that cyber-attacks involving nuclear facilities may be the highest risk type of state-sponsored terrorist attack.

Conclusion

The consequences of a successful nuclear terrorist attack could be catastrophic, which is to be balanced against a relatively low likelihood that such an attack would occur. Understanding why and how states might seek to sponsor nuclear terrorism is important, especially if their assistance could increase the capacity and willingness of terrorists to engage in such attacks. This chapter suggests that, while not entirely inconceivable, governments are highly unlikely to provide terrorist organizations with support in acquiring nuclear weapons for use in attacks.

State sponsors are much more likely to support terrorist organizations in targeting nuclear facilities for physical attacks or cyber-attacks. The nuclear security community's focus on the risks of terrorists denotating nuclear weapons overlooks a potentially greater risk when it comes to attacks against nuclear infrastructure. The Fukushima Daiichi nuclear disaster demonstrated that even well-constructed nuclear facilities are susceptible to catastrophic failure and that a large-scale incident involving a nuclear facility can cause an international crisis. Moreover, there have been dozens of successful physical attacks against nuclear facilities in the past by states[49] and terrorist actors.[50] Israel's use of state-sponsored cyber and physical espionage against Iran's nuclear programme demonstrates that governments are willing to target their adversaries' nuclear programmes by clandestine means. If governments are willing to carry out such attacks overtly and covertly themselves, sponsoring terrorists to conduct such attacks may be an attractive alternative tactic in some cases. More scholarship is needed to understand the circumstances in which states may seek to leverage this type of terrorism, as it poses an under-appreciated and under-studied risk. In particular, scholarship that focuses on the cyber-dimension of this threat is particularly important.

The latter is the most likely scenario, as there would be greater difficulty implicating sponsors in such attacks, and retribution for cyber-attacks generally is not as severe as for physical attacks. While further empirical research is needed to better understand how and under what circumstances such attacks occur, the lack of transparency involving such cases makes that challenging. Studying only publicly known or successful cases of cyber-attacks against nuclear facilities could lead to false inferences about the prevalence of such attacks taking place and their characteristics.

Policymakers have tended to focus on unstable and insecure nuclear weapons possessors as posing the greatest source of nuclear terrorism risks. Once physical and cyber-attacks on nuclear infrastructure are taken into account, the pool of states that potentially pose nuclear terrorism risks, and that face heightened risks of being a target of nuclear terrorism, may be greater than commonly appreciated by those who focus mainly on the risks of terrorists detonating a nuclear weapon. Indeed, a state grows more vulnerable to being targeted with nuclear terrorism the more it expands its civil or military nuclear infrastructure.[51] Securing nuclear infrastructure from both external and internal threats should thus be a key priority for all countries that currently possess or plan to enhance nuclear programmes in the future.

Notes

1. Graham Allison, *Nuclear Terrorism: The Ultimate Preventable Catastrophe* (New York: Owl Books, 2004).
2. For example, Bob Graham and James Talent, *World at Risk: The Report of the Commission on the Prevention of WMD Proliferation and Terrorism* (New York: Vintage Books, 2008).
3. John Mueller, *Atomic Obsession: Nuclear Alarmism from Hiroshima to Al-Qaeda* (Oxford: Oxford University Press, 2009).
4. Matthew Bunn, Martin B. Malin, Nickolas Roth, and William H. Tobey, *Preventing Nuclear Terrorism: Continuous Improvement or Dangerous Decline?* (Cambridge, MA: Harvard Kennedy School: Belfer Center for Science and International Affairs, 2016).
5. Daniel Byman, 'Understanding, and Misunderstanding, State Sponsorship of Terrorism', *Studies in Conflict & Terrorism*, 45/12 (2020), 1–19.
6. Ibid. Passive state sponsorship also includes when a state knowingly does nothing to prevent its citizens or individual members of its government from providing support to terrorists.
7. This is derived from Jeremy Berkowitz's work but does not restrict itself to non-domestic terrorist groups and incorporates Berkowitz's later stipulation that the assistance must provide concrete advantages and cannot merely be rhetorical support in public venues. See Jeremy Berkowitz, *Dangerous Delegation: Explaining the Rationales and Outcomes of State Sponsorship of Terrorism through the Principal–Agent Framework* (Binghamton, NY: State University of New York at Binghamton, 2018), 10.
8. Byman, 'Understanding, and Misunderstanding, State Sponsorship of Terrorism', p. 3.
9. Berkowitz, *Dangerous Delegation*.
10. Ibid.; Navin Bapat, 'Understanding State Sponsorship of Militant Groups', *British Journal of Political Science*, 42/1 (2012), 1–29; Daniel Byman and Sarah Kreps, 'Agents of Destruction? Applying Principal–Agent Analysis to State-Sponsored Terrorism', *International Studies Perspectives*, 11/1 (2010), 1–18.
11. Berkowitz, *Dangerous Delegation*; Brittnee Carter, Maya Van Nuys, and Cagil Albayrak, 'State Sponsorship of Religiously Motivated Terrorism: A Deadly Combination', *Democracy and Security*, 17/2 (2021), 181–209.

12. Berkowitz, *Dangerous Delegation*.
13. Ibid.
14. Charles Ferguson and William Potter, *The Four Faces of Nuclear Terrorism* (New York: Routledge, 2005).
15. US Department of Defense, 20 December 1993, 'Nuclear Accident and Incident Public Affairs (PA) Guidance', Directive Number 5230.16.
16. Especially from the terrorist side this might not always be the case, but assessment of terrorist capabilities lies outside the topic of state sponsorship.
17. Mark Bell, 'Defending the "Acquisition-Use Presumption" in Assessing the Likelihood of Nuclear Terrorism', *International Studies Quarterly*, 63/3 (2019), 774–8.
18. William C. Potter and Gary Ackerman, 'Catastrophic Nuclear Terrorism: A Preventable Peril', in Nick Bostrom and Milan M. Circkovic (eds.), *Global Catastrophic Risks* (Oxford: Oxford University Press, 2008), 402–49.
19. Bell, 'Defending the "Acquisition-Use Presumption"'.
20. Gary Ackerman, Markus K. Binder, Lauren E. Pinson, and John Sawyer, *Profiles of Incidents Involving CBRN And Non-State Actors (POICN) Database* (College Park, MD: National Consortium for the Study of Terrorism and Responses to Terrorism, 2017).
21. Keir Lieber and Daryl Press, 'Why States Won't Give Nuclear Weapons to Terrorists', *International Security*, 38/1 (2013), 80–104.
22. Robert S. Litwak, 'Recalibrating Deterrence to Prevent Nuclear Terrorism', *The Washington Quarterly*, 40/1 (2017), 55–70.
23. Quoted in Graham Allison, 'Nuclear Terrorism: Did We Beat the Odds or Change Them?', *Prism*, 7/3 (2018), 12.
24. Michael A. Levi, *Deterring State Sponsorship of Nuclear Terrorism, Council Special Report*. No. 29 (New York: Council on Foreign Relations Press, 2008), 12.
25. Berkowitz, *Dangerous Delegation*.
26. Carter et al., 'State Sponsorship of Religiously Motivated Terrorism'.
27. Caitlin Talmadge, 'Deterring a Nuclear 9/11', *The Washington Quarterly*, 30/2 (2007), 21–34.
28. Lieber and Press, 'Why States Won't Give Nuclear Weapons to Terrorists'.
29. Michael Kristo and Scott Tumey, 'The State of Nuclear Forensics', *Nuclear Instruments and Methods in Physics Research Section B*, 294 (2013), 656–61; Wyn Bowen, Matthew Cottee, and Christopher Hobbs, 'Multilateral Cooperation and the Prevention of Nuclear Terrorism: Pragmatism over Idealism', *International Affairs*, 88/2 (2012), 349–68.
30. Talmadge, 'Deterring a Nuclear 9/11'; Lieber and Press, 'Why States Won't Give Nuclear Weapons to Terrorists'.
31. Lieber and Press, 'Why States Won't Give Nuclear Weapons to Terrorists'.
32. Marianne O. Larivee, 'Legitimate Use of Military Force Against State-Sponsored International Terrorism', *Denver Journal of International Law & Policy*, 20/2 (1992), 433–6.
33. Bowen et al., 'Multilateral Cooperation and the Prevention of Nuclear Terrorism'; Talmadge, 'Deterring a Nuclear 9/11'; Daniel Byman, 'Iran, Terrorism, and Weapons of Mass Destruction', *Studies in Conflict & Terrorism*, 31/3 (2008), 169–81.
34. Litwak, 'Recalibrating Deterrence to Prevent Nuclear Terrorism'; James R. Van De Velde, 'The Impossible Challenge of Deterring "Nuclear Terrorism" by Al Qaeda', *Studies in Conflict & Terrorism*, 33/8 (2010), 682–99.
35. Byman, 'Iran, Terrorism, and Weapons of Mass Destruction'.

36. Agency loss can also occur as a result of incompetence, for example, if the terrorist group detonated the weapon in a neighboring state to the sponsor in sufficient proximity or under sufficient meteorological conditions such that fallout ended up contaminating the sponsor's own territory as well.
37. Lieber and Press, 'Why States Won't Give Nuclear Weapons to Terrorists'.
38. Ibid.; Christopher McIntosh and Ian Storey, 'Between Acquisition and Use: Assessing the Likelihood of Nuclear Terrorism', *International Studies Quarterly*, 62/2 (2018), 289–300.
39. Ibid., 294, footnote 16.
40. If the state provides fissile materials and the expertise and equipment to construct a nuclear weapon, this falls more along the lines of Threat Scenario 1.
41. A state helping a terrorist group to attack a nuclear facility to enable the terrorist group to gain access to fissile materials for use in an IND falls under Threat Scenario 2.
42. Gary Ackerman and James Halverson, 'Attacking Nuclear Facilities: Hype or Genuine Threat?', in Brecht Volders and Tom Sauer, eds., *Nuclear Terrorism: Countering the Threat* (New York: Routledge, 2016), 111–41.
43. Lieber and Press, 'Why States Won't Give Nuclear Weapons to Terrorists'.
44. Charles Q. Choi, 'Nuclear Cybersecurity Woefully Inadequate', *IEEE Spectrum* [online], 7 October 2015. https://spectrum.ieee.org/nuclear-cybersecurity-woefully-inadequate; Page Stoutland, 'Growing Threat: Cyber and Nuclear Weapons Systems', *Bulletin of the Atomic Scientists* [online], 18 October 2017. https://thebulletin.org/2017/10/growing-threat-cyber-and-nuclear-weapons-systems/
45. The NTI report finds that 24 per cent of countries (12) receive a zero score for cyber security in the nuclear sector, while only 24 per cent receive a high score and only 4 per cent receive a full score (pp. 40, 47). See NTI, *The NTI Nuclear Security Index: Losing Focus in a Disordered World*, 5th edn (Washington, DC: Nuclear Threat Initiative, 2020).
46. Patrick Tucker, 'Hacking into Future Nuclear Weapons: The US Military's Next Worry'. *Defense One* [online], 29 December 2016. https://www.defenseone.com/technology/2016/12/hacking-future-nuclear-weapons-us-militarys-next-worry/134237; Anthony Barrett, *False Alarms, True Dangers?* (Washington, DC: RAND Corporation, 2016). https://www.rand.org/pubs/perspectives/PE191.html
47. The lowest and highest values of the scale are included given the sometimes extreme probabilities or values when dealing with nuclear weapons. Although the authors are interested primarily in comparing the relative magnitude of likelihood across options, rather than the absolute probability, and there is extensive debate surrounding the use of terms of estimative probability, they have in mind the following rough equivalences between their values and probability ranges: Extremely unlikely: <3 per cent; Very Unlikely: 3–10 per cent; Unlikely: 10–35 per cent; Moderate: 35–65 per cent; Likely: 65–90 per cent; Very likely: 90–97 per cent; Extremely likely: >97 per cent.
48. Richard Lugar, *The Lugar Survey on Proliferation Threats and Responses* (Washington, DC: Senate Foreign Relations Committee, 2005); Lieber and Press, 'Why States Won't Give Nuclear Weapons to Terrorists'.
49. Matthew Fuhrmann and Sarah Kreps, 'Targeting Nuclear Programs in War and Peace: A Quantitative Empirical Analysis, 1941–2000', *Journal of Conflict Resolution*, 54/6 (2010), 831–59.
50. National Consortium for the Study of Terrorism and Responses to Terrorism, *Nuclear Facilities Attack Database* [online] (College Park, MD: National Consortium for the Study

of Terrorism and Responses to Terrorism, 2016. https://www.start.umd.edu/data-tools/nuclear-facility-attack-database

51. Bryan Early, Matthew Fuhrmann, and Quan Li, 'Atoms for Terror? Nuclear Programs and Non-Catastrophic Nuclear and Radiological Terrorism', *British Journal of Political Science*, 43/4 (2013), 915–36.

Suggested Reading

Bell, Mark, (2019), 'Defending the "Acquisition-Use Presumption" in Assessing the Likelihood of Nuclear Terrorism', *International Studies Quarterly*, 63/3, 774–8.

Byman, Daniel and Kreps, Sarah, (2010), 'Agents of Destruction? Applying Principal–Agent Analysis to State-Sponsored Terrorism', *International Studies Perspectives*, 11/1, 1–18.

Ferguson, Charles and Potter, William, (2005), *The Four Faces of Nuclear Terrorism* (New York: Routledge).

Lieber, Keir and Press, Daryl, (2013), 'Why States Won't Give Nuclear Weapons to Terrorists', *International Security*, 38/1, 80–104.

Stoutland, Page, (2017), 'Growing Threat: Cyber and Nuclear Weapons Systems', *Bulletin of the Atomic Scientists* [online], 18 October. https://thebulletin.org/2017/10/growing-threat-cyber-and-nuclear-weapons-systems/

CHAPTER 6

CYBER THREATS AND NUCLEAR VULNERABILITIES

ANDREW FUTTER AND JASON T. HARRIS

Introduction

In 1993, David Arquilla and David Ronfeldt warned that 'cyberwar is coming' and that the world had entered a new digital era of conflict and security threats.[1] In truth, evidence suggests that, by the early 1990s, a certain type of computer, information, or net-centric war had already arrived, although not as yet known by the 'cyber' moniker. Three decades later we live in a world far removed from the late twentieth century, where computers, networks, and digital technology have become all -encompassing, and almost every part of our everyday lives are impacted by what has come to be known colloquially as 'cyber'. Perhaps nowhere has this dynamic become more acute than in the field of nuclear security. Securing civilian and military nuclear infrastructure is critical in preventing a radiological release or, in a worst-case scenario, an accidental or deliberate nuclear explosion. Historically, nuclear security, both civilian and military, has focused on physical threats, such as theft of nuclear materials, sabotage of facilities, or direct attacks with kinetic weapons, but today it is arguably of most concern that there exists the additional risk that hackers could gain access to nuclear material, cause a radiation leak, or interfere with the safe and secure management of nuclear weapons. This new and growing threat presents significant challenges to facility operators, national authorities, and international security more broadly.

The urgency and importance of working to minimize the risks posed to nuclear security from a range of cyber threats has spurned a diverse and rapidly growing literature.[2] This nascent but fast-growing body of work includes important insights from a broad range of stakeholders ranging from academia, the professional, think tank and non-governmental organization (NGO) worlds, as well as from governments and international organizations. It also reflects the fact that this is an issue that spans numerous traditional disciplinary boundaries from international relations and strategic studies to psychology, computing, and information technology, and an issue that spans both public and private sectors of society. The aim of the chapter is first, to review this literature and second, to build on it, seeking to understand the intersection of the innovative field of information technology with the much more established literature on nuclear security (both civilian and military).

The chapter begins with an explanation of the multifaceted and often opaque nature of the cyber threat, unpacking the different types of challenges and operations that fall under this broad moniker, before looking in detail at 1) the threats posed to civilian nuclear systems and infrastructure, particularly power plants and research facilities, and 2) the threats posed to military nuclear weapons systems, including the possibility of cyber operations or attacks against weapons and nuclear command, control, and communications, and the broader challenges posed by an increasingly complex digitized nuclear information space. The chapter continues by investigating international, national, and facility approaches that identify criteria reflecting the basic, minimum security measures and regulatory requirements necessary to protect vulnerable nuclear systems from cyber-attacks. The chapter finishes with a discussion of some of the measures that can be pursued and adopted to minimize risks in the cyber nuclear space.

Unpacking the Cyber Threat Spectrum

Before any analysis of the cyber threat to nuclear security can be fully understood, it is important to unpack what is meant by the phrase 'cyber', and to outline the many different dynamics that might fall under this moniker. Indeed, cyber has become a very widely used term that means many different things to many different people, with the result that experts, organizations, and governments talk past each other when it comes to computer- and information technology-related risks. This lack of an agreed language, semantics, and definitions has unquestionably hindered finding solutions to cyber nuclear risks and it is clear that shared understanding on the key questions, dynamics, and challenges must come before any serious attempt can be made to address and mitigate these.

To start, it must be accepted that the label 'cyber' is a fundamentally contested and nebulous concept,[3] and began life as a phrase that had little to do with computers, let alone nuclear security.[4] Indeed, it became popularized as part of the 1980s 'cyberpunk' movement rather than anything specifically to do with security.[5] As a result, using the phrase 'cyber' can sometimes be unhelpful because the label is often used to refer to many different things with vastly different implications. The same is true for the label 'cyber-attack' (which might, for example, refer to a piece of malware that steals personal information about employees of a private company, a malicious piece of coding that causes a particular software programme on a personal computer to temporarily 'crash', or a sophisticated and deliberate attack on the computer systems that support a national power grid).

The term is also used differently by different actors, which creates a major barrier to resolving tensions and finding agreement. An example of this is when then-United States (US) presidential nominee Donald Trump referred to 'the cyber' as a major challenge, and 'something that we must do better'.[6] But Trump is by no means the only person to confuse cyber as a noun rather than a prefix, or to treat the concept of 'cyber' as some sort of monolith. It is therefore essential to be clear and transparent about how the term is being used, even if common language cannot be agreed upon. This is arguably the most important first step in mitigating the sometimes-misplaced hype due to misunderstanding, and avoiding loaded, misleading, and often imprecise terminology such as phrases like 'cyber 9–11' or 'cyber-Pearl Harbour'.[7]

Cyber challenges and threats in the civilian and military nuclear realms are incredibly diverse. The cyber nuclear threat spectrum ranges from nuisance, espionage, and nuclear intellectual property theft at one end (for example, stealing design secrets, mapping out systems, 'hoovering up' masses of data in the hope of finding something useful), through to indirect attacks against, for example, human nuclear operators and personnel (via phishing scams or social engineering), sensitive data (that control key system processes), or the wider information space (through manipulation of social media for example), up to operations that could cause physical damage and disruption to a nuclear facility or command and control system. Arguably the most serious threats are cyber operations intended to cause a massive release of radiation, steal fissile material, or cause a nuclear launch or explosion. But consideration of the entire threat spectrum is important because many of the challenges are linked.

'Cyber threats' also range across a spectrum in terms of seriousness and sophistication. Thus, there is a big difference between state and non-state threats: different actors will have different intentions, capabilities, and vulnerabilities in this space. As a general rule, states are likely to be more cyber-capable than a terrorist or non-state organization, and more likely to have the time and resources necessary to carry out a sophisticated cyber-attack against a nuclear target, the best example of which is the Stuxnet attack against the uranium enrichment facility at Natanz in Iran discovered in 2010, where hackers gained access to the control systems that managed the centrifuges.[8] At the other end of the spectrum, non-state actors may have quicker and easier methods of achieving their aims, or utilize more indirect methods, such as stealing passwords or social engineering that facilitate non-cyber interference. Thus, we should not assume 'cyber' will be the preferred tool for certain operations against nuclear systems just because it seems to dominate the current debate. Consideration of to 'whom' these systems might be vulnerable, and particularly how intentions and capabilities will differ between actors, is therefore essential.

When it comes to the nexus of nuclear and cyber threats, a plethora of nuclear facilities, assets, and related operations exist as 'targets'. As mentioned previously, these targets can be divided between military and civilian assets. Military assets include nuclear command, control, and communications systems (NC3), nuclear weapons and their delivery vehicles, and the facilities that are used to produce, house, launch, and decommission them. These assets, depending on the country, can also include research and production laboratories and reactors, and nuclear-powered craft, such as submarines and surface ships. Civilian assets include nuclear power plants (NPPs), both operational and decommissioned, new reactors, including small modular reactors (SMRs), advanced reactors, and floating nuclear power plants (FNPPs), research reactors, fuel fabrication facilities, storage sites, and transportation. If the definition of 'nuclear' is extended to include all radioactive materials, the list of assets expands greatly. To name a few: hospitals, universities and research institutes, radiopharmaceutical companies, and irradiation facilities would also be included.

Regardless of the asset type, many experts believe that the consequences associated with the risks of cyber nuclear interaction come from five scenarios.[9] These five scenarios are: 1) the theft of, access to, or control of an intact nuclear weapon; 2) the theft/diversion of separated plutonium (Pu), highly enriched uranium (HEU), or mixed oxide (MOX) fuel; 3) an accident or sabotage resulting in a radiation release from a civilian or military nuclear system or facility; 4) public response or reputational damage from an actual, potential, or false cyber event against nuclear infrastructure; and 5) business interruption.[10] Other incidents involving the compromise of all sorts of sensitive nuclear information (about systems, processes, operations, weapons designs, people, etc.) could also be added to this list.

Another complexity of the 'cyber nuclear' challenge is determining how to go about classifying threat. The field of IT has been evaluating cyber risks since the 1980s and there are several well-established methodologies.[11] For example, the spoofing, tampering, repudiation, information disclosure, denial of service and elevation of privilege (STRIDE) and operationally critical threat, asset, and vulnerability evaluation (OCTAVE) models are used extensively for threat classification.

However, nuclear infrastructure, including NPPs, has only recently legislated regulations and implemented security controls against cyber risks. For example, the US Nuclear Regulatory Commission provided guidance in 2010, and centralized oversight of the agency's activities related to cyber security did not occur until 2013.[12] Other countries, including China and South Africa, only have generic nuclear laws dealing broadly with issues relating to the implementation of cyber security at nuclear facilities.[13] As a result, attempting to directly apply established IT classification methods for nuclear infrastructure is not feasible. One reason is because availability—which is the assertion that a system is available or accessible by an authorized user whenever it is needed—is considered important in nuclear infrastructure, whereas confidentiality is considered important in the IT field. Therefore, modifications are required to ensure the application of 'cyber nuclear threat classification' is appropriate.

What this all suggests is that the cyber nuclear security challenge is more diverse than perhaps first meets the eye. For sure, this also means that certain operations are potentially more worrying than others, or at least have far more severe possible implications. Importantly, it means that any attempt to manage these risks will need to be multifaceted and will likely involve many different and complementary mechanisms.

Civilian and Military Vulnerabilities

The threat from cyber-attacks against nuclear systems is increasingly perceived as a problem of national and international security given the growth in attack number and sophistication. Actors behind these attacks are no longer only private hackers or organized criminals but also nation states. Likewise, attacks once confined to networks and business computer systems have now been extended to instrumentation and control (I&C) systems, with all the implications and potential consequences such attacks may carry.

Nuclear assets, whether they are civilian or military, have progressively become more reliant on digital instrumentation, digital control systems, or computer-based information systems.[14] In early implementation, digitization was necessary to supplement or replace existing analogue systems for improved control and diagnostics, reconfiguration, and remote access capability. The digitalization of these systems has, however, increased their vulnerability to cyber-attacks. Intrusion by an external or internal malicious attacker could violate the confidentiality, integrity, or availability of data. Potentially worse, an attacker may have the capability of taking control of a process or system, leading to a failure event of significant consequences.

Power Plants, Civilian Infrastructure, and Nuclear Material

Cyber-attacks against civilian nuclear energy infrastructure have gone from the possible to now inevitable. Such attacks have the potential to bring down critical infrastructures, which are so vital to the country that their incapacitation would have debilitating effects on national security, public health, or safety. Notable incidents include attacks against the Bowman Avenue Dam in New York in 2013, the Ukrainian electric grid in 2015, and the Korea Hydro & Nuclear Power Co. Ltd in 2014.[15] The Bowman Avenue Dam attack has been attributed to Iranian nationals who were members of the Islamic Revolutionary Guards Corps, mainly targeting financial institutions and companies like the New York Stock Exchange and AT&T. The other two attacks have been linked to nation states: Russia in the case of the Ukrainian electric grid, and North Korea targeting the nuclear power plant in South Korea. These attacks against critical infrastructure, whether they were from sophisticated non-state actors or nation states, have demonstrated a changing landscape of adversarial attributes and capabilities.

The threats to nuclear facilities are no different. Nuclear facilities responsible for power generation, enrichment, and storage are complex computing environments comprised of hundreds to thousands of individual devices. These devices, and the computer systems that manage them, are built from a combination of common, off-the-shelf computing technologies and custom, one-of-a-kind hardware, software, and networking protocols. Contributing to the complexity is that fact that digitization has become necessary to replace broken or outdated analogue components in many systems, especially in NPPs.

Additionally, in order to extend the lifetime of a domestic NPP fleet, modernization upgrades to digital instrumentation and control systems are required. Balancing the need to digitize while trying to reduce the vulnerabilities that come with it is a tremendous challenge.

The vulnerability of nuclear facilities to cyber-attacks comes from the fact that digitization has led to increased complexity of systems needed for secure and reliable operation. The interconnected nature of these systems to external networks exposes them to cyber-attacks. Digital systems have historically been standalone with no need to interface with outside systems, as their sole purpose was to automatically regulate a physical process. As such, early systems were designed with no cyber security measures in place that would protect them from modern day cyber threats due to an implicit assumption that they would be isolated from the internet and other outside network influences.[16] However, digital isolation is becoming increasingly difficult with wireless networking (e.g. Wi-Fi, Bluetooth, cellular data networks) and the ubiquitous 'internet of things' (IoT), where virtually any technological device has some wireless connectivity.[17]

Digital isolation loss increases the vulnerability of systems to a number of different types of attacks. Well-known attacks are usually blocked automatically by signature-based detection methods or vaccines. In signature-based detection methods, the unique patterns of malicious codes are registered in advance to find similar patterns in the network to block them or the URLs containing them. However, these defence mechanisms are not able to respond to newly created malicious codes or intelligent mutant codes in real time. Any analysis of these codes depends on the system, equipment, and users in place, which may or may not be able to successfully defend an attack. Sophisticated attacks, such as an advanced persistent threat (APT), make it even more difficult to defend because these attacks persistently target a specific object over a long period of time. Since this kind of attack is not carried out based on the latest unknown vulnerability or with a routine attack method, it is not easy to respond to it using the existing pattern matching method based on the signatures. The use of artificial intelligence (AI), machine learning (ML), and big data analysis methods are now being developed to successfully defend against the ever-growing number and sophistication of attacks.

It should be evident now that the implementation of digital systems introduces new vulnerabilities, and these new vulnerabilities must be assessed. For civilian public infrastructure such as hospitals and academic institutions, traditional IT methods are probably mostly 'adequate' given their lack of I&C systems. Business (financial) and customer information at these organizations are more likely the targets of attacks and should be addressed accordingly by the IT security department. This is not to say that equipment, such as radiation-delivering medical devices, should not be managed as well. In fact, the IT security departments at these types of facilities may be at a disadvantage because they do not know the vulnerabilities in these systems. It is vital that these organizations have the support of their vendors, regulators, and competent authorities to fully understand the threats they face.

In the case of NPPs, the situation is more complicated. Digital I&C systems must be resilient to cyber threats, support safe operation of the plant, and provide the operator with clear and prioritized information to make distinctions between safety events and cyber-based instrument manipulation. In most countries, tools and operating procedures do not currently provide adequate support to differentiate fault conditions due to a cyber event from those due to component or system failure, regardless of the age of the plant. Traditional vulnerability assessment methodologies only analysed the physical systems associated with the plant. However, with the introduction of digital control systems and computer networks, these methodologies must be expanded to include these fundamentally different systems and vulnerability concerns.

Civilian nuclear facilities face significant challenges when it comes to modernizing and making updates to their operation, policies, equipment, and systems. These challenges include working with tight margins and small budgets, maintaining compatibility with one-of-a-kind technologies, meeting regulatory requirements, and limiting service interruption, all while ensuring safe operation before, during, and after a change is made. These challenges create significant hurdles when trying to keep pace with the ever-changing cyber threat environment.

Military Cyber Nuclear Vulnerabilities

Many similar dynamics are at play when we look at the challenges posed to military nuclear systems. To begin with, there is a growing recognition, and even acceptance, that cyber-attacks and computer network operations will play a significant (and possibly central) role in future warfare and strategic missions. This is not to say that such capabilities will be the only tool used, but rather that they will be used to complement and augment other kinetic forms of military power. It also seems likely that such capabilities will be used early in a crisis or conflict: one could envision attacks on critical national infrastructure or various systems used for the command, control, and communication of conventional military forces (some of which may play a—possibly unknown to the attacker—dual role in nuclear operations) being a key role for cyber operators. The cyber threat in the military nuclear realm therefore has different components: first there is the risk of escalation, unintended or otherwise, from the civilian or conventional military domain to the nuclear level; second, there is the risk of hackers seeking to cause the detonation or activation/launch of a nuclear weapon, either directly or indirectly; third is the possibility of breaking into nuclear weapons and support systems so that they do not work, or at least so that confidence in them is greatly reduced.

While the possibility of cyber interference causing a nuclear explosion or nuclear war is the most sensational risk, and one that has featured as the plotline of many moves (most notably the 1983 film *WarGames*), is it also probably the least likely and most difficult. This is because, thankfully, a single direct route to cause nuclear weapon use through a cyber-attack probably does not exist (to the best of the authors' knowledge). As such, perhaps the most pressing major concern is that sophisticated hackers manage to break into nuclear command and control or/and weapons systems in order to stop them working as planned. This has the potential to undermine confidence in a state's nuclear deterrent and in the ability to retaliate after a strike, which sits at the heat of stability through mutual vulnerability (or mutual assured destruction; MAD). The result might be that a state that feared its nuclear systems might be breached and compromised would seek to act pre-emptively.

At the time of writing, the most worrying application of this is in 'left of launch' operations. These are military offensive cyber operations designed to infiltrate an adversary's most sensitive weapons and control systems to prevent weapons, including nuclear weapons, from being used as part of a 'full spectrum missile defence' mission. The idea is that 'left of launch' operations with cyber or electronic warfare tools used before launch will complement 'right of launch' operations that utilize kinetic interception (shooting down a missile with a missile) once a missile has been launched.[18] It is possible, though unconfirmed, that US hackers were responsible for a series of failed North Korean missile tests in the mid-2010s.[19]

Notwithstanding this type of acute risk, we should be careful in seeking to view 'cyber' operations and risks in the nuclear sphere in isolation. It will be the intersection of cyber operations with different technologies across different domains, perhaps more than any one capability or risks limited purely to cyberspace, that will create new types of nuclear dangers, and both deliberate and inadvertent pathways to escalation and nuclear use. This will also clearly have implications for nuclear stability, deterrence, and arms control.

Cyber operations against nuclear weapons systems and interference within the broader information space within which nuclear weapons decisions are made are potentially far more worrying, dangerous, and destabilizing during a crisis. The possibility of indirect attacks designed to 'spoof' early warning systems, the use of mis- and disinformation, or interference with key data or personnel, combined with a general fear and uncertainty about an adversary's intentions and capabilities, seem almost certain to heighten pressures on decision makers and create destabilizing pressures.[20] This is especially the case for actors with weapons held at high states of alert, where it is at least conceivable that cyber operations could be used to convince operators to make precipitous decisions based on worst-case scenario planning and false information. Therefore, building trust and confidence between nuclear-armed states in the cyber nuclear world is essential.

Lastly, modernizing and improving the functionality of nuclear weapons and support systems might be considered a double-edged sword.[21] Basically, the more complex, capable, and interconnected these nuclear and other non-nuclear systems become, the more vulnerable they will be to attackers or unintended outcomes. As such, simplicity and separation across the military nuclear realm is arguably better than rushing towards greater capability when it comes to managing the cyber challenge to nuclear weapons systems. Unfortunately, a certain technological determinism seems to be driving military organizations ever forward toward a world of more complex, interconnected, and sophisticated nuclear weapons systems. As a result, taking stock and thinking through both the positive and negative implications of rapid technological change in this space are essential to manage the challenges going forward.

Risk Management: Security, Defence, and Regulation

Given the myriad and diverse nature of the threats to civilian and military nuclear systems, there is no one-size-fits-all remedy or means of managing and mitigating the cyber risk. Instead, what is required is a broad mixture of complementary but also quite different mechanisms that include different actors across different levels of the nuclear enterprise. These range from 'cyber hygiene' and better nuclear security culture, through enhanced cyber defence and cyber security practices, up to more ambitious transnational initiatives and normative cyber frameworks. That said, it is important that cyber defence/security, and broader research in this area for that matter, be focused on distinguishing cyber-attacks from other fault-induced events, such as safety events in nuclear facilities.

Cyber Nuclear Security Framework

There are many different approaches that can be taken to reduce cyber vulnerabilities. At the core of any approach is to utilize a defence-in-depth model for cyber security. The model includes various tiers of cyber security needed to address cyber threats. These tiers extend from the international/national level down to the facility. Figure 6.1 shows an example of a defence-in depth model for cyber security.

Using the defence-in-depth-model as a base, a cyber nuclear security framework can then be constructed. The standard approach is to begin with international guidance, such as that provided by the Nuclear Security Series (NSS) of the International Atomic Energy Agency (IAEA), including the Fundamentals document NSS 20, which sets out the objective and essential elements of an appropriate and effective national nuclear security regime.[22] This approach is broader than is needed purely for cyber; but most essential elements can play a role when assessing a nation state in terms of cyber nuclear readiness, such as national legislation, regulatory framework, licensing, and education and training. More specific guidance, such as the IAEA document 42-G, 'Computer Security for Nuclear Security', provides detailed information about developing, implementing, and integrating computer security as a key component of nuclear security.[23]

At the national level, legislation should ideally reflect a contemporary approach to nuclear security, that is, reflecting the rapid evolution the field has seen since the 11 September 2001 terrorist attacks on the US and among others, incorporating

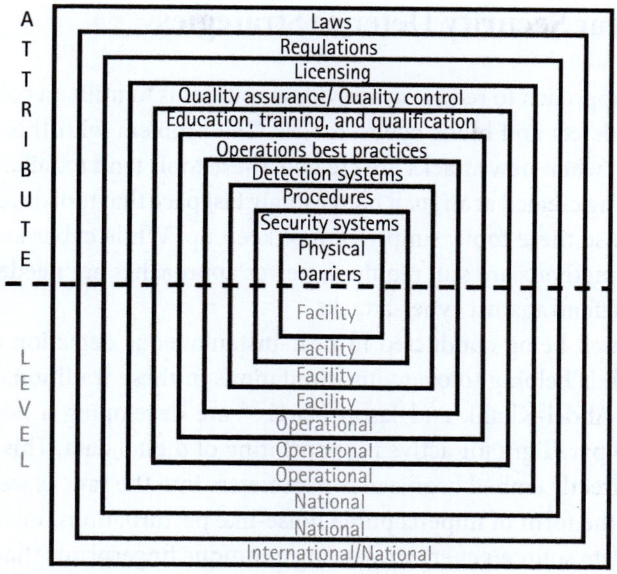

FIGURE 6.1. Defence-in depth model for cyber nuclear security.

Diagram adapted from figure by the Institute for Security and Safety, 2015.

concepts expressed in the 2005 Amendment to the Convention on the Physical Protection of Nuclear Material (CPPNM) as well as including or referring to the security of information as one of the key elements of nuclear security. Cyber threats should also be incorporated into the design basis threat (DBT) framework. This has proven to be difficult though, as infrastructure deficits and financial limitations in many countries have already limited the practical application of the DBT. Revising the threat assessment to include cyber is likely to worsen the current unsatisfactory situation.

From a regulatory perspective, specific authority details that are bound to evolve rapidly, like technology, should be avoided. Instead, the approach should focus on correct operation of the regulatory authority, on its ability to write and enforce regulations, and on the criminalization and prosecution of relevant crimes. Regulations also should provide guidance to licensees and industry and be able to adapt and respond quickly to change. For this to happen the process of regulation approval and modification itself needs to change. This approach truly involves a paradigm shift given how slowly the legislation process proceeds in most nation states.

Moving down to the licensee and operator, ideally cyber nuclear security should be embedded into the design of facilities themselves and their associated security plans. For this to happen, the licensing process and enforcement of policies need to be integrated throughout the entire lifecycle of the facility. A key part of this integration comes through the establishment of a strong cyber nuclear security culture. This is accomplished through successful implementation of education and training to ensure the necessary competence of the workforce.

Cyber Nuclear Security Defence Strategies

The traditional approach to reduce cyber vulnerabilities is to utilize a collection of tools that attempt to detect and block cyber-attacks. The problem with this 'attack-centric' approach is that, when new attacks are created, these tools tend to fail. Also, new types of attacks are being created at an such increasingly fast pace that tool developers, and the customers that use these tools, simply cannot keep up. While cyber experts maintain that traditional methods are still required, newer approaches are needed to better detect, deter, and defend against cyber-attacks.

Current research being conducted in near-instantaneous detection and mitigation of a cyber-attack is helping to overcome limitations of these traditional methods. For example, Hany Abdel-Khalik and Jason Harris[24] are developing a novel cognizance (self-awareness) paradigm for active fingerprinting of digital data. This new paradigm is designed to directly embed cognizance parameters into the raw, as well as processed, data streams in the form of imperceptible noise-like perturbations, inserted as close as possible to the data source/generator, serving as unique fingerprints that can be used to create verifiable and detailed records of operation. This technique has no impact on operation and is immune to self-learning via adversarial ML techniques, thereby ensuring resilience to tampering.

Other novel technical approaches to combat cyber-attacks include human reliability analysis and the use of AI, ML, and Monte Carlo (MC) techniques, the latter of which are simulations used to model the probability of different outcomes in a process that cannot easily be predicted due to the intervention of random variables. This technique is used to understand the impact of risk and uncertainty in prediction and forecasting models. For example, Hee Eun Kim and colleagues explain that cyber-attacks on non-safety systems can escalate into plant safety threats by inducing incorrect operator actions. Their work focuses on using conventional fault-tree (FT) analyses and cyber-attack propagation scenarios of operator actions that lead to the unavailability of safety systems and cause the failure of accident mitigation.[25] Wei Wang and colleagues use MC simulations and AI to identify the most vulnerable components of a cyber-attack.[26] Their approach is based on safety margins estimation, using an object-oriented model embedded within a MC-driven engine that injects different types of cyber-attacks at random times and magnitudes. Other methods, such as the use of Dynamic Bayesian Networks (DBN), which relate variables to each other over time, are also being explored to better distinguish abnormal events (safety events or cyber-attacks).

Another approach, introduced by George Chamales, is to prevent successful attacks by proactively addressing computer vulnerabilities.[27] This vulnerability-centric security approach is based on three fundamental principles that guide how security is selected, deployed, and managed. The first approach is to increase security by decreasing vulnerabilities. Here, facility operators focus on addressing a limited set of exploitable vulnerabilities in their systems instead of the ever-increasing number of attacks. The rationale is that, by eliminating a vulnerability, all attacks against that vulnerability are eliminated. The second approach is to decrease vulnerabilities using deterministic systems. Tools and strategies are applied to decrease vulnerabilities that ensure the systems do only what they are intended to do. This approach eliminates the need for deploying expensive, hard-to-manage, attack-detection technologies. The third approach holds that security should enhance operations. Facility operators manage their own defences using tools and techniques that increase their system's reliability. Following this framework avoids requiring dedicated security technologies that are only useful when under attack.

Regulation and Control

Ultimately, nation states and the nuclear facilities within their borders must adopt a strong cyber security culture. Several international and national organizations have published guidance in this direction, ranging from high-level guidance documents[28] to detailed technical guidance.[29] From these documents and existing best practices in the nuclear and other sectors, it is possible to model a preferred scenario of how such a strong cyber security culture can be established.

Adoption of a strong cyber security culture begins by accepting 'cyber norms'. Christopher Spirito from the US Idaho National Laboratory has outlined three proposed cyber norms for nuclear facilities.[30] First, personnel should consider whether observed events within the business and control system networks may be the result of a cyber campaign or attack. Next, incident response personnel should be routinely exchanging information within their vetted community of practitioners and exercising their cyber incident response communication plan with regional and/or national authorities. Finally, facility cyber defenders should consider implementing non-traditional capabilities to shift attack identification earlier in the cyber-attack lifecycle and improve their ability to actively engage with adversaries.

Standing back from this, it is also important to consider what role (if any) there is for international agreements and norm building in this space. This is especially important as there are no recognized international standards on how the national infrastructure for the cyber security of nuclear facilities should be organized. To some extent we have begun to see the international community begin to wrestle with what is undoubtedly both a global as well as local problem: The Council of Europe's Convention on Cybercrime,[31] known as the Budapest Convention, and the ongoing UN Group of Governmental Experts (GGE) on Developments in the Field of Telecommunications in the Context of International Security are good examples of this.[32] However, neither of these have really begun to touch on issues pertaining specifically to *nuclear* security.

There is some discussion about how agreement could be reached around prohibiting attacks on nuclear facilities as components of critical national infrastructure and talk about the possibility for a moratorium between states on cyber-attacks against NC3 systems.[33] But these have proved difficult to operationalize in practice, partly because the idea of cyber arms control is difficult and will probably look quite different to previous notions of arms control[34]—especially arms control in the nuclear and conventional military realms. But it is also because of the extra secrecy, mistrust, and intangibility that such endeavours face when the intersection with nuclear weapons is included. This is not helped by the lack of transparency, confidence building, and trust within the nuclear weapons space in general.

More broadly still there could be a role for traditional notions of deterrence in managing cyber nuclear security risks, although, given the highly heterogenous and diverse nature of the challenge, this could only really address a small, and the most acute, parts of the threat spectrum. A linked concept but perhaps one with wider applicability is 'cyber resilience',[35] though this concept potentially suffers from being rather nebulous and all-encompassing.

Conclusion

Nuclear security covers a wide range of different dynamics, threats, and risks, but arguably nowhere is the challenge more acute, growing, and less understood than in the cyber dimension. While cyber challenges in the nuclear realm are not as new as is sometimes portrayed in the mainstream global media, the nature and seriousness of cyber threats right across the nuclear enterprise are growing, diversifying, and becoming potentially more dangerous. It probably was not possible to cause a nuclear power plant to suffer a meltdown or somehow cause the launch of a nuclear weapon a generation ago through what we have come to think of as hacking, but today it is at least conceivable. This challenge is only likely to become more acute in the years ahead as modernization and further digitization across both civilian and military nuclear sectors creates more complexity and vulnerabilities for attackers to exploit.

What is striking about the cyber nuclear challenge is the number of similarities between the types of risks facing both the civilian and military sector; these range from the importance of culture and hygiene in guarding against intrusion to broader thinking about how to establish norms of control and governance. Indeed, many of the principal responses to managing cyber nuclear risks are not really about technology, but rather about people, practices, and education. Traditionally, human operators have been one of the easiest targets in terms of attack vectors and have played a role—wittingly or not—in many of the cyber-attacks seen thus far. Perhaps one of the major takeaways from this chapter therefore is that dialogue and the sharing of best practice between civilian and military nuclear security experts is absolutely essential (it is unclear if this currently happens).

Finally, cyber risks to nuclear security should not be seen in isolation. Whether this means understanding the link between compromising the security system at a particular research facility to facilitate a physical break-in, or how cyber operations might be used to prepare the battlefield and facilitate a strategic attack against a nuclear weapons command and control facility, the principle holds true. Neither should we immediately assume that all the cyber challenges are inherently new. Some, given the relatively new nature of certain digital systems and vulnerabilities, might be, but most will either look like older problems or take advantage of established pathways and thinking.

Notes

1. John Arquilla and David Ronfeldt, 'Cyberwar Is Coming!', *Comparative Strategy*, 12/2 (1993), 141–65.
2. This literature includes (but is not limited to): Andrew Futter, *Hacking the Bomb* (Washington, DC: Georgetown University Press, 2018); Herb Lyn, *Cyber Threats and Nuclear Weapons* (Stanford: Stanford University Press, 2021); Andrew Jenkinson, *Stuxnet to Sunburst: 20 Years of Digital Exploitation and Cyber Warfare* (Boca Raton, FL: CRC Press, 2021); Caroline Baylon, Roger Brunt, and David Livingstone, *Cyber Security at Civil Nuclear Facilities: Understanding the Risks* (London: Royal Institute of International Affairs, 2016); James M. Acton, 'Cyber Warfare and Inadvertent Escalation', *Daedalus*, 149/2 (2020), 133–49; Erik Gartzke and Jon Lindsay, 'Thermonuclear Cyberwar', *Journal of Cybersecurity*, 3/1 (2017), 37–48.
3. Andrew Futter, 'Cyber-Semantics: Why We Should Retire the Latest Buzzword in Security Studies', *Journal of Cyber Policy*, 3/2 (2018), 201–16.
4. Nobert Weiner, *Cybernetics: Or Control and Communication in the Animal and the Machine* (Cambridge, MA: MIT Press, 1961).
5. For example, William Gibson, *Neuromancer* (New York: Ace, 1984).
6. For example, Adrienne LaFrance, 'Trump's Incoherent Ideas About "the Cyber"', *The Atlantic* [online], 27 September 2016. https://www.theatlantic.com/technology/archive/2016/09/trumps-incoherent-ideas-about-the-cyber/501839
7. For example, Elisabeth Bumiller and Thom Shanker, 'Panetta Warns of Dire Threat of Cyberattack on US', *New York Times* [online], 11 October 2012. https://www.nytimes.com/2012/10/12/world/panetta-warns-of-dire-threat-of-cyberattack.html
8. Kim Zetter, *Countdown to Zero Day: Stuxnet and the Launch of the World's First Digital Weapon* (New York: Broadway Books, 2014).
9. Debra Decker, Kathryn Rauhut, Sara Kutchesfahani, and Erin Connolly. 'Nuclear Cybersecurity: Risks and Remedies'. Fissile Materials Working Group/Stimson Center Nuclear-Cybersecurity Workshop, 29–30 October 2018, Vienna, 2019.
10. Wilfred Wan, Andraz Kastelic, and Eleanor Krabill. 2021. 'The Cyber-Nuclear Nexus: Interactions and Risks'. United Nations Institute for Disarmament Research (UNIDIR).
11. Sang Min Han, Chanyoung Lee, Young Ho Chae, and Poong Hyun Seong, 'A Study on Classification of the Security Controls for the Effective Implementation to Nuclear Power Plant', *Nuclear Engineering and Technology*, 54/4 (2021), 1245–52. https://doi.org/10.1016/j.net.2021.10.009
12. US Nuclear Regulatory Commission, 'Cyber Security Programs for Nuclear Facilities', Regulatory Guide 5.71 (Washington, DC: USNRC, 2010).
13. Guido Gluschke, et al., *Cyber Security at Nuclear Facilities: National Approaches*, Institute for Security and Safety (Brandenburg: Brandenburg University of Applied Sciences, 2015).
14. Ibid.
15. John Peterson, Michael Haney, and R. A. Borrelli, 'An Overview of Methodologies for Cybersecurity Vulnerability Assessments Conducted in Nuclear Power Plants', *Nuclear Engineering and Design*, 346 (2019), 75–84. https://doi.org/10.1016/j.nucengdes.2019.02.025

16. D. Watts, 'Security & Vulnerability in Electric Power Systems', 35th North American Power Symposium, 20–21 October 2003, University of Missouri-Rolla.
17. Susan Adams, Nicole Murchison, and Robert Bruneau, 'Investigating Cyber Threats in a Nuclear Power Plant', *Chemical Engineering Transactions*, 77 (2019), 889–94.
18. See, for example, Andrew Futter, 'The Dangers of Using Cyberattacks to Counter Nuclear Threats', *Arms Control Today* [online], August 2016. https://www.armscontrol.org/act/2016-07/features/dangers-using-cyberattacks-counter-nuclear-threats
19. David E. Sanger and William Broad, 'Trump Inherits a Secret Cyberwar Against North Korean Missiles', *New York Times* [online], 4 March 2017. https://www.nytimes.com/2017/03/04/world/asia/north-korea-missile-program-sabotage.html
20. See Harold A. Trinkunas, Herbert Lin, and Benjamin Loehrke, eds., *Three Tweets to Midnight: Effects of the Global Information Ecosystem on the Risk of Nuclear Conflict* (Stanford, CA: Hoover Institution Press, 2020).
21. Andrew Futter, 'The Double-Edged Sword: US Nuclear Command and Control Modernization', *Bulletin of the Atomic Scientists* [online], 29 June 2016. https://thebulletin.org/2016/06/the-double-edged-sword-us-nuclear-command-and-control-modernization
22. IAEA, *Objective and Essential Elements of a State's Nuclear Security Regime*, IAEA Nuclear Security Series, No. 20 (Vienna: IAEA, 2013). https://www.iaea.org/publications/10353/objective-and-essential-elements-of-a-states-nuclear-security-regime
23. IAEA, *Computer Security for Nuclear Security*, IAEA Nuclear Security Series, No. 42-G (Vienna: IAEA, 2021). https://www.iaea.org/publications/13629/computer-security-for-nuclear-security
24. Hany Abdel-Khalik and Jason Harris, 8 September 2021. Personal communication with authors.
25. Hee Eun Kim, Han Seong Son, Jonghyun Kim, and Hyun Gook Kang, 'Systematic Development of Scenarios Caused by Cyber-Attack-Induced Human Errors in Nuclear Power Plants', *Reliability Engineering and System Safety*, 167/C (2017), 290–301.
26. Wei Wang, Antonio Cammi, Francesco Di Maio, Stefano Lorenzi, and Enrico Zio, 'A Monte Carlo-Based Exploration Framework for Identifying Components Vulnerable to Cyber Threats in Nuclear Power Plants', *Reliability Engineering and System Safety*, 175 (2018), 24–37.
27. George Chamales, *A New Approach to Nuclear Computer Security* (Washington, DC: Nuclear Threat Initiative, 2015).
28. IAEA, *Computer Security at Nuclear Facilities*, IAEA Nuclear Security Series, No. 17 (Vienna: IAEA, 2011). https://www.iaea.org/publications/8691/computer-security-at-nuclear-facilities; World Institute for Nuclear Security, *Security of IT and IC Systems at Nuclear Facilities – A WINS International Best Practice Guide*, Version 4.3 (Vienna: WINS, 2019). https://www.wins.org/document/4-3-security-of-it-and-ic-systems-at-nuclear-facilities
29. See International Society of Automation, 2007, 'Security for Industrial Automation and Control Systems', ANSI/ISA 99.00.01.
30. Christopher Spirito, Cyber Norms for Civilian Nuclear Power Plants, INL CON-16-39323 (Idaho Falls, ID: Idaho National Laboratory, 2016). https://www.osti.gov/servlets/purl/1358386
31. The Council of Europe, 23 November 2001, Convention on Cybercrime. https://www.coe.int/en/web/conventions/full-list/-/conventions/treaty/185

32. See https://digitallibrary.un.org/record/799853?ln=en
33. For a discussion of this and other policy recommendations, see Page O. Stoutland and Samantha Pitts-Kiefer, *Nuclear Weapons in the New Cyber Age* (Washington, DC: Nuclear Threat Initiative, 2018). https://media.nti.org/documents/Cyber_report_finalsmall.pdf
34. Andrew Futter, 'What Does Cyber Arms Control Look Like?' *European Leadership Network Policy Brief* [online], June 2020. https://www.europeanleadershipnetwork.org/wp-content/uploads/2020/06/Cyber-arms-control.pdf
35. Beyza Unal and Yasmin Afina, 'How to Deter Cyberattacks on Nuclear Weapons Systems', *Chatham House* [online], 18 December 2020. https://www.chathamhouse.org/2020/12/how-deter-cyberattacks-nuclear-weapons-systems; Yasmin Afina, Calum Inverarity, and Beya Unal. 'Ensuring Cyber Resilience in NATO's Command, Control and Communications Systems', Chatham House [online], 17 July 2020. https://www.chathamhouse.org/2020/07/ensuring-cyber-resilience-natos-command-control-and-communication-systems

Suggested Reading

Baylon, Caroline, Brunt, Roger, and Livingstone, David, (2016), *Cyber Security at Civil Nuclear Facilities: Understanding the Risks*. London: Royal Institute of International Affairs.

Chamales, George, (2015), *A New Approach to Nuclear Computer Security* (Washington, DC: Nuclear Threat Initiative).

Futter, Andrew, (2018), *Hacking the Bomb* (Washington, DC: Georgetown University Press).

Gluschke, Guido, et al., (2015), *Cyber Security at Nuclear Facilities: National Approaches* (Institute for Security and Safety. Brandenburg: Brandenburg University of Applied Sciences).

Jenkinson, Andrew, (2021), *Stuxnet to Sunburst: 20 Years of Digital Exploitation and Cyber Warfare* (Boca Raton, FL: CRC Press).

Lyn, Herb, (2021), *Cyber Threats and Nuclear Weapons* (Stanford,CA: Stanford University Press).

Spirito, Christopher, (2016), *Cyber Norms for Civilian Nuclear Power Plants*, INL CON-16-39323 (Idaho Falls, ID: Idaho National Laboratory). https://www.osti.gov/servlets/purl/1358386

CHAPTER 7

INSIDER THREATS TO NUCLEAR SECURITY

MATTHEW BUNN

INTRODUCTION

THREATS from insiders—individuals with authorized access to protected locations, materials, and information—represent the most serious threats to nuclear organizations around the world.[1] Insider threats are rare but can be devastating. Nearly all the cases of nuclear theft in which the circumstances of the theft are known were perpetrated either by insiders or with the help of insiders, as were nearly all the known cases of sabotage of nuclear facilities.[2] Much the same is true of non-nuclear operations: a 2014 Sandia National Laboratory report on cases of large-scale, multi-million dollar 'perfect heists' around the world found that over half of them involved an insider—often a coerced employee, but sometimes a planted or recruited criminal.[3]

Insiders pose substantial threats because they have access inside many of the layers of an organization's security; they may understand the security system and its vulnerabilities; they are able to plan, collect intelligence from inside the facility, and recruit for months at a time; and they are known and trusted individuals, which often makes the organization slow to detect and react to possible indicators of a threat. As noted by the physical protection recommendations of the International Atomic Energy Agency (IAEA), insiders 'could take advantage of their access rights, complemented by their authority and knowledge, to bypass dedicated physical protection elements or other provisions, such as safety procedures'.[4] Nuclear organizations must find ways to protect against insider threats while building internal trust and cooperation needed for their success.

The chapter first describes several illustrative cases of devastating insider threats to highlight the ongoing problem and some of its characteristics. Second, it outlines some of the categories of insider actions and motivations that insiders may have for taking them. Third, it describes some of the organizational challenges of mitigating the insider threat. Fourth, it offers a brief summary of terrorist interest in nuclear and radiological terrorism, as these interests relate to the insider threat. Fifth, it highlights some of the recent steps the international community has taken to address the insider issue. Sixth, it outlines good practices organizations are implementing to address the problem. Finally, it offers conclusions and areas for further research.

Illustrative Cases

To highlight the reality of the insider threat to nuclear operations, it is worth briefly considering several real cases that have occurred over the years.

Theft of Highly Enriched Uranium, Russia, 1992

One of the first well-documented thefts of highly enriched uranium (HEU) occurred at the Luch Production Association in Podolsk, near Moscow, in 1992. The facility specialized in research, development, and production of rare earth metals and its alloys used in the nuclear industry, including nuclear fuel made from HEU. Leonid Smirnov had been a loyal employee for more than 25 years and had passed all relevant security reviews. However, hyperinflation relating to the collapse of the Soviet Union had destroyed the purchasing power of his salary. Knowing that if output was within a couple of per cent of input the difference was considered normal losses to waste, Smirnov began stealing 25–70 grams of HEU at a time, never setting off an alarm. Over a period of months, he stole 1.5 kilograms (kg) of 90 per cent enriched HEU. Smirnov was in fact arrested by accident, because he was with neighbours who the police suspected of stealing batteries.[5] This was one of numerous thefts of HEU or plutonium that occurred in the former Soviet states in the 1990s, nearly all of them apparently perpetrated by insiders or with insider help.[6]

Explosives on Reactor Pressure Vessels, South Africa, 1982

In 1982, Rodney Wilkinson, an employee at the Koeberg nuclear power plant in South Africa, planted four limpet mines inside the facility, causing US$159 million in damages and the plant's commissioning to be delayed by 18 months.[7] Wilkinson had earlier made contact with the African National Congress (ANC), at the time an anti-apartheid opposition group. The ANC helped Wilkinson bring the mines into the plant despite a procedure requiring him to change clothes before entering the plant. Wilkinson placed the mines on the supports for the steel pressure vessels; on electrical cabling for the plant; and in one of the control rooms. Wilkinson detonated the mines on a timer, at a time when the plant would be deserted so that no one was hurt. The plant was still under construction and not loaded with nuclear fuel, so there was no danger of a radioactive release.[8]

Shooting of Guards, India, 2014

In October 2014, Head Constable Vijay Pratap Singh of the Central Industrial Security Force (CISF)—the force that guards India's nuclear facilities and other major facilities—carried out a mass shooting at the Kalpakkam nuclear plant in Tamil Nadu. While at his posting, Singh removed an automatic weapon from the facility's armoury in the early hours of the morning, went to his supervisor who was sleeping, and killed him with a controlled burst of fire. He then killed two other CISF personnel and wounded two more before being captured. Reportedly, multiple red flags had been ignored—including Singh's explosive temper and a complaint to a doctor about stress and exhaustion that led to a temporary removal from weapons duties.[9] Singh was convicted and sentenced to three life terms in 2018.

Destruction of a Reactor's Turbine, Belgium, 2014

In 2014, a still-unidentified insider at the Doel-4 nuclear reactor in Belgium opened a locked valve and allowed all the lubricant for the reactor's turbine to drain out. The turbine overheated and destroyed itself. The cost of replacing the turbine and of the loss of electricity output while the reactor was down amounted to US$100–200 million, making this one of the largest economic sabotages ever.[10] The incident occurred outside the nuclear island and as such did not pose a threat of a radioactive release. The motive of the sabotage remains unknown.[11]

Categories of Insiders

As these examples suggest, insiders can play many different roles. Security managers have to worry about workplace violence, petty theft, minor violations of security rules—and rare but potentially truly catastrophic events such as sabotage intended to cause a major radioactive release, or theft of plutonium or HEU. Insider threats are typically divided into several categories, as outlined in the IAEA guidance:[12]

- **passive insiders** who may provide crucial information to adversaries, but do not otherwise take an active part in actions against the organization;
- **active insiders** who carry out the key actions (such as Smirnov and Wilkinson); and
- **active violent insiders** who are willing to commit acts of violence to achieve their objectives (such as Singh).

Insiders may be conscious, or they may be inadvertent—unaware of the damage their actions may do to the organization. Insiders are often characterized by their levels of access to protected materials, information, and locations; authority over operations or security arrangements in the organization; and knowledge, particularly of the protected items and of the security system and its vulnerabilities. Clearly, insiders with more access, authority, and knowledge pose greater dangers. As one senior Russian security manager put it, members of the guard force can be 'the most dangerous internal adversaries', given their knowledge of the security system, access to key locations, and authority to carry weapons and to give a variety of directives to staff.[13] Yet, turnover among guard forces is often very high,[14] presenting a significant risk, given their privileged security knowledge and access.

Insiders may act alone (as appears to have been the case for Smirnov and Singh), they may conspire with outsiders (as Wilkinson did), or they may conspire with other insiders. Conspiracies involving multiple insiders, familiar with the weaknesses of the security system (and in some cases including guards or managers), are among the most difficult threats for security systems to defeat. Because of the difficulty of defending against multiple insiders, nuclear security systems often include only a single insider in the threats they are designed to protect against. Yet insider conspiracies routinely occur. The previously cited Sandia study found that, in their database of crimes, 'multiple-insider heists are actually more common than single-insider heists'.[15] As an example from the nuclear industry, an insider conspiracy at one of Russia's largest nuclear weapons facilities was behind an attempt in 1998 to steal 18.5 kg of HEU—potentially enough for a nuclear bomb.[16]

Insider Motivations

What motivates insiders to carry out hostile actions? A standard list is money, ideology, compromise (or coercion), and ego (or entitlement), giving the handy acronym MICE.[17] In the selection of incidents described, Smirnov was motivated by money and Wilkinson by ideology. The most loyal employees can be subject to coercion. For example, the Northern Bank in Northern Ireland had a security system that required two senior officials to turn their keys in order to open the vault—but in 2004 a gang held the families of two senior officials hostage and, to protect their families, they not only opened the vault but carried the loot to the robbers' van, and the gang made off with millions of dollars.[18] Ego—feeling that he was smarter than those trying to catch him and had never been adequately recognized—clearly played a role in the spying of Robert Hanssen, perhaps the most devastating mole in FBI history.[19]

But the MICE categories do not convey the full suite of motivations that can lead to insider actions against the interests of an organization. Mental health crises can be a key factor, particularly in violent insider incidents, as appears to have been the case with Singh's murders. In situations ranging from retail theft to information technology (IT) sabotage, disgruntlement has also been found to be a key driver of insider threats. For example, one important study of IT sabotage found that 92 per cent of the cases examined occurred 'following a negative work-related event such as termination, dispute with a current or former employer, demotion, or transfer.' Well over half of the insiders in these cases were already perceived by others in the organization to be disgruntled.[20]

Chelsea (then known as Bradley) Manning's decision to provide a vast trove of classified documents to Wikileaks from 2009 to 2010 is a classic example of the importance of disgruntlement. On deployment in Iraq, as a biological man dealing with gender issues in the days of 'Don't Ask, Don't Tell', Manning reportedly felt isolated and alone. After being told at a 'counselling' session that she would lose her one day off a week as a result of persistent lateness, Manning reportedly flipped over a table, damaging the computer that was on it, and had to be restrained from going for the gun rack in the room. Three weeks later, she began systematically downloading classified documents.[21]

Singh's murders appear to have been related not only to mental health, but also to disgruntlement driven by a poor relationship with the supervisor he murdered. Bullying bosses are perhaps the biggest source of employee disgruntlement; if organizations can deal with them, this can significantly mitigate the insider threat.[22] Inadvertent insiders—people who create vulnerabilities or contribute to conspiracies without meaning to—can also be a key problem, particularly for cyber security. Most cyber breaches involve an insider creating a vulnerability by clicking on a link or opening a document they should not have—or worse, hooking up an insecure computer or storage device to the system.

The escape of two convicted killers in 2015 from the Clinton Correctional Facility in Dannemora, New York, provides a compelling example of the dangers a non-malicious insider can pose. One of the insiders who helped the prisoners escape, Joyce Mitchell, a prison worker, did so consciously, in part because of nascent sexual relationships she allegedly had with them.[23] But another of the insiders who helped them claims he did so inadvertently. Gene Palmer, a corrections officer, reportedly gave the prisoners paint, a screwdriver, and access to the catwalk they later used to escape in return for paintings by the inmates and tips on the activities of other prisoners—never realizing that he was contributing to an escape plan.[24]

Nuclear Terrorism and Insider Threats

Clearly, the dangers posed by both hostile and inadvertent insider actions at nuclear facilities are very real. Terrorist interest in nuclear and radiological terrorism is also very real. While most terrorist groups have no interest in a nuclear level of violence, a few do. Both the Japanese terror cult Aum Shinrikyo and *jihadist* militant group al-Qaeda had active programmes to attempt to get nuclear weapons. Chechen terrorists stole radiological sources, plotted attacks on reactors, and Russian officials and a Russian state newspaper have reported that Chechen terrorists carried out reconnaissance at nuclear weapon storage sites and on nuclear weapon transport trains.[25]

For decades, Western counter-terrorism efforts have focused primarily on violent Islamist extremists. This has led to concern about Islamist extremists being employed at nuclear facilities, which happened in the cases of Sharif Mobley (a former New Jersey nuclear worker convicted of killing a prison guard) and Ilyass Boughalab (a former employee of the Doel plant, and killed in Syria). In recent years, however, many advanced democracies have faced increasing threats from a different kind of extremist: violent (and often racist) nationalists. This threat was highlighted in the 2011 attack by the Norwegian far-right extremist Anders Breivik, behind the massacre at the Workers Youth League summer camp and bombing of the prime minister's office building, and the violent effort on 6 January 2021 in the United States (US) to prevent Congress from certifying the democratic transfer of power.

The rise of violent nationalist extremists poses a particular insider threat problem for nuclear organizations, for several reasons. First, extreme views which *could* lead to violence are increasingly widespread in the populations of many advanced democracies. Second, in most democracies, holding such views and even being a member of extremist groups is constitutionally protected. An organization cannot fire a guard or staff member simply for believing his or her country should be reserved for white people and that its current government is illegitimate—yet those beliefs, and membership in groups that espouse them, can readily lead to violence. Third, right-wing extremists appear to have a particular fascination with the possibility of nuclear terrorism. *The Turner Diaries*, a 1978 novel sometimes referred to as 'the bible of the racist right' envisions bringing down the government with nuclear weapons on American cities, to build a new, racially pure society. Breivik's bizarre and lengthy manifesto offered detailed descriptions of various types of nuclear terrorism. One US violent extremist group even gave itself a nuclear name, 'Atomwaffen Division', German for the nuclear weapons division.[26] Fourth, this right-wing extremist ideology appears to appeal to some current and retired military and police personnel—groups that nuclear organizations draw heavily on for their guard forces. Ashli Babbitt, the woman killed by US Capitol Police during the violent effort on 6 January 2021 to reverse the US presidential election, was a retired Air Force security officer who had been a guard at two nuclear power plants.[27]

Nevertheless, to date there is little publicly available evidence of connections between nuclear insiders and terrorist groups. In one study, scholars compiled both a list of known incidents of hostile actions at nuclear facilities and a list of discussions of nuclear terrorism possibilities in documents and online postings from Islamist terrorist groups. They found very few cases of *jihadis* discussing nuclear thefts or attacks at all and, within those, virtually no cases where they explored the nuclear insider option—that is, attempting to infiltrate one or more of their people into the staff of a nuclear facility or find ways to coerce or recruit existing staff.[28] On the other hand, al-Qaeda's cooperation with Pakistani nuclear experts in the Ummah Tameer-e-Nau (UTN) militant organization and reported attempts to get help from A. Q. Khan's network are instances of working to recruit help from insiders.[29] The lack of internet postings is indicative, but it may not be definitive; one would not expect *jihadis* to post about a secret plan to carry out such an action before it occurred. Moreover, insiders at nuclear facilities may become radicalized without terrorist groups actively seeking to recruit them—as appears to have occurred in the Boughalab case. And nuclear insiders could potentially steal nuclear material and provide it to terrorists for their own reasons—including profit—without *jihadis* seeking to recruit them.

Challenges of Mitigating the Insider Threat

A range of challenges makes it difficult for organizations to mitigate insider threats. First, insiders may have knowledge of the security system and its weaknesses. Hanssen, for example, was an FBI counter-intelligence officer who was able to monitor the ongoing search for a spy in the organization and modify his behaviour accordingly. Smirnov knew, and took advantage of, the limits of the nuclear material accounting system at his facility and the absence of portal monitors to set off an alarm as he carried the material out. Second, insiders can observe, plan, and in some cases recruit others to help, for months or even years at a time. Wilkinson, for example, worked out a method for smuggling in the explosives, despite having to completely change clothes before starting his shift—and practised by smuggling in bottles of vodka. Third, insiders have authorized access to pass through many of the layers of the security system. They need to defeat only the comparatively modest number of security measures that their access does not allow them to bypass. Fourth, insiders are trusted by other employees. Few imagine that their colleagues, whom they have known and worked with for years, might pose a danger. In high-security organizations, these colleagues have typically also been through a formal process of review and been officially approved as trustworthy, removing them further from suspicion. Fifth, a wide range of cognitive and organizational biases can lead managers to downplay the threats insiders pose to their organizations, facilities, and operations. Managers should beware of the 'halo effect', in which well-liked employees are assumed to be trustworthy—a special case of *affect bias*, the tendency we all have to assume that something we like for a particular reason has other positive qualities as well.[30]

Moreover, in most organizations there are few positive incentives for being alert to insider threats. Being the only person complaining about a possible security issue is not likely to be career enhancing; as the saying goes, 'the nail that sticks up gets pounded down'. One employee at the Y-12 National Security Complex in the US told the author about having reported unusual behaviour by a colleague he was working with who had special expertise in machining HEU; the colleague had his access to HEU temporarily revoked while an investigation proceeded. The investigation found reasonable explanations for the behaviour and the colleague resumed his work with HEU. The supervisor, rather than complimenting the employee for identifying a potential security issue and following the organization's reporting rules, complained that not having the colleague's special HEU machining skills for several weeks would put them behind schedule. That response is unlikely to provide much incentive for the employee to report unusual behaviour in the future.[31]

Organizations in general have biases that tend to downplay the low-probability, high-consequence risks of what they are doing.[32] Long periods without a major security incident—as occur in almost all nuclear organizations—breed complacency and lack of concern. Moreover, responsibility and authority for dealing with potential insider problems are often dispersed among managers responsible for operations, for security, for legal affairs, for human resources, and other areas. The existence of a problem might be obvious if all the relevant information was in one place, but often that is not the case.

An examination of past cases makes clear that people sometimes fail to report, or organizations fail to act on, even the most extreme 'red flags'. Bruce Ivins, for example, the likely perpetrator of the 2001 anthrax attacks in the US, was thought by his first therapist to be the most dangerous patient she had ever seen (though the organization never bothered to look at his medical records), and set off countless red flags (including frightening a staffer so much she went to his supervisor; Ivins also spent many unexplained nights after hours alone in the hot cell with the anthrax).[33] But no one in the organization took action—in part because his strangeness over many years led colleagues to write off concerning behaviour as just 'Bruce being Bruce'.[34] Similarly, Major Nidal Hasan, who ultimately gunned down many of his colleagues at Fort Hood in 2009, had openly expressed support for violent *jihad* and was known to be emailing Anwar al-Awlaki, one of al-Qaeda's most prolific recruiters—and nothing was done to stop him. Many of his superiors were disciplined but, as Amy Zegart notes, when so many fail to act, the problem is with the system, not the individuals.[35]

Cooperation, teamwork, and trust are essential for an organization to function successfully. Managers are more likely to be focused on building teamwork than on encouraging employees to keep an eye out for the possibility that their colleagues are betraying the organization. Finally, organizational response is constrained by limited learning from experience. Insider incidents are typically shrouded in secrecy; no comprehensive set of incidents and lessons learned from them exists. Otto von Bismarck once said that only a fool learns from his mistakes; a wise man learns from the mistakes of others.[36] Leaders of nuclear organizations, however, have few opportunities to do so.

International Action on Mitigating the Insider Threat

Fortunately, nuclear organizations around the world are now taking action to try to reduce the dangers posed by insider threats. Most nuclear regulators require operators to take specific actions to protect against threats from both outsiders and insiders, and many have strengthened their requirements in recent years. Bilateral nuclear security cooperation programmes put significant focus on strengthening protection against insider threats. Organizations such as the IAEA's Division of Nuclear Security and the World Institute for Nuclear Security (WINS) offer guidance on what organizations should do to protect nuclear operations against insider threats.[37]

At the Nuclear Security Summits from 2010 to 2016, the participants repeatedly emphasized the importance of protecting against the insider threat. At the 2016 summit, one of the major group commitments or 'gift baskets' was an initiative focused on protecting against insider threats. Like other initiatives from the summits this has now been opened to all states, and subsequently distributed as an IAEA Information Circular, INFCIRC/908.[38] As of early 2021, 31 states were participating (along with Interpol as an observer), though unfortunately major nuclear players such as Russia, France, China, India, and Pakistan were not yet participating.[39] The participants have created an international steering committee on insider threats, held an international symposium on the topic in Belgium in 2019, and are pursuing a variety of ongoing activities.

Good Practices at Nuclear Organizations

Protecting against insider threats while maintaining a culture of teamwork and trust necessary for an organization to be successful is an inherently difficult challenge. Organizations must balance between operational and security imperatives, and between responding rapidly and effectively to real threats, while avoiding ruining people's careers over false accusations. Nuclear organizations have different items to protect and need to design their approaches within the context of their national and organizational laws, policies, and cultures. Hence, there is no one-size-fits-all approach to mitigating insider threats. Moreover, only very limited hard data are available to assess which elements of insider mitigation programmes for nuclear organizations are most important and effective. Nonetheless, a few broad categories of effort are likely to be essential to an effective programme.[40]

A truly effective nuclear security system must be able to protect against sophisticated individuals, multiple insiders working together, and insiders working with outsiders. This requires a comprehensive, multi-layered approach, rather than relying on only one or two protections—because any single measure can be defeated by a clever insider.[41] The goal should be to maximize the scale and complexity of the challenges insiders would have to overcome to achieve their aims. Such an approach should include action to focus the organization on mitigating the insider threat; to screen and monitor staff; to train and motivate staff; to control, monitor, and account for protected materials, information, and areas (including effective protection against cyber insiders); to limit and monitor insider access to those protected materials, information, and areas; to conduct investigations when needed; to minimize opportunities available to insiders; and to assess, test, and learn over time.

Focusing the Organization on Insider Threats

To combat insider threats effectively, organizations must build both high-performance and high-vigilance organizational cultures. This requires strong buy-in and focused effort from the top leaders of the organization. As the US government's National Insider Threat Task Force (NIITF) puts it, 'successful insider threat programmes receive strong, personal, and visible support from the agency head'.[42] Offering compelling stories of the damage to organizations caused by insiders in the past, and plausible scenarios for what could happen within their own organization, may help focus leaders' attention. Protecting against insider threats requires many parts of an organization to work together effectively. The US Department of Energy (DOE), for example, requires its sites to establish 'Local Insider Threat Working Groups' that involve representatives from security, counter-intelligence, information assurance, human resources, and the legal department, among others. Ultimately, however, the organization's leaders need to designate someone to take charge of the insider threat mitigation programme and to make decisions among competing recommendations. That responsible official and the organization's leaders will have to ensure that the working group has access to all the disparate sources of information it needs, and the resources required to implement the programme.

Identifying and prioritizing the items to be protected and the nuclear organization's potential vulnerabilities are, of course, crucial parts of protecting against insider threats, as for all security. In today's highly digitized environment, identifying key systems and information that might be vulnerable to a cyber intrusion is particularly key—including information that might be leaked or corrupted, alarms and detectors that might be disabled, systems that could be sabotaged, and more. Beyond the leaders of the organization and of the insider threat programme, mitigation is likely to require support and alertness from the entire staff, as part of building a strong security culture overall.[43] As the NIITF puts it, 'A highly aware workforce is key to the early detection and prevention of malicious insider threat conduct'.[44] Organizations should provide training for the whole workforce with real stories of disasters that alert staff could have prevented.

Screening and Monitoring Staff

A primary element of an insider programme is an effective system for screening employees for trustworthiness before giving them access to highly sensitive information, areas, or materials. Screening programmes should have the ability to check individuals' criminal and financial records, as well as access to any lists of individuals suspected of terrorist linkages that the authorities might maintain. Such programmes should be designed, however, with a reasonable level of speed and with fairness in mind; long delays, with no explanation or opportunity for the individual to appear when a clearance is denied (as sometimes occur in the US), are not good practice. While such background checks are important, managers should not rely on them too heavily, because 1) such checks are highly imperfect; 2) people change after the initial screening (as in the cases of radicalization within a few months); and 3) even trustworthy employees can be coerced. Hence, organizations should have ongoing programmes for monitoring employees and encouraging staff to report concerning behaviour. Such programmes should have access to, and the ability to integrate, information from a wide range of sources. As the NIITF puts it, 'maximum value is gained only when disparate information points are fused together to identify intricate patterns of conduct'.[45]

Getting such ongoing monitoring programmes to work effectively is difficult. Employees often resist reporting on friends and colleagues, even if they are acting in a suspicious manner; meanwhile, each individual incident, not combined with others, often appears too harmless to report. Moreover, in many organizations, employees often choose not to report problems because of the headaches involved, or they pass troublesome employees off to someone else to deal with. Indeed, US Defense Department research suggests that indicators of insider security problems are systematically underreported.[46] One study of IT sabotage found that almost all of the insider perpetrators had exhibited concerning behaviour, which the organization had ignored.[47]

Putting in place processes to increase the chances of noticing and acting on such signals is crucial to reducing insider dangers. Organizations should make the process of reporting easy and known to all employees (including the possibility of anonymous reporting). They should take steps to give employees incentives to report and to prevent retaliation against those who do. It is crucial to establish a process for responding to reports that employees understand and that is considered fair and reasonable. Training should include the possibility that reporting could lead to troubled individuals getting the help they need. As is the case with the US Personnel Reliability Program (PRP) for access to nuclear weapons and some weapons-usable materials, organizations should make it relatively easy and routine for employees to be excluded (or exclude themselves) from the most sensitive work if they are facing unusual stress or emotional issues, and relatively easy and routine for them to return to that work when those issues are resolved.

Another fundamental element of monitoring staff is making sure that no one is bringing contraband into or out of the facility—for example, bringing guns or explosives into the facility, or bringing nuclear material out. This typically requires searches and monitoring equipment as people are entering and exiting. Organizations need to think through all the potential pathways that something could be brought in or out, such as being passed through a window or sent through a pipe. Wilkinson smuggled the mines in by pushing them through a diaphragm pump adjacent to the airlock as he was stripping and donning protective clothing, and then picking them up on the other side. Even emergency evacuations should be planned to make it possible to ensure that no nuclear material is removed in the emergency.

Training and Motivating Staff

Nuclear organizations need to build organizational cultures in which all staff take nuclear security seriously, including the insider threat, and are always on the lookout for potential issues and vulnerabilities. As with safety, the focus must be on constant vigilance and continuous improvement, in a never-ending quest for excellence. The culture should be one in which everyone understands that security is everyone's responsibility, not something only the security team has to worry about.[48] Establishing clear incentives that make employees understand that they will be rewarded for good security performance is one key element of building such a culture, and of making clear the priority that management places on security.

Training should make the reality of the threat clear and vivid—and, as noted, should include real stories of insiders. The security systems and rules should be clearly designed to address the threat, and training should make clear why each key element of the security system is important and should not be ignored or bypassed. Workshops in which staff imagine how insiders might accomplish various types of thefts or attacks and then envision better ways to prevent such events represent one important approach to training on the insider threat; these highlight the threat and the relevance of security measures for addressing it, make staff feel included and involved, and help motivate staff to suggest improvements. The US State Department's Partnership for Nuclear Security has sponsored such workshops in several countries.

Employee satisfaction is another critical aspect of organizational culture. As noted previously, disgruntled employees are far more likely to pose insider threats. Nuclear managers should strive to build performance-orientated cultures in which employees believe that they are respected and treated well, in which they have avenues for their complaints and ideas to be heard, and in which they expect the organization to be helpful, rather than punishing, when issues of mental illness or emotional difficulties arise. Fortunately, organizations have found that it is not very difficult or expensive to combat employee disgruntlement. Providing complaint and ombudsman processes that are perceived to result in actions to address the issues, complimenting and rewarding employees for good work, and addressing the problem of bullying bosses all go a long way towards reducing disgruntlement and its contribution to the insider threat.[49]

Controlling, Monitoring, and Accounting for Protected Material and Areas

Keeping a constant watch and effective control over the items and areas to be protected is, of course, another crucial element of an effective insider protection programme. Nuclear material access areas, vaults, and nuclear facility vital areas should be continuously monitored with security cameras and alarm systems. In the case of nuclear material, the full suite of nuclear material control and accounting (MC&A) approaches can be used to help ensure that material does not go missing without being detected. MC&A for international safeguards and for protecting against insider threats, however, are not identical. For international safeguards, for example, it might not matter much how many people had access to a vault, whether the vault was monitored when people were inside, or whether nuclear material in use was left out in working spaces at night rather than being returned to the vault; for protection against insider theft, all of those factors might be important.[50]

Organizations should work to keep access to the protected items to a minimum—whether these are nuclear materials, equipment, locations, or information—and closely monitor whatever interactions with these items do occur. IAEA recommendations call for 'constant surveillance' whenever anyone enters an 'inner area' of a nuclear facility—for example, by use of the two-person rule.[51] In the case of weapons-usable nuclear materials, constant surveillance whenever there is human access to the material areas should be integrated with effective control and accounting of the material in an overall nuclear material accounting and control (NMAC) programme.

Conducting Investigations

Monitoring and reporting mechanisms will only have value if the organization is able to respond effectively when they raise a concern. Hence, organizations need to have effective processes for conducting investigations to clarify whether there is a real problem or not. These processes have to be seen by employees as fair and reasonable, or employees will be unlikely to report on issues they observe. Such investigations also have to respect relevant privacy laws and policies and provide protections for whistle-blowers. Many nuclear organizations have a first level of investigation that they perform internally, and then another level that is referred to others (such as law enforcement agencies) if needed.

Minimizing Opportunities for Insider Adversaries to Succeed

A variety of steps can minimize the opportunities insiders have to succeed in a theft or sabotage attempt.[52] Increased reliance on passive safety, for example, makes it more difficult for either accidental events or intentional actions to cause a major radioactive release. Locating backup safety systems far from the systems they are backing up makes it more difficult for either an accident or a saboteur to disable both systems. Minimizing the number of places where plutonium or HEU exist makes it possible to achieve higher security at lower cost by defending fewer locations and reduces the number of groups of insiders with potential access to the material or who might be able to assist in a theft attempt.

Minimizing large-scale bulk processing of nuclear materials—and ensuring the highest standards of security, control, and accounting for it wherever it occurs—is particularly important, as nearly all the known cases of theft of plutonium or HEU have been bulk materials, presumably in most cases stolen from bulk processing facilities. Activities such as reprocessing of plutonium and fabrication of fuels from plutonium or HEU inevitably create increased insider risks. At a minimum, such bulk processing should only proceed under the very highest attainable standards of security, control, and accounting. Processing systems that are highly automated and involve little or no direct human access to the material can also reduce the opportunities for insider adversaries to remove material. Keeping material in difficult-to-steal forms—such as large and heavy fuel assemblies containing only a modest fraction of weapon-usable material by weight—can also help minimize insider opportunities.

Assessment, Testing, and Learning

Finally, an effective insider protection programme requires an ongoing effort to assess and test the programme's effectiveness, to make corrections, and to learn from experience.[53] Rather than leaving their security system largely static until some crisis occurs, organizations need to find ways to learn and adapt as they go. As with safety, each incident or newly found issue should be examined for root causes and lessons learned and treated as an opportunity to improve. Performance tests should go well beyond simply testing whether a particular security component such as a camera is functioning as intended, to exploring the ability of intelligent insiders to find ways to defeat the security system.

The same is even more true of vulnerability assessments. Security managers need to find creative people with a hacker's mindset to come up with a wide range of ways that insiders might try to beat the security system—and then develop security measures that will be effective against a broad range of possibilities. A security system adequate to defend against the first few pathways thought of by an unimaginative committee is not likely to be good enough against the real threat. Such uncreative vulnerability assessments were the target for Roger Johnston's 'Thanks for Nothin' maxim: 'Any vulnerability assessment which finds no vulnerabilities or only a few is worthless and wrong.'[54]

Conclusion

Protecting against insider threats is the most important—and most challenging—element of nuclear security. Nuclear organizations need comprehensive, multi-layered approaches to mitigating insider threats; additionally, they need to always be looking for ways to improve them. Insider protection in nuclear organizations is clearly improving. Many countries have adopted new requirements and approaches in the past decade. Cyber protections for nuclear facilities—including against insider cyber threats—are strengthening. The international cooperation described in this chapter is helping to spread good practices around the world. But given the daunting nature of the challenge, much more still needs to be done.

This chapter has only skimmed the surface of what nuclear organizations are doing to cope with insider threats. More research on what measures are most effective, under what circumstances, is sorely needed. Past research has shown that some common insider-protection practices are not as effective as might be hoped: the second person in the two-person rule often does not notice an insider action, for example, and many tamper-indicating seals can be defeated easily.[55] Past work has highlighted lessons of what *not* to do, drawn from past insider incidents.[56] More research is needed, however, on what *does* work (and what does not). For example, several companies now offer software that can allegedly identify insider activity and that can be used for monitoring the computer activities of staff. How promising are such approaches in practice? Better approaches for assessing and modelling insider protection arrangements are also needed, although there will never be a replacement for creative human vulnerability assessors able to think like an adversary.

In some other fields (such as cyber security) insider issues occur more often than they do at nuclear organizations, offering more data. At the same time, increasing numbers of companies have urgent needs, creating a larger market for understanding effective practices; this drives a larger body of research and faster learning.[57] Nuclear agencies should consider financing comparable research on good practices to cope with insider threats in the nuclear sector. Only by learning and improving can nuclear organizations hope to respond effectively to the challenges posed by insider threats.

Notes

1. This chapter draws on Matthew Bunn and Scott D. Sagan, eds., *Insider Threats* (Ithaca, NY: Cornell University Press, 2017); Matthew Bunn, 'Scenarios of Insider Threats to Japan's Nuclear Facilities and Materials—and Steps to Strengthen Protection', *Napset Special Report* (Berkeley, CA: Nautilus Institute, 2017).
2. Matthew Bunn and Scott D. Sagan, *A Worst Practices Guide to Insider Threats: Lessons from Past Mistakes* (Cambridge, MA: American Academy of Arts & Sciences, 2014).
3. J. M. Lafleur, L. K. Purvis, A. W. Roesler, and P. Westland, 2014, 'The Perfect Heist: Recipes From Around the World', Paper No. SAND-2014-1790, Albuquerque, NM: Sandia National Laboratories.
4. IAEA, 'Nuclear Security Recommendations on Physical Protection of Nuclear Material and Nuclear Facilities', Recommendations, IAEA Nuclear Security Series, No. 13. Vienna, 2011.
5. Public Broadcasting System (PBS), 1996, 'Loose Nukes: Interviews'. Frontline; Noah Pope and Christopher Hobbs, *Insider Threat Case Studies at Radiological and Nuclear Facilities*, Technical Report No. LA-UR-15-22642 (London: Los Alamos National Laboratory and King's College London, 2015).
6. Lyudmila Zaitseva and Friedrich Steinhauser, 2014, 'Nuclear Trafficking Issues in the Black Sea Region', Non-Proliferation Paper, No. 39, SIPRI.
7. Rebecca Davis, 'The Man Who Blew Up Koeberg', *Daily Maverick*, 31 October 2020; David Beresford, *Truth is a Strange Fruit: A Personal Journey Through the Apartheid War* (Johannesburg: Jacana Media, 2010); Pope and Hobbs, 'Insider Case Threat Studies'.

8. Beresford, *Strange Fruit.*
9. Adrian Levy and R. Jeffrey Smith, 'India's Nuclear Explosive Materials are Vulnerable to Theft, U.S. Officials and Experts Say', *The Center for Public Integrity* [online], 17 December 2015. https://publicintegrity.org/national-security/indias-nuclear-explosive-materials-are-vulnerable-to-theft-u-s-officials-and-experts-say
10. Matthew Bunn, 'Belgium Highlights the Nuclear Terrorism Threat and Security Measures to Stop It', *Nuclear Security Matters* [Blog], 29 March 2016. https://www.belfercenter.org/publication/belgium-highlights-nuclear-terrorism-threat-and-security-measures-stop-it
11. Pope and Hobbs, 'Insider Case Threat Studies'; Erik Raspoet, 2015, 'Wie is De Saboteur van Doel 4?' ['Who is the saboteur of Doel-4?'].
12. IAEA, *Preventive and Protective Measures Against Insider Threats, Implementing Guide*, IAEA Nuclear Security Series, No. 8 (Vienna: IAEA, 2008); World Institute for Nuclear Security (WINS), *Managing Internal Threats: WINS International Best Practice Guide*, No. 3.4 (Vienna: WINS, 2015).
13. Igor Goloskokov, 'Refomirovanie Voisk MVD po Okhrane Yadernikh Obektov Rossii' ['Reforming MVD troops to guard Russian nuclear facilities'], *Yaderny Kontrol*, 9/4 (2003), 39–50.
14. Roger G. Johnston, *Mitigating the Insider Threat (And Other Security Issues)* (Lemont, IL: Argonne National Laboratory, 2011).
15. Lafleur et al., 'The Perfect Heist'.
16. Dmitry Litovkin, 'Interview: Victor Yerastov: MINATOM Has All Conditions for Providing Safety and Security of Nuclear Material', *Yaderny Kontrol Digest*, 5/1 (2000), 19–20.
17. The term is now widely used but appears to have been coined by Stanislav Levchenko, a former Russian KGB major who defected to the US in 1979.
18. C. Moore, 'Anatomy of a £26.5 Million Heist', *Sunday Life*, 21 May 2006.
19. David A. Vise, *The Bureau and the Mole: The Unmasking of Robert Hanssen, the Most Dangerous Double Agent in FBI History* (New York: Atlantic Monthly Press, 2002).
20. Andrew P. Moore, Dawn M. Cappelli, and Randall F. Trzeciak, 'The "Big Picture" of Insider IT Sabotage Across U.S. Critical Infrastructures', in Salvatore J. Stolfo, Steven M. Bellovin, Angelos D. Keromytis, Shlomo Hershkop, Sean W. Smith, and Sara Sinclair, eds., *Insider Attack and Cyber Security: Beyond the Hacker* (New York: Springer, 2012), 17–52.
21. Paul Lewis, 'Bradley Manning Flipped a Table During Counseling, Defence Tells Hearing', *The Guardian*, 12 August 2013.
22. Johnston, *Mitigating the Insider Threat.*
23. Karma Allen, 'Joyce Mitchell Released from Prison 5 Years after Helping Convicted Killers Escape', *ABC News*, 6 February 2020.
24. Associated Press, 'New York Prison Worker Who Helped in Escape Receives Seven-Year Sentence', *The Guardian*, 28 September 2015; M. Bunn, M. B. Malin, N. Roth, and W. H. Tobey, *Preventing Nuclear Terrorism: Continuous Improvement or Dangerous Decline?*, Project on Managing the Atom (Cambridge, MA: Belfer Center for Science and International Affairs, Harvard Kennedy School, 2016).
25. Matthew Bunn, Yuri Morozov, Rolf Mowatt-Larssen, Simon Saradzhyan, William Tobey, Viktor I. Yesin, and Pavel S. Zolotarev, *The U.S.–Russia Joint Threat Assessment of Nuclear Terrorism* (Cambridge, MA: Belfer Center for Science and International Affairs, Harvard Kennedy School, 2011).
26. The Soufan Center, *The Atomwaffen Division: The Evolution of the White Supremacy Threat* (New York: The Soufan Center, 2020; Rebecca L. Earnhardt, Brendan Hyatt, and

Nickolas Roth, 'A Threat to Confront: Far-Right Extremists and Nuclear Terrorism', *Bulletin of the Atomic Scientists* [online], 14 January 2021. https://thebulletin.org/2021/01/a-threat-to-confront-far-right-extremists-and-nuclear-terrorism/

27. Earnhardt et al., 'A Threat to Confront'.
28. Thomas Hegghammer and Andreas H. Daehli, 'Insiders and Outsiders: A Survey of Terrorist Threats to Nuclear Facilities', in Bunn and Sagan *Insider Threats*, 10–41.
29. Rolf Mowatt-Larssen, *Al-Qaeda Weapons of Mass Destruction Threat: Hype or Reality?* (Cambridge, MA: Belfer Center for Science and International Affairs, Harvard Kennedy School, 2010).
30. R. E. Nisbett and T. D. Wilson, 'The Halo Effect: Evidence for Unconscious Alteration of Judgments', *Journal of Personality and Social Psychology*, 35/6 (1977), 250–6; Daniel Kahneman, Paul Slovic, and Amos Tversky, eds., *Judgment Under Uncertainty: Heuristics and Biases* (Cambridge: Cambridge University Press, 1982).
31. Author interview with unnamed Y-12 employee, 2015.
32. Max H. Bazerman and Michael D. Watkins, *Predictable Surprises: The Disasters You Should Have Seen Coming, and How to Prevent Them* (Cambridge: Harvard Business School Publishing, 2003).
33. Jessica Stern and Ronald Schouten, 'Lessons From the Anthrax Letters', in Bunn and Sagan, 74–102.
34. Ibid.
35. Amy B. Zegart, 'The Fort Hood Terrorist Attack: An Organizational Post-Mortem of Army and FBI Deficiencies', in Bunn and Sagan *Insider Threats*, 42–73.
36. Goodreads, 2019. 'Otto von Bismarck', Goodreads Quotes.
37. IAEA, *Preventive and Protective Measures.*
38. IAEA, 9 January 2017, 'Communication Dated 22 December 2016 Received From the Permanent Mission of the United States of America Concerning a Joint Statement on Mitigating Insider Threats'.
39. IAEA, 2021, 'IAEA Information Circular (INFCIRC) 908: Mitigating Insider Threats', Fact Sheet, INFCIRC/908 International Working Group.
40. Bunn, 'Scenarios of Insider Threats'.
41. Bunn and Sagan, *Insider Threats.*
42. National Insider Threat Task Force (NIITF), *Insider Threat Guide: A Compendium of Best Practices to Accompany the National Insider Threat Minimum Standards* (Washington, DC: NIITF, 2017).
43. World Institute for Nuclear Security (WINS), *Nuclear Security Culture: WINS International Best Practice Guide*, No. 1.4 (Vienna: WINS, 2016).
44. NIITF, *Insider Threat Guide.*
45. Ibid.
46. Suzanne Wood and Joanne C. Marshall-Mies, *Improving Supervisor and Co-Worker Reporting of Information of Security Concern* (Monterey, CA: Defense Personnel Security Research Center, 2003).
47. Moore et al., 'The "Big Picture"'.
48. WINS, *Nuclear Security Culture.*
49. Johnston, *Mitigating the Insider Threat.*
50. IAEA, 2015, 'Use of Nuclear Material Accounting and Control for Nuclear Security Purposes at Facilities', Implementing Guide, IAEA Nuclear Security Series, No. 25-G. Vienna.

51. IAEA, 'Nuclear Security Recommendations on Physical Protection of Nuclear Material and Nuclear Facilities'.
52. IAEA, *Preventive and Protective Measures*.
53. Ibid.
54. Roger G. Johnston, *Security Maxims* (Lemont, IL: Argonne National Laboratory, 2018).
55. Roger G. Johnston, 'Tamper-Indicating Seals: Practices, Problems, and Standards', Paper presented at the World Customs Organization Security Meeting, Brussels, 11–14 February 2003.
56. Bunn and Sagan, *Insider Threats*.
57. CERT National Insider Threat Center, *Common Sense Guide to Mitigating Insider Threats* (Pittsburgh, PA: Software Engineering Institute, Carnegie Mellon University, 2018).

Suggested Reading

Bunn, Matthew and Sagan, Scott D., (2014), *A Worst Practices Guide to Insider Threats: Lessons from Past Mistakes* (Cambridge, MA: American Academy of Arts & Sciences).

Bunn, Matthew, and Sagan, Scott D., eds., (2017), *Insider Threats* (Ithaca, NY: Cornell University Press).

IAEA, (2008), *Preventive and Protective Measures Against Insider Threats*, Implementing Guide, IAEA Nuclear Security Series, No. 8. (Vienna: IAEA).

Pope, Noah, and Hobbs, Christopher, (2015), *Insider Threat Case Studies at Radiological and Nuclear Facilities*, Technical Report No. LA-UR-15-22642 (London: Los Alamos National Laboratory/King's College London).

World Institute for Nuclear Security (WINS), (2015), *Managing Internal Threats*, WINS International Best Practice Guide, No. 3.4. (Vienna, WINS).

PART III

INTERNATIONAL NUCLEAR SECURITY REGIME

CHAPTER 8

THE INTERNATIONAL LEGAL BASIS OF NUCLEAR SECURITY

YASEMIN BALCI

INTRODUCTION

STATES, companies, regulators, and individuals across the world rely on the international basis of nuclear security to organize and harmonize their activities at the national level. The international legal basis of nuclear security provides the framework in which national activities are regulated and establishes the foundation for cooperation across national borders. International law regarding nuclear security ultimately affects operational reality through the adoption of national laws that meet international obligations. This chapter focuses on the international legal sources that together form the foundation of those national legal frameworks.

Where can the international legal basis of, or the international law on, nuclear security be found, and who creates it? International lawyers will point to Article 38 of the Statute of the International Court of Justice,[1] according to which the sources of international law are treaties, custom, and general principles of law, as well as judicial decisions and the works of established scholars as a subsidiary means of identifying international law. With regard to nuclear security, treaties (which are developed by states) are the main source of international law on the subject. The beginning part of this chapter is therefore dedicated to treaties that are solely (the first section) or partially (the second section) related to nuclear security.

International rules that are not legally binding, known as 'soft law', act as a layer of support to the international legal basis of nuclear security. Soft law can be created by states, but also by other actors such as intergovernmental and non-governmental organizations. Soft law can be more readily accepted by states due to its non-binding nature, although it often develops into hard law over time. The third section addresses its role in nuclear security. The fourth section explores decisions of judges and shows how these are a possible source of law for nuclear security as well as a form of enforcement of national law meant to implement international law. The chapter ends with reflections on how the international legal basis of nuclear security might be improved.

Finally, it should be noted from the outset that there is no international legal definition of nuclear security. This chapter follows the Nuclear Security Fundamentals document 'Objective and Essential Elements of a State's Nuclear Security Regime', which can be considered a soft law instrument of the International Atomic Energy Agency (IAEA). It states that 'nuclear security focuses on the prevention of, detection of, and response to, criminal or intentional unauthorized acts involving or directed at nuclear material, other radioactive material, associated facilities, or associated activities'.[2] Nuclear security therefore covers both nuclear and other radioactive material, although the scope of a particular international legally binding or non-binding instrument can define the term in a more limited matter. For example, the Convention on the Physical Protection of Nuclear Material (CPPNM) only covers nuclear material.

Nuclear Security Treaties

The international legal basis of nuclear security consists mostly of treaties. Treaties are written, legally binding international agreements between states.[3] In a multilateral context, states will negotiate the content of treaties, adopt a final text, and usually open it for signature. However, states will not be legally bound by it, until they have expressed their 'consent to be bound'. If they have signed the treaty, such consent is normally expressed by depositing an instrument of ratification with the treaty's depositary (which is the person or office responsible for administering a treaty). If they have not signed the treaty, this is usually expressed through an instrument of accession. Most states will complete a national governmental and parliamentary or congressional approval procedure before they deposit such an instrument. Once bound by a treaty, a state is expected to comply with its newly acquired international obligations. This is captured by a general principle of law that is embodied in the Latin maxim *pacta sunt servanda*, that is, agreements must be kept. Often, national legislation will be required to implement and carry out treaty obligations. Treaty provisions may generally require the adoption of legislative measures to implement the treaty or specifically oblige states parties to, for example, establish criminal offences.

When looking at the international legal basis of nuclear security, its most noticeable feature is that there is no single treaty on the subject.[4] Instead, there are three treaties: the CPPNM,[5] the Amendment to the Convention on the Physical Protection of Nuclear Material (the 2005 Amendment),[6] which together are known as the 'amended CPPNM' or 'A/CPPNM', and the International Convention for the Suppression of Acts of Nuclear Terrorism (ICSANT).[7] Both the 2005 Amendment and ICSANT were adopted in 2005, but these treaties are the product of different multilateral forums, namely the IAEA and United Nations (UN), respectively.

At the time of writing, the CPPNM, which entered into force on 8 February 1987, has 164 states parties and its 2005 Amendment, which entered into force on 8 May 2016, has 135 states parties.[8] ICSANT entered into force on 7 July 2007 and has 122 states parties.[9] It took many more years for the 2005 Amendment to enter into force, as it required two-thirds of the CPPNM states parties to express their consent to be bound for the amendment to enter into force, compared with 22 parties needed to bring ICSANT into force (Article 20, CPPNM; Article 25, ICSANT). The lack of a single international treaty on nuclear security means states themselves have to consolidate their obligations, which are scattered over a number of different instruments, and in doing so, they may be faced with different definitions for the same terms or overlapping criminal offences.

Moreover, while these three treaties constitute the main legal framework on nuclear security, none use the term 'nuclear security'. Instead, these treaties are limited in their scope to certain aspects of nuclear security. As can be derived from its title, the focus of the amended CPPNM is on 'physical protection', although this term is not defined by the amended CPPNM. The IAEA's 'Nuclear Security Recommendations on Physical Protection of Nuclear Material and Nuclear Facilities', known as INFCIRC/225, which is a soft law standard that predates the CPPNM, uses the term 'physical protection' interchangeably with the 'nuclear security of nuclear material and nuclear facilities'.[10] However, 'physical protection' can also be understood as a narrower term than nuclear security, that is, excluding non-physical measures such as accounting and cyber security.[11]

ICSANT does not use the term 'nuclear security' either. Similarly, even though ICSANT mentions 'nuclear terrorism' in its title and in a reference to its title in the preamble, the articles of the convention do not use this term and no terrorist motivation is required for the offences it establishes. This is due to a lack of agreement between states on a legal definition of the term 'terrorism'.[12] Nevertheless, Article 6 of ICSANT underscores that there can be no justification of offences established by the convention on the basis of political or other similar considerations when an offence is carried out with the intention to provoke a state of terror.

There is overlap between the amended CPPNM and ICSANT, but their scope also varies. The latter explicitly refers to the CPPNM in its preamble, although not to the 2005 Amendment, which had yet to be adopted when ICSANT was drafted.[13] The definitions for 'nuclear material' are the same in both the amended CPPNM and ICSANT. The term 'nuclear facility' was added to the CPPNM by its amendment. ICSANT also includes this term but defines it more broadly. Unlike the amended CPPNM, ICSANT covers and defines 'radioactive material', which includes 'nuclear material and other radioactive substances'. Moreover, ICSANT does not make a distinction between radioactive material (and nuclear facilities) for peaceful or military purposes, but the amended CPPNM does not apply to nuclear material for military purposes or to a nuclear facility containing such material (Article 2(5)).[14] Unlike ICSANT, the amended CPPNM includes a definition for sabotage.

The overlap as well as differences are visible in the treaties' criminal provisions, too. Both the amended CPPNM and ICSANT require their respective states parties to establish criminal offences (Article 7, amended CPPNM; Articles 2 and 5, ICSANT) and apply related measures regarding jurisdiction, extradition, prosecution, and mutual legal cooperation (Articles 8 to 13, amended CPPNM; Articles 9 to 19, ICSANT). There are differences in the exact behaviour that is criminalized, however, and states parties to both the amended CPPNM and ICSANT will therefore need to carefully consolidate these in their national criminal laws. This requires detangling and accommodating the differences in definitions and criminal provisions. For example, Article 7(1)(e) of the amended CPPNM and Article 2(1)(b) of ICSANT both concern acts against a nuclear facility, but there are differences between the exact behaviour that is prohibited as well as the mindset of the perpetrator at the time of the offence. Article 7(1)(e) of the amended CPPNM requires making the following an offence in national legislation:

> The intentional commission of [...] an act directed against a nuclear facility, or an act interfering with the operation of a nuclear facility, where the offender intentionally causes, or where he knows that the act is likely to cause, death or serious injury to any person or substantial damage to property or to the environment by exposure to radiation or release of radioactive substances, unless the act is undertaken in conformity with the national law of the State Party in the territory of which the nuclear facility is situated.

Compare this with Article 2(1)(b) of ICSANT which states:

> Any person commits an offence within the meaning of this Convention if that person unlawfully and intentionally [...] uses or damages a nuclear facility in a manner which releases or risks the release of radioactive material: (i) With the intent to cause death or serious bodily injury; or (ii) With the intent to cause substantial damage to property or to the environment; or (iii) With the intent to compel a natural or legal person, an international organization or a State to do or refrain from doing an act.

Bearing in mind that ICSANT defines 'nuclear facility' more broadly than the amended CPPNM, this too needs to be taken into account when establishing these offences in national criminal law. Furthermore, both the amended CPPNM and ICSANT require protection measures for material and facilities. The amended CPPNM does so in Article 2A, providing a list of 12 fundamental principles of physical protection of nuclear material and nuclear facilities, while ICSANT simply requires its states parties to 'make every effort to adopt appropriate measures to ensure the protection of radioactive material taking into account relevant recommendations and functions of the International Atomic Energy Agency' (Article 8). Note that ICSANT does not limit this requirement to 'physical' protection. ICSANT specifically requires its states parties to 'have regard to' the IAEA's physical protection recommendations when seizing or taking control of radioactive material, devices, or nuclear facilities in the aftermath of an offence (Article 18).

As for legislative implementation requirements in addition to the criminal ones discussed previously, both the amended CPPNM and ICSANT include provisions that explicitly require national implementation measures. Article 2A(2)(a) of the amended CPPNM requires its states parties to 'establish and maintain a legislative and regulatory framework to govern physical protection'. There are further specific legislative implementation requirements with regard to the international transport of nuclear material (Article 3), the international transfer of nuclear material (Article 4), and the protection of confidential information (Article 6). The amended CPPNM also requires its states parties to inform the IAEA of the laws and regulations that implement the convention (Article 14). ICSANT explicitly requires legislative changes to enable cooperation in the prevention and countering of preparations of offences (Article 7(1)(a)).

Finally, the amended CPPNM and ICSANT differ in their governance mechanisms. The amended CPPNM provides for a conference of states parties (Article 16). This allows all states parties to the amended CPPNM to gather, review the status of the convention, and adopt further decisions if necessary. Such a mechanism enables the updating of understandings and agreements among states parties over time, thus making it possible for the amended CPPNM to keep up with technical, scientific, and other relevant developments.

The first conference of states parties for A/CPPNM took place in March 2022 and was heralded by the IAEA as 'a major milestone in the development of the international legal framework for nuclear security'.[15] It clarified, for example, that 'physical protection' is a 'key element of nuclear security' and that 'physical protection is linked or [...] interconnected with other areas of nuclear security, such as [...] computer security'. In addition, during the conference, states parties 'exchanged on good practices, lessons learned, and challenges' in implementing A/CPPNM. ICSANT, however, does not have such a governance structure. It should be noted that neither A/CPPNM nor ICSANT contain international verification mechanisms, which means the onus is solely on states to ensure that they have an effective national nuclear security system.[16]

The scattered international legal approach to nuclear security means that states will have varying international obligations regarding nuclear security, depending on whether they are party to the CPPNM, its 2005 Amendment, and ICSANT. This has led to calls from some nuclear security experts to establish a single convention: a Convention on Nuclear Security.[17] Such a convention could help close gaps in the international legal framework on nuclear security.

A key aspect of the proposed convention is to establish common security standards by making binding the IAEA's 'Objective and Essential Elements of a State's Nuclear Security Regime'. This is an example of proposing to turn soft law into hard law. However, the proposed convention does not intend to replace the existing nuclear treaty treaties, as such an exercise entails complications that may not be readily addressed due to varying number of parties to the multiple treaties across two distinct forums (the IAEA and UN). A single convention would therefore in effect constitute the fourth addition to the patchwork of treaties regarding nuclear security. This proposal was first put forward by a non-governmental group of experts in 2015, but has since not gained traction with states.

While it is no means of improving the international legal framework on nuclear security per se, working towards existing treaties' universality is a way to ensure that all states at least have the same international obligations on nuclear security. Almost two decades since the adoption of the 2005 Amendment and ICSANT, universality is not yet in sight. Currently, while the amended CPPNM has more states parties than ICSANT, roughly one-third of states still need to adhere to both treaties. Between 2018 and 2022, only around three states per year have joined the 2005 Amendment and even fewer, namely one or two states per year, have joined ICSANT. Contrast this with an average of around eight states that joined the 2005 Amendment and an average of around nine states that joined ICSANT yearly from 2006 to 2017. For A/CPPNM, there is a clear spike of 16 states joining in 2016, the year of the final Nuclear Security Summit, but this is not the case for ICSANT, which gained seven states parties that year.

It should be noted that the international obligations that states accept by joining these treaties stay confined to the international level and do not affect operational reality at the national level, unless national legislation allows for their application or implementation. To ensure the international legal framework on nuclear security has relevance in practice, states should take care to update their national legislation in conformity with their international nuclear security obligations.

Related Treaties

With the amended CPPNM and ICSANT, the international legal basis of nuclear security is already scattered across three treaties. However, these are not the only treaties that contain nuclear security-related obligations. There are other related treaties that pertain to nuclear security, which states will have to take into account when they are implementing their nuclear security obligations and establishing and maintaining a coherent nuclear security framework at the national level. The treaties discussed in this chapter are global in scope and do not include region-specific nuclear-weapon-free-zone treaties, some of which also contain nuclear security obligations.

Firstly, the treaty establishing the United Nations is the UN Charter.[18] Under Article 25 of the UN Charter, members of the UN 'agree to accept and carry out the decisions of the Security Council'. These decisions are therefore legally binding, and if there is a conflict with another treaty obligation, the UN obligation will prevail by virtue of Article 103 of the UN Charter. For the purposes of nuclear security, UN Security Council Resolution 1540 (UNSCR 1540) is an important resolution that was adopted unanimously in April 2004, only a year before the adoption of ICSANT in April 2005 and the 2005 Amendment in July 2005. Resolution 1540 acknowledges CPPNM and the Code of Conduct on the Safety and Security of Radioactive Sources in its preamble.[19] This resolution concerning the non-proliferation of nuclear, biological, and chemical weapons to non-state actors contains a number of decisions that are directly relevant to nuclear security.

These decisions require states to refrain from supporting non-state actors seeking such weapons (operative paragraph 1); adopting and enforcing laws that prohibit non-state actors to 'manufacture, acquire, possess, develop, transport, transfer or use nuclear, chemical or biological weapons and their means of delivery, in particular for terrorist purposes' (operative paragraph 2); and establishing national controls over related materials, including physical protection and transfer control measures (operative paragraph 3). Resolution 1540 provides definitions for key terms such as non-state actor and related materials. Importantly for nuclear security purposes, UNSCR 1540 does not cover radiological weapons and, by extension, radioactive material that is not nuclear material. Because the resolution requires states to adopt legislative measures of a general application (i.e. not limited to a specific time, place, person(s), or situation), the resolution is known as a 'legislative' resolution, of which there are only a few.[20] As a result, UNSCR 1540 has created obligations relevant to nuclear security for states whether or not they have joined any of the nuclear security treaties.

Secondly, there are a number of terrorism treaties that contain nuclear security-related obligations. The earliest treaty is the Terrorist Bombings Convention,[21] which establishes offences and related measures involving 'explosive and other lethal devices' (Article 2(1)). A lethal device is defined as including 'a weapon or device that is designed, or has the capability, to cause death, serious bodily injury or substantial material damage through the release, dissemination or impact of toxic chemicals, biological agents or toxins or similar substances or radiation or radioactive material' (Article 1(3)(b)). The Terrorist Financing Convention[22] establishes offences with regard to the funding of acts that are considered offences in the CPPNM and the Terrorist Bombings Convention (Article 2(1)(a)).

Besides these two terrorism conventions, there are three treaties specific to maritime and aviation security that include nuclear security obligations. For maritime security, these are the Protocol to the Convention for the Suppression of Unlawful Acts against the Safety of Maritime Navigation (SUA 2005)[23] and the Protocol to the Protocol for the Suppression of Unlawful Acts against the Safety of Fixed Platforms Located on the Continental Shelf (SUA PROT 2005).[24]

Both protocols establish offences and related measures with regard to biological, chemical, and nuclear (BCN) weapons, which include 'nuclear weapons and other nuclear explosive devices' (Article 1(1)(d)(iii)). SUA 2005 involves offences regarding the use of radioactive material or a BCN weapon against or on a ship, their discharge from, or transport on board a ship (Article 3bis(1)(a)(i) and (b)(i)–(ii)). SUA PROT 2005 establishes similar offences regarding the use of radioactive material or a BCN weapon against or on a fixed platform and their discharge from a fixed platform (Article 2bis(a)). For aviation security, the Convention on the Suppression of Unlawful Acts Relating to International Civil Aviation[25] (the Beijing Convention) uses the same terms of a BCN weapon (Article 2(h)), but (unlike SUA 2005 and SUA PROT 2005) also defines 'radioactive material' in line with ICSANT (Article 2 (e)). It establishes similar offences regarding the use of a BCN weapon or radioactive material against or on board an aircraft in service (Article 1(1)(h)), their release or discharge from an aircraft in service (Article 1(1)(g)), or their transport on board an aircraft (Article 1(1)(i)(1)–(2)).

Thirdly, nuclear security obligations can be considered within the context of nuclear non-proliferation, disarmament, and safeguards law. The focus of this area is on non-proliferation among states, but its rules to control nuclear material help achieve the aims of nuclear security as well. For example, nuclear accounting is a safeguards measure that could also help to determine any theft or sabotage by non-state actors. The Treaty on the Non-Proliferation of Nuclear Weapons[26] (NPT) requires non-nuclear weapon states to conclude a safeguards agreement with the IAEA (Article III(1)). This type of agreement is known as a comprehensive safeguards agreement (CSA) and is based on the model in INFCIRC/153 (Corr.).[27] Moreover, an additional protocol to such a CSA can be concluded between a state and the IAEA, on the basis of the model in INFCIRC/540 (Corr.). Unlike a CSA, an additional protocol provides the IAEA with verification tools to help detect undeclared nuclear material.[28]

Safeguards agreements aim to prevent the diversion of nuclear energy from peaceful uses. Safeguards agreements are also required as part of the Treaty on the Prohibition of Nuclear Weapons[29] (TPNW), which entered into force on 22 January 2021 (Articles 3 and 4). Moreover, the TPNW requires the adoption of legal and other measures to 'prevent and suppress' activities that are prohibited to a state party, mainly in Article 1, which include the development, testing, production, manufacturing, acquisition, possession, stockpiling, transfer, receipt, control over, and (threat of) use of nuclear weapons or other nuclear explosive devices.

Finally, treaties on response to nuclear accidents and radiological emergencies, more commonly associated with nuclear safety, are relevant to nuclear security. The Early Notification Convention[30] focuses on the sharing of information, and the Assistance Convention[31] on assistance between states, in order to deal with the effects of such emergencies. However, these conventions can also apply to deliberate acts, for example, terrorist acts.[32]

Soft Law

As outlined previously, the international legal basis of nuclear security consists mostly of treaties. Strictly speaking, 'soft law' or non-legally binding measures do not form part of the international law on nuclear security. However, soft law is an important layer of rules that supports, supplements, and sometimes develops into legally binding 'hard law'. Soft law can be created by states but, since it is not legally binding, this is an area in which other actors such as non-governmental organizations and intergovernmental organizations can take the initiative. For nuclear security, the IAEA has played a leading role in this regard by developing and adopting a number of instruments.

Covering safety aspects and specifically developed for radioactive sources, the IAEA created the Code of Conduct on the Safety and Security of Radioactive Sources (Code of Conduct). It has supplemented this with the Guidance on the Import and Export of Radioactive Sources and the Guidance on the Management of Disused Radioactive Sources.[33] According to the Code, a radioactive source is 'radioactive material that is permanently sealed in a capsule or closely bonded, in a solid form and which is not exempt from regulatory control. [...]' (Paragraph I.1). The Code of Conduct provides for further definitions of key terms and legislative and regulatory measures that are necessary to control radioactive sources. Member states can express their political commitment to the Code and 149 have done so. Similarly, at the time of writing, 133 and 63, respectively, have expressed their political commitment to the two Guidance documents.[34]

While the Code of Conduct and the two Guidance documents are focused on security (and safety) of radioactive sources, the IAEA's full range of standards on nuclear security can be found in its Nuclear Security Series. The series consists of four tiers: Fundamentals; Recommendations; Implementing Guides; and Technical Guidance. While the Fundamentals provide the key aspects of nuclear security, the Recommendations provide for their application, and the Implementing Guides and Technical Guidance give detailed information on the implementation of the Recommendations.[35]

Of these, the 'Nuclear Security Recommendations on Physical Protection of Nuclear Material and Nuclear Facilities' were first published in 1972, incorporated in the IAEA's INFCIRC series as INFCIRC/225 in 1975, and most recently revised in 2011 as INFCIRC/225/Rev.5.[36] The fifth revision in 2011 was carried out to reflect new legal developments, including the adoption of the 2005 Amendment, since the previous revision of INFCIRC/225 in 1998.[37] The amended CPPNM makes a general reference to INFCIRC/225 by 'recognizing that there are internationally formulated physical protection recommendations that are updated from time to time' in its preamble. Similarly, ICSANT refers to the 'relevant recommendations' (Article 8) and 'physical protection recommendations' (Article 18(1)(c)) of the IAEA, which can be taken to refer to INFCIRC/225.[38]

Conversely, parts of these soft law instruments have developed into hard law. INFCIRC/225 predates the CPPNM and has informed the content of the CPPNM. For example, 'an important feature of INFCIRC/225 is its categorization of nuclear material by type and quantity based on the security significance of the specific material'.[39] The CPPNM adopted this categorization by including it in its Annex II and it continues to form part of the amended CPPNM.[40] In addition, the draft language as proposed by a non-governmental group of experts[41] for a Convention on Nuclear Security includes the Nuclear Security Fundamentals document, 'Objective and Essential Elements of a State's Nuclear Security Regime', as an Annex to the convention.

Judicial Decisions

Judicial decisions, including rulings of both international and national courts, are considered a subsidiary source of international law.[42] Such rulings can serve as authoritative determinations and help further develop international law. Besides developing international law, judicial decisions can also be seen as a form of enforcement of current international law. For example, when states implement international obligations through national legislation, the enforcement of that legislation through the courts is a way to ensure compliance with national legislation, and ultimately with international law.

There is no comprehensive official or academic record of any international or national judicial decisions concerning nuclear security. The IAEA maintains the Incident and Trafficking Database (ITDB), which tracks nuclear security incidents and helps facilitate information sharing between IAEA member states. However, only an annual factsheet summarizing the contents of the database is made publicly available and it does not mention judicial decisions as a result of investigation and prosecution of incidents.[43] Similarly, the Geiger database of the International Criminal Police Organization (Interpol) compiles data on incidents involving radioactive material from its member states' police forces and provides analysis from the database to the member states to assist their investigation.[44] Those data are not publicly available. In order to be able to make an assessment of judicial decisions' contribution to the development of international law on nuclear security, a record or database of such decisions across various jurisdictions would be a useful first step. The few national cases elaborated on here only serve as examples of factual and/or legal situations relevant to nuclear security.

One type of relevant judicial decision is that related to sabotage. For example, there have been situations where anti-nuclear weapons activists have entered nuclear facilities or environmental activists have damaged property at nuclear facilities.[45] A notable case in the United States (US) involved an 82-year-old nun and two veterans entering the Y-12 nuclear facility, which stored enriched uranium, in 2012. The three were initially convicted and sentenced under Section 2155(a) of Title 18 of the US Code (18 USC 2155(a)), which prohibits injuring, destroying, contaminating, or infecting national-defence material, national-defence premises, or national-defence utilities. They were also convicted of injuring government property. A federal appeals court later overturned the first conviction, however, as it determined the defendants lacked the required intent to commit the sabotage offence.[46]

While the facts in this case are relevant to ICSANT, the prosecution has relied on the sabotage offence in 18 USC 2155(a), which is not an offence specific to nuclear facilities. This may be related to the fact that the case predated the enactment of implementing legislation of the US for ICSANT in 2015,[47] but it is also the prosecutor's discretion to make charging decisions on the basis of consideration of all facts, the availability of evidence, and the prospect of a successful conviction.

Another type of criminal case where legislation implementing the amended CPPNM and ICSANT has been used are those that involve malicious acts with radioactive material in a personal setting. For example, in October 2018, a man from the US state of Wisconsin tried to obtain polonium on the dark web for suicide purposes.[48] He was charged with two offences, but a plea agreement led to a sentence of two years' imprisonment for the second charge only. The first charge had been knowingly and unlawfully attempting to possess radioactive material with intent to cause death or serious bodily injury under 18 USC 2332i(a)(1) and (3), carrying a maximum penalty of life imprisonment and a fine. The second charge was intentionally attempting, without lawful authority, to receive and possess any nuclear material or nuclear by-product material, where circumstances have been represented to the defendant to exist, that are likely to cause the death of or serious bodily injury to any person, under 18 USC 831(a)(1)(B), carrying a maximum penalty of 20 years of imprisonment and a fine.

The first charge is an offence that implements Articles 2(1)(a), 2(3) and 5 of ICSANT in US legislation. The second charge is an offence that implements Article 7(1)(a) of the amended CPPNM. Given the plea agreement, there was no trial, and hence, no judicial decision on the substantive aspects of the case, which may have contributed to the interpretation or development of international law on this subject. However, the case is a good example of how international agreements concluded at the international level ultimately capture conduct on the ground. Moreover, the court documents for this case mention that the offences are rooted in treaty obligations and that the prosecution and defence have argued as to how they apply in this case. The defence even notes that 'the statute is written in a very peculiar way. It's the product of a treaty [...]. We've only found one indictment under it. Now, [the prosecution] has told me there's others, but we haven't found them, and there is no good case law fleshing this out.'[49] This observation applies not only to the US, but also to all states parties of the amended CPPNM and ICSANT. To better understand and apply the offences these treaties create, it is important to be able to track any court cases and judicial decisions on these offences.

As for international judicial decisions, it was noted previously that there are no such decisions specifically about nuclear security. However, it has been suggested in academic literature to allow for this, for example, by amending the Rome Statute of the International Criminal Court (the Rome Statute) to include the most serious crimes of transferring nuclear or other radioactive material.[50] The Rome Statute does not currently cover any nuclear weapons or nuclear material-related crimes. Moreover, there has even been a call from at least one state to establish a dedicated international tribunal on the issue. During the Nuclear Security Summit in 2010, which was the first of four meetings of world leaders specifically on the subject of nuclear security, France's President Nicolas Sarkozy proposed amending the Rome Statute or establishing an international tribunal under the auspices of the UN.[51] However, this idea does not seem to have gained traction since then.

Conclusion

The rules-based international order is made up of treaties, custom, general principles of law as well as judicial decisions and the works of eminent scholars. The international legal basis of nuclear security is firmly rooted in the first category, namely, treaty law. While the CPPNM had been the sole treaty on the subject since the 1980s, the 2000s saw a flurry of activity to strengthen the international legal basis by the adoption of, first and foremost, the 2005 Amendment to the CPPNM and the adoption of ICSANT, but also of related instruments such as UNSCR 1540, SUA 2005, SUA PROT 2005, and the Beijing Convention. The fact that there is no single comprehensive treaty on nuclear security has led to calls from some experts to establish one, but this proposal has not been advocated by states themselves yet.

Such a treaty is not proposed as a replacement of the current treaties and would therefore only add to the patchwork of treaties. However, it would provide an opportunity to improve the international legal system on nuclear security by making certain non-binding nuclear security measures binding. Such 'soft law', while not part of the international law on nuclear security, therefore has a role to play, either to supplement treaty obligations or to help develop them. The IAEA's soft law on nuclear security, primarily INFCIRC/225, has been used to draft treaty obligations and other instruments such as the IAEA's Fundamentals series and it continues to supplement existing treaty obligations or inspire the drafting of new ones. France has once suggested creating new international treaty obligations to establish the ability to try major violations of nuclear security obligations at the international level, but this idea has not been further developed.

Regardless of whether states agree to new treaty obligations, they will need to work towards the strengthening of the international legal framework, for which the first conference of states parties to the amended CPPNM provided an opportunity, and towards universality of the current (and any new) treaties to ensure states have the same international obligations on nuclear security. Once states join the main treaties on nuclear security, they are bound by them, as underpinned by the general principle of law, *pacta sunt servanda*, that is, states have to perform their treaty obligations in good faith. One aspect of this obligation is to draft and adopt national legislation in conformity with international obligations. Such national legislation allows for an effective set of rules on nuclear security, the violation of which allows courts to enforce the rules.

Judicial decisions can therefore be seen as a form of enforcement of national legislation and ultimately international law, and they can also be a source of international law in themselves. While the IAEA and Interpol have a database to track nuclear security incidents, there is no similar record of enforcement of such incidents through the courts. To analyse judicial decisions in the area of nuclear security and their possible contribution to the interpretation and development of the international law on the subject would therefore require a record of cases, preferably across a variety of jurisdictions.

Notes

1. Statute of the International Court of Justice (opened for signature on 26 June 1945; entered into force on 24 October 1945).
2. IAEA, *Objective and Essential Elements of a State's Nuclear Security Regime*, Fundamentals, IAEA Nuclear Security Series, No. 20 (Vienna: IAEA, 2013), Paragraph 1.1.
3. Other subjects of international law, such as intergovernmental organizations, can also conclude treaties. However, treaties between states are the most relevant for nuclear security purposes.

4. This is not exclusive to nuclear security. Other areas of nuclear law, such as nuclear safety and safeguards, also consist of multiple treaties. See Anthony C. Wetherall, 'Strengthening the International Legal Framework for Nuclear Security: Better Sooner Rather than Later', *Nuclear Law Bulletin*, 2/98 (2016), 25–26; Carlton Stoiber, 'Nuclear Security: Legal Aspects of Physical Protection, Combating Illicit Trafficking and Nuclear Terrorism', in International Nuclear Law: History, Evolution and Outlook (Paris: OECD, 2010), 219–42; Sonia Drobysz, 'Nuclear Security: Physical Protection, Illicit Trafficking and Nuclear Terrorism', in Kimberly Sexton Nick and Stephen G. Burns, eds., *Principles and Practice of International Nuclear Law* (Paris: OECD, 2022), 339–56.
5. Convention on the Physical Protection of Nuclear Material 1456 UNTS 101 (adopted on 26 October 1979; opened for signature on 3 March 1980; entered into force on 8 February 1987).
6. Amendment to the Convention on the Physical Protection of Nuclear Material (adopted on 8 July 2005; entered into force on 8 May 2016).
7. International Convention for the Suppression of Acts of Nuclear Terrorism 2445 UNTS 89 (adopted on 13 April 2005; opened for signature on 14 September 2005; entered into force on 7 July 2007).
8. IAEA, 'Convention on the Physical Protection of Nuclear Material', (CPPNM). https://www.iaea.org/publications/documents/conventions/convention-physical-protection-nuclear-material-and-its-amendment
9. UN Treaty Collection, '15. International Convention for the Suppression of Acts of Nuclear Terrorism', Status of Treaties. https://treaties.un.org/Pages/ViewDetailsIII.aspx?src=TREATY&mtdsg_no=XVIII-15&chapter=18&Temp=mtdsg3&clang=_en
10. Carlton Stoiber, Alec Baer, Norbert Pelzer, and Wolfram Tonhauser, *Handbook on Nuclear Law* (Vienna: IAEA, 2003), 146; IAEA, 'Nuclear Security Glossary: Terminology Used in IAEA Nuclear Security Guidance', Draft, August 2020, p. 1. https://www.iaea.org/sites/default/files/21/06/nuclear_security_glossary_august_2020.pdf
11. George Vardulakis, 'New Wine in Old Bottles? Cyber Security Coverage within the CPPNM/A and Its Implementing Guidance', *Trust & Verify*, 169 (2021), 1. https://www.vertic.org/2021/12/trust-verify-no-169-winter-2021
12. Odette Jankowitsch-Prevor, 'International Convention for the Suppression of Acts of Nuclear Terrorism', *Nuclear Law Bulletin*, 76 (2005), 7, 8.
13. Walter Gehr, 'The Universal Legal Framework Against Nuclear Terrorism', *Nuclear Law Bulletin*, 79 (2007), 5, 12.
14. IAEA, *The International Legal Framework for Nuclear Security*, IAEA International Law Series, No. 4 (Vienna: IAEA, 2011), 9–10, 12.
15. IAEA, Conference of the Parties to the Amendment to the Convention on the Physical Protection of Nuclear Material, 28 March–1 April (Vienna: IAEA, 2022), 4. https://www.iaea.org/events/acppnm2022
16. Ibid.
17. Kenneth C. Brill and John H. Bernhard, 'A Convention on Nuclear Security: A Needed Step Against Nuclear Terrorism', *Arms Control Today*, 45 (2015), 16–21; John H. Bernhard, Kenneth C. Brill, Anita Nilsson, and Shin Chang-Hoon, *International Convention on Nuclear Security* (Washington, DC: Nuclear Security Governance Experts Group, 2015). http://www.nsgeg.org/ICNSReport315.pdf
18. Charter of the United Nations (opened for signature on 26 June 1945; entered into force on 24 October 1945).
19. Drobysz, 'Nuclear Security', 300.

20. Stoiber, 'Nuclear Security', 92; Stefan Talmon, 'The Security Council as World Legislature', *The American Journal of International Law*, 99/1 (2005), 175–6.
21. International Convention for the Suppression of Terrorist Bombings, 2149 UNTS 256 (adopted on 15 December 1997; opened for signature on 12 January 1998; entered into force on 23 May 2001).
22. International Convention for the Suppression of the Financing of Terrorism, 2178 UNTS 197 (adopted on 9 December 1999; opened for signature on 10 January 2000; entered into force on 10 April 2002).
23. Protocol of 2005 to the Convention for the Suppression of Unlawful Acts Against the Safety of Maritime Navigation, LEG/ CONF.15/ 21 (adopted on 14 October 2005; opened for signature on 14 February 2006; entered into force on 28 July 2010).
24. Protocol of 2005 to the Protocol for the Suppression of Unlawful Acts Against the Safety of Fixed Platforms Located on the Continental Shelf, LEG/CONF.15/22 (adopted on 14 October 2005, opened for signature on 14 February 2006; entered into force on 28 July 2010).
25. Convention on the Suppression of Unlawful Acts Relating to International Civil Aviation, ICAO DOC 9960 (adopted on 10 September 2010; opened for signature on 10 September 2010; entered into force on 1 July 2018).
26. Treaty on the Non-Proliferation of Nuclear Weapons, 729 UNTS 161 (opened for signature on 1 July 1968; entered into force on 5 March 1970).
27. IAEA, *Safeguards Legal Framework* (Vienna: IAEA, 2022). https://www.iaea.org/topics/safeguards-legal-framework
28. IAEA, *Additional Protocol* (Vienna: IAEA, 2022). https://www.iaea.org/topics/additional-protocol
29. Treaty on the Prohibition of Nuclear Weapons, A/ CONF.229/ 2017/ 8 (adopted on 7 July 2017; opened for signature on 20 September 2017; entered into force on 22 January 2021).
30. Convention on Early Notification of a Nuclear Accident, 1439 UNTS 275 (opened for signature on 26 September 1986; entered into force on 27 October 1986).
31. Convention on Assistance in the Case of a Nuclear Accident or Radiological Emergency, 1457 UNTS 133 (opened for signature on 26 September 1986; entered into force on 27 February 1987).
32. IAEA, '*The International Legal Framework*', 4.
33. IAEA, *Code of Conduct on the Safety and Security of Radioactive Sources*, Non-serial Publications (Vienna: IAEA, 2004); IAEA, *Guidance on the Import and Export of Radioactive Sources*, Non-serial Publications (Vienna: IAEA, 2012); IAEA, *Guidance on the Management of Disused Radioactive Sources*, Non-serial Publications (Vienna: IAEA, 2018)
34. IAEA, 'List of States Expressing a Political Commitment', 9 November 2023. https://nucleus-new.iaea.org/sites/ns/code-of-conduct-radioactive-sources/Documents/Status_list%2015%20September%20%202021.pdf
35. IAEA, *Nuclear Security Recommendations on Physical Protection of Nuclear Material and Nuclear Facilities (INFCIRC/225/Revision 5)*, IAEA Nuclear Security Series, No. 13 (Vienna: IAEA, 2011), 1.
36. Ibid.
37. Ibid., Foreword.
38. Jankowitsch-Prevor, 'Suppression of Acts of Nuclear Terrorism', 20.
39. Stoiber, 'Nuclear Security', 222.
40. Ibid., 223.
41. The Nuclear Security Governance Experts Group (NSGEG) 'is a multi-sector coalition of experts with diverse nuclear experience that has developed more than 50 recommendations

for improving nuclear security. The NSGEG is a project of the Asan Institute for Policy Studies, Partnership for Global Security, and the Stanley Foundation.' Bernhard et al., '*International Convention on Nuclear Security*', 25.

42. Statute of the International Court of Justice, art. 38; Michael N. Shaw, *International Law*, 9th edn (Cambridge: Cambridge University Press, 2021), 92.
43. IAEA, 2022, Incident and Trafficking Database (ITDB). https://www.iaea.org/resources/databases/itdb
44. INTERPOL, 2022, 'Geiger Analytical Database'. https://www.interpol.int/en/Crimes/Terrorism/Radiological-and-Nuclear-terrorism/Our-response-to-radiological-and-nuclear-terrorism
45. Greenpeace, 'Greenpeace schildert grote scheur op koepel', *Kernenergie in Nederland* [online], 18 May 2005. https://www.kernenergieinnederland.nl/node/524
46. *United States v Walli*, No. 14-5220/5221/5222 (6th Cir. 2015), 2.
47. Leanna Burnard, 'Former US Congressional Candidate Attempts to Buy Radioactive Substance on the Dark Web', *Trust & Verify*, 163 (2019), 9. https://www.vertic.org/wp-content/uploads/2019/04/TV163.pdf
48. Yasemin Balci, 2021, 'Case Study III: A Criminal Case in the United States', *EU Non-Proliferation and Disarmament Law*, eLearning course, No. 17. https://nonproliferation-elearning.eu/learningunits/disarmament-law
49. *United States v Ryan*, 20 December 2019, Stenographic Transcript of Status Conference, No. 318-cr-00146.
50. The Rome Statute of the International Criminal Court, 2187 UNTS 3 (adopted and opened for signature on 17 July 1998; entered into force on 1 July 2002); Leonard S. Spector, 'Nuclear Material and Commodity Trafficking as International Crimes: Current Status, Gaps in Coverage, and Potential Steps Forward', *Proceedings of the Annual Meeting of the American Society of International Law*, 105 (2011), 133, 137.
51. Ibid., 137; Maria Sanminiatelli, 'Sarkozy Proposes Trying Countries Selling Nukes', *Associated Press* [online], 14 April 2010. https://www.arabnews.com/node/342254; Reuters, 'Nicolas Sarkozy propose de juger les fournisseurs des "groupes terroristes"', France24 [online], 13 April 2010. https://www.france24.com/fr/20100413-washington-sommet-prevention-nucleaire-sarkozy-propose-juger-fournisseurs-groupes-terroristes

Suggested Reading

Drobysz, Sonia, (2022), 'Nuclear Security: Physical Protection, Illicit Trafficking and Nuclear Terrorism', in Kimberly Sexton Nick and Stephen G. Burns eds., *Principles and Practice of International Nuclear Law* (Paris: OECD), 339–56.

IAEA, (2011), *The International Legal Framework for Nuclear Security*, IAEA International Law Series, No. 4 (Vienna: IAEA).

Stoiber, Carlton, Baer, Alec, Pelzer, Norbert, and Tonhauser, Wolfram, (2003), *Handbook on Nuclear Law* (Vienna: IAEA).

Stoiber, Carlton, (2010), 'Nuclear Security: Legal Aspects of Physical Protection, Combating Illicit Trafficking and Nuclear Terrorism', in *International Nuclear Law: History, Evolution and Outlook* (Paris: OECD), 219–42.

Wetherall, Anthony C., (2016), 'Strengthening the International Legal Framework for Nuclear Security: Better Sooner Rather than Later', *Nuclear Law Bulletin*, 2/98, 25–6.

CHAPTER 9

THE EVOLUTION OF GLOBAL NUCLEAR SECURITY GOVERNANCE

WYN Q. BOWEN, MATTHEW COTTEE, AND SARAH TZINIERIS

INTRODUCTION

THE architecture of global nuclear security governance comprises a complex set of treaties, international agreements, policy instruments, and guidance documents—some of which are formal and legally binding, and others that are informal and dependent on state-level commitment for their application. These instruments are broadly designed to prevent, deter, and respond to non-state actor acquisition of nuclear material for malign purposes. This chapter provides an overview of the evolution of global nuclear security governance. Employing a chronological approach, the chapter charts the formulation of international instruments on nuclear security and seeks to shed light on the international negotiations and policy debates that preceded important developments in the governance architecture. As will become clear, the evolution of global nuclear security governance has often been driven by policy responses to perceived threats or gaps in existing structures.

While the importance of nuclear security has been recognized since the discovery of atomic energy, the concept only became embedded in international governance structures from the 1970s. Today, nuclear security is accepted around the world as a fundamental requisite for the peaceful uses of nuclear energy. Over the years this understanding became legally anchored in various binding treaties and resolutions, most prominently those of the United Nations Security Council (UNSC) and the International Atomic Energy Agency (IAEA). Such efforts have created a distinct set of governance structures for nuclear security, although there remains resistance from many states to more permanent or extensive instruments.

Governance structures for nuclear security extend wider, however. Governance refers to 'the process of collective decision-making and policy implementation' and reflects 'norms and processes relating to the delivery of public goods'.[1] The concept of norms explains how heightened awareness among states encourages them to abide by certain standards; while some obligations on nuclear security are enshrined in international law, many others are discretionary and reflect a sense of shared responsibility, or 'oughtness'.[2] This chapter argues that the United States (US) has played a key role as a 'norm entrepreneur' in persuading states to join initiatives and pursue universalization.

Dawn of the Atomic Age, 1940s–1960s

Screwdriver Report

The genesis of the need to secure nuclear and radiological materials came with the dawn of the atomic age, even if the concept of nuclear security itself was formalized much later. Early work on nuclear fission research was driven by the Manhattan Project, which produced the first nuclear weapons, used in 1945 at the end of the Second World War. The recognition that nuclear fission could also be exploited as a cheap and potentially infinite energy source led to the rapid development of a civilian nuclear industry in the post-war period. It was clear from the outset that, whatever their application, the mishandling of fissile materials would pose major risks to human health and the environment—and stringent protections were necessary.

Even in these early years, the potential for what was later termed 'nuclear terrorism' was demonstrated during a closed Congressional hearing in 1946, when Manhattan Project architect Robert Oppenheimer warned about a scenario of hostile operatives smuggling a bomb into New York City. Detecting such a device, he advised, would require a 'screwdriver' to open every crate entering the city, thereby signalling that such a scenario was likely impossible to prevent. At the time, Oppenheimer's testimony appeared alarmist, but it was a catalyst of early thinking about an unconventional nuclear attack, set against the backdrop of the emerging Cold War. The US Atomic Energy Commission (AEC) subsequently commissioned a secret project on how to detect a clandestine nuclear device.[3] Informally called the 'Screwdriver Report', it marked a shift in perceptions about the threat posed by nuclear materials.

Rapid Growth of Civilian Nuclear Power

President Dwight D. Eisenhower's 1953 'Atoms for Peace' initiative sought to improve the image of atomic science as a non-military technology, and drove global growth in the industry by facilitating the export of US nuclear technology, materials, and expertise. However, an unintended consequence was the industry's unrestrained expansion.

Various incidents indicated that national regulators were ill-equipped to protect nuclear materials. A fire at Britain's Windscale plutonium plant in 1957 released radioactivity across northern Europe, underscoring the vital need to improve nuclear safety standards.[4]

In addition to safety concerns, there was growing unease among Western governments about the adequacy of nuclear safeguards, and in 1957 the IAEA was born. However, in 1965, the AEC discovered that 100 kilograms (kg) of highly enriched uranium (HEU) had been unaccounted for at a Pennsylvania fuel processing and fabrication plant 'over a period of years'.[5] The weakness of early American safeguards and the Windscale fire were a trigger for the development of additional domestic protection measures in the US and the UK. Nevertheless, the emphasis during this period was on nuclear safety standards and nuclear material accounting and control (NMAC), rather than nuclear security.

Growing concerns about security led the AEC to commission an independent review. The panel, led by Ralph Lumb, concluded in its 1967 report that measures should 'be designed in recognition of the problem of terrorist or criminal groups clandestinely acquiring nuclear weapons or materials useful therein'.[6] This was the first formal acknowledgement by any authority that nuclear terrorism posed a threat, and the first time security measures were formally advised. The Lumb panel recommended '[t]he AEC, in cooperation with its licensees, should develop minimum physical protection standards'.[7]

Recognition of the Threat, 1970s–1980s

Emergence of International Terrorism

By the 1970s, the need to safeguard civilian nuclear facilities from the potential diversion of nuclear materials to military use was well established, although the actual threat posed by non-state actors was less understood.[8] But as the decade progressed there was a significant expansion in international terrorist activity. Not only were international terrorist attacks increasing in frequency, but also they were becoming more indiscriminate and audacious—epitomized by the 1972 Munich Olympics hostage massacre. The immediate impact of Munich was the implementation of counterterrorism measures in several Western countries. The next logical step for terrorists, it was feared by some, might be to attempt to acquire nuclear materials. Scholars such as Thomas Schelling questioned whether 'future terrorists, for motives either personal or political, will hold cities hostage, rather than just airplanes full of people? It has to mean that they could'.[9]

Concerns about unrestrained nuclear proliferation in the 1960s resulted in the Treaty on the Non-Proliferation of Nuclear Weapons (NPT) which entered into force in 1970. Supported by both superpowers and regarded as the cornerstone of the international community's non-proliferation architecture, the NPT required state parties to control civil nuclear materials under a safeguards regime verified by the IAEA. It was within this context that a period of policy entrepreneurship on nuclear security ensued.

By now nuclear security was not only the preoccupation of a small number of countries with their own indigenous nuclear facilities. Aided by Eisenhower's Atoms for Peace initiative, civil nuclear energy and research programmes had sprung up globally and many scientific documents on nuclear applications had been declassified. While these developments removed the exclusivity that had characterized the early atomic age, the accessibility of sensitive technology and information gave rise to proliferation risks. The movement of nuclear personnel between countries further disseminated knowledge and skills around the world.[10] No longer was an advanced level of indigenous knowledge a prerequisite for building a nuclear programme.[11]

During the following decade, further atrocities—including bombings by Irish republican paramilitaries, a spate of aeroplane hijackings, and the 1988 bombing of a PanAm plane over Lockerbie—brought into sharp focus the lengths that terrorists were prepared to go to maximize publicity.[12] The upsurge in international terrorism fuelled concerns about weak regulatory practices accompanying the rapid global growth of the civil nuclear sector. There was growing evidence of poor guarding of fissile material, even in established nuclear sectors like the UK, where inadequate arrangements around nuclear waste had been exposed.[13] While terrorism was becoming a growing problem internationally, the threat of terrorist groups targeting nuclear materials was still rather abstract. Several incidents, however, suggested that nuclear terrorism presented a risk, such as the 1982 sabotage attacks on the Koeberg Nuclear Power Plant in South Africa sponsored by the African National Congress (ANC).[14]

First Legal Instrument

By the mid-1970s, there was growing international consensus that collective action was needed on nuclear security and, decisively, this was supported by both superpowers. In 1972, the IAEA had published a booklet by a group of experts titled *Recommendations for the Physical Protection of Nuclear Material*. This text became the basis of subsequent international efforts to formalize the requirement for states to secure nuclear materials. In 1976, the text was published as an IAEA guidance document, 'Information Circular 225 (INFCIRC/225)—The Physical Protection of Nuclear Material'. INFCIRC/225 underwent five further revisions—in 1977 (Rev.1), 1989 (Rev.2), 1993 (Rev.3), 1999 (Rev.4), and 2011 (Rev.5).

While not legally binding, INFCIRC/225 represented a key milestone in the evolution of global nuclear security governance. It provided recommendations on best practices in physical protection, and, for the first time, nuclear materials were categorized against levels of protection.[15] American 'norm entrepreneurship' was catalytic in the establishment of these early foundations of nuclear security governance. Washington was keen to improve global physical protection standards and sought to establish a close working relationship with the IAEA, amid political sensitivities about imposing additional nuclear governance requirements on other states in a top-down approach.

INFCIRC/225 paved the way for the first legal instrument on nuclear security. Adopted in 1979, the Convention on the Physical Protection of Nuclear Material (CPPNM) is a multilateral treaty with provisions for the physical protection of nuclear material. The CPPNM, which entered into force in 1987, was originally conceived with wide-ranging conditions on all non-military nuclear materials, associated facilities, and their transport. However, the scope of the final agreement was narrowed in order to gain consensus and covered only civilian nuclear material in international transit—widely understood as 'the weakest link in the fuel cycle from the standpoint of vulnerability to theft and diversion'.[16]

Despite this emergence of a 'nuclear security regime', the CPPNM was of limited scope and lacked verification mechanisms. Perceptions of the threat also varied between states. Compared with the established non-proliferation regime—built on the NPT and globally recognized safeguards—nuclear security lacked the same concern internationally. As would prove with later efforts, the difficulty in achieving consensus meant nuclear security initiatives have tended to be either non-binding or deficient in substance. Notwithstanding its shortcomings, the Convention still provides a baseline international guide in the area of physical protection and has the highest number of states parties for a legally binding nuclear security agreement. The CPPNM also retains a close symbiotic relationship with INFCIRC/225; the CPPNM ensures states protect nuclear material to a certain standard while INFCIRC/225 provides the necessary set of prescriptive recommendations against which states can be judged.[17] INFCIRC/225 and the CPPNM can thus be considered nuclear security's normative genesis.

Two subsequent nuclear accidents highlighted the linkage between nuclear safety and nuclear security: the 1979 partial meltdown of the Three Mile Island nuclear reactor in the US; and the 1986 meltdown and radiation release at the Chernobyl power plant in Ukraine (then part of the Soviet Union). The two disasters ultimately led to two international conventions on ensuring prompt notification of a nuclear incident and facilitating emergency assistance for any country facing a nuclear disaster: the Convention on Early Notification of a Nuclear Accident and the Convention on Assistance in the Case of a Nuclear Accident or Radiological Emergency, both adopted in 1986. These legally binding agreements also apply to deliberate acts, and thus can be considered as contributing to the consolidation of nuclear security governance.

Nuclear Leakage and the Advent of 'New' Terrorism, 1990s

Post-Cold War Proliferation Risks

With the end of the Cold War in the late 1980s, the strategic environment changed dramatically. Although the risk of an East–West nuclear exchange abated, the new security environment was far from stable, with the Soviet Union's dissolution leaving behind a sprawling complex of military and civilian nuclear facilities. Almost overnight, the newly independent governments of Russia, Ukraine, Kazakhstan, and Belarus inherited nuclear research, production, and storage facilities, as well as so-called 'secret cities' housing thousands of nuclear personnel. Alongside civilian infrastructure was a vast arsenal of nuclear weapons. Prior to its collapse, the Soviet Union had over 27,000 nuclear weapons and enough weapons-grade plutonium and HEU to triple that number.[18]

The Soviet Union's dissolution also triggered a severe economic crisis. Amid deep cuts to public spending, nuclear infrastructure fell into disrepair while regulatory bodies were severely compromised in their monitoring and enforcement capacities. Safeguarding of high-grade nuclear materials and the security controls at nuclear facilities lapsed. The profound economic and political turbulence also triggered rampant corruption in public institutions and a surge in organized crime. Hitherto well-paid nuclear workers were laid off or faced months-long delays to their wages.[19]

Inevitably, international concerns grew about poorly secured nuclear weapons being potentially traded on the black market (so-called 'loose nukes'). The actual market for such material was unclear but it was initially thought to include aspiring nuclear weapons states, particularly North Korea, Libya, and Iran.[20] With the benefit of hindsight it appears that the loose nukes threat was somewhat exaggerated. Nevertheless, the potential diversion of civilian nuclear materials, technology, and sensitive information was no longer abstract. The vast borders of the former Soviet Union (FSU) had become relatively porous, while lapsed security controls meant any stolen nuclear materials were unlikely to return to regulatory control.

The threat was characterized by an incident involving a nuclear engineer, Leonid Smirnov, who smuggled 1.5 kg of HEU out of a state research laboratory near Moscow. Caught in October 1993 before finding a buyer, Smirnov had siphoned out minuscule amounts over several months to avoid detection.[21] A series of nuclear smuggling incidents subsequently came to light, including the case of a Russian naval officer illegally appropriating 4.5 kg of HEU in Murmansk.[22] It was in this context of fear over new routes to nuclear proliferation that nuclear security became a more pressing priority for the international community. Progress was again primarily driven by US norm entrepreneurship, both in terms of identifying and publicizing the dangers emanating from the FSU and taking steps to counter them.

Cooperative Threat Reduction

Fears about nuclear leakage from the FSU to rogue states led to an array of policy initiatives, aimed at both securing chemical, biological, radiological, and nuclear (CBRN) materials and stabilizing the Soviet military during a time of transition. Washington funded and implemented the majority of these initiatives, although progress would not have been possible without Russia's collaboration. In December 1991, the George H. Bush administration signed the Soviet Nuclear Threat Reduction Act. The programme of work mostly focused on non-proliferation assistance and threat reduction—the most well-known being 'Cooperative Threat Reduction' (CTR).[23] Of direct relevance to nuclear security was the emphasis in many of the programmes on improving physical protection.

The various post-Cold War initiatives in the FSU were largely successful in reducing nuclear weapons arsenals and strengthening nuclear security at civilian facilities. Nevertheless, not all objectives were fulfilled because of apparent mismanagement and bureaucratic inertia within the FSU. One report even argued the assistance programmes may have eased Russian budgetary woes and enabled defence modernization.[24] Yet, against the odds, the US and Russia utilized a narrow window of opportunity to forge an unexpected partnership over nuclear security that shaped the international regime over the coming decade.

Dawn of 'New Terrorism'

By the 1990s, fears were growing in the international community that terrorists might seek to acquire CBRN materials. While terrorists had already proved themselves capable of launching mass-casualty attacks and operating across borders, this so-called 'new terrorism' was characterized by an apparent willingness to employ non-conventional weapons to achieve apocalyptic ends.[25] No longer focused on traditional goals of national liberation or revolution, these new terrorists—most prominently jihadist militant movements nascent in the Middle East—were driven by existential desires related to ideology and religion; the risk of alienating their supporters through indiscriminate killing had apparently ceased to be a deterrent. The arrival of the internet also meant sensitive CBRN information was potentially accessible to would-be terrorists.

A sarin attack in March 1995 by a religious doomsday cult in Japan, Aum Shinrikyo, highlighted the threat from non-state actors and unconventional weapons. In coordinated attacks, Aum adherents punctured packets of sarin with umbrellas to disperse the nerve agent across the Tokyo metro system, killing 13 and injuring hundreds more.[26] The use of a nerve agent broke an implicit taboo about the use of CBRN weapons. It later emerged that Aum Shinrikyo had produced the sarin itself and had experimented with developing botulinum and anthrax.[27]

Following the terrorist attacks on two US embassies in East Africa in 1998 and on the *USS Cole* in 2000, global attention turned to Osama bin Laden and his jihadist militant group, al-Qaeda. US intelligence reports indicated that al-Qaeda harboured an interest in CBRN weapons, leading the US to broaden its FSU activities from narrow non-proliferation efforts to encompass CBRN security. More generally, the attacks thrust the issue of unconventional terrorism into the policy spotlight. Compared to the tangible concerns about nuclear attacks and mutually assured destruction that characterized the Cold War, the risk of a terrorist nuclear attack was much harder to quantify—and therefore more difficult to defend against.

IAEA Headway on Nuclear Security Governance

Against this backdrop, the IAEA expanded its nuclear security role during the 1990s. A notable development was updating INFCIRC/225/Rev.2 to reflect the security environment and evolving policy architecture. It had become clear that vast implementation differences remained between states, despite INFCIRC/225 being the global reference point for physical protection. INFCIRC/225/Rev.3, published in 1993, was a significant development as it now included guidance on appropriate regulatory systems. It also went beyond the CPPNM by incorporating the concept of in-depth or layered protection aligned to the category of nuclear material.[28]

At the behest of the US, INFCIRC/225/Rev.4, published in 1999, incorporated for the first time the concept of the 'design basis threat' (DBT).[29] The state was still responsible for the evaluation of the threat, but the US had successfully institutionalized the DBT via the IAEA. The recognition that nuclear facilities required a specific threat assessment demonstrated the importance of US leadership in setting global standards. There were also additional provisions in INFCIRC/225/Rev.4 for preventing nuclear sabotage and, for the first time, for facilities specifically. In the wake of Chernobyl, there was also an extra clause on the 'release of radioactive substances'.[30] These developments reflected heightened fears about radiological terrorism and sabotage of nuclear reactors.

In 1995, the IAEA created the Illicit Trafficking Database—later renamed the Incident and Trafficking Database (ITDB). The ITDB records incidents of illicit trafficking of nuclear and radioactive materials dating from 1993 onwards and has made a significant contribution to the monitoring of nuclear smuggling and materials outside of regulatory control. Also in 1995, the IAEA formed the International Physical Protection Advisory Service (IPPAS). Performed around the world at the request of states, IPPAS missions provide an international peer review of the physical protection arrangements at nuclear facilities, identifying weaknesses, and assisting in rectifying them. The aim is to facilitate a more standardized approach to the domestic implementation of INFCIRC/225.

The earliest terrorism treaty containing nuclear security-related obligations was the International Convention for the Suppression of Terrorist Bombings, adopted by the UN General Assembly in 1998. However, by the end of the millennium it was evident there were gaps in the international policy architecture regarding terrorism, and that new instruments were required. The international community was committed to nuclear security, but the focus was still largely on the safeguards regime as the foremost tool to prevent illicit trafficking and improve NMAC.

Post-9/11 Period, 2000s

'War on Terror'

The events of 11 September 2001 ('9/11') profoundly influenced the global security environment and gave new impetus to concerns about nuclear terrorism. Al-Qaeda had proven its ability to circumvent what were previously considered robust civil aviation security measures in the continental US to launch synchronized attacks using commercial aircraft. During the 1990s, the militant jihadist group had indicated interest in CBRN materials. Now concerns were heightened about this 'potential conjunction of terrorism and WMD [weapons of mass destruction]', elevating risk calculations.[31] Yet without hard evidence, others argued such narratives were alarmist.[32]

The 9/11 attacks prompted a flurry of policy making, giving rise to new international norms to complement pre-existing standards for physical protection and nuclear safety. Most notably, the UN Security Council adopted two binding Resolutions: 1373 and 1540. Resolution 1373, agreed unanimously two weeks after the attacks, set out unprecedented measures on countering terrorism and, for the first time, included a focus on nuclear materials. Specifically, 1373 noted 'the close connection between international terrorism and transnational organized crime, illicit drugs, money-laundering, illegal arms-trafficking, and illegal movement of nuclear, chemical, biological and other potentially deadly materials'.[33] While there was no direct intelligence indicating an imminent CBRN terrorist threat, statements of intent by al-Qaeda to acquire WMD no longer seemed far-fetched. The A. Q. Khan network, centred in Pakistan, also highlighted the role of non-state actors in WMD proliferation as both suppliers and recipients of weapons and technologies.[34]

The existing non-proliferation regime appeared incapable of addressing these emerging proliferation risks, and a new means of tackling the threat was required. In this context, the Security Council unanimously adopted Resolution 1540 in April 2004. This was a comprehensive resolution requiring states to implement appropriate measures to ensure the security of 'nuclear, chemical or biological weapons', with specific provisions on physical protection, material accountancy and control, and border controls.[35] Of note, radiological materials were not explicitly referenced.[36]

Adopted under Chapter VII of the UN Charter, 1540 includes a mandate for a committee to oversee its implementation. However, the 1540 Committee largely relies on voluntary reporting by states and, due to an absence of verification or effective enforcement provisions in the resolution, lacks authority to carry out an effective oversight role. Still, 1540 does much to mitigate the threat of non-state actors acquiring WMD or the means to develop them and has become a cornerstone of nuclear security efforts. Mandate of 1540 has been extended several times by further resolutions: UNSCR 1673 (2006), UNSCR 1810 (2008), and UNSCR 1977 (2011).

Perceptions of the CBRN terrorist threat were exacerbated by the delivery of anonymous letters containing the biological agent anthrax to various locations across the US just a week after the 9/11 attacks. Posted to senators, media outlets, and other targets, the anthrax letters killed five people and injured another 17. The subsequent investigation represented the most expensive in American history and led biodefence funding to surge the following year to US$4 billion, up from US$633 million.[37]

In the wake of 9/11, governments across the world, but particularly in North America and Europe, strengthened domestic security legislation and enhanced intelligence capabilities and border security. Established supplier groups of nuclear material also adjusted their control lists and adopted language aimed at preventing illicit acquisition of WMD-related materials.[38] In the US, objective concerns about nuclear terrorism were sensationally packaged as part of the broader 'War on Terror' and played into Washington's justification for the 2003 invasion of Iraq. Arguably, this environment of heightened anxiety prompted the most innovative period in the evolution of nuclear security governance.

Catalyst for IAEA Initiatives

In this environment, the IAEA was able to elevate itself as an institutional actor in the field of nuclear security—playing the role of a non-partisan, expert-led organization, and thereby carrying legitimacy amid perennial tensions between the nuclear and non-nuclear countries. In 2002, the IAEA launched several initiatives. First, to support its work in assisting Member States, the IAEA created a Nuclear Security Plan (NSP), to be implemented by a newly formed Office of Nuclear Security. Renewed every four years, the NSP sets out the IAEA's future priorities and implementation strategy. Second, the IAEA developed Integrated Nuclear Security Support Plans (INSSPs). INSSPs provide ad hoc support to individual states by outlining improvements required in the area of physical protection. And third, to finance its expanded activities, the IAEA created a new funding mechanism, the Nuclear Security Fund (NSF), based on voluntary donations from states.

Another example of the IAEA's expanding role was its promotion of radiological source security. Awareness of the CBRN threat was increasing internationally. In 2004, the IAEA published the Code of Conduct on the Safety and Security of Radioactive Sources ('the Code'). The Code sets out guidelines for national authorities to ensure radiological sources are adequately protected. Despite the majority of states being signatories, the Code is not a legally binding agreement but represents 'soft law' where parties are simply expected to observe good conduct. Arguably, there was greater incentive for most states to push the radiological source security agenda because it is more universal than nuclear terrorism, given almost all countries used radiological sources in medicine, industry, and agriculture.

Another IAEA initiative was its fifth revision of INFCIRC/225. Published in January 2011, this was 'an evolutionary, not revolutionary' update intended to harmonize INFCIRC/225 with the CPPNM and other Nuclear Security Series documents.[39] While changes incorporated in this version were less substantial than those made in previous ones, INFCIRC/225/Rev.5 includes revised guidance on how to categorize self-protecting nuclear material when applying physical protection measures. This was an acknowledgment that adversaries may be willing to receive damaging or even lethal doses of radiation in order to accomplish their mission—a nod to the events of 9/11.

The IAEA's expanded engagement helped raise awareness about nuclear security, with its reach going beyond the smaller groupings of like-minded states on which the

Bush administration was focusing diplomatic efforts. The IAEA was also instrumental in issuing recommendations about best practice, particularly in the area of physical protection. Yet, the IAEA's potential to socialize norms and encourage states to develop binding commitments was hindered by still significant resistance in some quarters. While initial pledges to the NSF had been promising, then-IAEA director general Mohamed ElBaradei revealed these did not translate into received donations, leaving the US to underwrite the majority of initiatives.[40]

Global Partnership and East–West Diplomacy

In the wake of 9/11, Washington was highly successful in engaging like-minded states in plurilateral non-binding arrangements to achieve a specific set of policy goals related to the CBRN threat. These were designed to fill gaps in the existing nuclear security architecture, albeit an increasingly complex architecture characterized by overlapping objectives. Most prominent among such arrangements was the Global Partnership Against the Spread of Weapons and Materials of Mass Destruction ('Global Partnership'). Launched in 2002 at the G8 Summit, the Global Partnership aimed 'to prevent terrorists, or those that harbour them, from acquiring or developing nuclear, chemical, radiological and biological weapons; missiles; and related materials, equipment and technology'.[41]

The scale of funding and technical expertise committed to the Global Partnership was unprecedented. The US, as principal sponsor, provided a total of US$10 billion for the initial 10-year mandate. Urged by Washington, other G8 members, several other states, and the EU collectively pledged a further US$10 billion.[42] The Global Partnership built on pre-9/11 threat reduction efforts, first focusing on the considerable inventory of CBRN materials in Russia and other parts of the FSU, then expanding to include wider countries of interest. Underscoring its importance, the Global Partnership was renewed beyond its original 10-year mandate. However, Russia stopped participating following its suspension in March 2014 from the G8, in response to 'Russia's violation of Ukraine's sovereignty and territorial integrity'.[43]

The Global Partnership undoubtedly contributed to international security by decreasing CBRN stockpiles, implementing physical security measures, and promoting security culture. Nevertheless, as a non-legally binding, plurilateral initiative, and reliant on access agreements between funders and recipient states, the Global Partnership suffered from lack of coordination. It was not always clear if programmes were based on the greatest need and would deliver long-term solutions.

Shared nuclear security concerns led the US and Russia to announce a new joint initiative in July 2006, the Global Initiative to Combat Nuclear Terrorism (GICNT),[44] which was not a formal treaty organization, but rather one designed as an overarching programme to raise international awareness of the now elaborate nuclear security architecture. It called on governments to commit to implementing existing nuclear security legislation to mitigate the threat of nuclear terrorism. In effect, GICNT relied on norms to encourage states to consent voluntarily to nuclear security standards despite not being legally bound to these.

In May 2003, the US launched the Proliferation Security Initiative (PSI)—'a response to the growing challenge posed by the proliferation of weapons of mass destruction (WMD), their delivery systems, and related materials worldwide'.[45] The PSI was developed in response to an incident the previous December when the crew of a Spanish warship off the Yemeni coastline intercepted a shipment of scud missiles and chemical propellant purportedly from North Korea. Despite the implications for WMD proliferation, no breach of any law had been committed.[46]

Under the PSI, governments commit to non-binding Statement of Interdiction Principles, which set out a coordinated approach to interdicting WMD, delivery systems, and related materials being transferred to or from state or non-state actors of proliferation concern. The PSI has been adopted by more than half of the world's governments. However, implementation relies on participating states employing domestic tools, which poses 'myriad potential political, legal, operational, and informational hurdles'.[47] Some states—most notably China—have disputed the legality of the proposed interdictions.

Legally Binding Entrepreneurship

In this industrious policy-making period, the international nuclear security architecture evolved into a complex web of legally binding conventions, with gaps filled by an array of informal initiatives. Even before 9/11, it was clear that existing nuclear security agreements required updating—most obviously the CPPNM—to reflect both technical developments and the post-Cold War security environment. In June 2001, the IAEA had established a group of legal and technical experts to draft an Amendment to the CPPNM ('A/CPPNM'). But despite the group convening six times, the draft failed to generate consensus.[48] In response, 25 states parties to the CPPNM subsequently produced their own amendment.

Negotiations remained thorny not least because it was feared an additional protocol to the original convention would undermine the existing legislative framework and create parallel or incompatible regimes.[49] Washington also wanted to go further by using the mandatory acceptance of INFCIRC/225 as a baseline of physical protection standards and obligations for states to self-report to the IAEA on physical protection measures.[50] However, some states, including the UK, possessed large stockpiles of civilian plutonium that would fall under this proposal, and resisted. In this context, several governments advocated reinforcing the CPPNM by enlarging its membership and strengthening its application through existing signatories.[51] For other states, including Russia, the prevailing legislation was insufficient and unable to prevent nuclear terrorism in all its manifestations.

Ultimately, the A/CPPNM was adopted by all parties in July 2005. The A/CPPNM broadened the scope of the original convention text, making it a legal obligation for states to protect the domestic use, storage, and transportation of civilian nuclear material. The A/CPPNM also advanced cooperation among states regarding rapid measures to locate and recover smuggled or stolen material and to mitigate the radiological consequences of sabotage. Furthermore, once a suitable legislative framework is established in a state, it required implementation by a responsible national authority. The Convention and its amendment remain the only legally binding international instruments on the physical protection of nuclear and radioactive materials.

Another milestone in the evolution of nuclear security governance was the adoption in April 2005 of the International Convention for the Suppression of Acts of Nuclear Terrorism (ICSANT), which entered into force in July 2007 and was designed to criminalize acts of nuclear terrorism and encourage international coordination that would deter, investigate, and prosecute such acts. ICSANT had its roots in debates during the 1990s when policy makers were exploring additional methods to prevent unconventional activities by non-state actors.[52] In 1996, Russia drafted a document that formed the basis of the subsequent ICSANT agreement. However, negotiations over the proposal were complicated, in part as it was feared another initiative would undermine and dilute the existing nuclear security architecture. There were also concerns over respect for state sovereignty, and whether the document's wording might be interpreted as legitimizing the use of nuclear weapons.[53]

By incorporating 'the broadest possible definition of terrorist acts related to the use, or threat of use, of nuclear components,' ICSANT was designed to counter threats by individuals or organizations, regardless of the target.[54] Some states had already amended domestic legislation following 9/11. But ICSANT required all parties to make certain acts criminal offences in domestic law, to establish jurisdiction over such offences, and to prosecute or extradite persons alleged to have committed the defined criminal offences.[55] The requirement for domestic legal changes helped grow the uptake of norms associated with nuclear security. There was also now greater crossover with counter-terrorism policy. Provisions in the previous conventions had only applied to nuclear materials being transported across borders, or being used, stored, or transported in a given state; thus ICSANT helped to plug the gaps and complemented UNSCR 1540. ICSANT also broadened 'ownership' of the issues by being introduced at the UN, where all states could negotiate over the scope of the text.

Nuclear Summitry and Beyond, 2010s–Present

Prague Agenda and the NPT Pillars

Barack Obama's electoral victory in November 2008 gave huge momentum to international nuclear security efforts. Early in his presidency, the US president declared a vision of global nuclear disarmament during a speech in Prague.[56] The 'Prague speech', as it became known, emphasized that nuclear security was at the top of America's foreign policy agenda, with Obama warning nuclear terrorism was an 'immediate and extreme threat' that required a global response.[57] He proposed a 'new international effort to secure all vulnerable nuclear material around the world within four years.'[58] Obama also announced renewed efforts to disrupt nuclear smuggling and intercept sensitive materials at borders.[59] The Prague speech provided the origin for the Nuclear Security Summits, a series of four inter-governmental conferences on nuclear security held from 2010 to 2016.

Prior to the inaugural 2010 summit, international differences over how nuclear security fitted with the non-proliferation regime came to a head. The US and UK sought to reinvent nuclear security as a 'fourth pillar' of the NPT—equivalent to the three existing pillars of disarmament, non-proliferation, and peaceful uses of nuclear energy.[60] It was envisaged that insulating nuclear security from broader nuclear weapons politics by reconstituting it as a separate pillar would lead to heightened engagement by those states critical of the slow pace of disarmament. However, this idea backfired with the Non-Aligned Movement (NAM), which contained several nuclear newcomer states, objecting to what they perceived as additional obligations being imposed on their inalienable right to nuclear energy.[61] The fourth pillar concept was subsequently abandoned by the US and UK. In the run-up to the summit, emphasis was placed instead on state-level responsibility and the centrality of the IAEA. It was also recognized that progress could only be achieved by the international community working together on the basis of common interests.

Nuclear Summitry

The four Nuclear Security Summits aimed to mitigate the threat of nuclear terrorism and enhance the security of nuclear and radioactive materials around the world. The 2010 summit was hosted by Washington; the 2012 summit was hosted by Seoul, South Korea; the 2014 was hosted by The Hague, Netherlands; and the concluding 2016 summit returned to Washington. Underscoring their significance, the summits represented the largest gathering of world leaders since the UN Conference on International Organization in 1945. While the legacy of the summits is mixed, they helped to consolidate a global consensus over the threat of nuclear terrorism and provided a catalyst for governments to take concrete steps, thereby strengthening international nuclear security governance.

The overarching policy agenda remained the same throughout the summits but there were several notable developments. Reflecting Obama's preoccupation with nuclear terrorism, the summit process initially focused on the scenarios of extremists acquiring fissile materials to launch mass-casualty attacks; at the later summits a broader range of scenarios featured, including the unauthorized acquisition of radiological sources; during the 2012 summit, the safety–security interface came to the fore, reflecting international concern over the 2011 Fukushima nuclear disaster; at the 2016 summit, Russia did not participate amid tensions with Western governments after its annexation of

Crimea. A key trend to emerge over the course of the summits was the progressive expansion in commitments pledged by states, totalling more than 935 individual voluntary state actions.[62]

Indeed, the summit process can be said to have pioneered a novel implementation mechanism where states were motivated to make ambitious—albeit non-binding—commitments in the form of (national) 'house gifts' or (multilateral) 'gift baskets'. These were voluntary commitments but there was an implicit expectation that progress would be analysed by the international community at each subsequent summit, thereby motivating implementation.[63] With a greater number of states adopting nuclear security agreements, the corresponding standards became embedded into national regulations and laws.[64] The summits thus acted as a catalyst in consolidating international nuclear security norms.

A key milestone was the uptick in states approving the A/CPPNM, which reached the threshold to enter into force in May 2016.[65] Similarly, there was a substantial increase in new ratifications of ICSANT. Meanwhile, the number of countries that possessed weapons-usable material (defined as holding 1 kg or more) went from 35 when the process began to 24 when it ended.[66] States also took action to down-blend a substantial amount of HEU to low enriched uranium (LEU) and convert separated plutonium to mixed oxide (MOX) fuel.[67] Other achievements included the creation of new nuclear security training centres, increased contributions to the NSF, and Russia's cessation of plutonium production. Another landmark was achieved by the joint proposal 'Strengthening Nuclear Security Implementation' (SNSI), which obliged signatories to commit to implementing IAEA recommendations. SNSI was subsequently adopted by the IAEA as INFCIRC/869. By giving the force of law to these non-binding recommendations, signatories were contributing to norms development and helping to strengthen international nuclear security governance.

As the process wore on, tangible progress was becoming harder to achieve, while the overlap with non-proliferation politics remained a source of contention. Meanwhile, those states not invited to the summits were increasingly vocal.[68] There were also no formal mechanisms established to evaluate the implementation of commitments once the process ended. Above all, actions by states remained predominantly motivated by national interests, making any ambitious future agreements unlikely, such as the comprehensive 'International Convention on Nuclear Security' touted in 2015 by the Nuclear Security Governance Experts Group (NSGEG).[69] At the final summit, the emphasis was on enhancing the normative elements of nuclear security by communicating expected behaviour and best practice.

Beyond the Summit Process

Following the summits, in parallel with actions by states the agenda was integrated into five separate 'Action Plans' to be implemented by the IAEA, UN, Global Partnership, GICNT, and Interpol. It was envisaged that these successor schemes would provide a governance framework for sustaining the summits' ambitions. However, the Action Plans required neither a formal reporting process nor did they identify funding arrangements—resulting in no guarantee of long-term sustainability. Moreover, the momentum achieved through the enactment of progressive and ambitious commitments—on which governments would be nominally answerable—was not replicated in the Action Plans or any post-summit initiative.

Perhaps inevitably, the pace of policy development on nuclear security slowed after the summits ended. The IAEA's triennial International Conference on Nuclear Security (ICONS) sought to fill this gap. Together with the IAEA's annual General Conferences, these events are now the principal forums for government-level engagement on nuclear security by bringing together ministers, experts, and international organizations. Other mechanisms continue to shape nuclear security governance such as the Nuclear Security Contact Group (NSCG), which works to implement the summit commitments. Responding to the lack of verifiable policy implementation, the US-based Nuclear Threat Initiative (NTI) created a Nuclear Materials Security Index designed to encourage states to improve their ranking, according to several criteria, and generally improve transparency.

The most recent development for global nuclear security governance was the Review Conference of the A/CPPNM in March–April 2022. Pursuant to Article 16.1, the conference had to take place within five years of A/CPPNM coming into force (albeit being a year late due to the Covid-19 pandemic). The conference reviewed implementation of the legislation and discussed its continuing adequacy in light of the security environment and new technologies. There is potential for this conference structure to be maintained on a regular basis, akin to the NPT review conferences held every five years.[70] Such a development would formalize the only international forum for a nuclear security instrument with a legal mandate.

Despite such persistent efforts, however, the plethora of post-summit initiatives has arguably resulted in a refragmentation of the policy landscape. The number of new actors and organizations working on nuclear security increased but crucially absent in the post-summit period is an ongoing mechanism for political commitment at the inter-government level. The Nuclear Security Summit process illustrated that close engagement by states is crucial for making a tangible impact.

Conclusion

The business of strengthening nuclear security governance is far from complete. This chapter demonstrates how major governance gaps remain and how the international architecture is characterized by complex and overlapping structures. There are still no comprehensive international standards or best practices that all states must follow, implementation of legal instruments is not universal, verification and monitoring mechanisms are largely absent, and the current system does not extend to the vast majority of the world's weapons-usable nuclear material that is under military control.

Historically, major world events have triggered reactive policy making on nuclear security. Yet while broad international engagement suggests nuclear security has acquired a normative element, the embedding of governance structures into an international system remains fundamentally dependent on shifting national appetites. US leadership, as a 'norm entrepreneur', has helped strengthen governance structures and unequivocally solidified international consensus. Nevertheless, while broad uptake of a binding agreement might enhance the normative aspect of nuclear security, this does not automatically equate to implementation where it matters most at the facility-level.

Ambitions for future binding agreements seem distant at the current juncture, primarily because the construction of nuclear security as a state-level responsibility is so entrenched. Perceptions of the threats diverge profoundly around the world and many nuclear newcomer states still perceive nuclear security initiatives as further bureaucratic hurdles. Nuclear security is also inherently complicated by debates around disarmament. In many ways, the international architecture on nuclear security still represents an uneasy compromise between the converging preferences of states.[71] The extent to which nuclear security norms are truly internalized by states will thus remain limited, absent a major shock such as a nuclear terrorist event. But it is not just terrorist groups that can exploit civil nuclear infrastructure in pursuit of their objectives. Russia's exploitation of nuclear facilities for tactical advantage in its war of aggression against Ukraine—most alarmingly at the Zaporizhzhia Nuclear Power Plant—has demonstrated the willingness of one of the world's most significant civil and military nuclear powers to entirely discard safety and security norms for its own gains. It remains to be seen whether the war in Ukraine will strengthen or undermine international cooperative efforts to strengthen nuclear security going forward.

Notes

1. Definition by the *Concise Oxford Dictionary of Politics*, (Oxford: Oxford University Press, 2009).
2. Martha Finnemore and Kathryn Sikkink, 'International Norm Dynamics and Political Change', *International Organization*, 52/4 (1998), 887–917.
3. Kai Bird and Martin J. Sherwin, *American Prometheus* (New York: Knopf, 2005).
4. Lorna Arnold, *Windscale 1957: Anatomy of a Nuclear Accident* (Basingstoke: Palgrave Macmillan, 1995), 154; Richard Wakeford, 'The Windscale Reactor Accident—years on', *Journal of Radiological Protection*, 27 (2007), 211–15.
5. Mason Willrich and Theodore B. Taylor, 'Nuclear Theft: Risks and Safeguards', *Political Science Quarterly*, 90/1 (1975), 79.
6. US Atomic Energy Commission, *Report to the Atomic Energy Commission by the Ad Hoc Advisory Panel on Safeguarding Special Nuclear Material* (Washington, DC: AEC, 1967), 4.
7. Ibid.
8. Bruce G. Blair and Gary D. Brewer, 'The Terrorist Threat to World Nuclear Programs', *The Journal of Conflict Resolution*, 21/3 (1977), 379.
9. Thomas Schelling, 'Who Will Have the Bomb?', *MIT Press*, 1/1 (1976), 83.
10. Willrich and Taylor, 'Nuclear Theft', 80.

11. Ibid.
12. Brian Jenkins, 'International Terrorism: A New Mode of Conflict', in David Carlton and Carlo Schaerf, eds., *International Terrorism and World Security* (London: Croom Helm, 1975).
13. John Hannam, 'Atomic Energy Authority (Special Constables) Bill', *Hansard (House of Commons)*, 906, 26 February 1976, Cols 701–55. http://hansard.millbanksystems.com/commons/1976/feb/26/atomic-energy-authority-special
14. Jo-Ansie van Wyk, 'The ANC's Operation Mac and the attack on the Koeberg Nuclear Power Station in South Africa', *Historia*, 60/2 (2015). http://dx.doi.org/10.17159/2309-8392/2015/V60N2A3
15. Christopher Price. 'Development of the IAEA Nuclear Security Recommendations on Physical Protection of Nuclear Materials and Nuclear Facilities (INFCIRC/225/Rev.5)', *Journal of Nuclear Materials Management*, XXXIX/3 (2011), 11.
16. Samuel J. Walker, 'Regulating against Nuclear Terrorism', *Technology and Culture*, 42/1 (2001), 115.
17. Craig Everton, Stephan Bayer, and John Carlson, 'Developments in the IAEA's Nuclear Security Series and Physical Protection Guidance Document INFCIRC/225', Paper presented to the Annual Meeting of the INMM, Baltimore, 11–15 July 2010.
18. Council on Foreign Relations, 2006, 'Loose Nukes'. https://www.cfr.org/backgrounder/loose-nukes
19. William Langewiesche, 'How to Get a Nuclear Bomb', *The Atlantic*, December 2006.
20. Alex P. Schmid, 'Terrorism and The Use of Weapons of Mass Destruction', *Terrorism and Political Violence*, 11/4 (2007), 116.
21. PBS, 1996, Interview with Leonid Smirnov.
22. Jane Perlez, 'Tracing a Nuclear Risk: Stolen Enriched Uranium', *The New York Times* [online], 15 February 1995. https://www.nytimes.com/1995/02/15/world/tracing-a-nuclear-risk-stolen-enriched-uranium.html
23. Defense Threat Reduction Agency, 2006, 'Cooperative Threat Reduction'.
24. Rich Kelly, 'The Nunn-Lugar Act: A Wasteful and Dangerous Illusion'. Foreign Policy Briefing, No. 39. *Cato Institute* [online], 18 March 1996. https://www.cato.org/foreign-policy-briefing/nunn-lugar-act-wasteful-dangerous-illusion#
25. Martha Crenshaw, 'The Psychology of Terrorism: An Agenda for the 21st Century', *Political Psychology*, 21/2 (2000), 405–20.
26. Schmid, 'Terrorism and The Use of Weapons of Mass Destruction', 126.
27. Richard Danzig, et al., *Aum Shinrikyo: Insights Into How Terrorists Develop Biological and Chemical Weapons* (Washington, DC: CNAS, 2012).
28. Bonnie D. Jenkins, 'Establishing International Standards for Physical Protection of Nuclear Material', *The Nonproliferation Review*, 5/3 (1998), 100.
29. Author correspondence with former government official involved in the drafting of INFCIRC/225/Rev.4, September 2014.
30. IAEA, *The Physical Protection of Nuclear Materials and Nuclear Facilities (INFCIRC/225/Rev.4)* (Vienna: IAEA, 1999), 2.
31. Harald Müller, Marco Fey, and Carsten Rauch, 'Winds of Change: Exogenous Events and Trends as Norm Triggers (or Norm Killers)', in Harald Müller and Carsten Wunderlich, eds., *Norm Dynamics in Multilateral Arms Control* (Athens: University of Georgia Press, (2013), 151.
32. John Mueller, *Atomic Obsession: Nuclear Alarmism from Hiroshima to Al-Qaeda* (Oxford: Oxford University Press, 2010).

33. UN Security Council, 28 September 2001, Resolution 1373.
34. Peter Crail, 'Implementing UN Security Council Resolution 1540', *The Nonproliferation Review*, 13/2 (2006), 355–99.
35. UN Security Council, 28 April 2004, Resolution 1540.
36. Ibid.
37. Aidan McCarty, 'Changes in U.S. Biosecurity Following the 2001 Anthrax Attacks', *Journal of Bioterrorism & Biodefense* [online], 9/2 (2018). DOI: 10.4172/2157-2526.1000163
38. Todd Perry, 'The Growing Role of Customs Organizations in International Strategic Trade Controls', in James Doyle, ed., *Nuclear Safeguards, Security and Nonproliferation* (Oxford: Elsevier, 2008), 550.
39. Everton et al., 'Developments in the IAEA's Nuclear Security Series and Physical Protection Guidance Document INFCIRC/225'.
40. ElBaradei, *The Age of Deception*, 165. https://www.amazon.co.uk/Age-Deception-Nuclear-Diplomacy-Treacherous/dp/1408822245.
41. Group of Eight Summit, 2002, 'Statement by G8 Leaders: The G8 Partnership Against the Spread of Weapons and Materials of Mass Destruction'.
42. Alan Heyes, Wyn Q Bowen, and Hugh Chalmers. *The Global Partnership Against WMD: Success and Shortcomings of G8 Threat Reduction since 9/11*, RUSI Whitehall Paper Series (Abingdon: Routledge, 2012), 2.
43. Group of Seven Summit, Joint G7 Statement from the Hague on Ukraine (London: Prime Minister's Office, 2014).
44. George W. Bush and Vladimir V. Putin, *Announcing the Global Initiative to Combat Nuclear Terrorism* (Washington, DC: US Department of State Archive, 2006).
45. The White House, *Proliferation Security Initiative: Statement of Interdiction Principles* (Washington, DC: The White House, Office of the Press Secretary, 2003).
46. Paul Kerr, 'U.S. Stops Then Releases Shipment of N. Korean Missiles', *Arms Control Today* [online], January 2003. https://www.armscontrol.org/act/2003-01/news/us-stops-then-releases-shipment-n-korean-missiles
47. Andrew C. Winner, 'The Proliferation Security Initiative', *The Washington Quarterly*, 28/2 (2005), 132.
48. Council of the European Union, 2005, 'Proposal for a Council Decision Issuing Directives to the Commission for the Negotiation of the Amendments to the Convention on the Physical Protection of Nuclear Material', Ref 9264/05.
49. Ibid.
50. George Bunn, 'Raising International Standards for Protecting Nuclear Materials from Theft and Sabotage', *The Nonproliferation Review*, 7/ 2 (2008), 152.
51. UN Press Release L2584, 'UN Ad Hoc Committee Begins Considering Draft Convention to Suppress Nuclear Terrorism Acts', 17 February 1998.
52. Rohan Perera, *International Convention for the Suppression of Acts of Nuclear Terrorism* (New York: Audiovisual Library of International Law, 2005).
53. Sam Kane and Kingston Reif, *Fact Sheet: The 2005 Amendment to the Convention on the Physical Protection of Nuclear Material (CPPNM) and the International Convention on the Suppression of Acts of Nuclear Terrorism (ICSANT)* (Washington, DC: Center for Arms Control and Non-Proliferation, 2013).
54. UN Press Release L2584, 'UN Ad Hoc Committee Begins Considering Draft Convention'.
55. David P. Fidler, 'International Convention for the Suppression of Acts of Nuclear Terrorism Enters into Force', *Insights*, 11/18 (2007), n.p.

56. Barack Obama, 'Remarks by President Barack Obama in Prague as Delivered, Hradcany Square, Prague, Czech Republic, 5 April' (Washington, DC: The White House, Office of the Press Secretary, 2009). https://obamawhitehouse.archives.gov/the-press-office/remarks-president-barack-obama-prague-delivered
57. Ibid.
58. Ibid.
59. Ibid.
60. Hillary Clinton, 'Remarks by Secretary of State Hillary Rodham Clinton at the United States Institute of Peace, Renaissance Mayflower Hotel', Washington, DC, 21 October (Washington, DC: Archive of US Department of State, 2009; Cabinet Office, *The Road to 2010: Addressing the Nuclear Question in the Twenty First Century* (London: HMSO, 2009), 7.
61. Non-Aligned Movement, 2010, '2010 Review Conference of the Parties to the Treaty on the Non-Proliferation of Nuclear Weapons'.
62. Sara Z. Kutchesfahani, Kelsey Davenport and Erin Connolly, *The Nuclear Security Summits: An Overview of State Actions to Curb Nuclear Terrorism 2010–2016*, Arms Control Association and Fissile Material Working Group Report (Washington, DC: ACA, 2018).
63. Michelle Cann, Kelsey Davenport, and Jenna Parker, The Nuclear Security Summit (Washington, DC: Arms Control Association and Partnership for Global Security, 2016).
64. Kutchesfahani et al., *The Nuclear Security Summits*.
65. Nuclear Security Summits, 2016, 'Statement by National Security Advisor Susan E. Rice on the Entry into Force of the 2005 Amendment to the Convention on Physical Protection of Nuclear Materials'.
66. Houses of Parliament, 2016, 'Nuclear Security', Post Note. UK Parliamentary Office of Science and Technology.
67. Nuclear Security Summit, 2014, 'Communiqué'.
68. Leah Matchett, 'The Controversial Legacy of the Nuclear Security Summit', *Bulletin of the Atomic Scientists* [online], 4 October 2018. https://thebulletin.org/2018/10/the-controversial-legacy-of-the-nuclear-security-summit/
69. John Bernhard, Kenneth C. Brill, Anita Nilsson, and Shin Chang-Hoon, *International Convention on Nuclear Security* (Muscatine, IA: NSGEG, 2015).
70. Jonathan Herbach and Samantha Pitts-Kiefer, 'More Work to Do: A Pathway for Future Progress on Strengthening Nuclear Security', *Arms Control* Today [online], October 2015. https://www.armscontrol.org/ACT/2015_10/Features/More-Work-to-Do-A-Pathway-for-Future-Progress-on-Strengthening-Nuclear-Security
71. Harald Müller, 'Where It All Began', in Harald Müller and Carmen Wunderlich, eds., *Norm Dynamics in Multilateral Arms Control* (Athens, GA: University of Georgia Press (2014), 1.

Suggested Reading

Barrow, Alex, Bowen, Wyn, Cottee, Matthew, Hobbs, Christopher, Lentini, Luca, Moran, Matthew, and Tzinieris, Sarah, (2022), *Nuclear Security Briefing Book*, 2022 Edition. (London: King's College London).

Bowen, Wyn, Cottee, Matthew, and Hobbs, Christopher, (2012), 'Multilateral Cooperation and the Prevention of Nuclear Terrorism: Pragmatism over Idealism', *International Affairs*, 88/2, 349–68.

Bunn, Matthew, Roth, Nickolas, and Tobey, William H., (2019), *Revitalizing Nuclear Security in an Era of Uncertainty* (Cambridge, MA: Harvard Kennedy School: Belfer Center for Science and International Affairs).

Kutchesfahani, Sara Z., Davenport, Kelsey, and Connolly, Erin, (2018), *The Nuclear Security Summits: An Overview of State Actions to Curb Nuclear Terrorism 2010-2016*, Arms Control Association and Fissile Material Working Group Report (Washington, DC: ACA)

Schelling, Thomas, (1976), 'Who Will Have the Bomb?', *MIT Press*, 1/1, 77–91.

CHAPTER 10

THE UNITED NATIONS SECURITY COUNCIL AND THE 1540 COMMITTEE

BENJAMIN KIENZLE AND DANIEL SALISBURY

INTRODUCTION

RESOLUTION 1540 of the United Nations (UN) Security Council is a milestone in international efforts to prevent the proliferation of weapons of mass destruction (WMD) to non-state actors. When it was adopted on 24 April 2004, Resolution 1540 became the first international instrument to impose legally binding measures to combat WMD terrorism on UN member states. Its scope is very broad and covers nuclear, chemical, and biological weapons, as well as their means of delivery. Nuclear security is only one—albeit an important—part of the resolution.[1] Notwithstanding the broad nature of the resolution, it has become a key element of the international nuclear security architecture.

Resolution 1540 refers specifically to other elements of that architecture, in particular the Convention on the Physical Protection of Nuclear Material (CPPNM) and the Code of Conduct on the Safety and Security of Radioactive Sources of the International Atomic Energy Agency (IAEA). The resolution also established a Committee of the Security Council, the '1540 Committee', to oversee the resolution's implementation in UN member states. However, the adoption and implementation of Resolution 1540 has not been without controversies and challenges, including its legally binding character, the perceived legitimacy of some of its measures, the lack of resources, and the wide variety of stakeholders that require engagement.

This chapter provides an overview of Resolution 1540 and the work of the 1540 Committee in the area of nuclear security. It explores the resolution's key provisions in the wider context of global nuclear security governance, and the place of the nuclear security aspects of the resolution in the broader scope of the 1540 agenda. More specifically, the chapter examines four main elements: the genesis and content of the resolution; its institutional infrastructure; major issues and controversies; and its mid- and long-term future.

The UN Security Council and Resolution 1540

The Security Council is widely seen as the most powerful UN organ, as it has 'primary responsibility for the maintenance of international peace and security'[2] and is able to adopt relevant measures that are legally binding for all UN member states.[3] Most notably, under Chapter VII of the UN Charter, the Security Council can authorize the use of sanctions and the use of force,[4] if it determines 'the existence of any threat to the peace, breach of the peace, or act of aggression.'[5] In other words, the Security Council has (in principle) wide-ranging powers to enforce the maintenance of international peace and security.

In the area of nuclear weapons, the UN Charter specified in particular that '[i]n order to promote the establishment and maintenance of international peace and security with the least diversion for armaments of the world's human and economic resources, the Security Council shall be responsible for formulating [...] plans to be submitted to the Members of the United Nations for the establishment of a system for the regulation of armaments.'[6] As such, the Council was endowed with specific powers regarding the control of nuclear weapons. Some of the earliest resolutions, in particular Resolutions 18 and 20 (1947), were about the international control of atomic energy.[7]

In practice, however, the Security Council had very little impact on the evolving role of nuclear weapons in the decades after the end of the Second World War. Very rarely, it has adopted nuclear weapons-related resolutions, for instance, Resolution 255 (1968), in which the nuclear weapon states promised to not use nuclear weapons against non-nuclear weapon states—the so-called negative security assurances. However, the issues of nuclear security and nuclear terrorism more broadly did not play any role in the work of the Security Council until the end of the Cold War. In this regard, the lack of action largely reflected Cold War inertia of the Council due to the paralysing superpower rivalry between the United States (US) and the Soviet Union.[8] Unsurprisingly, in Resolution 1540 there is no reference to the work of the Security Council prior to the end of the Cold War.

In the wake of the collapse of the Soviet Union and easing East–West tensions, the Security Council was able to invigorate its role in international affairs and became more active—though not always successfully—in areas such as peacekeeping, peacebuilding, conflict prevention, and human rights. It also strayed into the area of nuclear weapons, in particular with a Statement of the President of the Security Council on behalf of the members of the Council (meeting at the level of Heads of State and Government) that identified WMD proliferation as a threat to international peace and security and emphasized international commitment to the non-proliferation of WMD. While this is an important precedent for Resolution 1540—the first one to which it refers specifically—the statement did not include explicit references to nuclear security or nuclear terrorism.[9]

In the early 2000s, the Security Council turned its attention increasingly to transnational issues including terrorism, piracy, and the AIDS epidemic, demonstrating a high degree of convergence among the interests of its members.[10] In this context, the Security Council adopted two resolutions that became important precursors of Resolution 1540, both of which are referenced in the resolution. First, Resolution 1267 (1999) addressed international terrorism and condemned, in particular, the Taliban for harbouring Osama bin Laden. Most notably, the resolution established a committee on which the 1540 Committee could be modelled. Second, Resolution 1373 (2001), which was adopted shortly after the terrorist attacks of 11 September 2001 ('9/11'), was a wide-ranging counter-terrorism resolution. It also established another committee on which the 1540 Committee could be modelled. Most importantly, Resolution 1373 set an important precedent for the broad legislative nature of Resolution 1540, as it established for the first time general rules—in this case on counter-terrorism—rather than responding to a specific situation. In the words of Masahiko Asada, a legal scholar, Resolution 1373 was the 'first international legislation by the Security Council'.[11]

While the adoption of Resolution 1540 needs to be examined in light of developments in the Security Council after the end of the Cold War, the origins of the resolution can be traced back to three concrete issues: the terrorist attacks of 9/11; the discovery of an illicit nuclear trade network run by Pakistani nuclear scientist A. Q. Khan; and new concerns about nuclear proliferation surrounding Iran and North Korea.[12] Although the attacks of 9/11 did not involve acts of nuclear terrorism, their tremendous scale demonstrated both the willingness and capability of non-state actors to commit acts of mass impact terrorism. In other words, 9/11 made it easy to contemplate that terrorist groups are keen to use WMD, in particular a nuclear device, in a terrorist campaign. Unsurprisingly, nuclear terrorism became a key security concern for policy-makers in the 2000s. A famous op-ed in the *Wall Street Journal* declared that 'In today's war waged on world order by terrorists, nuclear weapons are the ultimate means of mass devastation.'[13]

While not all experts agree that non-state actors would be able to use WMD for mass impact terrorism,[14] the discovery of the A. Q. Khan network demonstrated that illicit proliferation activities involving networks of non-state actors existed on a hitherto unknown scale, which potentially could allow terrorist groups to acquire their own nuclear device. The A. Q. Khan network revelations also showed that the existing international non-proliferation instruments, including the various IAEA instruments and the export controls of the Nuclear Suppliers Group (NSG), had significant loopholes. In 2002 and 2003, these concerns were underpinned by the exposure of clandestine nuclear proliferation activities in North Korea, Iran, and, what some believed at the time, Iraq. In sum, in the early 2000s, many governments, above all the US administration under George W. Bush, were very alarmed by a combination of a triple proliferation threat and a set of what it perceived to be highly inadequate international non-proliferation measures.

The Bush Administration was determined to establish a set of new counter-proliferation instruments to address the perceived shortcomings of the established non-proliferation instruments and bodies such as the Nuclear Non-Proliferation Treaty (NPT), the IAEA, or the 1997 International Convention for the Suppression of Terrorist Bombings. However, their approach deviated from the established instruments insofar as they were both more informal and more coercive, as exemplified by the 2003 Proliferation Security Initiative—an informal group of states that pledged to impede and interdict trade in WMD and related materials.[15] Although Resolution 1540 was not an informal counter-proliferation instrument, it was an unusual instrument and fitted within the broader approach of the US administration at the time. That is, it circumvented established relevant international organizations such as the IAEA and avoided a time-consuming treaty process to create a *de facto* legislative framework reminiscent of a treaty. At the same time, it opened up the possibility—at least in principle—of forceful actions against the proliferation of WMD under Chapter VII of the UN Charter. Unsurprisingly, and as discussed in the next section, the adoption of Resolution 1540 was controversial.

However, the US administration was able to muster the necessary support in the Security Council with patient diplomacy and some compromises. Three of the other four permanent members—France, Russia, and the United Kingdom (UK)—supported Resolution 1540 from the start. After all, the resolution reflected some of the ideas and language in the 2002 G8 Statement that established the G8 Global Partnership against the Spread of Weapons and Materials of Mass Destruction.[16] China was more reluctant, as it feared—particularly in the aftermath of the 2003 Iraq War—that the US would use the resolution as a pretext to justify the use of force, in particular against North Korea. But once these concerns were addressed, China also supported the resolution.

In general, the resolution turned out to be much less coercive than its adoption under Chapter VII would suggest, and relied heavily on 'soft modes of governance'.[17] In essence, Resolution 1540 established a legislative framework that attempts to prevent the proliferation of WMD to non-state actors by adhering to a wide variety of other relevant international instruments, such as the CPPNM; outlawing proliferation support by nation states; criminalizing proliferation-related activities; and establishing the necessary controls and enforcement mechanisms against illicit trade and trafficking, including the physical protection of relevant material, material accountancy and control, and export and border controls. While the resolution is only four sides of paper, it encompasses in total over 330 relevant measures that states should take to counter WMD terrorism. The resolution does not differentiate specifically between nuclear security and other measures, but virtually all parts of the resolution are relevant for nuclear security. Over the years, it has been renewed and updated seven times, though its basic approach has stayed the same.[18]

The Resolution 1540 Infrastructure

The implementation of Resolution 1540 suffers from an inherent contradiction in the resolution itself. While its adoption under Chapter VII has created the strongest mandate possible in international law, the resolution has not established any forceful implementation mechanisms or powerful institutional infrastructure. Although the UK floated the idea of a well-resourced Secretariat during the negotiations of the resolution, the implementation of the resolution needs to rely on the 1540 Committee and the voluntary commitments of UN member states.[19]

The 1540 Committee consists of representatives of the 15 Security Council members, with the role of the chair and the three vice-chairs rotating on a yearly basis.[20] The Committee also has four Working Groups, which reflect the four main areas of work of the Committee: monitoring and national implementation; assistance; cooperation with international organizations; and transparency and media outreach. The Committee holds both informal meetings at the Working Group-level and formal Committee meetings several times per year to examine the work of the different Working Groups, discuss reports from members of the Group of Experts, or listen to the view of external experts.[21] However, it is worth noting that there is no systematic or specific focus on nuclear security. Nuclear security-related issues are only part of the Committee's consideration as part of the wider work of the Committee Working Groups and the Group of Experts.

Like other committees of the Security Council, the 1540 Committee is bound by an unanimity rule, which makes the Committee's decision making more complicated and typically leads to decisions that reflect the lowest common denominator between the Security Council member states, in particular the permanent five members (P5). Even the arguably 'technical' work of the Working Groups and the Group of Experts are marred by differences between the Security Council member states, and, 'as a political rather than technical body, its decisions are often politicized, reflecting the interests in particular of the veto-wielding Security Council members'.[22] During the first seven years of the existence of Resolution 1540, the political nature of the 1540 Committee was exacerbated by its short mandate periods, which put the continuity of the Committee on shaky grounds and made long-term planning difficult. Only after three short-term prolongations in the 2000s did the Security Council decided in 2011 to grant a 10-year extension of the Committee.

In contrast to other comparable committees of the Security Council, the 1540 Committee suffers from another important structural shortcoming, that is it enjoys very little formal support.[23] For instance, it does not have a 40-person strong Executive Directorate, as the Counter-Terrorism Committee, nor an Analytical Support and Sanctions Monitoring Team, as the Security Council Committee pursuant to resolutions 1267 (1999), 1989 (2011), and 2253 (2015). Apart from some technical support by a small number of officials in the UN Secretariat, in particular the Office for Disarmament Affairs and the Department of Political and Peacebuilding Affairs, the Committee's main support mechanism is the previously mentioned Group of Experts, which consisted initially of eight (and, since 2012 of nine) national experts.

Neither the Committee itself nor the Group of Experts have been given the authority to assess, investigate, or sanction. That is, the 1540 Committee does not have the powers of a Security Council Sanctions Committee.[24] Likewise, the 1540 Committee has a small budget, especially in comparison with the wide array of tasks it is supposed to carry out (see below). A former member of the Group of Experts pointed out that '[t]he direct budget for the 1540 Committee [...] covering UN Secretariat staffing, fees for experts, official travel expenses, and various operational costs, started at roughly $2.1 million in 2006 (the first full year for which experts were on staff), and grew to $3.1 million in 2011'.[25] Although the 1540 Committee has also received significant additional funding from member states and other international organizations (e.g. the European Union, which has donated over four million euros),[26] this type of funding is entirely voluntary and, thus, less reliable. The ad hoc nature of this funding makes long-term planning for the 1540 Committee more difficult, while the funding priorities reflect those of the donors rather than those of the 1540 Committee.

The overall weakness of the infrastructure underpinning the 1540 Committee is a reflection of the controversial nature of the adoption of the resolution as *de facto* counter-proliferation legislation.[27] In other words, the Security Council opted for a soft implementation infrastructure to alleviate the fears of UN member states that the resolution would either put an undue burden on them—a point made especially by developing countries—or that they would be singled out for their lack of implementation of the resolution. These kinds of initial fears by UN member states are particularly reflected in the kind of work the Committee is allowed to carry out. According to Resolution 1540, the Committee's role was merely to 'report to the Security Council for its examination, on the implementation of this resolution'. Although subsequent resolutions—in particular Resolutions 1673 (2006), 1810 (2008), and 1977 (2011)—expanded slightly the competences of the Committee,[28] its work remains limited to the four areas of the Working Groups.

First, the 1540 Committee examines the national reports that UN member states send about their implementation measures regarding Resolution 1540. UN member states can also prepare, on a voluntary basis, National Implementation Action Plans to set out future measures in support of Resolution 1540. The Committee, and in particular the Group of Experts, supports UN member states by visiting them to raise awareness of Resolution 1540 and the facilitation of a peer-review process, which allows UN member states to learn from each other's measures and best practices. As soon as the Committee receives a national report, members of the Group of Experts put the information provided in the national report in a 'matrix' (complemented with official information from governments and/or international organizations), which offers an overview of the different measures taken by UN member states.

The overall trend of the implementation of Resolution 1540 is positive.[29] As the Committee highlights on its website, 'The average number of measures per State for the data fields concerned rose to 159 in 2016 from 134 in 2011 and from 93 in 2008'.[30] However, these are only initial trends and important variations between regions and regarding types of measures remain.[31] Furthermore, the matrices reveal little about the quality of the implemented national measures and potential loopholes. To some extent, detractors could argue that the matrices are merely a box ticking exercise.

Second, the 1540 Committee provides technical assistance to those states that require support with the implementation of Resolution 1540, in particular acting as a 'clearinghouse'—a targeted mechanism that matches member states that require assistance with international organizations and other states that can offer assistance. The website of the 1540 Committee lists both offers and requests, as well as provides relevant, points of contact. Nuclear security-specific offers and requests, however, are rare. For instance, Tajikistan has requested 'Assistance with regards to the acquisition of equipment covering NBC [nuclear, biological and chemical] fields for the Nuclear and Radiation Safety Agency (NRSA)' and Zambia has sought 'Advice and support to Zambia to strengthen its nuclear security regime.' Yet, most requests reflect assistance pertaining to the resolution across the board.

Third, over time, the 1540 Committee has established relations with nearly 50 international organizations at the global, regional, and sub-regional level to 'orchestrate' the implementation of Resolution 1540 by other organizations. In other words, the Committee collaborates with other organizations to compensate its own shortcomings using the expertise and resources of other organizations. Again, there is no systematic or specific focus on nuclear security, but the work with institutions such as the IAEA and the NSG falls typically within the realm of nuclear security. Yet, the collaboration lacks a clear pattern and continuity beyond a few ad hoc meetings and conferences, and it is hampered by an ineffective information exchange between the 1540 Committee and other international organizations.[32]

Finally, one of the key elements of the work of the 1540 Committee is transparency and outreach. The Committee maintains an easily accessible website, where it provides a wide variety of detailed information of its work—from summaries of its various meetings to national reports about the implementation of Resolution 1540. This high degree of transparency is certainly unusual for an international institution in the area of international security. Furthermore, the Committee and its Group of Experts is regularly engaged in outreach activities, such as interviews with the media and the organization of and participation in relevant international conferences and workshops, including with non-governmental organizations, academia, and representatives from industry. The extent and depth of the outreach activities, however, have been criticized for its 'strategic blindness'.[33] In other words, there is no coherent approach to engage civil society actors and industry in a consistent and effective way.

Controversies and Challenges Surrounding the Resolution

Resolution 1540 has faced a number of types of controversies and challenges. As highlighted previously, controversies were seen in the earlier years of the resolution's life, especially over the use of a Chapter VII resolution by the Security Council. The passage of 1540 in this manner led to some questioning the resolution's political legitimacy, with the Security

Council in some sense legislating for all UN members. The Security Council, despite two-thirds of the seats being elected by the General Assembly, is not a democratic body. The 15 members of the Security Council—less than 8 per cent of UN member states—effectively decided on measures that would require implementation by all member states.[34] Some states, for example, noted concern at the broadening and growing powers of the Security Council to 'legislate' for the international community.[35] Other states noted a perceived lack of justification for use of Chapter VII powers.[36] Others expressed concern over the imposition of 'externally prescribed norms or standards'.[37] These concerns surrounding legitimacy, however, largely faded in the months and years after the resolution's passage.

Indeed, in the longer term, the resolution has arguably worked to grow interest in the issues encompassed within the 1540 agenda around the world, helping to further universalize these policy tools. Traditionally, policy tools such as export controls, while used by more advanced supplier economies to counter proliferation, have been viewed differently by states of the Global South that are largely denied access to these technologies. In the area of nuclear technology, some have argued that supplier groups such as the NSG, which have worked to harmonize nuclear export controls, have acted in contravention of the peaceful uses pillar of the NPT.[38]

More broadly, as export controls have been applied to a wider range of dual-use technologies—items like advanced machine tools, materials such as carbon fibre and alloys, as well as high-grade electronic components—these policy tools have the potential to impact more broadly on states' economic development.[39] Resolution 1540 has arguably helped to overcome some of these perceptions surrounding export controls, working to further universalize the non-proliferation agenda largely driven by the US and a handful of more developed supplier states.

Nuclear security is arguably less politically contentious than the export control and non-proliferation-related aspects of the 1540 agenda. There are some aspects of nuclear security that are contested and more political—particularly threat perceptions—that can affect how states prioritize the issue. The burden of implementing nuclear security measures, however, clearly falls heaviest on states with nuclear assets and facilities. While nuclear security measures and the associated costs could be viewed as a barrier to those seeking to develop a nuclear programme for the first time, they are a minor dimension of a hugely costly endeavour. Export controls have arguably placed barriers preventing the transfer of technology between the 'haves' and the 'have nots', but nuclear security has only marginally raised the costs of entry.

Beyond the broader challenges of political legitimacy, challenges that have endured throughout the near 20-year lifespan of the 1540 agenda are more practical in nature. The greatest practical challenge surrounds ensuring states' implementation of the various measures included in the resolution—that is, ensuring compliance with it. As discussed previously, there is no means of enforcing the resolution or responding to 'noncompliance'. Implementation of many of the measures forming the 1540 agenda requires significant government resources: human in terms of staff time and expertise, as well as financial costs of putting in place relevant systems, processes, and hardware.

Implementation also requires political will. Many countries—especially those in the developing world—struggle to commit resources, especially when they often do not place such a high priority on countering WMD terrorism or proliferation as those states involved in putting the resolution together. These implementation challenges are perhaps less acute for the nuclear security measures of the resolution that fall heaviest on states holding nuclear materials and hosting nuclear facilities, rather than the broader spectrum of measures of 1540 that all states have to implement in theory (export controls, border controls, etc.).

Much of the progress in the resolution's implementation has come as a result of bilateral and multilateral cooperation—whether branded under the 1540 banner, or forming part of many of the global threat reduction and capacity-building programmes largely funded by the US and partners sharing similar interests. As noted, regional and sub-regional organizations, and international organizations such as the IAEA, Organization for the Prohibition of Chemical Weapons (OPCW), and others have also played important roles assisting states. Demand for assistance clearly exceeds supply, although there are also often political barriers to states requesting it.

The main burdens of the nuclear security aspects of the resolution mostly fall on a smaller sub-section of states that hold nuclear materials. While the security of nuclear materials is the responsibility of states, many of the nuclear security obligations listed in Resolution 1540 on these states also form a part of other international obligations and best practices—for example, those stemming from IAEA safeguards and nuclear security 'recommendations'.[40] The 2005 CPPNM Amendment, which entered into force in 2016, also places some of the first legally binding requirements on signatory states for securing their nuclear facilities and materials during domestic transport.[41] The resolution places further rationale to implement these types of measures through a legally binding resolution, rather than placing additional or different nuclear security requirements on states.

One question that has hung over the resolution regards the status of radiological materials within it. For many states—particularly those with limited nuclear assets and holdings—radiological security forms a core part of their broader nuclear security efforts, and a topic that they are keen to discuss in the 1540 context. International outreach and capacity-building efforts have also seen a lot of resources devoted to radiological risks, with states keen to showcase their progress in the context of Resolution 1540. Indeed, some states merely view radiological as the greater risk, one of much higher likelihood and—while not as catastrophic—still potentially hugely impactful.

The text of Resolution 1540 refers to 'nuclear, chemical or biological weapons', 'means of delivery', and 'related materials'.[42] Radiological materials are not explicitly included in the text, although they could arguably fall under the 'related materials'.[43] The preamble of the resolution also refers to the IAEA Code of Conduct on the Safety and Security of Radioactive Sources. Radiological sources are widely used in industry, health care, and research, and often are found in less-regulated, and more customer-facing, environments. When radiological materials are viewed as a part of the 1540 agenda, a much wider range of countries are affected by the physical protection-related obligations of the resolution.

A final challenge often discussed in relation to 1540 is engaging a wide range of stakeholders. The resolution calls on states to 'develop appropriate ways to work with and inform industry and the public regarding their obligations under such laws.'[44] This is potentially a broad category with some ambiguity for states in their implementation. Nuclear material holdings are often fairly centralized—largely based in the nuclear industry, as well as smaller quantities in research and academic institutions. Outreach on the nuclear security-related aspects of 1540 is more focused and less ambiguous, especially when compared with outreach required to the nuclear, chemical, biological, and dual-use supply chains on

Table 10.1. Nuclear security-related measures included in the approved 1540 Committee matrix.

Matrix Section	Measures	
I. OP 1 and related matters from OP 5, OP 8 (a), (b), (c), and OP10	Adherence to legally binding instruments, membership of organizations, participation in arrangements, and statements made	3. Convention for the Suppression of Acts of Nuclear Terrorism 4. Convention on the Physical Protection of Nuclear Material (CPPNM) 5. 2005 Amendment to the CPPNM 16. International Atomic Energy Agency (IAEA) 18. Statement on non-provision of WMD and related materials to non-state actors
II. OP 2—Nuclear Weapons (NW), Chemical Weapons (CW), and Biological Weapons (BW)	National legislation that prohibits persons or entities to engage in one of the following activities and its enforcement	1. Manufacture 2. Acquire 3. Possess 4. Develop 5. Transport 6. Transfer 7. Use 8. Attempt to engage in above-mentioned activities 9. Participate as an accomplice in above-mentioned activities 10. Assist in above-mentioned activities 11. Finance above-mentioned activities
III. OP 3 (a) and (b)—Account for/Secure/Physically protect NW, CW, and BW, including Related Materials 2	Measures to establish domestic controls to prevent the proliferation of NW, CW, BW, and their means of delivery; controls over related materials	1. Measures to account for production 2. Measures to account for use 3. Measures to account for storage 4. Measures to account for transport 5. Measures to secure production 6. Measures to secure use 7. Measures to secure storage 8. Measures to secure transport 9. Physical protection measures 10. Personnel reliability

Table 10.1. Continued

Matrix Section	Measures	
IV. OP 3 (a) and (b)—Account for/Secure/Physically protect NW, including Related Materials (NW specific)	Measures to establish domestic controls to prevent the proliferation of NW, and their means of delivery; controls over related materials	1. National regulatory authority 2. Licensing of nuclear installations/entities/use of materials 3. IAEA Safeguards Agreements 4. IAEA Code of Conduct on Safety and Security of Radioactive Sources 5. Supplementary Guidance on the Import and Export of Radioactive Sources of the Code of Conduct on the Safety and Security of Radioactive Sources 6. IAEA Incident and Trafficking Database (ITDB) 7. Integrated Nuclear Security Support Plan (INSSP)/International Physical Protection Advisory Service (IPPAS) 8. Applying the physical protection recommendations in INFCIRC/225/Rev.5 9. Other agreements related to IAEA 10. National legislation and regulations related to nuclear material, including CPPNM
VII. OP 3 (c) and (d) and related matters from OP 6—Controls of NW, CW, and BW, including Related Materials	Border controls and export and trans-shipment controls to prevent the proliferation of NW, CW, and BW and their means of delivery including related materials—national legal framework and enforcement	1. Border control to detect, deter, prevent, and combat illicit trafficking 2. Law enforcement to detect, deter, prevent, and combat illicit trafficking 3. Border control detection measures 5. Export control legislation in place 6. Licensing provisions and authority 7. Control lists of materials, equipment, and technology 8. Intangible technology transfers 12. Transit control 13. Trans-shipment control 14. Re-export control
VIII. OP 7 and 8 (d)—Assistance, Work with, and inform Industry and Public, and other Information		1. Assistance offered 2. Assistance point of contact (for assistance providers only) 3. Assistance requested 4. Action taken to work with and inform industry 5. Action taken to work with and inform the public 6. Point of contact 7. Voluntary National Implementation Action Plan (NAP) 8. 1540 Committee visits to states

Table courtesy of the authors.

export controls, for example. There still remain questions surrounding what activities could usefully 'inform' the public on various nuclear security-related laws (Table 10.1).

The Future of Resolution 1540

The resolution and its agenda have been extended by the Security Council several times since 2004.[45] After 2016, the work of the 1540 Committee slowed, as a result of a number of factors, more recently compounded by the Covid-19 pandemic and the Russian invasion of Ukraine. Work did not stop entirely—for example, the 1540 Experts produced for around 180 countries new matrices, which were approved in December 2020.[46] Two interim resolutions were used to extend the mandate of the 1540 Committee. In 2021 the Security Council extended the mandate for 10 months, until February 2022, and a further resolution in February 2022 extended the mandate an additional nine months, to 30 November 2022.[47] Most recently, on the day of the November deadline, the mandate of the Committee was extended by a further decade, to 2032.[48]

Resolution 1977 (2011), which extended the 1540 Committee's mandate by a decade, mandated a Comprehensive Review prior to further extension of the resolution. The final report would be delivered, and the Committee Chair and members would draft a new resolution to extend the mandate beyond 2021.[49] The 2021 Comprehensive Review was postponed in August 2020, due to 'delays caused by the coronavirus pandemic'.[50] The extension of the 1540 Committee's mandate into 2022 provided extra time, although the Covid-19 pandemic continued to create barriers. The lack of in-person meetings over almost two years was particularly problematic, with differing views among Committee members and governments about how business should be conducted. The lack of in-person meetings also delayed the Committee's 'open consultations' with UN member states and other organizations.

The 2021 Comprehensive Review was planned to be undertaken through four working groups, looking at the following topics: implementation trends; the assistance mechanism; collaboration with international and regional organizations; and outreach to other stakeholders.[51] As time passed, the build-up of business to be conducted by the UN in the aftermath of the pandemic meant that 1540 has needed to compete for resources and meeting space. 'Open consultations' on 1540 took place in May 2022, with the Comprehensive Review planned to report by November 2022. At the November 2022 Security Council meeting that extended the Committee's mandate, a number of countries congratulated the Mexican Chairperson for successfully concluding the review.[52]

The extension of the 1540 Committee's mandate required the support of all P5 members of the Security Council, which although likely, since the resolution's agenda has received unanimous support in the UNSC in recent years, was not a given, especially in light of fractures in the Security Council following Russia's illegal invasion of Ukraine.[53] The new resolution, beyond extending the Committee's mandate by a decade, also requires reviews to be conducted at the five-year point, as well as before the resolution is up for further extension in 2032. However, the new resolution largely reiterates the substance of previous resolutions, rather than encompassing new requirements or recommendations.

As the 1540 agenda moves towards a third decade in 2024, while it is difficult for the 1540 Experts and Committee members to talk publicly about priorities, a number of issues have seen discussion. Some of these are familiar, surrounding the implementation challenges discussed previously. Others are novel. Some of these priority areas match those listed in Resolution 2325 (2016) and reiterated in Resolution 2663 (2022), where the Security Council noted 'the need for more attention on': 'enforcement measures; measures relating to biological, chemical and nuclear weapons; proliferation finance measures; accounting for and securing related materials; and national export and transshipment controls'.[54]

Regarding implementation—one of the greatest challenges for the 1540 agenda—the 2016 Comprehensive Review noted the need to explore new approaches to implementation. The 2016 Review final document reads:

> The Committee should continue to explore and develop an approach with regard to implementation and reporting that takes due account of the specificity of States with respect to their ability to manufacture and export related materials, with a view to reducing unnecessary compliance burdens and prioritizing effort and resources towards where they are most needed, without affecting the comprehensive and balanced implementation of the resolution.[55]

The hint towards a need for a more tailored approach is also reflected in Resolution 2325, which followed the issue of the review. The resolution '*Urges* the 1540 Committee to continue to explore and develop an approach'.[56] This is an issue that has been under consideration by the Committee since 2016, although conclusion is pending. Developments in this area are unlikely to have a direct impact on nuclear security, since the 'manufacture and export' mentioned leans more towards non-proliferation and export controls. Any movement towards tailored reporting and implementation in these areas, however, could potentially have implications for nuclear security further down the line, perhaps with implementation and reporting requirements tailored on the basis of nuclear material holdings and facilities.

The resolution and the Committee need to maintain their relevance by adapting to consider new and emerging threats on the WMD landscape, for example developments in additive manufacturing (3D printing) and synthetic biology. While the resolution does not list specific technologies—rather leaving that down to the control lists—these technologies could potentially fall under the 'related materials' language of the resolution. This is a challenge for Resolution 1540, but also for international organizations, states, and industry in efforts to control these technologies.

The Covid-19 pandemic has also seen a renewed interest in biological threats. The experience has renewed interest in the bio-safety and bio-security of dual-use research, and tools that could benefit the prevention of bio-terrorism.[57] International tensions over investigations into the virus's origins are unlikely to subside. These two areas—disruptive technology and biological weapons—were highlighted by the UN High

Representative for Disarmament Affairs alongside capability as three priorities coming out of the 2016 Comprehensive Review.[58] Bio-security and nuclear security are areas with some synergies that may learn from each other's challenges and successes.

Some aspects of the 1540 agenda have received attention among a wider range of governments, international organizations, and non-governmental organizations in recent years, largely because they are areas where work was underdeveloped, or confined to a number of more developed states. One of these is proliferation financing (PF): Resolution 1540 required states to put in place and enforce 'appropriate effective laws' prohibiting financing any non-state actor WMD-related activities.[59] Since the original resolution, much work has been conducted on financial aspects by organizations such as the Financial Action Task Force, and a number of key states supporting the resolution have been pushing the PF agenda.

Another area, perhaps more directly relevant to nuclear security as well as proliferation, relates to controls on intangibles—sensitive information, research data, knowledge, and skills that could be relevant to the development, manufacture, or use of WMD. Resolution 2325, for example, included an operative paragraph *encouraging* states to control intangibles, an issue that was not explicitly mentioned in Resolution 1540, and was only referenced in the preamble of Resolution 1977 (2011) and Resolution 2663 (2022). However, it is unclear whether issues such as PF and intangibles could make it more into the formal work of the 1540 Committee in due course.

For the nuclear security aspects of the 1540 agenda, the main challenge lies in finding momentum for nuclear security following the conclusion of the Nuclear Security Summits. The series of four Summits was pushed by the US during the Obama presidency, concluding in 2016. This process added urgency and momentum to the nuclear security agenda, but also only included a fraction of the full list of UN members that have obligations under 1540. The CPPNM Amendment's entry into force in 2016, and the relatively successful review conference held in March–April 2022, perhaps represents a further avenue to drive the nuclear security agenda, and those aspects contained in Resolution 1540, forward in this regard.

Broader events will also have some implications for the future fortunes of the Resolution 1540 agenda, as well as broader cooperative non-proliferation and nuclear security efforts. A challenging international political climate, including fractures in the Security Council, tensions over related issues such as chemical weapons use in Syria, and the collapse of US–Russia cooperative threat reduction, set the tone for questions surrounding the resolution's eventual renewal. Russia's illegal invasion of Ukraine and false narratives and disinformation surrounding Ukraine's alleged development of WMD also complicate potential for collaboration in related areas.[60] Meanwhile, continually evolving WMD proliferation challenges posed by North Korea and Iran affect the threat landscape. More broadly, the lack of an agreed final document at the postponed 2020 NPT Review Conference in August 2022 also shapes this landscape. Persistent tensions between the non-nuclear weapons states and the nuclear weapons states potentially affect the former's willingness to further engage on non-proliferation tools such as Resolution 1540. In sum, the current international environment is not conducive to cooperative efforts to prevent WMD terrorism.

Conclusion

Resolution 1540 of the UN Security Council is an important element of efforts to prevent WMD terrorism, in particular nuclear terrorism. This chapter has provided an overview of Resolution 1540 and the work of the 1540 Committee in the area of nuclear security. It has explored the key provisions of the resolution, its place in the broader international nuclear security framework, as well as the status of nuclear security in the resolution and the broader 1540 agenda. More specifically, the chapter has examined the genesis and content of the resolution, related institutional infrastructure, major issues and controversies surrounding the resolution, and its mid- and long-term future.

Resolution 1540 has seen great progress made in global efforts to prevent WMD terrorism and proliferation. States around the world have taken many steps at the national level in order to comply with the resolution and implement the 1540 agenda, resulting in a transformation of the non-proliferation and counter-WMD-terrorism landscape over the past 20 years. There have been many challenges along the way—notably, those surrounding resources and buy-in for implementation around the world. These implementation-related challenges are unlikely to go away as the resolution places so many requirements on all UN member states.

Unfortunately, the future of the 1540 agenda is currently uncertain. Inertia in the 1540 Committee, the Covid-19 pandemic, and challenges across the broader international political landscape meant that the extension of the Committee's mandate in 2022 could not be taken for granted. Thankfully, the mandate of the Committee was extended by a decade, which is some good news for the non-proliferation regime. However, tensions in the international political environment will continue to have negative effects and create uncertainties for the future of cooperative non-proliferation and nuclear security efforts. In particular, the illegal invasion of Ukraine by Russia—a state holding the largest stocks of weapons-usable material—could cause problems in cooperative efforts to implement the 1540 agenda. Although great progress has been made on implementation of measures to counter WMD terrorism and proliferation through the 1540 agenda, future successes continue to be fragile and uncertain.

Notes

1. It should be noted that 'nuclear security' as a concept does not appear in the resolution.
2. UN Charter, Article 24.
3. Ibid., Article 25.
4. Ibid., Articles 41 and 42.

5. Ibid., Article 39.
6. Ibid., Article 26.
7. UN Security Council, n.d. 'Resolutions adopted by the Security Council in 1947'. https://www.un.org/securitycouncil/content/resolutions-adopted-security-council-1947
8. David M. Malone, 'Security Council', in Sam Daws and Thomas G. Weiss, eds., *The Oxford Handbook on the United Nations* (Oxford: Oxford University Press, 2018), Ch.7.
9. UN Security Council, 31 January 1992, 'Note by the President of the Security Council', S/23500. https://www.securitycouncilreport.org/atf/cf/%7B65BFCF9B-6D27-4E9C-8CD3-CF6E4FF96FF9%7D/PKO%20S%2023500.pdf
10. Malone, 'Security Council'.
11. Masahiko Asada, 'Security Council Resolution 1540 to Combat WMD Terrorism: Effectiveness and Legitimacy in International Legislation', *Journal of Conflict & Security Law*, 13/3 (2009), 303–32.
12. William H. Tobey, 'A History of United Nations Security Council Resolution 1540', in Daniel Salisbury, Ian J. Stewart, and Andrea Viski, eds., *Preventing the Proliferation of WMDs: Measuring the Success of UN Security Council Resolution 1540* (Cham: Palgrave Macmillan, 2018), 13–32.
13. George P. Shultz, William J. Perry, Henry A. Kissinger, and Sam Nunn, 'A World Free of Nuclear Weapons', *The Wall Street Journal* [online], 4 January 2007. https://www.wsj.com/articles/SB116787515251566636
14. For example, John Mueller.
15. Oliver Meier and Christoph Daase, eds., *Arms Control in the 21st Century: Between Coercion and Cooperation* (Abingdon: Routledge, 2013).
16. Tobey, 'A History of United Nations Security Council Resolution 1540'.
17. Monika Heupel, 'Combining Hierarchical and Soft Modes of Governance: The UN Security Council's Approach to Terrorism and Weapons of Mass Destruction Proliferation after 9/11', *Cooperation and Conflict*, 43/1 (2008), 7–29.
18. UN Security Council Resolutions 1673 (2006), 1810 (2008), 1977 (2011), 2055 (2012), 2325 (2016), 2572 (2021), and 2663 (2022).
19. Tobey, 'A History of United Nations Security Council Resolution 1540'.
20. The only exception is the UK, which has held the role of vice-chair since the inception of the 1540 Committee.
21. For the full list of meetings, see '1540 Committee' on the UN website. https://www.un.org/en/sc/1540/transparency-and-outreach/general-information.shtml
22. Ian J. Stewart, 'Preventing WMD Proliferation: The Future of UNSCR 1540', in Daniel Salisbury, Ian J. Stewart, and Andrea Viski, eds., *Preventing the Proliferation of WMDs*, 105–26.
23. Volker Beck, 'The tools of the 1540 Committee', in Nicolas Kasprzyk, Mothepa Shadung, and Noël Stott, eds., *Towards the 2016 Comprehensive Review: Former Experts Assess UNSC Resolution 1540* (Pretoria: Institute for Security Studies, 2015), 1–6.
24. See UN Security Council, 'Sanctions and Other Committees'. https://www.un.org/securitycouncil/content/repertoire/sanctions-and-other-committees
25. Beck, 'The Tools of the 1540 Committee'.
26. See Council Joint Action 2006/419/CFSP; Council Joint Action 2008/368/CFSP; Council Decision 2013/391/CFSP; Council Decision (CFSP) 2017/809.
27. Sarah Shirazyan, 'Building A Universal Counter-Proliferation Regime: The Institutional Limits of United Nations Security Council Resolution 1540', *Journal of National Security Law and Policy*, 10/1 (2019), 125–70.

28. Daniel Salisbury, 'UNSCR 1540 Implementation: Challenges Past and Present', in Daniel Salisbury, Ian J. Stewart, and Andrea Viski, eds., *Preventing the Proliferation of WMDs* 81–103.
29. Andrea Viski, 'UNSCR 1540: Implementation Trends', in Daniel Salisbury, Ian J. Stewart, and Andrea Viski, eds., Preventing the Proliferation of WMDs, 33–54.
30. 1540 Committee, n.d. 'General Information'. https://www.un.org/en/sc/1540/national-implementation/general-information.shtml; at the time of writing, the data for 2021 has not been published yet.
31. For an in-depth analysis, see Viski, 'UNSCR 1540'.
32. Benjamin Kienzle, 'Effective Orchestration? The 1540 Committee and the WMD Terrorism Regime Complex', *Global Policy*, 10/4 (2019), 486–96.
33. Dana Perkins, 'The 1540 Committee and Civil Society: Strategic Blindness and Ways to Overcome It', in Nicolas Kasprzyk, Mothepa Shadung, and Noël Stott, eds., *Towards the 2016 Comprehensive Review: Former Experts Assess UNSC Resolution 1540* (Pretoria: Institute for Security Studies, 2015), 101–14.
34. Johan Bergenäs, 'The Slippery Slope of Rational Inaction', *Nonproliferation Review*, 15/2 (2008), 380.
35. See for example, the statement of India: UN Security Council, 22 April 2004, 'Security Council, 4950th meeting, Thursday, 22 April 2004', S/PV.4950, 23. http://www.securitycouncilreport.org/atf/cf/%7B65BFCF9B-6D27-4E9C-8CD3-CF6E4FF96FF9%7D/Chap%20VII%20SPV%204950.pdf
36. Statement of Pakistan, Ibid., p. 15.
37. UN Security Council, 1 February 2005, 'National report submitted by the Republic of South Africa to the Security Council Committee established pursuant to Security Council resolution 1540 (2004): Implementation of operative paragraphs 1 to 3 and 6 to 10 of resolution 1540 (2004)', S/AC.44/2004/(02)/102, 2.
38. See for example, Mohamed Shaker, *The Nuclear Non-Proliferation Treaty: Origins and Implementation, 1959–1979*, Vol. 1 (New York: Oceana Publications, 1980), 332.
39. Krishnaswamy Subrahmanyam, 'Export Controls and the North–South Controversy', *Washington Quarterly*, 16/2 (1993), 135.
40. For example, the vast majority of states with nuclear material holdings have concluded a safeguards agreement with the IAEA, which places requirements on them to undertake material accountancy. The IAEA also has non-legally binding nuclear security 'recommendations' in the Nuclear Security Series.
41. Previously the CPPNM only covered international transportation of nuclear material.
42. UN Security Council, 28 April 2004, Resolution 1540, 1.
43. Defined in the resolution as 'materials, equipment and technology covered by relevant multilateral treaties and arrangements, or included on national control lists, which could be used for the design, development, production or use of nuclear, chemical and biological weapons and their means of delivery'. Radiological materials are included on many national control lists and are included in the NSG 'Guidelines Part 2', the 'dual-use' list.
44. Resolution 1540, OP8d.
45. Resolution 1540 was extended for two years by Resolution 1673 (2006), a further three years under Resolution 1810 (2008), a further decade by Resolution 1977 (2011), and, more recently, a further decade by Resolution 2663 (2022).

46. 1540 Committee, n.d. 'Committee Approved Matrices'. https://www.un.org/en/sc/1540/national-implementation/1540-matrices/committee-approved-matrices.shtml
47. UN Security Council, 22 April 2021, Resolution 2572; UN Security Council, 25 February 2022, Resolution 2622.
48. UN Security Council, 30 November 2022, Resolution 2663.
49. UN Security Council, 20 April 2011, Resolution 1977, OP5.
50. 1540 Committee, 'Press Release by the Chair of the 1540 Committee on the Postponement of the 2021 Comprehensive Review of the Status of Implementation of Resolution 1540 (2004)', UN Press Release, 27 August 2020. https://www.un.org/press/en/2020/sc14287.doc.htm
51. 1540 Committee, '2021 Comprehensive Review of the Status of Implementation of Resolution 1540 Modalities Paper', 2020. https://www.un.org/en/sc/1540/documents/2021%20CR%20Modalities%20Paper.pdf
52. UN Security Council, 'Security Council Extends Mandate of Committee Monitoring Nuclear, Biological, Chemical Weapons for 10 Years, Unanimously Adopting Resolution 2663 (2022)', *UN Press Release*, 30 November 2022. https://press.un.org/en/2022/sc15123.doc.htm
53. For example, all 15 UNSC members voted for Resolution 2572 (2021), Resolution 2325 (2016), and Resolution 1977 (2011).
54. UN Security Council, 15 December 2016, Resolution 2325, OP.12.
55. UN Security Council, 9 December 2016, 'Report of the Security Council Committee established pursuant to resolution 1540 (2004)', S/2016/1038, 40. https://undocs.org/S/2016/1038
56. Resolution 2325, OP.11.
57. Gregory D. Koblentz and Filippa Lentzos, 'Whether Covid Came from a Leak or Not, It's Time to Talk about Lab Safety', *The Guardian* [online], 15 June 2021. https://www.theguardian.com/commentisfree/2021/jun/15/covid-escaped-lab-or-not-biosecurity-laboratories
58. UN Security Council, 15 December 2016, 'Security Council Seventy-First Year 7837th meeting Thursday, 15 December 2016, 10 a.m. New York', S/PV.7837, 4. https://undocs.org/S/PV.7837
59. Resolution 1540, OPs 2 and 4d.
60. Abigail Stowe-Thurston, 'Russia's non-proliferation disinformation campaign', *Bulletin of the Atomic Scientists* [online], 22 March 2022. https://thebulletin.org/2022/03/russias-non-proliferation-disinformation-campaign

Suggested Reading

Asada, Masahiko, (2009), 'Security Council Resolution 1540 to Combat WMD Terrorism: Effectiveness and Legitimacy in International Legislation', *Journal of Conflict & Security Law*, 13/3, 303–32.

Kienzle, Benjamin, (2019), 'Effective Orchestration? The 1540 Committee and the WMD Terrorism Regime Complex', *Global Policy*, 10/4, 486–96.

Heupel, Monika, (2008), 'Combining Hierarchical and Soft Modes of Governance: The UN Security Council's Approach to Terrorism and Weapons of Mass Destruction Proliferation after 9/11', *Cooperation and Conflict*, 43/1, 7–29.

Salisbury, Daniel, Stewart, Ian. J., and Viski, Andrea, eds., (2018), *Preventing the Proliferation of WMDs: Measuring the Success of UN Security Council Resolution 1540* (Cham: Palgrave Macmillan).

Shirazyan, Sarah, (2019), 'Building a Universal Counter-Proliferation Regime: The Institutional Limits of United Nations Security Council Resolution 1540', *Journal of National Security Law and Policy*, 10/1, 125–70.

Website of the 1540 Committee. https://www.un.org/en/sc/1540

CHAPTER 11

THE IAEA'S CRITICAL ROLE IN NUCLEAR SECURITY

TREVOR FINDLAY

Introduction

The chapter begins by tracing the origins of the role of the International Atomic Energy Agency (IAEA) in nuclear security—from early concern with the physical protection of nuclear material in the 1970s to a much wider remit encompassing nuclear security, as it has come to be known since the 2000s. The chapter then outlines and assesses the Agency's current nuclear security programme and its notable achievements, including providing a global platform for exchanges, debate, and negotiation; guidance and assistance to member states; training, education, and information-sharing; and emergency preparedness and response. It examines planning, management, and funding of the IAEA programme, along with an assessment of the Agency's continuing challenges and shortcomings in the nuclear security realm. The conclusion includes recommendations for enhancing the future role of the Agency in bolstering global nuclear security.

Mandate Evolution: From Minimal Involvement to Physical Protection to Nuclear Security

Founded in 1957, the IAEA's original mission was to facilitate the peaceful uses of nuclear energy while avoiding the proliferation of military applications, notably nuclear weapons. Although there was early recognition that nuclear materials had to be physically secure to prevent unauthorized access, there was little concern in the 1950s and

1960s that terrorists would be able to produce nuclear weapons, even if they could acquire fissionable material. The IAEA Statute therefore contains no reference to nuclear security (a term that gained widespread use only this century) or even to the traditional notion of the physical protection of nuclear material. Governments appeared to believe that securing nuclear material was 'a problem of criminal justice, to be dealt with by national authorities responsible for internal security, and not by international agreement'.[1] Even today there is strong belief, constantly reiterated by its member states and repeated *ad nauseum* in IAEA documentation, that sole responsibility for nuclear security rests with the state.

With the advent of Comprehensive Safeguards Agreements, pursuant to the 1968 Non-Proliferation Treaty (NPT), the IAEA Secretariat began to realize that its role in promoting physical protection flowed naturally from the requirement that states ensure 'control of all nuclear material subject to safeguards'.[2] A fillip for the IAEA's role came with the negotiation, partly at the Secretariat's instigation, of the 1979 Convention on the Physical Protection of Nuclear Material (CPPNM). After an increase in illicit trafficking of nuclear material from former Soviet nuclear facilities in the early 1990s the Secretariat's focus turned to nuclear smuggling.[3]

The terrorist attacks in the United States (US) on 11 September 2001 ('9/11'), despite being non-nuclear, demonstrated the danger that suicide bombers and other well-organized terrorists posed to nuclear materials and facilities—and the need for enhanced international cooperation to thwart them. In 2002 the IAEA Board of Governors approved the Agency's first integrated, comprehensive action plan, focused not just on physical protection of material but 'protection against nuclear terrorism', while the Director General established an Advisory Group on Nuclear Security (AdSec) and a Nuclear Security Fund (NSF).[4]

At the initiative of US President Barack Obama, four nuclear security summits, held from 2010 to 2016, drew high-level attention to nuclear security for the first time since 9/11. Initial fears that the Agency's role would be overshadowed and in the longer term diminished proved unfounded. Each summit paid obeisance to the IAEA's 'central coordinating role' and boosted political support for the IAEA's mission. The final summit, held in Washington, DC in 2016, essentially passed the baton to the Agency, lauding it as 'crucial for the continuing delivery of outcomes and actions from the nuclear security summits'.[5] A seven-page Action Plan in Support of the IAEA was agreed, although it promised no new authorities or funding. Still, the summits increased member state participation in IAEA activities, stimulated extra voluntary funding, and accelerated entry into force of the amended CPPNM (A/CPPNM) in 2016.

Such is the evolution of the Agency's role in nuclear security from practically zero to its eventual broad acceptance as a vital component of global nuclear governance that it cannot be covered comprehensively in this chapter. Readers are encouraged to consult IAEA Annual Reports, the annual Nuclear Security Report, the biennial Programme and Budget, and the annual Financial Statements.

The IAEA as the 'Global Platform' for Nuclear Security

With its membership of 175 at the time of writing,[6] the IAEA is the only international body, apart from the United Nations (UN) General Assembly, where nuclear security can be debated by the majority of states. Its convening power is an essential component of its self-described role as a 'global platform' for nuclear security.[7] In addition to the annual IAEA General Conference, which has adopted increasingly detailed consensus resolutions on nuclear security, and the 35-member Board of Governors, which makes broad policy and budgetary decisions about nuclear security, the Agency also hosts International Conferences on Nuclear Security (ICONS) approximately every three years. ICONS—the most recent in 2020; the next planned for 2024—have attracted rising attendance and support from member states and other organizations. A one-day ministerial meeting (attended by 53 ministers in 2020) is intended to replicate, to some extent, the high-level representation at the summits.[8] Although the resulting ministerial declarations have not been ground breaking, the meetings have kept nuclear security on the agenda of ministers. ICONS have also partly taken on the 'gift basket' culture of the summits by encouraging states to compete in presenting new initiatives. As well as ICONS, the Agency also holds technical conferences on specific nuclear security topics.

The Secretariat has sought to make its conferences more interactive, with electronic posters, specialized technical sessions, and side events, in an attempt to counter the traditional mind-numbing UN conference culture and member state preference for formal speeches, rather than true debate. A new social media-conscious cohort in the Secretariat is experimenting with panel sessions chaired by professional media personalities, instant e-polling on critical issues, scoping exercises to map out long-term policy challenges, and even musical entertainment during young professional essay award ceremonies. The Agency's collaboration with other organizations, notably the World Institute for Nuclear Security (WINS), is also injecting novel approaches and ideas into the proceedings.

The Agency's convening powers are building on one of the true legacies of the summits—the nurturing of a community of dedicated nuclear security experts, both inside and outside government. Not only is this community increasingly aware of nuclear security risks and appropriate responses, but its members have a growing personal vested interest in securing governments' commitment to and resources for their chosen field. Over time, this expanding international cadre is embedding nuclear security permanently in the global and domestic nuclear governance architecture and building greater awareness of and support for the IAEA's role.

An institutional legacy of the final summit was the Nuclear Security Contact Group (NSCG), devised by 42 state participants as torchbearer for 'sustained action and ambition on nuclear security'.[9] At the time of writing, the group's membership comprises 48 states and four international organizations: the European Union, the International Criminal Police Organization (Interpol), the UN, and the IAEA itself. A 'Statement of Principles' was issued by the group in October 2016, reproduced as INFCIRC/899.[10]

Although it is not an official IAEA group, the NSCG normally meets in the margins of the IAEA General Conference, as well as at other times and locations as needed.

One of the NSGC's goals was to encourage implementation of the IAEA Action Plan agreed at the final summit and follow up on commitments made at ICONS. In addition, NSCG was intended to nurture the links to industry and civil society that had been forged successfully during the summits. The NSCG styles itself as a discussion group rather than a decision-making forum and its meetings are confidential to ensure frankness. The US Government Accounting Office (GAO) reported in 2019 that the NSCG is useful for formulating and refining ideas before exposing them to the general IAEA membership.[11] The NSCG website, run by the US State Department, was out of date at the time of writing, but by design provides no details of NSCG meetings or activities and only replicates information found elsewhere. The proposed outreach to non-governmental organizations and industry has not eventuated, and the IAEA Secretariat has reportedly not engaged actively with the group.[12] It is difficult for outsiders to assess what added value the NSCG brings to the IAEA's plethora of meetings and networks on nuclear security.

One of the nuclear security summits' novel diplomatic techniques bequeathed to the IAEA that shows more promise was to encourage 'house gifts' by individual states and 'gift baskets' by groups of states, comprising commitments or initiatives to advance nuclear security. There seems to be growing expectation that ministerial attendance at ICONS requires the announcement of a new initiative, or at least the fulfilment of previously announced promises. Some summit 'gifts' have been encapsulated in IAEA information circulars (INFCIRCs), deftly prolonging their life beyond the summits and permitting additional states, including non-summit participants, to subscribe.[13]

One of the most significant gift baskets, launched at the 2014 Nuclear Security Summit in The Hague, was the 'Strengthening Nuclear Security Initiative', advanced by the US, the Netherlands, and South Korea, which became INFCIRC/869.[14] Intended to overcome the difficulty of reaching consensus on more ambitious commitments, it attracted 35 initial subscribers (37 at the time of writing). The subscribing states pledged to promote continuous improvement and best practice in nuclear security, including by conducting internal assessments and periodic peer reviews and acting on the resulting recommendations. They also committed themselves to ensuring the IAEA's voluntary guidelines are reflected in, or exceeded by, their own national regulations. Finally, they undertook to ensure those responsible for nuclear security are professionally certified. The Nuclear Threat Initiative (NTI) has produced an 'Implementation Guide' to assist states in carrying out their commitments under INFCIRC/869.[15]

The IAEA as Provider of Nuclear Security Guidance

One of the earliest, most important, widely supported, and enduring roles of the IAEA in nuclear security has been the preparation and promotion of nuclear security guidance and recommendations. The inaugural 'Grey Book' was first revised in 1975 and issued as INFCIRC/225.[16] This established as a 'basis for concern' the possibility that 'the theft of plutonium, highly enriched uranium or uranium-233 could lead to the construction of a nuclear explosive device by a technically competent group'. INFCIRC/225 recognized that such materials could be used as 'radioactive contaminants' in what would later become known as a radiological weapon (or 'dirty bomb') and the possibility of sabotage against a nuclear facility or a shipment of nuclear material. INFCIRC/225 also recognized that its recommendations should be reviewed and updated periodically to 'reflect advances made in the state of the art or introduction of new types of facilities'.[17] Later versions emphasized the need for revisions to take into account 'recent developments and new international legal instruments'.[18]

The IAEA has reviewed and revised INFCIRC/225 five times: in 1987, 1989, 1993, 1998, and 2010. The process of preparing the revisions has evolved along with the complexity of the issues covered. Today a Nuclear Security Guidance Committee (NSGC), a standing body established in 2012 and open to all member states, reviews and approves the draft before opening it to member states' comments.[19] In September 2001, coincidentally just three days after the 9/11 terrorist attacks, the IAEA Board of Governors and the General Conference supplemented INFCIRC/225 by endorsing 'Physical Protection Objectives and Fundamental Principles'.[20] These included, in addition to the 'responsibility of the state' mantra, the need for a legislative and regulatory framework and establishment of a competent national authority; the key responsibility of licence holders for physical protection; the importance of security culture; the value of a threat-based, graded approach, deploying defence-in-depth, quality assurance, and contingency plans; and the importance of confidentiality.

In 2010, responding to further developments in the nuclear security landscape, the Secretariat conducted an extensive consultation process with member states which resulted in INFCIRC/225/Revision 5.[21] Hailed as 'a significant improvement upon its predecessors', it recommended for the first time protecting digital systems against cyberattack; defending against insider threats; prioritizing security culture; conducting regular force-on-force exercises; and ensuring sustainability of nuclear security systems.[22] Nonetheless, eminent critics pointed to the document's flaws and the need for further improvements.[23] Work has begun on a new revision, but major changes are apparently not expected, partly because since 2010 the Agency has prepared several implementing and technical guides tackling specific topics, including on design basis threat (DBT) development, mitigating insider threats, computer security, and security culture. In 2017 the Agency also published, for the first time, an implementing guide for INFCIRC/225.[24]

Today the IAEA publishes a Nuclear Security Series, a hierarchically arranged collection comprising Fundamentals, Recommendations, Implementing Guides, and Technical Guidance.[25] All remain non-legally binding, but have expanded in scope to become comprehensive advisory documents for member states in designing their legal and regulatory nuclear security architectures. Given fast-moving developments, such as in cyber security, the Secretariat has sought to establish a continuous review process, although it has been constrained by the lengthy consultations among member states needed to reach consensus on revisions.

Nurturing the CPPNM and Its Amendment

The amended Convention on the Physical Protection of Nuclear Material (A/CPPNM) is the key international agreement relating to nuclear security.[26] In its original form the treaty was restricted to what was then seen as the priority—the protection of nuclear material in international transport. States still insisted that physical protection of nuclear material within their borders was their own exclusive responsibility. The convention also did not cover military or radiological material. Opened for signature in October 1979, it took almost seven years before acquiring the necessary 21 ratifications to enter into force, an indication that governments did not give it high priority.

After 9/11, pressure grew for extending the convention beyond international transit to nuclear material in domestic use, storage, and transport, as well as protecting nuclear materials and facilities from sabotage. After extensive negotiations, an amendment along these lines was adopted in 2005. The Amendment also criminalized offences for illicit trafficking and sabotage of nuclear material or nuclear facilities and provided for strengthened international cooperation, including assistance and information sharing. The categorization of nuclear materials and appropriate levels of protection to be applied to them drew on the contents of INFCIRC/225, but there is no mention of that document or the need to comply with its recommendations. The Amendment did incorporate the so-called Fundamentals, the 'Principles of Physical Protection of Nuclear Material and Nuclear Facilities' agreed in 2001.

There was continuing opposition to a more radical transformation of the treaty to include implementation reports by states parties; peer review mechanisms; mandatory application of INFICRC/225; compulsory international oversight of physical protection measures; and the inclusion of nuclear material and facilities in military use. After much campaigning by the IAEA, some member states, and civil society, the amended treaty entered into force in 2016 (but only for states parties that had signed and ratified the amendment). As it did for the original convention, the Agency actively promotes universality for the amended version.

Despite the significant role played by the IAEA in its genesis and amendment, and despite being widely perceived as an 'IAEA Convention', the only specific role assigned to the Agency is as a 'post office' for states parties to notify each other of their central authority and point of contact and in case of breaches or incidents 'where appropriate'.[27]

As the depositary, the Agency also assumes the role of organizing review conferences. In 1992 it convened the rather desultory first review conference, five years after the convention entered into force. It took the events of 9/11 to generate interest in the implementation of the treaty and its revision. The first review conference for the A/CPPNM was held in Vienna in March–April 2022. The Agency assisted materially with preparations for the meeting. The conference Outcome Document acknowledged the work of the IAEA in promoting the objectives of the convention and its universalization, but it did not single it out for commendation or identify it as having a central role as a global platform for international cooperation in the field, which it surely does. This low-key consensus 'outcome' partly reflects the churlish attitude to the IAEA by some member states, which fret over their sovereign prerogatives, and partly the views of those that take a legalistic view that seeks to distinguish between states parties to a stand-alone international treaty and membership of an international organization, however vital that body is to the treaty's implementation.

Assisting Member States

In the absence of a bespoke international organization for the CPPNM in its original or amended versions, the IAEA has by default assumed the tasks of assisting states parties in carrying out their treaty obligations, providing them with technical guidance, advice, and assistance, including in drafting national implementation legislation, and increasingly acting as the focal point for international collaboration between them. It has simply added these roles to its existing assistance and advice to its member states, whether they are CPPNM parties or not. The range, volume, and sophistication of advice and assistance to all member states in the nuclear security field has surged since 2001, as detailed in the 38-page-long annual IAEA Nuclear Security Report of 2021.[28]

The Secretariat reports that it gives high priority to assisting states develop and implement Integrated Nuclear Security Support Plans (INSSPs), enabling them to take a comprehensive approach to enhancing their domestic nuclear security regimes. The plans, which are drafted and approved by states with the assistance and advice of the Secretariat, are specially designed to increase coordination between the Agency, the state, and potential donors. At the time of writing, 92 INSSPs have been approved and are operable, and 16 have been completed and await state acceptance.[29]

Paradoxically, states with sophisticated nuclear enterprises do not see the need for IAEA involvement in planning their nuclear security regimes, believing they have everything under control, while states with modest nuclear capabilities see no need for planning because they have so little to account for. The Secretariat has made scrubbed data about INSSPs available to member states to demonstrate the success of the programme and encourage more funding.

Meanwhile, demand for IAEA advisory and peer review missions is growing, although the Covid-19 pandemic interrupted on-site visits and a shortage of qualified experts hampers their expansion of the programme.[30] International Physical Protection Advisory Services (IPPAS), created by the IAEA in 1995 as a natural extension of support provided to former Soviet states, offer peer advice, at a state's request, on implementing international instruments and Agency guidance. IPPAS missions are coordinated by the Agency and conducted by teams of international nuclear security experts. Their scope is determined by the requesting state and the findings reflected in confidential mission reports. Missions may be complemented by a follow-up visit and IAEA assistance.

At the time of writing, the IAEA has conducted 92 missions since 1996, including follow-up missions, in 55 states, still less than one-third of the Agency's membership.[31] Since 2016, when a record six missions were carried out, the annual number of missions has declined to just two in 2020–2021, partly due to the shortage of experts and partly due to the Covid-19 pandemic. To spread the benefits of IPPAS to states that have not yet hosted a mission, the Agency has established an IPPAS Good Practices Database. Regional workshops are also convened to provide information to potential mission recipients on preparing for and conducting missions. Regrettably, no norm has emerged that would make IPPAS missions customary practice. Some larger states continue to decline to participate, presumably because they feel confident in their ability to manage their own nuclear security or due to (probably unwarranted) national security concerns.

International Nuclear Security Advisory Service (INSServ) missions, also conducted by international teams of experts, evaluate a state's nuclear security legal and regulatory regime, using the IAEA Nuclear Security Series as a basis.[32] The expert team provides a confidential report, including suggested improvements. The programme is not nearly as robust or active as IPPAS and appears to overlap with it in some respects. INSServ missions were suspended in 2016 while new guidelines were prepared to ensure that they were compatible with IPPAS missions.[33] The Covid-19 pandemic prevented the planned resumption of missions in 2020. A continuing difficulty is a shortage of expert participants, which the Agency is seeking to remedy through training workshops. There may be a case for amalgamating IPPAS and INSServ, not least to simplify the advisory process for member states with little national capacity to understand the finer points of nuclear security.

The IAEA has increasingly paid attention to the long-neglected security of radioactive sources. This is vital, not only for its potentially substantive impact, but also because it promises to draw all states, some of which have no nuclear material but notable radiological sources in medical or agricultural use, into the nuclear security enterprise and give them a stake in its success. The Agency's work encompasses guidance development; training courses and expert advice; the fostering of dialogue between states; and support for voluntary codes of conduct. At the time of writing, 140 states had committed to implement the 2004 'Code of Conduct on the Safety and Security of Radioactive Sources'.[34] Of these, 123 had notified the Director General of their intention to follow the Code's supplementary 'Guidance on the Import and Export of Radioactive Sources'.[35] Only 42 had given a similar notification regarding the 2017 supplementary 'Guidance on the Management of Disused Radioactive Sources'.[36] Despite all countries having, at a minimum, a hospital that uses radiological sources for X-rays, the number of states committed to the various voluntary measures is disappointing. Clearly more work needs to be done to achieve global coverage.

There is increasing recognition by the IAEA that nuclear safety and security are inextricably linked and member states should be assisted in acting on this nexus. Agency initiatives include the Global Nuclear Safety and Security Network (GNSSN), which is 'both a human network and a web platform allowing its members to share nuclear safety and security knowledge and services',[37] and the International Regulators Network (RegNet), a portal for web-based international cooperation among nuclear safety and security regulators.[38] The Agency's own Advisory Group on Nuclear Security (AdSec) and the International Nuclear Safety Advisory Group (INSAG) have also begun collaborating on joint projects. A Coordinating Committee on Safety Standards and Nuclear Security Series Publications and an Interface Group, comprising the chairs of the Safety Standards Committees and the Nuclear Security Guidance Committee, seek to ensure that the safety–security nexus is covered in all IAEA publications. A challenge for the Agency in inducing member states to consider safety and security holistically is that they have a variety of domestic arrangements to handle such matters. Some combine safety and security (and sometimes even safeguards) in the same office, while others strictly separate safety and security, often applying much stricter confidentiality to the latter. Some languages do not differentiate between safety and security. The Secretariat needs to be deft in handling these differences.

One IAEA service for member states not mentioned in the Agency's Nuclear Security Reports is its support for repatriating highly enriched uranium (HEU) from nuclear research reactors to the material's country of origin.[39] The two main IAEA roles of central coordinator and provider of support, provided by the Division of Nuclear Fuel Cycle and Waste Technology, have been described by US officials as indispensable.[40] Agency assistance with the recent repatriation of HEU from Ghana and Nigeria has rendered Africa free of HEU-fuelled research reactors.[41]

The IAEA as Trainer, Educator, and Information Disseminator

Many IAEA member states rely on the Agency as their sole source of information, education, and training in nuclear security. The demand for training, whether in-house or outsourced to member states, is far outstripping demand. Since 2010, with the assistance of member states, the Secretariat has supplemented in-person training by developing an extensive e-learning capacity. This served it well during the Covid-19 pandemic. Around 23,000 modules have been completed by around 9,900 users from 175 states.[42] The Secretariat is aware, however, of the pitfalls of e-learning, including low completion rates, uncertain knowledge retention, and questionable sustainability. It cautions that e-learning should supplement, but not replace, face-to-face learning, although it can be successfully used as a prerequisite for classroom learning. Aware that some states approach training unsystematically, the Secretariat has developed a 'systematic approach to training (SAT)' module designed to help states better identify their training needs and priorities.[43] The IAEA is trying to incorporate new technology into training, including an interactive three-dimensional model of a hypothetical Shapash Nuclear Research Institute, used in its course on Preventive and Protective Measures against Insider Threats.[44]

Recognizing that there is a worldwide dearth of nuclear security education at the tertiary level, the Agency has established the International Nuclear Security Education Network (INSEN). Comprising 198 institutions from 66 states the time of writing, INSEN has been successful in assisting member states to establish and enhance educational programmes based on international guidance and recommendations.[45] The Agency has also collaborated since 2011 with the Abdus Salam International Centre for Theoretical Physics in Trieste, Italy, in conducting the International School on Nuclear Security. Another significant initiative by the IAEA is the establishment of a dedicated nuclear security training and demonstration facility at the Agency's Seibersdorf site outside Vienna. In operating the facility, the Secretariat plans to follow its recommendations to member states on how to set up their national support centres.[46] This is a welcome development, as the Agency has not always followed, even at its own facilities, the recommendations it makes to member states.

With IAEA encouragement, states have begun pursuing an idea that arose from the nuclear security summits by establishing National Nuclear Security Support Centres (NSSCs) or Centres of Excellence (CoEs). These facilities provide training to states' own nationals and sometimes also to neighbouring states' personnel. In 2012 the Agency inaugurated a Nuclear Security Support Centres Network to encourage coordination and collaboration. It now has 66 participating states and holds annual meetings, enabling participants to visit other NSSCs and share best practice. The Asia–Pacific region has been particularly active in establishing such centres, with China, India, Indonesia, South Korea, and Japan each hosting one. Other regions have not been so successful. The Agency could help rectify this disparity by fostering regional centres where individual countries are unable to set up their own.

One of the critical roles that the IAEA can play as a global platform is to foster information dissemination and exchange. A key mechanism is the Nuclear Security Information Management System (NUSIMS), which provides member states with online access to reams of information. Another is the Incident and Trafficking Database (ITDB). Established in 1995, the ITDB is one of the Agency's oldest nuclear security activities. It records incidents of illicit trafficking in nuclear and other radioactive material reported by member states, as well as nuclear and other radioactive material out of regulatory control.

The data encompass all types of nuclear material (uranium, plutonium, and thorium), naturally occurring and artificially produced radioisotopes, and radioactively contaminated material such as scrap metal. States are also encouraged to report scams or hoaxes where material is purported to be nuclear or otherwise radioactive.[47] The number of reported trafficking and malicious use incidents has declined slightly in recent years. Incidents that involved kilogram quantities of HEU were always few and have not occurred since the 1990s.[48] The majority of thefts now involve sources in industrial or medical applications. Communication with states participating in the ITDB is maintained through a network of national Points of Contact (POC). The ITDB's biggest drawback is that participation is voluntary, and most states do not participate.

Preparing for Nuclear or Radiological Emergencies

The Agency has an irreplaceable function as a credible purveyor of information to the global public and as a provider of advice and assistance during a nuclear emergency. Although the IAEA's incident emergency preparedness and response (EPR) activities were originally prompted by concerns about nuclear safety, they are relevant to any nuclear or radiological incident or emergency 'independent of the triggering event'.[49] The Agency's mandate for such activities flows from its Statute and from two nuclear accident conventions negotiated after the 1986 Chernobyl accident: the Convention on Early Notification of a Nuclear Accident (CENNA) and the Convention on Assistance in the Case of a Nuclear Accident or Radiological Emergency (CACNARE). The challenge for the Agency is ensuring continuing support for such mechanisms that it is hoped will never have to be used. Nonetheless, the Agency persists in seeking to prepare itself for nuclear or radiological emergencies to the extent that resources permit, with a striking number of ongoing, often little-known, initiatives.

The Agency is, for instance, responsible for developing and maintaining the 2004 Joint Radiation Emergency Management Plan of the International Organizations (JPLAN) and the convening of the Inter-Agency Committee on Radiological and Nuclear Emergencies (IACRNE). The current system is based on the 2004 International Action Plan for Strengthening the International Preparedness and Response System for Nuclear and Radiological Emergencies.[50] The Agency publishes an Emergency Preparedness and Response series, in addition to its Safety Standards that include recommendations on emergency preparedness and response.

A key feature of the Agency's role is the Incident and Emergency Response Centre (IEC), which maintains a skeleton staff 24/7, but which can ramp up to full capacity should an incident occur. The IEC was utilized effectively during and after the 2011 Fukushima accident and has since been strengthened.[51]

The Secretariat seeks to advise and assist member states through the Emergency Preparedness and Response Information Management System (EPRIMS), an interactive web-based tool that permits self-assessment of EPR arrangements and shares the results with the IAEA and member states. The Secretariat offers states Emergency Preparedness Review (EPREV) missions, including follow-up visits, although the take-up rate is low.[52] The Agency also conducts ConvEx (Convention Exercises) for implementing CACNARE should the need arise, using the Unified System for Information Exchange in

Incidents and Emergencies (USIE) in exercise mode. In addition, the Agency manages the International Nuclear and Radiological Event Scale (INES), designed to communicate to the public and the media the severity of events at nuclear facilities, using a seven-level scale ranging from 'anomaly' to a 'major accident'. It can equally be used for nuclear security incidents as for accidents.

Finally, the Agency manages a Response and Assistance Network (RANET), intended to link recipients and potential providers of assistance by securing pledges in advance, and an International Radiation Monitoring Information System (IRMIS), accessible to all states. RANET enabled the Agency to respond to a request in 2020 from Lebanon for assistance in determining whether the explosion at the port of Beirut on 4 August had any nuclear or radiological implications.[53]

The Russian invasion of Ukraine and the capture by Russian forces of the defunct but contaminated Chernobyl nuclear facility and the Zaporizhzhia nuclear power plant, the largest in Europe, have put the Agency's emergency preparedness arrangements to the test. On 24 February 2022, as soon as the nuclear security and safety threat became clear, the Agency activated its Incident and Emergency Response Centre and has been monitoring developments closely.[54] In addition to providing its own advice and technical assistance to Ukraine, the Agency used RANET to engender emergency preparedness and response assistance from at least 11 member states. Despite longstanding scepticism about the need for RANET and a low-key response to appeals for pre-pledging of assistance in the absence of a crisis, it has now proved its worth.

In response to the Ukraine emergency, the Agency also dispatched combined safety, security, and safeguards missions to the country, sometimes led by Director General Rafael Mariano Grossi. As the need for a permanent IAEA presence at Ukraine's vulnerable nuclear power plants became increasingly obvious, the Agency in September 2022 established the IAEA Support and Assistance Mission to Zaporizhzhia (ISAMZ) and later similar missions at Chernobyl and the other Ukrainian nuclear power plant complexes (Rivne, South Ukraine, and Khmelnitsky).[55] Acting as both a deterrent to further military attacks, as an early warning mechanism, and to provide advice, support, and assistance in case of emergencies, a standing IAEA presence for such purposes is unprecedented (the IAEA has had permanent on-site deployments of safeguards inspectors, such as currently in Iran).

The IAEA's Ukraine response has also been novel in explicitly acknowledging the often-overlooked nexus between safety, security, and safeguards. The Director General's 'seven indispensable pillars' for ensuring nuclear safety and security during an armed conflict, while drawing on international law and existing IAEA principles and recommendations, were an inspired effort, drafted overnight, to encapsulate in digestible form the international norms for protecting nuclear power plants during wartime.[56] The IAEA's timely and creative response to Russia's attacks on Ukraine's nuclear facilities is a welcome assertion of the vital role of the Agency in nuclear security (and safety). It has also enhanced the political standing of the position of IAEA Director General in efforts to ameliorate future nuclear crises.

Organization, Planning, Management, and Budget

As nuclear security has grown in political salience, the Secretariat's organizational capacity has expanded—albeit with a time lag—to meet the need. An Office of Physical Protection and Material Security was established in 1999 to coordinate activities, especially the preparation of documentation and the organization of expert group meetings. In 2002 the office was moved from the Department of Safeguards to the Department of Nuclear Safety and renamed the Office of Nuclear Security (ONS). Despite the obvious nexus between non-proliferation and nuclear security, the move was useful bureaucratically, given the 'Cinderella' status of nuclear security in the IAEA's largest, best-funded, and most politically sensitive department. Housing it with nuclear safety did not entirely solve this problem, but nuclear security was helpfully united with radiological security, which had hitherto been considered primarily a safety concern.

In 2013, partly due to the high-level attention drawn to nuclear security by the summit process, the ONS was upgraded to a division within the renamed Department of Safety and Security. This enhanced its bureaucratic status and signalled its growing importance to member states. The division comprises four sections corresponding to the four nuclear security sub-programmes: Information Management; Nuclear Security of Materials and Facilities; Nuclear Security of Material Outside of Regulatory Control; and Programme Development and International Cooperation. There is also an Office of Safety and Security Coordination to ensure attention is paid to the nexus between safety and security.

The Secretariat's activities for the coming four years are meant to be guided by Nuclear Security Plans, incorporating views of member states expressed by the General Conference, the Board of Governors and ICONS. The latest Nuclear Security Plan is for 2022–2025.[57] A comparison of the plans since 2001 reveals increasing length and detail, but arguably declining utility. Having to incorporate the views of all member states, the plan now essentially describes all activities currently undertaken, adds new ones, and projects them vaguely into the future, without any prioritization. The plans do not envisage any reform, streamlining, or rationalization of the existing programme. Metrics for measuring success are also absent, challenging the Secretariat's alleged results-based management approach.

In its 2019 report on the IAEA's role in nuclear security, the GAO concluded that the Secretariat 'does not prioritize activities or fully measure or report on program performance'.[58] It especially criticized the absence of baselines or targets for each performance measure, meaning that the IAEA's ability to demonstrate results for its programme was limited.[59] Secretariat officials acknowledged this was deliberate, due to nuclear security being a national responsibility, the sensitivity of states about confidential information, and the limitations of the IAEA's nuclear security mandate. They explained that, instead of prioritizing activities, they respond to requests from member states as they come in and, to the extent that resources are available and donors' funding conditions permit, carry out the requested activities.[60]

The Agency's external auditors agreed with the GAO, recommending an interconnected risk management process at Agency-wide and project level to allow sub-programme management (like nuclear security) to improve operational risk identification, and risk mitigation planning in a visible and trackable manner.[61] The Agency agreed with the recommendation and undertook to implement it. One genuine challenge is that member states are often reluctant to provide information on the impact of Agency projects, perhaps due to embarrassment at the mixed results or (probably ill-conceived) national security concerns. The GAO investigators responded that many of the performance measures recommended were for the Agency's own activities, and not those of member states.[62]

Donor states need to know not only how many IPPAS missions have been conducted, how many radiation monitors have been installed at border crossings, or how many states have a recommended DBT in place, but also what the real impact has been on nuclear security. More transparency about the Secretariat's management and funding challenges could strengthen interactions with donors and ensure that member states and the Secretariat can better implement the programme.

The Agency's nuclear security work is, above all, constrained by its unusual financing arrangement. Unlike other Agency programmes, the regular budget allocation for nuclear security is largely devoted to management of the Division of Nuclear Security, not to substantive projects. Regular budget funding for the Division has increased almost every year (reaching €60.3 million in 2021), despite the Agency operating under zero nominal budgetary growth. But around 80 per cent of the substantive work is funded by the NSF, sourced from voluntary, extra-budgetary contributions from member states (amounting to €50 million in 2020).[63] This arrangement is a legacy of a time when there was disagreement among member states (albeit since largely resolved) about whether nuclear security was a core function of the Agency.

Although the NSF has been generously funded and is being regularly replenished, donor fatigue can imperil the Agency's work at short notice. Donors also tend to place conditions on their contributions, directing them to pet activities, projects, states, or regions—complicating planning, management, personnel recruitment and retention, and long-term risk assessment. Staff funded by the NSF are limited to two-year contracts, resulting in an absence of long-term commitment, a rapid staff turnover, and loss of institutional memory and experience. Cost-free experts provided by member states are there at the whim of the government concerned, again constraining long-term planning. This author has proposed a grand bargain for the IAEA's finances, in which funding for nuclear security (a Western preoccupation) and technical cooperation (a developing country concern) are both brought into the IAEA regular budget and that assessed contributions by all member states to that budget are allocated according to the standard formula.[64]

Conclusion

The IAEA is demonstrably growing and strengthening its role in enhancing nuclear security. Since 2016, it has become more visible as the principal forum for international dialogue on nuclear security matters, expanding its educational and training role and seeking to meet increasing demand from states for its advice and services. It has adopted innovative approaches in several areas, including e-learning, university-level nuclear security education, and global networking, along with its new training facility at Seibersdorf. Importantly, it is contributing to the steady expansion of a permanent, informed global nuclear security community. The IAEA's role in Ukraine has been a practical demonstration of the vital part it can play in bolstering nuclear security during a multi-dimensional nuclear crisis.

Yet, the IAEA could achieve much more if its mandate was less equivocal. The Board of Governors could take the initiative to update the IAEA's nuclear security mandate to something like, 'work to enhance global nuclear security in a comprehensive and sustainable fashion and do everything that derives from that mandate.' More cohesive and logical strategic planning could flow from such a development. Even without a clarified mandate, the Agency should adopt enhanced strategic planning, risk assessment, and prioritization to its work, as well as encouraging feedback from recipients of its advice and technical assistance. It should draw the appropriate lessons from its necessarily improvised but to date successful response to the Ukraine crisis.

Currently, the bureaucratic and budgetary complexity of the Agency's nuclear security undertakings and a lack of transparency makes it difficult to draw definitive conclusions about the Agency's overall, day-to-day contribution to nuclear security on the ground. This in turn poses challenges for promoting the programme's value to member states and other stakeholders. A comprehensive external evaluation of the sprawling IAEA nuclear security programme seems warranted to help it establish proper prioritization and performance measurement. Altogether indispensable, the IAEA's role in nuclear security needs to be enhanced and much more widely promoted and appreciated.

Notes

1. David Fischer, *History of the International Atomic Energy Agency: The First Forty Years* (Vienna: IAEA 1997), 229.
2. IAEA, 22 August 1974, 'The Standard Text of Safeguards Agreements in Connection with the Treaty on the Non-Proliferation of Nuclear Weapons', GOV/INF/276, Annex A, INFCIRC 153 (Corrected), para. 7.
3. Piet De Klerk, 'How Nuclear Security Arrived at the IAEA', in Joseph F. Pilat, ed., *The International Atomic Energy Agency: Historical Reflections, Current Challenges, and Future Prospects* (London: Routledge, 2022), 196.
4. IAEA, 2002, 'General Conference: Measures to Improve the Security of Nuclear Materials and other Radioactive Materials', GC(45)/1; IAEA, 12 August 2002, 'Nuclear Security—Progress on Measures to Protect Against Nuclear Terrorism: Report by the Director General', GOV/INF/2002/11-GC(46)/14; IAEA, 2022, 'Nuclear Security Plan.' https://www.iaea.org/topics/security-of-nuclear-and-other-radioactive-material/nuclear-security-plan
5. Nuclear Security Summit, 'Action Plan in Support of the International Atomic Energy Agency', Nuclear Security Summit, Washington, DC, 1 April 2016. http://static1.squaresp

ace.com/static/568be36505f8e2af8023adf7/t/56feeeb94d088e7781f9e41c/1459547833689/Action+Plan+-+IAEA_FINAL.pdf

6. IAEA, 2 March 2022, 'List of Member States'. https://www.iaea.org/about/governance/list-of-member-states
7. The current official language is: 'The responsibility for nuclear security within a State rests entirely with that State. At the same time, Member States have consistently recognized the central role of the IAEA in strengthening the nuclear security framework globally and in coordinating international cooperation in nuclear security.' IAEA, 'The Agency's Programme and Budget 2020–2021', GC(63)/2 (Vienna: IAEA, 2019), 117.
8. Samantha Neakrase, 'ICONS 2020: the Good, the Bad, and the Ugly … Actually, Mostly Good', *Atomic Pulse* [online], 5 March 2020. https://www.nti.org/atomic-pulse/icons-2020-good-bad-and-ugly-actually-mostly-good
9. NSCG, 'Joint Statement on Sustaining Action to Strengthen Global Nuclear Security Architecture', Nuclear Security Summit, Washington, DC [online], 5 April 2016. http://www.nscontactgroup.org/joint-statement.php
10. IAEA, 2 November 2016, 'Communication dated 24 October 2016 received from the Permanent Mission of Canada concerning the Statement of Principles of the Nuclear Security Contact Group', INFCIRC/899.
11. US Government Accountability Office (GAO), July 2019, 'Nuclear Security: The International Atomic Energy Agency Could Improve Priority Setting, Performance Measures, and Funding Stabilization', GAO-19-429 (Washington, DC: GAO), 43.
12. Ibid., 35.
13. These include 'Nuclear and Radiological Terrorism' (South Korea), INFCIRC/904; 'National Nuclear Detection Architectures' (Finland), INFCIRC/905; 'Insider Threats' (US), INFCIRC/908; 'Transportation Security' (Japan), INFCIRC/909; 'Strengthening the Security of High Activity Sealed Radioactive Sources' (France), INFCIRC/910; 'Minimization and Elimination the Use of Highly Enriched Uranium in Civilian Applications' (Norway), INFCIRC/912; 'Forensics in Nuclear Security' (Australia), INFCIRC/917; 'Countering Nuclear Smuggling' (Jordan), INFCIRC/918.
14. IAEA, 'Joint Statement on Strengthening Nuclear Security Implementation, INFCIRC/869 (Vienna: IAEA, 22 October 2014). https://www.iaea.org/sites/default/files/publications/documents/infcircs/infcirc869.pdf
15. NTI, 'Joint Statement on 'Strengthening Nuclear Security Implementation' (INFCIRC/869): Guide for Implementation and Confidence Building' (Washington, DC: NTI, 2016). https://www.nti.org/wp content/uploads/2021/09/INFCIRC869_Implementation_Guide.pdf
16. IAEA, September 1975, 'The Physical Protection of Nuclear Material (INFCIRC/225)'; a corrected version was issued in 1976 as 'INFCIRC/225 (Corrected)'.
17. IAEA, February 1976, 'The Physical Protection of Nuclear Material (INFCIRC/225 (Corrected))', 1.
18. IAEA, *Nuclear Security Recommendations on Physical Protection of Nuclear Materials and Nuclear Facilities (INFCIRC/225/Revision 5)* (Vienna: IAEA, 2011), 7–8. https://www-pub.iaea.org/MTCD/Publications/PDF/Pub1481_web.pdf
19. IAEA, 'IAEA Fact Sheet: Nuclear Safety and Security: The IAEA Nuclear Security Series', (Vienna: IAEA, 2019). https://www.iaea.org/sites/default/files/19/05/nuclear-security-series.pdf
20. IAEA, *General Conference: Measures to Improve the Security of Nuclear Materials and Other Radioactive Materials*, GC(45)/INF/14, 14 September (Vienna: IAEA, 2001).
21. IAEA, *Nuclear Security Recommendations*, Foreword.

22. Matthew Bunn, Laura Holgate, Dmitry Kovchegin, William H. Tobey, and Nicholas Roth, *IAEA Nuclear Security Recommendations (INFCIRC/225): The Next Generation* (Washington, DC: Henry L. Stimson Center, 2020). https://www.stimson.org/2020/iaea-nuclear-security-recommendations-infcirc-225-the-next-generation
23. Ibid., 4–9.
24. IAEA, *Physical Protection of Nuclear Material and Nuclear Facilities (Implementation of INFCIRC/225/Revision 5), Implementing Guide*, IAEA Nuclear Security Series, No. 27-G (Vienna: IAEA, 2018).
25. For a complete list, see the IAEA Nuclear Security Series. https://www.iaea.org/publications/search/type/nuclear-security-series
26. The other major treaty in the field is the 2005 International Convention for the Suppression of Acts of Nuclear Terrorism (ICSANT), negotiated by the UN General Assembly.
27. IAEA, 1982, Convention on the Physical Protection of Nuclear Material (CPPNM), Articles 5.1 and 5.2.
28. IAEA, 30 July 2021, 'Nuclear Security Report 2021', GOV/2021/35-GC(65/10). https://www.iaea.org/sites/default/files/gc/gc65-10.pdf
29. Information provided by the IAEA, 19 July 2022.
30. IAEA, *IAEA Annual Report 2020* (Vienna: IAEA, 2021), 19. https://www.iaea.org/sites/default/files/publications/reports/2020/gc65-5.pdf
31. IAEA, 'Nuclear Security Report 2021', 16.
32. IAEA, *International Nuclear Security Advisory Services (INSServ) Guidelines*, IAEA Services Series, No. 13 (Vienna: IAEA, 2019). https://www-pub.iaea.org/MTCD/Publications/PDF/SVS39web.pdf
33. GAO, 15.
34. IAEA, 2004, 'Code of Conduct on the Safety and Security of Radioactive Sources'. https://www-pub.iaea.org/MTCD/publications/PDF/Code-2004_web.pdf; see also IAEA, 15 September 2018, 'Status List of Countries'. http://www-ns.iaea.org/downloads/rw/imp-export/status-list.pdf; for 2021 figures, see 'Nuclear Security Report 2021', 18.
35. IAEA, 2012, 'Guidance on the Import and Export of Radioactive Sources'. https://www-pub.iaea.org/MTCD/Publications/PDF/8901_web.pdf
36. IAEA, 2018, 'Guidance on the Management of Disused Radioactive Sources'. https://www-pub.iaea.org/MTCD/Publications/PDF/Guidance_on_the_Management_web.pdf
37. IAEA, 2022, 'Global Nuclear Safety and Security Network'. https://gnssn.iaea.org/main/pages/default.aspx
38. IAEA, 2022, 'International Regulatory Network'. https://www.iaea.org/resources/databases/international-regulatory-network
39. Laura Gil, 'Countries Move to Low Enriched Uranium to Fuel Their Research Reactors', *IAEA Bulletin* [online], 21 February 2020. https://www.iaea.org/newscenter/news/countries-move-towards-low-enriched-uranium-to-fuel-their-research-reactors
40. A. Atkins, US Department of Energy, Presentation to main panel session 'CPPNM/A. International Cooperation for Physical Protection of Nuclear Material and Nuclear Facilities', in *International Conference on Physical Protection of Nuclear Material and Nuclear Facilities: Summary of an International Conference*, 13–17 November 2017, Vienna, (Vienna: IAEA, 2017).
41. Petr Chakrov and Thomas Hanlon, 'Nigeria Converts Its HEU Research Reactor to LEU fuel', *IAEA Press Release*, 20 December 2018. https://www.iaea.org/newscenter/news/nigeria-converts-its-research-reactor-from-heu-to-leu-fuel
42. IAEA, 'Nuclear Security Report 2021', 27.

43. Ibid., 28.
44. See A. Askin, 4 June 2018, 'Hypothetical Nuclear Facility Interactive Training Tool to Support Practical Application of Nuclear Security Skills', LLNL-CONF-752430, Lawrence Livermore National Laboratory; IAEA, 'Nuclear Security Report', 16.
45. IAEA, 'Nuclear Security Report 2021', 28; The author is a member and has been impressed by the organization and continued evolution of INSEN.
46. IAEA, *Establishing and Operating a National Nuclear Security Support Centre*, IAEA-TDL-010 (Vienna: IAEA, 2020).
47. IAEA, 2021, '2020 Fact Sheet', Incident and Trafficking Database (ITDB). https://www.iaea.org/sites/default/files/20/02/itdb-factsheet-2020.pdf
48. IAEA, 'Nuclear Security Report 2021', 11.
49. IAEA, 'Annual Report 2020', 87.
50. IAEA, n.d. 'International Action Plan for Strengthening the International Preparedness and Response System for Nuclear and Radiological Emergencies'. http://www-ns.iaea.org/downloads/rw/action-plans/ers-action-plan.pdf
51. See Trevor Findlay, *Unleashing the Nuclear Watchdog* (Waterloo: Centre for International Governance Innovation (CIGI), 2012), 27–8.
52. IAEA, The Agency's Financial Statements for 2020, GC(65)/4 (Vienna: IAEA, 2021), 182.
53. See 'Case Study: Agency Assistance to Lebanon in the Aftermath of the Beirut Blast', IAEA Annual Report 2020', 89–90.
54. IAEA, 'The Safety, Security and Safeguards Implications of the Situation in Ukraine', Oral report to the Board of Governors by the Director General, 9 June 2022. https://www.iaea.org/newscenter/statements/the-safety-security-and-safeguards-implications-of-the-situation-in-ukraine
55. IAEA, 'Update 149—IAEA Director General Statement on Situation in Ukraine', 2 March 2023. https://www.iaea.org/newscenter/pressreleases/update-149-iaea-director-general-statement-on-situation-in-ukraine
56. IAEA, 'Nuclear Safety, Security and Safeguards in Ukraine: 2nd Summary Report by the Director General', 28 April-5 September 2022 (Vienna: IAEA, 2022), p. 7.
57. IAEA, *Nuclear Security Plan 2022-2025*, GC(65)/24 (Vienna: IAEA, 2021).
58. GAO, 19.
59. GAO, 22.
60. GAO, 19.
61. See IAEA, *The Agency's Accounts for 2017*, GC(62)/5 (Vienna: IAEA, 2018), 160.
62. GAO, 22, 27.
63. IAEA, 'The Agency's Programme and Budget 2020–2021'.
64. Trevor Findlay, *What Price Nuclear Governance? Funding the International Atomic Energy Agency*, Project on Managing the Atom (Cambridge, MA: Harvard Kennedy School: Belfer Center for Science and International Affairs, 2016).

Suggested Reading

De Klerk, Piet, (2021), 'How Nuclear Security Arrived at the IAEA', in Joseph F. Pilat, ed., *The International Atomic Energy Agency: Historical Reflections, Current Challenges and Future Prospects* (London: Routledge), 189–201.

Findlay, Trevor, (2012), *Unleashing the Nuclear Watchdog: Strengthening and Reform of the IAEA* (Waterloo, ON: Centre for International Governance Innovation).

Findlay, Trevor, (2016), 'Nuclear Security Diplomacy Beyond Summitry', in Brecht Volders and Tom Sauer, eds., *Nuclear Terrorism: Countering the Threat* (London: Routledge), 232–48.

IAEA, (2021), 'Nuclear Security Plan 2022-2025: Report by the Director General', 15 September, GC(65)/24 (Vienna: IAEA).

IAEA, (2021), 'Nuclear Security Report 2021: Report by the Director General', 30 July, GOV/2021/35-CG(65)/10 (Vienna: IAEA).

CHAPTER 12

US–RUSSIAN COOPERATIVE THREAT REDUCTION TO SECURE NUCLEAR WARHEADS

GEOFFREY CHAPMAN AND WILLIAM M. MOON

Introduction

Most studies on nuclear security focus on commercial and civilian stocks of nuclear material used for energy, industry, and medical applications. Some studies have examined the world's stockpiles of military materials, which, according to the latest data available, comprise approximately 83 per cent of the total estimated nuclear materials,[1] but fewer still have examined the security of nuclear warheads that pose the most powerful nuclear threat. For an in-depth examination of nuclear warhead security practices, the 1992–2013 jointly executed United States (US) and Russian Cooperative Threat Reduction (CTR) programme provides a unique case study. This chapter makes extensive use of first-hand accounts from William Moon, a leading figure in the programme.

Following the dissolution of the Soviet Union in December 1991, Russia underwent a period of national upheaval. International concern inspired two American Senators, Sam Nunn and Richard Lugar, to initiate a programme to work cooperatively with Russia to secure the two countries' nuclear weapons, overcoming years of Cold War rivalry. The Nunn–Lugar CTR programme was a pioneering concept. Instead of providing funding to the Russian government like traditional assistance programmes had done, CTR established contracts with US and Russian companies to provide nuclear security assistance and support eliminating nuclear, chemical, and biological weapons and infrastructure.

The US and Russia signed the CTR agreement in 1992, paving the way for cooperation on the elimination of strategic delivery systems, including ballistic missiles, silos, launchers, and submarines. These efforts proceeded quickly—helped by previous experience negotiating and implementing arms control agreements. Cooperation on nuclear material security was also initiated and grew from previous international efforts.

Discussions on securing nuclear warheads, however, were much more sensitive and proved extremely difficult to initiate and make progress on.

The Russian NTV programme *Top Secret* reported that the situation of the 12th Main Directorate was 'catastrophic' in the mid-1990s, and that ' … the wages of officers at the facilities are no more than $70 per month and their wives are forced to look for jobs in the small military units remote from the towns'. One Officer assigned to the 12th Main Directorate reported: 'On a 10-point scale I would rate the protection of nuclear weapons as two out of ten. Physical protection, meaning protection against terrorists, does not exist'.[2]

The US and other Western states took such concerns over the security of the Russian nuclear warheads very seriously, making improvement in this area a high priority for cooperation under CTR. However, developing cooperation on nuclear warhead security between erstwhile adversaries had no precedent. The Russians also harboured concerns that assistance might cast doubt on their legitimacy as a nuclear power. The CTR programme worked hard to overcome such insecurities and took extraordinary steps to develop the trust necessary for nuclear rivals to work together to enhance the security of their most sensitive nuclear assets. This took years of technical engagements and bold demonstrations of respect and commitment before the partners agreed to the necessary site access arrangements.

In this regard, the CTR's efforts were successful as they produced 'heightened levels of trust between U.S. and Russian officials, military officers, and scientists … [that were] a unique result of this work'.[3] Although the role and mechanisms by which trust acts at an emotional and rational level in international relations remains 'slippery', this chapter demonstrates that it was a prerequisite in explaining the programme's success in warhead security, its most sensitive work area.[4] The chapter recounts how the CTR effort built a partnership on nuclear security through a focus on shared interests and a deliberate, step-by-step process involving 'patience and persistence'.[5] This success would be replicated elsewhere as efforts were expanded to work on reducing the risks posed by chemical and biological agents, and then other CBRN threats, worldwide.[6]

The trust established via bilateral CTR cooperation continues to have an important legacy. Indeed, it is credited by Hanna Notte as helping to facilitate Syrian chemical weapons disarmament in 2013, even as the Russian–US relationship deteriorated and joint warhead security work ceased.[7] Therefore, this case study has wider implications for ongoing CTR and arms control work, demonstrating that cooperation is possible between rivals and that threat reduction work can provide a foundation for stronger partnerships between nuclear weapons states.

Phase One: Preliminary Engagements (1992–1995)

Although the Soviet Nuclear Threat Reduction Act of 1991 had authorized the CTR programme, little progress was made initially between 1991 and 1993, partially the result of US budgetary issues. Funding for the programme derived from the budget of the Department of Defense (DoD) so its reallocation to CTR work would come at the expense of the US armed services. An additional factor was the lack of political initiative; although passing the threat reduction act had been a bi-partisan effort, Nunn and Lugar had limited input after the legislation was passed.[8] It was only after William Perry was appointed Deputy Secretary of Defense in March 1993 that 'major reform' was brought to the 'unfocused way of handling the CTR programs'.[9] This became manifest when the US Congress appropriated dedicated DoD funding to the programme, eventually growing to US$400 million per year. The early programmes focused on assisting Ukraine, Belarus, and Kazakhstan to denuclearize in adherence with the Non-Proliferation Treaty (NPT). Russia was reluctant to welcome DoD technical experts for discussions on their weapons of mass destruction (WMD) programmes but accepted US offers of equipment such as armoured blankets that were used to protect warheads as they were shipped from the former Soviet states back to Russia.

The CTR effort to engage with Russia was bolstered by the appointment of Harold P. Smith as US Assistant to the Secretary of Defense in June 1993, with a mandate from Perry to make sure the '[CTR] program ... [would] move, and ... move quickly'.[10] Smith enjoyed a close personal relationship with both Perry and Secretary of Defense Les Aspin and was given 'carte blanche' for CTR efforts.[11] Perry institutionalized this arrangement by holding formal bi-weekly meetings for Smith to report CTR progress. According to Smith, this 'sent a powerful message throughout the bureaucracy'.[12] The continued ability of Smith and his subordinates to cite their political support, their sole dedication to CTR work, and the resources available to them convinced their Russian counterparts of their seriousness.[13]

Cooperation on nuclear warheads could only start in earnest after trust had been established. Despite issues with warhead security being evident, the greatest barrier to cooperation was concern of American ulterior motives (such as intelligence gathering or efforts to denigrate Russian's sovereign capability to manage their nuclear arsenal), combined with distrust of a former adversary. General Evgeniy Maslin, Chief of the Russian Ministry of Defence's (MOD's) 12th Main Directorate (GUMO) responsible for the security of the Russian nuclear warhead arsenal, summarized the prevailing Russian view that 'free cheese only comes in mousetraps'.[14] Further systemic barriers to cooperation remained, including 'bureaucratic inertia, legal restrictions, and simple pride'.[15] While these issues were shared across all CTR efforts, the security implications of work involving nuclear warheads made these issues acutely sensitive.

In 1993, both the Pentagon and Smith recognized the need to build trust with their Russian colleagues. However, the initial problem confronting the US side was identifying who to build a relationship with.[16] US intelligence identified General Maslin as the key figure in charge of non-alert weapons. The first step to initiating CTR work on nuclear warheads was courting this relationship.[17] That so much depended on Maslin highlighted a systemic difference in bureaucratic systems; while multiple individuals on the American side dealt with Maslin, he was the only senior Russian point of contact for the Americans working on warhead security.[18]

To advance CTR efforts, Smith insisted on arranging a meeting with Maslin. Their first meeting was arranged through the military attaché's office in the American Embassy in Moscow. Both Smith and Maslin formed a positive first impression of each other. This was despite (or on Smith's reflection, because of) all their communication being filtered through their translator, Irene Nehonov.[19] According to Smith, Nehonov would continue to play a pivotal role throughout early Russian–US CTR efforts due to her considerable Russian language skills, extensive contacts, and her ability to smooth any building tension in negotiations.[20] This initial meeting was followed by a series of cultural exchanges in Russia, organized by Maslin. These included visits to the Hermitage, the opera, ballet, and Borodino. Throughout these exchanges, Smith 'was certain' Maslin was judging his reactions to see if he could be trusted.[21]

In return for these exchanges in Russia, Smith was conscious of the need to reciprocate and invited Maslin on a private tour of the Smithsonian in Washington, DC, followed by a serendipitous meeting with Justice Breyer.[22] A casual meeting between a Russian general and a Justice of the Supreme Court demonstrated Smith's influence within the US government, encouraging Maslin to continue increasing his trust and developing the relationship. Both Smith and Major General (ret.) Roland Lajoie, who was assigned as the first director of the CTR programme, believed that Maslin's first visit to the US was pivotal in convincing Maslin to work with Smith; according to Smith it was 'from that point on … pretty easy sledding'. Despite the importance of these exchanges, little 'business' was discussed at these meetings.[23] However, it was only after they had occurred that Smith was able to convince Maslin to support technical negotiations on nuclear security projects. By 1995, Lajoie and Maslin would meet regularly as part of the US–Russian Nuclear Weapons Security Group to further CTR efforts.[24]

While removed from the technical matters of warhead security, this initial building of trust was essential given that Maslin would continue to act as an advocate on the Russian side for further CTR efforts.[25] The US side continued to engage with Maslin as he was perceived to be trustworthy by the Americans, due to his competency, management, and lack of corruption. With Maslin's cooperation, work to improve Russian warhead security would secure the 'greatest achievement of the CTR programme', largely unheralded due to a lack of 'photo ops'.[26] Beyond the immediate CTR work, this relationship would continue to pay dividends.

Phase Two: Promoting Trust 'Step-by-Step' (1995–1997)

Once Maslin and Smith opened the door to cooperation, Lajoie and the DoD technical team needed to demonstrate the benefits of such cooperation. Until then, the only deliveries of nuclear security equipment provided to the Russian MOD were armoured blankets that provided ballistic protection to warheads during transportation and storage. The first deliveries were from US Army surplus stockpiles shipped to Moscow in 1992; by June 1993, 4,000 armoured blankets had been provided.[27] Later deliveries included newly manufactured ballistic and fire protection blankets developed under a DoD contract.[28] These were shipped to the Russian Ministry of Atomic Energy, the Executive Agent for the CTR agreement, who turned them over to the MOD—but US DoD and Russian MOD officials never met on this project.

When DoD and MOD teams finally met in person, they had strict orders from Smith and Maslin to identify projects and equipment that would provide immediate nuclear security benefits without requiring Russian security officials to reveal sensitive information on their warhead stockpile. Even though the technical teams developed a strong rapport based on a mutual appreciation of nuclear security requirements, the Russians were still concerned that any equipment provided by the US might include imbedded listening devices, so options for assistance were limited.[29]

The first plan developed by the technical teams was for the US to provide enhanced safety and security equipment for Russian cargo cars that were used to ship warheads from storage sites to dismantlement facilities. The project was a logical follow-on to the provision of armoured blankets, and it would enhance security for the great number of warhead shipments returning from Ukraine, Belarus, and Kazakhstan.[30] However, technical challenges of designing and integrating such systems for Russian railcars created delays.[31] To do the design work, the Russians agreed to have a railcar shipped to Albuquerque, New Mexico, where the nuclear security experts from Sandia National Laboratory could design and install intrusion detection and fire protection systems in a prototype. The project took more than a year before the results could be delivered to Russia; after the deliveries, the Russians took considerable time to inspect the railcar for listening devices. Once the Russians were satisfied that the Americans were genuinely seeking to enhance security, confidence in the cooperation grew and 100 cargo security kits were transferred by October 1994. Nevertheless, a better and faster way to deliver enhancements was needed.[32]

The DoD technical team was motivated to demonstrate their sincere desire to provide nuclear security enhancements and further build trust. The Russian side shared their concern that, after years of relying on the secrecy of their site locations to provide for nuclear security, physical security now needed to be enhanced, too. As Russian sites lacked the kind of security fencing used to protect DoD warhead facilities, the DoD technical team sought to meet this requirement. Knowing the impossibility of being granted site access and that requests could expose basic security vulnerabilities, the DoD agreed to purchase security fencing from Eleron, a Russian company originally founded by MinAtom. The Russians were impressed with the US willingness to establish direct contracts with a Russian company, which enhanced cooperation.[33]

The technical teams dubbed the effort to rapidly procure and deliver fencing as a 'quick fix', and agreed that the fencing should be triple layered with clear zones, microwave detectors, and an integrated fence disturbance detection system.[34] The DoD established a major contract with Eleron for 50 km of fencing to provide coverage around Russian storage bunkers. The contract was for over US$25 million, and the fencing was delivered rapidly from September to December 1997.[35] The considerable funding demonstrated the DoD's commitment, and the rapid procurement conducted without questioning the quantity of equipment engendered trust. The Russians saw that the DoD was willing to work to meet their needs without demanding sensitive information. DoD contracting procedures required the technical team to verify that Eleron was fulfilling the contract, so the technical teams established a process to inspect the equipment in person at a delivery facility in Sergiev Posad in order to approve payments.[36]

From that point on, the technical teams worked feverishly to define additional nuclear security equipment requirements, negotiate major contracts, and develop procedures to verify and transfer equipment deliveries. For example, Motorola radios were quickly purchased and delivered to assist the guard forces patrolling bunkers. Talks ensued on nuclear security inventory management systems, emergency response, and additional guard force support.[37] While the equipment purchases and technical talks continued at an expedited pace, the DoD teams assigned additional military and civilian personnel to the CTR team full time, and CTR leadership in the Pentagon and Congressional leaders continued to increase the budgets to meet demands, further emboldening the relationship.[38]

As these relationships grew, in 1996 the Russian side proposed further cooperation by requesting assistance to establish a major training centre. Both sides could then work together to design comprehensive security enhancements for nuclear warhead storage sites, and personnel could be trained to operate and maintain those systems.[39] In 1997, the US side agreed, and the Security Assessment and Training Center (SATC) was established in February 1998.[40] The SATC became the centrepiece of the nuclear security programme, providing a place for technical experts to discuss and develop advanced nuclear security capabilities. The effort was conducted by a US prime contractor that established subcontracts with multiple Russian companies to provide construction, installation, and equipment purchases on-site.[41] For the first time in the CTR nuclear security programme, US contractor personnel were hired to live and work in Russia to support ongoing cooperation.

The institutionalization of US–Russian nuclear security cooperation through the foundation of the SATC enabled technical discussions on issues that would have previously been taboo.[42] This included insider threats—the topic could not be broached earlier because it would imply the Russians had personnel reliability issues.[43] In the context of the SATC, the Russian team was provided with briefings by the DoD Personnel Reliability Program (PRP) and the DOE/National Nuclear Security Administration (NNSA) Human Reliability Program, and they were also invited to visit US drug and alcohol testing labs.[44] A PRP drug and alcohol testing lab was subsequently established at the SATC.[45]

The work of the SATC also allowed for a broadening and deepening of US–Russian nuclear security cooperation by establishing a nuclear warhead inventory management programme.[46] At first, the DoD provided a small quantity of computers and fax machines to assist with some inventory management tasks. Technical talks led to the provision of additional software and hardware, and the SATC complex provided a convenient and accessible place to design and construct a prototype inventory management field facility.[47] This led to the production and delivery of modular inventory management facilities across MOD storage sites.[48]

In terms of deepening already-established nuclear security cooperation, the SATC conducted some of its most important work through the development and design of comprehensive nuclear security systems to be installed at MOD warhead storage facilities. A comprehensive suite of integrated nuclear security equipment was jointly developed, including access control systems; a state-of-the-art central processing facility; entry and exit portals; guard force surveillance stands; hand-held metal detectors; and modern proximity card badging stations. In addition to the work of the SATC, a parallel effort focused on improving transportation security by supplying the MOD with 'super containers' that provided warheads with fire and ballistic protection during transportation. The United Kingdom (UK) worked in conjunction with MOD technical experts, delivering 150 super containers.[49] The French provided an additional 100 super containers that were specially designed to accommodate the world's largest warheads. The US delivered 150 super containers from 1995 to 1997.[50]

The other major warhead transportation security programme was the design and development of Emergency Response Modules used to respond to any accident or incident involving a nuclear warhead. The modules were designed to be transported by truck, train, or aircraft for quick response. The Russians explained that the roads and railways used for warhead transportation were in disrepair and could lead to accidents, especially since the pace of transportation operations had been accelerated to accommodate warheads returned from Belarus, Ukraine, and Kazakhstan. The modules were stocked with a mix of US-and Russian-supplied equipment used to cut into trucks and railcars if their cargo was inaccessible, towing capabilities in case a vehicle was forced off a bridge over a river, lifting equipment to recover a truck to an upright position, and sensors to detect potential warhead damage.[51]

Phase Three: Increased Technical Engagement and Reciprocal Visits (1998–2000)

When Maslin retired from his post in 1997, cooperative efforts came to a sudden halt as General Igor Valynkin replaced him as head of the 12th GUMO. That a singular change of a leading official disrupted nuclear security cooperation illustrates the importance of personal relationships in incrementally building trust. To help rebuild this trust, Maslin introduced Valynkin to Smith. Valynkin's initial meeting with Smith was an exception to his previous refusals to meet with Americans.[52] When Valynkin was appointed as the Chief of the 12th GUMO, CTR assistance was being rapidly delivered, but he was nonetheless suspicious of American cooperation. More importantly, Valynkin was motivated to establish himself and the 12th GUMO as the elite organization it had been under the Soviet Union. The 12th GUMO was not only responsible for the safety and security of the Soviet nuclear arsenal, the crown jewels of the USSR, but also it owned command responsibilities over all the storage sites. The Chief of the 12th GUMO reported directly to the President, much like the DoD Chairman of the Joint Chiefs, and until the CTR agreement was signed, its very existence and its personnel were classified as state secrets. Valynkin would not be beholden to American assistance or cooperation.

The CTR team learned from his staff that Valynkin was conducting command visits to each of his sites and was particularly interested in seeing how the quick fix fencing was being installed and operating. Valynkin's apparent displeasure with conditions at the storage sites led him to visit the SATC repeatedly. He reviewed every piece of equipment at the site, how it worked, and how it needed to be installed. His staff showed him that most of the equipment being installed at the site was built by Russian companies and that the Americans were motivated to find Russian vendors to meet nuclear security standards.[53] These visits must have impressed Valynkin because afterwards activity at the SATC to purchase and test components was accelerated.

Technical engagements to design, develop, and test a comprehensive nuclear security system to be installed at all Russian short-range, tactical, and strategic warhead storage sites required careful negotiations. Constant attention was paid to ensure that the discussions would not reveal any potential vulnerabilities.[54] The CTR team knew that it would not be able to visit the sites to conduct assessments of current or future vulnerabilities. Instead, the CTR team worked with the nuclear security experts in the US Office of the Secretary of Defense Office for Nuclear Matters and provided the MOD with a declassified and cleared version of the DoD Manual for Nuclear Security, 5210. Using the manual, the teams established an overall standard for 'probability of detection' for the system, that it must exceed a 95 per cent chance of warning of unauthorized activity. For purposes of designing the comprehensive security system, the teams also established a standard that the system must be capable of responding to an attack force of 10 heavily armed terrorists.[55]

CTR also provided an unclassified and cleared version of a vulnerability assessment model called ASSESS which was developed by Sandia National Laboratory in the US. The model enabled the 12th GUMO experts to conduct their own assessments of their sites as well as to conduct independent evaluations of the jointly developed comprehensive security system.[56] The CTR programme purchased three or more equipment items for each component of the system. The technical teams tested the items side by side for their performance, reliability, and sustainability. American and Russian technologies were tested, and the best performing equipment was selected to be part of the comprehensive system to be standardized for Russian storage sites. The final system was configured on-site, including fencing, perimeter sensors, access control systems, central processing equipment, video cameras, guard towers, and the doors, gates, and non-lethal smoke and sound systems for the bunkers.[57] Once the system was defined and installed at the SATC, the 12th GUMO assigned troops to conduct an operational test of the system that included an exercise with armed intruders attempting to break into the facility. The CTR team helped set up the operational exercise but left the site during the exercise to allow the Russians to gather their own data.[58]

Once the comprehensive security system was designed and defined, the programme reached one of its most important milestones. The next step should have been to procure, deliver, and install systems at all the warhead storage sites. However, installing and calibrating such systems for all the nuclear warhead sites across Russia would require more resources and technical expertise than the 12th GUMO was prepared to supply. Due to procurement laws associated with the validation of work completed under governmental contracts, the CTR team was not willing to procure suites of equipment without having the confidence that they would be installed, having now provided nearly all the equipment and support it could without a site access agreement. The CTR team offered to provide delivery and installation support, but just like the first quick fix deliveries, it would need to confirm that the work was accomplished properly to make contract payments. While the Russians acknowledged the need for installation support, they were not willing or able to approve site access. The prospects for this seemed unlikely, especially with Valynkin in charge.[59]

Concurrently, the CTR team learned that after Valynkin visited his warhead storage facilities he relieved 16 of the site commanders from duty. He was apparently not satisfied that the commanders were meeting nuclear security standards and were not correctly installing and utilizing the Russian equipment made available through CTR cooperation.[60] This was an extraordinary development in terms of nuclear warhead security; it would be equivalent of the Chairman of the Joint Chiefs firing half of all the base commanders in the US at once.[61] This demonstrated that Valynkin was motivated to enhance security at the sites and was willing to fire anyone who couldn't fulfil that need. However, he also determined that the 12th GUMO did not have the resources or personnel capable of implementing the comprehensive security systems designed at the SATC.

To resolve this issue, Valynkin started encouraging his staff to engage with the CTR technical teams to convince them to provide the security systems and installation services without requiring site access. The Russian side offered to take pictures, but that would not satisfy DoD contracting requirements, especially for a programme that would require US$20–40 million per site. Still, Valynkin would not accept any invitations to meet with the CTR technical teams or CTR leadership.[62] The CTR team knew that they needed to further cooperation with Valynkin, so the 12th GUMO staff was asked for suggestions. With support from Smith and Lajoie, the teams jointly developed a plan to invite Valynkin to visit a US nuclear weapon storage site. Earlier under Maslin, senior US leaders had visited a Russian storage site and went inside a bunker. Now the plan was to bring Valynkin to the US on a reciprocal visit.[63]

Inviting a Russian three-star general to a US nuclear weapons storage site required high-level support and extensive planning. The visit was supported by Under Secretary of Defense Ash Carter and Secretary of Defense William Cohen, who had dedicated the SATC during its opening ceremony in February 1998.[64] So, in January 2000, CTR Director, Brigadier General (ret.) Tom Kuenning invited Valynkin to F. E. Warren Air

Force Base.[65] The visit in July 2000 turned out to be a watershed for the programme. Valynkin met with US DoD three- and four-star generals during the visit as well as Under Secretary of Defense Jacques Gansler in the Pentagon. Thereafter, Valynkin regularly discussed the procurement and installation of the comprehensive security systems with the CTR teams. During these engagements, the CTR team consistently demonstrated to Valynkin that their only motivation in developing potential site visits was to ensure that the contract requirements were met and that the only way this could be done was through on-site visits. Valynkin continued to press the Americans to consider alternatives as he was frequently called away to meet with President Vladimir Putin on the state of nuclear warhead security.[66]

Phase Four: Agreement to Allow Site Access (2001–2003)

Few senior US leaders thought the Russians would ever allow Americans site access to the 12th GUMO's nuclear warhead storage sites when the programme began. After his visit to a US site, however, Valynkin began advocating for increased CTR cooperation, but he did not have the authority to approve site access. The technical teams continued to discuss requirements and explored potential procedures that would enable CTR to pay for construction and on-site installation. The 12th GUMO agreed to establish pre-arranged telephone calls twice a month between Valynkin's Deputy, General Major Frolov and the Defense Threat Reduction Agency (DTRA) CTR Program Manager, William Moon. Nearly every week, there were technical meetings in Moscow, and US contractors were on-site at the SATC daily. Both sides knew that the decision to grant access would not come easily or quickly, but it would not be for lack of effort on behalf of the technical teams.

Valynkin's ardent desire to bolster the 12th GUMO's capabilities and international reputation, his motivation to enhance nuclear warhead security, and his willingness to cooperate with the US CTR programme were critical during Russian deliberations over site access. After seven years of technical negotiations on equipment and installation requirements, Frolov broke the news to the CTR team in July 2002 that President Putin had approved Valynkin's request to allow negotiations on limited site access to the 12th GUMO's nuclear warhead storage sites.[67]

Once the decision to allow limited access was approved by President Putin, negotiations on site access progressed quickly during intensive negotiating sessions.[68] The CTR technical teams focused on explaining what they would need to see to make contract payments and carefully designed procedures tied to payment requirements.[69] The sides developed a methodology that would allow for three visits to each site: one prior to construction to observe the state of the facility and lay out the plan; the second mid-way through construction to verify the work was on schedule and to make an interim payment for construction, procurement, and delivery of equipment; and the third to verify all work was complete and equipment was correctly installed and operational, so that the final payments could be issued. With this understanding in place, site access procedures, including the number of personnel on the team (four), and the duration of the time on-site (four hours) were agreed.[70] Based on site diagrams that were built according to the equipment installation plans and layout, the sides were able to negotiate the locations of the four stops allowed for each site.[71]

Phase Five: Implementing a 'Defense-in-Depth' Comprehensive Security System (2003–2008)

In February 2003, the US DoD and Russian MOD signed three documents governing site access.[72] This included a Protocol on Protection of Sensitive Information that was designed so that the CTR teams would not release specific data on the sites gathered during visits; a Protocol on the Limited Access allowing for the visits; and Administrative Arrangements to the Protocol on the Limited Access that contained the specific rules for conducting the site visits.[73] With a site access agreement in place, the DoD set out to negotiate and sign contracts with US contractors, who then began procuring the systems and hiring the Russian subcontractors to perform the work on site.[74] Unlike the SATC work, the US contractors were not allowed on-site until the official site visits were conducted. The total effort required a major increase in funding and staff to manage the contracts.[75] The cost for procurement and installation at each site ranged from US$25–40 million, depending on the size, condition, and location of the site.[76]

Despite the dedicated efforts of the DoD and MOD personnel, construction and installation efforts required over a year to complete per site. Progress was limited by the production and installation capacity of the Russian subcontractors cleared to do the work on site. Valynkin pressured his staff and contractor teams to hasten the effort. Having invested his leadership and credibility to enable access to the sites, he was anxious to realize the results.[77] The first site completed with a full suite of security equipment was a Rail Transfer Point near Saratov. Rail transfer sites were prioritized due to their vulnerability during warhead operations. Since the site did not store warheads on a permanent basis and was one of the smaller sites, it was completed early.[78] But other large sites were not completed until 2005–2008.[79]

Valynkin also did his part to press the teams to implement the nuclear security enhancements as quickly as possible. He shared stories demonstrating the urgent need for the work to be done quickly. In June 2005, Valynkin called a press conference to report that armed terrorists had been caught conducting surveillance at two different nuclear warhead storage sites in 2002 and 2003.[80] He also told the CTR technical teams that warhead transport trucks were old and in disrepair, highlighting that one truck had stalled on a railway crossing and needed to be pushed off the tracks by the escort teams. The message was clear: the security enhancements were urgently required.[81] The urgency of the installation efforts attracted presidential attention. When presidents Bush and Putin met for a summit at Bratislava in February 2005, they agreed to accelerate efforts to enhance security at warhead storage sites.[82] They set an ambitious goal for both DoD and DOE teams working on upgrades that they needed to complete their installations by the end of 2008.[83]

The CTR teams conducted site visits as soon as arrangements could be made to see work that was ready to be verified for contract payments. Over the next three years, the security enhancements were completed on time, with the last site visit conducted by the CTR Director, Rear Admiral (ret.) John Byrd on New Year's Eve 2008.[84] Once the installation work was completed, there was considerable debate by the DTRA implementers and policy makers in the Pentagon on how much additional work and cooperation should continue. Given the pace of the installation work, all agreed that some training would be required to fully implement operations and sustain the security systems.

As training on operations and maintenance was carried out, it became clear to the technical teams that long-term sustainment of the nuclear security capabilities would require additional infrastructure. The technical teams and policy makers recognized the value of continued engagement with the 12th GUMO and were interested in exercising the DoD's rights to conduct audits and examinations of the sites to monitor that the security systems were being used as intended. So, with political and budgetary support in Congress, Russian and American experts worked together to identify and develop a set of projects constituting a comprehensive sustainment programme.[85]

The sustainment projects were developed and designed to transfer operational and maintenance responsibilities to the MOD. These included expanding the SATC to include a dormitory for trainees and expanded training rooms. US and Russian teams worked together to develop a centralized operations management system and establish a Unified Operations Center to manage procurement of spare parts and support maintenance, repair, and training programmes. Mobile repair vehicles were outfitted and procured, and on-site repair facilities were designed and built to conduct security system repairs at multiple sites.

Other projects expanded, including providing equipment for guard forces to conduct live fire and simulator training. The Automated Inventory Control and Management System (AICMS) was expanded to include additional facilities to be located at Strategic Rocket Forces (SRF) bases where the 12th GUMO maintained warhead storage bunkers. Additional cargo and guard force railcars were designed and procured as well, with the Russians agreeing to eliminate two railcars for every new one purchased by CTR. CTR even provided a metallurgy laboratory that the Russian teams used to assess the tensile strength of warhead handling equipment such as overhead cranes to ensure warhead operational safety. Previous efforts to support shipments of warheads to central storage and disassembly sites continued to encourage consolidation and reduction in the overall warhead inventory.[86]

During this time frame, one of the most long-lasting benefits of the nuclear security cooperation began to emerge. For each of these efforts, the US prime contractor established subcontracts with Russian firms that were newly established by former officers retiring from the 12th GUMO. In effect, the cooperation helped produce a Russian contracting community trained and equipped to support nuclear security. This contracting community went beyond support for operations, repair, and maintenance; it was also capable of supporting future nuclear security enhancements and would persist beyond the termination of US–Russian bilateral nuclear warhead security cooperation in 2013.

Conclusion

The CTR nuclear security cooperation between the US DoD and the Russian MOD's 12th GUMO demonstrates that nuclear security cooperation can provide additional value to overall stable relations. Although these activities were enabled by geopolitical conditions following the end of the Cold War and were subject to the decline of US–Russian bilateral relations, they serve as an outstanding model of how progress can be made by generating trust between agents if activities are conducted on a reciprocal basis. In the face of continuously emerging threats, the nuclear security community should continue to explore future opportunities for rivals to engage in nuclear security cooperation. This would enhance their mutual interests in the safety and security of nuclear warheads and military materials stockpiles.

Notes

1. Des Browne, Richard Lugar, and Sam Nunn, *Bridging the Military Nuclear Materials Gap* (Washington, DC: Nuclear Threat Initiative, 2015). https://media.nti.org/pdfs/NTI_report_2015_e_version.pdf
2. William M. Moon, 'The Story Behind U.S. Access to Russian Nuclear Warhead Storage Sites', 4 February (Washington, DC: Stimson Center, 2021). https://www.stimson.org/2021/the-story-behind-u-s-access-to-russian-nuclear-warhead-storage-sites
3. Kenneth N. Luongo and William E. Hoehn, III, Reform and Expansion of Cooperative Threat Reduction, *Arms Control Today* [online], June 2003. https://www.armscontrol.org/act/2003-06/features/reform-expansion-cooperative-threat-reduction
4. Tuomas Antero Forsberg, 'Taking Stock of the Trust Study in International Relations', in Hiski Haukkala, Carina van de Wetering, and Johanna Vuorelma, eds., *Trust in International Relations: Rationalist, Constructivist, and Psychological Approaches* (Milton Park: Taylor & Francis, 2018), 158–68.
5. Joseph P. Harahan, *With Courage and Persistence, Defense Threat Reduction Agency* (Washington, DC: US Department of Defense, 2014). https://www.gospanews.net/wp-content/uploads/2022/05/With-Courage-and-Persistence-CTR.pdf
6. Sharon K. Weiner, 'The Evolution of Cooperative Threat Reduction', *The Nonproliferation Review*, 16/2 (2009), 211–35.

7. Hanna Notte, 'The United States, Russia, and Syria's chemical weapons: a tale of cooperation and its unravelling', *The Nonproliferation Review*, 27/1–3 (2020), 206–9.
8. Harold Smith in interview with authors, April 2021.
9. Harahan, *With Courage and Persistence*, 117.
10. Ibid., 118.
11. Harold Smith in interview with authors, April 2021.
12. Ibid.
13. Ibid.
14. Evgeniy Maslin, Программа Совместного Уменьшения Угрозы И Интересы Национальной Безопасности России, *PIR Center Scientific Notes*, 13 (2000), 5.
15. Harold P. Smith, *American Access to Russian Weapons Storage* (Berkeley: University of California, 2003), 16. https://www.hsdl.org/?view&did=441987
16. Harold Smith in interview with authors, April 2021.
17. Ibid.
18. Ibid.
19. Ibid.
20. Ibid.
21. Ibid.
22. Ibid.
23. Ibid.
24. Harahan, *With Courage and Persistence*, 266.
25. Harold Smith in interview with authors, April 2021.
26. Ibid.
27. US General Accounting Office, 'Weapons of Mass Destruction, Reducing the Threat From the Former Soviet Union: An Update', Report to Congressional Requesters (1995), 25. https://www.gao.gov/products/nsiad-95-165
28. Harahan, *With Courage and Persistence*, 33.
29. Based on recollections of William Moon, the programme manager for these projects.
30. US General Accounting Office, 'Weapons of Mass Destruction, Reducing the Threat From the Former Soviet Union: An Update', 26.
31. US General Accounting Office, 'Weapons of Mass Destruction, Reducing the Threat From the Former Soviet Union', Report to Congressional Requesters (1994). https://www.gao.gov/products/nsiad-95-7
32. Office of the Coordinator of U.S. Assistance to the NIS, 'Government Assistance to and Cooperative Activities with the New Independent States of the Former Soviet Union', FY 1998 Annual Report, Part 2(1999), 157. https://1997-2001.state.gov/regions/nis/rpt_fy98_nis_assist2.pdf
33. Based on recollections of William Moon, the programme manager for these projects.
34. Office of the Coordinator of U.S. Assistance to the NIS, 'Government Assistance', 157–8.
35. Ibid.
36. Based on recollections of William Moon, the technical team lead for these discussions.
37. Based on recollections of William Moon, the programme manager for these projects.
38. US General Accounting Office, 'Weapons of Mass Destruction, Reducing the Threat', 26.
39. US General Accounting Office, 'Weapons of Mass Destruction: Status of the Cooperative Threat Reduction Program', Report to Congressional Requesters (1996), 11. https://www.gao.gov/products/nsiad-96-222
40. Harahan, *With Courage and Persistence*, 271.

41. US General Accounting Office, 'Weapons of Mass Destruction, Reducing the Threat', 157.
42. Harahan, *With Courage and Persistence*, 270.
43. See Harahan, *With Courage and Persistence*, 270–1, for a description of the SATC's development based on interviews conducted by Harahan with General Maslin in 2005 and William Moon in 2004.
44. Based on technical team meetings led by William Moon, June 1996 and July 1997.
45. See William Moon, 'Russian-American Cooperation on Personnel Reliability' (Washington, DC: Stimson Center, 2021). https://www.stimson.org/2021/russian-american-cooperation-on-personnel-reliability
46. Harahan, *With Courage and Persistence*, 270.
47. See William Moon, 'Inventory Management Cooperation: A Core Function of Nuclear Security' (Washington, DC: Stimson Center, 2021). https://www.stimson.org/2021/inventory-management-cooperation-a-core-function-of-nuclear-security
48. Ibid.
49. Office of the Coordinator of U.S. Assistance to the NIS, 'Government Assistance to and Cooperative Activities with the New Independent States of the Former Soviet Union', 157.
50. Harahan, *With Courage and Persistence*, 265.
51. William Moon, 'Russian Ministry of Defense Nuclear Security Exercise 2002'(Washington, DC: Stimson Center, 2020). https://www.stimson.org/2020/2002-russian-ministry-of-defense-nuclear-security-exercise
52. Harold Smith in interview with authors, April 2021.
53. Insights into Valynkin's activities were shared by Frolov and his staff during technical meetings with William Moon and his team in Russia in April 1998, May 1999, July 1999, and October 1999.
54. Harahan, *With Courage and Persistence*, 273.
55. Based on technical team meetings led by William Moon in Russia, in April 1998, May 1999, July 1999, and October 1999.
56. Harahan, *With Courage and Persistence*, 272.
57. Ibid., 273.
58. Based on recollections of William Moon, the technical team lead for these discussions.
59. See Harahan, *With Courage and Persistence*, 273–4, for a description of the impasse between the DoD's requirements to verify work and the MOD's position that access to actual storage sites would be prohibited.
60. See Moon, 'The Story Behind U.S. Access'; additional details based on 12 technical meetings conducted by William Moon in Russia, 1998–2000.
61. William Moon, '2000 Russian Ministry of Defense Visit to Francis E. Warren Air Force Base' (Washington, DC: Stimson Center, 2020). https://www.stimson.org/2020/2000-russian-ministry-of-defense-visit-to-francis-e-warren-air-force-base
62. Harahan, *With Courage and Persistence*, 273–4.
63. The plan to invite Valynkin to the US was discussed during bi-monthly phone calls between William Moon and Frolov and referenced in the author's notes from those calls.
64. Harahan, *With Courage and Persistence*, 271.
65. William Moon, '2000 Russian Ministry of Defense Visit'.
66. Based on personal notes of William Moon, who escorted Valynkin to all meetings during his visit.
67. Moon, 'The Story Behind U.S. Access'.
68. Harahan, *With Courage and Persistence*, 275.

69. Based on technical discussions on site access conducted 23–28 September 2001 with William Moon in attendance.
70. Provisions were included in 'Protocol on the Limited Access by United States Representatives to Ministry of Defense Work Site Exhibitions of Physical Protection Systems Enhancements', February 2003.
71. Moon, 'The Story Behind U.S. Access'.
72. Amy F. Woolf, 'Nonproliferation and Threat Reduction Assistance: U.S. Programs in the Former Soviet Union', CRS Report for Congress, RL31957 (2004). https://www.everycrsreport.com/files/20040617_RL31957_32e3c0b2937893c015c5ea3f0fbfcc7c1919de26.pdf
73. US House Committee on Armed Services, 'U.S.–Russian Cooperative Threat Reduction and Non-Proliferation Programs' (2003), 6. https://www.hsdl.org/?view&did=35337
74. Woolf, 'Nonproliferation and Threat Reduction Assistance'. The sides signed two agreements: a protocol on Limited Access and Administrative Arrangements to the Protocol, in February 2003.
75. Based on personal experience of William Moon, who led the DTRA projects.
76. See Harahan, *With Courage and Persistence*, 275. The initial total cost estimate for identified sites was US$573.4 million.
77. Based on technical discussions and quarterly reviews led by William Moon with US and Russian contracting teams, 2003–2008.
78. William Moon led the first site visit to the Saratov RTP in July 2003, and he oversaw completion of the work.
79. George W. Bush, 'Statement by President Bush on 15th Anniversary of the Cooperative Threat Reduction Program', 29 August (Washington, DC: The White House, 2007). https://georgewbush-whitehouse.archives.gov/news/releases/2007/08/20070829.html; this statement references the Bratislava Agreement to complete site security upgrades at Russian nuclear warhead storage sites; also see Harahan, *With Courage and Persistence*, 317.
80. Andrew Osborn, 'Chechen militants have tried to steal nuclear weapons twice, says Russian security chief', *The Independent* [online], 24 June 2005. https://www.independent.co.uk/environment/chechen-militants-have-tried-to-steal-nuclear-weapons-twice-says-russian-security-chief-294371.html
81. Remarks to the DoD delegation by Valynkin during Executive Review meetings in Moscow, February 2004.
82. See n. 79.
83. Woolf, 'Nonproliferation and Threat Reduction Assistance'; George W. Bush and Vladimir Putin, 'Joint Statement by President Bush and President Putin on Nuclear Security Cooperation' (Washington, DC: The White House, 2005). https://georgewbush-whitehouse.archives.gov/news/releases/2005/02/20050224-8.html
84. Based on recollections of William Moon, the programme manager for these projects.
85. Based on numerous meetings between William Moon, DTRA, and Jim Reid, DoD CTR. This resulted in the addition of a 'Sustainment Funding' line to the CTR Global Nuclear Security budget line.
86. See Harahan, *With Courage and Persistence*, 356, for a description of the sustainment programme.

Suggested Reading

Bidgood, Sarah and Potter, William C., eds., 'End of an Era: The United States, Russia, and Nuclear Nonproliferation', *James Martin Center for Nonproliferation Studies* [online], 23 August 2021. https://nonproliferation.org/end-of-an-era-the-united-states-russia-and-nuclear-nonproliferation

Harahan, Joseph P., (2014), *With Courage and Persistence*. Defense Threat Reduction Agency (Washington, DC: US Department of Defense). https://www.gospanews.net/wp-content/uploads/2022/05/With-Courage-and-Persistence-CTR.pdf

Moon, William, (2021), 'What It's Like to Visit a Russian Nuclear Warhead Storage Site' (Washington, DC: Stimson Center). https://www.stimson.org/2021/what-its-like-to-visit-a-russian-nuclear-warhead-storage-site

Walker, Paul F., (2016), 'Cooperative Threat Reduction in the Former Soviet States: Legislative History, Implementation, and Lessons Learned', *The Nonproliferation Review*, 23/1–2, 115–29.

Weiner, Sharon K., (2009), 'The Evolution of Cooperative Threat Reduction', *The Nonproliferation Review*, 16/2, 211–35.

CHAPTER 13

THE FOUR NUCLEAR SECURITY SUMMITS

AMANDEEP GILL

Introduction

The Nuclear Security Summits were a timely and tailored response to the danger of non-state actors, for example, terrorist groups, accessing nuclear materials, facilities, and information to cause death and destruction. The summits forced a change of pace on national governments and international organizations through a combination of factors, in particular, the direct involvement of leaders, concerted international efforts, and a 'whole of government' approach to nuclear security. A small circle of countries—chiefly the Group of Eight (G8; later the Group of Seven)—dealing with the protection of nuclear material, facilities, and expertise against unauthorized access was expanded to more than 50 countries, many of which had been traditionally suspicious of nuclear security efforts. The reasons for this reluctance to engage on nuclear security were varied. Some nuclear weapon states saw it as an attempt to prise open their secrets or put pressure on their policies and programmes; non-nuclear weapon states perceived it as pretext to limit their choices on civil nuclear energy.

The Nuclear Security Summits were able to address these concerns and build more diverse support for nuclear security. Nuclear security funding and attention subsequently gravitated towards the International Atomic Energy Agency (IAEA), which began to coordinate its work more with the United Nations (UN), the International Criminal Police Organization (Interpol), plurilateral forums such as the Global Initiative to Combat Nuclear Terrorism (GICNT), industry, and a network of national 'centres of excellence' around the world.

Despite this progress, a made-to-measure process focused exclusively on nuclear security was not a foregone conclusion. For one, nuclear orthodoxy demands that primary attention be placed on nuclear disarmament, non-proliferation, and civilian uses of nuclear energy—the so-called three pillars of the Treaty on the Non-Proliferation of Nuclear Weapons (NPT).[1] For the traditionalists, nuclear security could at best be a distraction, or at worst a deliberate ploy to divert attention from the more pressing work on the three pillars. Further, nuclear security already had a home of sorts in the IAEA where it featured in an annual resolution of the General Conference and where a small team within the Secretariat provided advisory services to member states. Progress might have been slow but almost every country that mattered was present in the room.

The principal reason a Nuclear Security Summit was convened in Washington, DC in April 2010 and a new multilateral process took shape outside of traditional nuclear forums was United States (US) leadership in the shape of then-president Barack Obama. As a first-time US senator, Obama had accompanied senior Senator Richard Lugar in 2005 to nuclear and other weapons of mass destruction (WMD) sites in Russia, Ukraine, and Azerbaijan.[2] This rekindled Obama's interest in nuclear security and nuclear arms control, which he had been curious about as a senior student at Columbia University.[3] He later joined legislative efforts pioneered by the Nunn–Lugar Cooperative Threat Reduction (CTR) approach to prevent nuclear terrorism through US-led global cooperation.[4] Nuclear security featured in Obama's presidential campaign speeches and became a part of his election platform. In his speech to the Chicago Council of Global Affairs on 23 April 2007, he called nuclear terrorism 'a threat that rises above all others in urgency' and said that as president he would lead 'a global effort to secure all nuclear weapons and material at vulnerable sites within four years—the most effective way to prevent terrorists from acquiring a bomb'.[5]

The 2007 Chicago speech made it clear that President Obama wanted to lock down vulnerable material susceptible to be used for nuclear weapons by terrorists or others within a specific time frame. Moreover, Obama wanted this to happen across the globe and not just in the former Soviet Union territory, which had been on the policy radar of his nuclear security mentors in the US Senate. Notably, he did not express interest in radioactive material that could be used in dirty bombs rigged with conventional explosives. Terrorism, including WMD terrorism, had of course been a top priority for Obama's predecessor George W. Bush, whose time in office was marked by the attacks of 11 September 2001 ('9/11'). Where Obama differed was with his emphasis on turning ad hoc efforts such as the Proliferation Security initiative and the GICNT from his predecessor's term into durable processes and institutions.[6]

The idea of a Nuclear Security Summit matured in the year prior to Obama's election; the campaign's non-proliferation team led by Ivo Daalder proposed a nuclear security speech that included a reference to a summit of the 'P5 and other interested countries'.[7] That speech was delivered on 5 April 2009 in Prague, capital of the Czech Republic. In the speech, Obama announced a new international effort to secure all vulnerable nuclear material around the world within four years and offered to host a 'Global Summit on Nuclear Security' within the next year to start the effort.[8] A one-year dash leading up to one of the largest gathering of leaders on US soil started with this announcement. All summits need to be scaled with the help of a team. Borrowing from the practice of designated preparatory teams employed in the G8 summits, senior officials with access to leaders ('Sherpas') were entrusted with the preparations. They were assisted by technical teams led by 'Sous-Sherpas'.

The Road to the First Nuclear Security Summit

To develop the summit, a team was put together at the White House in July 2009. The key figures were Gary Samore, Laura Holgate, and Joyce Connery. Samore, who joined as the White House Coordinator on Arms Control and Weapons of Mass Destruction in 2009, served as US Sherpa for the 2010 and 2012 Nuclear Security Summits, after which Holgate took charge as the US Sherpa for the 2016 summit. Both had a background in non-proliferation; for instance, Samore played a role at the US State Department in negotiating the 1994 nuclear agreement with the Democratic People's Republic of Korea (DPRK), while Holgate had worked on the CTR programme at the Department of Defences in 1995–1998. They also shared common intellectual roots in Harvard University's Belfer Center, which had done pioneering work on nuclear terrorism.[9]

The US State Department and the Department of Energy (DOE) were partners in this effort led from the White House. An initial list of 25 or so countries eventually grew to 47 participants by the time the summit took place in April 2010; diverse representation and geopolitical considerations were factors in this choice as were practical considerations of potential contributions to nuclear security.[10] As well as the P5 (China, France, Russia, the UK, and the US), nuclear-armed states not party to the NPT (India, Israel, and Pakistan) were invited, as were the UN and the IAEA. Most countries that held vulnerable nuclear material and facilities or had plans to use nuclear material or build nuclear facilities were invited with a few exceptions, most notably Iran and the DPRK. Iran's exclusion can be debated. However, a separate policy track, the EU3 + 3 talks on the Iranian nuclear programme, was in place and without a political breakthrough, eventually the Joint Plan of Action agreed with Iran in July 2015, an invitation to the summit might have sent a different signal. In fact, President Obama ended up using the 2010 Summit to increase pressure on Iran.[11]

The US sent out an invitation letter in September 2009 from President Obama to 37 countries, the European Union (EU), UN Secretary-General Ban Ki-moon and IAEA Director General Yukiya Amano. The accompanying information note clarified that the Nuclear Security Summit to be held in April 2010 would have as its goal 'a common understanding of the threat posed by nuclear terrorism and to recognize that nuclear material, whether in civilian or military use, should not be vulnerable to that threat'.[12] Participants should therefore commit, as a 'responsible steward of nuclear material, to reduce the stocks of nuclear material where possible and to protect the remainder of that material to the highest standards'.[13] This was the 'theory of change' to realize Obama's call for an international effort to lock down vulnerable nuclear materials within four years.

Obama's invitation letter also requested participating countries and organizations to designate a Sherpa for the summit. Very few invitees had the time to set up such a team by the time a first organizing meeting was held in Vienna on 15 September 2009 with US Sherpa Samore in the Chair. But by the time the Sous-Sherpas met in Washington on 4–5 November 2009 at the US State Department under Holgate's leadership, a community of arms control experts, nuclear policy specialists, and others had started to come together in a two-tiered preparatory process. The Vienna meeting and the bilateral consultations that the US had begun to pursue with key protagonists provided some feedback on the initial slate of ideas. More concrete suggestions came in response to a draft 'Possible Summit Outcomes' paper circulated by the US ahead of the Washington meeting. This helped the hosts prepare a draft 'Work Plan Elements' document (separate from a draft summit 'Communiqué', which the Sherpas were to consider at their meeting in Tokyo in December 2009).

A key substantive issue was the extent to which nuclear security efforts should rely on existing mechanisms and institutions such as the IAEA or the G8's Global Partnership. Samore had clarified in Vienna that the US was not proposing a new mechanism, but rather would seek to reinforce existing ones.[14] This implied, for example at the normative level, more countries committing to the Amended Convention on the Physical Protection of Nuclear Material (A/CPPNM), the main international legal instrument in the area of nuclear security, so that it could enter into force,[15] or joining the International Convention for the Suppression of Acts of Nuclear Terrorism (ICSANT). It also implied clarity on the meaning of existing obligations on nuclear security such as the requirement in UN Security Council Resolution 1540 that called for 'appropriate effective' physical protection measures.[16] At an institutional level, it meant developing a new shared understanding of the role of the IAEA in nuclear security and the IAEA's relationship with other organizations.

Another important issue was the standards to use for benchmarking nuclear security measures if such measures were to remain a national responsibility. There were pre-existing 'Nuclear Security Guidelines' in the form of the Recommendations on Physical Protection of Nuclear Material and Facilities (INFCIRC/225), or the more narrowly prescribed Plutonium Management Guidelines (INFCIRC/549) adopted by nine countries (Belgium, China, France, Germany, Japan, Russian Federation, Switzerland, the UK, and the US) voluntarily whereby they report on their civilian plutonium holdings and plutonium management policies to the IAEA. However, standards with regard to the design and construction of facilities as well as for personnel reliability and cyber security were weak or absent. Implementation too was far from universal. A related question was the manner of assessing adherence to and implementation of existing standards, and whether states should be obliged to accept existing mechanisms such as the IAEA's International Physical Protection Advisory Service (IPPAS) missions.

A third issue was the materials themselves. Should the summit focus only on nuclear material directly usable in nuclear weapons, highly enriched uranium (HEU), and separated plutonium, or as argued by a few participants, should it extend the scope to radioactive material for dirty bombs? From the perspective of Germany, the EU, and the

Republic of Korea, leaving out radiological sources such as caesium-137 used in medical treatment meant that a crucial part of the threat spectrum was being ignored. Another sensitive topic was whether some of the existing materials such as HEU used in research reactors, including for the production of medical isotopes, might be minimized and/or returned to countries that had originally provided it. Highlighting the separation of plutonium through reprocessing as a potential source of nuclear insecurity made countries, such as France, India, and Japan, wedded to closed nuclear fuel cycle approaches uncomfortable.[17]

For many participants, a key question was whether material in civilian use should be the exclusive focus of the summit, or if the approach should be 'comprehensive' and cover military materials in the possession of a few states through new international measures. While it was true that civilian material was more dispersed around the world and often not as highly protected as it ought to be, military nuclear material vastly outweighed such material in quantity, and even small volumes could be highly dangerous in the wrong hands. China, France, India, Russia, and Pakistan argued against any references to nuclear weapons or military material as outside the scope of the summit; others would not see the summit as credible without somehow covering the 85 per cent of the material in military possession.[18] An additional complication was how to refer to the nuclear-armedstates, with Egypt in particular sensitive to any international recognition of the status of Israel's nuclear programme.

There were two other important political issues that permeated the technical discussions. First, was the status of nuclear security vis-à-vis the so-called three pillars of the NPT, in particular, the pillars of nuclear disarmament and peaceful uses of nuclear energy. The Information paper of September 2009 stated that the summit was not directly related to the Review Conference of the NPT scheduled for 2010. However, it also stated, 'pursuit of nuclear energy requires diligence and good stewardship of that (nuclear) material'.[19] This seemed to undercut the 'right' to civil nuclear energy that some countries such as Egypt saw as part of the NPT bargain. Additionally, Egypt wanted an explicit reference to nuclear disarmament in the summit's Action Plan.

Second, the idea of a briefing on the threat posed by nuclear terrorism, vital to Obama's goal of a common understanding on nuclear security, looked like finger pointing to some countries. Pakistan was particularly sensitive about the issue given the history of proliferation by the nuclear scientist A. Q. Khan and the interest in WMD technologies and experts shown by al-Qaeda. Australia, which had volunteered to give a threat briefing at the summit, made an initial presentation at the Sherpa meeting in Tokyo. However, Pakistan questioned the right of a participating state to stand in judgement over others and called for a neutral organization such as the UN or the IAEA to make a presentation on the threat using only recognized and verified information sources. The idea of the threat briefing also elicited criticism from China, Russia, and South Africa.[20]

The Tokyo meetings of the Sherpas and Sous-Sherpas on 2–3 December 2009 were crucial to delineating the preceding fault lines and assessing potential compromises. The next round of meetings at The Hague on 9–11 February 2010 offered the opportunity to test some of these compromises. A draft Communiqué text was circulated by the hosts on 30 January prior to that meeting. Given the importance of defusing the NPT overlap issue, the US also sent out a non-paper distinguishing the summit process from the NPT Review Conference being held within weeks of the summit. The meeting at The Hague and bilateral conversations thereafter on a few pending issues, for instance, the reference to military material, helped the hosts finalize the Communiqué and the Work Plan by March 2010 in time for adoption at the summit.[21]

Reaching Agreement about the Summit Process

The final compromise on the status of nuclear security in the hierarchy of nuclear issues was a Communiqué reference to the three pillars as 'shared goals', with nuclear security described as a 'shared objective' additional to the goals.[22] Explicit references to nuclear disarmament were dropped, thus preserving the narrow focus that Obama's team had set for the summit. Participants welcomed the move and joined Obama's call to secure all vulnerable nuclear material within four years.

The preamble to the Communiqué termed nuclear terrorism as one of the most challenging threats to international security. The attempt to describe the threat of nuclear terrorism through a regular report or briefing was dropped.[23] In its operative part, the Communiqué succeeded in defining broadly what the global nuclear security architecture of existing or emerging nuclear security instruments and mechanisms could look like.[24] The A/CPPNM and ICSANT were listed as among the essential (legally binding) elements of this architecture. The Work Plan added detailed references to a third legally binding element, UN Security Council Resolution 1540.[25] While the IAEA's 'essential role' was affirmed, the roles and responsibilities of other organizations such as the UN, GICNT, and the G8 Global Partnership were recognized within their respective mandates and memberships. The nature of the Work Plan, a matter of considerable debate in the preparatory meetings, was clarified as 'guidance' to 'national and international action including through cooperation within the context of relevant international fora'.[26]

On standards, agreement was reached in the Work Plan on treating the latest version of the IAEA recommendations on physical protection (INFCIRC/225) as the 'gold standard' for implementing the nuclear security aspects of UNSC Resolution 1540. Much to Germany's disappointment there was no specific reference to the IAEA Code of Conduct on the Safety and Security of Radioactive Sources as a standard for the security of radiological material.[27] A *de minimus* design basis threat (DBT) standard to ward off insider and outsider threats to nuclear facilities was brought into the Work Plan through a compromise that recognized the primacy of national standards, with the IAEA's Implementing Guide on the subject taken into account.[28] As for benchmarking, the value of IAEA's advisory and support services including the voluntary IPPAS peer review missions was recognized.

With regard to materials, the US and Germany worked out a general reference to the security of radioactive material in paragraph 12 of the Communiqué without explicitly making those materials part of the scope (although this would change at the 2012 Summit in Seoul). Further, France, India, Japan, and the US agreed a compromise text on securing and accounting separated plutonium, taking into account the fact that all forms of plutonium were not usable in a nuclear explosive device. This was understandable from a technological perspective, as trying to use plutonium for nuclear weapons would be much more difficult than HEU for a non-state actor. However, political perceptions were another story. Read along with the appeal for HEU minimization, this meant that countries with HEU stocks or continued HEU use in civilian programmes (such as South Africa) felt unfairly singled out.

As for the sensitive matter of military material, the Communiqué in its very first paragraph reaffirmed the fundamental responsibility of states, consistent with their respective international obligations, to maintain effective security of all nuclear materials, including nuclear materials used in nuclear weapons. This avoided singling out nuclear-armed states or opening a debate about NPT nuclear weapon states and non-NPT states. An important set of substantive outcomes focused on the day-to-day work of nuclear security: capacity building, fostering a nuclear security culture in the industry, consolidation of sites where nuclear material is stored, cooperation to prevent illicit nuclear-trafficking, exchange of information on nuclear forensics and law enforcement, and new technologies.

The Washington, DC Nuclear Security Summit, 2010

The summit itself was an outcome of substantive significance. There were 38 heads of state or government in attendance, and the hosts basked in what some called the 'nuclear spring'.[29] A pre-summit dinner on 12 April 2010 was the occasion for a discussion among the leaders present on the nuclear threat. What could not be discussed by the Sherpas was broached by French President Nicolas Sarkozy when he called for an international mechanism to try officials or leaders who assist terrorists with nuclear material and technology along the lines of the ad hoc tribunals created to try pirates off the coast of Somalia.[30]

On the margins of the formal meetings, the summit was an occasion for participants to announce national commitments or 'house gifts' as they came to be known.[31] For instance, Canada, Chile, Kazakhstan, Mexico, Russia, Ukraine, and the US committed to removing and eliminating some of their nuclear material. Chile, Kazakhstan, Mexico, and Vietnam committed to converting their HEU-fuelled research reactors to alternative fuels. Russia announced the closure of its last plutonium production reactor. A few countries committed to hosting the IAEA's IPPAS nuclear security peer review missions. Even though the summit was not a pledging conference, several countries announced funding support to the IAEA's nuclear security-related activities. China, France, Italy, India, Japan, Kazakhstan, the US, and the UK announced the creation of national centres of excellence for nuclear security training and capacity-development. Armenia, Egypt, and Malaysia announced new export controls and nuclear security-related regulations. Argentina, Italy, and the United Arab Emirates (UAE) announced measures to improve infrastructure at ports to prevent nuclear trafficking. Argentina, Philippines, Thailand, and Vietnam committed to joining the GICNT.[32]

The Seoul Nuclear Security Summit, 2012

Consolidating and implementing the commitments made at the first summit was the first order of business for the second Nuclear Security Summit. Exploration of new themes to be added to the Washington Work Plan and outreach to those not present at the first summit were two other priorities identified by the US during summer 2010. How to assess progress on translating the Washington political commitments into concrete actions became another issue of debate as soon as the Sous-Sherpas gathered again in Buenos Aires in November 2010. A food-for-thought paper compiled by the US in September 2010 ahead of the meeting proposed a 'tracking mechanism' based on national reporting on the actions taken.[33] Standardized reports could, for instance, be uploaded to a secure website and compiled in a database that placed national actions into categories derived from the Washington documents. Alternatively, the reports could be compiled by the host of a summit and circulated along with a chairman's summary.

In a classic display of the exploitation versus exploration dynamic in learning,[34] the US proposed a few additional areas of focus that promised a high return on investment. These included establishing counter-nuclear smuggling teams, sharing actionable information through an independent international entity on illicit trafficking, agreeing a global norm on HEU minimization, addressing insider threats through concrete measures, strengthening information security, and addressing nuclear material that was no longer needed.[35]

The host, the Republic of Korea, and a few other summit participants supported the US approach. Others, in particular Egypt, Russia, and Pakistan, however, resisted the attempt to turn the Washington political agreement into trackable sub-components on which they would have to report regularly as if these were legally binding commitments. The issue was not resolved until the New Delhi meeting of the Sherpas in January 2012. Once an overly prescriptive approach was dropped by the co-chairs (the Republic of Korea and the US), a simple format for voluntary 'National Progress Reports' proposed by Algeria became the basis for a compromise. As per this format at the Seoul Summit,

46 countries reported voluntarily, thus institutionalizing a soft implementation compliance mechanism.[36]

Half-way through the four-year timeline set by President Obama, the HEU minimization issue assumed critical importance. Attention turned to HEU targets for radioisotope production in addition to the HEU fuel for research reactors. This had global commercial implications since even Canada, Belgium, France, and the Netherlands used such targets to produce medical isotopes such as molybdenum-99. A peculiar problem was the sovereignty-related concerns of South Africa, which was loath to give up its HEU stocks even though it had given up its apartheid-era nuclear weapons.[37] Progress was made nonetheless with 18 countries converting their research reactors from HEU to low-enriched uranium (LEU) by the time of the summit, while 530 kilograms of HEU sufficient for dozens of nuclear weapons was removed from eight countries, including a total clean-out from Ukraine and Mexico.

The summit's Communiqué added calls *inter alia* for transparency on HEU policies and civilian HEU stocks to the Washington commitments and urged countries to take specific action to minimize HEU use by 2013.[38] An attempt spearheaded by France and with US backing to craft new language on guidelines for HEU management was abandoned in the face of opposition from South Africa and others. South Africa applied counter pressure of its own by proposing language on nuclear disarmament for the preamble of the Communiqué at the New Delhi meeting; the impasse could be broken only at Seoul.[39]

In the midst of the preparations for Seoul, the nuclear accident caused by a tsunami at the Fukushima Daiichi nuclear power plant in Japan in March 2011 led to renewed public interest in nuclear safety and the place of nuclear energy in modern lives. There were synergies between nuclear security (keeping bad actors out) and nuclear safety (keeping bad stuff in), but it was becoming increasingly obvious that the communities of practice and the regulatory frameworks often worked in silos. The Seoul Summit debated the interface, with the Communiqué highlighting capacity building for emergency response, the proper management of spent fuel and radioactive waste, and an integrated approach to facility design and operations that minimized both types of risks.[40]

A key advance was security of radiological sources, which the US had not foreseen as part of the scope of the summit process. Paragraph 6 of the summit's Communiqué brought in a specific reference to the need to reflect the IAEA Code of Conduct and other IAEA guidance on the safety and security of radioactive sources in national practice, and to establish national registers of such sources. In a sign of the maturation of the field of practice of nuclear security, the Seoul Summit outcomes expanded the Washington references in a number of areas of practical significance. These included nuclear forensics, information and cyber security, sharing of information including through the IAEA's Illicit Trafficking Database (later renamed the Incident and Trafficking Database; ITDB), and working with Interpol and the World Customs Organization to prevent illicit trafficking. Interpol joined the summit for the first time, as did six more countries to take the total of participants to 53.

A key process innovation at Seoul was the 'gift baskets'. This was a step up from what Samore had termed 'house gifts' brought by participants to the 2010 Washington

Summit. Small groups of like-minded enthusiasts came together to present joint commitments at Seoul on areas such as 'Transport Security', 'Nuclear Training Centers', 'Nuclear Information Security', and 'Security of Radioactive Sources'.[41] This allowed progress at a faster pace in smaller groups and brought more dynamism to the consensual track led by the Sherpas. However, despite an attempt to interject more interactivity, written statements and pre-selected discussants remained the bedrock of the summit discussions. The next summit would break fresh ground with a more engaging format.

The Hague Nuclear Security Summit, 2014

The 2014 summit at The Hague was initially thought to be the last of a series. The Dutch hosts approached the preparations strategically in terms of both substance and process. At the very first meeting of the Sherpas in Istanbul in November 2012, they presented a list of deliverables and ideas for an innovative summit format. For instance, on vulnerable material (HEU and plutonium), they suggested that participants go beyond the mantra of material minimization to commit specifically to less production, fewer sites, fewer stocks, and fewer countries with the material. Military material of various types was sought to be brought into the ambit of the summit commitments as part of a comprehensive approach to nuclear security. A second suggestion was greater reliance on IAEA reviews and services to align nuclear security practices across the globe. Another powerful idea, which continued to be discussed even beyond the summit at The Hague, was providing 'assurances' to domestic and international audiences on the robustness of nuclear security measures through increased transparency.

Other ideas that fared less well and dropped off by the time of the summit included a greater role for industry in nuclear security governance, bringing international legal instruments such as the Amended CPPNM into effect by the time of the summit, and turning existing guidance on security of radiological sources into binding obligations. Initially, the Dutch also proposed discussing the future of the summit process in order to maintain a high-level focus on the nuclear security mission beyond 2014. However, when it became clear that the US would host another summit, the idea was put on the back burner.[42] It also became clear that regardless of how the summit process ended, the IAEA, which would host its first Nuclear Security Conference in 2013 at the ministerial level, would play the central role in the future nuclear security architecture.

The Dutch Prime Minister Mark Rutte stated that he did not want a 'boring' summit.[43] Leaders played a 'game' by reacting to a hypothetical nuclear security threat presented to them through short videos. The choices they had to submit on computer tablets made them think about having plans in place and rehearsing them, securing nuclear material, communicating with the public to avoid panic, and building habits of international cooperation in advance. What seemed obvious was not necessarily so when gaming a crisis response. This experience inspired a bilateral nuclear security exercise between Argentina and Chile in August 2014.[44] Another example of innovation was a leaders-only retreat within the summit where the participants broached some sensitive issues such as the security of nuclear material and the need to stop the production of fissile material for nuclear weapons through a new treaty.

Just the new titles that the Dutch proposed for their first Communiqué draft underlines their determination to push the boundaries of the consensus reached at Washington and Seoul. When the Sherpas met at Ottawa in October 2013, the text before them had sub-sections titled 'Nuclear security as part of a broader agenda', 'Building international confidence', 'Strengthened international nuclear security architecture', and 'Fissile material'. Those who thought that the Washington formulation of 'all nuclear material' was comprehensive enough and that there was nothing to discuss at the summit (since this was a matter of national responsibility), faced calls to give an account of the security of their arsenals and materials for military purposes. The US and the Netherlands encouraged this reopening of the Washington consensus but had to step back when the proponents of a comprehensive approach—including Brazil, Egypt, Mexico, South Africa, and Switzerland—pushed for explicit references to nuclear disarmament. Eventually, the Washington compromise survived, and the supporters of a more comprehensive approach had to content themselves with a gift basket.[45]

The gift basket also became a vehicle to capture the ambition on the issue of international assurances, which was essentially the familiar issue of how to track progress without having a treaty-like compliance regime. Proponents could argue that confidentiality and national responsibility had not come in the way of transparency in aviation and shipping. There were existing commitments, such as UNSC Resolution 1540 or the CPPNM, to which these assurances could be linked. The opponents did not budge from their view that nuclear security was a national responsibility. Eventually, 36 countries signed up to their own version of an implementation matrix based on existing IAEA guidance and terminology and presented it as the 'excellence paper' at the summit.[46] The summit barely avoided the geopolitical fallout of the crisis between Russia and the West over Ukraine. On the eve of the summit, the G7 withdrew from the G8, effectively expelling Russia, while Russia downgraded its participation to the level of foreign minister at the summit. While the hope in 2010 had been that Russia might actually host a Nuclear Security Summit, now the participation itself of a co-constructor of the idea of nuclear security was coming under doubt.

The Washington, DC Nuclear Security Summit, 2016

Russia withdrew from the preparations on the eve of the first meeting of the Sherpas in Washington in October 2014. Russia explained its decision on three grounds, through a note circulated to the participants. First, it claimed that the tasks of the summit process had been fulfilled and there would be no added value from another summit. Second, it disagreed with the US approach of convening informal working groups to decide post-2016 follow-up action to be pursued in the UN, the IAEA, Interpol, GICNT, and the Global Partnership; according to Russia, this amounted to interference in the working of international organizations that had their own mandates and memberships. Finally, it opposed the privileged rights for the summit hosts (the US, the Republic of Korea, and the Netherlands) in steering the process. Despite its withdrawal from the summit, Russia would continue to pursue what it called 'Physical Nuclear Security' (PNS) through the IAEA, which would hold its second ministerial conference in 2016. In fact, the US Sherpa kept her Russian counterpart copied into the summit emails until 2015, when Russia formally announced its decision not to participate.[47]

The unease about the process question was shared by other participants. The US wanted to do away with the negotiations on a Communiqué and instead have informal working groups (IWGs), chaired by pre-selected participants, which would come up with specific Work Plans for the five organizations[48] most relevant to nuclear security. This fitted the two-part vision that the US Sherpa described in her first communication to her counterparts: continue to strengthen national implementation through 'house gifts' and 'gift baskets', and build an enduring global nuclear security architecture comprised of 'binding legal commitments, multilateral institutions, voluntary collectives, and national bodies responsible for executing nuclear security'.[49] The concern of the doubters was that common political commitment channelled thus far through the summit Communiqués and a unitary Sherpa process would dissipate.

The Action Plans were negotiated in a series of meetings and brought to the Sherpas' meetings for approval. The Sherpas also negotiated a short nine-paragraph Communiqué, which allowed the leaders to look back on progress since 2010, reaffirm their commitments, and pledge to make nuclear security an enduring priority.[50] The five Action Plans were linked to the Communiqué through a paragraph that described them as reflecting the political will of participating states, to be implemented in support of the five concerned organizations and initiatives on a voluntary basis consistent with national laws and respective international obligations.

The needle on substance did not move much between 2014 and 2016. A fresh attempt was made to persuade nuclear weapons possessors to agree to share information on nuclear security measures for their arsenals and military materials. Within the Sherpas meetings, this was led by Switzerland, Australia, Chile, Sweden, and the Netherlands, while on the margins and bilaterally, the US tried quietly to persuade China, France, India, Pakistan, and the UK to agree to a 'gift basket'-type of statement that would underline a shared commitment to the highest level of security for military materials and facilities. The reference to the security of military material in the US National Progress Report of 2014 was cited by the US as a potential exemplar for reporting. However, the attempt foundered due to lack of mutual trust among the nuclear possessors, differing approaches to nuclear security and transparency (even within the group of three

Western nuclear weapon states), and concerns about such measures opening an additional avenue of pressure on nuclear disarmament.[51]

A much-needed rebalancing of focus happened due to external events. Despite the participation of Interpol since 2012, and despite the inter-disciplinary expertise in the summit process, the nuclear non-proliferation and arms control community still dominated the discussions. Their natural tendency was to focus on the safeguarding of nuclear materials against misuse by governments through non-proliferation controls such as inspections by the IAEA. Traditionally, nuclear security had been about the 'material', the 'facilities' which held such material, and (to some extent) the 'people' given the focus on illicit trafficking and terrorism. However, the sudden growth of the Islamic State in Iraq and the Levant (ISIL) network in the West—with their use of digital platforms and technology to radicalize and recruit, their apparent interest in WMD, and the terror attacks in Paris and Brussels—brought home the necessity of intelligence co-operation, counter-terrorism, staying up to date with the use of disruptive technology by non-state actors, and law enforcement efforts. The scenario-based discussion among the leaders on 1 April 2016 reflected these important considerations.[52]

Among the gift baskets announced at the concluding summit, a few stand out. Governments from 35 countries, Interpol, and the UN committed through the 'Joint Statement on Sustaining Action to Strengthen Global Nuclear Security' to set up a Nuclear Security Contact Group. This was the idea that President Obama expressed at the 2014 Summit of an advisory group to provide guidance and keep the process going by regular meetings on the margins of the IAEA and other forums. Crucially, the Contact Group, which would be open to all countries that subscribed to the goals set out in the Joint Statement, could recommend holding a future summit.

The 2014 'excellence paper', now issued as an IAEA document (INFCIRC/869), gained two prominent new participants, China and India. Two other significant gift baskets were the 'Statement of Activity and Cooperation to Counter Nuclear Smuggling' subscribed to by 35 countries and Interpol and the 'In Larger Security: Looking Ahead', where the advocates of nuclear disarmament laid out an alternative vision of future work on nuclear security with measures that were mutually reinforcing across nuclear disarmament, non-proliferation, peaceful uses of nuclear energy, and nuclear security.[53] Thus, through the Action Plans and the gift baskets, the last summit succeeded in putting nuclear security in a broader institutional context—which might be described as a global architecture—and channelling collective international will into areas most critical to preventing nuclear terrorism.

Conclusion

What is the legacy of the Nuclear Security Summits today? Pandemics and dramatic climate change—two themes evoked in 2007 in Obama's Chicago speech—are currently at the forefront of global consciousness. The threat of catastrophic terrorism appears to have receded from the minds of political leaders even though a former sanctuary for the perpetrators of the 9/11 attacks, Afghanistan, is again under the thumb of the protectors and allies of al-Qaeda. Even the nuclear threat is framed more in classical terms today with worries centred on a new Cold War as a resurgent China squares off against the US, and as nuclear arms control engagement between Russia and the US remains hobbled by their political differences.[54] Digital vulnerabilities have also risen sharply in the public consciousness in the five years since the last summit. The Cambridge Analytica scandal demonstrated new existential vulnerabilities, and supply chains of everything from chips to ingredients for pharmaceuticals are seen today in stark national security terms. When it comes to national security, the attention of leaders is naturally on these new threats and realities.

Nuclear security might not figure in today's headlines, but the routinization of the nuclear security function in several international organizations and in a large number of countries is a welcome development. It is a direct legacy of the Nuclear Security Summit process, as is the creation of a network of experts across disciplines, countries, and organizational boundaries. The summit community of Sherpas and Sous-Sherpas meets regularly in a Nuclear Security Contact Group on the margins of the IAEA in Vienna as well as at Interpol and other forums. The idea of nuclear security comes alive in their routine interactions in these follow-on initiatives. The impact of the Nuclear Security summit recommendations is felt widely, for instance, in the uptick in the IAEA's IPPAS missions, which jumped from a total of 73 in 2016 to 96 by the end of 2021.[55] Above all, the enduring legacy of the summits is their instructive value for accelerating global cooperation and coordinated action on a shared threat.

Notes

1. Nina Tannenwald, 'Justice and Fairness in the Nuclear Nonproliferation Regime', *Ethics and International Affairs*, 27/3 (2013), 299–317.
2. Jeff Zeleny, 'A Foreign Classroom for a Junior Senator', *Chicago Tribune*, 23 September 2005.
3. Jaime Sneider, 'Obama's Missing "Thesis" ', *NBC News*, 24 July 2008.
4. Congress of the United States, 'To provide for sustained United States leadership in a cooperative global effort to prevent nuclear terrorism, reduce global nuclear arsenals, stop the spread of nuclear weapons and related material and technology, and support the responsible and peaceful use of nuclear technology', Bill S, 1977 (110th), Washington, DC, 2007.
5. Barack Obama, 'Remarks to the Chicago Council on Global Affairs', 23 April 2007, The American Presidency Project (Santa Barbara, CA: University of California).
6. For details of membership and activities, see www.gicnt.org and www.psi-online.info
7. Amandeep S. Gill, *Nuclear Security Summits: A History* (London: Palgrave Macmillan, 2020).
8. Barack Obama, '*Remarks by President Barack Obama in Prague as Delivered, Hradcany Square, Prague, Czech Republic*, 5 April' (Washington, DC: The White House, Office of the Press Secretary, 2009). https://obamawhitehouse.archives.gov/the-press-office/remarks-president-barack-obama-prague-delivered

9. During his 2005 trip, Obama read a book by the centre's Graham Allison on nuclear terrorism.
10. Gill, *Nuclear Security Summits.*
11. Ibid.
12. Information paper on the Nuclear Security Summits circulated by the US in September 2009.
13. Ibid.
14. Embassy UN Vienna Cable, 'The Nuclear Security Summit Sherpa Meeting in Vienna', 24 September 2009, Wikileaks Cable: 09UNVIEVIENNA442.
15. The Nuclear Security Summits accelerated the process, and the goal was realized in 2016.
16. United Nations Security Council Resolution, 28 April 2004, S/RES/1540 (2004).
17. As opposed to open fuel cycles, where nuclear waste from reactors is stored in repositories rather than being reprocessed.
18. Government of South Africa, 'Statement by the President of South Africa', 2016 Nuclear Security Summit, 1 April 2016, Washington, DC.
19. US Information Paper of September 2009, reproduced in Gill, *Nuclear Security Summits.*
20. Author interview with a Sous-Sherpa who attended the meeting.
21. Nuclear Security Summits, 'Pre-decisional Ad Ref draft', 2010 Nuclear Security Summit, 4 March 2010, Washington, DC.
22. Nuclear Security Summits, 'Communiqué', 2010 Nuclear Security Summit, 13 April 2010, Washington, DC, Preamble.
23. Australian Prime Minister Kevin Rudd made a general presentation on the threat at the 2010 summit.
24. Nuclear Security Summits, 'Communiqué', 2010 Nuclear Security Summit, 13 April 2010, Washington, DC, paras. 6–8.
25. Nuclear Security Summits, 'Work Plan', 2010 Nuclear Security Summit, 13 April 2010, Washington, DC.
26. Ibid.; Nuclear Security Summits, 'Communiqué', 2010 Nuclear Security Summit.
27. Author interview with a Sous-Sherpa who participated in the meeting.
28. Ibid., Nuclear Security Summits, 'Work Plan', 2010 Nuclear Security Summit, 13 April 2010, Washington, DC.
29. This reference in *The New Yorker* was inspired by the signing of the New START treaty between Russia and the US in Prague on 8 April 2010 as well as the release of a new US Nuclear Posture Review.
30. Ben Rhodes, Gary Samore, and Laura Holgate, *Transcript of the Post-Summit Press Conference, The American Presidency Project* (Santa Barbara, CA: University of California, 2010).
31. Ibid.
32. Nuclear Security Summits, 'National Statements and Highlights of National Commitments', 2010 Nuclear Security Summit, 12–13 April, Washington, DC.
33. US non-paper, September 2010. See Gill, *Nuclear Security Summits.*
34. Mary M. Crossan, Henry W. Lane, and Roderick E. White, 'An Organizational Learning Framework: From Intuition to Institution', *Academy of Management Review*, 24/3 (1999), 522–53.
35. Ibid.
36. Nuclear Security Summits, 'Progress Reports', 2012 Nuclear Security Summit, Seoul. http://www.nss2016.org/2012/progressreports

37. Douglas Birch and R. Jeffrey Smith, 'US Unease about Nuclear-Weapons Fuel Takes Aim at a South African Vault', *The Washington* Post, 14 March 2015.
38. Nuclear Security Summits, 'Communiqué', 2012 Nuclear Security Summit, Seoul, para. 5.
39. Gill, *Nuclear Security Summits*.
40. Nuclear Security Summits, 'Communiqué', 2012 Nuclear Security Summit, para. 7.
41. Nuclear Security Summits, 'Joint Statements for the 2012 Nuclear Security Summit', 2012 Nuclear Security Summit, Seoul. http://www.nss2016.org/2012/jointstatements
42. President Obama announced the hosting of a Nuclear Security Summit in 2016 in his June 2013 Berlin speech.
43. Author interview with a member of the Dutch preparatory team.
44. Gill, *Nuclear Security Summits*.
45. Nuclear Security Summit, 'In Larger Security: A Comprehensive Approach to Nuclear Security', Statement by Brazil on behalf of 14 countries, 2014 Nuclear Security Summit, 24 March 2014, The Hague.
46. The matrix comprised of five sets of action: subscribing to the IAEA's Nuclear Security Fundamentals document (Nuclear Security Series, No. 20); committing to meet or exceed the intent of the recommendations in three IAEA guidance documents (Nuclear Security Series, No. 13, 14, and 15); ensuring effectiveness of nuclear security regimes and operators' systems; ensuring the demonstrable competence of personnel accountable for nuclear security; and ensuring continuous improvement of nuclear security through one or more of listed steps. See IAEA, 'Communication Received from the Netherlands Concerning the Strengthening of Nuclear Security Implementation', INFCIRC/869, 20 October 2014, Vienna.
47. Gill, *Nuclear Security Summits*.
48. As aforementioned, these were the IAEA, UN, the IAEA, Interpol, GICNT, and the Global Partnership.
49. US Sherpa, 16 September 2014. 'Nuclear Security Summit 2016: Invitation to Washington, DC', Communication on Sherpa meetings.
50. Nuclear Security Summits, 'Communiqué', 2016 Nuclear Security Summit, 1 April 2016, Washington, DC.
51. There was growing momentum by this time on a Treaty on the Prohibition of Nuclear Weapons (adopted in 2017), fuelled by a reference to humanitarian consequences of the use of nuclear weapons in the outcome of the 2010 NPT Review Conference.
52. Barack Obama, 'Remarks by President Obama at the Closing Session of the Nuclear Security Summit', 2016 Nuclear Security Summit, 1 April 2016, Washington, DC.
53. Nuclear Security Summits, '2016 Gift Baskets', 2016 Nuclear Security Summit, 1 April 2016, Washington, DC.
54. Joe Biden and Vladimir Putin, *US–Russia Presidential Joint Statement on Strategic Stability* (Washington, DC: US Department of State Archive, 16 June 2021).
55. IAEA, 'International Physical Protection Advisory Service: Twenty Years of Achievement', IAEA Press Release, 26 September 2016, Vienna. https://www.iaea.org/newscenter/news/international-physical-protection-advisory-service-twenty-years-of-achievement; IAEA, '25 Years of Strengthening Nuclear Security with Physical Protection Peer Advice', IAEA Press Release, 23 December 2021, Vienna. https://globalimpactnews.com/2021/12/24/25-years-of-strengthening-nuclear-security-with-physical-protection-peer-advice/?utm_source=rss&utm_medium=rss&utm_campaign=25-years-of-strengthening-nuclear-security-with-physical-protection-peer-advice

Suggested Reading

Allison, Graham T., (2004), *Nuclear Terrorism: The Ultimate Preventable Catastrophe* (New York: Times Books).

Bong-Geun, Jun, (2012), *Road to the 2012 Seoul Nuclear Security Summit* (Baltimore, MD: US–Korea Institute at SAIS).

Bowen, Wyn Q., Cottee, Matthew, and Hobbs, Christopher, (2012), 'Multilateral Cooperation and the Prevention of Nuclear Terrorism: Pragmatism over Idealism', *International Affairs*, 88/2, 349–68.

Cann, Michelle, Davenport, Kelsey, and Williams, Sarah, (2013), *The Nuclear Security Summit: Progress Report* (Washington, DC: Arms Control Association and Partnership for Global Security Report).

Carlson, John, (2016), 'Assessing Obama's Nuclear Security Summits', *Global Asia* [online], Vol. 11/2. https://www.globalasia.org/v11no2/focus/assessing-obamas-nuclear-security-summits_john-carlson

Gill, Amandeep S., (2020), *Nuclear Security Summits: A History* (London: Palgrave Macmillan).

Pomper, Miles and Warner, Meghan, (2012), *Progress Since the 2010 Washington Nuclear Security Summit: Successes, Shortcomings, and Options for the Future* (Washington, DC: James Martin Center for Nonproliferation Studies).

PART IV

SECURING NUCLEAR FACILITIES AND ASSETS

CHAPTER 14

THE FUNDAMENTALS OF PHYSICAL PROTECTION

MARTIN GRIGG AND JOHANNES H. STERBA

Introduction

Early civilizations, such as that in the Indus Valley from 3,300 BCE, fortified their cities and citadels with walls and bastions to protect the inner sanctum.[1] These were intended to shelter their people in times of danger, or to protect the ruler's most valuable assets. Over time, these were supplemented with added layers of physical defences to further deter and delay their adversaries. This and other historical approaches to securing valuable assets through identifying high-priority targets and developing multiple lines of defences are still replicated today in nuclear and other security systems. They have been further reinforced through security strategies that consider the interplay of adversary actions and security operations.

A defence strategy provides an understanding of vulnerabilities and develops a framework for reducing risk. This is crucial when dealing with evolving adversary capabilities and changing security technologies, which may serve to degrade or enhance the effectiveness of security systems. The past two decades have seen a significant shift in the threat to nuclear and other assets, evolving from largely state-based actors to encompass small groups of non-state actors that have access to new information and technology that can be utilized for malicious purposes.[2,3,4,5]

In the nuclear arena, international effort to strengthen physical protection systems (PPSs) have been spearheaded by the International Atomic Energy Agency (IAEA), which started providing physical protection training in the 1970s. This was underpinned by the development of international guidance. In 1975, the first version of INFCIRC/225,[6] 'Nuclear Security Recommendations on Physical Protection of Nuclear Material and Nuclear Facilities', was published by the IAEA. This has since been updated four more times, in order to reflect changes in the international security environment and remain a cornerstone of international efforts to protect nuclear materials.

In 1987 the 'Convention on the Physical Protection of Nuclear Material' (CPPNM), the first legally binding set of rules dedicated to physical protection, albeit focused on material in international transport, came into force.[7] Focused on the materials to be protected rather than how to protect them, this served to complement INFCIRC/225, which provides detailed information on measures to be taken to secure nuclear material. They have evolved together in line with ever-changing threats and fluctuating international focus on nuclear security. For example, in 2016, an Amendment to the CPPNM came into force, extending its scope to the protection of nuclear facilities and material in domestic use, storage, and transport.

The IAEA's role extends beyond training and the development of international guidance to advisory and other targeted services that can directly support individual states' nuclear security efforts. At the 2019 General Conference, the IAEA emphasized it would give 'high priority to the development and implementation of Integrated Nuclear Security Support Plans' (INSSPs).[8] An INSSP recognizes the security needs within a state, prioritizes them, and, importantly, highlights which organization is responsible for establishing strategies and implementation for each need. The typical functional areas within an INSSP are legislation and regulation, threat and risk assessment, physical protection, detection of unauthorized acts, response to unauthorized acts, and sustaining a nuclear security regime. This promotes an integrated approach to protection against nuclear theft, sabotage, terrorism, or protest, that calls on relevant agencies to work together to establish the threats and mitigate risks at the state level. At the operational level, integration is also essential whether addressing the protection of a building, the secure transportation of material, or mitigating the risks of sabotage at a facility.

Design Philosophy

Approaches to the design of PPSs in general follow five key design principles:

1. graded approach
2. layered protection system
3. defence-in-depth
4. adaptive security
5. security-by-design.

Graded Approach

A graded approach to physical protection is the application of security measures that are proportionate to the potential effects of a malicious act. In simple terms, each asset, material, or structure is assessed and graded to determine the number of control measures needed to protect it. INFCIRC/225 provides a table that categorizes nuclear material from the perspective of theft.[9] In general, the categorization is done by the amount and type of material, taking into consideration its potential use in an improvised nuclear device (IND); for example, all unirradiated plutonium that weighs 2 kilograms (kg) or more should be graded as Category I. Category II and III are used for smaller amounts of material. However, grading does not only apply to physical assets, but also to operations, activities, management, training, inspections, and procedures, among other aspects. Each will have altered characteristics based on the potential consequences of a failure. As the seriousness (the grade) increases, so will the number and type of resources applied to support it, as well as the controls that are applied. The graded approach is one of the fundamental principles of INFCIRC/225.[10]

Grading is based on consequences. The greater the impact of an uncontrolled event or security breach, the higher the grade becomes. Therefore it is essential to understand the facility or operation, and threats that it faces. A detailed understanding and analysis of the facility and its operation must be carried out to identify vital areas, assets, operations, and processes. The likelihood, the significance, and the magnitude of the risks and associated hazards must all be captured to determine the types and levels of controls necessary to mitigate the risk. The levels of control, or the number of resources applied to any asset or operation, may change over time, as threats and processes evolve. This means that the analysis of potential consequences based on a graded approach is not a 'do it once' exercise, but one that should be repeated regularly.

Grading must be systematic and ongoing to ensure consistency and to reduce subjectivity (see Figure 14.1). The rank and definition of each grade should be set using predefined criteria and be applied in a consistent manner. While the IAEA specifies three categories of nuclear material in INFCIRC/225, states may choose to apply four, or even five, grading levels. Often, grade four or five is the lowest criticality and grade one is the highest. However, in the interest of adaptability of a PPS, having one as the highest category is arguably the wrong way around. There are no logical numbers left should something be introduced with an even higher consequence. Threats are also evolving, and risks are increasing, as adversaries find new and more imaginative ways to achieve their goal. Therefore, having grade one as the lowest significant consequence, and working up from there, allows for the system to adapt as threats change.

To grade an asset, activity, or operation, and establish the appropriate levels of mitigation measures, it is necessary to carry out a risk analysis. The basic premise of risk is that it is a combination of consequence and likelihood. Consequence is the (negative) effect of an action, and it can range from a very little effect to a catastrophe. Likelihood is the probability of something happening. The likelihood of a malicious act is a combination of target attractiveness, the threat, and the vulnerability of the asset or activity in question. The consequences of an asset being compromised will be established through a close liaison with the facility operators and process experts. To estimate the proportionate risk associated with a nuclear asset, it is necessary to understand the magnitude of the negative effects should it be lost through theft or sabotage.

Not all assets are attractive to an adversary, and they may not contribute to a cause of a malicious act. Therefore, the attractiveness of an asset is the perception that an adversary

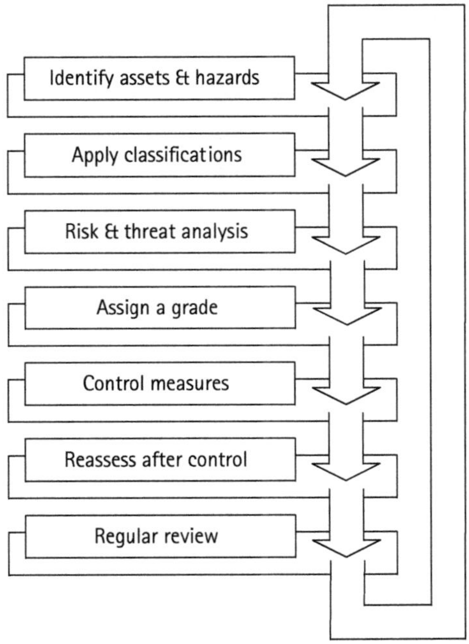

FIGURE 14.1. The graded approach as a simplified process.

Figure courtesy of the authors.

has of it. This may be influenced by any or a combination of the following: its economic value; the physical, collateral, political, or reputational damage that it may cause; and its ability to be used elsewhere—potentially for a greater malicious act.

Any aspect of a facility or activity that can be exploited by an adversary to execute their intentions is a vulnerability. This may be a physical weakness, a procedural flaw, or an operational fault. Having a good understanding of the potential attack paths allows for the identification of vulnerabilities in existing systems, operations, and procedures. Vulnerabilities are the potential opportunities for an adversary. The more metrics that can be applied to calculating likelihood and consequence, the better because this will reduce the subjective element of the findings and produce a more accurate understanding of the risk.

The degree of effort required to protect an asset includes the technology, the processes, procedures, and resources taken to provide effective mitigation measures that lower the risk of compromise to a practicable level. The outputs of a graded approach are assets categorized by risk level. Grading is a management process to determine the degree of effort required to protect an asset. As a minimum, the grading process includes the following: identification of the consequence, likelihood, and mitigation measures; determination of the depth and extent of mitigation; and documentation for the basis for the graded approach.

When multiple asset assessments are clustered, a risk profile starts to emerge for the whole facility. The individual threats can be prioritized and grouped so that high-risk/high-priority issues can be graded as 'critical' from a security point of view. A graded approach is a process that works through every aspect of the classification of security issues

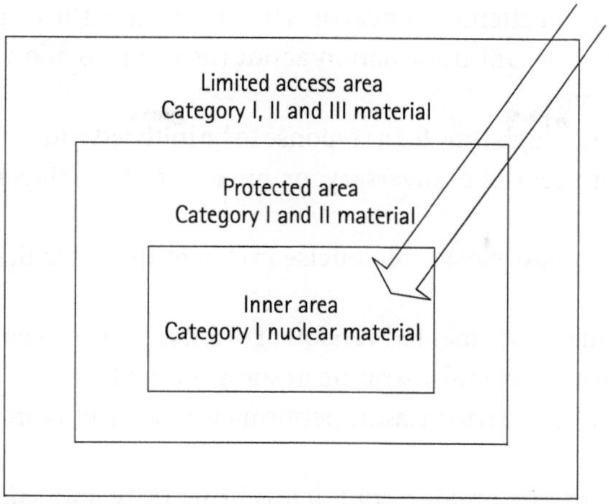

FIGURE 14.2. Layered physical protection as the grade increases.

Figure courtesy of the authors.

and hazards at a facility and, with thorough examination and analysis, determines the appropriate level of mitigation measures and controls that are required to reduce the risks.

Layered Protection System

When discussing PPSs, control measures usually take the form of structures, technology, people, and procedures. These are often arranged in incrementing layers or concentric shapes around an asset, or a 'vital area' (see Figure 14.2). One of the strategies of crime prevention through environmental design is to use multiple layers of incremental protection that start at the perimeter and increase at buildings and then doors within the buildings, etc. In a nuclear facility, under the guidance on INFCIRC/225, Category I, II, and III nuclear materials should be stored in a 'limited area'. Category I and II material should be stored in a 'protected area', and Category I material should be stored in an 'inner area'. This naturally creates a layered approach to physical protection where the number of physical layers increases with the criticality of the nuclear asset, that is, the higher the categorization or grade of the asset, the more controls that are put in place.

Each layer of security that protects an area will provide control measures to either restrict access to the next layer or provide a delay to allow time for a response team to intercept the event. Structures, technology, people, and procedures are deployed at each layer to provide security functions. The following measures are typically implemented:

- **Demarcate**: to clearly define the boundary of a restricted area so that people on the non-secure side of the boundary understand that they would need authorization to enter.
- **Deter**: to provide enough discouragement to potential adversaries to avoid an attempt to steal or sabotage.

- **Detect**: to identify an attempt to steal or sabotage at the earliest opportunity.
- **Assess**: to gather relevant information about the event to allow an informed and appropriate response.
- **Delay**: to provide enough time for a response to be initiated and to intercept the event.
- **Respond**: to intercept the adversary or an event that triggered the detection mechanism.
- **Communicate**: to pass clear and concise information to facilitate an appropriate response.
- **Record**: to provide a documented version of events for post-event analysis.
- **Recover**: to return to normal operation as soon as possible.
- **Review**: post-event analysis to measure performance and identify areas of improvement.

Layered protection systems provide confidence in security and safety because of the increased protection that they provide, and the inherent redundancy of multiple layers. A single control measure on a physical layer is unlikely to mitigate all vulnerabilities. Hence, it is necessary to layer the control measures as well. For example, an electronic access control system that only grants access to somebody with the correct card or fingerprint does not stop another person following an authorized person through a door. In this case, it might be necessary to include a turnstile, or an airlock, to ensure that only the authorized person enters. The addition of a closed-circuit television (CCTV) camera will assess an alarm condition from the access control system; it will provide video verification of the event and it will record the evidence for later review. Adding people and procedures builds up the layers of physical protection.

A facility is unlikely to only have one vital area or one critical asset. It is common for a facility to have multiple sets of layered physical protection within an outer boundary (see Figure 14.3). Physical protection layers do not determine who can go where in a facility. A person having access to one inner area might not have access to another. A controlled area can be described as an area where only a set of authorized people can gain access. There may be several areas with the same level of protection, or in the same security zone, but the people who are allowed in one area may not be allowed in another.

Defence-in-Depth

A layered PPS provides strength and confidence, but it can be improved upon through integration. Integrating layers and reducing the linear nature of the PPS creates a more effective security system, albeit more complicated. Defence-in-depth applies to a complex web of technology, people, procedures, structures, and communications to create a PPS that will provide uncertainty for an adversary. An adversary must defeat or avoid several protective mechanisms to be successful. This integrated and layered approach is often depicted in a model of concentric circles (see Figure 14.4).

This model holds well for a threat approaching from the outside without internal knowledge or access. In a scenario where employees, former employees, or other persons

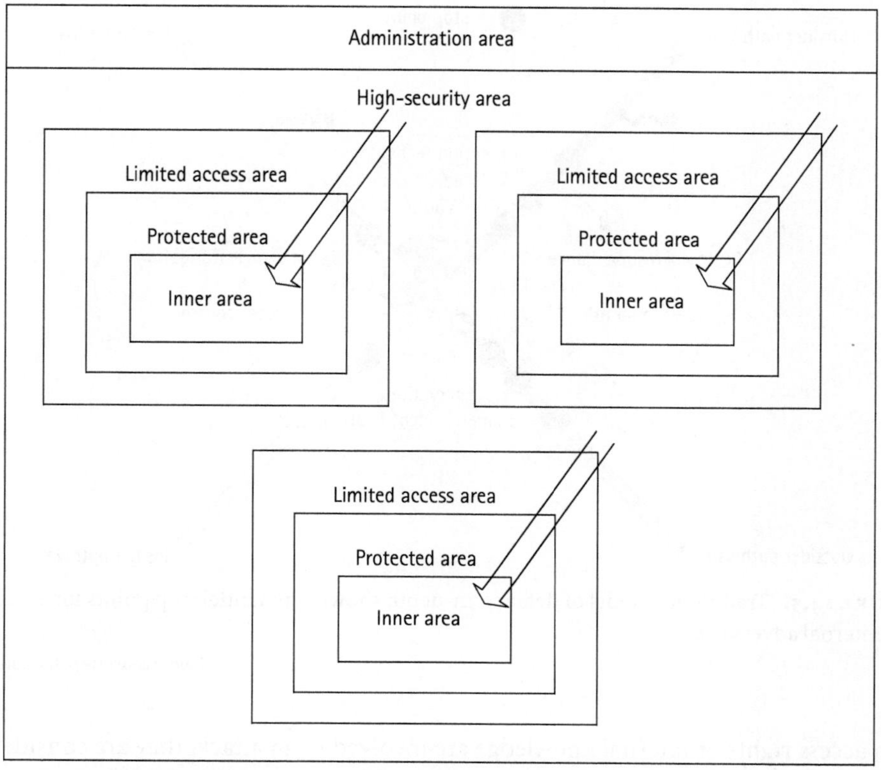

FIGURE 14.3. Multiple areas of layered physical protection in a facility.

Figure courtesy of the authors.

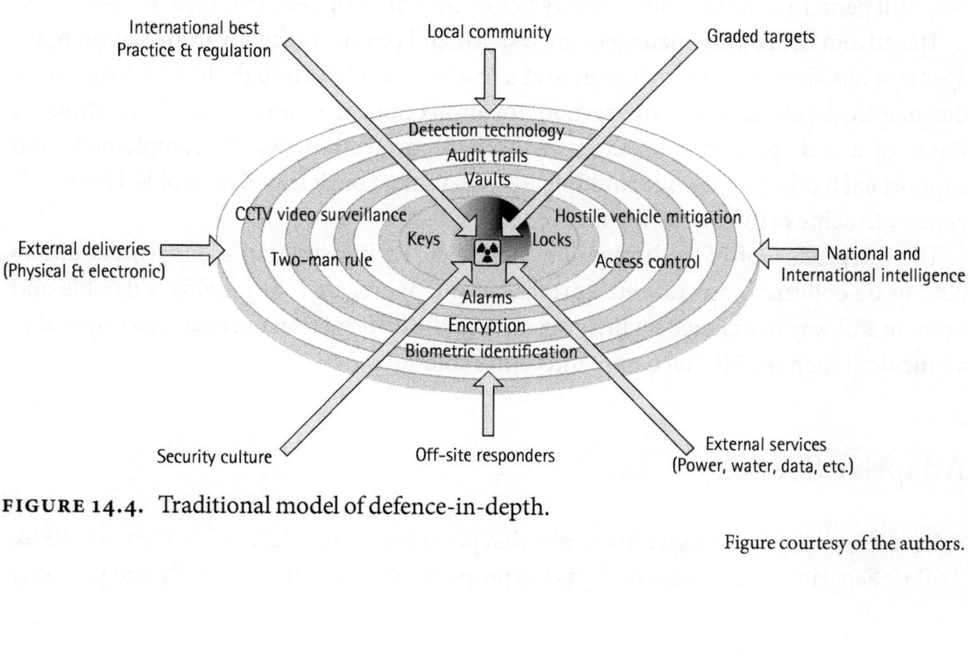

FIGURE 14.4. Traditional model of defence-in-depth.

Figure courtesy of the authors.

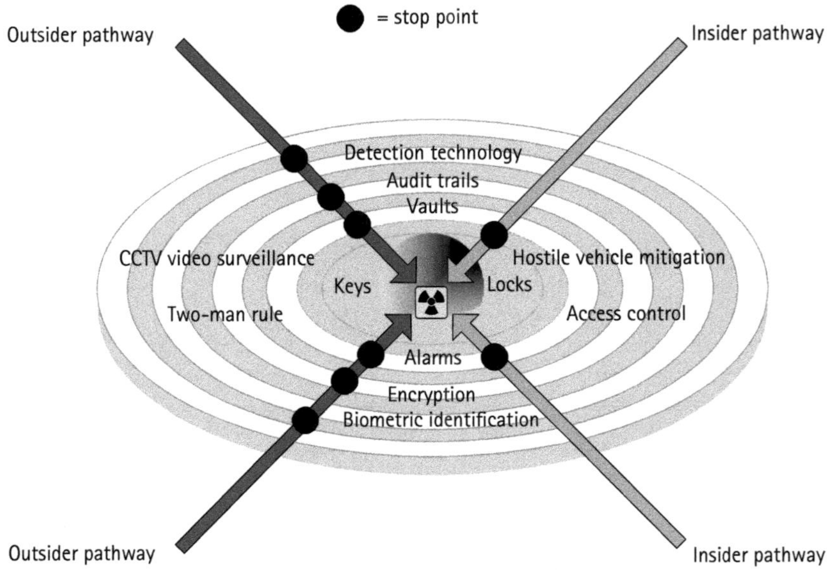

FIGURE 14.5. Traditional model of defence-in-depth showing potential stop points for external and internal adversaries.

Figure courtesy of the authors.

with access rights or internal knowledge are involved in an attack, they are considered an 'insider threat'. This insider threat is a true test of defence-in-depth because with trust and authority, an insider can bypass many defence mechanisms. Policies such as a graded approach, a two-man rule, a need to know, welfare support, and a good security culture all combine to make it very difficult for an insider to be successful. But an insider may still penetrate deeper into a facility before they are stopped, as shown in Figure 14.5.

The advent of security management systems and physical security information management has allowed PPSs to integrate at a machine level. Although the IAEA describes defence-in-depth as layers of methods of protection, it is now possible to think of them as a web of interconnected methods of protection, which complement and support each other to provide multiple PPSs that must be defeated or avoided by an adversary to achieve their goal (see Figure 14.6).[11]

The integration of PPSs through physical security information management allows data to be collected, centralized, correlated, and aggregated, to provide a flexible and more useful way to manage security events and the information necessary to respond to an incident than traditional command centre solutions.

Adaptive Security

Adaptive security is a design philosophy that provides another layer of resilience to PPSs. As Rafe Sagarin writes, 'adaptability is the property of being able to actively and passively

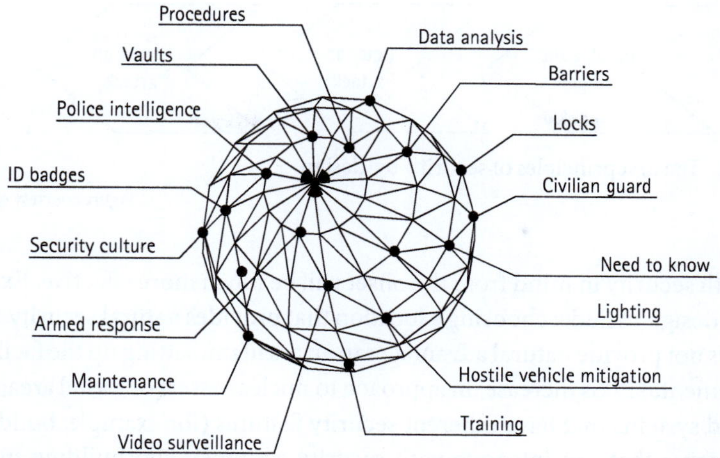

FIGURE 14.6. Representation of physical protection measures with multiple interconnections.

Figure courtesy of the authors.

alter structures, behaviours, and interactions with other organisms and the non-living world in response to selective pressures.'[12] Adaptive security is the ability to predict, respond to, and contain threats by dynamically adjusting to and adapting to new threats before they occur. Adaptive security requires a combination of trend analysis in historic data and real-time analysis of physical security information to detect patterns of behaviour that could predict an upcoming event. Having advanced notification allows control measures to be evaluated and adjusted to match the threat. This capability requires a degree of artificial intelligence (AI) and an adaptable physical system architecture. This kind of adaptation is used in maximum-security prisons, where intelligence is gathered and analysed to initiate change in physical and procedural security measures.

Security-by-Design

Building on the principles of a graded approach, layered protection, defence-in-depth, and adaptability, security-by-design aims to integrate good practice into a facility or operation at the earliest opportunity. Security-by-design is a risk-informed approach that requires multi-disciplinary teamwork and a clear security strategy. Implementing security-by-design requires that all stakeholders agree on how to reduce security risks during the design process—and long before construction begins. It is about considering security from the outset. The phrase 'build security in, not on' refers to a facility being far more robust if PPSs are designed into the fabric and processes of the facility, rather than added in as an afterthought.

Although initial design costs might be higher, introducing security throughout the entire design process will help to reduce the cost of consistently delivering, or upgrading, nuclear security throughout the lifetime of a facility. Consequently, a system

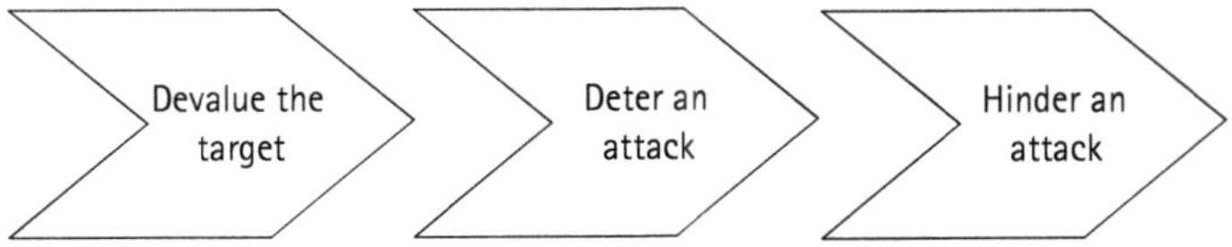

FIGURE 14.7. The first principles of security-by-design.

Figure courtesy of the authors.

designed with security in mind from the onset will be much more effective. Examples of security-by-design include: choosing a location that provides natural security, or at least one that does not provide natural advantages to an assailant; setting up the facility layout so that incremental PPSs increase on approach to nuclear assets or a vital area; choosing products and systems that have inherent security features (for example, building management systems that can integrate with security systems); and building in one-way systems and mantraps (airlocks) to control access to authorized personnel only and to enable lock-down to control an event.

The first principles of security-by-design are to devalue the target, deter an attack and, if necessary, hinder an attack (see Figure 14.7). Devaluing the target is about changing the facility design so that less attractive materials are used, most obviously by converting from highly enriched uranium (HEU) to low enriched uranium (LEU). Deterring an attack requires a careful combination of physical protection measures that portray a safe and secure environment without being oppressive or overbearing. However, if an assailant is determined they will attack regardless of deterrence. In this case, it is essential to detect the event at the earliest opportunity and delay the attacker so that an appropriate response can be deployed.

The Functions of a PPS

The major functions of a PPS are to protect people, property, and assets by bringing together technology, structures, people, and procedures to *deter*, *detect*, *delay*, and *respond* to a malicious human attack whether by external actors, internal threats, or a combination of both.

Deterrent

Designing a PPS with high fences topped with razor wire, plenty of overt video surveillance cameras, and regular armed patrols will send a message to adversaries that the facility is well protected, and its security cannot be breached. Nevertheless, the projection of hard security can lead to over-confidence, which might negatively impact security culture at the facility.

Detection

Detection is the process of discovering, assessing, and alerting an operator, or a response mechanism, to a specific event. The event may be a movement in a controlled area, a door being opened, vibration induced into a structure, contraband material being identified, or any number of security issues that can activate a sensor or be noticed by a person. Detection systems use intruder-sensing devices such as movement sensors, vibration sensors, video content analysis, and a whole array of scanning technology, including human presence.

Detection must operate at the earliest opportunity to provide enough time to analyse the event, communicate relevant information, and prepare a response. Until detection occurs, an adversary can proceed unhindered with the attack. A continuous line of detection should be available across all attack paths, and it must operate before the critical detection point (CDP), which is defined as the last opportunity to detect an event to allow time for a response to intercept and deal with it before the adversary achieves their goal.

The effectiveness of a detection system is also influenced by the nuisance alarm rate (NAR). The NAR is the number of detected events that have been assessed incorrectly and therefore did not need to be communicated. A nuisance alarm is not a false alarm. False alarms are rare and happen when an alarm is raised without any stimulus of the sensor. A nuisance alarm is likely to be caused by environmental conditions such as adverse weather, vegetation, debris blowing in the wind, or wildlife. It is believed that false (nuisance) alarms leading up to the 2012 break-in at the Y-12 National Security Complex may have led to complacency of the Protective Force and diminished facility security.[13]

Delay

The delay function in a PPS allows time for the detection system to sense and assess and to allow time for a response to be deployed and intercept the adversary. Delay is achieved by inserting physical barriers, structures, or a protective guard force along the adversary path. Products that provide physical delay—such as doors, window grills, hatches, and fences—can be tested and rated based on the amount of delay they provide depending on what tools or equipment the adversary has with them.

There are various standards available, and they all tend to have a range of delay ratings that vary with time and attack methods. Different delay mechanisms can be layered along an attack path so that each delay requires a different set of tools or skills to defeat. This layered approach provides strength to the design philosophy of defence-in-depth. In an integrated PPS, the delay performance is measured from the point of detection. Prior to this, the delay period is irrelevant because a response will not be initiated.

Response

The response is initiated by a detected event. The response may be fully or partly automated by switching on lights and directing video surveillance cameras to the location of the event. The response may be to deploy an armed guard force to intercept the adversary. Whatever form the response might take, information is required to establish an appropriate reaction; time is also needed to deploy the response.

Competing Timelines

The competing timelines between an adversary and the PPS is possibly the most fundamental and important principle of a physical security system. The gap in time between the detection of an adversary and the time it takes to interrupt the attack is critical, that is, it should take a long time for an adversary to succeed, and a short time for a response mechanism to interrupt the event. The balance of the competing timelines can be shifted in favour of the response team by detecting an event at the earliest opportunity and then delaying the adversary as much as possible with physical measures. However, it is important to ensure that the delay mechanisms only delay the adversary and do not delay the response team.

Design Foundations

Design Basis Threat

A key part of the development of Integrated Nuclear Security Support Plans (INSSP) is the development of a national-level threat and risk assessment, referred to as the design basis threat (DBT), which is a definition of the attributes and characteristics of potential insider and/or external adversaries who might attempt unauthorized removal of nuclear material or sabotage a facility, against which a PPS is designed and evaluated. The concept of a DBT was introduced by the IAEA in its fourth revision of INFCIRC/225 in 1999. A DBT is used by operators to facilitate and develop appropriate, proportionate, and affordable PPSs by understanding adversaries and their capabilities. A DBT is also a valuable tool for intelligence-led early detection of an attack. An understanding of adversary motivations, intentions, and capabilities will allow national agencies to intercept and monitor potentially malicious developments with a view to intervention and neutralizing the threat as early as possible.

It is important to bear in mind that the DBT[14] is defined by the state, based on a threat assessment.[15] A DBT collects intelligence data to identify a robust set of threats, and it includes each threat objective, and the various capabilities associated with it. A DBT develops a profile of the type, composition, and capabilities of a potential adversary. A threat could be an insider, an external adversary, or a combination of both, that might carry out sabotage or the unauthorized removal of nuclear or other radioactive material.

A DBT assessment requires agencies and competent authorities—national, international, civil, and military—to come together to collect, collate, and analyse intelligence data and other publicly available information, with a view to establishing a robust and credible list of adversaries, threats, likelihood of action, and the associated adversary capabilities (see Figure 14.8).

A DBT addresses all the relevant legal and regulatory constraints, the competencies of integrated agencies, operator competency, as well as the issues associated with identifying appropriate, proportionate, and affordable PPSs. A DBT is not intended to be a statement about actual, prevailing threats. It is used by the state to achieve appropriate allocations of resources for the protection of nuclear materials and facilities against malicious attack. A state-level DBT provides a national risk strategy that acts as the basis for an agreement between the state and the operator. It establishes a robust list of credible threats that can be used for the design and evaluation of PPSs, and it places accountability on the operator to ensure that those PPSs are appropriate and proportionate to the identified threat scenarios. At the same time, there is no legal obligation for the operator to protect against threats outside of the DBT.

Although a state-level DBT is considered classified, it can be disseminated by the state to trusted parties and used to develop individual DBTs to address security issues relating to specific material, transportation, facilities, and activities. These individual

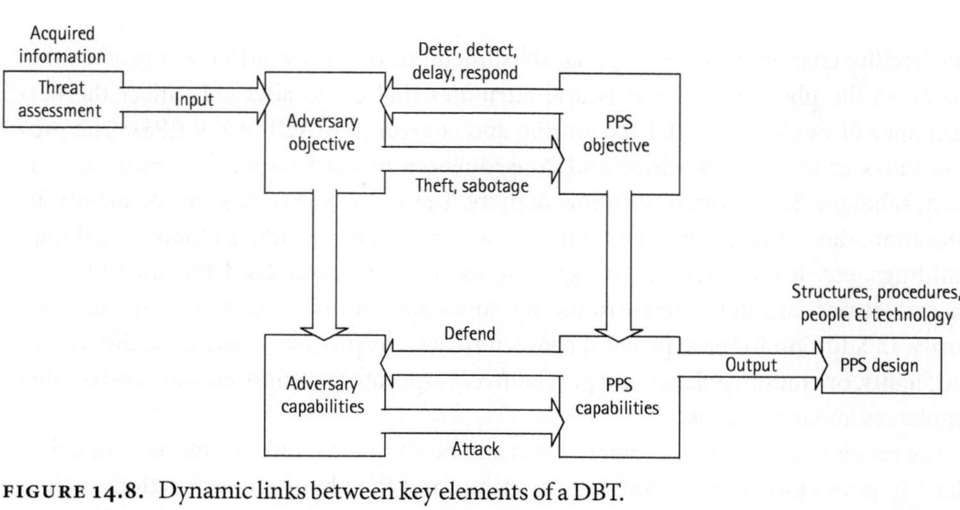

FIGURE 14.8. Dynamic links between key elements of a DBT.

Figure courtesy of the authors.

DBTs will influence the design of PPSs from cyber security to airborne threats. They set up the basis for security hardening, efficient deployment of resources, and system evaluation. A DBT should be regularly reviewed against emerging threats and new intelligence. Changes should be disseminated to operators and relevant agencies without delay.

Design Objectives

A PPS has a set of functional and non-functional objectives. These are the outputs of the system. Ultimately, the functional output of the system design is to prevent an act of theft or sabotage. However, the design objectives can be separated into individual functions and arranged in an order of priority to make the design more efficient. An efficient design will reduce risk at the earliest opportunity. ERIC-PD (eliminate, reduce, isolate, control–protect, discipline) is an abbreviation that is often used when considering safety and hazard control. It also applies to security design and builds on the concepts of devalue, deter, and hinder.

These functional design objectives reduce risk, counter sabotage, and prevent the unauthorized removal of material. They provide resilience and create defence-in-depth. The non-functional outputs of the design are for it to be adaptable, sustainable, and cost-effective. An adaptable PPS will allow fast and cost-effective changes to adapt to new or changing threats. A sustainable design is important for the environment and the security culture at the facility. A cost-effective design is essential to support the wider economics of the facility or operation.

Facility Characterization

The 'facility characterization' of an establishment, or of an operation, is a process that identifies the physical and procedural attributes that could adversely affect the performance of nuclear material accounting and control (NMAC) or the PPSs. The process looks at the threats, risks, and consequences posed by specific events such as theft, sabotage, terrorism, or criminal activity. The process also looks at the facility, its operation, and the assets—equipment, raw materials, product, tools, machines, buildings, etc. It considers the geographic location to understand the site's physical characteristics and how the surrounding landscape and environment may affect security. In addition to these physical characteristics, the process assesses the lifecycle of the facility, operation, policies and procedures, corporate ambition, as well as associated regulatory and legal issues.

The process for facility characterization can be simplified and summarized in a flow chart (Figure 14.9). The site context, its setting, and the physical construction, such as buildings, fences, pedestrian and vehicle access points, and open spaces in between buildings, are the basics of facility characterization.

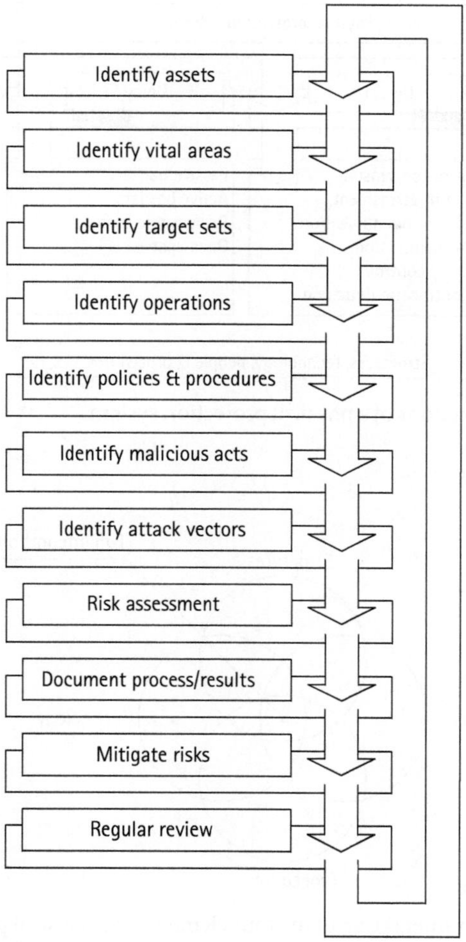

FIGURE 14.9. Facility characterization process.

Figure courtesy of the authors.

INTEGRATING A PPS

Structures, Technology, Procedures, and People

The functions of a PPS can be simplified as *deter*, *detect*, *delay*, and *respond*. Each function has many facilitators. For example, a well-maintained system that is clean and obviously working is a good deterrent. Movement sensors or scanners provide detection, and locks, doors, and window grills provide delay. The facilitators for deter, detect, delay, and respond can be grouped into four categories. These categories form the basis of a model for integrated physical security. These categories are *structures*, *technology*, *people*, and *procedures* (see Figure 14.10).

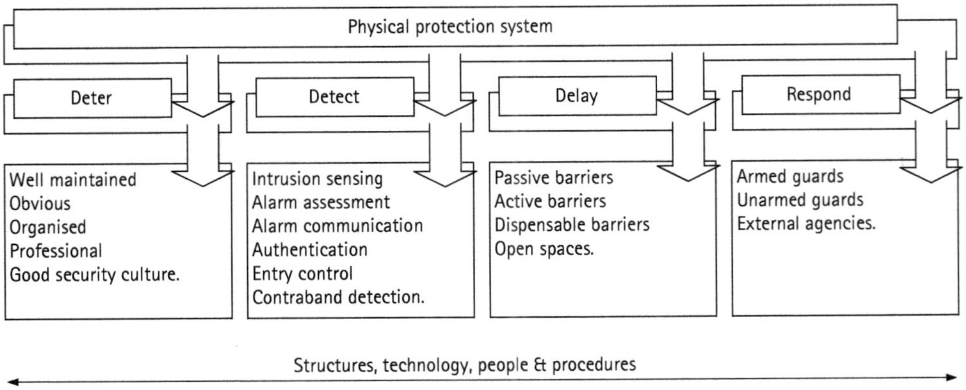

FIGURE 14.10. Basic functions of a physical protection system.

Figure courtesy of the authors.

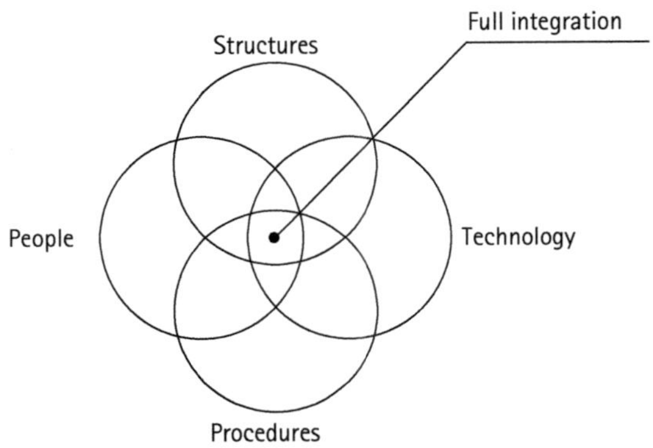

FIGURE 14.11. Model for integration of the four elements of a physical protection system.

Figure courtesy of the authors.

Integration in a PPS is about complementary systems supporting each other. The categories of structures, technology, people, and procedures can be developed into a model that shows how each element interacts with another (see Figure 14.11).

Synergies can be created by aligning people with procedures, procedures with the available technology, and technology with the surrounding structures. Each element can support and interact with the other to achieve a result that is greater than the sum of the individual parts. An example of this is a procedure that describes a situation where remote surveillance that is required is only effective if the right technology is in place to carry out that surveillance; people must also be able to access the procedures and have sufficient resources to implement them. The concept of a fully integrated security scheme is that every interaction between the four elements should be optimized for efficiency. This concept not only creates synergies, but also builds defence-in-depth.

Benefits of Integration

Integrating structures, technology, people, and procedures builds synergy in a PPS. Each element can operate as a stand-alone unit but when supported by complementary systems, the combined effect is greater than the sum of the individual components. When designing a security scheme, it is important to consider how one element of protection might affect another, so that they can be combined to complement each other, rather than cause a conflict. Advancements in communication technology and data processing mean that more data can be aggregated and analysed to produce meaningful results. 'Physical security information management' systems compare data sets and allow operators to make intelligent security decisions. Using this type of technology reduces the risk of human error. Physical security information management also brings synergy to the overall scheme because it allows operators to benefit from the following conditions that would not normally be present without integration:

- improved situational awareness and holistic analysis
- manage multiple systems (such as thousands of sensors)
- reduced operator training and costs
- Enhanced system intelligence
- instant initiation of adjacent technology
- clear, concise, relevant communication
- automated presentation of standard operating procedures
- 24/7 central device monitoring
- detect a corridor of opportunity
- analysis of operator actions
- event trend analysis
- defence-in-depth.

Technology as a Tool

There is a vast range of technology available to help enhance PPSs and provide solutions to problems that people cannot solve alone. From video surveillance to detecting contraband or minute environmental changes caused by an intruder, technology complements people, procedures, and structures. By focusing on the basic functions of a PPS, different technologies can be categorized into the functions of deter, detect, delay, and respond. Figure 14.12 lists a few of the major technical systems, but many more could be included.

Figure 14.12 illustrates how each item, or a type of technology, has a purpose in the model of deter, detect, delay, and respond. However, there are many items in each category. This is because technology is a tool to aid the deterrent factor, or to aid detection, delay, or response. Any one or any combination of the listed technologies may help in any of the categories.

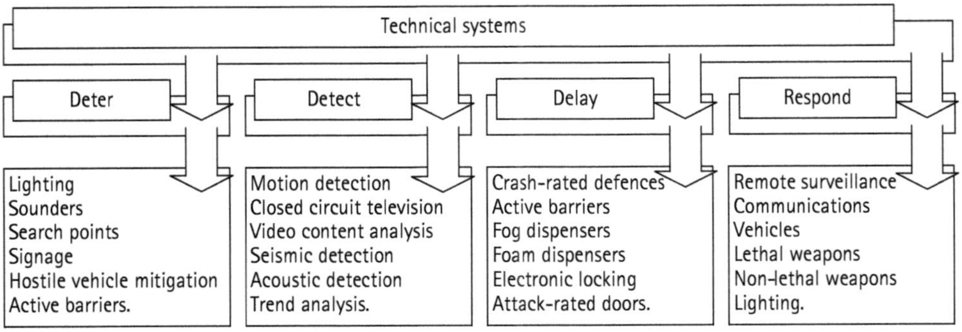

FIGURE 14.12. Basic functions of technology in a physical protection system.

Figure courtesy of the authors.

In the integrated model of structures, technology, people, and procedures, technology is only one part of the PPS. It would be an error to think that technology is a complete solution to a security problem because technology needs structures to support it, people to use it, and procedures to ensure that it is effective. The advent of AI and physical security information management systems has made technology a very powerful tool, but it still does not replace the people in the integrated model. People have insight and intuition at a level that machines do not, are innovative and creative, and are also far more flexible and adaptable than machines. As an example, a surveillance system may well be able to detect movement or even specify the size of objects moving. However, a human made aware of the detected movement is much better suited to assess the detection than an algorithm. Thus, rather than a solution, technology should be viewed as a tool that can be used to help solve the problem of securing a facility or a particular operation.

Increased Efficiency

Technology increases efficiency in a system by detecting environmental changes imperceptible to human senses. Radio frequency can be used to detect different metallic objects or communicate large amounts of data—and very quickly over a wide area. X-rays can be used to detect different compounds inside a sealed container. These are only two examples of how technology can be used as a tool to help mitigate security risks, but there are many others. For example, AI and probabilistic programming mean that technology can make more efficient and valid assessments of what is happening in the environment—that is, actions that might be either hidden or deduced at a speed that humans cannot compete with. For this reason, AI is often deployed in crowded places, such as airports, to identify a wanted person; or used in high-security applications to compensate for the problems associated with operator lethargy and change blindness. This level of integration enhances the efficiency of the PPS by detecting trends and presenting information that humans cannot ordinarily see, or at least, much faster than humans could detect it.

Interaction of PPS with Other Security Measures

PPSs interact with other safety and security measures that are implemented in nuclear facilities. However, in most cases, interfaces between the PPS and other systems are minimal and of little significance. In the following three cases, interfaces need to be considered and planned accordingly.

1. **Contraband detection**: Since PPSs require access control, it is advisable to set up systems in a way that they can also be used for contraband detection. It is important to note that contraband detection should be implemented in both directions, while access control usually only controls entry. Separation systems that are needed for access control can be used for contraband detection if set up accordingly. Sufficient space and a station to perform manual searches increase efficiency.
2. **Accountancy and control**: While NMAC is usually considered to be a tool for nuclear non-proliferation, reliable accountancy and a material balance clearly supports the goals of nuclear security in general and the prevention of theft in particular. The interface between NMAC with the PPS is mostly a matter of the exchange of information.
3. **Cyber security:** With the advent of more and more PPSs relying on IT infrastructure, awareness of and interfacing with cyber security has become an important issue. A successful cyber-attack could potentially move an external threat scenario into an insider threat scenario, therefore increasing the necessary level of protection needed.

Conclusion

International guidance and national efforts to protect nuclear assets are based on the key principles of a graded approach to security and the creation of defence-in-depth. These allow for appropriate and proportionate resources to be applied to nuclear facilities and other assets. A PPS that considers specific high-risk areas can apply multiple layers of incremental protection that increase as a potential adversary gets closer to their target. However, it is likely that an attack on a nuclear facility or an operation involving nuclear or radiological material is going to involve the use of technology, a physical attack, or insider assistance. What types of threats to consider, as well as what adversaries to protect against, is part of the development and revision of the DBT. Changes in the threat landscape will most likely also necessitate changes in the PPS.

The ideal time to start designing a PPS is before the facility is built. Understanding the threat origins and their capabilities and designing security into a facility at such an early stage will allow for threat reduction through design, by either devaluing the targets or building in such a way that an attack is either deterred or hindered to allow for a response to intercept the adversary and neutralize the threat. The competing timelines of the adversary and the response are critical to a timely resolution to an attack. Building in early detection and maximum delay are two of the basic principles of good security design. There are many facets for design consideration, including understanding the threats, integrating and evaluating, and verifying that the result is fit for purpose, proportionate, and appropriate. A PPS is more than just fences, walls, and CCTV. It is a cohesive scheme that brings together structures, technology, people, and procedures. A PPS also has a lifecycle of updating technology and adapting to changing threats.

Notes

1. Michiel Petersen, *Aggressive Architecture: Fortifications of the Indus Valles in the Mature Harappan Phase*, Master's thesis, Leiden University, 2012.
2. Beatrice de Graaf, E.J. Van der Heide, Sabine Wanmaker, and Daan Weggemans, 'The Anders Behring Breivik Trial: Performing Justice, Defending Democracy', *Terrorism and Counter-Terrorism Studies* [online] 4/6 (2013). http://dx.doi.org/10.19165/2013.1.06
3. Aysel Morin, 'Typifying Terror: The Pattern of Storytelling in Terror News and the Case of San Bernardino Shooting', *Global Media Journal*, 9/17 (2018), 33–52.
4. Raymond Archee, 'The Technology of Experience: Narrative vs Media Analysis of Bastille Day, Nice, France, 2016', *PEOPLE: International Journal of Social Sciences*, 4/3 (2018), 1590–1609.
5. Bankmycell, 2021, 'How Many Smartphones are there in the World' [blog]. https://www.bankmycell.com/blog/how-many-phones-are-in-the-world
6. IAEA, *Nuclear Security Recommendations on Physical Protection of Nuclear Material and Nuclear Facilities (INFCIRC/225/Revision 5)*, Recommendations, IAEA Nuclear Security Series, No. 13 (Vienna, IAEA, 2011).
7. Craig Everton, Stephan Bayer, and John Carlson, 'Developments in the IAEA's Nuclear Security Series and Physical Protection Guidance Document INFCIRC/225', Paper presented to the Annual Meeting of the Institute of Nuclear Materials Management, Baltimore, Maryland, 11–15 July 2010.
8. IAEA, 'Nuclear Security Report 2019', GOV/2019/31/Rev.1-GC(63)/10/Rev.1, Vienna, 6 September 2019.
9. IAEA, *Physical Protection of Nuclear Material and Nuclear Facilities*, Table 1, 20.
10. Physical protection requirements should be based on a graded approach, taking into account the current evaluation of the threat, the relative attractiveness, the nature of the nuclear material, and potential consequences associated with the unauthorized removal of nuclear material and with the sabotage against nuclear material or nuclear facilities. See 'Fundamental Principle H: Graded Approach' in IAEA, *Nuclear Security Recommendations on Physical Protection of Nuclear Material and Nuclear Facilities*.
11. IAEA, *Physical Protection of Nuclear Material and Nuclear Facilities*.
12. Rafe Sagarin, *Learning from the Octopus: How Secrets from Nature Can Help Us Fight Terrorist Attacks, Natural Disasters, and Disease* (New York: Basic Books, 2012).
13. Office of Audits and Inspections, 'Special Report: Inquiry into the Security Breach at the National Nuclear Security Administration's Y-12 National Security Complex', US Department of Energy Office of Inspector General, August 2012.

14. IAEA, *Objective and Essential Elements of a State's Nuclear Security Regime*, Fundamentals. IAEA Nuclear Security Series, No. 20 (Vienna, IAEA, 2013).
15. IAEA, 'Design Basis Threat (DBT)', 2022. https://www.iaea.org/topics/security-of-nuclear-and-other-radioactive-material/design-basis-threat

Suggested Reading

Barnard, Robert, (1988), *Intrusion Detection Systems*, 2nd edn (Burlington, MA: Butterworth-Heinemann).

Cumming, Neil, (1992), *Security: A Guide to Security System Design and Equipment Selection and Installation*, 2nd edn. (Burlington, MA: Butterworth-Heinemann).

Garcia, Mary Lynn, (2008), *The Design and Evaluation of Physical Protection Systems* (Burlington, MA: Butterworth-Heinemann).

IAEA, (2021), *Handbook on the Design of Physical Protection Systems for Nuclear Material and Nuclear Facilities*, IAEA Nuclear Security Series, No. 40-T. Vienna.

Sagarin, Rafe, (2012), *Learning from the Octopus: How Secrets from Nature Can Help Us Fight Terrorist Attacks, Natural Disasters, and Disease* (New York: Basic Books).

CHAPTER 15

NUCLEAR SECURITY CULTURE

JAMES R. HOLMES AND IGOR KHRIPUNOV

Introduction

NUCLEAR facilities depend on human performance to avoid security incidents involving adverse consequences. Such incidents include deliberate acts and unintentional personal errors. Inadequate organizational procedures and processes, management failures, and ergonomic issues related to the design and layout of software or hardware hamper efforts to prevent or mitigate security incidents. In the past, facility leaders focused chiefly on designing system hardware and software to prevent and mitigate events with adverse consequences. Experts and practitioners now realize the importance of the interface between security technologies and operators and the need for effective input from the human element of this system. They also acknowledge that weaknesses in organizational processes and cultural values contribute significantly more to the occurrence of security events than do individual mistakes. However, addressing the human component of security has proven as challenging and complicated as it is essential. No security regime succeeds if left to incompetent, apathetic, or malicious personnel. After all, the finest equipment is no better than its user.

Past incidents demonstrate that the performance of complex socio-technical systems is contingent on interactions among human, organizational, technical, social, and environmental factors.[1] The consensus view that has evolved since the late 1980s is that accidents, incidents, and similar events are consequences not only of technical failures and human behaviour but also of organizational cultures inhospitable to nuclear security. In other words, an organization-wide cultural malaise creates the context for security failures. The International Atomic Energy Agency (IAEA) report on the 2011 Fukushima nuclear accident in Japan attributed the disaster not just to a natural cataclysm or a failure of technology but also to human shortcomings. The report concluded that the interaction of human, organizational, and technical factors across all stakeholder organizations and between different levels within each organization must be evaluated and understood for each phase of a nuclear installation's lifecycle.[2]

'Security culture' is a subset of organizational culture, and it shares many similarities with its counterpart 'safety culture'. An institution with a robust safety culture takes safety seriously, knows how to ensure it, and enforces safety measures—all as a matter of corporate and individual habit. The same goes for security culture. The IAEA defines nuclear security culture as the 'assembly of characteristics, attitudes and behaviour of individuals, organizations and institutions which serves as a means to support and enhance nuclear security.'[3] The Agency eventually added 'sustain' as a third function of culture in the context of nuclear security. In other words, cultural characteristics, attitudes, and behaviour foster and perpetuate nuclear security throughout an installation's service life.

The bedrock of nuclear security culture is the recognition and the conviction—by those who have a role in regulating, managing, or operating nuclear facilities or activities, and by those who could be affected by these activities—that a credible threat of a security breach exists and that nuclear security is important. Given the expanding scope of nuclear security, as demonstrated by IAEA publications and the agendas of the Nuclear Security Summits (2010–2016), nuclear security culture covers a wide playing field. In addition to physical protection and material accounting and control, it may include—explicitly or implicitly—cargo inspections, customs and border security, export controls, cooperation to identify and interdict shipments, interception of illicit trafficking, and personnel-reliability screening and training, among other aspects.

A robust nuclear security culture and its regular assessment bestow multiple benefits on an organization. Such an approach encourages management to reflect upon nuclear security far more than normal, while at the same time alerting them to cultural strengths and weaknesses. Furthermore, consistently tending to security culture plays a significant role by broadening the range of thinking among staff about what constitutes a solid and sustainable foundation for security. More s pecifically, the benefits of a healthy culture could include better IT security, protection of trade secrets, an improved interface between security and safety, reduced theft and diversion of nuclear materials, reduced risk of vandalism and sabotage, improved mechanisms for control during emergency situations, and—in some cases—less need for cumbersome auditing procedures.

Security culture assessment plays a key role in developing and maintaining an awareness of the strengths and weaknesses of organizational culture, and of 'nuclear security culture' as its subset. By focusing on perceptions, views, and attitudes at all levels of the organization, culture assessment and enhancement programmes help management understand the organization's patterns of behaviour in certain circumstances and devise more effective security arrangements. As safety and security cultures coexist in an organization, it is important to foster an approach that integrates them in a mutually supporting manner.

Security culture should embrace the entire workforce, encouraging all employees and contractors—whether or not they are directly involved in security—to remain vigilant, question irregularities, carry out their work diligently, and maintain high standards of personal and collective accountability. To describe security culture in more practical terms, it would be appropriate to go back several centuries and quote Benjamin Franklin, the American founder who said that 'an ounce of prevention is worth a pound of cure'.[4] It is better to take precautions now than to suffer severe consequences later. Security culture is not a panacea, but it can prevent major security incidents that result in severe consequences.

Edgar Schein's Concept and the IAEA Model

The IAEA security culture design is based on the organizational-culture model developed by Professor Edgar Schein of the Massachusetts Institute of Technology. Schein's approach was successfully used to develop a methodology for nuclear safety culture following the 1986 Chernobyl accident in Ukraine, which exposed serious gaps in safety compliance and a disastrous failure of the human factor in the Soviet nuclear complex.[5] Given the many synergies between safety and security and the fact that the two endeavours overlap as part of overall organizational culture, reviewing Schein's safety-culture methodology provides a ready-made analytical and practical framework for exploring and modelling nuclear security culture. Schein defined culture as

> A pattern of shared basic assumptions that was learned by a group as it solved its problems of external adaptation and internal integration, that has worked well enough to be considered valid and, therefore, to be taught to new members as the correct way to perceive, think, and feel in relation to those problems.[6]

What group members regard as self-evident constitutes the institutional culture.

Applied to the field of security, Schein's concept helps elucidate the essence of nuclear security culture. It is a set of jointly learned and relevant values, beliefs, and assumptions that become shared and taken for granted as a nuclear facility strives to operate at acceptable levels of risk and compliance. Schein proposes that culture in organizations can be considered in layers comprised of *underlying assumptions*, *espoused values*, and *artefacts*. Some layers are directly observable, while others are invisible and must be deduced from what can be observed in the organization. Assumptions ultimately manifest themselves in observable forms such as documents, procedures, and behaviour. It is the leaders and managers who guide the way as role models in this process, while others typically accept security-related precepts unconsciously and seldom discuss them. Espoused values are the principles in which the leadership says it believes and wants the organization to display in action. Artefacts include the group's visible products, such as the physical environment, technology, style, patterns of behaviour, procedures, and documents. Any group's culture can be studied at these three levels.

Using Edgar Schein's three layers of culture, the IAEA model for nuclear security culture divides the artefacts of culture into three parts, giving a total of five elements:

1. Beliefs and attitudes (corresponding to Schein's 'underlying assumptions').
2. Principles for guiding decisions and behaviour (corresponding to Schein's 'espoused values').
3. Leadership behaviour (specific patterns of behaviour and actions designed to foster an effective nuclear security culture).
4. Management systems (processes, procedures, and regulations in the organization that make security a priority and have an important impact on security culture).
5. Personnel behaviour (the desired outcome of the leadership's efforts and the operation of the management system).

The IAEA model consists of 37 characteristics within these five elements, along with 321 metrics of culture within the model's management and behaviour segments. The ultimate objective is to achieve the five key characteristics of personnel behaviour described by their associated indicators: professional conduct, personal accountability, adherence to procedures, teamwork and cooperation, and vigilance.[7] The IAEA methodology recognizes that even a structured series of characteristics cannot be comprehensive, nor can it apply to all circumstances. The objective of identifying these characteristics is to encourage organizations and individuals to undertake self-scrutiny—to stimulate thought, rather than to prescribe. Furthermore, it is evident that history, traditions, and established management practices often leave a distinct imprint on national security culture as it is observed in different regions and countries. In other words, examples of the characteristics of security culture in the IAEA methodology can be adapted for specific situations and requirements.

Security culture indicators can be assigned to the characteristics and help gauge the state of the culture and identify practical ways to improve the organization's performance. The main purpose of using indicators is to stimulate thought. Indicators suggest a way to evaluate relevant characteristics and erect a framework under which to facilitate change and development while promoting desirable behaviour and discouraging undesirable behaviour. They are designed to encourage managers to reflect upon security culture and increase their awareness of the role played by the human factor in security. A thorough review of these indicators reveals the state of nuclear security in an organization, identifies human factor-related gaps in its security regime, and points the way to corrective measures—even if management does not undertake a full-scope self-assessment of security culture. Such simple self-reflection does not, however, preclude a full self-assessment. It is advisable to verify whether the original cultural diagnosis was correct, whether remedial measures taken were effective, and whether the organization is on the right track towards upgrading its nuclear security culture.

Self-Assessment to Identify Strengths and Weaknesses

The purpose of a security culture self-assessment is to support competent security performance in anticipation of or in response to all risks–including insider threats—by painting a clear picture of the influence of the human factor on the organization's security regime. This involves evaluating the key characteristics of security culture and comparing what the culture *is* to what it ideally *should be*. Performance indicators supply the benchmarks for such a self-appraisal.

Security culture self-assessment plays a pivotal role as the leadership attunes itself to institutional strengths and weaknesses. Because of their heavy focus on perceptions, views, and behaviour at all levels, regularly held assessments help one understand the rationale behind an organization's patterns of behaviour in certain circumstances. They also optimize the cost of security arrangements and predict how the workforce may react in the face of the unknown. In this sense they contrast with audit-type assessments, in which the focus tends to fall more on technical issues and much less on intangibles. In other words, launching a self-assessment requires conscious efforts to think in terms of the human system (the dynamic interaction of individuals and teams within an organization) as part of a complex interplay among the organization's cultural traits. At the same time, a security culture self-assessment generally will not yield clear-cut and easily actionable results. Instead it bolsters understanding of why different security-related issues emerge and what their root causes may be, while also furnishing insights into what can be done to enhance nuclear security. The six-stage process of conducting self-assessment of nuclear security culture is described in the IAEA Technical Guidance, 'Self-assessment of Nuclear Security Culture in Facilities and Activities'.[8]

Security culture self-assessment is an important learning curve to traverse, both for those directly involved in security and for the entire workforce, as it provides an opportunity for leaders to learn how culture influences security performance and for staff members to perceive the role of security in their organization. An effective self-assessment increases ownership of the results while facilitating decisions taken to foster continued improvement. Examples of specific benefits produced through self-assessment include:

- Improved understanding of the human factor and nuclear security culture.
- Improved understanding of employee concerns, needs, aspirations, and motives.
- Identification of motivations to improve security performance, along with barriers to improvement.
- Identification of motivations for change, along with sources of resistance to change.
- Clarification of employee opinions about important security-related topics.
- Ability to assess the organization's security regime, either by itself or in contrast to comparable organizations, to monitor progress.

- Ability to set priorities on further moves to strengthen the overall organizational culture in areas like human resource management and internal communications.

An assessment of nuclear security culture concentrates on patterns of behaviour among the staff and management in the context of expected and new threats, both internal and external. It ideally complements the evaluation methodology currently used to detect vulnerabilities and highlight requirements for physical protection, thus helping the facility achieve optimal security arrangements. While the IAEA's Nuclear Security Series on evaluating vulnerabilities and physical protection has been published, the Agency has issued no guidelines explaining how to assess how the human factor shapes the effectiveness of detection, delay, and response—the three main pillars of nuclear security against outside and insider threats. Employing the assessment methodology for security culture suggested here will make a realistic, rational, risk-based approach to detection, delay, and response possible. Practitioners would benefit from a well-designed means for designing and maintaining a security regime, and from currently missing data indicating the state of security consciousness and skill among the workforce.

Culture change is a long-term project.[9] Its success depends largely on continuity and long-term commitment on the part of management and staff. Since security culture needs to be periodically assessed to trace how it has evolved and adjust relevant programmes to nurture it, senior leadership should institutionalize this activity inside the organization. An apparatus for reviewing and reforming the culture could include appointing a senior manager in charge; periodically disseminating information on the status of security culture; and assembling a core group of staff members who are motivated to improve their assessment-related skills and participate in subsequent assessment projects. And as a side benefit, highly visible administrative arrangements would act as a deterrent to insiders contemplating committing malicious acts.

A crucial task before starting a self-assessment is to invest time in building commitment throughout the organization. It is commonplace for a self-assessment to run into problems owing to misconceptions and lack of support. The following five steps are in order:

1. *Establish a Self-Assessment Team.* A self-assessment team should consist of staff members who represent different departments and are familiar with the procedures for conducting a security culture self-assessment. A staff member with practical experience in appraising nuclear safety culture would be helpful as a member of the team.
2. *Draft a Self-Assessment Plan and Prepare to Implement It.* The team and senior management work together to develop a self-assessment plan that covers the entire duration of the process, with due regard for the need to limit its cost and disruptions to the institution's operations. Methods for executing the plan depend on several variables, such as the time allocated for the exercise, the availability of team members to perform their assessment functions, budget considerations, and kindred administrative constraints. These methods fall into two categories:

a. Non-interactive methods such as document reviews, surveys, and observations, and
b. Interactive methods such as focus-group discussions and individual interviews. All these methods have their strengths and weaknesses. Hence it is best to start with a survey followed by on-site interviews to ferret out gaps in understanding, clarify ambiguities, and generate qualitative data.

3. *Launch the Data Collection Phase: Survey, Interviews, Supplementary Research.* The team first explains to the facility staff the self-assessment's purposes, mainly improved adherence to procedures and enhanced vigilance, and points out that it focuses on attitudes and behaviour rather than matériel or bureaucratic procedures. The team next launches the assessment through surveys employing relevant performance indicators, and then conducts interviews. Collecting data and ensuring their consistency is the goal of this initial phase. Concurrently the team may search for supplementary information by reviewing documents or conducting on-site observations. The team compiles raw information about the state of nuclear security culture at the facility for evaluation against the performance indicators selected by facility leaders.
4. *Analyse Data and Consolidate Assessment Results.* The team analyses the raw data collected and integrates it with the results from the survey and interviews. While surveys yield quantitative data, interviews are better at capturing the quality of human interactions and experiences. The team compares across the quantitative and qualitative data sets in order to draw conclusions and identify themes. Results that contradict one another are double-checked and clarified during this phase.
5. *Discuss Preliminary Results throughout the Organization, Submit Final Report, and Support the Development of a Follow-up Action Plan.* The team communicates its findings to the organization, requests feedback, revises its draft report if necessary, and submits its final report. Management invites some members of the team to join the effort to develop an action plan. In developing the plan, it is important to go beyond visible artefacts of culture to address its deeper, intangible tiers. By identifying inconsistencies and conflicts between behaviour, practices, and policies and principles, beliefs, and attitudes, drafters will be able to address the elemental causes of deficiencies and problems. This approach will enable the organization to enrich nuclear security culture during the post-assessment period.[10]

Self-assessment is an important learning opportunity, helping an organization estimate the reliability of the human factor and manage culture to influence security performance. The self-assessment process conscripts a large number of staff members and contractors at all levels and, as a result, increases ownership of its outcome. In turn, the security-related behaviour of the workforce tends to improve. The earlier the process starts, the better. There is often a time gap between the development of weaknesses and the occurrence of an event entailing significant security consequences. Weaknesses tend to interact synergistically, creating a volatile security environment that renders an organization vulnerable. By identifying early-warning signs through regular self-assessments, corrective actions and enhancements can be taken in sufficient time to prevent security breaches.

Corrective Measures and Enhancement of Culture

Dysfunctional security culture, on the other hand, risks leading to a security breach fraught with potentially catastrophic consequences. One such case happened in July 2012, when three anti-nuclear activists cut through a fence and sneaked through the multi-layered security perimeter protecting the Y-12 National Security Complex in Oak Ridge, Tennessee in the United States (US). Undetected by guards securing the perimeter, the activists spray-painted anti-nuclear slogans along the exterior walls of the highly enriched uranium (HEU) facility, a US$500 million structure that houses most of the weapons-grade uranium in the US. The intruders wielded sledgehammers, spread human blood on the walls, and sang religious hymns until finally apprehended. The follow-up investigation concluded that the break-in was the logical result of mismanagement, complacency, deficient culture, low morale, and inadequate standards of professionalism plaguing the facility.[11]

What needs to be done in such situations to improve security culture and human performance? Before developing and launching a programme to enhance nuclear security culture, it is important to select specific issues besetting the organization and determine whether the prevailing culture aids or hinders efforts to solve them. Such issues are identified through self-assessments, routine observations by the management team, reviews of past security breaches, and reports from personnel of abnormal security-related occurrences.

Typical issues include:

- Nuclear security is not a clearly recognized value within the institution.
- Workers are apathetic about complying with security procedures.
- Lines of communication between security and non-security personnel are tenuous.[12]

The vehicle to achieve culture enhancement is an action plan that sets in motion training, briefings, workshops, information dissemination, and other measures. While developing an action plan it is necessary to remember that culture is a multi-faceted phenomenon combining both weaknesses and strengths. Even if some elements of the culture look dysfunctional, they may coexist with elements of strength that can be used. It is a mistake to concentrate exclusively on weaknesses.

Action plans may be structured differently but generally include the following:[13]

- Specific goals for the enhancement programme.
- Specific actions to be taken to achieve those goals.
- Personnel responsible for these actions.
- The time frame in which the actions should be completed.
- The resources needed to complete the actions.
- Potential barriers to completing the actions.
- Steps to be taken to surmount barriers and complete the actions.
- Expected results of the actions.

Leadership and management allocate resources to implement the action plan, issue an official document to inform the staff about the enhancement programme, and stress the need for cooperation. With sustained effort and by employing incentives and disincentives, they establish patterns of desirable behaviour. Executing the plan may even alter the physical environment. Leadership and management lead by example. Through their own behaviour they demonstrate their commitment to nuclear security and, in so doing, play an important role in promoting security culture.

It is important to understand that enhancement does not mean instilling an entirely new culture that the workforce may be reluctant to embrace. Leaders have several tools available for achieving their enhancement goals, including: paying visible attention to security-friendly values, beliefs, and priorities; reacting to crises the way they expect subordinates to respond in similar circumstances; acting as role models, exhibiting patterns of behaviour for subordinates to emulate; allocating rewards ranging from praise to promotions to instil desired behaviour and encourage desired outcomes; and setting criteria for hiring, firing, and other professional-development milestones.

Even so, cultural reform does not always come easy. There is always a risk that such efforts will encounter misunderstanding and resistance inside the organization. One reason is that few organizational cultures are homogeneous—subcultures may exist that see things differently. For example, individuals who are not part of the facility's security contingent may think security is someone else's responsibility, and that security successes and failures have little to do with anything they do or fail to do. A mismatch in attitudes between these two subcultures risks undercutting the facility's capacity to prevent, detect, delay, and respond to security problems. Bridging the gap between security and non-security subcultures should be a focal point of any cultural enhancement plan.

Yet another challenge is how to coordinate enhancement with off-site stakeholders. After all, their perceptions, beliefs, and attitudes are central to effective on-site security culture and teamwork among all players. These stakeholders include regulators, law-enforcement agencies, off-site response forces, emergency services, trade unions, and local communities. The action plan's success often hinges on the extent to which the organization has a culture compatible with those of off-site players.

Executing a security culture action plan imparts a shared understanding of threats to nuclear and other radioactive materials, as well as associated facilities and activities.

Most importantly, it creates a common awareness of security issues at all levels and improves coordination among various stakeholders. The process gives rise to an ethos of teamwork that is a prerequisite for robust nuclear security culture.

Safety–Security Interface

Nuclear safety and security have the same ultimate goal, namely, to protect individuals, the public, and the environment from the harmful effects of ionizing radiation. A fundamental mutual goal of safety culture and security culture is to instil among all personnel a general awareness of the hazards resulting from radioactive materials and associated facilities, as well as a sense of joint responsibility for minimizing the risk of nuclear incidents. However, there are significant differences between the two. While both nuclear safety and security address the dangers of inadvertent human error, nuclear security places more emphasis on deliberate acts meant to cause harm. Accordingly, security culture, unlike safety culture, puts a premium on attitudes and behaviours, for example, confidentiality, as well as efforts to deter malicious acts.

Security culture also entertains broader objectives such as instilling awareness of the threat and predisposing all employees and contractors to adopt a strict and prudent approach to security, be vigilant, take a questioning attitude, and react quickly and correctly when the need arises. Security, like safety, is everyone's business—or should be. Despite this, nuclear security culture has historically lagged behind nuclear safety culture, due to an absence of major forcing events.

In terms of nuclear safety there have been three major incidents over the past 50 years that have clearly demonstrated the importance of human factors in nuclear safety systems and driven the development of safety culture within the global nuclear industry:

- The 1979 accident at the Three Mile Island reactor in the US highlighted the influence of the human factor on nuclear safety.[14]
- The 1986 accident at the Chernobyl reactors in the USSR illustrated the impact of organizational and cultural factors.
- The 2011 tsunami and nuclear disaster at the Fukushima Daiichi nuclear power plant in Japan served as a reminder of how safety culture contributes to facility operations.

It is likewise important to understand national cultures and subcultures when considering how to manage the safety–security culture interface. After all, the national culture encompasses all stakeholders in any nuclear complex and shapes how they conduct their affairs. Consequently, each facility's leadership should take an all-encompassing view when designing measures to guard against inadvertent and hostile acts. The leadership should also take a common approach to evaluating the effects of safety and security events. Communiqués to the staff should stress that the causes of safety and security incidents may be quite different, but their repercussions may be similar if not identical. The Fukushima Daiichi nuclear accident is a case in point. An earthquake and tsunami did the damage at Fukushima, but a man-made attack would have been just as capable of crippling Japan's economy and power supply. This being the case, it is of utmost importance to bring safety culture into harmony with security culture.

A significant milestone in the evolution of the safety–security interface came when the IAEA affirmed that nuclear safety principles should also govern security when an endeavour involves *both* safety and security.[15] The Agency's guidance applies to such efforts as designing and constructing nuclear installations; controlling access to sites to prevent the loss, unauthorized removal, possession, transfer, or use of radioactive materials; mitigating the consequences of accidents and failures that have a bearing on security; and safeguarding radioactive sources and radioactive materials.[16]

Promoting nuclear safety and security jointly is part of an efficient cultural programme that ensures that personnel view safety and security as co-equal elements of nuclear-related operations. The beliefs and attitudes necessary to garner success are the same for both safety and security culture. Everyone involved must internalize the importance of culture and believe in the need for close coordination. Everyone must believe that threats are real and credible and deserve abundant attention. A facility where the staff takes these precepts as axiomatic is a facility primed for success.

Integration or Harmonization

Clearly, then, there is a nexus between nuclear security and safety culture. Even so, it is important to acknowledge that certain contradictions make it impractical—and indeed undesirable—to combine safety and security programmes into a single grand programme. Security culture is sufficiently distinct in its approaches to justify its status as a separate field. The common objectives of security and safety efforts are best achieved through disparate measures, while certain aspects of safety stand in tension with security (and vice versa). Below are several examples:

- While safety is the principal responsibility of the facility operator, measures to ensure security involve off-site players.
- Security and safety experts often come from different backgrounds, use different professional terminology, and have distinct mindsets.
- Safety experts typically call for building greater redundancy into at-risk systems, while security experts point out that increased redundancy might render these systems more difficult to protect.
- Safety experts expect engineers and managers to pay frequent visits to operating spaces at a facility, or to any other safety-critical area, while security personnel may prefer to restrict access on security grounds.

- Linguistic difficulties sometimes inhibit a merger. Sometimes one word has multiple meanings within a language. Trying to talk about safety and security as distinct concepts in these languages requires the use of qualifying adjectives. Concepts cannot be easily conveyed with single words. The need for elaborate phrasing tends to pose communication difficulties, especially for the public, a major stakeholder. In such cases it is advisable to keep safety and security separate.

On the other hand, as with safety, the ultimate success of a nuclear safety culture depends upon the human element, including leadership, commitment, and accountability. Such factors are crucial to introducing cultural change, which can be a challenging and time-consuming endeavour. Since safety culture is better recognized by, and thus more acceptable to, constituents within the organization, exploring the interface between these two cultural programmes offers a vehicle to generate additional support for security culture by association. Invoking well-accepted safety tenets helps constituents transpose concepts to security efforts, provided these efforts enjoy strong support from consolidated management.

Several areas are most ripe for harmonization. Mission statements and plans of action should clearly outline the benefits of harmonization in terms of value, strategy, and time scheduling. Training and qualification apparatuses should contribute by holding integrated safety and security training sessions and conducting joint briefings to raise safety–security awareness. Self-assessment and enhancement benefit from including both safety and security assessors. Combined self-assessments should help top leadership understand both safety and security culture and identify obstacles in the way of bringing them in harmony. Corrective action plans should drive improvement and support implementation based on best practices. Dissemination of lessons learned and case studies makes use of the experience gleaned from a project, operation, or assignment to improve performance in future endeavours. Under emergency response plans, safety and security representatives should work together to prepare and upgrade emergency response plans, ensuring that both safety and security culture are afforded equal consideration. Options for emergency response planning might include integrating the regulatory process and addressing everyone who has a stake in safe, secure plant operations.

There is a great opportunity to improve the overall operational culture of a nuclear facility or activity by harmonizing safety and security ventures wherever possible. After a harmonization programme is established, facility overseers should closely monitor progress to ensure that the programme yields such results as streamlined operations, reduced costs, and enhanced safety and security culture.

Security Culture Drivers

Despite its recent provenance, nuclear security culture has taken on the trappings of a mature management practice that contributes to robust nuclear security. One driver that brought the concept to prominence in international law was the 12 'fundamental principles' of nuclear security compiled in the aftermath of the 11 September 2001 terrorist attacks on the US ('9/11') and later codified in the 2005 Amendment to the 1980 'Convention on the Physical Protection of Nuclear Material', which came into force in 2016. Principle F stipulates that 'all organizations involved in implementing physical protection should give due priority to the security culture, its development and maintenance necessary to ensure its effective implementation in the entire organization'.[17]

Although security culture is listed alongside other fundamental principles, for example, threat evaluation, the graded approach, defence-in-depth, and quality assurance—implying a co-equal status—it is clear that culture is an enabler and a driving force. Culture stands above the other principles. It is an overarching and integrating element without which most of the other fundamental principles cannot be successfully and fully implemented. Indeed, experts increasingly characterize culture as 'an obvious requirement throughout the nuclear infrastructure and across all initiatives'.[18]

The Nuclear Security Summits in Washington, DC (2010), Seoul (2012), The Hague (2014), and Washington, DC (2016) boosted the concept of nuclear security culture. The Hague Summit, for example, encouraged all nuclear stakeholders to build and sustain a strong security culture to combat nuclear terrorism and criminal threats. The summit communiqué listed nuclear security culture as one of the three pillars of nuclear security, the other two being physical protection and material accountability. The Washington Summit (2016), the last in the series, adopted five action plans for five international organizations and initiatives, which were to take over the nuclear security agenda. Three of them (the United Nations, IAEA, and Global Partnership against the Spread of Weapons and Materials of Mass Destruction) were assigned specific roles for nuclear security culture.[19]

Nor do security culture drivers stop there. The IAEA's 'Code of Conduct on the Safety and Security of Radioactive Sources' is a non-binding instrument that stipulates that every state should promote safety and security culture with respect to radioactive sources. Following the 9/11 acts of terrorism, the Code of Conduct—one of the first international instruments to mention both safety and security culture—was revised to strengthen several security-related provisions and to address malicious misuse of radioactive sources. A vast majority of the member states made political commitments to implement the code, signalling wide support for exploring prospects to upgrade it to binding status.

Furthermore, numerous non-binding documents have a direct or indirect impact on nuclear security culture. They include UN General Assembly and Security Council resolutions, UN declarations, IAEA technical recommendations, and many others, which upon codification or endorsement can contribute to guiding principles and standards for states or organizations and indirectly strengthen the nuclear security regime. For example, the IAEA publications in the Nuclear Security Series are designed to help member states meet their obligations and develop a robust security culture. The series has a four-tiered structure consisting of fundamentals, recommendations, an implementing guide, and technical guidance publications, of which three are dedicated to nuclear security culture.[20]

It is important not only to amass new technical knowledge, expertise, and culture in nuclear organizations, but also to preserve accumulated knowledge and ensure it is available when needed. The experience, skills, and knowledge acquired through operating a nuclear security system should be shared among stakeholders and actors within the nuclear community to achieve better performance and feedback. Knowledge management culture is an element that addresses the practices, behaviour, and attitudes that exist within a nuclear organization and demonstrates the value ascribed to knowledge. Trust, teamwork, and active collaboration are the main characteristics of a positive knowledge management culture. The key indicators associated with these are a no-blame policy for sharing knowledge, rewards for sharing knowledge, learning by example, and vibrant team relationships.

Knowledge management can be supported by constructing and using knowledge networks. Knowledge networks are made up of people familiar with nuclear security and culture. Together they serve as a platform to enable the analysis and sharing of expertise and experience in these matters, supplementing efforts to build national nuclear security capacity. Knowledge networks can be established at both national and international levels, with a variety of functions that promote security culture. For example, a pool of master's graduates in nuclear security might supply knowledgeable and capable individuals to serve as nuclear security officers at nuclear facilities and related national agencies. Nuclear Security Support Centres (NSSCs) are another example. This IAEA-initiated programme brings together national stakeholders, including regulatory bodies, nuclear operators, technical support organizations, law-enforcement agencies, and first responders.[21]

At the international level, knowledge networks are expected to share lessons learned and serve as channels for international cooperation. Knowledge networks can be established to connect states that experience common challenges in nuclear security and culture, allowing participants to learn from one another. The International Network for Nuclear Security Support Centres (NSSC Network) is another example of a knowledge network for building capacity in nuclear security. The network is sometimes referred as 'Centres of Excellence' and is designed to facilitate the efforts of participating states to build capacity in many areas, including nuclear security culture. The NSSC Network serves as a platform to standardize the quality of nuclear security training in the same way that the IAEA supports the development of training curricula and materials.

As to what drives individuals to excel at security culture, personal experience can shape a mindset favourable to the theory and practice of nuclear security culture. Other factors could be university instruction in nuclear security, workplace managers who acknowledge and nurture the human element, facility leaders who act as role models, continuous learning at the organizational level, IAEA pilot projects and training workshops, or participation in security-related national and international events. This list is not exhaustive, but it demonstrates the growing interaction between the sustainability of a nuclear security regime and culture as a cross-cutting element rooted in human behaviour.

Conclusion

The rationale behind developing a security-minded nuclear culture is that stopping an intelligent adversary from outwitting a security system is an intensely interactive process. As adversaries refine their capabilities and tactics and new scenarios come into view, security systems must evolve rapidly to keep pace. Focusing on people's values, beliefs, and attitudes as components of nuclear security culture is perhaps the most effective way to deal with human fallibility and prevent security incidents. Only personnel who have security culture imbued in them as second nature can constantly evaluate security systems, predict their performance against changing scenarios, and stay ahead of threats. A robust culture is key to long-term and sustainable success at achieving high standards of reliability within a nuclear facility's workforce. Concentration on performance, reductions in training errors, and improved practices is achievable in the short run as a vehicle to attain the long-term objective of an effective and sustainable culture. To this end, a security incident must be seen as a failure of multiple controls in the organization—a failure that demands interdisciplinary efforts to learn lessons and get to the root causes.

Notes

1. Christopher Hobbs and Matthew Moran, 'Exploring the Human Dimension of Nuclear Security: The History, Theory, and Practice of Security Culture', *The Nonproliferation Review* [online], 9 June 2022, 3–6. https://doi.org/10.1080/10736700.2020.1811532
2. IAEA, *IAEA Report on Human and Organizational Factors in Nuclear Safety in the Light of the Accident of the Fukushima Daiichi Nuclear Power Plant: International Experts Meeting*, Vienna, 21–24 May 2013 (Vienna: IAEA, 2014).
3. IAEA, *Nuclear Security Culture*, Implementing Guide, IAEA Nuclear Security Series, No. 7 (Vienna: IAEA), 3.
4. 'Idioms', The Free Dictionary by Farlex, 2022.
5. International Nuclear Safety Advisor Group (INSAG), *Summary Report on the Post-Accident Review Meeting on the Chernobyl Accident*, IAEA Safety Series, No. 75-INSAG-1 (Vienna: IAEA, 1986), 76.
6. Edgar Schein, *Organizational Culture and Leadership*, 3rd edn (San Francisco: Jossey-Bass, 2004), 17.
7. IAEA, *Nuclear Security Culture*, 19.
8. IAEA, *Self-Assessment of Nuclear Security Culture in Facilities and Activities*, IAEA Nuclear Security Series, No. 28-T (Vienna: IAEA, 2017).
9. This is true in normal times, at any rate. A drastic event such as a Chernobyl or a 9/11 can shatter a culture overnight, showing that it is defective and creating an opportunity to institute swift and sweeping remedial change. This chapter limits itself to cultural reforms in ordinary, steady-state times when no disaster has taken place.
10. IAEA, *Self-Assessment of Nuclear Security Culture*, 12–17.
11. US Department of Energy, Office of Inspector General, Office of Audits and Inspections, 'Special Inspection Report Inquiring into the Security Breach at the National Nuclear Security Administration's Y-12 National Security Complex', DOE/IO-0868, Washington, DC, August 2012.

12. IAEA, *Enhancing Nuclear Security Culture in Organizations Associated with Nuclear and Other Radioactive Material*, Technical Guidance, IAEA Nuclear Security Series, No. 38-T (Vienna: IAEA, 2021), 22.
13. Ibid., 28–36.
14. Mitchell Rogovin and George T. Frampton, *Three Mile Island: A Report to the Commissioners and to the Public*, Vol. 1 (Washington, DC: Nuclear Regulatory Commission Special Inquiry Group, US NRC, 1980), 122. https://www.osti.gov/servlets/purl/5395798
15. IAEA, *The Interface Between Safety and Security at Nuclear Power Plants*, International Nuclear Safety Group, INSAG-24 (Vienna: IAEA, 2010).
16. IAEA, *Fundamental Safety Principles*. IAEA Safety Standards Series, No. SF-1 (Vienna. IAEA, 2006), 6.
17. IAEA, *Nuclear Security–Measures to Protect Against Nuclear Terrorism: Amendment to the Convention on the Physical Protection of Nuclear Material*, Report by the Director General, No. GC(49) INF/6.2995 (Vienna: IAEA, 2006), 8.
18. Elizabeth Turpen, 'Global Lockdown: Moving the Needle on Nuclear Security', *Stanley Foundation* [online], Policy Analysis Brief, November 2010, note 32. https://stanleycenter.org/publications/global-lockdown-moving-the-needle-on-nuclear-security/
19. Nuclear Security Summits, 'Action Plan in Support of the United Nations', Washington Security Summit, 2016; Nuclear Security Summits, 'Action Plan in Support of the International Atomic Energy Agency', Washington Security Summit, 2016; Nuclear Security Summits, 'Action Plan in Support of the Global Partnership Against the Spread of Weapons and Materials of Mass Destruction', Washington Security Summit, 2016.
20. For a list of these with references to the application of nuclear security culture, see IAEA, 'Self-Assessment of Nuclear Security Culture', 4–5.
21. For a list of IAEA programmes that focus on nuclear security culture, see Igor Khripunov, 'Nuclear Security Culture: The State of Play' (Vienna: IAEA/ INSEN, 2018), Ch. 4.

Suggested Reading

Cameron, Kim S. and Quinn, Robert E., (2006), *Diagnosing and Changing Organizational Culture* (San Francisco, CA: Jossey-Bass).

Geert Hofstede, Hofstede, Gert Jan, and Minkov, Michael, (2010), *Cultures and Organizations: Software of the Mind*, 3rd edn (New York: McGraw-Hill).

Hobbs, Christopher and Moran, Matthew, (2022), 'Exploring the Human Dimension of Nuclear Security: The History, Theory, and Practice of Security Culture', *The Nonproliferation Review* [online], 9 June. https://doi.org/10.1080/10736700.2020.1811532

IAEA, (2008), *Nuclear Security Culture*, Implementing Guide, IAEA Nuclear Security Series, No. 7 (Vienna: IAEA).

IAEA, (2017), *Self-Assessment of Nuclear Security Culture in Facilities and Activities*, Technical Guidance, IAEA Nuclear Security Series, No. 28-T (Vienna: IAEA).

IAEA, (2021), *Enhancing Nuclear Security Culture in Organizations Associated with Nuclear and Other Radioactive Material*, Technical Guidance, IAEA Nuclear Security Series, No. 38-T (Vienna: IAEA).

Schein, Edgar, (2004), *The Corporate Culture and Leadership*, 3rd edn (San Francisco, CA: Jossey-Bass).

CHAPTER 16

NUCLEAR MATERIAL ACCOUNTING AND CONTROL

SUKESH K. AGHARA AND MARCO MARZO

INTRODUCTION

THE military dimensions of the use of nuclear materials and technology were recognized at the beginning of the nuclear era. In the subsequent decades all governments in possession of nuclear material began applying tight access control to the materials and the technology being developed. The United States (US) Atomic Energy Act[1] and the Baruch Plan,[2] both proposed in 1946, were broadly aimed at restricting the dissemination of nuclear technology information and preventing any supply of nuclear material. The US had already developed a simple accounting system, similar to bank ledgers, for its domestic nuclear material accounting, which was loaned to scientists and organizations working with nuclear material. With the subsequent development of nuclear fuel cycles for civilian applications, when large amounts of nuclear material came to be used by private and state industries, the US Atomic Energy Commission (AEC) developed an elaborate system of accounting and control to improve record keeping.[3]

Nuclear material accounting and control (NMAC) systems were later implemented internationally, following the 1953 Atoms for Peace Programme and 1954 Atomic Energy Act, which sought to transfer nuclear technology for civilian applications to partner countries.[4] This was later brought under the auspices of the International Atomic Energy Agency (IAEA), following its establishment in July 1957, in an effort to ensure that nuclear materials and facilities transferred from one member state to another in the context of cooperation agreements were not diverted to military use. These early IAEA-administered measures, applied under what was known as a safeguards agreement, used accounting and control measures similar to those already being implemented by the AEC. This process was formalized following the entry into force of the Treaty on the Non-Proliferation of Nuclear Weapons (NPT) in 1970, which sought to promote cooperation in the peaceful uses of nuclear energy and prevent the spread of nuclear weapons and weapons technology. Under the NPT, non-nuclear weapon state signatories were required to establish IAEA safeguards on their nuclear activities, with NMAC programmes a core part of these agreements.

Over time, several cases of theft of nuclear material from nuclear facilities, by non-state actors, both by external and internal adversaries, have highlighted the importance of applying NMAC measures in support of nuclear security. This is recognized in Resolution 1540 of the United Nations (UN) Security Council regarding the non-proliferation of weapons of mass destruction (WMD). Adopted in 2004 in the wake of the 11 September 2001 terrorist attacks, Resolution 1540 established obligations for all UN member states to, inter alia, 'Develop and maintain appropriate effective measures to account for and secure such items [nuclear material] in production, use, storage or transport.'[5] The IAEA also recognizes the importance of NMAC systems as a means of controlling access to nuclear material, defining it as

> An integrated set of measures designed to provide information on, control of and assurance of the presence of nuclear material, including those systems necessary to establish and track nuclear material inventories, control access to and detect loss or diversion of nuclear material, and ensure the integrity of those systems and measures.[6]

An effective NMAC system implemented by a nuclear facility operator is essential to maintain knowledge of quantities, type, location, use, and movement of nuclear materials in the facility.[7] When properly implemented, NMAC can detect timely unauthorized abrupt and protracted removals of nuclear material by outsiders or insiders. If a facility security system identifies a theft of nuclear material, NMAC is the only means available to provide complete, accurate, and timely information on the stolen nuclear material, and that can also support a complete material recovery.[8] Despite its importance, NMAC for nuclear security was not fully recognized until almost 60 years after its first uses in nuclear safeguards, when in 2015 the IAEA published its Implementing Guide, 'Use of Nuclear Material Accounting and Control for Nuclear Security Purposes at Facilities'.[9] As a consequence, many nuclear facility operators are still implementing NMAC only to comply with international safeguards commitments and are not fully exploiting the synergies between nuclear safeguards and security systems.

Many of the NMAC elements are common to both nuclear safeguards and security, although both systems have different objectives. Nuclear safeguards aim to detect in a timely manner the diversion of significant quantities of nuclear material for the production of nuclear weapons or other nuclear explosive devices by states; detection times are of the order of months and goals on material quantities are of the order of many kilograms (kg). Nuclear security, meanwhile, aims to detect in the shortest possible time the removal of any quantity of nuclear material by non-state actors.[10] The chapter is structured in three main sections: NMAC systems at the facility level;

the establishment of a system of records and reports; and nuclear material control. Each section introduces the fundamentals of NMAC using examples and concludes with a discussion of synergies in the system that could be applied to safeguards and security.

NMAC Systems at the Facility Level

Accounting measures and statistical methods are used to maintain information about the nuclear material amount present in each area of a nuclear facility. These include physical inventory taking (PIT) to determine the real quantity of nuclear material present at a facility at a given time; quantifying all nuclear materials from the time of receipt up until their final disposition or shipment; and closing and evaluating a material balance period (MBP) between two consecutive PITs.[11] Control systems are used to detect nuclear material losses when they occur or as soon as possible after they occur. These include: a material containment system to establish physical integrity of an area; a material surveillance system aimed at detecting unauthorized activities; an access control system to prevent undetected access to, or movements of, nuclear material; and a detection and assessment system aimed to prevent unauthorized removals of nuclear materials within specified time limits.[12]

Basic Concepts of Accounting

Accounting consists of measuring the nuclear material inventories and the nuclear material flows in and out of an area of a facility, to determine any nuclear material loss. For both nuclear safeguards and security, the change in material balance of uranium, plutonium, and thorium is considered during an MBP, that is, the time between two consecutive PITs. A nuclear facility is divided into one or more material balance areas (MBA) that comply with two conditions: that the nuclear material quantity involved in each transfer to the area (input) or out of the area (output) can be determined; and that the physical inventory in the area can be determined, when necessary.[13]

MBAs are the basic accounting system for records at a facility, with information provided to the relevant national authority. For nuclear safeguards, the material that flows through an MBA's border is recorded as inventory changes, and the material present inside an MBA's border is recorded as the physical inventory.[14] Measurements of nuclear material flows and inventories are performed at key measurement points (KMPs), which are defined in an MBA. The KMP definition includes the entries (inputs) and exits (outputs), and storage areas in MBAs.[15] A KMP structure at a nuclear site was historically defined for nuclear safeguards purposes. However, a facility operator can use a different KMP structure for nuclear security purposes, although the two structures should be compatible. There are two main types of KMPs:[16]

1. Inventory KMPs: locations where nuclear material appears in a form that can be measured to determine inventory in an MBA (typically material storage locations).
2. Flow KMPs: locations where nuclear material is in flux and can be measured to determine nuclear material transfers (entries and exits) or any other relevant inventory change (nuclear loss or production, accidental gains, and losses, etc.).

The MBA concept is relevant to nuclear security as well as being the central accounting system for nuclear safeguards, but the definition of the MBA boundaries may be quite different. For safeguards purposes, a facility is divided into MBAs to comply with the requirements of a safeguards agreement,[17] with the facility operator annually declaring the facility's material balance. Most nuclear facilities (for example, nuclear reactors) are in fact only constituted by one MBA. Even in complex bulk facilities with large material flows and more intensive internal control (for example, conversion and fabrication plants, and enrichment plants), only two or three MBAs are adopted. This concept limits the need for material sampling and what is reported for nuclear safeguards purposes. However, the operator has much more detailed accounting information available to them, which is used to support routine operations.

For nuclear security purposes, the facility operator has the option to define a more granular MBA structure. More MBAs allow for better internal control and facilitate the localization of a material removal or loss. This could be particularly relevant in identifying and responding to insider threats, which are normally not considered in state-level diversion scenarios for safeguards. For example, through deploying multiple MBAs at a particular facility it may be possible to pinpoint in which area of a facility nuclear material was lost and consequently who had access to that material, helping to narrow down a list of potential suspects.

MBAs are established based on criteria such as material types, processes, and functions to allow material unaccounted for (MUF) and their causes to be identified and localized. If an MBA includes more than one geographical area, all areas are administratively controlled by the same person. Therefore, the following elements should be considered when implementing an MBA for nuclear security:[18]

- If the area contains nuclear material in both item and bulk form, it is advisable not to mix them in the same area.
- The MBA boundary should not cross a process area (as this can jeopardize the material flow measurement) or a controlled area (as different categories of nuclear material and custodians would be involved).
- The potential application of containment and surveillance means (seals, cameras, irradiation monitors) on the area borders should be considered to help ensure the completeness of flow measurements.
- A separation between the amounts involved in shipper–receiver difference (SRD) and that involved in MUF can increase control effectiveness, as SRDs are not considered in MUF determination.
- If the MBA structure at a nuclear facility is modified for security purposes, it should remain compatible with safeguards.

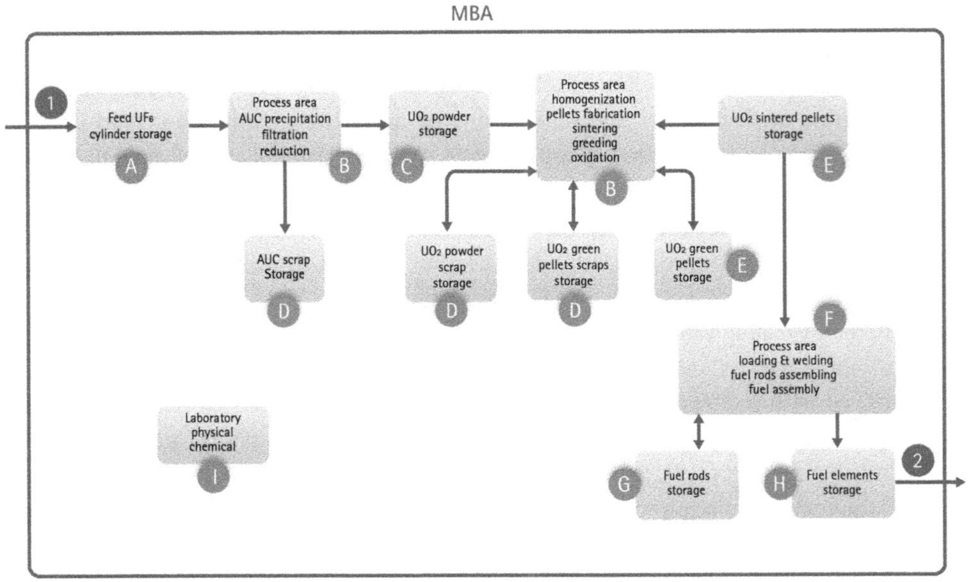

FIGURE 16.1. An MBA structure for a fuel fabrication plant consisting of only one MBA for safeguards purposes.

Figure courtesy of the authors. Also see IAEA, *International Safeguards in the Design of Fuel Fabrication Plants*, IAEA Nuclear Energy Series, No. NF-T-4.7 (Vienna: IAEA, 2017).

As an example of an MBA structure for nuclear security, consider a hypothetical scenario of a fuel fabrication plant. The plant receives uranium hexafluoride (UF_6) cylinders, which are then processed to produce uranium dioxide (UO_2) sintered pellets. These pellets are loaded to produce fuel rods, which are then used to produce fuel assemblies. Historically, the facility has only deployed one MBA for safeguards purposes. Figure 16.1 shows two flow KMPs (1 and 2) and nine inventory KMPs (A to I). It is now decided that this system should be modified so it can support the implementation of nuclear security. Here, one possible approach is to divide the facility into 10 MBAs (Figure 16.2). This 10-structure (security) MBA is compatible with the original (safeguards) MBA. However, the new structure will increase the number of accounting records, as transfers between MBAs must now be recorded. While this may seem like a significant increase in effort, for operational purposes the accounting data and source documents are already available; they just need to be logged and analysed from a security perspective.

Closing a Nuclear Material Balance

Closing a periodic nuclear material balance is essential to determine the material unaccounted for and therefore draw conclusions about material loss or diversion. Accounting practices for safeguards consider the material balance component by strata.[19] A stratum is a group of items or batches that have similar physical and chemical characteristics formed to facilitate statistical sampling.[20] For instance, in a typical

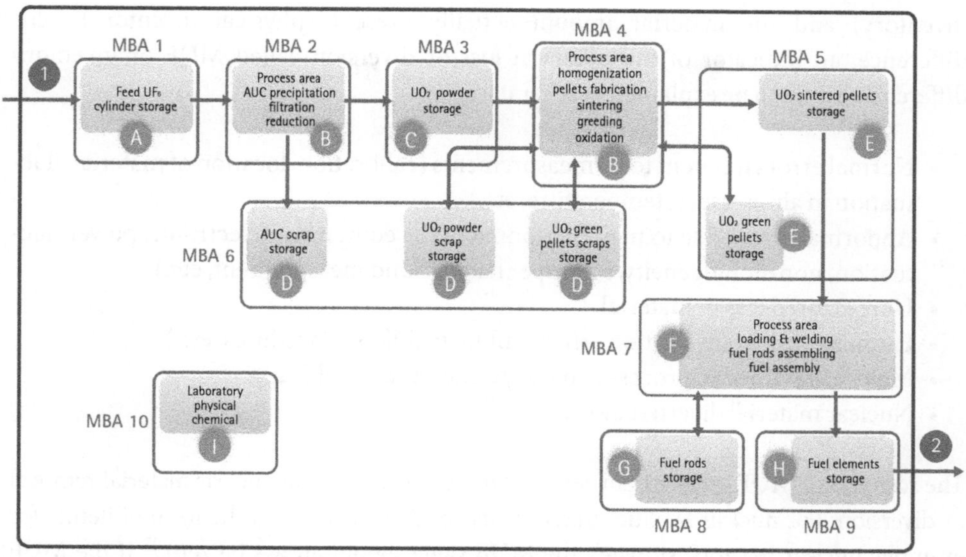

FIGURE 16.2. An MBA structure for a fuel fabrication plant consisting of 10 MBAs for security purposes.

Figure courtesy of the authors. Also see IAEA, *International Safeguards in the Design of Fuel Fabrication Plants*, IAEA Nuclear Energy Series, No. NF-T-4.7 (Vienna: IAEA, 2017).

fuel fabrication plant that receives low enriched UF_6 and produces fuel assemblies, the beginning and ending inventory strata would normally be UF_6 cylinders, drums of UO_2 powder, pellet boxes, fuel rods, fuel assemblies, drums of scrap, or drums of waste. The input stratum would be UF_6 cylinders, and the output stratum would be fuel assemblies.

An item is an identifiable unit that is maintained intact during its stay in the MBA.[21] An item has an identification, a weight or a volume declared, and the possibility to be measured by destructive analysis (DA) techniques or by non-destructive assay (NDA) techniques. Examples of items are a fuel assembly, a fuel rod, a pellet box, a UF_6 cylinder, or a drum of UO_2 powder. A batch is a nuclear material quantity treated as an accounting unit in the MBA and whose composition and quantity are defined by a common set of specifications or measurements (for example, element factor or enrichment). A batch can consist of one or many items. An example of a batch is a number of fuel rods with the same specification, or various drums of UO_2 powder produced in the same production line. These definitions are important, as normally it is impracticable to measure all available material, and NDA and DA measurements are performed in randomly selected items. The measurement results are extrapolated to the batch and to the strata of batches.

Evaluation of the Material Unaccounted For

All components of the nuclear material balance are focused on determining the difference between the material amount that should be present at a facility (book inventory) and the material amount actually present (physical inventory). This difference, an indicator of the potential loss or diversion called MUF or inventory difference (ID), can be explained in several ways:[22]

- Normal errors inherent to all measurements (calibration, location of material, fluctuation in air pressure, temperature, etc.).
- Abnormal errors due to malfunctioning of the equipment (electronic, power fluctuation, non-homogeneity, improper background measurement, etc.).
- Unmeasured loss of material.
- Human error (transcription errors, failure to follow procedures, etc.).
- Material retained in process and not yet measured (hold-up).
- Nuclear material diverted/stolen.

The scenario of MUF greater than zero could indicate a possible nuclear material removal or diversion. For nuclear facilities that only use nuclear material in the form of items (for example, nuclear reactors, storage), the MUF value should always be zero.[23] If the MUF value is not zero then an evaluation should be launched to establish whether the MUF value is acceptable. This will determine whether the MUF value can be explained plausibly (measurement errors, process losses, etc.), or if there is suspicion of possible nuclear material diversion or theft. For MUF evaluations, first the variance σ^2_{MUF} and its standard deviation σ_{MUF} are calculated. As all measurements have uncertainties—random and systematic errors—the measurement uncertainties must be propagated through accounting calculations. Advanced statistical methods are used to quantify expected random and systematic measurement uncertainty associated with closing a material balance.[24]

As measurement uncertainties may be either positive or negative, a confidence interval may be established within which the MUF value could be only due to measurement uncertainties. This confidence interval is defined by the limit of errors of MUF (LEMUF). The LEMUF is a confidence interval, meaning there is a probability of 95 per cent (confidence) that the limits of this interval contain the true value of the MUF.[25] The aim is to arrive at a LEMUF as small as possible that reflects an effective measurement system and good operational practices. Taking L and U as the lower and upper limit of LEMUF, respectively, the following functions can be identified:[26]

- If $L > 0$, the true MUF is positive.
- If $U < 0$, the true MUF is negative.
- If $L < 0 < U$, the true MUF is approximately zero, since the magnitude of LEMUF is small. If the magnitude of LEMUF is too big, no conclusion can be drawn, and the measurement system should be improved.
- Practically, if MUF > LEMUF, the MUF value should be declared significant and investigated.

The MUF evaluation procedure, as described, deals with a single MBP, and, therefore, is used to detect single large thefts that could occur in this area. Additional analysis must be performed to extend these statistical concepts to detect several small thefts, where these are each smaller than LEMUF and occur over a sequence of MBPs. Advanced techniques can be applied to the analysis of the MUF sequences.[27] The cumulative sum technique, or CUSUM, is normally used, with the analysis of a sequential sum of MUFs (CUMUF) enabling the identification of trends in MUF that can indicate a long-term diversion or protracted theft.[28] To reduce LEMUF at a facility, the operator has two options. One option is improving the measurement system. However, reduction of measurement uncertainties is limited by the available instruments and techniques. Another option is increasing the PIT frequency and reducing the MBA size. Normally, physical inventories of plutonium, uranium-233 (U-233), and highly enriched uranium (HEU) should be taken at least every two months, and low enriched uranium (LEU) physical inventories should be taken at least every six months.[29]

To highlight the importance of evaluating MUF, consider a real case of theft of 90 per cent HEU that occurred in 1992 following the Soviet Union's dissolution.[30] Leonid Smirnov, an engineer who had worked for over 25 years at the Luch Scientific Production Association near Moscow, diverted 1.5 kg of HEU as UO_2 powder. He performed about 30 diversions of 50 g HEU over five months (an average of 300 g of HEU removed a month). The HEU inventory in the facility was about 150 kg of HEU, meaning Smirnov diverted 1 per cent of the facility's entire inventory. The removal of such quantities was possible first due to the facility's poor measurement system, which allowed a very large measurement uncertainty at 3 per cent of the inventory (equivalent to 4.5 kg of HEU); this value is 10 times larger than the recommended measurement uncertainty for a fuel fabrication plant (0.3 per cent).[31] And second, the PIT frequency at the facility was too long; in the Smirnov case, a two-month PIT frequency would have been enough to detect the HEU removal.

Nuclear Material Measurement System

The MUF evaluation shows the importance of a good-quality facility measurement system to comply with both safeguards and security objectives, with this particularly critical for the latter. The measurement system must have the following attributes: adequate accuracy to measure nuclear material in both its physical and chemical form; viability for application in KMPs; ability to give a quick result, considering the required timeliness goal; and reasonable price point to make its application viable. A nuclear facility must implement and maintain a measurement programme that consists of well-documented procedures for controlling, calibrating, and maintaining measurement equipment to assure the credibility of the material records at the facility.

The NMAC unit should also periodically perform revisions independent from the operational area to evaluate the measurement system and the measurement control programme, and to verify conformity with international and national standards. The results of these revisions should be documented and maintained as records that indicate the measurement quality. Methods and techniques to calculate the propagation of uncertainties associated with the inventory change measurements and physical inventory measurements should be employed, and their results used to evaluate the contribution of the measurement uncertainties to the MUF.

Physical Inventory Taking (PIT)

Periodically, a facility operator must carry out a physical inventory of all nuclear material at a facility. PIT is the central point of the activity of nuclear material accounting. The primary purpose of the physical inventory is operational, meaning the operator is ensuring that nuclear material losses are within acceptable limits. But it is also an important requirement of both nuclear security and safeguards. The frequency of the PIT depends on the capacity of the facility in terms of processing nuclear material and of accumulated and stored inventories of nuclear material. In general, for safeguards purposes an annual PIT is required.[32] However, as discussed previously, from a nuclear security standpoint, a more frequent PIT is necessary. In the case of a suspected loss of material, a PIT should be immediately conducted to quantify this.

The national authority defines the procedures that the operator must follow to carry out the physical inventory. Typically, in MBAs that only handle nuclear material items, the physical inventory is simple and limited to counting and identifying all items present. Some randomly selected items can also be measured by NDA techniques. In plants that handle bulk material, where the material undergoes physical or chemical processes, carrying out the physical inventory is more complicated. In many cases, the operator must 'clean' the plant's process area, that is, transfer all nuclear material to the inventory KMPs. In large commercial plants this procedure is costly as it often requires stopping the plant, which is a time-consuming activity.[33]

Shipper–Receiver Difference (SRD)

SRD is the difference between the nuclear material quantity in a batch declared by the shipper MBA and the quantity measured by the receiver MBA. The receiver should initially record the batch received using the shipper's data. Later, once the receiver performs a batch measurement, it may find the values are not exactly the same due to difference in measurement instrument accuracy or techniques. Therefore, an SRD is always calculated and informed by the receiver. If the receiver does not perform any quantitative measurement in the received batch, there is no SRD, and the receiver takes on the responsibility for the shipper's declared values when the batch integrity is terminated (for example, the batch is loaded into the process area). When the receiver measures the received batch, the national authority must be informed of the SRD, even if it is zero.[34]

The SRD evaluation is an important element of NMAC, as it enables the detection of nuclear material diversion that could be covered up during transfers of nuclear material between two facilities or between two MBAs. For facilities operating for profit, there is a commercial interest of a shipper and receiver in measuring and recording transferred quantities accurately. As such, certain security and safeguards measures related to the determination of SRD present additional work to normal operations. The situation would become complicated in a scenario where there was collusion between the shipper and the receiver, which could enable material to go missing without being detected.

The most complex and important scenario of material diversion is where a shipper and/or a receiver inflate the measurement error to cover deviations within those errors. It is a scenario that cannot be detected just by examining the accounting statement, item counting, or applying NDA measures. To mitigate this, statistical methods should also be used to assess the size of the combined measurement errors of the shipper and receiver.[35] Although detecting deviations covered by transfers appears relatively easy, studies have shown that this is an important diversion scenario.[36] In certain cases, the shipper's measurement errors are unknown to safeguards, which hampers the evaluation process. As such, particular attention should be given to match material transfers. For instance, some cases have been reported on diversions or diversion attempts through the declaration of material shipments that never have been received at the receiver facilities.[37]

Establishing a System of Records and Reports

There is an internationally accepted format to record and report transactions of nuclear material, as set out in detail by the IAEA in its *Nuclear Material Accounting Handbook*.[38] Countries with comprehensive safeguards agreements with the IAEA are obliged to establish a record and report system following this format, as provided for in Code 10 of the IAEA's *Model of Subsidiary Arrangements*.[39]

Accounting Records

Accounting records consist of a set of documents that are kept at the facility and contain information on: the quantity of each type of nuclear material present at the facility at any time; its distribution within the facility; and the occurred quantity changes. In general, accounting records include the following elements:[40]

- **Book of accounting records**: These provide a summary of the inventory changes and the book inventory in any given period of time. The book inventory is the quantity of nuclear material that should be present at the facility. Each entry should be identifiable through a system of references that links to a support or origin document. The book of accounting records should be separated by nuclear material category (natural uranium, enriched uranium, depleted uranium, plutonium, and thorium).

- **Source documents**: These are the primary documents used for data recording at the operational places where nuclear material originated (the origin data). Normally these documents contain recorded data from measurements. The main data that should be recorded are an inventory change identification; the date of the inventory change; the type of inventory change; and the material description code.

Operational Records

Operational data are used to establish the changes in nuclear material quantities and compositions. They are data recorded at the place where the measurement is performed. For a nuclear reactor they are records that provide the necessary data for the calculation of the nuclear loss (uranium burnup) or nuclear production (plutonium production). Other examples of operational records include description of the actions performed to determine the value and causes of any nuclear material loss (accidental or unexpected).

Accounting Reports

In normal operations, to comply with safeguards agreements, a facility operator must prepare an inventory change report (ICR), a material balance report, and a physical inventory list. An ICR should be prepared and sent to the national authority in the time frame required (usually monthly), as per the safeguards agreement. The data necessary to prepare an ICR should be available in the inventory change documents and reflected in the book of accounting records. A material balance report (MBR) is a declaration of the nuclear material balance, and considers all inventory changes, adjustments, corrections, and physical inventories in a given MBP (normally yearly). A physical inventory list (PIL) is prepared at the end of an MBP, and this describes the nuclear material batches and their data present in the PIT. All data on nuclear materials should be contained in the facility measurement system. Information on the nuclear material inventory is generated through procedures for the PIT. The record can be done automatically or manually, depending on the sophistication level of a facility's control system. In some nuclear facilities that handle items, a system for item monitoring is also implemented. The objective is to assure the timely detection of item losses through the periodic verification of the presence and integrity of items.[41]

Near Real Time Accountancy (NRTA)

The development and implantation of NRTA is an important advancement of nuclear safeguards monitoring in facilities that process large throughput of nuclear material.

Uncertainties about process material measurement contribute in an important way to MUF. In some cases, the MUF is so large that using a conventional nuclear material accounting system is not enough to achieve safeguards objectives. As an example, consider a large reprocessing plant that processes 1,000 tons per year of spent fuel from light water reactors. About 1 per cent of this material is plutonium (Pu). Given the expected measurement uncertainty associated with closing the material balance in reprocessing facilities of 1 per cent,[42] the inventory measurement uncertainty is 100 kg Pu per year. Even performing a PIT once a month would produce a large and unacceptable MUF. In this case, the only solution is applying a NRTA approach that allows periodic (for instance, weekly) closing of the material balance at the facility. NRTA can also be an effective security tool for a facility operator to promptly detect the unauthorized removal of material during processing.

NRTA involves the ability to measures all inputs, outputs, and physical inventories without affecting normal plant operation. It requires performing a large number of bulk and analytical measurements and analysing and evaluating the large volume of accountancy data obtained. Such conditions can only be complied with by automated systems of nuclear measurements and data transmission. The capital cost to implement NRTA may be acceptable by a facility operator when the system is found to be essential for both safeguards and security.

Nuclear Material Control

At the facility level, accounting measures are essential to effectively quantify nuclear material losses or removals. However, accounting measures alone do not allow the immediate detection of an unauthorized removal of nuclear materials. For example, conclusions on the application of accounting measures might be drawn long after the unauthorized removal of nuclear materials have taken place. In this context, nuclear material control is important, as it maintains the continuity of knowledge about nuclear material status and location. This enables control over who has access to the material, in order to help detect any unauthorized handling or movements and identify irregularities that may have occurred. A nuclear material control system encompasses technical and administrative measures used to physically monitor the access and use of the nuclear materials. A comprehensive description of these control measures is provided in the IAEA's Technical Guidance, 'Establishing a System for Control of Nuclear Material for Nuclear Security Purposes at a Facility during Use, Storage and Movement'.[43]

Effective nuclear material control for security purposes should take into consideration all possible scenarios of removing nuclear materials from an MBA. This should include malicious insider actions and both the abrupt removal of a large amount of nuclear materials and the protracted removal of small quantities of nuclear materials, including items or samples. Special consideration should be given to difficult to measure materials, such as material in process or waste, as the associated large measurement uncertainty could be used to mask unauthorized removal.

The intensity of the control measures to be applied depends on the amount of nuclear material, its category, type, and form, and the type of process occurring in the MBA. The custodian, designated to each MBA, is responsible for nuclear material control and must be sufficiently senior in the organizational hierarchy to work with personnel across the facility in the design of an effective system. As many control measures may cut across safeguards, physical protection, nuclear safety, and facility operation, these should be planned and implemented in a consistent and synergistic way. For example, information from a subaquatic video camera put in place by the facility operator to identify spent fuel in pools for operational and safety purposes can be also used for security and safeguards purposes.

Some control measures, such as containment and monitoring, are applied by safeguards organizations as complementary measures to nuclear material accountancy. Normally these are used to maintain continuity of knowledge and reduce the need of material re-verification. For instance, safeguards applied to spent fuel stored at dry storages in nuclear power plants are based only in containment and monitoring measures. As such, the same control methods and equipment can be used for both security and safeguards at the facility level but with different objectives and context. The most important control measures are described in the following subsections.

Containment

Containment aims to maintain the continuity of knowledge on nuclear materials—impeding or making it more difficult to gain access to them, and recording any attempts made to gain unauthorized access. Containment is related to the structural features of a facility area, containers, or equipment used, in order to establish the physical integrity of an area. If the containment is integral, it is not possible to access the materials under containment or the access will be more difficult such that it can only be carried out with a large delay and likely trigger alarms at the facility.

Tamper indicating devices (TIDs) are employed in the movable physical barriers or in the open part of the containment, to assure the continuity of knowledge on the nuclear material inventory. TIDs consist of seals or other technologies used to provide evidence of unauthorized access to nuclear materials. TIDs do not prevent a removal of nuclear materials, but rather provide indication of an attempt to violate a containment. The most common TIDs are metallic seals, which are cheap and reliable. Metallic seals can be tampered with, so to fully confirm their integrity it is necessary to implement a well-established random procedure for the verification of the seal integrity. This can be a time-intensive procedure, which may delay the detection of an adversary.

More advanced systems of verification use passive fibre optic or active electronic radio frequency seals that can be promptly checked using a seal reader without needing to replace them in extreme environmental conditions. This method provides a substantial time reduction for checking seal integrity and near real time alarms.[44] However, these types of seal are very expensive, and are only used in specific situations to contain sensitive items or equipment or a large amount of nuclear materials. As yet, there is no cheap, reliable, and on-site verifiable seal type available, and efforts for its development are needed.[45] There are a number of containment technologies being developed that use advanced fibre optics and lasers.[46] A laser curtain for containment technology (LCCT) has started to be applied to contain large dry storage at nuclear power plants.[47] Different types of nuclear facilities require different types of containment due to obligations related to the nuclear material handling. Right at the design stage of new facility, a robust containment system should be incorporated to enhance nuclear material control.

Monitoring and Access Control

Monitoring measures help detect and deter unauthorized activities, with information collected by observation, measurements, and checks. Surveillance systems are applied for several reasons: to maintain the continuity of knowledge over nuclear materials; to identify nuclear material movements; to identify personnel access to an area; and to confirm the integrity of an area containment. They are also used to maintain continuity of knowledge over equipment in operation (e.g. a crane, a valve, a door) or to confirm an activity or operation in an area. Surveillance systems are very useful in an investigation to resolve an anomalous situation by recovering information about an area. NMAC systems can provide valuable information from receipts and shipments of nuclear materials. This includes checking for consistency in packaging documents and any change of physical location, for instance transfers between facility buildings.

Although closed circuit television (CCTV) and other video cameras can be easily procured and installed, they must be employed carefully. The review of images and checks of a camera's functioning normally result in a large workload for personnel. Surveillance systems are also faced with challenges related to operators, reliable power supply, unstable network, wiring problems, hardware (cameras, data storage, or monitor) failures, among others. Hence, nuclear facilities also use other surveillance measures such as radiation monitors, radio frequency tags, motion detectors, laser monitors, weight sensors, and heat sensors. Radiation monitors are used at entrances and exits for the detection of unauthorized introduction or removal of nuclear materials. Portal monitors are often used for this purpose. To avoid the introduction of shielded containers that can be used to blind a nuclear material removal, portal monitors should be used together with metal detectors. Normally, physical protection systems (PPS) control access to the facility, buildings, and protected areas. The integration of PPS with NMAC allows for the controlling of access to nuclear materials to prevent unauthorized or undetected access to, or movement of, nuclear material. Only persons who have a need to access nuclear materials should be granted access to nuclear materials and nuclear material accounting data and equipment.

Detection and Response

Any unexpected observation in material control measures, an 'irregularity', might indicate an unauthorized removal of nuclear materials. Therefore, every irregularity should immediately initiate an investigation to determine its cause and magnitude. As there is an extensive list of possible irregularities, the facility operator should define, in advance, the relevant irregularities for any given situation. In general, the irregularities can be divided into two groups. The first is directly related to nuclear material loss or removal (e.g. a missing item, a statistically significant value of MUF, a significant difference between the measured and recorded value of nuclear materials, a significant SRD, an overstatement of shipments, or an understatement of receipts). The second involves failures in administrative or technical control measures that might indicate an attempt to remove nuclear materials (e.g. a broken TID, a failure of a surveillance measure, a violation of the two-person rule, an unauthorized action involving nuclear materials/data/equipment).

When an irregularity is detected, the facility's management team (both those responsible for NMAC and physical protection management) should be informed immediately, whereupon they should start a coordinated response. The nature of the response will be reflective of the nuclear materials involved. Detailed procedures and guidelines should be developed, approved by the facility management, and maintained on record. Response actions may include a records audit and evaluation of possible back-up measures. In critical cases, an emergency PIT can be performed. The investigation should be well documented and a report describing the nature of the irregularity, all steps of the investigation, and the corrective actions adopted should be prepared and submitted to the facility management.[48]

Organization and Management

The responsibility for NMAC at the facility level should be attributed to a formal NMAC unit (department, division, etc.). The NMAC unit should be formed by a director and one or more qualified employees, and it is recommended that these functions are independent of the general manager or operations manager of the facility. The NMAC unit has responsibility for maintaining the records system used to document and track all nuclear materials at the facility, including inventories, receipts to, and shipments from all MBAs. Therefore, the NMAC unit must be informed about all nuclear material flows and inventory changes.

Safeguards requirements, whose functions and activities are subject to international inspections, have led to state support for NMAC systems and consistent safeguards personnel teams at nuclear facilities. However, the security system and personnel vary between facilities. Mandatory inspections and verification are characteristics of safeguards but there are no such requirements for nuclear security. There is also significant variation in organizational sub-cultures, which often leads to separate personnel being involved with safeguards and nuclear security. In general, nuclear security personnel lack awareness about NMAC and its application; likewise, safeguards personnel lack awareness about nuclear security. There are no international recommendations on developing an integrated nuclear safeguards and security system at nuclear facilities. As a result, an integrated safeguards and security culture involving all relevant personnel is missing at many facilities. Nevertheless, as discussed, there are benefits at a national level in integrating NMAC for nuclear safeguards and security, encouraging regular interaction between the safeguards and the security teams.

Conclusion

NMAC systems provide an essential means of reducing the risk of the theft or diversion of weapons-useable fissile materials by both states and non-state actors. This chapter highlights some of the key concepts of NMAC and their application to nuclear safeguards and security. Significant gains can be realized by adopting an integrated nuclear security and safeguards approach to NMAC. Nuclear operators, governments, and the international nuclear community should all pursue ways to leverage accounting and monitoring systems to this end. Integrated NMAC systems should include a clearly defined response plan, with independent authority for safeguards personnel to be able to coordinate effectively with the physical protection security force. The integration of NMAC and physical protection systems can bring efficiency and greater effectiveness of nuclear security at nuclear facilities.

Notes

1. Senate Special Committee on Atomic Energy, *The United States Atomic Energy Act of 1946, S 1717* (Washington, DC: US Government Printing Office, 1946), 1–9.
2. Bernard Baruch, 'Statement before the UN Atomic Energy Commission', Hunter College, New York, 14 June 1946.
3. George T. Mazuzan and J. Samuel Walker, *Controlling the Atom: The Beginnings of Nuclear Regulation 1946–1962*, NUREG-1610 (Washington, DC: US Nuclear Regulatory Commission, 1997).
4. Ibid.
5. UN Security Council, *UN Security Council Resolution 1540 (2004)*, S/RES/1540 (New York: United Nations, 2004).
6. IAEA, *Use of Nuclear Material Accounting and Control for Nuclear Security Purposes at Facilities*, Implementing Guide, IAEA Nuclear Security Series, No. 25-G (Vienna: IAEA, 2015).
7. Ibid.
8. Ibid.

9. Ibid.
10. IAEA, *IAEA Nuclear Security Glossary: Terminology Used in IAEA Nuclear Security Guidance*, Draft, 2020 edn (Vienna: IAEA, 2019).
11. IAEA, *Nuclear Material Accounting Handbook*, IAEA Services Series, No. 15 (Vienna: IAEA, 2008).
12. IAEA, *Establishing a System for Control of Nuclear Material for Nuclear Security Purposes at a Facility during Use, Storage and Movement: Technical Guidance*, IAEA Nuclear Security Series, No. 32-T (Vienna: IAEA, 2019).
13. IAEA, *IAEA Safeguards Glossary*, 2001 Edition, International Nuclear Verification Series, No. 3 (Rev. 1) (Vienna: IAEA, 2022).
14. Ibid.
15. Ibid.
16. Ibid.
17. Ibid.
18. R. K. Larsen, C. E. Crawford, and P. W. Gibbs, 'Comparison between Nuclear Material Accounting and Control for Nuclear Security and a State System of Accounting and Control for Safeguards' in *Proceedings of the IAEA International Conference on Physical Protection of Nuclear Material and Nuclear Facilities*, 13–17 November 2017 (Vienna: IAEA, 2017), 2471–7
19. IAEA, *IAEA Safeguards Glossary*, 44.
20. Ibid.
21. Ibid., 28.
22. IAEA, *IAEA Safeguards Technical Manual Part F: Statistical Concepts and Technique*, Vol. 3 IAEA-TECDOC-26 (Vienna: IAEA, 1982).
23. Ibid.
24. Ibid.
25. John Jaech, *Statistical Methods in Nuclear Material Control*, TID-26298 (Washington, DC: US Atomic Energy Commission, Technical Information Center, 1973).
26. Ibid.
27. W. Michael Bowen and Carl A. Bennett, *Statistical Methods for Nuclear Material Management*, PNL-5849 (Richland, WA: Pacific Northwest Laboratory, 1986).
28. T. Burr and M. S. Hamada, 'Revisiting Statistical Aspects of Nuclear Material Accounting', *Science and Technology of Nuclear Installations*, 2 (2013), Article ID 961360.
29. IAEA, *Nuclear Material Accounting Handbook*.
30. Noah Gale Pope and Christopher Hobbs, *Insider Threat Case Studies at Radiological and Nuclear Facilities*, IAEA Technical Report No. LA-UR-15-22642 (London: Los Alamos National Laboratory (LANL) and King's College London: 2015).
31. IAEA, *Quantifying Uncertainty in Nuclear Analytical Measurements*, IAEA-TECDOC-1401 (Vienna: IAEA, 2004).
32. IAEA, *Safeguards Implementation Practices Guide on Provision of Information to the IAEA*, IAEA Services Series, No. 33 (Vienna: IAEA, 2016).
33. IAEA, *Use of Nuclear Material Accounting and Control for Nuclear Security Purposes at Facilities*, Implementing Guide, IAEA Nuclear Security Series, No. 25-G (Vienna: IAEA, 2015).
34. IAEA, *Safeguards Implementation Practices Guide on Provision of Information to the IAEA*.
35. A. Rota, 'The Use of a Shipper-Receiver Difference in the Facility Accountancy and the Safeguards Evaluation', *ESARDA Bulletin*, No. 13 (1997), 22–8.

36. Ibid.; R. Schenkel and W. Kloeckner, 1993, 'The Role of Non-Quantifiable Aspects in Nuclear Safeguards', in Proceedings of the 34th Annual Meeting of the Institute of Nuclear Materials Management, 18–21 July 1993, Scottdale, AZ.
37. Schenkel and Kloeckner, 'The Role of Non-Quantifiable Aspects in Nuclear Safeguards'.
38. IAEA, *Nuclear Material Accounting Handbook*.
39. IAEA, *Model Subsidiary Arrangements Code 10 (Labelled)*, SG-FM-1172 (Vienna: IAEA, 2017).
40. IAEA, *Nuclear Material Accounting Handbook*.
41. K. K. S. Pillay, *Fundamentals of Materials Accounting for Nuclear Safeguards*, LA-11569-M (Los Alamos, NM: Los Alamos National Laboratory, 1989).
42. IAEA, *IAEA Safeguards Glossary*.
43. IAEA, *Establishing a System for Control of Nuclear Material for Nuclear Security Purposes at a Facility during Use, Storage and Movement*.
44. IAEA, *Safeguards Techniques and Equipment*, International Nuclear Verification Series, No. 1 (Vienna: IAEA, 2003).
45. Ahn Jihyun, Park Junsung, Sim Hayoung, An Geunyoeng, and Seo Hee, 'Design Requirements of Safeguards Sealing System for Real-Time Verification', *Nuclear Technology & Radiation Protection*, 36/4 (2021), 376–83.
46. IAEA, *Development and Implementation Support Programme for Nuclear Verification 2022–2023*, IAEA Safeguards, STR-400 (Vienna: IAEA, 2022).
47. Vitor Sequeira, et al., 'LASER Curtain for Containment and Tracking', in *Proceedings of the INMM & ESARDA Joint Virtual Annual Meeting*, August 23–26 & August 30–September 1, 2021 [online].
48. IAEA, *Use of Nuclear Material Accounting and Control for Nuclear Security Purposes at Facilities*.

Suggested Reading

Doyle, James E., (2019), *Nuclear Safeguards, Security and Nonproliferation: Achieving Security with Technology and Policy*, 2nd edn (Kidlington: Elsevier).

Eldridge, Christopher, (2006), *Protection, Control, and Accounting of Nuclear Materials* (Washington, DC: National Research Council; National Academies Press).

Siegel, Jonas, Gallagher, Nancy, and Steinbruner, John, (2014), *Comprehensive Nuclear Material Accounting: A Proposal to Reduce Global Nuclear Risk* (College Park, MD: Center for International Security Studies).

IAEA, (2015), *Use of Nuclear Material Accounting and Control for Nuclear Security Purposes at Facilities*, Implementing Guide, IAEA Nuclear Security Series, No. 25-G (Vienna: IAEA).

IAEA, (2019), *Establishing a System for Control of Nuclear Material for Nuclear Security Purposes at a Facility during Use, Storage and Movement*, Technical Guidance, IAEA Nuclear Security Series, No. 32-T (Vienna: IAEA).

CHAPTER 17

SECURING NUCLEAR MATERIALS DURING TRANSPORT

KARL DEWEY, BOB OFFICER, AND BEN WHITTARD

Introduction

Shipments of nuclear and radioactive material take place daily all over the world, via road, rail, air, inland waterway, and sea. Most transportations are conducted domestically by national transporters, which largely transport material by road. However, to cater for international transportations, there are a small number of specialist international transport companies, for example, the UK's Nuclear Transport Solutions (NTS) or France's Orano Nuclear Packages and Services (Orano NPS), that are licensed to transport nuclear materials of up to Category I, which refers to the categorization system of the International Atomic Energy Agency (IAEA).

Despite being an everyday occurrence, the provisions undertaken to transport nuclear and radioactive materials safely and securely are generally not well understood, and there is a perception that transportation is the weakest link within the global civil nuclear estate.[1] In part, this is because transportation is considered a niche area of endeavour and expertise, conducted by relatively few licensed operators (although the intricacies of mobility itself and the need to work with multiple stakeholders also increase the complexities of any operation). Despite this perception, the transportation industry remains a highly regulated environment that has numerous measures in place to ensure the safety and security of nuclear and radiological material when in transport.

To unpick the workings of effective transport security, this chapter adopts a practical industry-informed perspective. It first outlines the international framework that governs the transport of nuclear and radiological material, and the key bodies that regulate the various modes of transport options. Building on this, the chapter shows that, with the application of risk-based approaches, the use of intelligence, and the design and execution of effective physical and cyber security solutions, nuclear carriers can maintain an edge over potential adversaries. The chapter also provides further detail on how fundamental principles of nuclear security such as 'graded approaches' and 'defence-in-depth' are applied to transport security, before outlining additional security measures such as guarding, escort, and the role of armed response forces. The chapter then discusses the importance of security contingency measures and how contingency planning for 'nuclear security events' (NSE) during transport complements wider emergency preparedness and response paradigms. Finally, the conclusion draws these aspects together to show how extensive planning and preparation result in the successful execution of transport operations.

Although the planning and execution of secure transport operations is complex, the chapter argues that international standards clearly delineate the roles and responsibilities of states, competent authorities, and operators. When enacted, these form a comprehensive system that contributes to an effective nuclear security regime. Moreover, through the systematic consideration of various threat landscapes, the use of intelligence and risk-based practices, and the use of effective physical protection systems (PPS) and cyber protection systems (CPS), nuclear carriers can ensure the safe and secure transportation of nuclear and radiological material.

International Framework: Treaties, Guidance, and Coordinating Mechanisms

The main international legal instrument governing the security of nuclear material is the 1979 Convention on the Physical Protection of Nuclear Material (CPPNM) and its 2005 Amendment (A/CPPNM). The CPPNM entered into force in 1987 and establishes legal obligations for states parties to protect civil nuclear material during transport. The CPPNM also requires states parties to criminalize certain offences involving nuclear material and to cooperate internationally in certain circumstances, for example, the theft of nuclear material.[2] The 2005 Amendment entered into force in 2016 and extends the CPPNM's original scope to also cover nuclear facilities and nuclear material that are used for peaceful purposes. This includes when nuclear materials are in domestic use, storage, and transport. States parties are obliged to 'establish, implement, maintain and sustain an effective and appropriate nuclear security regime to prevent, detect and respond to such nuclear security events', and must implement 'an appropriate legislative and regulatory framework for physical protection including against sabotage'.[3] For example, in the United Kingdom (UK), the national-level regulatory mechanism is the 2003 Nuclear Industries Security Regulations.[4]

To support the application of A/CPPNM by approved carriers, in 2011 the IAEA updated its Nuclear Security Series (NSS) document 'Nuclear Security Recommendations on Physical Protection of Nuclear Material and Nuclear Facilities' (INFCIRC/225/Revision 5).[5] INFCIRC/225 does not provide detailed guidance on the implementation and compliance responsibilities for the transport of nuclear material, but instead works in combination with the NSS Implementing Guide 'Security of Nuclear Material in Transport' (NSS 26-G) to assist carriers in fulfilling their physical protection responsibilities in the transport of nuclear materials.[6] Although the A/CPPNM is legally binding on states parties, the guidance offered in the NSS is not, and is dependent on states implementing these principles nationally. For example, INFCIRC/225 recommends that the transportation of different categories of nuclear materials are protected with security measures aligned to the material's categorization. This is reflected in the UK's approach and outlined by the UK Competent Authority's Technical Assessment Guide.[7]

Modes of Transport Considerations

Nuclear and other radioactive materials are generally conveyed using five different modes of transport: road, rail, maritime (oceanic), inland waters (i.e. lakes, rivers, canals), and air. Transport operations that use multiple modes are called 'multimodal'. The mode(s) of transport chosen depend on a wide range of factors. These vary between states but may include factors such as the operating environment and distances to be traversed, public support for certain modes of transport, national geopolitics and threat profiles, and the responsible state's overall approach to balancing these competing needs.[8] The concept of threat-based risk informed approaches is also captured by the IAEA's Implementing Guide, 'National Nuclear Security Threat Assessment, Design Basis Threats and Representative Threat Statements', which recommends that national threat statements should take into account policy factors before wider dissemination to key stakeholders.[9]

Internationally, the United Nations (UN) has sought to harmonize standards of regulations for the transport of dangerous goods. In doing so, the UN's 'Recommendations on the Transport of Dangerous Goods Model Regulations' (commonly referred to as the 'Orange Book'), codifies nine classes of dangerous goods in transport.[10] According to the Orange Book, nuclear and radioactive materials are 'Class 7' goods, for which the Orange Book recommends distinguishing between 'basic' levels of security and 'enhanced' levels of security, with the latter to be undertaken for cargoes deemed to be of higher consequence.[11]

Although such high-level guidance applies to all modes of transport, each modality is further regulated by specific guidelines and expectations, with compliance overseen by distinct regulatory bodies. The International Civil Aviation Organization (ICAO) is the international modal authority for air transport. The ICAO has authored several relevant Technical Instructions (TIs) that incorporate Orange Book recommendations. Similarly, the

International Air Transport Association (IATA) acts as the trade association for the world's airlines and issues the only industry-recognized standard for the shipping of dangerous goods by air. For sea conveyances, the International Maritime Organization (IMO) is the international modal authority. The IMO authored the 'International Maritime Dangerous Goods (IMDG) Code', which includes the 'International Code for the Safe Carriage of Packaged Irradiated Nuclear Fuel, Plutonium and High-Level Radioactive Wastes on board Ships' (INF Code),[12] as well as the 'International Ship and Port Facility Security Code' (ISPS Code).[13] Regarding the regional modal authority for road and inland waterways transport, the Committee on Inland Transport of the UN Economic Commission for Europe (UNECE) is responsible for two main agreements, namely, the 'International Carriage of Dangerous Goods by Road' (ADR) and the 'European Agreement on the International Carriage of Dangerous Goods by Inland Waterways' (ADN). International rail transport is subject to oversight by the Intergovernmental Organisation for International Carriage by Rail (OTIF). OTIF has global reach, developing the 'International Carriage of Dangerous Goods by Rail' to promote, improve, and facilitate international traffic by rail and as the global authority for rail transport.

Depending on the desired mode(s) of transport, operators can refer to the aforementioned documents and determine whether the materials intended to be transported present a significant threat to people, property, or the environment. For radioactive material, tables clearly outlining radioactive activity threshold levels for isotopes—both in isolation and in aggregation—also help inform whether basic or enhanced levels of protection are required. The IAEA's NSS Implementing Guide 'Security of Radioactive Material in Transport' offers guidance on what this might look like and aids carriers in meeting their security obligations.[14]

Although the ultimate responsibility for nuclear security rests with the state,[15] during transportation ownership of risk chiefly lies with the carrier. Because of this, and the fact that the impacts of any incidents would be unlikely to be limited to any single country, several initiatives have sought to increase the transparency of nuclear transport operations and associated risks. For example, in the late 1990s, many coastal states began requesting greater levels of information on the material transiting their waters. This led to the formation of the 'Coastal and Shipping States' informal engagement forum within the IAEA.[16] Similarly, bodies such as the IAEA's Incident and Emergency Centre (IEC) have used periodic exercises to facilitate information sharing between different actors. For example, in 2017, 24 participants from across governments, regulatory bodies, and nuclear carriers took part in an IEC exercise centred on a fictional scenario of a fire on a ship transporting nuclear waste (Figure 17.1).[17]

Where applicable, carriers have also taken the initiative to extol their track records, both through their own communication portals, but also via the Global Acceptance programme. This unique collaborative industry effort includes carriers such as NTS and Orano NPS, and a consortium of Japanese utilities working in a harmonized fashion to educate and build relationships at the diplomatic level. The programme seeks to assure states on such topics as safety, security, and emergency response regarding nuclear and radioactive material transport (Figure 17.2).

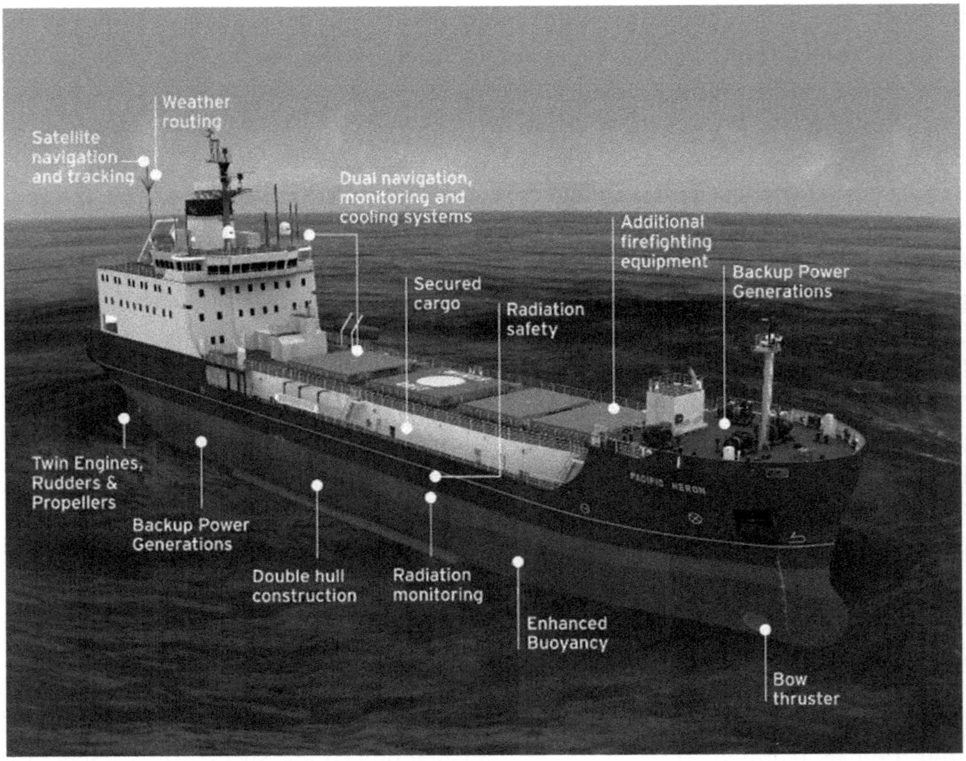

FIGURE 17.1. Cutaway model of a ship transporting radioactive material.

Image courtesy of Nuclear Transport Solutions (NTS).

FIGURE 17.2. Image of a loading operation involving a ship, train, and flask.

Image courtesy of Nuclear Transport Solutions (NTS). See also NTS, 'NTS completes successful repatriation of intermediate-level waste to Australia', NTS [online], March 2022. https://nucleartransportsolutions.com/2022/03/ntscompletes-successful-repatriation-of-intermediate-level- waste-to-australia

Defining Threats within the Operating Environment

Because carriers face a range of potential threats, no single measure can offer total security. Instead, a risk-based and graded approach offers pragmatism, value for money, and proportionality.[18] Risk is often defined as a function of threat, consequence, and vulnerability. Here, threat is the product of a malicious actor's capability and intent, and in practice carriers must work closely with state authorities to understand the threats faced for each shipment. This may be done either generically (perhaps via an annual nuclear sector threat assessment) or upon request with transport-specific threat assessments. In the latter case, assessments will include factors such as the attractiveness of the material, its potential to create an unacceptable radiological consequence (URC), and the graded approach taken to protect the material.

In addition to these threat assessments, other resources—such as state-produced design basis threat (DBT) or representative threat statements (RTS)—can provide information relevant to planning assumptions, although only in cases where information therein is made available to carriers.[19] Such resources might include further details of various actors' attributes and characteristics such as motivations, strength, training, relevant skills, competency, tactics, support, likely weapon systems/explosives, cyber tools, and level of access, both from a cyber and physical perspective.

The Importance of Intelligent Customers

As well as information provided by state authorities, carriers may gain additional insight from looking at comparable industries—for example, from critical national infrastructure or those employed in high-risk areas, such as theatres of military operations. The latter are particularly relevant where they are close to areas of transit. Although not directly applicable, extrapolating lessons from horizon scanning exercises can enlighten the design and delivery of protection solutions and help keep carriers abreast of changing threats. To that end, continuous threat monitoring forms part of the duties of an 'intelligent customer' and aids a carrier's intelligence collection plan, including specific requests for information (RFIs) to state authorities to explore the credibility of actors and attack vectors within the nuclear transport domain. Carriers with an 'intelligent customer' functionality may identify knowledge gaps and other information relevant to security operations. Using this information, a competent security team will be able to design a business-centric collection plan before applying reporting, comment, and assessment in a nuanced fashion to each bespoke transport conundrum.

Although the use of intelligence in planning transport operations is often overlooked by external commentators, for carriers the role of intelligence is critical in understanding how potentially malicious actors may threaten the security of individual shipments.[20] This is especially important because, regardless of any actual radioactive release, attacks against nuclear transports may also evoke considerable psychological impact, such as from 'dreaded risk'. These are 'risks that are viewed as being invisible and involuntary with large-scale, lethal, and long-term effects and indiscriminant [sic], inequitable, and uncontrollable consequences'.[21]

If used properly, intelligence is an enabling capability that allows for the focused application of security efforts or mitigatory solutions. However, intelligence alone cannot lead to secure operations and bringing various strands of information together to provide timely, accurate, and actionable intelligence products requires carriers to use a threat-based intelligence management system. Such tools convert intelligence into practicable information and help support core business operations, for example, by aiding the design of both PPS and CPS and apprising all key security stakeholders of emergent threats. Therefore, through a combination of key stakeholder engagement; tasking of relevant sources and agencies (SANDAs); fusion of local area knowledge from subsidiaries, partners, or academia; and employment of open-source monitoring, carriers can use intelligence to inform their operational decision making. Such an approach maximizes preparations against anticipated adversaries while minimizing risks to a shipment, even before disembarkation.[22]

Responsibility for developing malicious actor attack scenarios varies from state to state and on whether national and regulatory bodies adopt prescriptive- or outcomes-focused approaches to their planning. Nonetheless, attack scenarios form an essential consideration in assessing adversarial pathways, as does the operating environment itself. Assessing the operational environment promotes situational awareness and highlights future requirements. This is an essential planning tool that allows carriers to make an informed decision on how—or indeed whether—a shipment of nuclear or radiological material should be conducted. Assessments of an operational environment will also feed into the design, and finessing, of any PPS/CPS captured within the Transport Security Plan (TSP) and concomitant Security Contingency Plan (SCP).

According to the UK's Joint Doctrine Publication,[23] a well-structured estimate will:

- Define the operating environment. [This will differ with the modality chosen.]
- Describe the environment's effects on the conduct of the shipment. [Consider limited infrastructure, weather, etc.]
- Analyse the threat actors or intended targets. [These should be mode specific and include RFIs for SANDAs.]
- Identify possible risks to operations. [Consider attack scenarios for each shipment.]
- Identify areas where intelligence collection assets must monitor or detect threats
- Identify opportunities where response forces can influence outcomes. [Posturing.]
- Identify decision points when the escort commander must act to influence the outcome of the shipment. [These are usually captured within the carrier and escort force's Concept of Operations.]

- Analyse the escort commander's and malicious actor's centres of gravity. [De-risk the shipment through avoidance of high-risk areas.]

The 2019 Attack at Angra dos Reis

Although such plans may sound routine, and attacks on nuclear transport operations are exceedingly rare, some have nonetheless been documented. For example, on 20 March 2019 a fresh fuel convoy supplying the Angra Nuclear Power Plant in Brazil was attacked while travelling from Resende Nuclear Fuel Factory—a journey of approximately 145 km.[24] On the day of the attack, the convoy appears to have entered an area of the Angra dos Reis municipality where different armed gangs were vying for control and fighting each other. The convoy itself came under attack, although it later arrived intact at its destination.

Eletronuclear, Brazil's state-owned nuclear operator, later downplayed the significance of the attack, saying that the convoy had not been intentionally targeted and that the assailants had been scared off by the convoy's strong armed escort.[25] However, the eventual safe return of the transport is less relevant because overcoming the attack was more a result of luck than judgement. The risk to the transport could have been avoided if the right intelligence had been provided.

As such, the Angra dos Reis example illustrates the importance of considering all actors on the threat spectrum and the need to consider a range of sources—including local partners, police, and open sources (such as social media)—to ensure decision makers are informed with timely and actionable information. This is because the consequences of even an incidental attack cannot be dismissed. Had the cargo been of a different nature and form with a greater toxicity or radioactivity level, the attack at Angra dos Reis might have resulted in a URC, requiring local, regional, and—most likely—national responses.

Graded Approach and Defence-in-Depth during Transportation

Transport does not benefit from the physical protection advantages afforded to fixed nuclear sites and facilities such as layers of security fences, closed-circuit television (CCTV) cameras, and perimeter detection systems. Similarly, the weight of such mitigatory measures within a mobile environment may preclude their incorporation. Moreover, a shipment may travel through multiple regions and different types of environments, all of which combine to make the threats posed to them far more dynamic. Nonetheless, the security principles used to protect nuclear materials in transport do not differ greatly from those applied to fixed facilities, with many predicated on ensuring effective measures are adopted that are proportionate to the consequence of theft/sabotage.

Common Requirements

INFCIRC/225 outlines several common requirements for the transport of nuclear material.[26] Referred to as the 'fundamental transport principles,' common IAEA requirements include:

- Minimizing the total time during which the nuclear material remains in transport. [This is a key requirement, underscoring the importance of 'time at risk' when material leaves protected facilities.]
- Minimizing the number and duration of nuclear material transfers, that is, transfer from one conveyance to another, transfer to and from temporary storage, and temporary storage while awaiting the arrival of a conveyance, etc.
- Protecting nuclear material during transport and in temporary storage in a manner consistent with the category of that nuclear material. [This is critical to ensure the continuity of protection, regardless of the material's location.]
- Avoiding the use of predictable movement schedules by varying times and routes where possible. [This can often be difficult and is subject to infrastructure restrictions, such as major motorways or rail lines.]
- Requiring predetermination of the trustworthiness of individuals involved during transport of nuclear material. [This involves gaining assurance that those involved are honest, reliable, and trustworthy.]
- Limiting advance knowledge of transport information to the minimum number of persons necessary.
- Using a material transport system with passive and/or active physical protection measures appropriate for the threat assessment or DBT.
- Using routes that avoid areas of natural disaster, civil disorder, or a known threat.
- Ensuring that packages and/or conveyances are not left unattended for any longer than is necessary. [This includes at intermodal points and locations of rest.]

These are common principles and are considered good practice. However, on occasion, these principles may conflict with one another. For example, minimizing the duration of a shipment may not be possible if a carrier is taking a mindful decision to avoid a known area of conflict or threat, via a less direct route. Another challenging area relates to limiting advance knowledge of transportation, as the transport sector relies heavily on its supply chain. For example, a vessel will need to work with agents, port authorities, lifting companies, border control, freight forwarders, etc., which increases the risk of information being compromised. Additional mitigation is therefore required to minimize this risk to a tolerable level. Mitigation measures may include compartmentalization of tasks and operational information to reduce the total number of personnel in possession of all the details relating to the transport. Similarly, protectively marking critical elements of the transport security plan, conducting comprehensive background checks, and fostering a robust security culture all combine to engender corporate confidence.

Defence-in-Depth

INFCIRC/225 lists a number of 'Fundamental Principles', with Fundamental Principle I focusing on 'defence-in-depth'. Defence-in-depth involves creating several layers and protective methods that have to be overcome or circumvented by an adversary. These may be structural, technical, personnel, and/or organizational in nature. Defence-in-depth is also used in the transport of nuclear materials, although its application will vary by transport mode. Nonetheless, it will often involve a mixture of hardware (such as security devices including detection systems and GPS tracking), procedures (such as security instructions on what to do when stopping at overnight rest points), and transport design (involving physical barriers such as containers and packages) (Figure 17.3).

FIGURE 17.3. Image showing a lifting mechanism being attached to a flask.

Image courtesy of Nuclear Transport Solutions (NTS). See also NTS, 'NTS completes successful repatriation of intermediate- level waste to Australia', NTS [online], March 2022. https://nucleartransportsolutions.com/2022/03/ntscompletes-successful-repatriation-of-intermediate-level- waste-to-australia

Categorization

Like materials stored at licensed facilities, when determining which security measures are necessary, carriers must first determine the category of the material being transported. Both theft and sabotage scenarios are considered. For example, categorization for theft is based on the type, form, and quantity of material to be transported, and its potential use in nuclear proliferation—to create a nuclear explosive device or a nuclear dispersal device. Plutonium and highly enriched uranium (HEU) therefore attract higher categorization levels and far more stringent security requirements.[27] Categorization against sabotage is determined by assessing the potential radiological consequences should the material in transport be sabotaged. This considers the impact on the public but relies heavily upon states defining what is considered an 'unacceptable radiological release'. Interestingly, materials that attract higher categorizations for theft may be lower for sabotage, and vice versa. This can create challenges for carriers. On nuclear sites, this measurement is often taken over 24 hours at the perimeter of the site (i.e. the point at which the public is impacted); however, in transport this takes place at the point of release (i.e. the package) as transport takes place in the public domain. As such, the threshold for nuclear transport is often far lower than that of nuclear facilities.[28]

The shipment of nuclear material can, on occasion, also involve the transport of various types, forms, and quantities of materials in the same conveyance. This can make the categorization of shipments more complex, and sometimes the categorization of a shipment changes throughout its journey. In this instance, nuclear material should be aggregated to ensure an appropriate categorization level is concluded. There are various mathematical approaches to achieve this, many of which are outlined in the IAEA's Implementing Guide NSS 26-G. These may be then adopted through national guidance, such as the UK's Nuclear Security Technical Assessment Guide, 'Categorization For Theft', published by the Office for Nuclear Regulation (ONR).[29]

Security Provisions and Vulnerability Assessments

NSS 26-G outlines security provisions that should be applied to all transportations of Category I–III nuclear material. These reflect international good practice and are advised for all categories of nuclear material across all modes of transport. Moreover, NSS 26-G provides additional detail on how these arrangements should be applied and become more stringent for Category I and II nuclear materials. For example, the provisions for guards during transport differ significantly for Category III nuclear materials because 'for lower category nuclear material, it may not be necessary for guards to accompany the shipment'. In contrast, NSS 26-G states that for Category I, 'adequately equipped and trained guards should accompany each shipment to protect nuclear material' and that if those guards are not armed, 'compensatory measures should be applied, such as adding delay barriers and the use of non-lethal incapacitating agents'.[30]

NSS 26-G includes additional provisions according to the mode of transport. This acknowledges that the security risks associated with each mode can differ significantly. Again, these provisions are broken down into provisions covering Category I–III and additional provisions for Category I and II nuclear materials. Nuclear carriers are also required to conduct a vulnerability assessment (VA) to ensure physical protection measures are effective. A VA is the method of identifying, categorizing, and mitigating, tolerating, or transferring known security vulnerabilities. For nuclear transport, the VA normally focuses on assessing whether physical protection measures are sufficient to protect against theft and/or sabotage.

Creating a Transport Security Plan

In most instances, states will require that a TSP be written for each shipment of Category I–III nuclear material. Depending on the state's requirements, this TSP may also need to be submitted for approval prior to transport. The content of the TSP will vary between states but should provide a sufficient written overview of the PPS to be used during transport, as well as (broader) indirect security arrangements. The TSP may also refer to other regional, national, or international contingency plans, for example, those coordinated by policing agencies involved in any national response. The TSP should therefore be aligned with wider organizational and stakeholder plans to ensure a continuity of security proportionate to the category of the material in transport. A carrier may also wish to share elements of, or the entire TSP, with key stakeholders (for example, security forces) to ensure a unified approach is adopted and that clarity on roles and responsibilities is achieved. For the avoidance of doubt, Appendix I of NSS 26-G includes a detailed example structure of a TSP. This is broken down into three core areas—administrative requirements and information, shipment security, and response planning—and provides an outline for state carriers.[31]

Guarding, Escort, and Armed Response Forces

Planning and preparation confer numerous advantages to a carrier over potential adversaries. Nevertheless, should security threats arise, the responsibility of meeting them falls to the guarding, escort, and armed response forces included in the TSP. The roles of these forces are defined in advance and are informed by the underpinning national and international regulatory tenets of the materials that need to be transported, namely, that security plans should be specific to the nature and form of the material (characterization and categorization) being transported and should assess the threat (from malicious actors) posed to it. Only then can carriers determine what a bespoke and proportionate force protection provision might look like, and how its provision will complement the holistic physical protection solution, mindful of the graded approach and potential cost implications that such a solution will deliver.

In short, the main objective is to detect an NSE and delay it sufficiently to allow the formulation of comprehensive response to it. As required by INFCIRC/225, 'Arrangements should be made to provide sufficient guards and/or response forces to deal with nuclear security events consistent with the category of nuclear material being transported'. Moreover, security plans should also consider contingencies where threats are greater than anticipated so that 'physical protection measures should include communication from the conveyance capable of summoning appropriate responders'.[32]

In calculating required levels of force protection, carriers must also contend with several layers of complexity, such as the mode of transport and the operating environment being traversed. For example, a multimodal transport operation that includes road, maritime, air, and rail elements will employ very different types of force protection for each leg of the journey. Numbers of required personnel will vary across each journey leg, while the training and ability of personnel to operate 4 x 4 vehicles, speedboats, bespoke weapon systems, etc., will also factor heavily in the types of solution offered. Conversely, the environment may work in the favour of force protection, namely, if it is hostile, or difficult terrain to traverse. These are in addition to the consistent pressures of ensuring appropriate training and the cost of supplying modal-specific assets, such as modified escort carriages for rail transport, or intelligence, surveillance, target acquisition, and reconnaissance (ISTAR) systems suitable for a maritime environment of sea-based transport.

In addition, SCPs must not only operate in tandem with the TSP, but indeed be integral to them—particularly for the more dangerous and attractive cargoes being transported. Subjecting transport security plans to both internal and external rigour prior to shipments commencing provides for great confidence in the robustness of their drafting, as does awareness of how each plan interoperates and interconnects with the various operational, tactical, and strategic levels, and the nuances therein.

Complex Operations

One factor contributing to the perception that the transportation of nuclear or radiological material represents a higher area of risk is the fact that maintaining the security of conveyances differs quite markedly from that required at a site or facility. Although both environments require a multi-stakeholder approach, transport—especially international transport, and/or of higher category or consequence shipments—requires a truly cohesive response from law enforcement and emergency responders (end-to-end) who are invariably not well versed in this area. The main reason for this is the relative lack of exposure to such transport conundrums compared to fixed facilities, which will have a regularly proven and mature security plan that meets competent authority expectation.

To overcome this, carriers must use concise command, control, and communication (C3) protocols to ensure clear and non-conflicting communications. Optimally, these will be disseminated via a bespoke central transport control centre (TCC) in tandem with clearly defined rules of engagement (ROE) for the force protection teams. The IAEA's Implementing Guide 'Security of Nuclear Material in Transport' reinforces this, stating a 'clear chain of command should be established to respond to a nuclear security event during transport, with appropriate lines of communication'.[33] Moreover, to improve familiarity '[c]oordination between guards that may accompany shipments and response forces should be regularly exercised. Transport personnel should be trained and prepared to act in full coordination with the guards, response forces and other response teams in accordance with the contingency plan'.[34]

This guidance could not be more apt given the challenges faced by carriers. Taking maritime lanes of communication as an example, researcher Chris Rahman highlights the international politics of combatting piracy in South-East Asia. Rahman emphasizes that, although there is a common interest among regional states to improve sea lines of communication and energy security, diverging strategic interests and varying political cultural hamper an effective security response.[35] As disappointing as this might sound, Rahman's assessment reflects the reality of the paradigm within which many multi-stakeholder operations occur. In addition to diverging state interests and capabilities, cross-border activities pose numerous practical challenges that have the potential to reduce a security team's effectiveness. Here, variances in language, national risk appetites, corporate agendas, cultural, and organizational differences all feature.

The Role of Rules of Engagement

In addition to clear delineations of responsibilities, escorts, guards, and response forces need clear instructions on when and how they may engage in the use of force to protect the transport of nuclear and radioactive materials. These are defined by the ROE. As noted by Alan Cole and colleagues, ROE 'are issued by competent authorities and assist in the delineation of the circumstances and limitations within which military forces may be employed to achieve their objectives'.[36] More substantially however, ROE also 'provide authorisation for and/or limits on, among other things, the use of force, the positioning and posturing of forces, and the employment of certain specific capabilities'.[37]

As such, ROE provide, *inter alia*, guidance to those on the ground delivering operational security effect but equally doctrine support for strategic commanders and tactical parameters for the command, control, communication, and coordination of incidents, most especially concerning the use of lethal force. It follows that the measured response or conflict management model employed by the force protection teams protecting the cargo must reflect, and accord with, the ROE. These cannot exist in isolation and will be informed by, and be consistent with, the hierarchy of related documentation. Together, this will span guidance ranging from the concept of operations (CONOPs) through to the standard operating procedures (SOPs) and/or tactics, techniques, and procedures (TTPs) at the granular level.

However, such prescriptions are perhaps easier said than done, and ambiguity of intent may hamper an effective security response. For example, in October 2000 the United States (US) Navy guided missile destroyer *USS Cole* was attacked by two suicide bombers who approached the *Cole* in a small fibreglass boat laden with explosives. The terrorist attack killed 17 sailors and injured a further 37. In the subsequent investigation, the US House Armed Services Committee Staff noted how the ambiguity in determining the attackers' hostile intent prior to the attack and insufficient attention to force protection measures by the ship's commanding officer were contributing factors to the successful attack.[38] To remove ambiguity, ROE must be interpreted through the lens of up-to-date intelligence, with the ROE clearly articulated in writing, and possessed by all stakeholders with a security responsibility within the C3 model/chain of command. Above all else, ROE must be proportionate through the adoption of a graded approach to reflect all relevant threats.

International Cross-Border Considerations and Handover of Responsibilities

Most shipments of nuclear and radioactive material are of lower-category radioactivity and will be conveyed by road within the borders of a single state. However, some shipments will be international, wherein the crossing of international borders adds a layer of complexity for consideration, most especially for high-consequence cargoes or those more attractive to malicious actors. As noted, states themselves are responsible for the safety and security of all nuclear and radiological material within their jurisdiction.

Consistent with this, Article 4 of the CPPNM obliges states parties to refuse the import or export of nuclear material 'unless the State Party has received assurances that such material will be protected during the international nuclear transport at the levels described in Annex I';[39] similarly, the IAEA advises that for the shipment of radiological material 'international transport, shippers and/or carriers should ensure in advance that any state by state variations in security requirements are applied and should determine the point at which the responsibility for security is transferred'.[40] As required by the IAEA, such notifications should take place by 'prior arrangement . . . specifying time, place and procedures for transferring transport responsibility'.[41] For those cargoes that attract special arrangements, such as an armed escort, the formalization of intergovernmental (non-legally binding agreements) are called records of discussion (ROD). These instruments serve to harmonize planning and preparations for each shipment taking stock of the complexity of the operation to identify areas of overlap, uncertainty, and potential gaps in the arrangements as the responsibility is transferred from one state's jurisdiction to another.

FIGURE 17.4. Image of a spent fuel transportation via rail.

Image courtesy of Nuclear Transport Solutions (NTS).

By its very nature the ROD is high level and requires follow-up meetings to be scheduled on the strength of ROD consensus. Hierarchically, these next stage tactical-level meetings require senior practitioners in positions of management or command from the carrier, consignor, consignee, escort, and—in some instances—armed forces, to meet and exchange details at a more granular level on how the transhipment points will be controlled. This all culminates in the formation of multi-stakeholder coordination groups, or multi-agency command centres (MACCs), to oversee the departure from the consigning, and arrival at the receiving facilities, respectively (Figure 17.4).

Security Contingency Measures

As noted, rigorous planning is imperative to securing nuclear and radiological transportation and—given the gravity of the consequences of failure—it is only appropriate that carriers deliver exemplary standards in response to an NSE. A popular maxim within the regulatory space is for legally responsible operators to 'provide evidence over assertion' that such efficacy exists. In short, carriers must demonstrate not only that their PPS can defeat the full spectrum of threats that are outlined in a state's DBT or RTS, but that any SCP will also meet the objective of denuding unauthorized removal or sabotage against cargoes under a carrier's charge. If they are to be deemed truly effective, these plans must be predicated by credible attack scenarios and evaluated within a pressurized environment that reflects a live scenario.

These basic requirements are captured within the A/CPPNM's Fundamental Principle K (Contingency Plans) and Fundamental Principle J (Quality Assurance).[42] However, Essential Elements 11 and 12 within the IAEA's Fundamentals document, 'Objective and Essential Elements of a State's Nuclear Security Regime', provide greater clarity on what quality assurance programmes expect for emergency preparedness and response (EP&R) at local, national, and international levels to an NSE, and ideally how SCPs should be stress tested (type, difficulty, frequency, and oversight).[43] Such exercises help engender confidence not only within the teams responsible for delivery, but also for other corporate and governmental stakeholders, including a state's competent authority for nuclear security.

Quality assurance can be provided by several avenues. Internally, plans and exercises can be rigorously evaluated by an operator's own intelligent customer utility. This internal assurance has numerous advantages, including the provision of an unequivocally clear understanding of a carrier's nuclear security capability and capacity by ensuring relevant good security practice is exercised. Internal assurance also gives insight into other key areas: how operational capability (OC) is maintained; equipment availability (EA) is examined and inspected; evidential data for the competent authority (CA) is provided; and corporate risk is managed.

That said, despite the benefits of internal assurance in helping to build a body of evidence to support carrier claims of sustainability, the potential for subjectivity inevitability hampers the credibility of such programmes. To bolster this, peer reviews from external partners, such as other responsible carriers/operators, add yet more weight and generate learning points from which corrective alterations and finessing can be fed back into a constantly evolving plan. Even with such measures of assurance, carriers must ultimately demonstrate their levels of competence to their CA. These are summative demonstrations that are graded as pass/fail to provide a 'litmus test' of a carrier's security effect; they should be conducted on a periodic basis, or in advance of bespoke shipments, to both meet legal requirements and provide confidence to state authorities of a carrier's quality assurance processes.

As the risk owner, carriers must build on their internal assurance programme with a holistic EP&R training and exercise regimen to adequately prepare their force protection elements and C3 teams. These must act in unison within a pressurized but safe environment, to deter, detect, and defeat those threat actors that they might reasonably be expected to counter. During exercises and assessments, threat-based scenarios draw upon prevailing sector threat assessments, DBT, and TTPs, and form part of the criteria against which response drills are assessed. Here, proportionality is key, requiring credible scenarios to be crafted based on the graded approach, the attractiveness of the material, and/or the environmental impact of sabotage. This is to not overburden the carrier with prohibitive costs that would also ultimately make a regulatory relationship unsustainable.

To that end, demonstrations (in practice) at one end of the assurance spectrum can consist of little more than exercising sound C3 arrangements against, for example, a protestor threat and the commensurate mobilization of external security stakeholders, or managing a road transport traffic accident (a highly plausible scenario) and the security implications therein. At the other end of the spectrum, carriers may be required to demonstrate the ability to control a full blown 'force-on-force' scenario, where they defeat dedicated and malicious threat actors.

Conclusion

Nuclear and radiological materials are essential to the global economy and used for a variety of civilian purposes. Although it is through logistics and transportation that such materials are moved to where they need to be, the provision of nuclear security during transportation is generally not well understood and there is a perception that transportation is the 'weak link' of the global civilian nuclear estate. Compared to nuclear sites and facilities, the provision of security to nuclear transport operations is inherently more complex. Indeed, although the majority of conveyances take place via roadways and within a single state, the need for multi-stakeholder consultation, use of intelligence, and modal-specific security considerations make such shipments complex affairs. The complexities only increase with international shipments because of varying national regulations, competent authorities, and threat environments.

Nevertheless, transportation remains a highly regulated environment with specific roles and responsibilities clearly defined and delineated by the international security regime. Moreover, carriers can systematically consider various threat landscapes, and use intelligence-led threat solutions that include combinations of key stakeholder engagement, SANDA tasking, and local area knowledge that is supplemented with open-source monitoring. This allows carriers to pursue risk-based practices and develop effective protection solutions to maintain the safe and secure transportation of nuclear and radiological material. Such solutions undergo rigorous scrutiny and audit through internal and peer review, while also being subject to national regulatory approval. When compared to radioactive materials, nuclear materials only form only a small part of global conveyances. However, as governments increasingly view nuclear energy as a means of generating low-carbon electricity, the importance of safely and securely transporting civilian nuclear material is likely to grow in both the volume of shipments, but also in politicization.

Notes

1. Office of Nuclear Energy, '5 Common Myths About Transporting Spent Nuclear Fuel. Office of Nuclear Energy', US Department of Energy [online], 26 May 2020. https://www.energy.gov/ne/articles/5-common-myths-about-transporting-spent-nuclear-fuel

2. IAEA, *The Convention on the Physical Protection of Nuclear Material*, INFCIRC/274/Rev.1 (Vienna: IAEA, 1980). https://www.iaea.org/sites/default/files/infcirc274r1.pdf
3. IAEA, *Objective and Essential Elements of a State's Nuclear Security Regime*, Fundamentals. IAEA Nuclear Security Series, No. 20 (Vienna: IAEA, 2013), 4.
4. UK Office for Nuclear Regulation (ONR), *Nuclear Industries Security Regulations 2003*, ONR Guidance Document, CNSS-SEC-GD-002, Version 4.1 (London: ONR, 2013).
5. IAEA, *Nuclear Security Recommendations on Physical Protection of Nuclear Material and Nuclear Facilities (INFCIRC/225/Revision 5)*, Recommendations, IAEA Nuclear Security Series, No. 13 (Vienna: IAEA, 2011).
6. IAEA, *Security of Nuclear Material in Transport*, Implementing Guide, IAEA Nuclear Security Series, No. 26-G (Vienna: IAEA, 2015).
7. UK Office for Nuclear Regulation (ONR), *Physical Protection Systems*, ONR Security Technical Inspection Guide, CNS-INSP-GD-6.0, Revision 1 (London: ONR, 2018).
8. World Institute for Nuclear Security (WINS), *Nuclear Transport Security, Good Practice Guide*, Version 2.0 (Vienna: WINS, 2022).
9. IAEA, *National Nuclear Security Threat Assessment, Design Basis Threats and Representative Threat Statements*, Implementing Guide, IAEA Nuclear Security Series, No. 10-G (Rev. 1), Vienna: (IAEA, 2021).
10. UN Economic Commission for Europe (UNECE), *UN Recommendations on the Transport of Dangerous Goods: Model Regulations*, Vol. 1, rev. 21st edn (New York: United Nations, 2019).
11. Ibid.
12. International Maritime Organization (IMO), 'International Code for the Safe Carriage of Packaged Irradiated Nuclear Fuel, Plutonium and High-Level Radioactive Wastes on Board Ships (INF Code)', IMO [online], n.d. https://www.imo.org/en/OurWork/Safety/Pages/INF-Code.aspx
13. UK Maritime and Coastguard Agency, 'Guidance: Ship security', UK Government [online], 11 June 2021. https://www.gov.uk/guidance/maritime-security
14. IAEA, *Security of Radioactive Material in Transport*, Nuclear Security Series, No. 9-G (Rev. 1) (Vienna: IAEA, 2020).
15. IAEA, *Objective and Essential Elements.*
16. Alessandra Scalia, 'Coastal and Shipping States Conduct Exercise Simulating Maritime Transport Emergency', *IAEA News* [online], 19 June 2015. https://www.iaea.org/newscenter/news/coastal-and-shipping-states-conduct-exercise-simulating-maritime-transport-emergency
17. Sinead Harvey, 'Nuclear Emergencies at Sea: Experts Practice Emergency Communication Procedures', *IAEA News* [online], 5 July 2017. https://www.iaea.org/newscenter/news/nuclear-emergencies-at-sea-experts-practice-emergency-communication-procedures
18. IAEA, '*National Nuclear Security Threat Assessment, Design Basis Threats and Representative Threat Statements*'.
19. IAEA, *Nuclear Security Recommendations on Physical Protection.*
20. IAEA, *National Nuclear Security Threat Assessment, Design Basis Threats and Representative Threat Statements.*
21. Gregory D. Koblentz, 'Predicting Peril or the Peril of Prediction? Assessing the Risk of CBRN Terrorism', *Terrorism and Political Violence*, 23/4 (2011), 501–20.
22. IAEA, *National Nuclear Security Threat Assessment, Design Basis Threats and Representative Threat Statements.*

23. UK Ministry of Defence, 2011, *Understanding and Intelligence Support to Joint Operations*, JDP 2-00, London.
24. *BBC News*, 'Brazil gunmen shoot at convoy carrying nuclear fuel in Angra dos Reis', [online], 20 March 2019. https://www.bbc.co.uk/news/world-latin-america-47635706
25. Dom Phillips, 'Brazilian Drug Gang Opens Fire on Convoy of Trucks Carrying Nuclear Fuel', *The Guardian* [online], 20 March 2019. https://www.theguardian.com/world/2019/mar/20/brazilian-drug-gang-opens-fire-on-convoy-of-trucks-carrying-nuclear-fuel
26. IAEA, *Objective and Essential Elements*, 39.
27. IAEA, *Nuclear Security Recommendations on Physical Protection*.
28. Ibid.
29. UK Office for Nuclear Regulation (ONR), *Categorisation For Theft*, ONR Nuclear Security Technical Assessment Guide, CNS-TAST-GD-6.1 Revision 1 (London: ONR, 2020).
30. IAEA, *Security of Nuclear Material in Transport* 59.
31. IAEA, *Security of Nuclear Material in Transport*.
32. IAEA, *Nuclear Security Recommendations on Physical Protection*, 41.
33. IAEA, *Security of Nuclear Material in Transport*.
34. Ibid., 26.
35. Chris Rahman, 'Multilateral Approaches to Combating Piracy at Sea', in Peter Lehr, ed., *Violence at Sea: Piracy in the Age of Global Terrorism* (New York: Routledge, 2006), 183–98.
36. Alan Cole, Phillip Drew, Rob McLaughlin, and Dennis Mandsager, *Sanremo Handbook on Rules of Engagement—Prepared under the auspices of the International Institute of Humanitarian Law* (Sanremo: International Institute of Humanitarian Law, 2009), 1.
37. Ibid.
38. US Congress, 2001. *The Investigation into The Attack on the U.S.S. Cole*, Report of the House Armed Services Committee Staff, Washington, DC. https://www.bits.de/public/documents/US_Terrorist_Attacks/HASC-colereport0501.pdf
39. IAEA, *The Convention on the Physical Protection of Nuclear Material*, 2.
40. IAEA, *Nuclear Security Recommendations on Physical Protection*, 22.
41. IAEA, *Amendment to the Convention of the Physical Protection of Nuclear Material*, INFCIRC274/Rev.1/Mod.1 (Vienna: IAEA, 2016), 16. https://www.iaea.org/sites/default/files/publications/documents/infcircs/1979/infcirc274r1m1c.pdf
42. IAEA, *Amendment to the Convention of the Physical Protection of Nuclear Material*.
43. IAEA, *Objective and Essential Elements*.

Suggested Reading

IAEA, (2015), *Security of Nuclear Material in Transport*, Nuclear Security Series, No. 26-G (Vienna: IAEA).

IAEA, (2020), *Security of Radioactive Material in Transport*, Nuclear Security Series, No. 9-G (Rev. 1) (Vienna: IAEA).

Lange, F., Mairs, J., and Niel, C., (1997), *Risk Associated with the Transport of Radioactive Materials in the Fuel Cycle* (Vienna: IAEA).

World Institute for Nuclear Security (WINS), (2022), *Nuclear Transport Security. Good Practice Guide*, Version 2.0 (Vienna: WINS).

World Nuclear Association (WNA), 'Transport of Radioactive Material', WNA [online], September 2022. https://world-nuclear.org/information-library/nuclear-fuel-cycle/transport-of-nuclear-materials/transport-of-radioactive-materials.aspx

CHAPTER 18

CYBER SECURITY FOR NUCLEAR FACILITIES

SUKESH K. AGHARA AND CHRISTOPHER SPIRITO

Introduction

Nuclear facilities are large industrial facilities with complex systems of operations. Through international cooperation between the nuclear industry, governments, and state regulators, nuclear materials and facilities have operated for decades with high standards in safety, security, and material control. Since the late twentieth century, the application of digital controls in the nuclear industry has led to an important change in the design of engineering systems and their management, operations, and maintenance, but this has posed a new set of security challenges. In response, cyber security efforts have been increasingly promoted through various international bodies, most notably the International Atomic Energy Agency (IAEA), the World Nuclear Association (WNA), and the World Association of Nuclear Operators (WANO). In just the past decade, the 2007–2010 Stuxnet attacks at the uranium enrichment facility in Iran, the 2014 cyber hacking at nuclear power plants (NPPs) in South Korea, and 2016 detection of malware at German NPPs are some examples of the growing cyber threat profile faced by nuclear facilities. The nuclear sector has effectively used quality management programmes to maintain high standards in safety and physical protection systems. The sector must accept the inevitable penetration of digital systems at nuclear facilities and apply its management programmes to mitigate the security risks posed by cyber-attacks.

This chapter describes the key cyber security challenges for the sector. First, the potential attack surfaces at nuclear facilities are defined. The following section identifies best practices to build a cyber security architecture using the graded approach. Using some hypothetical examples, the chapter also highlights some known capabilities of cyber threat actors that should be considered by the industry to build resiliency against cyber vulnerabilities. In doing so, the chapter refers to the MITRE ATT&CK framework.[1] The chapter then goes on to examine risk management strategies, followed by a discussion about defence in operation and incidence response—again, drawing upon MITRE strategies. Nuclear facilities are complex and there are inherent challenges for security designs. Nevertheless, the most effective security is based on simplicity of the system; protecting nuclear facilities against this evolving threat requires a review of designs and regulations to generate a simplified cyber security solution for a multi-functional, digitally networked system of thousands of components.

Understanding the Attack Surface

The modernization of NPPs has included replacing analogue systems with digital systems, offering enhanced functions and capabilities, and perceivably lowering the overall cost of ownership.[2] These enhanced functions and capabilities include improvements in component level monitoring to support lifecycle maintenance, graphical user interfaces for interacting with the reactor and balance of plant (BOP) systems, and network stacks for connectivity between digital systems. From the perspective of cost and operational efficiency, it is logical that the current fleet of NPPs and new-generation advanced reactors are fitted with digital systems, but with this choice comes a responsibility to understand how these affect the 'attack surface'. An 'attack surface' is the total number of all possible components or features of a digital system that could be exploited due to a vulnerability. Digital systems include hardware, operating systems, applications, and network protocol stacks, and each of these is part of the digital attack surface that a cyber-attacker may leverage to deliver effects onto the target systems.[3]

Digital systems are built upon components that allow for a function or capability to be fielded. Examples of digital systems within an NPP include programmable logic controllers (PLCs) that implement ladder logic (a programming language) for control loops; engineering workstations used to manage PLCs locally; relational databases that provide access control decisions throughout the site; and time series databases responsible for recording control system actions and events. The systems responsible for connecting digital systems with one another and enforcing security policies are themselves digital systems. Each digital system and digital system interconnects present an opportunity for an adversary to subvert NPP functions and capabilities, should they be capable of identifying a technical or procedural vulnerability. A research reactor may have hundreds to thousands of digital systems or assets. An NPP may have tens of thousands of digital systems—a collection of heterogenous systems with operational and managerial independence distributed over a large geographical space.

To identify vulnerabilities in digital systems, the security team must put themselves in the mind of a potential adversary through delineating the digital system into its respective components and then analysing the process by which the system at an NPP was designed, built, tested, deployed, and operated.[4] An adversary would evaluate digital system functions by attack vector, looking for instances where the system had been misconfigured, or had a kernel or design flaw. They would look for vulnerabilities related to memory management, such as buffer overflows, and flaws in how binaries, operating systems, and file systems interacted, allowing for race conditions and file descriptor attacks.[5] They also would consider each digital system user at an NPP as a target, where access to one system may allow access to all systems such as through spear phishing and social engineering attacks. The complexity of these human and digital system interconnects makes the attack surface complicated to analyse. One approach to understanding this problem is to segment the analysis into three areas: (1) analysis of digital system components; (2) analysis of security guidance and best practices; and (3) analysis of past incidents.[6] A comprehensive assessment of the digital system attack surface should include all three areas. The following examples demonstrate the application of this approach.

The attack surface can initially be derived from an analysis of the system hardware, operating system, and software. However, this is a complex task with the results dependent upon the knowledge and assessment skills of the engineer(s), which may vary widely depending upon the individual(s')'s background and experience.[7] Using a hypothetical example, an engineer wants to carry out an assessment of a USB device, to be used to transfer media from a nuclear vendor to a trusted system. In this case, there might be three possible scenarios. First, a typical protocol would be a review by a security engineer of the documentation to ensure that the USB device was designed and manufactured by a trusted party, and that the functions on the USB device are limited to the storing of digital files and interfacing with Windows operating systems. Additionally, a function validation test might be performed to check if, when plugged into a non-Windows system, the device is recognized or rejected and whether an event is recorded in the audit log. Second, a more experienced engineer might also examine the security controls implemented on the Windows system and carry out a functional test of the software performing the data transfer between the USB device and the system. The audit log verification might also be performed, including a check on the security privilege that prevents read/write by non-privileged users. Third, an even more experienced engineer, with knowledge of hacking techniques, would carry out additional analysis of the USB device to determine whether it contained a hardware modification from the design specification. They might examine the USB driver, the file system driver, and the configuration of the media scanner to assess whether there were any known exploitation pathways that could be leveraged by a skilled adversary working with a trusted or unwitting insider. Each of these three digital system analysis methods is reasonable, but each will produce a different assessment of the attack surface based upon the knowledge and skills of the individual security engineer.

Similar issues may occur when applying security guidance, since how well an engineer's knowledge of the attack surface is tightly coupled to how well they understand the security guidance provided by the vendor. Using another hypothetical example, a PLC vendor provides security configuration guidance and best practices for an instrumentation and control (I&C) engineer. This guidance is based upon the vendor's knowledge of the digital systems attack surface for their product and includes a set of recommended parameters for securing the PLC from network-based classes of attack.

While this vendor may have reported past vulnerabilities with the core ladder logic implementation, they might not have reported issues related to the control system protocol network interface or the administrative web interface.

A less-experienced engineer might read through the security guidance and best practices and apply them using a checklist approach, such that they can assure the system owner that they have adhered to the security recommendations and integrated those recommendations into the site security policy. A more-experienced engineer might perform a validation of the industrial control system (ICS) protocols that are enabled, and confirm that the disabled protocols are indeed non-responsive to requests. The ICS protocol analysis includes confirmation that the cryptographic functions have been correctly implemented and that fuzzing techniques applied to the web interface do not produce any application faults or anomalous behaviour on the PLC.[8] Consequently, the identified digital attack surface will vary based upon the engineer performing the analysis and the knowledge and skills of the NPP vendor security teams.

Analysis of past incidents is another valuable method for understanding potential attack surface issues. A security team at an NPP might utilize incident reports related to the operation of different digital systems. As the engineer reads through the report, they may understand the general mechanics of the cyber-attacks and then seek to negate these by confirming that the NPP operating system and application libraries have been updated to non-vulnerable version levels. They may extract from the report observables in the form of Yara or network IDS signatures (i.e. detection of attacks by looking for specific patterns), which can be implemented within the NPP network sensor suite that supports the detection and response capability. The engineer involved in installing or maintaining the system may proactively reach out to the security staff member responsible for capturing the prior security event to discuss the likelihood that the same type of attack would be successful on their plant's digital systems.[9] Results from an assessment of the digital attack surface can be incorporated into an implementable risk management plan that addresses the gap between the potential cyber security vulnerabilities and the available technical and procedural mitigations. This can then be utilized as part of a graded protection approach to securing digital systems within the NPP from potential cyber-attacks.

Graded Protection and Nuclear Facility Architectures

One of the challenges that NPP security teams must address is where the complexity of the architecture and number of components exceeds the ability of the team to implement comprehensive security controls across each digital system and the interconnects. Here, the use of a graded approach for the protection of digital systems—which prioritizes the application of security resources based on the consequences of a successful attack—is essential to provide an appropriate level of protection informed by the skills and capabilities of identified threat actors.

The IAEA, the Nuclear Regulatory Commission (NRC), National Institute of Standards and Technology (NIST), and several other organizations have developed guidance documents that nuclear facilities operators can consider to manage cyber security risks at NPP.[10] The general processes and procedures that incorporate a graded approach to assessing and protecting against the consequences of cyber-attacks on NPPs involve performing an analysis of the operational environment relative to cyber security threats. This is followed by the identification of 'critical digital assets' (CDAs) to use as a guide for investment prioritization.[11] Graded mechanisms are then applied within the existing NPP defence-in-depth strategy before security controls informed by a graded systems analysis are applied. This enables a continuous monitoring and assessment infrastructure to support the incorporation of changes to the NPP threat model over time. Finally, a change control process is enacted to support the overall graded approach, including a periodic review of the cyber security programme to ensure this remains effective against evolving cyber threats.

A successful graded approach to cyber security at NPPs starts by establishing a cyber security team to perform the assessment and make recommendations for updates to the defence-in-depth strategy and application of security controls.[12] The cyber security team will need to include subject matter experts (SMEs) in the fields of NPP engineering, I&C, and cyber security. The NPP engineering SMEs are responsible for ensuring that the I&C and cyber security engineers are informed on the operational boundaries and functions of the reactor and BOP systems. The I&C engineers are responsible for ensuring that the NPP engineers and cyber security engineers understand how they have implemented the digital systems and digital system interconnects. The cyber security engineers are responsible for communicating the threat actor capabilities to the NPP and I&C engineers who, in turn, can provide impact assessments that will be used to determine which digital systems will be protected with which security controls.

To support the graded approach, the cyber security team will perform an assessment of digital systems and identify which systems and system components would be considered CDAs.[13] Given the number of digital systems within NPPs, this process will ultimately be constrained by the cyber security resources available for implementation and monitoring of security controls. In a sense, this process is like the assessment of the attack surface, as it is dependent upon the knowledge and skills of the cyber security team. Generally, this type of assessment can be conducted at a high level where every system that contains a safety, security, or emergency preparedness (SSEP) function would be classified as a CDA. However, in environments where the assessment engineers are more comfortable with lower levels of analysis, the CDA assessment process may be bounded at the function and component level. A CDA assessment generally finds that systems such as the reactor protection system (RPS), post-accident monitoring system, safety network, and radiation monitoring system are considered CDAs—and thus, require additional attention during the application of security controls within the graded approach.[14]

After the cyber security team has identified the CDAs, it will perform an assessment of these within the defence-in-depth strategy. This includes an assessment of the CDA architecture with a focus on the protection of each system and component within the deployed or architected security level. The assessment may also include cross-boundary data communication that is implemented to conduit data between critical and information processing systems. Defence-in-depth analysis includes understanding whether there is component and function resiliency through the implementation of redundant connection and tiered approaches to application services. Defence-in-depth involves layers of security controls such that a skilled adversary is deterred or contained for as long as possible during the attack lifecycle. Implementing these concepts within the defence-in-depth approach is essential for successful cyber defence.

The graded approach to cyber security also includes the application of security controls defined and updated during the CDA enumeration and defence-in-depth analysis process.[15] These controls would normally include the following: implementation of access enforcement and least privileges within the systems architecture; implementation of a host-based detection system to complement the network intrusion detection system; and implementation of a comprehensive incident handling and response function.[16] Each of these security controls maps back through the CDA assessment process such that there is traceability for the implementation team as well as data available to support the review of the cyber security programme. Even for those systems not classified as CDAs, it does not mean as part of the graded approach that they receive no attention with regards to the application of security best practices. Those interconnected systems and sub-systems will be configured based upon vendor and site best practices, and monitoring will include a set of rules consistent with the applied security policies. The difference between these systems and the CDAs derived from the classification process is that the additional analysis and application of security controls are based upon their attractiveness to threat actors and consequences from possible compromise or subversion of system or component functions.[17]

Implementation of the graded approach within the operational environment requires the inclusion of a continuous monitoring and assessment capability. Many experts agree that once a system is accredited and transferred into operational mode, the operational states that were not predicted often impact the set of security controls that were implemented.[18] Consequently, monitoring and assessment are essential to identify those gaps as well as to detect anomalous behaviour consistent with cyber-attacks and the attempted delivery of cyber effects onto these targets. The continuous monitoring and assessment system should consist of a graded set of actions based upon the perceived vulnerability of the target asset such that changes in operational state are available for analysis in a timely way to minimize the risk of function compromise.

The change control capability should provide recommended remedial actions that can be applied when the continuous monitoring and assessment capability results in events that require change within the CDAs or associated systems. If an immediate solution is not available, a list of mitigation measures is generated to manage the risk. These measures are used until a permanent remediation to the vulnerability is identified. The steps taken should be integrated into a broader risk management plan, which should include periodic review of the cyber security plan to determine effectiveness of the graded approach and whether there are any systemic changes that need to be enacted.

Cyber Threat Actor Capabilities against Nuclear Facilities

This section discusses different techniques and tactics that a threat actor may use at multiple levels—defend, respond, resolve—to build up an understanding of the cyclic nature of the attack, as it pertains to cyber threats at nuclear facilities. A threat actor is likely to exploit existing capabilities across different industry domains to develop a credible cyber security threat at a nuclear facility. The following example explores this using hypothetical scenarios and fictitious threat actors at the NPP functional level. For example, threat actor LAZY DOG may be capable of disabling the turbine control system, preventing an NPP from generating energy, while a separate threat actor, BROWN COBRA, may be capable of creating a safety event that causes reactor shutdown. Threat actor capabilities against nuclear facilities are not limited to just an NPP. Consider threat actors NOISY KITTEN, capable of conducting an influence campaign, and STEALTHY BANDIT, capable of conducting a spear phishing campaign to exfiltrate engineering documents from a third-party contractor. These functional-level capabilities, which exist independently, could be exploited by a cyber threat actor to mount a cyber-attack at a nuclear facility.

The next level of enumeration of cyber threat actor capabilities involves analysing the impact of different attack scenarios. Here it is useful to refer to the MITRE ATT&CK framework for ICS—now widely in use across the industry—which is a unique compilation of adversary tactics and techniques based on real-world observations developed by the MITRE Corporation, a US not-for-profit organization.[19] The ATT&CK knowledge base, a free resource, could be used as a foundation for the development of specific threat models and methodologies by any person or organization. The ATT&CK database lists two techniques—"Damage to Property" (MITRE ID: T0879)[20] and "Manipulation of Control" (MITRE ID: T0831)[21]—that the threat actor LAZY DOG could use to disable the turbine control system (TCS).

The threat actor in this hypothetical case utilizes a series of other tactics, techniques, and procedures (TTPs) to cause physical damage to the TCS or remotely take over control of the TCS human–machine interface (HMI) in order to force the TCS into a degraded operational state. BROWN COBRA may choose to apply a "Loss of Safety" (T0880) technique to deliver the "Loss of Productivity and Revenue" (T0828) impact. To do so, BROWN COBRA applies a series of TTPs to achieve the safety event that causes the reactor to shut down. NOISY KITTEN, noticing an unplanned shutdown of NPP, would then deploy the "Loss of Control" (T0827) effect to build a disinformation influence campaign based on a momentarily deprived NPP operator controllability. STEALTHY BANDIT subsequently applies its capabilities to deliver "Theft of Operational Information" (T0882), which allows them to exfiltrate information from the third-party contractor systems, which they could then sell via a criminal forum, to LAZY DOG and BROWN COBRA, whose capabilities are enhanced in real time during delivery operations. As the mapping from functional-level capabilities to impact-focused capabilities is performed, the specific TTP that the threat actors employ, although difficult, can be enumerated.[22] These systems-level capabilities can be delivered against NPP personnel and the digital systems, system interconnects, and sub-systems that they manage and interact with.

The MITRE ICS ATT&CK framework sets out these capabilities, with detailed descriptions, in 11 tactical categories.[23] This framework can help identify tactics that LAZY DOG could use to achieve delivery of the "Damage to Property" and "Manipulation of Control" impacts. One pathway is to inject a malicious code into the TCS via a "Supply Chain Compromise" (T0862), which utilizes a vulnerability in one of the control system drivers to cause the PLC "Device to Restart" (T0816). The code also deploys "Manipulating the Supply, Use and Input/Output Table" (T0835) so that the PLC fails in a catastrophic way when it reboots. Each of these TTPs is a capability option which, when combined, enables the cyber threat actor to deliver impacts at the NPP functional level. NOISY KITTEN may choose to utilize its capability of exploiting "External Remote Services" (T0822) to gain access to the control system network so that it can observe how the safety system communicates, in terms of "Monitor Process State" (T0801), and then utilize an "Adversary-in-the-Middle Attack" (T0830) and "Alarm Suppression" (T0878) to momentarily prevent NPP operators from controlling the safety system.

In evaluating the capabilities of cyber threat actor, the maturity level at which they exist should also be considered when implementing a graded approach and building counter measures with adequate response pathways within the defensive capability suite of the security system. Mapping of defensive capabilities to effect delivery is also helpful in characterizing the effect on the cyber adversary. Here, Bodeau and Graubart offer a detailed description, with examples of six defensive action categories.[24] It is important for organizations to recognize that the cyber-attack and defence systems are extremely dynamic, and that capabilities evolve constantly, which influences both sides. A cyber security team must understand how the changes to defensive capabilities may affect the choices made by cyber threat actors, which in turn impact planning for future updates needed in cyber defence systems.

Inclusion of Cyber Security in the Risk Management Plan

Risk management is a complex process by which an organization weighs the costs and benefits of developing security and other capabilities with different operational environments. While there exist quantitative approaches for assessing the risk to which a system or function is exposed given the capabilities of a set of cyber threat actors, human performance on both sides makes this process inevitably imprecise. Thus, as cyber security in an organization's broader risk management plan is worked through, there will be areas where estimation errors will require agile response options to adjust the security environment. The plan should be built such that it can be adapted to changes in defensive and adversary cyber capabilities. Essentially, cyber security operations are more of a game than an engineering science with adjustments continuously needed to optimize risk coverage.

Although it is intrinsically challenging to implement an effective risk management plan, the steps that can be taken towards this goal are uncontroversial.[25] It starts with the preparation phase where a team is established, and the operational infrastructure needed to support the process is identified and fielded. This is followed by a categorization process where the analysis of the target environment is conducted, including documentation of system-and component-level functions with data flows and an associated impact analysis such that security controls can be subsequently selected. The CDA assessment described previously would then be conducted, followed by the selection phase in which the team would choose which security controls to implement, using a graded approach. These controls are then deployed and evaluated during the implementation and assessment phases. If these tests are successful, the risk manager or designated authority will authorize the system, which then leads to a final monitoring phase where the infrastructure is observed for anomalous behaviours and detections of policy violations.

During the preparation phase, the cyber security team starts with establishing communication with key stakeholders in each of the NPP teams that will be involved in the assessment process. The US National Institute of Standards and Technology (NIST) Risk Management Framework describes many of the definitive tasks such as identifying roles and evaluating privacy framework.[26] The two most important outcomes from the preparation phase are (1) ensuring that NPP staff members are aware of the process and expectations; and (2) senior leadership messaging support for this work and that management will provide the necessary resources.[27] The information derived during these types of assessments is often categorized as sensitive due to the identification of vulnerabilities and weaknesses in the systems, components, and processes. Understanding the human element of these types of findings, especially within the cultural context of the NPP where the assessment is being performed, will improve the outcomes of the process.[28]

During the categorization phase, the risk management team works in coordination with the functional teams to evaluate NPP systems and components against an agreed-upon framework. The challenge for the team performing the assessment is bridging the gap between proscriptive assessment methodologies. The assessment of the system and components using a proscriptive checklist can be assigned to less skilled team members. Cyber security experts are tasked to perform the assessment within the context of threat scenarios, availability of attack, and defence capabilities.

Using a hypothetical example, the team performs an assessment of a system and determines that it is a CDA for SSEP functions. The proscriptive checklist assessment indicates that this system will require security controls from several categories that impose higher lifecycle ownership costs but meet the requirements of these types of systems. The cyber security engineer performs their assessment and determines that the systems and network architecture used is such that, if the impact of an attack on this system would be high or significant, the probability or likelihood of an attack on this system would be low. Exceptions to these cases are where Trusted Functions A and B are exposed from Systems C and D, which are only accessible from Networks E and F. The security and systems architect may then decide that applying sensors (or controls) to Networks E and F and Systems C and D will lower the overall likelihood of a successful attack against this CDA due to moving the detection option further to the left on the cyber-attack lifecycle.[29]

The gap between the proscriptive tick-box thinkers and the more dynamic thinkers is often what separates good risk management practices from the exceptional ones. Effective selection of security controls is performed by starting with best practice recommendations. These are often based upon observations from past experiences and merging them with security controls lifecycle and incident response plans. This provides control maturity to be shifted based upon future incidents and informed by vendor announcements and security community research. Implementing a 'security control-incident response' cycle is one of the more challenging techniques at NPPs, given the constraints placed on the system operators with regards to changes to operational systems and sub-system components.[30]

Security control implementation and assessment can be considered a dry run for what the NPP cyber security team will experience once the systems are placed into an operational state. As each control is established, such as tracking of unsuccessful login attempts or unauthorized IP connections to access control systems, an assessment should be performed to validate each security control state.[31] In the examples, those states may include a successful or unsuccessful login attempt, and an authorized or unauthorized IP connection. These states may also be extended to encompass more complex security policies, such as three unsuccessful login attempts within 120 seconds or an unauthorized IP connection followed by brute-force login attempts. This is again the challenge related to the inclusion of cyber security within a risk management plan, since the actual control implementation and assessment will depend upon the knowledge, skills, experience, and creativity of the engineers responsible for its implementation.

Once the security controls are implemented and assessed, an evaluation is conducted, and a decision made to authorize the operation of the new capability. If the cyber security implications of this decision are based on the assessment findings where security controls of a sufficient maturity were not available for implementation, there exists a residual risk that needs to be accepted prior to putting the system in operation. It is important to note that the evaluation of residual risk occurs at a particular point in time, but the risk itself is not static in nature and will change based upon improvements in defensive skills and changes in cyber threat actor capabilities. The last phase of the risk management plan is the operation of a monitoring infrastructure that is central to the fielding of an incident response capability. This is the core of the defensive operations and incident response, as elucidated in more detail below.

Defensive Operations and Incident Response

Ensuring both the safety and security of NPP operations requires a cyber incident response capability that is baked into the plant's defensive operations plan. The first challenge for NPP staff is understanding that the maturity of any incident response capability is predicated on the inclusion of safety and security functions as part of the daily plant operations cycle. There are several strategies that an NPP could adapt. The MITRE publication '11 Strategies of a World-Class Cybersecurity Operations Center' is one of the most comprehensive publications on this subject.[32] The document states that the strategies should include knowing what is being protected and why, building a security operations centre (SOC) structure to match organizational needs, hiring and growing quality staff, and optimizing the incident response lifecycle. Implementing these strategies in full would exceed the operational budgets of most NPPs, and hence, will require a cost-impact analysis. This implies a challenge of prioritization of capability investment such that the risk to NPP operations is minimized through focusing on incident response capabilities that would have the highest impact on NPP digital systems. For new reactors, the security-by-design approach to digital systems would provide enhanced cyber resiliency with manageable cost increases.

Incident response capabilities can be described using seven high-level categories: incident triage, analysis, and response; cyber threat intelligence, hunting, and analytics; expanded SOC operations; vulnerability management; SOC tools, architecture, and engineering; situational awareness, communications, and training; and leadership and management.[33] Two priority capabilities are recommended for each of the seven categories. Historically, NPPs are low-visibility environments with regards to anomalous events that may indicate a cyber-attack is taking place, so real-time monitoring and triage would arguably not be as helpful as developing a depth of expertise in malware analysis.[34] The goal for NPPs should be to ensure that their incident analysis and investigation processes are well practised. A coordination plan should be ready such that a whole-of-plant response is available when an impactful event is observed. The incident response team should take time to identify expertise within their sector for artefact analysis such as national-level CERTs (computer emergency response team) or professional incident response teams, but the capability focus needs to be on maturity of the incident analysis and coordination functions.

One of the advantages that NPPs have is the extensive analysis that has been conducted on broader threats within the design basis threat (DBT) document. This document, while generally not focused on cyber security threats, does include threat actors and their capabilities. An additional layer of analysis can be performed to understand whether the capabilities can be deployed against NPP systems and sub-systems. Using this as a starting point, there are two capabilities for NPPs to invest in within the category of cyber threat intelligence, hunting, and analytics: sensor and analytics tuning, and threat hunting. Sensor and analytics tuning is aligned with the incident analysis capability, while threat hunting brings together NPP elements to identify where threat actors may be and what they may do within NPP systems. Choosing complementary capabilities in each of the categories helps to create a cohesive incident response that can strengthen synergistically over time.

NPPs should develop capabilities within expanded SOC operations, including insider threat detection and development of attack simulations.[35] Nuclear safety and security is predicated on a mature human reliability programme and an insider threat detection programme. For example, an insider threat mitigation expert may work with the cyber security team to design a scenario where an unwitting insider is recruited to insert a USB device into a sensitive digital asset, introducing malicious code into a trusted environment.[36] Attack simulations in a laboratory would allow for this scenario to be conducted and observations collected. This data would inform the implementation and improvement of security controls and deployment of signatures to be used during the incident analysis process.

NPPs are complex systems of systems infrastructure that are uniquely difficult to patch when a vulnerability is apparent. The primary challenge for patching NPP control systems is that they are designed and validated for use against an operational design basis and altering these systems is not possible without a consequence analysis and generally bringing these systems offline, whereas for information and communication technology (ICT) systems it is generally acceptable to patch and revert where necessary. For this reason, the focus is on two capabilities within the vulnerability management category for NPPs: asset mapping and vulnerability patching and mitigation.[37] The first capability area ensures that potential impact, mitigation, and remediation measures are available for use when vulnerabilities are announced. The second capability focuses on the implementation of mitigation process when a patching is not possible. Updated analysis and response procedures should be applied to ensure that there is minimal delay in response once indicators reveal an attack is underway.

NPP systems architecture best practices call for the use of levels and zones for separation and protection of functions and assets.[38] The defence-in-depth approach needs layers of security controls to detect and deny adversaries access to the delivery of cyber effects onto NPP targets. For NPP cyber security incident response, there are two capabilities under the SOC tools that take priority: operational technology security capability engineering and management, and sensing and network security capability engineering. The focus on operational technology should seem obvious, given the number of control systems that exist within NPPs, as well as the unique challenges that arise from configuration and monitoring of these systems in an operational state. Compared to ICT, the ICS protocols do not always have security mechanisms implemented. It is critical to ensure that there is a comprehensive connection management surveillance function that can support the enforcement of trusted system connection limitations. Subsequently, this identifies anomalous behaviours earlier in the attack lifecycle. A third capability in this category to consider is analytic platform engineering; the collection and analysis using advanced analytical platforms should yield operational efficiencies in the number of events or incidents handled.

In the situational awareness, communications, and training category, the two capabilities to focus on are: internal and external training and education, and the use of exercises. Education and training should be a high-priority area. This training might include short courses in cyber security awareness, longer courses on systems security architectures and risk management, and learn-as-you-go technical hacking platforms for developing and improving hands-on skills over time. Bringing together incident response stakeholders to test and evaluate incident-handling policies and procedures through use of exercises can support strengthening of emergency preparedness goals.

The leadership and management category has two areas that require focus: continuity of operations, and metrics. One of the primary goals of the incident response team is to analyse events and inform stakeholders to ensure the continued safe operation of the NPP. Teams that can shorten the analysis timeline while maintaining the quality of analysis will help the NPP towards this resiliency goal. Metrics are necessary to understand how investment in specific capabilities map onto managing the risk of operating the NPP with the threat from cyber-attacks. These metrics will also be used to allocate NPP resources towards further maturation of existing capabilities or investing in the development of new capabilities.

Conclusion

The application of digital controls in the nuclear industry has led to an important change in the design of engineering systems and their management, operations, and maintenance. With the increasing digital assets at NPPs, these facilities not only need cyber security experts, but also cyber security training for the wider multi-disciplinary workforce. Nuclear facilities have managed the digital transformation well, and thus far there has been no large-scale incident of public consequence. Nevertheless, the nuclear industry should continue to elevate cyber security efforts through international cooperation as cyber threats evolve and more digital systems are used at NPPs. The industry also needs to invest in designing tools to identify the attack surface at NPPs and develop design solutions to mitigate vulnerabilities. The nuclear industry has a strong history of success in applying management programmes to characterize and mitigate safety and security risks at NPPs. It should extend these programmes to integrate risk posed by cyber-attacks.

Notes

1. MITRE ATT&CK (2022), 'ATT&CK Matrix for Enterprise', MITRE Corporation [online]. https://attack.mitre.org
2. IAEA, *Core Knowledge on Instrumentation and Control Systems in Nuclear Power Plants.* IAEA Nuclear Energy Series, No. NP-T- 3.12 (Vienna: IAEA, 2011).
3. Ron Ross, Victoria Pillitteri, Richard Graubart, Deborah Bodeau, and Rosalie McQuaid, *Developing Cyber-Resilient Systems: A Systems Security Engineering Approach.* National Institute of Standards and Technology (NIST) Special Publication 800–160, Vol. 2 (McLean, VA: MITRE, 2021). https://doi.org/10.6028/NIST.SP.800-160v2r1
4. Joao Antunes, Nuno Neves, Miguel Correia, Paulo Verissimo, and Rui Neves, 'Vulnerability Discovery with Attack Injection', *IEEE Transactions on Software Engineering*, 36/ XX (2010), 357–70. doi: 10.1109/TSE.2009.91.
5. Ibid.
6. Stefan Authen and Jan-Erik Holmberg, 'Reliability Analysis of Digital Systems in a Probabilistic Risk Analysis for Nuclear Power Plants', *Nuclear Engineering and Technology*, 44/5 (2012), 824–84. doi: 10.5516/net.03.2012.707.
7. Larry J. Shuman, Mary Besterfield-Sacre, and Jack McGourty, 'The ABET "Professional Skills"—Can They Be Taught? Can They Be Assessed?', *Journal of Engineering Education*, 94/1 (2005), 41–55. doi: 10.1002/j.2168-9830.2005.tb00828.x.
8. Edward J. M. Colbert and Alexander Kott, *Cyber-Security of SCADA and Other Industrial Control Systems.* Advances in Information Security (Cham: Springer International Publishing, 2016).
9. Authen and Holmberg, 'Reliability Analysis'.
10. Han Bao, Hongbin Zhang, and Kenneth Thomas, *An Integrated Risk Assessment Process for Digital Instrumentation and Control Upgrades of Nuclear Power Plants* (Idaho Falls: Idaho National Laboratory, 2019). doi: 10.2172/1616252; US Nuclear Regulatory Commission (NRC), 'Protection of Digital Computer and Communication Systems and Networks', NRC Regulations Title 10, Code of Federal Regulations §73.54, USNRC [online], 24 March 2021. http://www.nrc.gov/reading-rm/doc-collections/cfr/part073/part073-0054.html; IAEA, Computer Security Techniques for Nuclear Facilities, Technical Guidance. IAEA Nuclear Security Series, No. 17-T (Rev.1) (Vienna: IAEA, 2021), and Ross et al., *Developing Cyber-Resilient Systems.*
11. Jonathan West, J. Hale, Mauricio Papa, and Peter Hawrylak, 'Automatic Identification of Critical Digital Assets', Second International Conference on Data Intelligence and Security (ICDIS), June 2019. doi: 10.1109/ICDIS.2019.00040.
12. Michael T. Rowland, Donald D. Dudenhoeffer, and J. Scott Purvis, 'Computer Security for I&C Systems at Nuclear Facilities', in *Proceedings of the International Conference on Nuclear Plant Instrumentation, Control, and Human-Machine Interface Technologies, San Francisco, 11–15 June 2017* (La Grange Park, IL: ANS, 2017), 1–10.
13. West et al., 'Automatic Identification of Critical Digital Assets'.
14. Jong Woo Park and Seung Jun Lee, 'Application of Probabilistic Risk Assessment to Cyber Security of a Nuclear Power Plant', Probabilistic Safety Assessment and Management Conference, 16–21 September 2018.
15. Z.-W. Guo, L. Zhu, and L. Zhou, *Research on Defense-In-Depth Standards for Information Security in NPP I&C System.* Lecture Notes, Nuclear Power Plants: Innovative Technologies

for Instrumentation and Control Systems (Cham: Springer, 2021). doi:10.1007/978-981-16-3456-7_50.

16. Park and Lee, 'Application of Probabilistic Risk Assessment'.
17. Ibid.
18. B. Stott, O. Alsac, and A. J. Monticelli, 'Security Analysis and Optimization', *Proceedings of the Institute of Electrical and Electronics Engineers (IEEE)*, 75/12 (1987), 1623–44. doi: 10.1109/PROC.1987.13931.
19. MITRE ATT&CK (2019), 'Impact', MITRE Corporation [online]. https://attack.mitre.org/tactics/TA0040
20. MITRE ATT&CK (2020), 'Damage to Property', MITRE Corporation [online]. https://attack.mitre.org/techniques/T0879
21. MITRE ATT&CK (2020), 'Manipulation of Control', MITRE Corporation [online]. https://attack.mitre.org/techniques/T0831
22. Tony UcedaVélez and Marco M. Moran, *Risk Centric Threat Modeling: Process for Attack Simulation and Threat Analysis* (Hoboken: John Wiley & Sons, 2015).
23. MITRE ATT&CK (2022), 'ICS Techniques', MITRE Corporation [online]. https://attack.mitre.org/techniques/ics/
24. Deborah Bodeau and Richard Graubart, *Characterizing Effects on the Cyber Adversary*. MTR130432 (Bedford, MA: The Mitre Corporation, 2013).
25. MITRE Corporation (n.d.), 'Systems Engineering: Risk Management Approach and Plan', MITRE Corporation [online]. https://www.mitre.org/our-impact/mitre-labs/systems-engineering-innovation-center/risk-management-approach-and-plan
26. National Institute of Standards and Technology (NIST), *Risk Management Framework for Information Systems and Organizations*. National Institute of Standards and Technology (NIST) Special Publication 800–37 (McLean, VA: MITRE, 2017), Rev. 2. https://doi.org/10.6028/NIST.SP.800-37r2
27. Ibid.
28. Rick Jacobs and Sonja Haber, 'Organizational Processes and Nuclear Power Plant Safety', *Reliability Engineering & System Safety*, 45/1–2 (1994), 75–83.
29. Yong Wang, Garhan Attebury, and Byrav Ramamurthy, 'A Survey of Security Issues in Wireless Sensor Networks', *CSE Journal Articles* (2006) No. 84. https://digitalcommons.unl.edu/csearticles/84/
30. Mislav Findrik, Ivo Friedberg, Ewa Piatkowska, Paul Smith, and Jae-Gu Song, 'Trustworthy Computer Security Incident Response for Nuclear Facilities', in *Proceedings of the Fourth International Symposium on Future Instrumentation and Control for Nuclear Power Plants, 26–30 November 2017* (Vienna: IAEA, 2017).
31. Vincent C. Hu, David F. Ferraiolo, and Rick Kuhn, 'Assessment of Access Control Systems', Interagency Report 7316. Computer Security Division (McLean, VA: MITRE, 2006).
32. Kathryn Knerler, Ingrid Parker, and Carson Zimmerman, *11 Strategies of a World-Class Cybersecurity Operations Center* (Bedford, MA: MITRE Corporation, 2022).
33. Ibid.; also see IAEA, *Computer Security Incident Response*. Non-serial Publications, IAEA-TDL-005 (Vienna: IAEA, 2016).
34. Balazs Peter Hamornik and Csaba Krasznay, 'Prerequisites of Virtual Teamwork in Security Operations Centers: Knowledge, Skills, Abilities and Other Characteristics', *Academic and Applied Research in Military and Public Management Science*, 16/3 (2017), 35–48. doi:10.32565/aarms.2017.3.5.

35. Knerler et al., *11 Strategies*.
36. CISA, 'Insider Threat Mitigation Guide'.
37. Morey J. Haber and Brad Hibbert, *Asset Attack Vectors: Building Effective Vulnerability Management Strategies to Protect Organizations* (New York: Apress, 2018).
38. US Nuclear Regulatory Commission (NRC), 'Physical Protection of Plants and Materials, Protection of Digital Computer and Communication Systems and Networks', NRC Regulations Title 10, Code of Federal Regulations §73.54. USNRC [online], 2 November 2015. http://www.nrc.gov/reading-rm/doc-collections/cfr/part073/part073-0054.html; IAEA, Computer Security Techniques for Nuclear Facilities; Ross et al., *Developing Cyber-Resilient Systems*.

Suggested Reading

Deborah Bodeau and Richard Graubart, (2013), *Characterizing Effects on the Cyber Adversary*. MTR130432 (Bedford, MA: The Mitre Corporation).

Cybersecurity and Infrastructure Security Agency (CISA), (2020), *Insider Threat Mitigation Guide* (Washington, DC: CISA).

Kathryn Knerler, Ingrid Parker, and Carson Zimmerman, (2022), *11 Strategies of a World-Class Cybersecurity Operations Center* (Bedford, MA: MITRE Corporation).

Mislav Findrik, Ivo Friedberg, Ewa Piatkowska, Paul Smith, and Jae-Gu Song, (2017), 'Trustworthy Computer Security Incident Response for Nuclear Facilities', in *Proceedings of the Fourth International Symposium on Future Instrumentation and Control for Nuclear Power Plants, 26–30 November 2017* (Vienna: IAEA).

MITRE ATT&CK, (2022), 'ATT&CK Matrix for Enterprise', MITRE Corporation [online]. https://attack.mitre.org

Michael T. Rowland, Donald D. Dudenhoeffer, and J. Scott Purvis, (2017), 'Computer Security for I&C Systems at Nuclear Facilities', in *Proceedings of the International Conference on Nuclear Plant Instrumentation, Control, and Human-Machine Interface Technologies, San Francisco, 11–15 June 2017* (La Grange Park, IL: ANS), 1–10.

CHAPTER 19

RADIOLOGICAL SOURCE SECURITY

IRMAN ABDUL RAHMAN, GEOFFREY CHAPMAN, AND FAIZAL MOHAMED

Introduction

While radiological sources are used in a wide array of research, industry, and healthcare applications worldwide and provide undoubted utility to their respective industries, ensuring that they are adequately secured poses several unique challenges due to their numerous applications and widespread use in public settings. Compared to nuclear security, many sources are retained in 'public facing' institutions, such as hospitals, which are not able to restrict public access. Sources may also be embedded in equipment, which by necessity is easily portable and routinely moved to open sites, such as in the case of nuclear density gauges used in construction. The prevalence of radiological sources means that not all staff may be sufficiently aware of their security obligations, while institutions may not devote adequate resources to their security. Previous incidents demonstrate that source misuse can occur, with tragic consequences to local communities. Striking the correct balance between access, safety, economic sustainability, and effective implementation of security are critical ongoing issues.

Although the increasingly widespread use of radiological sources worldwide has brought great benefits, such as the provision of blood transfusions and radiotherapy, the greater availability of radiological sources poses increased risks to their misuse when 'many of them [are] poorly secured and vulnerable to theft'.[1] Determining this trend is challenging as the precise number of radiological sources in use worldwide is unknown. What is clear is high-activity radiological sources that pose a security risk are present at 'thousands of sites in more than 100 countries'.[2] The only definitive statistic on the increase in Category I and II radiological sources comes from the United States (US).

Between 2008 and 2021 when the data were centrally collected by the Nuclear Regulatory Commission, there was a 30 per cent increase in their total number.[3] If the trend within the US is indicative, then the challenge of implementing radiological source security will only increase—more so in developing countries where there is greater scope to expand usage and where alternative technologies are less suitable.

This chapter surveys the current state of the threat, assessing the motivations and capabilities of non-state actors. Potential scenarios range in severity from the theft of a radiological source for economic benefit to mass casualty incidents resulting from the intentional dispersal of radiological material. The chapter also highlights the importance of security in preventing such incidents, covering the fundamentals of radiological source security as well as the challenges in implementation. The chapter then examines recent international measures to improve radiological source security, including formal and informal initiatives and efforts such as the adoption of alternative technologies. The chapter also examines differing national approaches to upholding radiological source security in a sustainable manner, through the inclusion of two case studies: the approach of the United Kingdom (UK) to radiological security regulation; and Malaysia's response to a radiological incident, with a focus on how it responded by improving its regulatory capacity. The chapter concludes by highlighting best practices and recommendations for radiological source security.

Threat Assessment

When considering the resources that should be devoted towards radiological source security, the potential impact of a misuse scenario must be considered. Lower-consequence scenarios include radiological material being stolen for economic or blackmail purposes; this could impact operations, cause reputational damage, and lead to inadvertent distribution of radiological material. Deliberate radiological incidents intended to inflict harm encompass a very wide range from 'hardly any casualties' up to 'mass casualty terrorism'.[4] Radiological attack methods include passively placing a device as a radiation emission device (RED), dispersing the material explosively as a radiological dispersal device (RDD), distributing particulate radiological matter through a radiological incendiary device (RID), or other methods causing inhalation, ingestion, and immersion of radioactive material in what James Acton and colleagues term 'I3 attacks'.[5]

The ability of each scenario to produce fatalities from deterministic effects would depend on the efficacy of dispersal, exposure times, and any internal contamination. The consequences would further vary depending on the quantity and type of radiological material used. The International Atomic Energy Agency (IAEA) provides a risk-based ranking through the categorization of radioactive sources. This categorization system considers the deterministic effects to health resulting from exposure and contamination in relation to the quantity of material involved. Small quantities of low-activity sources are correspondingly placed in lower-risk categories than large quantities of high-activity sources, and this assessment can be used for 'risk-informed decision making' relating to security.[6]

Radiological material can either be contained within 'sealed' sources (i.e. radiological material that is encapsulated in a solid material) or present in 'unsealed' sources (i.e. liquid, powder or gaseous radioactive material). Unsealed material is typically either present in small quantities, has a short half-life, or has lower activity. IAEA guidance has therefore historically focused on sealed sources to avoid a 'loss of focus on the most serious safety threat'.[7] However, the current IAEA guidance on the security of radiological material now also covers unsealed material where the quantity present exceeds thresholds of unsafe values.[8]

However, rather than direct health effects, the psychological impact of misusing radiological material means that it could alternatively be employed as a 'weapon of mass disruption'. An area exposed to radiological material would be disrupted until decontaminated and there could be secondary 'severe, unpredictable, long-lasting' political, social, and economic consequences to an attack.[9] Miles Pomper cites government estimates of recovery costs for major urban areas impacted by radiological terrorism ranging from US$25 billion per square kilometre to US$500 million per square kilometre.[10] In contrast, some planned terrorist activity involving radiological material has involved quantities insufficient to have anything other than a psychological impact. A UK-based terrorist plot in 2004 involving americium-241 harvested from smoke detectors would have required material from 20 million units to reach a thresholds equivalent of a Category II source, in comparison to the 10,000 envisaged to be used.[11]

Nonetheless, the ability of radiological sources to inflict harm and disruption has been proven repeatedly in unintentional accidents involving material outside of regulatory control. In 1985, a caesium-137 teletherapy unit was abandoned in a private clinic in Goiânia, Brazil.[12] The unidentified source assembly was salvaged for scrap in 1987 and the source capsule ruptured. As the caesium salt was soluble and glowed in the dark, it fascinated the local population and was being handled by several people, leading to prolonged exposure, dispersal, and ingestion. In the 15-day period between the removal of the source and the realization that a radiological incident was occurring, many people were exposed to radiation, leading to four deaths and a further 28 suffering radiation burns. The incident necessitated a substantial decontamination operation, involving the removal of contaminated topsoil in the area.

Around 11,2000 people were monitored for exposure, 'of whom 249 were contaminated either internally or externally'.[13] The 'disproportionate ... number of people who ... sought medical attention ... [compared to those] that actually needed it' indicates the potential psychological impact of an intentional radiation attack.[14] Peter Zimmerman and Cheryl Loeb note the incident 'shows that some forms of radiological attack could kill tens or hundreds of people and sicken hundreds or thousands'.[15] Similarly, a radiation accident that occurred in Samut Prakan, Thailand, in 2000 mimicked the impact of an RED. A cobalt-60 source was partially dismantled in a scrapyard and left *in situ*, exposing workers and locals to radiation, leading to three deaths and seven injuries, and 258 people who lived within 50 metres of the scrapyard were followed up by the Ministry of Public Health for long-term effects.[16]

Fortunately, few terrorist groups have attempted to misuse radiological material despite it being 'well within terrorists' ability to acquire radiological weapons'.[17] Chechen terrorists buried a caesium canister in a Moscow park in 1995 and an al-Qaeda-affiliated cell led by Dhiren Barot plotted in 2006 to use radiological material in an attack.[18] In 2014, a member of so-called Islamic State claimed the group possessed a 'dirty bomb' after the seizure of 40 kilograms of unsuitable low-grade uranium compounds from Mosul University.[19]

While no intentional mass casualty incidents involving radiological materials have occurred to date, the viability of radiological sources to inflict harm on individuals has been proven with the 2006 Litvinenko poisoning—where a Russian defector in London died by ingesting polonium-210—and several other less-well-documented incidents.[20] Nonetheless, given the availability of radiological material lost either within territory held by non-state actors, outside of regulatory control, or loosely secured within states, the discrepancy between the potential for a radiological attack and the lack of observed incidents is notable.

In view of the potential risks compared to the few recorded incidents, scholars Charles Ferguson and Michelle Smith attempt to explain the rarity of radiological terrorism by suggesting that conducting a radiological attack is more technically challenging than commonly assumed.[21] The ability to acquire, safely work, and effectively disperse radiological material poses additional challenges when compared to conventional terrorism, in an environment when knowledge acquisition is challenging and avoiding detection is paramount. For instance, Islamic State captured a cobalt-60 radiotherapy unit in Mosul Hospital in 2014 but was either unaware of the source's existence (despite the information being available online) or lacked the motivation or expertise to exploit it.[22] In addition, Wyn Bowen and Jasper Pandza note the role of pre- and post-event deterrence via denial strategies that can be employed by states, such as detection of radiological material at borders or impact mitigation measures, that further reduce the appeal of radiological terrorism.[23]

The Principles of Radiological Source Security

The chief means by which to prevent radiological material being misused is for its security to be improved. In practice, this means applying core concepts for radiological source security, such as providing defence-in-depth or a layered security approach. This is a strategy to build up a robust and redundant security system, aimed at radiological risk reduction. Any security protocol or system cannot guarantee 100 per cent protection, but the combined effect of multiple systems can provide a greater degree of protection than a single layer. An effectively implemented layered security system will ensure that an adversary attempting to acquire radiological material will be deterred, detected, and delayed. Even in the event of unauthorized attempted acquisition of radiological material, a sufficiently robust security system should allow the authorities sufficient time to respond, either by securing the material or to coordinate an emergency response and decontamination efforts. When combined, such security measures will minimize both the possibility and consequences of the misuse of radiological material. In turn, this will limit the appeal of misappropriating radiological material by malicious non-state actors.

The specific security measures to be applied to radiological sources will depend on the particular source involved—balancing local conditions, access, and the potential risks they could pose if misused. For instance, embedding a radiological source deep within a multi-tonne radiotherapy unit could sufficiently delay access to a source, but is an inapplicable solution for a mobile blood irradiator unit. The need to balance security with continuous public access for equipment such as blood irradiators or sources used in academic research means that unique approaches are required, and ideally these will integrate a range of human and technical measures.[24] The greater the quantity and radioactivity of the material, together with any on-site vulnerabilities, will determine the stringency with which layered defence-in-depth deter, detect, and delay measures should be implemented.

Aside from improving the security of radioactive sources on site, additional measures to mitigate the risks of radiological misuse include decreasing the use of certain types of very radioactive or dispersible materials, enhancing regulatory controls, improving export controls, increasing government cooperation in intelligence sharing about threats, deploying radiation detectors, rounding up orphan sources, and developing improved disposal and recycling pathways for sources. Reducing consequences involves developing and deploying better decontamination technologies, improving the training of emergency first responders, and increasing the capacity for effective international response to radiological incidents.[25] Effective measures that limit the opportunity to acquire radiological material, covertly transport it, and minimize any misuse scenario will further deter radiological misuse.

Of these measures, ensuring regulatory control of disused sources will remain a central challenge. The cases of Goiânia and Samut Prakan, and an incident in 2010 when an orphan cobalt-60 source in Mayapuri, India, resulted in one death and eight hospitalizations, indicate the potential for high-activity sources to continue to harm, especially if acquired by actors with malicious intent.[26] As commercial incentives end with the termination of use by an operator, governments and regulators must ensure sufficient measures are in place to fund the costly transport and disposal of radiological materials, including in situations where any of the commercial entities involved have become insolvent in the intervening time period.[27]

International Efforts

The effective implementation of these principles requires both private and state capacity to be devoted towards radiological source security. Given the global nature of sophisticated non-state actor threats, preventing radiological incidents 'requires co-operation on a near-global scale, and consistency of resolve'.[28] However, in contrast to nuclear materials for which there exists a relatively developed international nuclear security regime, underpinned by the legally binding Convention on the Physical Protection of Nuclear Material and its Amendment, international efforts to secure radiological materials have been far more limited.

The IAEA has developed the Code of Conduct on the Safety and Security of Radioactive Sources ('Code of Conduct'). The prelude to the Code of Conduct was an IAEA-sponsored conference on radiation safety in 1998 that established the framework on radiological source categorization. Following the 11 September 2001 ('9/11') terrorist attacks, nuclear security concerns rose to the fore.[29] The Code establishes principles and a framework for regulating the safety and security of radiological sources by prescribing legislative and regulatory objectives; these include establishing a national source registry, improving export and import controls, and enforcing regulatory security requirements. The IAEA subsequently produced two supplementary documents: 'Guidance on the Import and Export of Radioactive Sources', published in 2004; and 'Guidance on the Management of Disused Radioactive Sources', published in 2017. Additional implementation guidance on the security of radioactive sources is available via the IAEA's Nuclear Security Series: 'Security of Radioactive Sources', published in 2009; and 'Security of Radioactive Material in Use and Storage of Associated Facilities', published in 2019.

However, both the Code of Conduct and its supplementary Guidance are not legally binding, and engagement has arguably been uneven. At the time of writing, 141 countries have expressed political support for the Code of Conduct since 2004, but only 106 have returned an Importing and Exporting survey urged to be returned under the supplementary Guidance. Just 45 states have notified the IAEA of their commitment to implement the 2017 guidance on disused radioactive sources.[30] The task of creating the necessary laws and regulations and ensuring implementation on a state-by-state basis is further complicated due to the many competing demands on the resources of legislators, regulators, and operators responsible for controlling radioactive materials.

Recognizing this relative lack of international engagement on radiological source security issues, the international community has sought to raise awareness over potential radiological risks and champion initiatives intended to improve radiological source security around the world. Arguably, this has most prominently included the Nuclear Security Summit process, where over the course of four biannual summits (2010 to 2016), numerous countries pledged to implement improved security measures.

While primarily aimed at improving the security of nuclear material, the summit process was subsequently broadened to include radiological source security. The 2012 summit Communiqué stressed the need for all states to secure radiological material through adherence to measures such as the Code of Conduct and its supplemental guidance documents. Participating states pledged to create national registers of high-activity radioactive sources.[31] At the 2014 summit, 23 countries pledged to make specific advancements to secure their most dangerous radiological sources by 2016, with 22 countries making significant progress in that time frame. This included improved cooperation and notification mechanisms for loss of control of radiological material, transitioning to alternative technologies, and updating threat assessments.[32] At the 2016 summit, 27 states committed to take further action to prevent insider threats to nuclear and radiological material.[33]

An important legacy of the Nuclear Security Summits has been the institutionalization of progress for radiological source security in the foundation of IAEA INFCIRC/910, which commits signatories to improve and harmonize approaches to the security of high-activity sealed sources. However, only around 30 states that are signatories have subscribed to INFCIRC/910.[34] Further work is required on universalization and implementation of existing international measures, including the Code of Conduct.

Other mechanisms for international cooperation include the IAEA fostering bilateral cooperation on radiological security. To help enable this, the IAEA maintains a directory of national regulatory bodies.[35] Additional forums to improve radiological security include the Global Partnership against the Spread of Weapons and Materials of Mass Destruction, a G7 initiative launched after 9/11 to reduce chemical, biological, radiological, and nuclear (CBRN) risks. Their work most prominently included the removal of high-risk CBRN materials from the former Soviet Union.[36] The Global Partnership has since founded a Nuclear and Radiological Security sub-Working Group which 'match makes' expertise and pairs donor and recipient countries on projects intended to improve radiological security.[37]

Alternative Technologies

One of the major initiatives of the summit process was the push for the adoption of alternative technologies to mitigate the risk posed by high-activity radiological sources. The adoption of alternative technologies usually consists of four alternative approaches. These include replacing the application with an alternative radiological source that is less suitable to use in an RDD, replacing the radioisotope with other technology (such as a linear accelerator), changing the chemical or mechanical form of the source to be more dispersion resistant, and modifying the equipment to be more theft or tamper resistant.

If appropriate to local conditions, some applications of radiological sources can be replaced by particle accelerators, which include electron accelerators and cyclotrons. These accelerators can produce a range of secondary particles and can therefore fulfil several functions, such as sterilization, radiotherapy, material science, and research applications. This versatility and risk reduction, however, comes with increased capital and maintenance costs and greater technical complexity. Alternative technologies are also dependent on stable power supplies, reliable cooling systems, and highly qualified personnel.[38]

According to the latest Nuclear Threat Index (NTI) assessment on radiological sources, only 6 per cent of countries had committed to a policy of replacing high-activity sealed sources with alternative technologies.[39] While research has highlighted the success of promoting alternative technologies to replace high-activity sources in Japan, France, Norway, and the US, these countries likely represent a minority of countries where the prerequisite conditions are in place and increased costs represent a favourable exchange for the corresponding risk reduction.[40]

If alternative technologies are not appropriate, all states can still improve the security of high-risk sources deemed vulnerable by making modifications to improve their physical security. For instance, in medical facilities, sources within teletherapy and blood irradiator units should be secured in such a way in that they can only be removed by special tools and procedures. Unauthorized access attempts should trigger alarms both inside the facility and remotely. For applications that are mobile, such as blood irradiators, equipment should be tracked via GPS so the authorities can more easily locate the source if it is stolen. Whenever a large radiological source is being transported, alarms and tracking devices should be installed in the vehicle. If the vehicle departs from its planned route, the authorities can respond quickly and secure the material.[41]

Case Study: The UK and Radiological Source Security Regulation

Radiological source security for the UK's non-nuclear sites is regulated by the environmental agencies of the country's four constituent nations: England, Wales, Scotland, and Northern Ireland. Each agency maintains standards up to the equivalent national level, with regulatory powers authorized to England's Environment Agency and Natural Resources Body for Wales under The Environmental Permitting (England and Wales) Regulations 2016 (amended 2018), the Scottish Environment Protection Agency under the Environmental Authorisations (Scotland) Regulations 2018, and the Northern Ireland Environment Agency under the Radioactive Substances Act 1993 and High-activity Sealed Radioactive Sources and Orphan Sources Regulations 2005.[42] This section uses the UK as an example as it was recently judged by the IAEA's Integrated Regulatory Review Service (IRRS) as demonstrating 'good practice' on the 'interface between the environmental regulators and national counter terrorism security agencies for Category I to IV sources'.[43] Nonetheless, the UK has a complex regulatory system that has developed gradually over time and continues to evolve to account for the numerous government bodies involved in delivering radiological source security.

The UK's radiological security system is coordinated through regular meetings of the environmental regulators with police and transport security regulators to ensure consistent standards and coordination over overlapping competencies.[44] The UK's strategy for radiological source security encompasses all four strands of the UK's counter-terrorism CONTEST strategy, including measures to 'Prevent', 'Pursue', 'Protect', and 'Prepare' against acts of terrorism. This includes effective police counter-terrorism response in the case of incidents, but also efforts such as Project PORTOS and Cyclamen to prevent the importation of radiological material outside of regulatory control.[45] In line with general policing in the UK, there are no armed guards for sites housing high-activity sealed sources.

Regarding the development of a nuclear security regime in the UK, although radiological sources have long since been required to be secured against loss and theft, the 9/11 terrorist attacks demonstrated the need to significantly improve the regulation of radiological sources to protect them from armed hostile adversaries.[46] Regulatory powers over security were granted to the environmental agencies through 'The High-activity Sealed Radioactive Sources and Orphan Sources Regulations 2005'. Accordingly, specific duties for licensees and regulators are embedded in primary legislation.

Environmental agencies may initially appear as odd choices for security regulators. They are non-departmental government bodies with a range of responsibilities, from ensuring flooding protection to issuing fishing licences. However, the need to rapidly introduce an effective and fully authorized security regime following 9/11 meant that the environmental agencies were granted an extension to their existing licensing powers covering the safe use of radiological sources that had existed since the 1960s.[47] While the UK's environmental agencies retain a broad range of competencies, the regulation of security for high-activity sealed sources (HASS) remains an important part of their remit. As noted by the IAEA IRRS mission in 2019, the UK's environmental agencies have proven to be both effective and independent security regulators.[48]

In practice, the UK's environmental agencies work closely with local police Counter-Terrorism Security Advisors (CTSAs) to accomplish their security functions. CTSAs will often accompany environment agency staff on inspections as, although environment agency staff are knowledgeable as regulators, CTSAs are security specialists and they 'need to work together to get [the] best outcome'.[49] The CTSA parent organization, the National Counter Terrorism Security Office, sets the appropriate physical security standards for radiological sources in liaison with the UK's intelligence and security services, acting in conjunction with the Centre for the Protection of National Infrastructure.[50] These institutions ensure that security regulation is set in cooperation with law enforcement at both a local and national level.[51]

The security standards that must be upheld by licensees are encoded in a frequently updated restricted document, which specifies the required standards depending on the category and use of the source in question, in line with the national threat level. The UK's security requirements for radiological sources follow IAEA guidance set out in the Nuclear Security Series and are consistent with the source Categories I–IV, with additional requirements on some specific use cases of sources.[52] As the radioactivity of the sources increases, so do the required security arrangements, but these are maintained in proportion with operational and safety concerns.

From the perspective of licensees, compliance with the UK's radiological security regime requires early and persistent engagement. Even before being granted a licence, operators must already have security provisions in place and have demonstrated these during an inspection. Requirements include developing a site security plan, developing an information security plan, demonstrating the potential for security upgrade pathways, and implementing staff background checks.[53] Application for a licence in England costs approximately £4,750–£6,000, with annual fees of between £1,950 and £3,550 as of 2022.[54] Given the information needed to obtain the permits, the documentation is restricted to ensure information security. Licensees need to inform the regulator of any site change that could impact upon security.[55]

Compliance with regulations is ensured by frequent inspections at sites housing HASS. According to the IAEA 2019 IRRS, 'in general, the UK's environmental agencies inspect HASS every year, open sources and waste permits every second year and smaller sources every four years'.[56] Inspections can be pre-planned, in reaction to events, and either announced or unannounced.[57] The perception of frequent interaction with the regulator is shared among licensees who generally believe the environmental agencies are competent at maintaining regular contact.[58] If deficiencies are detected, the environmental agencies can act against licensees, with such sanctions including warnings, formal cautions, and prosecutions. No individual criminal cases have yet been brought as a result of deficiencies in fulfilling security regulations.[59] Sources have been lost and lesser penalties issued, but the permit holders, CTSAs, and Environment Agency staff work within a small community where their working relationship depends to an extent on good faith cooperation.

The transport of radiological material within the UK is regulated by the Office for Nuclear Regulation (ONR), which inspects duty holders' compliance to general criteria set out under international regulations (for example, the European Agreement Concerning the International Carriage of Dangerous Goods by Road–ADR), such as maintaining a security plan, establishing temporary storage areas, and drivers receiving security training.[60] Common issues encountered by ONR inspections include either an absence of or out-of-date security plans and a lack of awareness of the measures needed to secure high-activity sources.[61]

In terms of future areas for reform in the UK, further consideration could be given to improving the security of unsealed sources. However, the environment agencies recognize that the varied nature of these sources means their current prescriptive regulations would have to be altered towards a more goal-orientated approach.[62] Rather than setting rules that must be followed, a goal-orientated approach specifies outcomes that licensees must demonstrate that their security arrangements perform to. Some areas that could be reformed involve shifting regulations in line with existing improved practices. This is the case with securing radiological sources during transport within the UK.[63] Improved measures have been developed and deployed over time, but these standards are not yet specified in prescriptive or goal-orientated regulations.[64]

Unlike nuclear security, the UK's environmental agencies have no plans to adopt a goal-orientated approach for sealed sources that the ONR has used since 2017.[65] The environmental agencies maintain that the number of facilities, range of security assessment experience, and threats involved make such an approach impractical. For the most part, licensees will not have the same resources or capacity to dedicate to formulating their own security approaches compared to nuclear institutions.[66] It is also unlikely that the environmental agencies, ONR, and the Health and Safety Executive will be consolidated into a 'super' nuclear and radiological regulator for safety and security, as is the case in some other countries. Although the framework that governs the security of radiological material in the UK is complex, it is currently perceived to perform satisfactorily; any structural change would require changes to primary legislation and would erase complimentary cultures embodied by the different regulatory bodies.[67]

Case Study: Radiological Source Security in Malaysia

With more countries using ever-increasing amounts of radiological material for a variety of purposes, the threats posed by the loss of regulatory control of radiological material are becoming better understood worldwide. As radiological source security remains the responsibility of the state housing the material, how to improve radiological source security and learn lessons from incidents are of critical importance. This was demonstrated in Malaysia when on 9 February 2017, two sources from a gamma projector unit owned by an authorized industrial radiography company were stolen in the south-west city of Klang.

The two iridium-192 sources involved had an activity of 43.6 curie (Ci) and 35.8 Ci. The gamma projectors were stolen at around 03:30 from a vehicle parked at the company building. After becoming aware of the theft, the licensee reported the incident to the police and the Atomic Energy Licensing Board (AELB). Malaysia's Atomic Energy Licensing Act (Act 304) meant that, while the police investigated under their Criminal Procedure Code (CPC) to apprehend the perpetrators, the AELB also conducted their own investigation to recover the sources.

On 11 February 2017, a joint search operation between AELB and police found the projector at a scrap metal shop near the company building. After questioning the owner and workers in the scrap metal shop, the missing sources were recovered in the possession of a janitor within a nearby residential area, three days after the initial theft. Elevated radiation levels were detected from a distance of 500 metres from where the source was found. The Malaysian authorities immediately evacuated all 300 residents in the proximity and police set up roadblocks to prevent the public entering the area.[68] The cleaner and all residents in the cordoned area were provided with a medical check-up by the State Health Department. In addition to concerns over the direct effects of radiation, many residents experienced panic attacks and anxiety over their wellbeing. In response, the Department of Social Welfare provided counselling to them.

In order to improve the response to radiological incidents, the Malaysian authorities have made significant changes in the approach to institutional cooperation between the Royal Malaysian Police (RMP) and the AELB. Existing Malaysian authority standard operating procedures have now been incorporated for both the safety and security of radiological sources. The RMP has set up an internal task force to ensure that security forces can better handle the threat of nuclear and radiological terrorism. The AELB's authorization and inspection programme has also been updated to address the vulnerabilities identified in the 2017 Klang incident, ensuring that both the AELB and RMP now are better prepared to deal with nuclear and radiological security in Malaysia.[69] Further efforts have included a nationwide user and public awareness programme covering the benefits and security needs for radiological sources.

Civil society has also played a part in improving Malaysia's capacity in the area of nuclear and radiological security, with some universities offering relevant courses. These include training offered by Universiti Tenaga (UNITEN), Universiti Teknologi Malaysia (UTM), and Universiti Kebangsaan Malaysia (UKM). Other regional actors in Asia can also help in improving nuclear and radiological security training, such as universities in Japan and South Korea that offer courses in this area. Through research and cooperation across these institutions, all partners can learn from their respective experiences, disseminate best practices, and assist in generating policy to improve radiological source security across the region.

The Malaysian experience demonstrates that the increasing requirements of radiological and nuclear security have created new challenges, demanding the development of an overall state strategy to handle national implementation across all relevant stakeholders. To coordinate this effort, the AELB obtains assistance from institutions such as the National Nuclear Support Centre (NSSC). The role of the Malaysian NSSC is training authorized users, sustaining nuclear security devices, detection and response, providing expert support, optimizing national resources and best practices, and encouraging experience sharing. The establishment and effective use of the NSSC is intended to increase national capacity for radiological source security, promote international collaboration, and foster a developing network of experts.

Conclusion

Radiological sources continue to be used worldwide in numerous beneficial applications. However, some high-activity sources remain vulnerable to theft. Given that several non-state actor groups have demonstrated an interest in radiological misuse and past accidents have illustrated the potential consequences of a loss of regulatory control, ensuring the sufficient implementation of radiological source security remains a global priority. This chapter explained both the fundamental principles of radiological source security, such as defence-in-depth and applying a graded approach, as well as international initiatives aimed at universalizing the application of these measures.

Upholding these principles will remain challenging as it requires consistent commitment from a range of stakeholders, including manufacturers, regulators, operators, and governments. To lessen this burden, alternative technologies can be used where appropriate, while human capital can be used to better effect in delivering security through training and capacity-building. Radiological source security can also be further improved by sharing national approaches and best practices. In this regard, this chapter highlighted the UK's approach to radiological source security and Malaysia's efforts to ensure regulatory control. States and institutions must continue to cooperate in sharing their experiences to ensure that the momentum for continual improvement in radiological source security is maintained.

Notes

1. Andrew J. Bieniawski, Ioanna Iliopulos, and Michelle Nalabandian, *Security for Radiological Sources* (Washington, DC: Nuclear Threat Initiative (NTI), 2016), 3. https://www.belfercenter.org/publication/security-radiological-sources
2. Ibid., 4.
3. National Academies of Sciences, Engineering and Medicine, *Use of Radioactive Materials in Commercial Applications Has Increased by 30 Percent*, NAS [online], 14 June 2021. www.nationalacademies.org/news/2021/06/use-of-radioactive-materials-in-commercial-applications-has-increased-by-30-percent-government-should-improve-security-and-support-development-of-alternatives-says-new-report
4. Wyn Bowen and Jasper Pandza, 'Preventing Radiological Terrorism: Is There a Role for Deterrence?', in Alex Winer and Andreas Wenger, eds., *Deterring Terrorism: Theory and Practice* (Redwood City: Stanford University Press, 2012), 180–204.
5. James Acton, M. Brooke Rogers, and Peter Zimmerman, 'Beyond the Dirty Bomb: Rethinking Radiological Terror', *Survival*, 49/3 (2007), 151–68.
6. IAEA, *Categorization of Radioactive Sources*, IAEA Safety Guides, No. RS-G-1.9 (Vienna: IAEA, 2005). https://www-pub.iaea.org/MTCD/Publications/PDF/Pub1227_web.pdf
7. Steve McIntosh and Kristen Cutler, 'The Code of Conduct on the Safety and Security of Radioactive Sources', in *Safety and Security of Radioactive Sources: Maintaining Continuous Global Control of Sources throughout their Life Cycle: Proceedings of an International Conference*, Abu Dhabi, 27–31 October 2013 (Vienna: IAEA, 2015), 68.
8. IAEA, *Security of Radioactive Material in Use and Storage and of Associated Facilities*, Implementing Guide, IAEA Nuclear Series, No. 11-G (Rev.1) (Vienna: IAEA, 2019). https://www-pub.iaea.org/MTCD/Publications/PDF/PUB1840_web.pdf
9. Igor Khripunov, 'The Social And Psychological Impact Of Radiological Terrorism', *The Nonproliferation Review*, 13/2 (2006), 275–316.
10. Miles Pomper, *Mind the Gap: The Role of Liability and Insurance Regimes in Strengthening Radiological Security* (Monterey, CA: James Martin Center for Nonproliferation Studies, 2014). www.jstor.org/stable/resrep09883?seq=1
11. Raymond Murray and Keith Holbert, 'Biological Effects of Radiation', in *Nuclear Energy*, 8th edn (Oxford: Elsevier, 2019), 87–108; *The Guardian*, 'Mass Damage and Chaos',

[online], 7 November 2006. https://www.theguardian.com/uk/2006/nov/07/terrorism.usa

12. IAEA, *The Radiological Accident in Goiânia* (Vienna: IAEA, 1988). www-pub.iaea.org/mtcd/publications/pdf/pub815_web.pdf
13. Ibid., 2.
14. Bowen and Pandza, 'Preventing Radiological Terrorism'.
15. Peter Zimmerman and Cherly Loeb, 'Dirty Bombs, The Threat Revisited', *Defense Horizons*, 38 (2004), 1–11.
16. IAEA, *The Radiological Accident in Samut Prakarn* (Vienna: IAEA, 2002). https://www-pub.iaea.org/MTCD/publications/PDF/Pub1124_scr.pdf
17. Bowen and Pandza, 'Preventing Radiological Terrorism'.
18. Charles Ferguson and Michelle Smith, 'Assessing Radiological Weapons: Attack Methods and Estimated Effects', *Defence Against Terrorism Review*, 2/2 (2009), 15–34.
19. Geoffrey Chapman and Robert Downes, 'Dirty Business: Challenges Remain to Islamic State RDD Usage', *Jane's Intelligence Review*, 4 (2016), 46–52.
20. Charles Streeper, Marcie Lombardi, and Lee Cantrell, 'Nefarious Uses of Radioactive Materials', Session D, in *Proceedings of the 48th Annual Institute of Nuclear Management Meeting*, 13–17 July 2008 (Madison, WI: Omnipress, 2008).
21. Ferguson and Smith, 'Assessing Radiological Weapons', 16.
22. Chapman and Downes, 'Dirty Business'.
23. Bowen and Pandza, 'Preventing Radiological Terrorism'.
24. Mark J. Baumann, Douglas Adams, and Michal Kuca, 'Protection Concepts for Irradiators in a Laboratory Environment', IAEA-CN-269, in *Proceedings of the International Conference on the Security of Radioactive Material: The Way Forward for Prevention and Detection* (Vienna: IAEA, 2018), 1–11. https://www.osti.gov/servlets/purl/1594333
25. Gregory J. Van Tuyle, Tiffany L. Strub, Harold A. O'Brien, Caroline F. V. Mason, and Steven J. Gitomer, *Reducing RDD Concerns Related to Large Radiological Source Applications*, LA-UR-03-6664 (Los Alamos, NM: Los Alamos National Laboratory, 2003). https://media.nti.org/pdfs/27-doe.pdf
26. Rajoo Kumar, 'Lessons Learned from the Radiological Accident in Mayapuri, New Delhi', in *Safety and Security of Radioactive Sources: Maintaining Continuous Global Control of Sources throughout their Life Cycle: Proceedings of an International Conference*, Abu Dhabi, 27–31 October 2013 (Vienna: IAEA, 2015), 517.
27. Charles Ferguson, *Ensuring the Security of Radioactive Sources: National and Global Responsibilities* (Baltimore, MD: US–Korea Institute At Johns Hopkins University, 2012). https://www.jstor.org/stable/pdf/resrep11141.pdf
28. Michael Levi, *Preventing Nuclear and Radiological Terrorism* (Washington, DC: Brookings Institution, 2003. https://www.brookings.edu/wp-content/uploads/2016/06/levi20031024.pdf
29. McIntosh and Cutler, 'The Code of Conduct on the Safety and Security of Radioactive Sources'.
30. IAEA, 'List of States Expressing a Political Commitment', Code of Conduct on the Safety and Security of Radioactive Sources, IAEA [online], 18 May 2022. https://nucleus.iaea.org/sites/ns/code-of-conduct-radioactive-sources/Documents/Status_list%2018%20May%20%202022.pdf
31. Nuclear Security Summit, 'Communiqué', 2012 Nuclear Security Summit, Seoul, 2012.
32. Bieniawski et al., *Security for Radiological Sources*.

33. The White House, Office of the Press Secretary, 'Joint Statement on Insider Threat Mitigation', NSS [online], 5 April 2016. http://www.nss2016.org/document-center-docs/2016/4/1/joint-statement-on-insider-threat-mitigation-gb
34. Nuclear Threat Initiative (NTI), 'What Are Nuclear Security INFCIRCs?', NTI [online], n.d. https://www.ntiindex.org/story/what-are-nuclear-security-infcircs
35. IAEA, 'Directory of National Regulatory Bodies For the Control of Radiation Sources', IAEA [online], 7 October 2020. https://www.iaea.org/sites/default/files/18/02/coc-reg-auth-directory.pdf
36. Ferguson, *Ensuring the Security of Radioactive Sources*, 17.
37. Global Partnership Against the Spread of Weapons and Materials of Mass Destruction, 'Nuclear & Radiological Security', n.d. www.gpwmd.com/nrswg
38. Ingrid Kirsten, Noah Mayhew, and Anthony Stott, 'The Safe, Secure and Sustainable Application of Nuclear Technology in Industry and Agriculture', Vienna Center for Disarmament and Non-Proliferation [online], July 2021, 5. https://vcdnp.org/wp-content/uploads/2021/07/Nuclear-Applications-Industry-and-Agriculture-Case-Study_final.pdf
39. Nuclear Threat Initiative (NTI), 'Losing Focus in a Disordered World', 5th edn, NTI [online], July 2020. https://www.ntiindex.org/wp-content/uploads/2021/01/2020_NTI-Index_Rad-Excerpt_FINAL.pdf
40. Bieniawski et al., *Security for Radiological Sources.*
41. Tuyle et al., *Reducing RDD Concerns Related to Large Radiological Source Applications.*
42. Office for Nuclear Regulation, 'High-Activity Sealed Radioactive Sources (HASS) on Nuclear Licensed Sites', ONR [online], updated 9 August 2021. www.onr.org.uk/hass.htm; Department of Agriculture, Environment and Rural Affairs, 'Radiation – overview', DAERA [online], n.d. https://www.daera-ni.gov.uk/articles/radiation-overview
43. IAEA, IRRS Report to the United Kingdom of Great Britain and Northern Ireland, Rev. 1, IAEA Integrated Regulatory Review Service, April 2020. https://assets.publishing.service.gov.uk/government/uploads/system/uploads/attachment_data/file/899129/irrs-report-2020-to-uk.pdf
44. Author correspondence with Environment Agency, 2021.
45. Robert Downes, 'Alarming Cargo: Regulation and Control at the UK Border', *Terrorism and Political Violence*, 33/3 (2019), 549–71.
46. Author interview with former Environment Agency staff member, 2021.
47. Ibid.
48. IAEA, *IRRS Report to the UK*, 33.
49. Author interview with President of the Society for Radiological Protection, 2021.
50. UK Department for Business, Energy & Industrial Strategy, 'How We Regulate Radiological and Civil Nuclear Safety in the UK', UK Government [online], 20 April 2021. www.gov.uk/government/publications/how-we-regulate-radiological-and-civil-nuclear-safety-in-the-uk/how-we-regulate-radiological-and-civil-nuclear-safety-in-the-uk-webpage
51. Author correspondence with Environment Agency, 2021.
52. IAEA, *IRRS Report to the UK.*
53. Ibid.
54. Environment Agency, 'The Environment Agency (Environmental Permitting and Abstraction Licensing) (England) Charging Scheme 2022', UK Government [online], 2022. https://assets.publishing.service.gov.uk/government/uploads/system/uploads/attachment_data/file/1098117/Environment_Agency_EPR_and_Abstraction_Licensing_Charging_Scheme_2022.pdf

55. Author correspondence with Environment Agency, 2021.
56. IAEA, *IRRS Report to the UK*.
57. Ibid.
58. Author interview with President of the Society for Radiological Protection, 2021.
59. Author correspondence with Environment Agency, 2021.
60. Office for Nuclear Regulation, 'The Carriage of Dangerous Goods and Use of Transportable Pressure Equipment Regulations 2009—Inspection Of Transport Security Requirements', ONR [online], July 2019. https://www.onr.org.uk/operational/tech_insp_guides/ns-insp-gd-072.pdf
61. Nicola Jaynes, 'ONR's Regulation of Transport Duty Holders', presented at RWA and Radioactive Transport Update 2022, [Online], 16 February 2022, Institute of Physics and Engineering in Medicine. https://liveportal.ipemhosting.org.uk/Portals/0/Documents/Conferences/RWA%20Programme%20and%20Abstracts%202022.pdf?ver=2022-02-14-163801-430
62. Ibid.
63. UK Department for Business, Energy & Industrial Strategy, 'How we Regulate Radiological and Civil Nuclear Safety in the UK'.
64. Author interview with former Environment Agency staff member, 2021.
65. Office for Nuclear Regulation, 'Secretary of State Consent for ONR's SyAP Changes', ONR [online], 7 April 2022. https://news.onr.org.uk/2022/04/secretary-of-state-consent-for-changes-to-onrs-security-assessment-principles
66. Author correspondence with Environment Agency, 2021.
67. Author interview with former Environment Agency staff member, 2021.
68. Rozman Tahar and Noor Bakri, 'Nuclear Security Response Experience in Malaysia', n.d. https://www.amssnursecurityconference.org/sites/default/files/sessions/3_T7_Noor%20Fitriah%20Binti%20Bakri.pdf
69. Rozman Tahar and Noor Bakri, *Nuclear Security Awakening: Strengthening Response Mechanism* (Vienna: IAEA, 2018).

Suggested Reading

Acton, James, Rogers, M. Brooke, and Zimmerman, Peter, (2007), 'Beyond the Dirty Bomb: Re-thinking Radiological Terror', *Survival*, 49/3, 151–68.

Bieniawski, Andrew J., Iliopulos, Ioanna, and Nalabandian, Michelle, (2016), *Security for Radiological Sources* (Washington, DC: Nuclear Threat Initiative (NTI)).

Ferguson, Charles D. and Potter, William C., (2005), *The Four Faces of Nuclear Terrorism* (New York: CNS).

Iliopulos, Ioanna and Boyd, Christopher, (2019), *Preventing a Dirty Bomb: Case Studies and Lessons Learned* (Washington, DC: Nuclear Threat Initiative (NTI)).

Van Tuyle, Gregory J., Strub, Tiffany L., O'Brien, Harold A., Mason, Caroline F. V., and Gitomer, Steven J., (2003), *Reducing RDD Concerns Related to Large Radiological Source Applications*, LA-UR-03-6664 (Los Alamos, NM: Los Alamos National Laboratory).

Wirz, Christopher and Egger, Emmanuel, (2005), 'Use of Nuclear and Radiological Weapons by Terrorists?' *International Review of the Red Cross*, 87/859, 497–510.

CHAPTER 20

NUCLEAR SECURITY FOR NEXT-GENERATION REACTORS

SUKESH K. AGHARA AND ROSS PEEL

INTRODUCTION

THIS chapter explores the application of nuclear security to the next generation of nuclear power plants (NPPs) that are expected to begin deployment from the late 2020s. Significant new build activity over the past two decades, primarily in China, Russia, and other non-Organisation for Economic Co-operation and Development (OECD) countries, is adding new Generation III and III+ NPPs to the global fleet.[1] A new wave of NPP construction is also expected to begin during the current decade, with an anticipated diversification in technology towards a broad range of evolutionary and innovative designs (EIDs).[2] These EIDs include long-anticipated Generation IV reactor technologies, as well as more recently conceptualized small modular reactors (SMRs), among others. Such technologies present a range of novel attributes compared to established NPPs, leading to differences in the application of security.

This chapter first provides context to explain why many of the physical protection approaches developed from the late 1960s and into the 1970s are still informing security approaches today. It then gives a brief description of the various next-generation reactor technologies. The chapter presents some of the common security features across a wide range of next-generation reactors and discusses the associated security challenges, in the expectation that this will facilitate next-generation NPP developers to design and implement effective solutions. It also covers safeguards considerations where these have a security dimension. The chapter then briefly explores the legal and regulatory frameworks governing nuclear security as applied to next-generation NPPs. Finally, a brief conclusion draws together the main points of the chapter.

The Changing Approach to Nuclear Security Design

The rationale for nuclear security has shifted significantly since the early days of the nuclear era when the main concern was the prevention of potential espionage by foreign powers into nuclear weapons programmes. Over time, the scope of the security threat has grown to include sabotage, theft, and other potential threats to NPPs and fuel cycle facilities by non-state actors, with these first becoming a consideration in the late 1960s.[3] Since then, security systems have been updated as the threat has evolved. Nevertheless, the nuclear security measures taken from the late 1960s, such as fences and access control points, overseen by teams of armed or unarmed human security personnel, remain the dominant elements of physical protection. This approach is often described as 'guns, gates, and guards', and remains central to nuclear security even for today's Generation III/III+ NPPs. These NPPs are an advancement over Generation II designs, using new technologies to deliver improved safety and operational efficiencies. However, many of the underlying nuclear security approaches are akin to those employed by Generation II NPPs and represent an evolution of the past rather than being truly novel or revolutionary.

Generation II and III plants were not designed with physical protection in mind, and security was largely retrofitted onto completed facilities, rather than being designed into the NPP itself. Retrofitting of security measures onto existing plants can be a potentially costly activity,[4] creating additional non-revenue generating expenses for these NPPs.[5] In the United Kingdom (UK), nuclear security costs at Generation II and III plants are estimated to make up approximately 10 per cent of total operating budgets,[6] and have been suggested to be as high as 25 per cent in the United States (US),[7] with most costs being personnel related. For large NPPs, these costs are bearable, if undesirable, but for how much longer this will remain the case is uncertain; numerous plants in the US have been permanently shut down in recent years due to economic challenges.[8] NPPs dedicate approximately 66 per cent of their budgets to operation and maintenance (O&M) costs, which includes security. In comparison, natural gas combined cycle (NGCC) plants, with significantly smaller geographical footprints and fewer on-site personnel (80 compared to 300), only spend approximately 13 per cent on O&M costs. With relatively lower fuel costs (87 per cent of the expense for NGCC compared to 34 per cent for NPP), the economic pressures on NPPs have been a significant contributor to planned plant closures in the US.[9]

The construction, maintenance, and operational experience of the past is now informing the development of the next generation of NPPs, with many designs looking to integrate safety-, security-, and safeguards-by-design at the conceptual phase of the development cycle. Developers already follow safety-by-design approaches, where the NPP is designed with, for example, passive safety features that minimize the release of radioactive material without the requirement for operator intervention.[10]

This principle is also being extended to security, with the aim of integrating security as an intrinsic element of the NPP design, incorporating technological solutions, and reducing the need for costly retrofitted security that is reliant on large numbers of guards and/or response forces.[11]

Next-generation NPP designers are thus recommended to integrate security-by-design (SBD) approaches into their design activities from an early stage. SBD promotes the building of design features into the facility itself to mitigate or neutralize the security threat profile of the NPP, with the aim of reducing the number of essential personnel on site (including security personnel), and thereby reducing O&M costs. Some Generation III/III+ NPP designs have explored possible routes to apply SBD with some limited success, such as by designing facilities to minimize or eliminate the possibility of deliberate malicious operations by saboteurs, designing facilities to use less-attractive nuclear materials, and laying out buildings to optimize surveillance.[12]

The next generation of NPPs are currently undergoing conceptual design, and so the timing is ideal for developers to realize the benefits of SBD. Going one step further, the interactions between security, safety, safeguards, and operations are of particular interest for EIDs early in the design phase, in order to leverage synergies and mitigate conflicts between these four elements.[13] The demand for next-generation NPP designs to provide operational flexibility—such as non-electrical power generation applications and load following capability—while remaining economically competitive creates a range of new safety, security, and safeguards challenges. While safety is already central to NPP design, security and safeguards both need more attention as these designs are developed further. EIDs have a unique opportunity to shrink their safety and security envelope, leading to reduced capital and operational costs.[14]

Evolutionary and Innovative NPP Designs

Since the start of the nuclear power era, NPPs have steadily grown, with the largest NPP power ratings now exceeding 1,700 megawatts electric (MWe),[15] while also incorporating new safety features.[16] It is also increasingly common to collocate many reactors on one site, a trend informed by the economic challenges experienced from single-unit NPP sites in the US.[17] However, although benefitting from cost efficiencies per megawatt of installed capacity, larger NPPs require a more reliable energy grid, significant baseload demand, and a greater degree of capital expenditure. The appeal of large NPPs in future energy markets, particularly in developing countries, is thus limited.

Over the past decade or so, there has been increasing interest globally in moving towards smaller and/or more advanced nuclear power technologies. Growth in intermittent renewable generation capacity, distributed grids, and interests in decarbonization are driving new thinking in energy strategies.[18] To meet expected demand for energy programmes, many nuclear technology developers are preparing and marketing so-called evolutionary and innovative reactor designs for countries both with and without existing nuclear energy infrastructure or experience. EID developers are claiming a range of advantages for their technologies compared to large conventional NPPs. Many claim that they will be less expensive to build and operate, have a wider range of potential siting options and applications, be less prone to delays and cost overruns in construction, provide more flexible power generation while decarbonizing energy generation, and other benefits.[19] Overall, the authors estimate that approximately 100 EIDs are currently in development. Information on many of these can be found in the Advanced Reactor Information System database of the International Atomic Energy Agency (IAEA).[20]

EIDs can be divided into four broad groups, with some of these overlapping.

1. SMRs are generally evolutionary designs based on the miniaturization of large conventional Generation III+ NPP technologies, and as such are commonly light water reactors (LWRs).[21] They are usually defined as having an electrical power output of less than 300 MWe.
2. Generation IV reactors, also known within the UK primarily as advanced modular reactors (AMRs), are revolutionary designs based on nuclear technologies that have yet to be successfully commercialized. They are generally of one of six types as defined by the Generation IV International Forum (GIF).[22]
3. Micro reactors are a subset of SMRs and Generation IV reactors, defined by their particularly small size and are generally accepted to be reactors with a maximum power of 10 MWe.[23]
4. Transportable NPPs are designed to generate power from a mobile platform such as a barge or railcar.[24]

The timeline for deployment of EID technologies is variable. At the time of writing, prototype or first-of-a-kind units for four designs have recently entered operation or are under construction: the CAREM-25 SMR under construction in Argentina; the two KLT-40s units aboard the floating NPP in commercial operation in Russia; and the operating HTR-PM and under construction ACP100 in China.[25] These units are at least partially operated as technology testing and demonstration units, with the intention of building additional units later. Large-scale commercial deployment of SMRs is anticipated to begin around 2030, with Generation IV reactors to follow during the 2030–2040 period or later depending on the design. Some states, such as China and Russia, are seeking to commercialize EIDs more rapidly.[26]

Generation IV Reactor Types

The Generation IV International Forum has done significant work to assess the nuclear security of six major types of Generation IV reactors: gas-, lead- and sodium-cooled fast reactors; molten salt reactors; supercritical water reactors; and very high temperature reactors. These are summarized in Table 20.1.[27]

Table 20.1. The six Generation IV International Forum-selected advanced reactor technologies.[a] Note that not all AMR concepts fall neatly within these categories, and some may use novel elements or take elements from several technology types.[b]

Reactor Type	Neutron spectrum	Coolant	Moderator	Core outlet temperature (indicative) /°C
Gas-cooled fast reactor (GFR)	Fast	Carbon dioxide or helium	N/A	400–850
Sodium-cooled fast reactor (SFR)	Fast	Liquid sodium	N/A	475–550
Lead-cooled fast reactor (LFR)	Fast	Lead or lead-bismuth eutectic	N/A	390–650
Molten salt-cooled reactor (MSR)	Thermal or fast	Molten salt (e.g. lithium fluoride)	Graphite	650–750
Supercritical water-cooled reactor (SCWR)	Thermal or fast	Light water	Light water	500–560
Very high-temperature gas-cooled reactor (VHTR)	Thermal	Helium	Graphite	750–850

a Table courtesy of the authors.

b IAEA, 'Advanced Reactors Information System (ARIS)'; Edward Blandford, et al., 'Kairos Power Thermal Hydraulics Research and Development', *Nuclear Engineering and Design*, 364 (2020), 110636. doi: https://doi.org/10.1016/j.nucengdes.2020.110636.

From a nuclear security perspective, a notable difference of AMRs from other plant types is operation of these reactors in the fast neutron spectrum range. Reactors using fast neutrons can use fissile or fertile isotopes as fuel, extending the range of nuclear materials and fuel cycle options compared to those of thermal neutron spectrum reactors. It is important to note that, although the range of materials would be wider at these reactor sites, the expectation is that the design and form of materials will essentially reduce the security risks compared to the current fleet of reactors.[28]

Most current and planned reactors use solid fuel that is fixed and inaccessible during operation, with access only possible during closely monitored refuelling and maintenance outages. Fuel inventory monitoring also only takes place during these outages. Real-time inventory monitoring of fuel within the reactor can significantly reduce risks of diversion and protracted theft, enhancing security and nuclear safeguards of the reactors. However, these designs can also present challenges, such as those related to online fuel reprocessing, which would generate a wide range of radiological materials in various separated states at an operating site. Currently, the only site from which relevant industrial experience can be drawn for this type of operation is the large-scale reprocessing plant at Rokkasho, Japan.

Two types of Generation IV reactors that do not conform to the common position of fixed, inaccessible fuel are most molten salt-cooled reactors (MSRs) and some high-temperature gas-cooled reactors (HTGRs). MSRs can be operated in the fast or thermal neutron spectrum range. The fuel is dissolved within the coolant salt and flows throughout the whole primary coolant loop. Many designs opt to continuously process the flowing salt to remove fission products and other by-products while introducing fresh fuel in the reactor loop. If designed and operated correctly, MSRs could provide a range of operational and economic benefits over conventional batch refuelling.

HTGRs operate at much higher temperatures than other reactor types and some are fuelled with so-called TRISO pebble fuels, consisting of sub-millimetre fissile material particles dispersed within 5–10 cm graphite spheres, which are mobile within the core.[29] The reactor core is formed of potentially hundreds of thousands of these pebbles.[30] Some of the security benefits and challenges of MSR are also applicable to pebble-bed HTGRs, as these designs generally plan for the fuel pebbles to be continuously removed at the base of the reactor, automatically assessed, and then either reintroduced at the top of the reactor or designated as spent and removed. This flow of pebbles throughout the core will necessitate similar consideration of real-time inventory monitoring. The form of the fuel—small solid spheres—also makes it more readily amenable to diversion or theft than either bulk molten salt or very large fuel assemblies, although with only a few grams of fissile material per sphere, a very large number of spheres would need to be diverted to accumulate a significant quantity of nuclear material.[31]

Both MSRs and TRISO-fuelled reactors will require enhanced nuclear material accountancy and control (NMAC) measures, such as near-real-time accountancy (NRTA) methods to maintain accurate and credible inventory records of nuclear materials. NMAC approaches currently employed at NPPs will not be applicable to non-fixed fuels. New approaches will be required to ensure these materials can be adequately tracked and monitored throughout the nuclear fuel cycle and within nuclear facilities. This will impact nuclear security, safeguards, and inventory verification.

Challenges and Opportunities in EID Nuclear Security

This section discusses the main areas of difference between EIDs and large conventional NPPs and their security implications.

Reduced Power, Modular Manufacturing, and Construction

Being smaller in size than large conventional NPPs allows for the mass manufacture of many units of an EID on a production line, representing a departure from the previous economy of scale model towards an economy of volume through the production of many identical units. This means that units of a given EID design will be highly standardized, unlike large conventional nuclear power plants, where the same design often varies significantly between sites. The EID units will be loaded onto vehicles as a small number of finished modules, complete reactors, or even complete NPPs, and then transported to their final site for installation. Some designs will be transported in a fuelled state or use factory sealed cores of nuclear material for refuelling.

Factory sealed cores reduce many of the proliferation and security concerns associated with current fleets.[32] However, the designs will have to address any new threat vectors to the material in transport. Criticality accidents resulting from attack or sabotage during transport are of particular note, as the fuel will be moving in a configuration designed for criticality, rather than a sub-critical cargo configuration. Movement of nuclear material by road, rail, river, and sea is already a very common activity globally, with a huge amount of existing industry experience, and many of the practices currently employed will be equally applicable. However, cores and/or reactor vessels themselves must now include suitable protective measures in their designs that prevent access to material or sabotage during transport.

Turning to manufacturing facilities, any factory producing complete EIDs will present a greater appeal to adversaries compared to geographically distributed facilities producing individual components. These integrated manufacturing facilities will need a very different security profile, more akin to NPP sites themselves. Nuclear security risks for the current reactor fleet are limited to the nuclear facilities; however, with EIDs the threat is spread throughout the supply chain. Espionage to exploit common design weaknesses, hardware tampering, and other vulnerabilities will need to be addressed, which is not common practice within the nuclear industry today. Here, experience from other industries, such as aerospace and information technology, could be used to inform practices for the nuclear industry.

Reduced Capital Costs and Operating Budgets

The reduced size of EIDs promises a scaled reduction in total capital cost of construction and operation. However, smaller generation capacity translates to a higher effective cost of electricity per unit. The developers argue that increased operational efficiencies, factory manufacturing, and short time to complete modular construction would more than offset the increased cost per unit of electricity generated relative to large conventional NPPs.[33] This is expected to make EIDs more financially accessible compared to conventional NPPs, while also reducing the dependence of states on complex and/or risky foreign financial assistance arrangements, again making nuclear energy more widely accessible.

Successfully incorporating nuclear security within NPP designs through SBD will lead to significant cost reductions in O&M, including for the secure operation of EID units. Nuclear security is delivered today using measures such as numerous armed and unarmed guards, tall fences surrounding large sites, staffed access control points, and human-monitored detection systems such as cameras. These approaches represent a major cost for nuclear installations. Incorporating real-time process monitoring systems, digital process controls, and other innovative approaches could potentially replace some of these more traditional approaches to nuclear security, helping ensure safe, economically operable, and secure plants with reduced proliferation risks.[34] Plant designers are encouraged to think about security as an engineering challenge, and to address the challenge using engineering design best practices to address the specific needs of their technology.

Most security costs are associated with personnel, be that as salaries, costs for training and certification, transport to and from isolated plants, and so on.[35] As such, novel approaches in nuclear security, such as the use of automated monitoring systems, intelligent decision-making algorithms, plant design that eliminates or reduces the need for personnel access to vital areas, and others, are expected to allow for a reduction in the number of security personnel required to protect facilities.[36] This represents a shift in focus from deterrence through the show of force and rapid response to threats, towards the earlier detection and imposition of greater time delay by design. This could be achieved through, for instance, designing safety systems in such a way that sabotage will be clearly and immediately apparent, or designing storage buildings such that material quantities in a given location are low enough that a breach of one storage building will not result in unacceptable radiological consequences. Employing such an approach would allow interdiction to be carried out by either a centralized response force stationed off site and potentially covering multiple facilities, or by local law enforcement or military personnel not specifically linked to a nuclear facility.

Incorporating security into the fabric of an NPP, rather than relying on operators to deliver security through retrofitted systems, will provide both relative uniformity to the approach across all units of a given design and greater opportunities for making nuclear technology accessible to a broader global community, as it will not be necessary for inexperienced nuclear operators to develop complete security plans from scratch. Technology vendors should continue early engagement with states to adopt international best practice in safety and security. Technology vendors might engage potential target markets early in their design cycle to understand their unique security situation and design their technology with features to meet the threat. Developers might offer tailored security feature selections into the reactor system to be deployed based on the needs of the specific target market.

Increasing Automation and Remote Operation

Digital transformation is impacting all sectors of technology development. NPP designs, particularly EIDs, are well placed to take advantage of technology solutions to reduce total personnel on site.[37] With the adoption of automation and computer-assisted operation, the role of the operator is expected to evolve from operational delivery of all functions to a supervisory role monitoring the proper function of several automated operations built on logic controls.[38] A shift to operation hubs that monitor several geographically dispersed NPPs with agile operational and security personnel groups available for deployment from a central location is expected.[39] Besides delivering the physical protection function, on-site security staff would be responsible for access control for on-site O&M staff. Smaller NPP physical site footprints and lower numbers of O&M staff would reduce the security staffing necessary on site.

Reduced on-site staffing further translates to a lower number of potential insiders. This has an additional benefit of reducing costs associated with human reliability programmes. The automation of plants further supports this by reducing the ability of insiders to act, while increasing the chance that insider sabotage will be detected and the perpetrator caught.[40] With fewer areas requiring human access, plant designers can harden physical structures and reduce their size. For example, thicker walls without doors or windows can be used for structures that do not need human presence. Similarly, a smaller operations staff means the control rooms can be smaller in size with safety critical systems placed more to the interior of the plant, further reducing opportunities for external force-on-force attack scenarios, as well as vulnerability to stand-off projectiles.

NPP security is increasingly making use of semi-or fully automated nuclear security systems, such as uncrewed aerial vehicles (UAVs, or drones) and multi-sensor surveillance masts, managed by a small number of human operators. However, these systems are not a panacea, and their potential drawbacks and limitations must be given due consideration. Any proposed element of a security system must be well proven and sufficiently trusted, and the complete system must still be able to prove its performance as sufficiently effective. Technological solutions need not solely be used for threat detection and monitoring but may also take the form of active delay features, such as strobe lighting, white noise, slippery/adhesive agent sprays, and other systems to impede the effectiveness of threat actors. This might also include the use of remotely operated weapons systems, although these are more controversial and further from deployment. Technology may also form part of the decision-making process as to whether to call for human, or potentially in future robotic, response forces. In addition, systems thinking will be critical for developers to ensure the optimal use of both human and digital elements of a security system in an integrated manner.

In all cases, security features must be carefully selected and deployed based on their compatibility with nuclear safety, non-proliferation, and operational considerations. The civil nuclear sector has historically been relatively slow to adopt novel technologies until well proven. However, EID developers have consistently demonstrated a willingness to consider and implement novel technologies within their designs and, furthermore, may be more open to holistic design approaches considering security synergies with other functions. Overall, an appropriate balance will need to be found between caution with regards to unproven technology and the need for cost-effective security. An SBD approach will allow for this balance to be defined and maintained throughout the EID lifecycle.

The increasing use of digital systems will create a greater surface for cyber-attacks and increase the potential impact of such attacks. Digital systems are likely to be used not only in security, but throughout EIDs for safety, operational, and other purposes. Such systems are only just starting to be deployed in the civil nuclear sector, and so it is critical that cyber security is given appropriate priority during design activities, both for current and prospective future threats. Security systems for EIDs will have to develop cyber-physical testing environments and digital twins to enable increased digitization of plant systems for autonomous and remote operations. Deployment of such highly autonomous nuclear plant systems is at least several decades away.

Advanced Technologies, Fuel Materials, and Fuel Cycles

Despite their many differences, a common feature of most EIDs is the incorporation of a range of advanced technologies, such as passive safety systems, as well as novel nuclear fuels and fuel cycles. Many of these have previously been tested or at least theorized—often during the early stages of nuclear technology development in the 1950s and 1960s—but for various reasons these designs were not commercialized. EIDs generally use lower quantities of nuclear material than large NPPs, in accident-tolerant fuel forms, with potentially greatly extended in-core lifetimes.[41]

The large variety of novel fuels proposed for use in EIDs has the potential to create novel security challenges, and individual assessments will be required to address these. Of note from a security perspective are VHTR and MSR reactors, which will have fuel in 'flow' during operation. Several designs are also proposed that would use higher enrichments of uranium than today's commercial reactors, as well as fuels containing thorium or plutonium, significantly changing the threat profile of the fuel and the fuel cycle facilities. In particular, a number of designs propose the use of high-assay low-enriched uranium (HALEU) fuel of up to 20 per cent enrichment in uranium-235, meaning that under the current IAEA security guidance, such fuels are of a higher category of nuclear material, warranting significantly greater protection. Such material would require only a fraction of the feed and enrichment work to convert to a weapons-grade product compared to standard low enriched uranium (LEU) fuels of <5 per cent uranium-235. This is also the case for fuels including plutonium or uranium-233, derived from the irradiation of thorium.[42] In addition, there is less operating experience in working with these novel materials, meaning that current security systems and processes for uranium fuels may not be directly applicable.[43]

For the majority of EIDs the reduced unit capacity will potentially lead to reduced inventories of nuclear material at plant sites and thus reduced potential consequences of security incidents. However, this advantage will be negated if many units are collocated, particularly if their material is aggregated and stored in common facilities. A potential exception here is the several EIDs that are proposed to operate with greatly extended fuel cycle durations, with some planned to operate for a decade or more without refuelling. This will lead to relatively enlarged core sizes due to the need to have enough fuel for long-term operation, increasing the potential consequences of security incidents.

However, this will reduce the number of nuclear transport and refuelling operations, which should provide a net security benefit, assuming there is adequate protection of the site during operation.

Greater Flexibility in Siting and Deployment Options

Whereas large conventional NPPs are reliant on off-site power and active cooling, EIDs are generally designed to be self-reliant, using passive systems to ensure safe plant operation without access to off-site resources, power, or even operator action. The passive safety systems that many EID concepts feature will also deliver benefits in terms of nuclear security by eliminating several vulnerabilities present in the current fleet, for example, dependence on off-site power.[44] Many vendors are also looking to reduce plant emergency planning zones to the site boundary, allowing for much more compact sites.[45] Together with the reduced reliance on outside infrastructure and support, this allows for a much-enlarged range of plant siting scenarios while shrinking the safety and security profile of EIDs. It also opens deployment options to, for example, isolated research stations and communities in polar regions. However, all other things being equal, smaller site profiles will lead to more rapid adversary access to vital areas, effectively driving a need for more rapid response.

As outlined previously, well-applied SBD approaches will likely be required to instead detect and delay adversaries early, achieving security objectives while minimizing or eliminating the need for on-site security personnel. This might be achieved by, for instance, positioning sensors beyond the site boundary to detect unusual activity and potential threats early, with automated systems to identify potential threats in the site's vicinity. New detector architecture will have to be developed with the ability to discern threats against a background of normal activity beyond the site boundary. Overall, multiple independent, redundant detection and delay capabilities should be layered, following the defence-in-depth principle.[46]

One application of EIDs involves the installation of multiple units at a single site to provide large-scale energy demand. Such a scenario could aid security compared to large conventional NPPs, as adversaries would need to breach multiple smaller EID units to access the same quantity of radioactive materials or to cause a large-scale power outage. However, as outlined earlier, aggregation of material in common storage can negate these advantages, as can the sharing of security features, where a breach of one unit at a site is effectively a breach of all.

Some EID concepts are planned to be mounted onto and operated from mobile platforms, such as road/rail vehicles or barges. A barge-mounted plant is already in operation in Russia, and several concepts are in development.[47] These transportable NPPs offer a range of potential benefits over fixed NPPs, but also present a range of novel safety, security, and safeguards challenges that must be overcome before they can be acceptable sources of nuclear power, such as the potential for hijacking or sinking of completed NPPs. The IAEA and others are engaged in processes to examine and address these challenges, and developers are encouraged not to rush EID concepts to market before the necessary independent assessments can be completed by the international community of experts.[48] An update to a 2013 study by the IAEA was in progress at the time of writing.[49]

Floating NPPs, meanwhile, face novel threats from divers, torpedoes, and other water-based attack vectors. There is currently limited applicable international law, with significant gaps between maritime law and nuclear law and their applications across sovereign borders.[50] Removing mobility from plants during routine operations could allow the use of security features from traditional plants. The ability to relocate transportable NPPs in the face of threats can be a potential security and safety advantage, though it also raises questions as to the possibility of hijack or capture. Finally, transportable NPPs may lack a site boundary as the term is currently understood, and in some cases the boundary may become the outer edge of the vehicle or barge, reducing the potential for early detection or certain delay features.

Wider Range of Developers and Business Models

While some EIDs are being developed by established nuclear technology developers, many others are being designed by start-up companies or teams within academic institutions, which may have relatively little experience of delivering information security, managing export control considerations, and so on. Even groups performing research on specific sub-components need to be aware of the risks of technology integration by groups that might take information on several such components to produce a more significant item of nuclear technology or gain access to controlled knowledge. There are several business models that include developers only interested in the research and development of certain technologies to be sold or licensed to other EID stakeholders. There are other business models that would pursue designing complete plants for construction and operation by third parties. A large supply chain approach to EID designs could leave security as a matter for operators to address, leading to high costs or inconsistent security envelopes for the same design under different operators. Developing methods to quantify lifetime operating costs that include security costs could level the playing field for all developers and benefit the SBD approach.

Nuclear Security Frameworks

The international legal instruments on nuclear security, the Convention on the Physical Protection of Nuclear Material (CPPNM; in force 1987), its Amendment (A/CPPNM; in force 2016), and the International Convention for the Suppression of Acts of Nuclear

Terrorism (ICSANT; in force 2007) have been accepted by most IAEA Member States.[51] These define the general guiding principles and approaches for effective nuclear security, and these will likely be unchanged in EIDs. Existing approaches to nuclear security planning and delivery will therefore remain broadly applicable.

The international instruments and guidance related to nuclear security are broadly applicable and are not specific to individual reactor technologies. However, the derived legislation of many states was developed at a time when large conventional NPPs, particularly LWRs, were the dominant technology types globally. As such, the national laws and regulatory frameworks of individual states may have been unintentionally tailored towards specific reactor technologies, making them inflexible and thus presenting challenges in acceptability towards novel EIDs, which generate different security considerations and approaches. Legal and regulatory frameworks developed around specific reactor types are at risk of becoming overly prescriptive in terms of setting security requirements, resulting in the deployment of security solutions which are neither cost-effective nor well adapted to individual nuclear facilities. The development of EIDs also presents an opportunity to review and revise national legislation and regulatory requirements, and to develop new international guidance.

Conclusion

Overall, EIDs represent a significant step away from the large conventional NPPs in operation today. The variety of EIDs is vast, and this chapter has discussed security features that could be common to broad Generation IV categories, instead of specific reactor technologies. The chapter has highlighted opportunities for security to be integrated with safeguards and safety by considering the broad reactor features in terms of neutron spectrum, size, siting, digitalization of monitoring and control, economic models of EID development, and elements of nuclear fuel cycles.

The reduced size of many EIDs and their method of manufacture offer a range of potential advantages, including more flexible deployment, reduced capital and operating costs, improved nuclear safety, and more. However, nuclear security remains critical, and developers are encouraged to adopt the principle of SBD to build security into the fabric of facilities themselves. This must include consideration of all security dimensions; this chapter has primarily considered physical protection, but cyber security and insider threats remain. There is a particular need for enhanced cyber security considerations as EIDs will utilize a much greater degree of digital instrumentation and control systems, automation, and other cyber-vulnerable systems. Developers, operators, and buyers must be alert to security risks throughout a plant's design life and work together to find security solutions to alleviate operating costs, which will minimize the economic burden of security faced by the current reactor fleet.

Notes

1. IAEA, *Nuclear Power Reactors in the World*, 2021 Edition, IAEA Reference Data Series, No. 2 (Vienna: IAEA, 2021).
2. The term EID was adopted by the IAEA in 2021 to replace the previously used term 'novel advanced reactors'. Both terms aim to encompass the full set of novel, post-Generation III+ reactor technologies.
3. Amandeep S. Gill, 'The Priority of Control: 1950s–1960s', in *Nuclear Security Summits: A History* (Cham: Palgrave MacMillan, 2020), 20–2.
4. Lap-Yan Cheng and Robert A. Bari, 'Proliferation Resistance and Physical Protection (PR&PP) in Small Modular Reactors (SMRs)', in Daniel T. Ingersoll and Mario D. Carelli, eds., *Handbook of Small Modular Nuclear Reactors* (Cambridge: Woodhead Publishing, 2021), 221.
5. Vaibhav Yadav, et al., *Current Challenges, Constraints and Recommendations for Reducing Cost of Physical Security at US Commercial Nuclear Power Plants* (Washington, DC: US Department of Energy, 2019).
6. Ross Peel, George Foster, and Sukesh Aghara, *Nuclear Security and Safeguards Considerations for Novel Advanced Reactors* (London: King's College London, 2022).
7. Edwin Lyman, *Small Isn't Always Beautiful: Safety, Security, and Cost Concerns about Small Modular Reactors* (Cambridge, MA: Union of Concerned Scientists, 2013).
8. Yadav et al., *Current Challenges, Constraints and Recommendations.*
9. World Nuclear Association (WNA), *Nuclear Power Economics and Project Structuring*, 2017 edn (London: WNA, 2017), 4.
10. Joanne Liou, 'Safety by Design: How the New Generation of Nuclear Reactors Addresses Safety', *IAEA Bulletin*, 62/1 (2021), 18–19. http://inis.iaea.org/search/search.aspx?orig_q=RN:52035686
11. World Institute for Nuclear Security (WINS), *4.1: Implementing Security by Design at Nuclear Facilities*, 2.1 edn (Vienna: WINS, 2019).
12. Ibid.
13. Shahen Poghosyan, et al., 'Safety, Security and Safeguards (3S) Interfaces and Challenges for Innovative Reactors', in IAEA Symposium on International Safeguards [online], 3 February 2022. https://iaea.mediasite.com/Mediasite/Play/d196f38851a842c083a0b8d5f59f75341d
14. WINS, *4.1 Implementing Security by Design at Nuclear Facilities.*
15. S. Bilbao y León, J. H. Choi, J. Cleveland, and Intisar Khamis, 'Available and Advanced Nuclear Technologies for Nuclear Power Programs', in Agustín Alonso, ed., *Infrastructure and Methodologies for the Justification of Nuclear Power Programmes* (Cambridge: Woodhead Publishing, 2012), 261–93.
16. IAEA, *Technology Roadmap for Small Modular Reactor Development*, IAEA Nuclear Energy Series, NR-T-1.18 (Vienna: IAEA, 2021).
17. IAEA, *Nuclear Power Reactors in the World*; Yadav et al., *Current Challenges, Constraints and Recommendations.*
18. IAEA, *Technology Roadmap.*
19. Ibid.
20. IAEA, 'Advanced Reactors Information System (ARIS)', IAEA [online], 26 May 2021. https://aris.iaea.org/sites/SMR.html
21. Ibid.

22. Harold McFarlane, 'Special Section on Status of Generation IV Reactor Development', Special issue of *Progress in Nuclear Energy*, 77 (2014), 1–420.
23. IAEA, *Technology Roadmap*.
24. IAEA, *Legal and Institutional Issues of Transportable Nuclear Power Plants: A Preliminary Study*, IAEA Nuclear Energy Series, No. NG-T-3.5 (Vienna: IAEA, 2013).
25. Andreas Schaffrath, Andreas Wielenberg, Robert Kilger, and Armin Seubert, 'SMRs–Overview, International Developments, Safety Features and the GRS Simulation Chain', *Frontiers in Energy*, 15/4 (2021), 793–809. doi: 10.1007/s11708-021-0751-2; World Nuclear News (WNN), 'Foundation Completed for Chinese SMR Turbine Building', WNN [online], 6 May 2022. https://www.world-nuclear-news.org/Articles/Foundation-completed-for-Chinese-SMR-turbine-build
26. IAEA, *Technology Roadmap*.
27. Generation IV International Forum, 'Evaluation Methodology for Proliferation Resistance and Physical Protection of Generation IV Nuclear Energy Systems', Revision 6, The Proliferation Resistance and Physical Protection Evaluation Methodology Working Group and System Steering Committees of the Generation IV International Forum, 15 September 2011; Giacomo G. M. Cojazzi and Keiichiro Hori, 'GIF Proliferation Resistance and Physical Protection (PR&PP) Working Group Activities', Generation IV International Forum, 13th GIF–IAEA Interface Meeting, 18–19 March 2019, Vienna, Austria.
28. L.-Y. Cheng, et al., 'White Papers on Proliferation Resistance and Physical Protection Characteristics of the Six GEN IV Nuclear Energy Systems', in *Proceedings of the INMM & ESARDA Joint Virtual Annual Meeting*, [online] August 23–26 & August 30–September 1, 2021.
29. Charles Folsom, Changhu Xing, Colby Jensen, Heng Ban, and Douglas W. Marshall, 'Experimental Measurement and Numerical Modeling of the Effective Thermal Conductivity of TRISO Fuel Compacts', *Journal of Nuclear Materials*, 458 (2015), 198–205. doi: 10.1016/j.jnucmat.2014.12.042
30. Rachel L. Seibert, Brian C. Jolly, Mehdi Balooch, Daniel P. Schappel, and Kurt A. Terrani, 'Production and Characterization of TRISO Fuel Particles with Multilayered SiC', *Journal of Nuclear Materials*, 515 (2018), 215–26. doi: 10.1016/j.jnucmat.2018.12.024
31. Ibid.
32. Ross Peel and George Burnett, 'Transportation Security Considerations of Small Modular and Mobile Reactors', Canadian Nuclear Society (CNS) Fourth International Conference on Generation IV and Small Reactors (G4SR-4), 3–6 October 2002, Toronto, Ontario.
33. B. Mignacca and G. Locatelli, 'Economics and Finance of Small Modular Reactors: A Systematic Review and Research Agenda', *Renewable and Sustainable Energy Reviews* [online], (2020) 118. doi: 10.1016/j.rser.2019.109519; Benito Mignacca, Giorgio Locatelli, and Tristano Sainati, 'Deeds Not Words: Barriers and Remedies for Small Modular Nuclear Reactors', *Energy* [online], 206 (2020). doi: 10.1016/j.energy.2020.118137
34. Peel et al., *Nuclear Security and Safeguards Considerations*; Lyman, *Small Isn't Always Beautiful*.
35. Pralhad H. Burli and Vaibhav Yadav, *Economic Analysis of Physical Security at Nuclear Power Plants, Light Water Reactor Sustainability Program* (Idaho Falls, ID: Idaho National Laboratory, 2020).
36. Ibid.
37. Paul Miller, 'Small Modular Reactors Provide Opportunity to Rethink Automation for Nuclear Generation', *POWER* [online], 2 May 2022. https://www.powermag.com/

small-modular-reactors-provide-opportunity-to-rethink-automation-for-nuclear-gen eration

38. Ahmad Al Rashdan and Shawn St. Germain, *Automating Surveillance Activities in a Nuclear Power Plant*, Light Water Reactor Sustainability Program, INL/EXT-19-55620 (Idaho Falls, ID: Idaho National Laboratory, 2019).
39. World Institute for Nuclear Security (WINS), *Security of Small Modular Reactors (SMRs) Workshop Report* (Vienna: WINS, 2019). https://www.wins.org/wp-content/uploads/2018/11/2019-20-3-REPORT-WS-on-Security-of-SMRs-1-1.pdf
40. IAEA, *Preventive and Protective Measures against Insider Threats*, Implementing Guide. IAEA Nuclear Security Series, No. 8-G (Vienna: IAEA, 2020), 24.
41. IAEA, 'Advanced Reactors Information System (ARIS)'.
42. IAEA, *Physical Protection of Nuclear Material and Nuclear Facilities (Implementation of INFCIRC/225/Revision 5)*, Implementing Guide, IAEA Nuclear Security Series, No. 27-G (Vienna: IAEA, 2018).
43. Lyman, *Small Isn't Always Beautiful*; Generation IV International Forum, 'Evaluation Methodology'; Edwin Lyman, *'Advanced' Isn't Always Better: Assessing the Safety, Security, and Environmental Impacts of Non-Light-Water Nuclear Reactors* (Cambridge, MA: Union of Concerned Scientists; Global Nexus Initiative, 2021); 'Advancing Nuclear Innovation: Responding to Climate Change and Strengthening Global Security', Global Nexus Initiative, June 2019; World Institute for Nuclear Security (WINS), *Security of Advanced Reactors*, WINS Special Report Series (Vienna: WINS, 2020).
44. Francesco Di Maio, et al., 'Reliability Assessment of Passive Safety Systems for Nuclear Energy Applications: State-of-the-Art and Open Issues', *Energies* [online], 14/15 (2021). doi: 10.3390/en14154688.
45. Juan Carlos de la Rosa Blul, 'Determination of Emergency Planning Zones distances and scaling-based comparison criteria for downsized Nuclear Power Plants', *Nuclear Engineering and Design*, 382 (2021), 111367. doi: 10.1016/j.nucengdes.2021.111367.
46. Andrew J. Clark and Michael T. Rowland, *Safety and Security Defense-in-Depth for Nuclear Power Plants* (Albuquerque, NM: Sandia National Laboratories, 2021), doi:10.2172/1832309.
47. M. Hadid Subki, 'Floating Nuclear Power Plants With Small Modular Reactors', in Ehud Greenspan, ed., *Encyclopedia of Nuclear Energy* (Oxford: Elsevier, 2021), 711–20.
48. IAEA, *Legal and Institutional Issues of Transportable Nuclear Power Plants.*
49. Ibid.
50. Marc Richard Fialkoff, 'The Floating Chameleon: Floating Nuclear Power Plants and the Nexus of Maritime and Nuclear Security Law', *Journal of Maritime Law and Commerce*, 51/2 (2020), 103–47. https://www.osti.gov/pages/biblio/1649247
51. IAEA, 'Convention on the Physical Protection of Nuclear Material (CPPNM) and its Amendment', IAEA [online], n.d. https://www.iaea.org/publications/documents/conventions/convention-physical-protection-nuclear-material-and-its-amendment; United Nations Treaty Collection, Status of Treaties, International Convention for the Suppression of Acts of Nuclear Terrorism (adopted on 13 April 2005; entered into force on 7 July 2007), 2445 UNTS 089. https://treaties.un.org/pages/ViewDetailsIII.aspx?src=TREATY&mtdsg_no=XVIII-15&chapter=18&Temp=mtdsg3&clang=_en

Suggested Reading

Kuznetsov, Vladimir, (2008), 'Options for Small and Medium Sized Reactors (SMRs) to Overcome Loss of Economies of Scale and Incorporate Increased Proliferation Resistance and Energy Security', *Progress in Nuclear Energy*, 50/2, 242–50. doi: 10.1016/j.pnucene.2007.11.006

Lee, Manseok and Woo, Seung Min, (2018), 'A New Security Strategy for Small Medium Sized Reactor (SMR) Plants', *Progress in Nuclear Energy*, 108, 66–70. doi: 10.1016/j.pnucene.2018.05.012

Peel, Ross, Foster, George, and Aghara, Sukesh, (2022), *Nuclear Security and Safeguards Considerations for Novel Advanced Reactors* (London: King's College London).

Snell, Mark K. and Jaeger, Calvin Dell, (2014) 'Incorporating Security-by-Design in Both Planned and Operational Nuclear Facilities', in *Proceedings at the Institute of Nuclear Materials Management 55th Annual Meeting*, 20–24 July 2014, Atlanta, GA (Deerfield, IL: INMM), 552. https://www.osti.gov/biblio/1315235

World Institute for Nuclear Security (WINS), (2019), *4.1: Implementing Security by Design at Nuclear Facilities*, 2.1 edn (Vienna: IAEA).

CHAPTER 21

ADDRESSING THE CONFIDENCE GAP IN THE SECURITY OF MILITARY NUCLEAR MATERIALS

SCOTT ROECKER AND NICKOLAS ROTH

Introduction

All weapons-useable nuclear materials and their associated facilities, as well as all major nuclear facilities, should be protected to prevent unauthorized access and use of that material. To reduce the risk of theft or sabotage, countries have worked together for decades to strengthen nuclear security around the world. This work has been carried out through bilateral channels, multilateral groups, and international organizations. Over time, many countries have made significant progress on reducing the likelihood that nuclear materials would be stolen or that nuclear facilities would be sabotaged. Despite significant accomplishments, these efforts have primarily focused on nuclear materials used for civilian purposes, which represent a small subsection of global stocks of nuclear materials. International cooperation has proven less successful at securing any aspect of military nuclear materials—within nuclear weapons or programmes that develop, build, and dispose of them.

In contrast to security for civilian nuclear facilities, there is no internationally agreed upon guidance for the security of military materials.[1] Furthermore, little is known about military nuclear security in countries that possess these assets. Nuclear weapons are steeped in secrecy for political, cultural, and security reasons. Military nuclear materials are largely set apart from international laws and institutions and beyond the reach of existing mechanisms for international nuclear security cooperation. In recent years, geopolitical trends have made cooperation between countries that possess military nuclear materials even more difficult. This trend is likely to continue for the foreseeable future.

Strengthening international confidence in military materials security is a critical element of the nuclear non-proliferation regime. Confidence-building activities between states or on a multilateral basis like national reporting, peer review, best practice exchanges, and joint trainings and exercises can help to improve global security, deter potential adversaries, assure domestic and international stakeholders, and strengthen international credibility. This chapter explains what military materials are, describes the legal mechanisms and international institutions that support strengthening security for military materials around the world, identifies why confidence in military materials security is important, and provides recommendations for the further strengthening of security for military materials and confidence in that security.

What Are Military Nuclear Materials?

The vast majority of the world's stocks of nuclear weapons-useable materials exist in military programmes. This includes nuclear materials that fall into the following categories: inside active nuclear warheads; inside retired nuclear warheads awaiting dismantlement; inside non-civilian naval reactors and materials designated for non-civilian naval reserves; declared excess to military needs and awaiting down-blending or disposition by governments; and designated for other non-civilian purposes, including bulk storage.[2] As of 2021, there were nearly 13,000 nuclear weapons around the world located in at least 100 sites.[3] This is a significant reduction from the more than 70,000 nuclear weapons that once existed across the world in 1986.[4] Of these, approximately 90 per cent of nuclear weapons are located in the United States (US) and Russia.[5]

More than 80 per cent of the world's stocks of nuclear weapons-useable materials exist in military programmes.[6] Global stocks of weapons-useable nuclear material that could be used for military or potentially dual-use programmes are large and growing. According to a report published in July 2022 by the International Panel on Fissile Materials (IPFM), the total stockpile of civilian and military plutonium has grown by 46 tons since the end of 2014. The report also showed the total inventory of highly enriched uranium (HEU) has decreased by 35 tons for a net increase of 11 tons of weapons-useable nuclear material over the past seven years.[7]

As of 2020, public assessments compiled by the IPFM of Russian nuclear stocks estimated that there were approximately 88 metric tons of plutonium within the country's military stockpile, while the US possessed 34 metric tons of plutonium within its military stocks. In the same year, Russia possessed approximately 672 metric tons of HEU and the US possessed 361 metric tons available for military purposes.[8] Most US and Russian weapons are deployed or part of reserve stockpiles, with the remaining weapons retired or awaiting dismantlement. There has been a significant reduction in the number of locations with military-purpose materials over the past several decades. Over the next decade, the total number of US nuclear weapons is expected to continue to decline.[9] However, there are no current plans to further consolidate nuclear weapons locations in either Russia or the US.

While both the US and Russia continue to observe arms control treaties that require limits on total numbers of deployed nuclear weapons, long-term nuclear weapons modernization plans in both countries are moving forward. Russia is in the process of a multi-decade nuclear weapons modernization programme. Proponents of modernization plans in the US argue that these programmes enhance the safety and security of nuclear weapons. For example, the W87-1 warhead modernization programme includes the incorporation of insensitive high explosives within the warhead, which some argue 'allow a greater range of security tactics, techniques, and procedures to ensure secure transportation of the warhead ... over long distances from Air Force bases to the approximately 450 geographically dispersed ICBM [intercontinental ballistic missile] launch facilities'.[10] Others have argued that the W87-1 modernization programme is unnecessary and costly.[11]

The US, which is in the process of ramping up its nuclear weapons modernization, is also starting the process of resuming large-scale plutonium pit production for the first time in more than two decades.[12] There are more than a dozen naval training reactors and reactors related to defence activities that use HEU in the US.[13] The US Department of Energy (DOE), which is responsible for managing the US nuclear weapons complex, US navy reactors, and significant nuclear facilities, has eight sites with weapons-useable nuclear material: the Los Alamos National Laboratory, the Pantex Plant, the Y-12 National Security Complex, the Nevada National Security Site, the Hanford Site, the Idaho National Laboratory, the Savannah River Site, and the Oak Ridge National Laboratory. The US stores weapons-useable nuclear material in fewer locations than it once did. All Category I and II special nuclear material has been removed from Lawrence Livermore National Laboratory (LLNL), Sandia National Laboratory, Rocky Flats (which once produced plutonium pits for the US nuclear arsenal), and several other facilities throughout the US nuclear weapons complex.[14] Despite this, the US has not made any significant progress on consolidating its weapons-useable nuclear material since the announcement in 2012 that the LLNL had been cleaned out. Russia, meanwhile, has more nuclear weapons and weapons-usable nuclear materials located in more buildings than any other country. Russia operates nearly 60 HEU-fuelled research reactors, many of which are in national security-related facilities.[15]

Nuclear stocks are likely to grow in several countries in the coming years, with potential military implications. China, which possesses approximately 350 nuclear weapons and 17 tons of weapons-useable nuclear material, will likely be increasing its nuclear stocks.[16] China operates a pilot reprocessing plant at its Jiuquan Atomic Energy Complex and non-proliferation experts say the plant's production capacity will likely double soon.[17] While the Jiuquan Complex has been described as a civilian plant, its parent company, the China National Nuclear Corporation, was originally responsible for producing HEU and plutonium for Chinese nuclear weapons. Some have argued that if Chinese nuclear weapons stocks continue to grow, it is conceivable that the government could require the plant to produce plutonium for nuclear weapons.[18]

India, whose nuclear stocks are growing, operates four reprocessing plants, none of which are under International Atomic Energy Agency (IAEA) safeguards.[19] After Iran resumed its enrichment activities, the IAEA announced that the country had produced 42 kilograms of HEU, more than it ever had before.[20] The nuclear stockpile in the Democratic People's Republic of Korea (DPRK) continues to steadily grow, producing enough nuclear material for at least 20 nuclear weapons.[21] Pakistan's nuclear forces, which consist of 165 warheads, also continue to grow.[22]

Security Risks

Regardless of where they are located, all nuclear stocks, including those within military programmes, should be protected by strong and sustainable security systems capable of protecting against the full range of threats a facility might face. Such protective measures should include the ability to defeat outsider and insider adversaries that have been demonstrated within a country or region.[23] At a minimum, facilities should have the capacity to withstand: a knowledgeable, well-placed insider; a small group of well-trained and well-armed outside adversaries, capable of operating in multiple teams; or insider and outsider groups working in coordination.[24] Protective measures should also account for adversaries employing advanced technologies capable of disrupting digital systems or stand-off attacks (i.e. unmanned vehicles).

While there is evidence that nuclear weapons and weapons-useable materials within military programmes are surrounded by elaborate security systems, it is also clear there is a need for consolidating and improving security for weapons-useable nuclear materials. There have been numerous incidents in recent years where insiders, insurgents, protesters, and natural events have threatened or demonstrated inadequate security at sites with nuclear weapons or materials intended for nuclear weapons.[25] Some prominent examples include:

- In July 2016, at Incirlik Air Base in Turkey—where some 50 US nuclear weapons are reportedly stored—the local government cut off power to the base and arrested the base commander and others for alleged participation in a coup attempt. Just prior to the coup attempt, security concerns at the base were so high that family members of US personnel were ordered to leave. However, no plans are apparent at the time of writing to remove any of the nuclear weapons from this location.[26]
- In August 2015, the deputy director and chief engineer of the Elektrokkhimpribor Combine, one of Russia's two remaining nuclear weapon assembly–disassembly facilities, was arrested for bribery.[27] The incident demonstrated the challenge insider threats pose to nuclear facilities as well as the need for constant monitoring of employees.

- In April 2018, seven activists broke into the Kings Bay Naval Submarine Base in the US—which houses six ballistic missile submarines—carrying hammers and baby bottles of their own blood and smeared either red paint or blood on the walls of buildings. According to one account, they managed to access three locations on the base, including nuclear weapons storage bunkers.[28] Through their actions, the activists revealed that the base's system for detecting intruders in a timely way was inadequate.
- Environmental calamities have compromised security at nuclear sites with military materials. For example, the Cerro Grande wildfire threatened Los Alamos National Laboratory (LANL) in May 2000. The fire burned about a quarter of the 43-square mile area of the lab's property, causing hundreds of millions of dollars in damage to the facility, destroying research, equipment, and many structures, and forcing LANL to shut down for two weeks.[29] While none of the five LANL locations—where weapons-useable nuclear material was stored at the time—were destroyed, catastrophe was only narrowly averted. There was extensive damage to areas around nuclear storage facilities.[30] The fire burned close to the plutonium facility at Technical Area 55 and its supporting buildings and damaged land near the Critical Assembly Facility at Technical Area 18.[31] When employees returned to the site, 'no one knew the status of the nuclear materials held at LANL and several questions needed to be addressed by the MC&A [Material Control and Accountability] personnel before normal material control and accounting (MC&A) operations could be permitted to resume'.[32] It took weeks for material balance areas to resume normal MC&A operations.

While none of these incidents resulted in disaster, they each demonstrate in different ways that security at sites with nuclear weapons or materials intended for military purposes can be found wanting.

Nuclear Security Architecture for Military Materials

Countries with nuclear weapons and materials in weapons programmes are ultimately responsible for the security of those weapons and materials. They can, however, fulfil this responsibility more effectively through international cooperation that includes dissemination of information and best practices. Despite the continuing need to improve security for military-purpose stocks, there has been relatively little international cooperation focused on this area. Most international agreements and recommendations related to nuclear security apply only to civilian materials.

Developing mechanisms for building confidence in security for military materials is necessary for fulfilling several international commitments. UN Security Council

Resolution (UNSCR) 1540, approved in 2004 in the aftermath of the 11 September 2001 terrorist attacks, legally obligates *all* countries to provide 'appropriate effective' security and accounting for all nuclear weapons or related materials along with export and border controls, and, among other measures, to report to the Security Council on their efforts. Countries have not, however, developed a common definition of what 'appropriate effective' means.[33]

The four Nuclear Security Summits held from 2010 to 2016 all produced communiqués that referenced military materials and contained a pledge from participating countries to 'maintain at all times effective security ... including nuclear materials'.[34] The summit communiqués, however, did not address protecting or consolidating military stocks in any significant way due to a lack of agreement among some nuclear weapons states on whether or how to describe these activities. This lack of agreement also prevented efforts among a few countries to release a joint statement on the security of military nuclear materials, despite a concerted effort to do so. There has been some interest since the end of the Nuclear Security Summits in holding subsequent such gatherings of world leaders. For example, in 2016, Indian Prime Minister Narendra Modi committed India to hosting a summit on preventing weapons of mass destruction (WMD) terrorism;[35] this took place in 2018 as the India–Wiesbaden Conference.[36] Subsequent summits could provide an opportunity for countries with military nuclear materials to present on the security of these materials.

Additionally, successful implementation of the Treaty on the Non-Proliferation of Nuclear Weapons (NPT) will be difficult, if not impossible, without states parties prioritizing nuclear security. The NPT states that nuclear weapon states party to the treaty cannot transfer 'to any recipient whatsoever nuclear weapons or other nuclear explosive devices or control over such weapons or explosive devices directly, or indirectly'.[37] Implicit in this provision is the responsibility to prevent the proliferation of weapons-useable nuclear material to non-state actors. Preventing this type of proliferation requires states to have effective security systems that prevent the transfer of nuclear weapons via theft. As a practical matter, it will be difficult for nuclear weapon states to fulfil their nuclear disarmament obligation under Article VI of the NPT if they are concerned that other states or terrorist groups will steal their nuclear material.[38] Nuclear security has been addressed in a general sense within the NPT review process, but thus far it has not been a significant mechanism for encouraging progress or cooperation over military stocks.

The Convention on the Physical Protection of Nuclear Materials (CPPNM) includes requirements on criminalizing nuclear theft and smuggling and physically protecting material in international transport. In 2005, the convention was amended to address materials in domestic use, storage, and transport, and to cover sabotage of nuclear facilities as well as nuclear theft.[39] The 2005 Amendment to the CPPNM (A/CPPNM) also includes a list of 'fundamental principles of physical protection,' which were incorporated into the IAEA's 2013 'Fundamentals' publication, 'Objective and Essential Elements of a State's Nuclear Security Regime.'[40] While nuclear theft could include theft of military materials, the international provisions related to physical protection only address civilian nuclear facilities. This, however, leaves open the possibility that states could voluntarily report on the security of their military materials within the context of the A/CPPNM.

One of the provisions in the A/CPPNM requires states parties to convene a review five years after the convention entered into force to evaluate the implementation and adequacy of the convention, considering the 'then prevailing situation'.[41] This review conference took place in early 2022. While the review did not address military materials, one of its most important achievements was the endorsement, as per Article 16.2 of the convention, of another review. On 25 March 2022, just prior to the review conference, 50 countries endorsed a statement calling for the IAEA to 'convene the next conference of States Parties … no later than six years after the first Review Conference'.[42] The review's outcome document also endorsed another meeting. Subsequent reviews could potentially provide a forum for countries to report on the security of their military materials.

Multilateral fora like international treaty reviews and high-level summits also provide opportunities for dialogue on accidents and security incidents. National authorities can report non-sensitive lessons learned and corrective measures taken following nuclear security incidents. For example, the US DOE's Office of the Inspector General published a report (and there were other reports as well) identifying numerous deficiencies in communications, equipment, and organizational culture after a nun and two other activists gained access to the Y-12 site in 2012.[43] These findings were shared with international partners. There has been some progress advancing multilateral initiatives to increase security for military materials and confidence in that security, but to date, there is no systemic or comprehensive international initiative focused on peer review and confidence building for the security of all nuclear weapons.

International Frameworks for Nuclear Security Cooperation

Bilateral Nuclear Security Cooperation

For decades, the US has been the world leader in supporting international nuclear security cooperation. The US and Russia once cooperated extensively on security for nuclear weapons and weapons-usable materials for military purposes under the Cooperation Threat Reduction (CTR) initiative, but that cooperation effectively ended in 2014, along with several other important programmes aimed to increase nuclear security inside Russia. As the two countries with by far the most weapons-useable nuclear material, they share a special responsibility for nuclear safety and security.

The US continues to cooperate with many countries to improve nuclear security practices, including ongoing dialogues with both the United Kingdom (UK) and France.[44] The US once had broad cooperation with China on nuclear security, but tensions between the two countries have disincentivized cooperation.[45] There was also extensive nuclear security cooperation between Pakistan and the US, but this engagement has likewise tapered off. US nuclear security cooperation with India has been limited to technical exchanges, despite US interest in deeper engagement.[46] There are also examples of non-possessor states providing significant financial and technical support to nuclear security activities, most notably Canada, Norway, and the Netherlands.[47] These activities are currently focused on civilian materials, but there is clear interest among countries without nuclear weapons to play a larger role in securing military nuclear materials.

Unilateral Mechanisms for Nuclear Security Confidence Building

Occasionally countries unilaterally release information about the status of their military materials. The US, for example published a fact sheet for the 2016 Nuclear Security Summit titled, 'United States Military Nuclear Material Security.'[48] Among other details, the document revealed that US military materials either met or exceeded the IAEA's nuclear security guidance and included detailed information about all of the country's nuclear stocks, which had been shared previously as well. Other countries have also intermittently provided information on the quantities of nuclear materials in their possession. In 2006, the UK released a report on the use of HEU in its defence programmes.[49]

Unlike most countries, the US publishes information about the security of most of its military materials on an annual basis. Most US weapons-useable nuclear material related to US military programmes is managed by the National Nuclear Security Administration (NNSA), which is responsible for maintaining US nuclear weapons and overseeing the eight nuclear weapons laboratories and production facilities in the US. Every year, the NNSA publishes its Stockpile Stewardship Management Plan (SSMP) describing US programmes that maintain the US nuclear weapons stockpile.[50] The plan covers a range of issues, including: the strategic rationale for nuclear weapons; programmes and budgets to maintain US nuclear weapons; status of nuclear weapons production facilities; scientific research related to US nuclear weapons; and descriptions of the workforce. The SSMP also describes security for nuclear weapons, weapon components, special nuclear material in transport, and cybersecurity. It also provides information on various other issues, including the threats NNSA security systems are designed to protect against; vehicles used to transport weapons and materials; security technologies; training activities; performance testing, including force-on-force exercises and vulnerability assessments; and training activities.

Strikingly, the plan includes descriptions of security challenges that the NNSA faces and what it is doing to mitigate these challenges. For example, the 2022 SSMP identified that the NNSA needs to update its personnel vetting, stating, 'As the Government moves to a more real-time vetting and adjudication model, there is a need for DNS and DOE to adjust to meet the needs of the changing landscape.'[51] The plan also described how the pandemic has impacted security operations, identifying that, 'Due to the COVID-19 pandemic, the IT and Cybersecurity program had to adapt and rapidly pivot to accommodate a large portion of the workforce working from home and for extensive videoconferencing, including classified videoconferencing.'[52] This is a level of detail rarely seen in international discussions of security for nuclear materials within military programmes.

Despite many calls for increased assurances regarding security for military materials, the SSMP was not intended for that purpose.[53] Rather than functioning as a mechanism of international confidence building, it was a product of good government oversight. In 1992, the US declared a moratorium on nuclear testing. Over the next decade, it began developing the industrial and scientific capabilities, known as Stockpile Stewardship, to retain US nuclear weapons in the absence of explosive testing. During the development of Stockpile Stewardship, some members of Congress were concerned that President Bill Clinton's administration would not provide sufficient resources for maintaining US nuclear weapons. As a result, they began requiring reports on how US nuclear weapons were being maintained. Those reports were eventually consolidated into the SSMP, which included issues related to security. The SSMP could provide a model for what countries could share about security for their military materials without revealing sensitive information. This would be a significant step forward for confidence building around security for military materials.

The P5: A Multilateral Nuclear Security Framework

Bilateral, trilateral, and/or multilateral groups can develop mechanisms for sharing information on military stocks and security arrangements. One such framework for this cooperation could be the so-called 'P5 Process.' The P5 refers to the permanent members of the UN Security Council: China, France, Russia, the US, and the UK. Established in 2009, the P5 Process provides a forum for nuclear weapon states to regularly discuss their obligations under the nuclear NPT.[54] Current tensions between P5 members may make dialogue on nuclear disarmament extremely difficult, but this does not make the work any less important. The P5 should still find areas for cooperation and demonstrate leadership on security standards for military nuclear materials.[55]

One area of potential progress where some of the parties have cooperated in the past is on nuclear security. Security for military materials could be an extension of these discussions. An initial step could be for the P5 to agree to include discussions of security for military nuclear materials as part of the permanent agenda of the P5 Process. Over time, the P5 could work towards establishing a nuclear security working group featuring expert meetings hosted by participating countries. This working group could be a forum for the P5 to share information about the security of military stocks. Non-sensitive elements of this information could be submitted in a form on military materials to the UN Security Council as an element of UNSCR 1540.

The P5 framework also allows for different groupings of countries to collaborate, depending on who might be considered trusted actors. A P3 group of France, the UK, and the US could take the first steps to begin addressing the security of military nuclear materials. This could include the development of a system for peer review of each other's security arrangements. Such a system could compliment the IAEA's International Physical Protection Advisory Service (IPPAS), an important peer-review mechanism for disseminating best practices. The IPPAS teams review the physical protection measures at civilian nuclear sites and national physical protection regulations and approaches, and identify best practices and areas for improvement. With nearly 100 missions completed, IPPAS demonstrates the benefits of peer review for improving nuclear security. Bilateral, trilateral, and/or multilateral groups could develop mechanisms for sharing information on the security arrangements of military stocks.

The P3 could create the basis for further progress within the P5, reporting out and providing opportunities for all states with such materials to strengthen their security. This could also be a platform for countries to work with nuclear weapon states outside the P5 and report back about progress. While there are currently significant hurdles within this group, it remains an important platform for advancing nuclear security.

Shrouded in Secrecy

Given the large quantities of military nuclear materials and the security risks therein, one might assume that countries in possession of military nuclear materials have a strong incentive, at the very least, to cooperate with each other to provide assurances that these materials are secured to the highest level possible. Instead, the opposite is true. Aside from limited cooperation that took place between the US and Russia following the collapse of the Soviet Union, there has been little to no discussion between countries with military programmes.

There are multiple reasons for this lack of interaction. Historically, information on every aspect of a country's military programme has been closely held to the point where non-military agencies in the same country have no real understanding of what happens inside the institutions of their military inter-agency colleagues. The classification of certain information is an additional hurdle. In addition, some have the view that any information shared on military nuclear materials will make them inherently less secure. Finally, competition among some of these countries does not lend itself to a conducive environment for sharing information.

Conclusion

While it receives less attention in international non-proliferation fora, security for military materials is a key element of preventing nuclear theft and sabotage. Past incidents at nuclear facilities around the world illustrate the need for greater focus on strong and sustainable security for military materials. Most nuclear weapons-useable material falls within the category of military materials, which means that meaningful nuclear risk reduction will have to include non-civilian activities. Moreover, as countries move forward with modernizing and perhaps even increasing their nuclear weapons stockpiles, the importance and need for security will increase. This means that countries need to dedicate domestic and international programmes to strengthening the security of military materials. Military materials need to be protected at least as well as civilian materials and, most likely, need greater security than what is provided for civilian facilities.

Countries must also have confidence in one another's nuclear security. Governments should develop dedicated programmes and initiatives to build international confidence that facilities with military materials are protected against all probable threats they might face. This chapter has outlined several international initiatives that could be used for sharing non-sensitive information about security for military materials: the A/CPPNM treaty review process, ministerial or presidential-level nuclear security summits, and bilateral information sharing and peer review. Countries are, however, unlikely to take advantage of these opportunities without true leadership to break through the flawed conventional wisdom that any exchange of information on securing military nuclear materials will make those materials less secure. In an increasingly unpredictable world filled with rapidly evolving threats, this leadership is needed now more than ever.

NOTES

1. For civilian nuclear facilities, the IAEA has published nuclear security guidance. For example, see IAEA, *Nuclear Security Recommendations on Physical Protection of Nuclear Material and Nuclear Facilities (INFCIRC/225/Revision 5)*, Recommendations, IAEA Nuclear Security Series, No. 13 (Vienna: IAEA, 2011). http://www-pub.iaea.org/MTCD/publications/PDF/Pub1481_web.pdf
2. See Des Browne, Richard Lugar, and Sam Nunn, *Bridging the Military Nuclear Materials Gap* (Washington, DC: Nuclear Threat Initiative (NTI), 2015), 7. https://www.nti.org/analysis/articles/bridging-military-nuclear-materials-gap. Not all military materials are controlled by militaries. Some military materials are under civilian control and protection. For example, in the US, the National Nuclear Security Administration (NNSA) is responsible for managing most of the country's weapons-useable nuclear material, together with nuclear weapons being refurbished or awaiting dismantlement.
3. Hans M. Kristensen and Robert S. Norris, 3 September 2017, 'Worldwide deployments of nuclear weapons, 2017', *Bulletin of the Atomic Scientists* [online], 3 September 2017. https://thebulletin.org/2017/09/worldwide-deployments-of-nuclear-weapons-2017
4. Federation of American Scientists (FAS), 'Status of World Nuclear Forces: Who owns the world's nukes?', FAS [online], n.d. https://fas.org/issues/nuclear-weapons/status-world-nuclear-forces
5. Frank N. von Hippel and Masafumi Takubo, *Banning Plutonium Separation* (Princeton, NJ: International Panel on Fissile Materials (IPFM), 2022), 17. https://fissilematerials.org/library/rr20.pdf

6. International Panel on Fissile Materials (IPFM), 'Fissile material stocks', 2022. https://fissilematerials.org
7. International Panel on Fissile Materials (IPFM), *Global Fissile Material Report 2022. Fifty Years of the Nuclear Non-Proliferation Treaty: Nuclear Weapons, Fissile Materials, and Nuclear Energy. A Report of the International Panel on Fissile Materials* (Princeton, NJ: International Panel on Fissile Materials (IPFM), 2022), 4, 16, and 19. https://fissilematerials.org/publications/2022/07/global_fissile_material_r.html; International Panel on Fissile Materials (IPFM), *Global Fissile Material Report 2015* (Princeton, NJ: International Panel on Fissile Materials (IPFM), 2015), 11, 25, 19, and 29. https://fissilematerials.org/library/gfmr15.pdf
8. IPFM, *Global Fissile Material Report 2022*, 16–20.
9. Hans M. Kristensen and Matt Korda, 'Nuclear Notebook: How Many Nuclear Weapons Does the United States Have in 2022?', *Bulletin of the Atomic Scientists* [online], 10 May 2022. https://thebulletin.org/premium/2022-05/nuclear-notebook-how-many-nuclear-weapons-does-the-united-states-have-in-2022/#post-heading
10. Government Accountability Office (GAO), 'NNSA Should Further Develop Cost, Schedule, and Risk Information for the W87-1 Warhead Program', *GAO Highlights*, GAO-20-703, September 2020. https://www.gao.gov/assets/gao-20-703.pdf
11. Eryn Macdonald, 'New President, Same Wasteful Spending: Why Is the US Still Spending So Much on Nuclear Weapons?' *Union of Concerned Scientists* [blog], 25 June 2021. https://blog.ucsusa.org/emacdonald/biden-nuclear-weapons-spending/
12. Office of NEPA Policy and Compliance, *Plutonium Pit Production at Savannah River Site; Aiken, South Carolina* (Washington, DC: US Department of Energy, 2020). https://www.energy.gov/nepa/doeeis-0541-plutonium-pit-production-savannah-river-site-aiken-south-carolina
13. International Panel on Fissile Materials (IPFM), 'Research Reactors: United States', IPFM [online], 4 September 2021. https://fissilematerials.org/facilities/research_reactors/united_states.html
14. See International Panel on Fissile Materials (IPFM), 'Special Nuclear Material Removed from Livermore', IPFM Blog [online], 21 September 2012. https://fissilematerials.org/blog/2012/09/special_nuclear_material_.html; International Panel on Fissile Materials (IPFM), 'Most Sensitive Nuclear Material Removed from Sandia', IPFM blog [online], 28 February 2008. https://fissilematerials.org/blog/2008/02/most_sensitive_nuclear_ma.html
15. International Panel on Fissile Materials (IPFM), 'Research Reactors: Russia', IPFM [online], 4 September 2021. https://fissilematerials.org/facilities/research_reactors/russia.html
16. International Panel on Fissile Materials (IPFM), 'Countries: China', IPFM [online], 2 May 2022. https://fissilematerials.org/countries/china.html
17. Ibid.
18. Hippel and Takubo, *Banning Plutonium Separation*, 17–18.
19. Ibid., 18.
20. Arms Control Association, 'EU Attempts to Save Iran Negotiations', *Arms Control Today* [online], June 2022. https://www.armscontrol.org/act/2022-06/news/eu-attempts-save-iran-negotiations
21. FAS, 'Status of World Nuclear Forces'.
22. Hans M. Kristensen and Matt Korda, 'Pakistani Nuclear Weapons, 2021', *Bulletin of the Atomic Scientists*, 77/5 (2021), 265–78. https://www.tandfonline.com/doi/pdf/10.1080/00963402.2021.1964258?needAccess=true

23. Matthew Bunn and E. P. Maslin, 'All Stocks of Weapons-Usable Nuclear Materials Worldwide Must be Protected Against Global Terrorist Threats', *Journal of Nuclear Materials Management*, 39/2 (2011), 21–7.
24. For more on what threats security systems should protect against and on security for military materials overall, see Matthew Bunn, Nickolas Roth, and William H. Tobey, *Revitalizing Nuclear Security in an Era of Uncertainty* (Cambridge, MA: Belfer Center for Science and International Affairs, Harvard Kennedy School, 2019), 168. https://www.belfercenter.org/publication/revitalizing-nuclear-security-era-uncertainty
25. For more, see Bunn et al., *Revitalizing Nuclear Security*.
26. See, for example, Eric Schlosser, 'The H-Bombs in Turkey', *The New Yorker* [online], 17 July 2016. https://www.newyorker.com/news/news-desk/the-h-bombs-in-turkey
27. 'Two Top Managers of Elektrokhimpribor Accused of Bribe-Taking', *Nuclear.Ru* [online], 19 August 2015. www.nuclear.ru
28. Lindsey Bever, 'Activists Raid Nuclear Submarine Base with Hammers and "Baby Bottles of Their Own Blood"', *Washington Post* [online], 5 April 2018. https://www.washingtonpost.com/news/energy-environment/wp/2018/04/05/activists-raid-nuclear-submarine-base-with-hammers-and-baby-bottles-of-their-own-blood
29. M. D. Webb and K. Carpenter, *The Cerro Grande Fire, Los Alamos, New Mexico* (Los Alamos, NM: Los Alamos National Laboratory (LANL), 2001). https://permalink.lanl.gov/object/tr?what=info:lanl-repo/lareport/LA-UR-01-1630
30. William Earl Haag, *Material Control and Accountability (MC&A): Recovery from the Cerro Grande Fire at Los Alamos National Laboratory* (Los Alamos, NM: Los Alamos National Laboratory (LANL), 2001). https://www.osti.gov/servlets/purl/975592
31. Los Alamos National Laboratory (LANL), 'LANL Plutonium-Processing Facilities', TA-55 PF-4, National Nuclear Security Administration (NNSA), n.d. https://www.energy.gov/sites/prod/files/2018/02/f49/NNSA%202016_TA-55-factsheet.pdf
32. Haag, *Material Control and Accountability (MC&A)*.
33. See United Nations, '1540 Committee' [online], n.d. http://www.un.org/sc/1540
34. Nuclear Security Summits, 'Communiqué', 2016 Nuclear Security Summit, Washington, DC, 1 April 2016. https://obamawhitehouse.archives.gov/the-press-office/2016/04/01/nuclear-security-summit-2016-communiqu%C3%A9
35. US Department of State, 'Joint Statement: The United States and India: Enduring Global Partners in the 21st Century', U.S. Embassy & Consulates in India [online], 7 June 2016. https://in.usembassy.gov/joint-statement-united-states-india-enduring-global-partners-21st-century-june-7-2016
36. Office for Disarmament Affairs, 'India–Wiesbaden Conference calls for enhanced cooperation with industry in combating proliferation of WMD', United Nations, 23 April 2018. https://www.un.org/disarmament/update/india-wiesbaden-conference-calls-for-enhanced-cooperation-with-industry-in-combating-proliferation-of-wmd/
37. Treaty on the Non-Proliferation of Nuclear Weapons (opened for signature on 1 July 1968; entered into force on 5 March 1970). https://www.un.org/disarmament/wmd/nuclear/npt/text
38. See Matthew Bunn, 'Nuclear Security is the Foundation for the Three Pillars of the Nonproliferation Treaty', Nuclear Security Matters [blog], 14 March 2014. https://www.belfercenter.org/publication/nuclear-security-foundation-three-pillars-nonproliferation-treaty
39. IAEA, *Nuclear Security Recommendations on Physical Protection of Nuclear Material and Nuclear Facilities* (INFCIRC/225/Revision 5), Recommendations, IAEA Nuclear Security

Series, No. 13 (Vienna: IAEA, 2011). http://www-pub.iaea.org/MTCD/publications/PDF/Pub1481_web.pdf

40. IAEA, *Objective and Essential Elements of a State's Nuclear Security Regime, Fundamentals*, IAEA Nuclear Security Series, No. 20 (Vienna: IAEA, 2013). https://www-pub.iaea.org/MTCD/Publications/PDF/Pub1590_web.pdf
41. See IAEA, 'Amendment to the Convention on the Physical Protection of Nuclear Material' (Vienna: IAEA, 5 July 2022.). https://www-legacy.iaea.org/Publications/Documents/Conventions/cppnm_amend_status.pdf
42. Ibid.
43. See Office of the Inspector General, *Inquiry Into the Security Breach at the National Nuclear Security Administration's Y-12 National Security Complex*, DOE/IG-0868 (Washington, DC: US Department of Energy, 2012). http://energy.gov/sites/prod/files/IG-0868_0.pdf
44. For more, see National Nuclear Security Administration (NNSA), *Prevent, Counter, and Respond—A Strategic Plan to Reduce Global Nuclear Threats FY2022-FY-2026*. Report to Congress, December 2021. (Washington, DC: US Department of Energy, 2021). https://www.energy.gov/sites/default/files/2021-12/FY2022%20NNSA%20Prevent%2C%20Counter%2C%20and%20Respond%E2%80%94A%20Strategic%20Plan%20to%20Reduce%20Global%20Nuclear%20Threats.pdf
45. For example, see Miles Pomper and Ferenc Dalnoki-Veress, 'The Little Known Success Story of U.S.-China Nuclear Security Cooperation', Nuclear Threat Initiative (NTI) [online], 10 June 2020. https://www.nti.org/analysis/articles/little-known-success-story-us-china-nuclear-security-cooperation
46. NNSA, *Prevent, Counter, and Respond*, 2.3.
47. For example, Canada has provided US$28 million to the IAEA Nuclear Security Fund since 2003. See Government of Canada, 'Nuclear Disarmament and Non-Proliferation', Government of Canada [online], 24 October 2022. https://www.international.gc.ca/world-monde/issues_development-enjeux_developpement/peace_security-paix_securite/nuclear_radiological-nucleaire_radiologique.aspx?lang=eng
48. White House, 'FACT SHEET: United States Military Nuclear Material Security', White House Press Release, 31 March 2016. https://obamawhitehouse.archives.gov/the-press-office/2016/03/31/fact-sheet-united-states-military-nuclear-material-security
49. UK Ministry of Defence, 'Historical Accounting for UK Defence Highly Enriched Uranium', Nuclear Information Service [online], March 2006. https://www.nuclearinfo.org/wp-content/uploads/2020/09/Historical-Accounting-for-UK-Defence-Highly-Enriched-Uranium.pdf
50. See US Department of Energy (DOE), 'Stockpile Stewardship and Management Plan (SSMP)', USDOE [online], 16 March 2022. https://www.energy.gov/nnsa/articles/stockpile-stewardship-and-management-plan-ssmp
51. National Nuclear Security Administration (NNSA), *Fiscal Year 2022 Stockpile Stewardship and Management Plan: Report to Congress March 2022* (Washington, DC: US Department of Energy (DOE), 2022), 2–5. https://www.energy.gov/sites/default/files/2022-03/FY%202022%20SSMP%20March%202022.pdf
52. Ibid., 5–14.
53. Des Browne, Richard Lugar, and Sam Nunn, *Bridging the Military Nuclear Materials Gap* (Washington, DC: Nuclear Threat Initiative (NTI), 2015). https://www.nti.org/analysis/articles/bridging-military-nuclear-materials-gap
54. Andrea Berger and Malcolm Chalmers, *Great Expectations: The P5 Process and the Nonproliferation Treaty* (London: Royal United Services Institute, 2013). https://www.rusi.org/downloads/assets/WHR_3-13_Web.pdf

55. For more, see Fissile Materials Working Group (FMWG), 2015, *The Results We Need in 2016: Policy Recommendations for the Nuclear Security Summit*, 19–20.

Suggested Reading

Browne, Des, Lugar, Richard, and Nunn, Sam, (2015), *Bridging the Military Nuclear Materials Gap* (Washington, DC: Nuclear Threat Initiative (NTI)). https://www.nti.org/analysis/articles/bridging-military-nuclear-materials-gap

Bunn, Matthew and Maslin, E. P., (2011), 'All Stocks of Weapons-Usable Nuclear Materials Worldwide Must be Protected Against Global Terrorist Threats', *Journal of Nuclear Materials Management*, 39/2, 21–7.

Bunn, Matthew and Roth, Nickolas, 'The Effects of a Single Terrorist Nuclear Bomb', *Bulletin of the Atomic Scientists* [online], 28 September 2017. https://thebulletin.org/2017/09/the-effects-of-a-single-terrorist-nuclear-bomb

Bunn, Matthew, Roth, Nickolas, and Tobey, William H., (2019), *Revitalizing Nuclear Security in an Era of Uncertainty* (Cambridge, MA: Belfer Center for Science and International Affairs, Harvard Kennedy School). https://www.belfercenter.org/publication/revitalizing-nuclear-security-era-uncertainty

IAEA, (2011), *Nuclear Security Recommendations on Physical Protection of Nuclear Material and Nuclear Facilities (INFCIRC/225/Revision 5)*, Recommendations, IAEA Nuclear Security Series, No. 13 (Vienna: IAEA. http://www-pub.iaea.org/MTCD/publications/PDF/Pub1481_web.pdf

International Panel on Fissile Materials (IPFM), (2022), *Global Fissile Material Report 2022. Fifty Years of the Nuclear Non-Proliferation Treaty: Nuclear Weapons, Fissile Materials, and Nuclear Energy. A Report of the International Panel on Fissile Materials* (Princeton, NJ: International Panel on Fissile Materials (IPFM)). https://fissilematerials.org/publications/2022/07/global_fissile_material_r.html

National Nuclear Security Administration (NNSA), (2022), *Fiscal Year 2022 Stockpile Stewardship and Management Plan: Report to Congress March 2022* (Washington, DC: US Department of Energy (DOE)). https://www.energy.gov/sites/default/files/2022-03/FY%202022%20SSMP%20March%202022.pdf

PART V

NATIONAL APPROACHES TO NUCLEAR SECURITY

CHAPTER 22

NUCLEAR SECURITY FOR NEWCOMER COUNTRIES

ZENOBIA HOMAN, SARAH TZINIERIS, AND ŞEBNEM UDUM

Introduction

THE global nuclear energy sector is poised for expansion. As governments grapple with the intertwined challenges of climate change, dwindling fossil fuel supplies, rising energy demand, and energy reliance on potentially hostile states, there is increasing recognition that nuclear energy could hasten the global pathway to net zero.[1] This chapter examines how states develop nuclear security systems as they navigate the lengthy journey of building a new civilian nuclear sector. It outlines the steps set out by the International Atomic Energy Agency (IAEA), which states must follow sequentially to establish a civilian nuclear industry. In so doing, the chapter focuses on the hurdles states have needed to overcome to establish national regulatory and legal frameworks and implement physical protection systems for the first time, and also highlights where opportunities have arisen. Accordingly, the chapter considers the gap between the ideal and the reality, comparing the standards set by the IAEA with what individual states have been able to accomplish.

In order to highlight differences in national approaches, the chapter utilizes two case studies: Turkey and the United Arab Emirates (UAE). At the time of writing, Turkey is in the final stages of operationalizing its first nuclear power plant (NPP), while the UAE's first reactor went online in 2020. The development of the nuclear industry in each country is explored in order to understand how nuclear security systems might diverge. While implementation inevitably differs between states, nuclear security is an intrinsic and non-delineable aspect of the broader nuclear sector. In this process, what might be considered 'norms' of the global nuclear security regime are established, although perceptions about security might also evolve.

Nuclear Security for Newcomers

Who Are 'Nuclear Newcomers'?

'Nuclear newcomer' is not a formal term but is widely employed to describe those countries in the process of developing a new civilian nuclear sector; they are also known as 'embarking countries'.[2] A government will have expressed an interest in this area and taken its first steps towards developing the necessary infrastructure. The past two decades have seen a surge in new states expressing interest in developing a nuclear sector. At the time of writing, around 30 countries are considering, planning, or starting nuclear power programmes—and a further 20 countries have at some point expressed an interest in doing so. According to the World Nuclear Association, this ranges from 'sophisticated economies to developing nations'.[3]

States have varying motivations for pursuing a civilian nuclear programme, but the trend today is largely driven by growing demand for electricity amid concerns about fossil fuels.[4] Beyond these more obvious drivers, the development of such a complex and costly resource confers an element of 'prestige'.[5] Scholars have also drawn attention to 'nuclear latency', which entails the possession of some or all of the technical capabilities to develop a potential future nuclear weapon programme.[6]

Notwithstanding this flurry of interest, adding nuclear power to the national energy mix is no mean undertaking, given the eye-watering upfront costs and no guarantee of future commercial return. The process is also riddled with unexpected technology redesigns, safety and security risks, worker skills shortages, and regulatory obstacles, coupled with 'hidden costs', such as budget overruns and construction delays—plus eventual decommissioning. Thus, a number of states will find themselves stalling on the journey to go nuclear, or even backtracking. Indeed the World Nuclear Association notes, 'despite the large number of these emerging countries, they are not expected to contribute very much to the expansion of nuclear capacity in the foreseeable future—the main growth will come in countries where the technology is already well established'.[7] This may change over the longer term as the 'trend to urbanization in less-developed countries will greatly increase the demand for electricity'.[8]

The nuclear energy sector is growing most quickly in Asia, which accounts for two-thirds of reactors globally currently under construction, the majority of these in China.[9] The Middle East is also experiencing fast growth.[10] Many countries in these regions face acute insecurity, however. Even if newcomers are not affected directly, the transborder and 'spillover' nature of insecurity[11] can change the risk calculations of nuclear programmes abruptly. Whether national, regional, or local in origin, insecurity may threaten access roads, transport links, the national grid, cyber networks, and other critical infrastructures that are prerequisites before the nuclear facilities themselves are even built. Similarly, it has been highlighted how the trafficking of nuclear and radiological materials could become more active and dangerous as the industry continues to expand to states that have not before dealt with nuclear technology.[12]

International Norms for Nuclear Security

There are profound security implications for states seeking to establish nuclear programmes.[13] The decision to pursue a civilian nuclear capability is always taken at the state level, but, in reality, states cannot go beyond an articulation of interest before they are required to work closely with the IAEA, the primary body responsible for supporting states introducing (or re-introducing) a civilian nuclear sector. Any state pursuing nuclear activities must also adhere to stringent safeguards systems. Failing to do so would mean operating outside the global non-proliferation regime, governed by the Non-Proliferation Treaty (NPT).[14] Within these normative frameworks, there is an implicit expectation that newcomers will build nuclear safety, security, and safeguards measures into their programmes from the outset (even if historically this was not always the case).

Nuclear technology and materials are always sensitive owing to their dual-use potential for military application, and a select group of states has partnered to control the global flow of nuclear technology in an effort to curb proliferation. A basic requirement for a civilian nuclear programme is to formally recognize IAEA safeguards and the guidelines of the Nuclear Supplier Group (NSG). States also need to sign multiple agreements with international vendors to acquire components required in the fuel cycle. Consequently, states establishing a nuclear programme tend to forge new trading relationships—sometimes with governments where diplomatic relations barely existed before. Meanwhile, importation of nuclear materials and technology may lead to a dependency on other states. All these dynamics coalesce, with the risk that new nuclear programmes may inadvertently become entangled in complex geopolitical issues. Nuclear newcomers need to be mindful of how their nuclear programmes may alter existing local, national, or regional security dynamics.

Another important early consideration for newcomers is to determine in which international vending agreements they want to participate. Most prominent of these is Section 123 of the US Atomic Energy Act, which facilitates transfers of nuclear material or equipment as well as technical exchanges and scientific research.[15] Based on the principle of peaceful nuclear cooperation, a 123 Agreement provides nuclear newcomer states with access to US nuclear technology but also obliges states to forego any domestic enrichment and reprocessing facilities; indeed, a 123 Agreement tends to be more restrictive than equivalent arrangements with states such as Russia and China. By opting for a US-led regime on nuclear technology, states are in effect espousing American authority in global nuclear rules and norms. Some have even argued that states signing 123 Agreements are aligning their nuclear commercial interests with US national security goals.[16]

A Phased Approach to Nuclear Security

Security-by-Design

All states possessing a nuclear programme were once newcomers. The civilian nuclear industry sprang up after the Second World War, but it took many decades for so-called mature nuclear countries to develop effective safety, security, and safeguards regimes. An advantage to newcomers is that they do not have to retrofit nuclear security onto older systems and sites, thereby saving time and costs (and also avoiding historical mishaps). What is more, nuclear newcomers can leverage security lessons previously learned by mature nuclear states, for example, utilizing past security incidents to design better nuclear security systems from the outset, as well as integrate safety with security. Indeed, the modern concept of nuclear security has developed as a result of 'reactive' policy making, rather than in any strategic way—most obviously in response to, first, the dissolution of the Soviet Union in the early 1990s and, second, the 11 September 2011 terrorist attacks in the United States (US).[17]

Security-by-design is a risk-informed approach where security is incorporated from the outset into the design process of a facility. To be successful, physical protection systems must be sufficiently robust to withstand future changes in security requirements over the lifecycle of a project.[18] Such an approach offers obvious benefits in incurring lower operational impacts and project costs. However, security-by-design requires a clear security strategy and multi-discipline teamwork throughout the development of a project. Threats exist right from the outset of a new nuclear programme because malicious actors may seek to take advantage of perceived loopholes and vulnerabilities when security systems are undeveloped, as highlighted by the 1982 attack on South Africa's NPP at Koeberg during the construction phase.[19]

IAEA's Milestones Approach

To establish the essential elements of a new nuclear programme, the IAEA has developed the 'Milestones Approach'. First issued in 2007 and revised in 2015,[20] this sequential framework is formalized in the IAEA guidance document 'Milestones in the Development of a National Nuclear Infrastructure for Nuclear Power'.[21] The document sets out 'a phased comprehensive method to assist countries considering or planning a nuclear power programme'.[22] Newcomers are supported to prepare for the commitments and obligations associated with developing a safe, secure, and sustainable nuclear power programme; meanwhile, countries with existing programmes can assess their readiness for expansion. There is no formal obligation for a newcomer to adopt the

Milestones Approach, but most choose to do so, and no other equivalent frameworks exist. Notably, while the Milestones Approach includes nuclear security within its comprehensive framework, it does not offer specific guidance in this area.

The Milestones Approach identifies three phases of development—containing 19 infrastructure issues—as follows:

1. Phase 1: the three bodies integral to building an NPP (government, regulator, and owner/operator) are ready to make a commitment to an NPP. At this point, a broad 'pre-feasibility study' is undertaken. Regarding nuclear security, recommendations for national policy and strategy are prepared, as well as the legal framework. Nuclear security roles and responsibilities are also defined, and the country's human resource needs and capabilities are evaluated.
2. Phase 2: the country is ready to invite bids and negotiate contracts for the construction of a nuclear facility, following a policy decision. The three key bodies will have established the crucial and regulatory frameworks and conducted preparatory work. Regarding nuclear security, the graded approach for the nuclear security regime is now developed—covering areas such as physical protection, sensitive information, nuclear security culture, and emergency response.
3. Phase 3: the country is ready to commission and operate its first nuclear facility. Activities relating to contracting, licensing, and construction are undertaken. Regarding nuclear security, physical protection systems are tested and approved, coordination between safety, security, and safeguards are put in place, a contingency plan to counter sabotage and theft is set up, a national nuclear incident response plan is developed, and appropriate inspection and verification measures for facilities are implemented.

IAEA's Nuclear Security Guidance for Newcomers

The IAEA's Nuclear Security Series (NSS) provides international guidance to support states in fulfilling their nuclear security responsibilities. Launched in 2006, the NSS contains four sets of publications: Fundamentals, Recommendations, Implementing Guides, and Technical Guidance. These publications cover all aspects of nuclear security. The IAEA's Fundamentals document 'Objective and Essential Elements of a State's Nuclear Security Regime' contains specific nuclear security guidance for newcomers, and sets out 12 'Essential Elements'.[23] Intended to serve as 'an overarching document' in the NSS,[24] the document is aimed at the various national stakeholders involved in establishing a nuclear security regime.[25] Guidance is based on a synthesis of 'the experiences of Members States' and 'provisions from the many international instruments that contribute to defining the international legal framework'.[26] Notably, the IAEA emphasizes that the nuclear security regime 'is part of the State's overall security regime'.[27]

Further Support for Newcomers

Developing comprehensive nuclear security infrastructure is essential from the outset of a nuclear programme. Early on, the country needs to engage in sustained capacity-building and training. The IAEA provides dedicated support through Integrated Nuclear Security Support Plans (INSSPs). Developed jointly between the IAEA and Members States, INSSPs identify specific actions required to ensure the country's nuclear security systems are effective.[28] Typically, six functional areas are addressed: legal and regulatory framework; threat and risk assessment; physical protection; detection of unauthorized acts involving material out of regulatory control (MORC); response to unauthorized acts involving MORC; and sustainability. INSSPs are aligned with the IAEA's NSS guidance, providing a best practice starting point for newcomers.

Another valuable resource that newcomers can access is the IAEA's International Physical Protection Advisory Service (IPPAS). Upon request, IPPAS missions provide advice and peer review of physical protection arrangements, including a country's implementation of the relevant international legislation and NSS guidance. At the time of writing, 97 IPPAS missions have been completed (since the programme began in 1995),[29] with a sizeable number being newcomers. Following the Nuclear Security Summits (2010–2016), the IAEA has also provided organizational support to regional 'Centres of Excellence', including in newcomers Morocco, Indonesia, and Uzbekistan. In an effort to strengthen coordination between the growing number of these centres, the IAEA has established a 'Nuclear Security Training and Support Centre' (NSSC) network. Another useful resource is the IAEA's e-learning modules, 21 of which focus on nuclear security.[30] These are offered for free and are available online, so enabling any nuclear stakeholder to gather information before decisions are taken.

There are also several industry organizations that provide nuclear security support to nuclear newcomers, including the World Association of Nuclear Operators (WANO) and the Institute of Nuclear Power Operators (INPO). The World Institute for Nuclear Security (WINS) offers dedicated professional training on nuclear security. Cooperating and collaborating with 'mature' nuclear states is another a useful route to develop domestic nuclear security expertise. The US has been particularly active in this area through the provision of training, capacity-building, and professional development courses, under its Cooperative Threat Reduction (CTR) programme.[31]

Establishing the Essential Elements of Nuclear Security

National Authority

The state is unequivocally responsible for nuclear security (Essential Element 1) and primary work on implementation begins at the national level. A national authority governs all aspects of a state's nuclear security regime—identifying and defining the key responsibilities (Essential Element 2). Early on, a country will establish the competent authorities; normally these include a national authority, a regulatory body, and border control and law enforcement. These institutions should be granted both 'adequate legal authority' and 'sufficient financial, human and technical resources' to fulfil their assigned nuclear security responsibilities.[32] The state legislature also needs to adopt national laws that implement international obligations. Most countries complete a congressional or parliamentary approval procedure to deposit instruments of international law. This involves consolidating the key binding treaties on nuclear security—Convention on the Physical Protection of Nuclear Material (CPPNM), its Amendment (A/CPPNM), and the International Convention for the Suppression of Acts of Nuclear Terrorism (ICSANT)—as well as relevant obligations in United Nations Security Council Resolutions and other treaties.

The responsibility structure will be different in each country. The IAEA's 'Responsibilities and Functions of a Nuclear Energy Programme Implementing Organization' is a useful resource that includes case studies illustrating this divergence.[33] For example, in Kenya the Nuclear Electricity Board served until recently as the implementing organization, while in Belarus day-to-day activities are coordinated by the Nuclear Energy Department of the Ministry of Energy with a high-level 'Interdepartmental Commission for Nuclear Power Plant Construction' that monitors implementation.[34] Whatever structure is implemented, effective coordination and communication between the relevant authorities is essential. Indeed, Belarus has highlighted the expediency of strong government support for its Nuclear Energy Department, which in turn has helped cooperation among the various stakeholders.[35]

In practice, there is no 'one-size-fits-all' approach but there are some common challenges. Key ones include: human capital (which may be limited); overreliance on 'turnkey' projects (a delivery method where a single entity completes all project stages, which might be executed by a foreign company); and low acceptance of state-owned companies on the part of publics (for example, Spain, India, and South Africa have all experienced attacks on nuclear facilities).[36] In the case of Belarus, ensuring public support has been critical following the Chernobyl disaster in neighbouring Ukraine.[37] Other challenges may include national security realities that appear incompatible with the Milestones Approach. Furthermore, while it is possible to learn from the collective experience of other states in the IAEA guidance, this is often high-level, broad, and undetailed. Thus, countries need to evaluate how the guidance information is relevant to them–and how they might obtain further insights.

In developing its approach, the national authority needs to identify and assess potential threats to its future nuclear estate, employing a risk-informed approach (Essential Element 9). This is a complex exercise as the security landscape may be in a state of flux and nuclear facilities may not yet exist. However, general lessons can be learned from comparable domestic industries and the nuclear sectors of other countries. At the outset, the national authority should develop a 'design basis threat' (DBT) or 'national threat assessment', in line with the IAEA's physical protection recommendations contained in INFCIRC/225/Rev.5.[38] The DBT informs the threat assessment performed by nuclear operators to determine the scope of physical protection required at their site. The DBT is revised when security threats change but for newcomers it might need to be iterated more regularly as the nuclear programme establishes itself.

Regulatory Framework

Establishing a regulatory framework is how nuclear security laws and obligations translate into operational reality. To satisfy IAEA criteria, a country needs to maintain appropriate regulatory oversight of nuclear security for all nuclear material inventories at all times (Essential Element 3). In addition to providing adequate legal authority and resources, regulatory bodies require 'appropriate independence in their nuclear security decision making'.[39] The case of the 2011 Fukushima nuclear disaster highlighted the dangers of 'regulatory capture' where regulatory authorities were not independent of government and industry.[40] To support states, the IAEA has produced a *Handbook on Nuclear Law*.[41] It also offers an *Integrated Regulatory Review Service* to strengthen the effectiveness of regulatory infrastructure being developed by newcomers.[42]

Owing to their divergent historical trajectories, varying regulatory frameworks for nuclear security have been adopted around the world. Nevertheless, regulatory bodies should all share the common aim of exercising legal authority to establish a nuclear security regime within their jurisdiction. In recent years, regulatory frameworks have become known as either 'rules based' or 'outcome focused'. The rules-based (also called 'prescriptive') framework is by far the most common, and nuclear newcomers tend to select this model. Here, the regulator prescribes detailed security arrangements for operators, and then conducts regulator inspections to determine compliance. The outcome-focused framework is mostly associated with the United Kingdom (UK). In this model, operators are given more flexibility but take full responsibility for decision-making over security arrangements and are expected to prove the validity of these to the regulator through evidence.[43]

While newcomers have tended to work towards establishing a rules-based regulatory framework, the reality is more complex. In creating the framework, the regulator will liaise closely with nuclear operators in a continuous process of iteration. Thus, it may be that in the early stages of a nuclear programme the regulatory system is technically more akin to a performance-based framework as it is unfeasible to implement a fully prescriptive framework until all the relevant information is gathered. This approach provides authorities some flexibility to modify and enhance arrangements; in due course, the regulator can work towards a more uniform framework.

Nuclear Operators and the Workforce

The operator of a nuclear facility bears greatest responsibility for the day-to-day delivery of nuclear security, namely, identifying threats against potential onsite targets and assessing the potential consequences of a security event (Essential Elements 7 and 8); detection of nuclear security events (Essential Element 10); and preparedness for and response to a nuclear security event (Essential Element 11). In addition to the obvious requirement for physical protection, operators need to address the threat of a cyber-attack. Cyber risks are growing as facilities become increasingly reliant on advanced informed technology infrastructure. The IAEA recommends that operators take a risk-informed approach to developing and implementing nuclear security systems (Essential Element 9). By systematically considering the threats and risks, resources can be allocated more effectively and efficiently.[44] For example, evaluation of risk-informed approaches helped strengthen security at the Doel NPP in Belgium following a sabotage attack in 2014.[45]

One of the greatest challenges faced by newcomers is building sufficient human resources for nuclear security.[46] In the early stages of a programme, capacity-building and training of the workforce needs to be scaled up fast. Human resource capacity impacts all aspects of a nuclear programme, but especially the operator. As well as support from the IAEA, technical skills can be learned from responsible vendor states that are likely to have a 'mature' programme. The pressure to scale up human resources capacity might lead to operators seeking to recruit from abroad or from non-nuclear industries. Countries with strict immigration requirements, or a state-run industry such as in India, might face more obstacles in this regard. One solution for operators is to recruit temporary workers. However, there are practical issues in vetting such workers who tend to work at a facility for a short period and be more transient, complicating the process of conducting thorough background checks or personnel reliability programmes (PRPs).

Security Incident Responders

From the outset of a programme, a newcomer should consider the scenario of a nuclear security incident. This requires developing local and regional law enforcement, emergency response, and investigative capabilities (Essential Element 11). The state authority, regulator, emergency services, and operators all need to work together in this area. Verification and enforcement measures are also required to ensure compliance with applicable laws, regulations, and requirements. Newcomers already face the challenge of scaling up security personnel quickly, but this is even more complicated when it comes to recruiting emergency responders and investigators such as nuclear forensic scientists, given the specific skillsets required. In fact, many countries will opt not to develop national capabilities for performing full forensic characterization of nuclear material because it is logistically and financially more expedient to establish arrangements with foreign laboratories, such as JRC Karlsruhe in Europe.[47]

Another challenge is developing a multi-agency response for emergency scenarios involving nuclear and radiological materials. Individual emergency services may conduct regular exercises, but joint exercises involving multiple agencies is often not the norm. For example, lack of coordination between agencies was a contributing factor to the weak response to the 2008 terrorist attacks in Mumbai.[48] Newcomers should therefore ensure that coordination between first responders is in place from the outset. Countries might find it useful to tap into existing capabilities, such as whether their armed forces already have a chemical, biological, radiological, and nuclear (CBRN) unit.

International Cooperation

The IAEA recommends that states cooperate and assist one another in developing nuclear security, either directly or via the IAEA (Essential Element 6).[49] This includes providing timely responses to requests for nuclear security assistance from other states, offering timely information about criminal or intentional unauthorized acts, and cooperating and exchanging information. One obvious area for cooperation is border and export controls. With the IAEA's assistance, governments are increasingly working with their neighbours to counter the smuggling of nuclear and radioactive materials. For example, potential nuclear newcomers Malaysia and Thailand have conducted joint exercises to detect the illicit trafficking of MORC at their shared border.[50]

Another key consideration is transportation (Essential Element 4), not least as this area is governed by two of the three binding international agreements on nuclear security: the CPPNM and its Amendment. Although operators will be responsible for the logistics of transportation, in the initial stages of developing a new nuclear programme, state authorities need to consider issues such as domestic and regional security threats, the organization of major events, and a national response if materials go missing. For example, Morocco faces security risks from terrorist groups as well as political instability in the Sahel-Sahara region, with these affecting maritime shipment of radioactive sources.[51] In this context, newcomers might want to forge agreements with their neighbours over transborder assistance in the event of an incident.

Sustainability and Security Culture

The IAEA's 'Objective and Essential Elements' document places emphasis on sustainability (Essential Element 12). Sustainability is about maintaining nuclear security over time in a changing threat environment and ensuring continuous improvements at both the state and site level. As well as the more obvious requirements of system maintenance and quality assurance, sustainability involves security culture, management, and leadership.[52] Sustainability in nuclear security came to the fore in the early 2000s when the US realized its substantial CTR programme in Russia could not continue indefinitely and would require the host authorities to maintain progress.[53]

Sustainability is related to the concept of organizational culture with its emphasis on the human factor. A subset of organizational culture is 'security culture', defined by the IAEA as the 'assembly of characteristics, attitudes and behaviour of individuals, organizations and institutions which serves as a means to support and enhance nuclear security'.[54] Brand new security systems can be built but without a robust security culture they will be ineffective and may fail. For example, weak security culture contributed to protesters penetrating the Y-12 National Security Complex in the US in July 2012, despite the facility having been recently upgraded to house Y-12's uranium to the tune of $0.5 billion.[55]

A problem faced by newcomers is that while the concept of security culture may be understood and well-received, it can, because of the nebulous nature of culture and human behaviour, take considerable time to establish and become effective. It is therefore imperative that newcomers consider security culture from the outset, even before the commissioning and construction phases start; when a new project begins formal operations, appropriate behaviours and attitudes are not only understood but also can be embodied by leaders and regarded as normal conduct by plant personnel. This helps organizations avoid significant events in the first place.[56] Here again, security culture will differ between countries depending on local social and cultural contexts. Nevertheless, newcomers may find establishing an effective security culture is actually an easier process in comparison to more established countries, where historical safety events can lead to a 'safety first' mentality becoming ingrained at the expense of security.

Case Studies

To deepen understanding about the challenges faced by newcomers, this section discusses two countries with relatively new nuclear energy programmes: Turkey and the UAE.

Nuclear Security in Turkey

Turkey first became interested in establishing a civilian nuclear programme after US President Dwight D. Eisenhower's 1953 'Atoms for Peace' speech, leading to a nuclear cooperation agreement between Ankara and Washington. Turkey established its Atomic Energy Commission in 1956, before bringing two research reactors online: in 1962, a pool-type 1MW test reactor at the Çekmece Nuclear Research and Training Center (ÇNAEM); in 1979, a TRIGA Mark-II research reactor at Istanbul Technical University, still operational today.[57] Turkey embarked on nuclear power projects between the 1970s and 1990s, but these early attempts faced a series of setbacks. In the following decades, Turkey engaged in nuclear-related talks with Canada, France, Japan, the Republic of Korea, and the US, among others.[58]

Ankara's interest in nuclear energy is not only for diversification to alleviate dependence on natural gas and reduce the energy bill, but it also views the possession of an NPP as a sign of a 'developed country'. Turkey's NPP at Akkuyu on the Mediterranean coast began construction in April 2018. The first reactor is due to come online in 2023. When completed, Akkuyu's four reactors are projected to reach 4,800 MW and supply around 10 per cent of the country's energy needs,[59] with funding largely provided by the Russian state nuclear corporation Rosatom.[60] Turkey also signed an agreement with Japan in 2013 for the construction of a second NPP at Sinop by the Black Sea coast, but negotiations stalled due to economic concerns.[61] China has expressed interest in building the country's third NPP at İğneada in the Thrace region. Meanwhile, Turkey has uranium resources it might seek to exploit.

Turkey broadly adheres to global norms on nuclear security. It has ratified the key legal instruments—the CPPNM, its Amendment, and ICSANT—as well as other instruments with nuclear security elements such as the Convention on Early Notification of a Nuclear Accident. Regulatory decisions are led by the Nuclear Regulatory Authority (NDK), which also has some licensing functions.[62] Notably, the original regulator, the Turkish Atomic Energy Authority (TAEK) formed in 1982, is now exclusively focused on the management of radioactive waste.[63] Various technical regulations with nuclear security provisions have been implemented,[64] most prominently the 2012 Regulation on Physical Protection of Nuclear Materials and Nuclear Facilities', aligned to INFCIRC/225.[65] This includes provisions against theft, sabotage and insiders, among other risks, and also sets out detailed measures for protecting particular zones of an NPP.[66]

Turkey is situated in a turbulent neighbourhood. Threats are diverse but mostly centre on separatist or terrorist organizations that could target nuclear facilities as a symbol of the state.[67] Turkey shares borders with Syria, Iran, and Iraq, and in 2017 a terrorist attack linked to extremist Islamist groups based in Syria targeted the country's largest city, Istanbul, where the TRIGA reactor is located.[68] Nevertheless, this incident is not reflective of the more deep-rooted security threat that Turkey faces from domestic militant groups, especially the separatist Kurdistan Workers' Party (PKK) based in the country's south-east; owing to its geographical location, the Akkuyu facility is a potential target. As a consequence of its security environment, Turkey has decades of experience of combatting terrorism and has developed an advanced civil defence structure.

Various national authorities shape the national threat assessment, or DBT. Adhering to the 2012 regulations, a Nuclear Security Plan is devised by the NDK and evaluated by the Gendarmerie, Ministry of Interior, Security Department, National Intelligence, Coast Guard, and TAEK, as needed.[69] The 2012 regulations highlight that nuclear security culture is not only relevant for dedicated security personnel but also for the wider workforce, particularly those responsible for information and cyber security. The concept of security culture is being embedded in Turkey's nuclear sector, but challenges remain. For example, the words 'safety' and 'security' have the equivalent translation in Turkish, as is the case in other languages, which complicates awareness raising.[70]

The Gendarmerie General Command, under the Ministry of the Interior, is responsible for nuclear security at the site level, including the plant under construction at Akkuyu. Here, the plant footprint and the wider shore area are treated as an 'exclusive security zone'.[71] To mitigate against threats, the personal details of employees, contractors, and all those who require access to the zone, are shared with the authorities.[72] Since the 2016 coup attempt, the authorities have tightened up the vetting process to address the insider threat. Staff at Akkuyu have undergone training in Russia, including on nuclear security,[73] underlining close Russia–Turkey cooperation in the nuclear sphere.

Turkey has a track record at emergency preparedness and response gained from dealing with earthquakes. The Department of Disaster and Emergency Management is predominantly engaged with disaster relief and emergency response. In 2019, it coordinated a 'stakeholders emergency drill' to enact a multi-agency response to an emergency scenario.[74] Thanks to this existing skill base, Turkey is not overly dependent on external contractors for nuclear security, as is the case for many newcomers. However, Turkey's prior experience is confined to confronting physical dangers, and the country is still working towards bolstering its cyber defences and security culture. Most recently, Turkey is confronting the devastating earthquake of 6 February 2023 and its sequence of aftershocks. The Akkuyu site itself did not sustain damage and construction work is continuing. In addition to not being sited in the immediate proximity of an active fault line, the facility has been built to withstand an eight-magnitude earthquake and with instrumentation in place to automatically shut down in the event of a nine-magnitude quake.[75] Nevertheless, the magnitude of the quake has revived policy debates in Turkey about the quality of structural engineering.[76]

Nuclear Security in the UAE

In August 2020, the UAE became the first Arab country with an operational nuclear power reactor. The Barakah NPP in the Abu Dhabi emirate comprises four APR-1400 reactors: unit one, which began commercial operations in April 2021, unit two, which came online in August 2021, and a further two under construction. When completed, total capacity is expected to reach 5,380 MW and supply around 25 per cent of the country's energy needs.[77] Nuclear energy officially became a consideration for the UAE in 2006 when the Gulf Cooperation Council (GCC) commissioned a study on a regional nuclear programme for peaceful uses.[78] The UAE took this further, releasing a 'white paper' in 2008 that set out elements of establishing a regulatory and legislative framework.[79]

Unlike Turkey, the UAE did not hold existing experience with related nuclear technology, for example, research reactors. Still, its nuclear energy programme has accelerated fast, in relative terms. As part of preparatory efforts, the UAE followed the IAEA's Milestones Approach, which led to a predictable process—and, arguably, in many ways the UAE exemplifies an IAEA newcomer 'archetype'. The UAE positioned itself within norms of the global non-proliferation regime by signing the NPT in 1995 and a bilateral safeguards agreement with the IAEA in 2003; in 2009, the UAE signed the Additional Protocol to the NPT. But the crucial step came in 2009 when a 123 Agreement with the US entered into force. To date, the UAE is the only Gulf state to have signed such a deal.

As part of the First Phase, in 2009 the UAE established the Federal Authority for Nuclear Regulation (FANR), an independent nuclear regulatory agency, and the

Emirates Nuclear Energy Corporation (ENEC), to evaluate and implement the country's nuclear plans. The UAE also established a legal framework for the industry, including 'Federal Law No. 13' (2007), which establishes nuclear export control legislation, and 'Federal Law No. 6' (2009), which prohibits uranium enrichment or spent fuel reprocessing facilities within the UAE's borders. During the Second Phase, the UAE evaluated the various nuclear technologies on offer. Notably, the UAE diverged here from the traditional commercial path of a bidding process, specifications issuance, bid evaluation, and contract negotiation, which were deemed cumbersome.[80]

Instead, the UAE sought integrated bids. After considering competitive proposals from Japan, France, and Korea, the latter appeared most willing to adapt to the UAE's preferred approach. The Korean consortium also promised the shortest timeline for completion. In 2009, the UAE awarded the contract for the construction of four APR-1400 reactors—including the associated maintenance, fuel supply, and training and education—to a Korea Electric Power Company (KEPCO)-led consortium. The Third Phase, which commenced in 2011, involved further evaluation work (with additional assessments on account of the recent Fukushima nuclear disaster).[81] The UAE also fully integrated the IAEA's recommendations into nuclear security implementation. By 2015, all four reactors were under construction.

Like Turkey, the UAE adheres to global nuclear security norms. The UAE adopted the CPPNM in 2003, its Amendment in 2009, and ICSANT in 2008. Notably, the UAE was invited to the first Nuclear Security Summit in 2010 and was one of the original signatories to the 2014 NSS commitment 'Strengthening Nuclear Security Implementation' (INFCIRC/869).[82] Such agreements helped lay the groundwork for international cooperation over nuclear security even before the UAE began building physical infrastructure. Closely adhering to the Milestones Approach, the UAE has also publicly released its nuclear security regulations and annual nuclear security reports. In 2016, a IPPAS mission was conducted in the UAE.[83] The UAE responded to feedback such as amending legislation to reflect CPPNM provisions.[84] The UAE's proactive engagement with the international community has enabled fast advancement on nuclear security, even despite the relative infancy of its programme. The UAE has also benefitted from its close relationship with Korea, with regular high-level consultations providing a platform to discuss nuclear security challenges.[85]

Security remains an important consideration at the Barakah site. While the UAE is generally stable domestically, its location in a geopolitical hotspot inevitably poses a risk, which applies in respect to the ongoing war in Yemen, political instability in Iraq, and regional non-state actors linked to Iran. The total population of the UAE is around 10 million, but, notably, just 15 per cent are Emirati citizens, with the remainder expatriates. Due to its relative inexperience dealing with nuclear materials, combined with its population make-up, the UAE has relied heavily on expatriate personnel.[86] To compose a skilled indigenous workforce, the UAE has set up several effective educational initiatives. However, until sufficient numbers have been trained, the UAE will continue to rely on recruiting from abroad.

Such a transient workforce can pose an insider threat, although the UAE authorities appear to be proactively taking steps to address this. For example, employing over 40 nationalities, Nawah Energy Company has created a 'Nuclear English Development Center' at the Barakah plant.[87] This unique initiative addresses issues such as preventing miscommunication, ensuring the integrity of the workforce through conducting international background checks and frequent and repeated vetting, reporting suspicious behaviour to relevant authorities, and maintaining a high level of threat awareness.

Conclusion

Historically, the routes to a civilian nuclear energy programme have varied hugely between states. With the dissemination of the IAEA's Milestones Approach, states can now follow a prescribed pathway, which helps instil a best practice approach to nuclear security. Developing a comprehensive nuclear security system is essential from the outset as malicious actors might seek to take advantage of perceived weaknesses and loopholes in a new programme. States benefit from international assistance where they are willing to adhere to normative frameworks, such as the CPPNM and its Amendment, or to receive guidance from IPPAS missions and collaborate with responsible vendors.

The global interest in civilian nuclear power is only likely to accelerate in the face of growing pressures related to climate change. However, the development of nuclear security systems requires both upfront and ongoing operational costs, which could be viewed as a barrier to newcomer states, albeit a minor proportion of an overall hugely costly enterprise. It is incumbent on the wider international nuclear community that newcomers do not perceive these costs and other regulatory and legal imperatives as further hurdles to developing peaceful nuclear energy programmes.

Notes

1. See government statements in IAEA, *Nuclear Energy for a Net Zero World* (Vienna: IAEA, 2021).
2. See, for example, IAEA, *E-learning for Nuclear Newcomers* (Vienna: IAEA, 2021); IAEA, *A Decade of IAEA Integrated Support to Embarking Countries* (Vienna: IAEA, 2022).
3. The World Nuclear Association maintains an up-to-date list of emerging nuclear energy states. https://world-nuclear.org/information-library/country-profiles/others/emerging-nuclear-energy-countries.aspx
4. David Bodanksy, 'The Motivation for Nuclear Energy', in *Nuclear Energy: Principles, Practices, and Prospects* (New York: Springer, 2005).
5. Joseph F. Pilat, *Nuclear Latency and Hedging* (Washington, DC: Woodrow Wilson Center, 2019).
6. Ibid.; Rachel E. Whitlark and Rupal N. Mehta, 'Hedging Our Bets: Why Does Nuclear Latency Matter', *The Washington Quarterly*, 42/1 (2019), 41–52; Ariel E. Levite, 'Never Say Never Again: Nuclear Reversal Revisited', *International Security*, 27/4 (2003), 59–88.

7. World Nuclear Association, 'Emerging Nuclear Energy Countries', May 2022.
8. Ibid.
9. World Nuclear Association, 'Asia's Nuclear Energy Growth', December 2021; Global Nexus Initiative, 'Evolving Nuclear Governance for a New Era', Global Nexus Initiative Working Group Workshop, Washington, DC, 2017.
10. World Nuclear Association, 'Emerging Nuclear Countries'; Global Nexus Initiative, 'Evolving Nuclear Governance for a New Era'.
11. Kristian Coates Ulrichsen, 'The Geopolitics of Insecurity in the Horn of Africa and the Arabian Peninsula', *Middle East Policy*, XVIII/2 (2011), 120–35.
12. Justin V. Hastings, 'Consequences of the Nuclear Renaissance for Nonstate Nuclear Trafficking', in Adam N. Stulberg and Matthew Fuhrmann, eds., *The Nuclear Renaissance and International Security* (Stanford, CA: Stanford University Press, 2013), 223–49.
13. Laura El-Katiri, 'The GCC and the Nuclear Question', Oxford Institute for Energy Studies, December 2012.
14. Treaty on the Non-Proliferation of Nuclear Weapons, 729 UNTS 161 (opened for signature on 1 July 1968; entered into force on 5 March 1970).
15. National Nuclear Security Administration, '123 Agreements for Peaceful Cooperation', 2019.
16. Laura Holgate, 'Virtuous Circles: Linking Business and Nuclear Security', Paper presented at the Norwegian Nobel Institute High-Level Panel on Nuclear Security, 8–10 June 2017.
17. Nuclear Security Governance Experts Group, 'Improving Nuclear Security Regime Cohesion: Summary Report & Initial Policy Recommendations', The Stanley Foundation, 2012.
18. Mark K. Snell and Calvin D. Jaeger, *Incorporating Security-by-Design in both Planned and Operational Nuclear Facilities*, SAND2014- XXXXC (Albuquerque, NM: Sandia National Laboratories, 2014).
19. Christopher Hobbs and Matthew Moran, 'Exploring the Human Dimension of Nuclear Security: the History, Theory, and Practice of Security Culture', *Nonproliferation Review* [online], 9 June 2022. https://doi.org/10.1080/10736700.2020.1811532.
20. Elisabeth Dyck, 'IAEA Marks a Decade of Integrated Support to Nuclear Newcomers', IAEA Press Centre [online], 19 September 2018. https://www.iaea.org/newscenter/news/iaea-marks-a-decade-of-integrated-support-to-nuclear-newcomers
21. IAEA, *Milestones in the Development of a National Infrastructure for Nuclear Power*, IAEA Nuclear Energy Series, No. NG-G-3.1 (Rev. 1) (Vienna: IAEA, 2015).
22. IAEA, 'Milestones Approach', IAEA [online], n.d. https://www.iaea.org/topics/infrastructure-development/milestones-approach
23. IAEA, *Objective and Essential Elements of a State's Nuclear Security Regime*, Fundamentals, IAEA Nuclear Security Series, No. 20 (Vienna: IAEA, 2013), 4.
24. Matthew Bunn, Laura Holgate, Dmitry Kovchegin, Nickolas Roth, and William H. Tobey, 'IAEA Nuclear Security Recommendations (INFCIRC/225): The Next Generation', *Stimson Center* [online], July 2020. https://www.stimson.org/2020/iaea-nuclear-security-recommendations-infcirc-225-the-next-generation/
25. IAEA, *Objective and Essential Elements*, Foreword.
26. Ibid.
27. IAEA, *Objective and Essential Elements*, 4.
28. IAEA, 'Integrated Nuclear Security Support Plan (INSSP)', IAEA [online] n.d. https://www.iaea.org/topics/integrated-nuclear-security-support-plan-inssp

29. IAEA, 'IAEA Completes Nuclear Security Advisory Mission in Finland', *IAEA Press Service* [online], 17 June 2022. https://www.iaea.org/newscenter/pressreleases/iaea-completes-nuclear-security-advisory-mission-in-finland
30. IAEA, 'Nuclear Security E-Learning'.
31. US Department of State, 'Office of Cooperative Threat Reduction', n.d.
32. IAEA, *Objective and Essential Elements*, 5.
33. IAEA, *Responsibilities and Functions of a Nuclear Energy Programme Implementing Organization*, IAEA Nuclear Energy Series, No. NG-T-3.6 (Rev. 1) (Vienna: IAEA, 2019).
34. Ibid.
35. IAEA, 'Governing New Nuclear Programmes: Newcomer Success Stories: Belarus', IAEA [webinar], 11 November 2021. https://www.iaea.org/newcomers-success-stories-webinars/3-belarus-experience
36. In Spain, Basque groups attacked the Lemóniz NPP in the 1970s and 1980s; in India, protesters prevented construction work on the Kudankulam NPP in 2011; in South Africa, the Africa National Congress (ANC) supported an attack on the Koeberg NPP in 1982.
37. IAEA, 'Newcomer Success Stories: Belarus'.
38. IAEA, *National Nuclear Security Threat Assessment, Design Basis Threat and Representative Threat Statements*, Implementing Guide. IAEA Nuclear Security Series, No. 10-G (Rev. 1) (Vienna: IAEA, 2021); IAEA, 2010, *Nuclear Security Recommendations on Physical Protection of Nuclear Material and Nuclear Facilities (INFCIRC/225/Revision 5)*, Recommendations, IAEA Nuclear Security Series, No. 13 (Vienna: IAEA, 2010).
39. IAEA, *Objective and Essential Elements*, 5.
40. Charles D. Ferguson and Mark Jansson, *FAS Issue Brief: Regulating Japanese Nuclear Power in the Wake of the Fukushima Daiichi Accident* (Washington, DC: Federation of American Scientists, 2013).
41. Carlton Stoiber, Alec Baer, Norbert Pelzer, and Wolfram Tonhauser, *Handbook on Nuclear Law* (Vienna: IAEA, 2003).
42. IAEA, *Integrated Regulatory Review Service (IRRS)* (Vienna: IAEA, 2022).
43. Christopher Hobbs, 'INSF Country Update: United Kingdom Nuclear Security', Stimson Center [online], 27 October 2021. https://www.stimson.org/2021/insf-country-update-united-kingdom-nuclear-security/
44. IAEA, Risk Informed Approach for Nuclear Security Measures for Nuclear and Other Radioactive Material out of Regulatory Control, Implementing Guide, IAEA Nuclear Security Series, No. 24-G (Vienna: IAEA, 2022).
45. DeMorgan, 'Engie Electrabel vraagt bijkomend onderzoek naar de sabotage van Doel 4', *Gazet van Antwerpen* [online], 24 December 2016. https://www.demorgen.be/nieuws/engie-electrabel-vraagt-bijkomend-onderzoek-naar-de-sabotage-van-doel-4~b571fdbe/?referrer=https%3A%2F%2Fwww.google.com%2F
46. Paul Day, 'Finding a Workforce May Be Nuclear's Largest Challenge', *Reuters* [online], 3 October 2022. https://www.reuters.com/business/energy/finding-workforce-may-be-nuclears-largest-challenge-2022-10-03/
47. Vitaly Fedchenko, 'Nuclear Forensics as Nuclear Material Analysis for Security Purposes', Non-Proliferation Papers, 60/June (2017), 1–16.
48. For example, see Angel Rabasa, et al., *The Lessons of Mumbai*, Occasional Paper (Santa Monica, CA: RAND Corporation, 2009).
49. IAEA, *Objective and Essential Elements*, 7.

50. Government of Malaysia, 'National Progress Report', Nuclear Security Summit, 31 March 2016.
51. IAEA, 'The Gate to Africa Exercise Programme', Non-serial Publications, STI/PUB/1886 (Vienna: IAEA, 2020).
52. IAEA, *Nuclear Security Recommendations on Physical Protection of Nuclear Material and Nuclear Facilities (INFCIRC/225/Rev.5)*, IAEA Nuclear Security Series, No. 13. (Vienna: IAEA, 2011).
53. US Government Accountability Office (GAO), 'Nuclear Nonproliferation: Past U.S. Involvement Improved Russian Nuclear Material Security, but Little is Known about Current Conditions', Report to the Committee on Armed Services, US Senate, February 2020.
54. IAEA, *Nuclear Security Culture*, Implementing Guide, IAEA Nuclear Security Series, No. 7 (Vienna: IAEA, 2008), 3.
55. Eric Schlosser, 'Break-In at Y-12', *The New Yorker*, 1 March 2015.
56. World Association of Nuclear Operators (WANO), 'Nuclear Industry New Build/New Entrants', White Paper, December 2019.
57. TRIGA Mark-II Reaktörü, 'TRIGA Mark II Araştırma ve Eğitim Reaktörü', n.d.
58. World Nuclear Association, 'Nuclear Power in Turkey', World Nuclear Association [online], updated July 2022. https://world-nuclear.org/information-library/country-profiles/countries-t-z/turkey.aspx
59. World Nuclear News, 'Akkuyu 2's Reactor Vessel Is Installed', World Nuclear News [online], 22 September 2022. https://www.world-nuclear-news.org/Articles/Akkuyu-2%E2%80%99s-reactor-vessel-installed
60. GCR Staff, 'Russia, Turkey Launch Third Reactor at World's Biggest Nuclear Power Site', *Global Construction Review* [online], 12 March 2021. https://www.globalconstructionreview.com/russia-turkey-launch-third-reactor-worlds-biggest/
61. Murat Yetkin, 'Japonlarla Sinop nükleer santral projesi durdu: Erdoğan açıkladı', *Yetkin Report* [online], 27 June 2019. https://yetkinreport.com/2019/06/27/japonlarla-sinop-nukleer-santral-projesi-durdu-erdogan-acikladi
62. Nükleer Düzenleme Kurumu, 'NDK Hakkında', 2022.
63. Pelin Cengiz, 'Erdoğan Becomes Sole Authority over Turkey's Nuclear Energy', *Ahval News* [online], 16 July 2018. https://ahvalnews.com/nuclear-energy/erdogan-becomes-sole-authority-over-turkeys-nuclear-energy
64. Nükleer Düzenleme Kurumu, Teknik Yönetmelikler [Technical Regulations], n.d.
65. Government of Turkey, 'National Report', The 7th Review Meeting of the Convention on Nuclear Safety (Vienna, IAEA, 2016).
66. Government of Turkey, *The Regulation on the Physical Protection of Nuclear Facilities and Nuclear Material*, Official Gazette No. 28300, 2012.
67. Emin Daşkın, 'Justification of Violence by Terrorist Organisations: Comparing ISIS and PKK', *Journal of Intelligence and Terrorism Studies*, 1 (2016).
68. Ibid.
69. 'Nükleer Tesislerin ve Nükleer Maddelerin Emniyetine İlişkin Yönetmelik', *Resmi Gazete*, Sayı 31,207, 2020.
70. Z. S. Homan and S. Udum, 'The Influence of Language on Nuclear Security Education and Training', *Institute of Nuclear Materials Management Annual Meeting Proceedings*, 60 (2019), 1–11.
71. Jandarma Genel Komutanlığı, 'Jandarma Sorumluluk Alanları ve Korunan Tesisler', 2019; Jandarma Genel Komutanlığı, 'Faaliyet Raporu', 2020, 32.

72. Akkuyu Nükleer, 'Türkiye'nin İstemediği İçeri Adım Atamayacak', n.d.
73. Akkuyu Nüclear, 'The Turkish Specialists Participated in the Russian Seminar involving IAEA', 2012.
74. Sezgin Pancar, 'Akkuyu Nükleer Güç Santralinde "fırtına" tatbikatı', *Anadolu Ajansı* [online], 26 October 2019. https://www.aa.com.tr/tr/turkiye/akkuyu-nukleer-guc-santralinde-firtina-tatbikati/1626812#
75. Akkuyu Nükleer/Star Newspaper, 'Akkuyu Uçak Çarpmasına ve 8 Şiddetinde Depreme Dayanıklı (Akkuyu is Resilient to Plane Crash and a 8 magnitude earthquake)', 2020. http://akkunpp.com/akkuyu-ucak-carpmasina-ve-8-siddetinde-depreme-dayanikli
76. Yevgeny Kuklychev, 'Do Videos Show Tsunami, Nuclear Plant Explosion in Turkey After Earthquake?', *Newsweek* [online], 7 February 2023. https://www.newsweek.com/do-videos-show-tsunami-nuclear-plant-explosion-turkey-after-earthquake-1779476; Thomas Johnson, 'Turkey earthquake: Experts believe collapse of buildings was preventable', *New Civil Engineer* [online], 15 February 2013. https://www.newcivilengineer.com/latest/turkey-earthquake-experts-believe-collapse-of-buildings-was-preventable-15-02-2023/
77. ENEC, 'Barakah Nuclear Energy Plant', 2022.
78. Laura El-Katiri, *The GCC and the Nuclear Question* (Oxford: Oxford Institute for Energy Studies, 2012).
79. UAE Government, 'Policy of the United Arab Emirates on the Evaluation and Potential Development of Peaceful Nuclear Energy', April 2008.
80. Charles Ebinger, John Banks, Kevin Massy, and Govinda Avasarala, *Models for Aspirant Civil Nuclear Energy Nations in the Middle East*, Energy Security Initiative, Policy Brief 11-01 (Washington, DC: Brookings Institution, 2011), 6.
81. ENEC, *Report on Lessons Learned from Fukushima-Daiichi Accident To FANR* (Abu Dhabi: Emirates Nuclear Energy Cooperation, 2012).
82. IAEA, *Joint Statement on Strengthening Nuclear Security Implementation*, INFCIRC/869 (Vienna: IAEA, 2014).
83. IAEA, 'IAEA Completes Nuclear Security Review Mission in United Arab Emirates', IAEA Press Centre [online], 10 November 2016. https://www.iaea.org/newscenter/pressreleases/iaea-completes-nuclear-security-review-mission-in-united-arab-emirates
84. Osama Al Shehhi, et al., 'Feedback Experience from the IAEA Missions with Respect on Nuclear Security of Barakah NPP in the United Arab Emirates', presented at the IAEA International Conference on Nuclear Security (ICONS), 10–14 February 2020, Vienna, Austria.
85. FANR, 'Annual Report 2018'.
86. Hamad Al-Kaabi, 'United Arab Emirates and the Experience of a Nuclear Newcomer', *Bulletin of the Atomic Scientists*, 67/4 (2011), 52–9.
87. Technical Review Middle East, 'World's first nuclear language proficiency standards established in the UAE', 15 October 2018.

Suggested Reading

IAEA, (2013), *Objective and Essential Elements of a State's Nuclear Security Regime*, IAEA Nuclear Security Series, No. 20 (Vienna: IAEA).

IAEA, (2015), *Milestones in the Development of a National Infrastructure for Nuclear Power*, IAEA Nuclear Energy Series, No. NG-G-3.1 (Rev. 1) (Vienna: IAEA).

Perkovich, George and Ulgen, Sinan, (2015), *Turkey's Nuclear Future* (Washington, DC: Carnegie Endowment for International Peace).

Pilat, Joseph F., (2019), *Nuclear Latency and Hedging* (Washington, DC: Woodrow Wilson Center).

Stoiber, Carlton, Baer, Alec, Pelzer, Norbert, and Tonhauser, Wolfram, (2003), *Handbook on Nuclear Law* (Vienna: IAEA).

CHAPTER 23

RISK-INFORMED NUCLEAR SECURITY REGULATION IN THE UNITED STATES

GREGORY B. JACZKO

Introduction

The public is keenly aware of the safety and security of nuclear power plants and nuclear materials in the United States (US). This chapter discusses the adoption, evolution, and challenges of these approaches in the US. Two factors drove the evolution of those initial regulations: changes in the external terror threats and the shift from deterministic to risk-informed regulations. After discussing these factors, the chapter closes with a look at the emerging threats of cyber-attacks and an examination of the challenges of developing security regulations in a publicly transparent process.

Framework for Nuclear Security in the United States

The US established a comprehensive legal framework for governing the use of nuclear materials in 1946 with the passage of the Atomic Energy Act (AEA),[1] which established the Atomic Energy Commission (AEC). The AEC developed nuclear weapons and promoted and regulated civilian nuclear power and materials use. Due to safety concerns, Congress passed the Energy Reorganization Act of 1974 to separate weapons development and commercial, promotional responsibilities from the safety and security regulation function of the AEC. This act established the Nuclear Regulatory Commission (NRC), an independent federal agency.

The NRC's authority and structure have change minimally since its establishment in 1974. The five-member, presidentially appointed, senate-confirmed commission establishes policies for the agency through legal orders, regulations, and guidance statements. The commission chair serves as both the presiding member of the commission and the chief executive officer of the agency, a position that has evolved in response to major incidents, such as the Three Mile Island (TMI) accident in 1979. The agency's staff develop, propose, implement, and enforce the policies the commission establishes under the executive direction of the chairman. The jurisdiction of the agency covers all licensing and regulatory aspects of civilian uses of nuclear material derived from uranium and thorium, the seeds for most of the radiation specific activities in the US—which include ore processing, fuel fabrication, isotope use and production, nuclear power generation, and nuclear waste control. Congress extended authority of the agency over accelerator produced radioactive materials in 2005.

While the governing laws for the NRC give the agency exclusive authority to establish policies for all civilian use of nuclear materials, these laws allow individual states to implement and enforce policies for some areas of NRC authority, as long as the state programme is consistent with the standards established by the NRC. Policies covering radioactive materials used in medicine, industry, research, and some aspects of radioactive waste disposal may be implemented by states. However, nuclear security policies may not be implemented by states.

Effective nuclear regulation requires independence by the regulatory authorities. The NRC was created as a separate agency specifically to reduce conflicts between promotional and safety aspects of civilian uses of nuclear materials of the AEA. Congress created the NRC solely to regulate health and safety without regard to national efforts to promote civilian nuclear technology. The laws governing the NRC ensure independence because they give the agency full and exclusive authority to determine health and safety and security regulations. The President and Congress have some influence over the agency's authorized actions through the appointment of the agency commissioners, the designation of one commissioner as chair, and the approval of the agency's annual budget. Nonetheless, the agency possesses broad legal and practical independence.

The NRC's focus has been primarily on the safety of nuclear power plants since the agency's inception. The NRC regulates many areas besides power plants, but these facilities tend to dominate the focus and resources of the agency. Following the terrorist attacks on the US on 11 September 2001 ('9/11'), the emphasis shifted for many years to nuclear security. Nuclear security concerns raised the profile of the NRC's regulation of fuel fabrication and radioactive material use and production.

Although nuclear safety and nuclear security are often treated as and perceived to be separate programmes, in the US nuclear security is a subset of nuclear safety efforts. Nuclear security regulations are primarily differentiated by the cause for potential harm to people or the environment. Nuclear security regulation carves out cases of individuals, groups, or nations who intend damage or misuse civilian nuclear infrastructure and materials. Nuclear safety covers all other types of harms to people and the environment from the civilian use of nuclear materials under the NRC's jurisdiction.

Because the primary distinction for security events from safety events rests with the cause of the incident, all the safety analysis methods, science, regulations, and enforcement tools for nuclear safety apply. For example, there is no distinction within the NRC for a harmful exposure from a radiation source used by a terrorist or misused by a health professional. The health impacts are treated identically. Similarly, a specific nuclear plant failure triggered by a terrorist sabotage of a nuclear plant can be analysed the same way as such failure instigated by a natural disaster or human error. Additionally, the mitigation measures used for nuclear safety failures may be applicable to nuclear security events.

Nuclear security has some regulatory practices, however, that are not employed for nuclear safety programmes, such as testing security protection schemes through mock terrorist assaults on a facility. In addition, some tools for nuclear safety analysis, such as probabilistic risk assessment (PRA), may apply poorly to nuclear security events. But generally, nuclear security events are simply safety incidents with a unique trigger. Furthermore, to ensure nuclear security programmes do not adversely impact overall nuclear safety objectives, the NRC requires assessment of the impacts of nuclear security programmes on nuclear safety efforts.

The approach to nuclear safety and nuclear security is unique to the US. Other countries handle these issues differently. The regulatory system in the US is well positioned to address nuclear security for nuclear power plants, fuel fabrication facilities, and the use and production of radioactive isotopes. The broad legal mandate for nuclear safety and statutory independence of the agency provides the foundation for the agency to establish meaningful nuclear security programmes.

Evolution of Regulatory Approaches: Deterministic to Risk-Informed

Nuclear security regulations in the US are generally seen as a subset of existing nuclear safety approaches. The NRC interprets the statutory safety standard in the AEA as amended in 1957 to be 'reasonable assurance of adequate protection', but the act leaves the detailed implementation of this provision to the agency.[2] The primary basis for regulations in the US has been a deterministic approach, including a defence-in-depth supplement.[3] The deterministic, defence-in-depth framework is implemented as a series of specific engineering requirements that nuclear power plants must meet to obtain and maintain a licence to operate. These standards derive from a postulated accident scenario known as a 'design basis accident' (DBA).

The DBA provides the metric for the analysis of adverse operational conditions at a nuclear power plant. The reactor coolant system must be designed to function within prescribed limits in the event of a loss of coolant accident from the double-ended guillotine break of the largest cooling system pipe.[4] Actual accidents may deviate from this scenario, which occurred with the TMI accident in 1979, the Chernobyl accident in 1986, and the Fukushima accident in 2011.[5]

Initially, the DBA was to be the most challenging scenario for a reactor consistent with the reasonable assurance standard. Safety authorities and industry experts could postulate more extreme scenarios, but the NRC established the double-ended guillotine as the DBA. Following the TMI accident, many criticized the agency's deterministic, defence-in-depth framework as too rigid. The TMI accident occurred due to a failed valve, not a double-ended guillotine break. Operators then misinterpreted the plant instrumentation and responded incorrectly, worsening the situation.[6] The TMI accident showed that radiation releases could occur from plant damage less severe than the DBA. Independent reports following the accident recommended using more sophisticated computational risk tools to provide a better assessment of the safety needs of reactors.[7] This process culminated with the agency's adoption of a policy statement in 1995 on the use of risk-informed methods for nuclear safety.

Risk analysis methods use probabilistic computer simulations to determine negative reactor outcomes based on possible faults in the reactor safety system, plant operation, or external events. A risk analysis approach would utilize these consequence and likelihood metrics to establish a threshold for allowable operation. Risk regulations would replace the rigid engineering standards derived from deterministic, defence-in-depth goals based on the DBA with flexible metrics based on the results of computational risk analysis calculations. In principle, this approach would enhance safety by identifying previously unknown significant accident scenarios.

Proponents of these measures assumed that the DBA distracted from other more plausible scenarios. The goal was to better rank accident scenarios and (as the industry hoped) reduce the reliance on the existing DBA scenario, which the industry believed over-engineered the plants around a single event, unnecessarily increasing capital and operating costs while leaving vulnerabilities to other accident scenarios. Critics, primarily among nuclear safety advocates at non-governmental organizations (NGOs), believe a greater focus on risk decisions reduces safety requirements, and therefore operation costs.[8] Partially to address criticism, the NRC has adopted a risk-informed approach that couples the results of risk analysis with defence-in-depth principles instead of relying solely on the results of computations risk calculations. Proponents of enhanced risk-based safety standards believed the approach would identify unknown accident scenarios that evaded the simplicity of the deterministic DBA.

Challenges of Risk-Informed Regulations

The safety and security framework for controlling nuclear power plants in the US has evolved from a strictly deterministic framework to a mottling of deterministic and risk-informed regulations. These efforts have important implications for the nuclear security requirements. Replacing the long-standing deterministic regulations and standards, which led to the design and operational characteristics of the current operating fleet of nuclear reactors in the US, has been a difficult task. One reason for the challenge is the technical complexity of the existing regulations and the challenge of cost-effectively modifying existing plant structures. The NRC has approached this problem by providing optional risk-informed alternatives to existing deterministic regulations for some parts of its requirements. Replacing the double-ended guillotine DBA with a comprehensive risk-based approach to an accident has not happened. In the area of fire protection, for example, the industry and the agency developed an alternative standard that is risk based. Adoption, though voluntary, proved difficult because some plants required expensive upgrades. The industry expected the risk-informed rules would simplify fire protection and reduce costs.[9]

Additionally, adopting risk-informed regulations requires each plant to develop and maintain a sophisticated computer model of the plant. Owners have developed risk models with varying fidelity to the plant's design and performance. Each successive replacement of deterministic regulations with risk-informed regulations requires expensive computer model development. A complete risk analysis would include emergency response and detailed radiation release modelling. In practice, most models only simulate core melt. The corresponding regulatory limits consider the frequency of core melt as a proxy for public radiation exposure. This is generally conservative, but also limited.

Finally, natural phenomena and some equipment failures have well-established occurrence frequencies, which can be easily modelled in PRAs. However, many plant accident precursors do not. Human errors, which are crucial elements in almost all accidents, are difficult to model. Previously unknown material deficiencies can be difficult to postulate. In the US, for example, an undetected reactor pressure vessel leak at the Davis–Besse Nuclear Power Station in 2001 almost led to a full breach of the reactor vessel.[10] Assigning a probability of occurrence to this problem is difficult because the progression relied on the falsification of inspection records. For many plants, adopting a new risk-informed regulation hinged on the potential for operational enhancements or design changes that would reduce expenses. While risk-informed approaches may often enhance safety, the adoption decision rests on economic feasibility.

This model for nuclear safety applies to nuclear security requirements too with a few notable exceptions. Instead of a DBA to guide the plant engineering, security regulations rely on a design basis threat (DBT). Given the sensitivity of the security activities, most of the implementation of this DBT is protected as a form of classified information by the NRC, another deviation from traditional safety regulation. In addition, the nuclear security regulations adopt a higher threshold for compliance. Instead of a 'reasonable assurance' of safety, security requirements must meet a 'high assurance' of protection. Finally, developing the computer models and theoretical underpinnings for security risk analysis is challenging. Most, but not all, accident risk analyses have a well-known initiation event with good confidence in occurrence frequencies. Security events, however, may involve intentional actions for which experts struggle to agree on the likelihood of taking place. As discussed, security events are a unique set of initiating events for reactor accidents, yet these instigation actions are not easily calculated as risk elements.

Although the debate over regulations being deterministic or risk-informed has primarily focused on the nuclear power plant responsibilities of the NRC, efforts exist in other parts of the agency's jurisdiction. But those efforts are less sophisticated and have less-advanced computation risk tools than even the nuclear power plants sector. The control of radiation sources utilized risk measures to address security concerns following the 9/11 attacks. These tools, however, are much less sophisticated than the tools used for nuclear plants. Notably, the DBT concept has become the global standard for nuclear security threats.

Basic Elements of Commercial Nuclear Security Regulations

The primary security regulations for nuclear security exist in 10 CFR Part 73, 'Physical Protection of Plants and Materials'.[11] These regulations, which were adopted in their modern form in 1979,[12] establish the basic deterministic, performance standards for physical security of facilities and transportation of special number material 'which will have capabilities for the protection of special nuclear material at fixed sites and in transit and of plants in which special nuclear material is used'.[13] The primary nuclear security requirement is the 'high assurance' of protection against the security DBT.

This approach parallels the nuclear safety framework by establishing a postulated threat (accident in the case of safety) that aids the facilities in designing their physical security protection system and strategies. Two elements differentiate the security approach from the safety approach. First, security programmes must protect against the DBT with high assurance. This phrase is unique to the security programmes and differs from the 'reasonable assurance of adequate protection' standard generally accepted as the norm for safety regulations. The standard existed in the original formulation of the modern security regulations in 1979. Second, the security regulations, while deterministic in nature, primarily are stated through a broad performance standard. Instead of specific engineering parameters and security strategy parameters, the security performance is measured through 'high assurance' of protection against the DBTs (10 CFR 73.20).[14] Since their original formulation, the security regulations have adopted additional specific, deterministic requirements, but the primary protection requirement is adherence to the DBT.

The security regulations initially derived from concerns about theft of enriched uranium or other weapons useable material—'special nuclear material' in the NRC's governing laws. This issue manifests in one of the two DBTs, the DBT for theft and diversion. This hazard postulates adversaries attempting to steal strategic special nuclear material, which is special nuclear material of higher enrichment. Initially, this threat was the dominant concern. Only a few facilities regulated by the NRC are covered by this threat, because only a few facilities maintain accessible sources of strategic special nuclear material. They are the fuel fabrication facilities that produce fuel for nuclear fleets and, to a lesser degree, research reactors. The other (now more attention-grabbing) DBT is the threat against radiological sabotage, which postulates adversaries attempting to damage a nuclear facility to create a release of radiation.

Although both DBT threats could apply to both—enriched uranium fuel fabrication and power production facilities—the threats have become uniquely identified with only one class of facility. Since reactors deal with low enriched uranium fuel, they do not generally protect against the theft and diversion DBT. While nuclear power plants do contain strategic special nuclear material in spent reactor fuel, this material is considered too hazardous to allow successful theft. Any adversary attempting to steal highly radioactive spent nuclear fuel would be unable to do so because of their inability to maintain shielding; therefore, power plants do not protect against the theft and diversion scenario. Fuel fabrication facilities are not subject to the radiological sabotage scenario because they do not possess enough nuclear material to pose a significant health and safety hazard through sabotage of the facility. Unirradiated uranium fuel is low activity and presents minimal radiological health hazards.

Inspections and enforcement of security regulations follow the same pattern as safety regulations with one unique difference. Adherence to the prescriptive, non-performance standards in the security requirements is determined through inspections and enforcement in the agency reactor oversight process just like safety requirements. The performance standard of DBT compliance, however, is assessed through a series of mock adversary attacks. This programme, known as the force-on-force programme, is unique to security requirements. The force-on-force programme involves mock intruders attempting to infiltrate the facility and achieve goals or reach areas of the plant which, if successful, would lead to a radiation release or the theft of strategic special nuclear material. These exercises are performed by a dedicated group of independent security personnel acting as the adversary. Both the plant and adversary use laser equipped weapons to simulate ammunition.

Finally, since the commercial nuclear industry in the US was originally part of a broader nuclear weapons programme, information security has always been a crucial element of all aspects of the operational and regulatory framework. The nuclear security system respects this legacy with strict control of information—another unique characteristic of the security approach in the US. The details of the DBT, the results of performance and standard inspections, the operational security plans, and regulatory efforts are often withheld from public to protect sensitivity security strategies or vulnerabilities. Controversy surrounds this approach.

Evolution of Security Requirements for Power Plants Following 9/11

The DBT is paramount within the US nuclear security approach. Changes in the security environment in the US and globally have challenged the adequacy of the existing

DBT. Following major security events like 9/11, the public and many policy makers expect adjustments to the DBT, and, in some cases, the agency has agreed. While written generically in public regulations, the DBT relies on specific details of the postulated adversary in protected guidance documents, the so-called adversary characteristic description (ACD). Following 9/11, many suspected elements of the DBT were challenged by nuclear safety NGOs. In particular, the lack of specific protections against intentional missile attacks designed to cause radiological releases stand out as an obvious omission. The period after the 9/11 attacks was an active time for reform and modernization of the DBT. Much of those actions stemmed from the agency's own actions and much resulted from Congressional requirements.

Following the terrorist attacks on New York City and Washington, DC on 11 September 2001, the government was concerned about attacks on energy infrastructure. Nuclear power plants attracted public and government attention. About five months after the attack, in his 2002 State of the Union address then–president George W. Bush tied al-Qaeda to nuclear plant threats when he said that US forces 'found diagrams of American nuclear power plants'.[15]

During this time, the NRC worked to understand the effects that aircraft impacts could have on nuclear power plants. Although not specifically an element of the radiological sabotage DBT, the agency took this threat seriously. The agency issued a major, mostly classified, security order (a fast, mandatory requirement process that bypasses lengthy public regulation development) that imposed many new security restrictions on nuclear power plant owners.[16] The main element of this programme, called 'b5b' for its location within the February 2020 order, was eventually released to the public and required plans for, procurement of, and maintenance of equipment to deal with large fires and explosions in the plant.[17]

The NRC also issued orders to facilities subject to a DBT to update the characteristics of the DBT in 2003. The basic elements of this order are also classified, but both the b5b provision and the generic aspects of the DBT orders became public regulations several year later. Despite the agency acting to address an element of the terrorist attacks through the interim compensatory measures order and the DBT order, the confidential nature of the content made security concerns a persistent public issue for the nuclear industry.

Aircraft impacts continued to be a concern, especially since the nuclear industry began to consider new nuclear power plant construction in the mid-2000s. These plans coincided with the public debate over the commission's response to the 9/11 attacks. The aircraft impact attacks were particularly controversial. Following confidential risk assessments, the NRC declared that the 'likelihood of both damaging the reactor core and releasing radioactivity that could affect public health and safety is low'.[18] This statement left open the possibility that, while possible, catastrophic impacts from directed aircraft strikes were unlikely. More importantly, the NRC had no specific requirements to protect against aircraft impacts for intentional purposes. Some facilities that were in the proximity of airports considered but never fully analysed these threats because the agency determined the likelihood of occurrence was too low.[19]

After several failed efforts, the agency required new plant designs to address aircraft impacts. This regulation led to modifications to or confirmation that all the designs under consideration in the US at that time would withstand an intentional aircraft impact. The detailed characteristics of the aircraft impact, including the aircraft type, velocity, fuel, and mass remain confidential. Unlike most security regulations, this regulation was included in the provision of the NRC's regulations covering safety design criteria, because the commission did not believe this feature merited the designation of a DBT or DBA. The proscription, while robust, did not rise to the level of a design basis characterization. The only two nuclear plants under construction at the time of writing comply with this regulation.

The new requirement is a hybrid of risk-informed and deterministic regulations. The rule does not prescribe specific engineering features to address aircraft impact. Similar to risk-informed regulations, it requires a high-level performance standard confirmed through computer simulation. The compliance threshold, however, is not risk based. Protection for existing reactors focused on mitigation of the effects of an aircraft impact. The interim compensatory orders issued in the years immediately after the 9/11 attacks required modifications to address this hazard. The most important element of the order is the b5b provision. Similar to the aircraft impact requirement, the b5b provision was codified in the safety element of the agency's regulations and relied on a high-level performance standard.

Nonetheless, the agency did make significant modifications to the security elements of its regulations following the 9/11 attacks. The primary change involved adoption of a new DBT definition. Just like the interim compensatory orders, the commission ordered changes to the DBT in the aftermath of the 9/11 attacks to provide an interim change to the DBT. In subsequent years, the NRC adopted and modified these provisions in new regulations. Although some public interest groups requested inclusion of an aircraft attack in the DBT, the agency addressed that issue in another manner.[20] A major driving force for the development of the new DBT regulation was the Congressional requirement to review 12 specific factors in evaluation of the new rule.[21] These factors addressed the number and capabilities of the 9/11 hijackers, and specific attack modes such as aircraft impacts, employee sabotage, and response planning.

While the NRC did not acknowledge the specific elements for the postulated adversary adopted after review of the Congressional requirements, public interest groups believed some of these known capabilities of the 9/11 attackers and other terrorists were not included in the protected ACD details that accompany the public DBT regulations.[22] Despite the controversy surrounding the adoption of the new DBT, the new standard upgraded the protection requirements in general terms.

The final evolution of the security approach to reactors is to utilize the risk analysis methods that have become more prevalent for specific nuclear safety elements, replacing some of the deterministic standards with a voluntary risk-informed approach.[23] In the approach usually considered by the NRC for such applications, the risk-informed approach to security would involve developing advanced probabilistic models for assessing the triplet of: What could go wrong? How likely is it? What are the consequences?[24] Nonetheless, risk-informed approaches in the US suffer from two fundamental weaknesses: plants are not required to have nor maintain full PRA models; and models used to assess security threats may be unrealistic because the determination of event frequencies is difficult—if not impossible—for security-initiated events.

Solving the first issue is a question of resources and requirements. The NRC has never required power plants to maintain and utilize PRAs, but it could. The NRC requires PRAs with minimal capabilities if a plant chooses to comply with a regulation using PRA methods. As a result, PRA models to determine core damage frequency have been developed for many plants for a limited set of initiating events, but full PRAs that include off-site consequences exist for only a few US nuclear power plants. There is in principle no obstacle to developing these models, except resources. Due to the high cost, the nuclear industry in the US has resisted efforts by the NRC to develop them.

The second issue continues to be a major challenge for the use of risk tools in all security events. The crucial difference between these types of initiating events and natural hazard events is that the frequency of occurrence is dependent on factors that may not be subject to some type of probabilistic distribution, namely, the interest and motive of the attacker. For most other events, failure probabilities or occurrence frequencies can be determined at least in a relative manner to provide a basis for comparison of impacts under a risk assessment approach.

In the years following the 9/11 terrorist attacks, the NRC made considerable changes to the security framework. These requirements relied on generic performance standards instead of proscriptive detailed engineering requirements. Some details, such as the size of security forces and passive defence systems, were delineated in regulations. Despite the general effort to increase the use of risk-informed regulations, the new security modifications primarily eschewed risk-informed requirements.

Strengthening Nuclear Security Beyond Power Plants

Just as the 9/11 terrorist attacks increased attention on the security of nuclear facilities, the incident sparked interest in the security of radioactive sources used for medical, industrial, and research applications. These materials are primarily regulated by the NRC, and while some aspects of these regulations are implemented by individual states, security regulations are exclusively implemented at the federal level. After the 9/11 terrorist attacks, the threats identified for radioactive sources involved the use of radioactive materials in a dispersion device, commonly referred to as a 'dirty bomb', and the use of radioactive materials in an exposure device. Dirty bombs were the most significant concern identified by the US.

Since radioactive sources are used throughout the world in medical, industrial, and research applications, the International Atomic Energy Agency (IAEA) convened a series of meetings following the 9/11 attacks to review and revise guidance for the control of radioactive sources. Those efforts produced an updated version of the IAEA Code of Conduct on the Safety and Security of Radioactive Sources. The new code identified five risk categories for sources; the top three categories represent the riskiest sources. Member states were expected, but not obligated, to develop or enhance security practices for the manufacture, use, and distribution of these sources commensurate with the risk. The US used the Code of Conduct to update its regulations for the control of sources.

Following the adoption of the Code of Conduct, which the NRC helped develop,[25] the NRC imposed two methods for implementing the code. The NRC ordered organizations that may possess Category I and II sources to adopt stricter access and monitoring controls on their facilities. The NRC also began developing a national source inventory database called the National Source Tracking System (NSTS). While not a real-time source tracking system, the NSTS provides ownership and location information for radioactive sources from the time of manufacture to the time of disposal. Since the NRC determined only Category I and II sources were risk significant enough to warrant orders, the NSTS only applied to those sources. Disagreements arose, however, over the risk significance of Category III sources.

Since the risk categorization of sources is based on activity level, aggregation of sufficient Category III sources could produce Category II quantities of material. Because Category III sources fall outside the security infrastructure of both the control orders and NSTS, some groups raised concern that a security vulnerability existed.[26] This concern was exacerbated when the Government Accountability Office (GAO), an oversight agency of the US Congress, conducted a clandestine investigation to obtain licences for Category III sources using fictious names.[27] The GAO was able to easily obtain licences for these sources. This exposed a weakness in the safety licensing of radioactive sources that could be exploited for a security threat.

Most critics of the restriction of the NSTS to Category I and II sources believed Category III sources should also be tracked through the NSTS. That way, if sources were sold to individuals with malevolent intent there would be an early identification. The NRC in the final requirements for the NSTS did not expand the NSTS to cover these Category III sources. The orders and subsequent regulations adopted for security sources and NSTS provided strong US support for the Code of Conduct, but they are limited in their effectiveness. The Code of Conduct ranked sources primarily based on activity level, but additional factors are significant in assessing the risk of radioactive sources.

Dirty bombs are the most prevalent concern for misuse of sources, but not all sources are easily used in dispersive devices. The US government investigated potential radionuclides to determine which sources could create the greatest impact if used in a dirty bomb. This approach provided a better assessment of risk than just a measure of activity level. Nonetheless, the NRC established regulations for source security based strictly around the rough risk categorization of the Code of Conduct.

Securing sources is more challenging than power plants. Unlike nuclear power plants—which are isolated, contained facilities—radiological sources are used throughout the world in diverse settings. Some places can be secured but not highly protected or they fail to meet their primary mission. A more complex risk assessment of sources would be cost prohibitive for most of these facilities and beyond their capability. A hospital may employ a blood irradiator but does not have full-time security staff or radiological security analysts to develop detailed risk methods. Thus, source security has relied primarily on the Code of Conduct.

Given these constraints and a recognition that some sources present significant risk based on confidential security studies, the US government and the NRC also focused attention on replacing commonly used sources. The government considered efforts to replace caesium-137 blood irradiators with alternative inactivation methods. These efforts are likely to provide the greatest reduction in risk by simply eliminating the threat. As of 2018, the US government has facilitated the replacement of 43 caesium-137 irradiators.[28] The Code of Conduct greatly enhanced the risk-informed nature of approaches to controlling radiological sources. The volume of such sources—the global distribution of them—means this will continue to be a security challenge, but significant improvements have been made.

Emerging Threats from Cyber-Attacks

The focus of security efforts at 9/11 primarily dealt with physical assaults and aircraft impacts. Although some nuclear safety stakeholders challenged the adequacy of the efforts, significant changes to physical security and plant design and operation were made to address these new threats. As the public focus on physical security and aircraft impacts wanes, a new class of threats is emerging: cyber threats against digital instrumentation and control systems. Simultaneously, nuclear plants in the US were evolving their control systems from analogue to digital systems. Because of their previous reliance on analogue control systems, most US nuclear power plants appear resistant to these evolving cyber threats. The move from analogue to digital control systems means this assumption is no longer valid. Unlike other security threats to nuclear power plants that can be dealt with almost exclusively through security measures at the plant, cyber-attacks can occur remotely and not even impact the nuclear power plant directly.

At the plant, cyber protection focuses on isolating systems from outside connections. This addresses direct efforts to initiate a nuclear accident through electronic sabotage of reactor operations. A broader class of threats, however, affects the entire bulk power systems. US plants are reliant on external power for normal operations. Loss of the grid due to cyber-attacks could—in isolation or as part of a physical assault—lead to a radiation release. This concern is especially challenging because the vulnerability extends beyond the plant's physical and managerial control. It is a challenge beyond the scope of the nuclear power sector.

Congress mandated broader protection of the electric power grid in 2005. It authorized the Federal Energy Regulatory Commission (FERC) to address cyber threats affecting the US power grid. Because the NRC has exclusive authority over nuclear power plant security, the FERC and NRC entered into an agreement to work cooperatively to address cyber threats to nuclear power plants initiated by attacks against vulnerable infrastructure within the national power grid.[29]

Adopting a risk framework for examining these issues is complex. Since the NRC's legal authority covers protection of the public and environment against nuclear radiation exposures, damage to the bulk power system lies outside the agency's jurisdiction. Operation issues only matter if they relate to safety of the power systems. For cyber risks, this narrow focus creates potential gaps. With the potential for new digital control systems increasing the risk of cyber intrusion at a nuclear power plant, they and the entire electricity grid will continue to be an important risk for public safety.

Limitations and Challenges of US Regulatory Programmes

The 9/11 terrorist attacks instigated massive changes to the US security programme for nuclear power plants and radioactive sources. The international community reacted and adopted aggressive risk categories for radioactive sources. Despite this significant progress, the modifications revealed several inherent challenges with security programmes for nuclear infrastructure. With an ever-increasing threat, nuclear power plants could be asked to adopt even more aggressive security measures. Following 9/11, some groups called for greater use of military technology, for instance. This raises significant societal questions about how much capability private enterprise should adopt.

In fact, the NRC advocated for changes to federal legislation to allow certain weapons to be used at nuclear power plants because state law barred those weapons.[30] In addition, some safety proponents called for the installation of anti-aircraft weapons at nuclear power plant sites.[31] The NRC periodically reviews the adversary characteristics of the DBT to ensure plants are protected against modern threats. Extensive public debate occurred around the tactics assumed by the adversary with some proponents asking the NRC to determine if adversaries could be armed with military weapons, for example, rocket-propelled grenades. The details of the DBT are protected information but, as the debate shows, as the adversary advances, the plants may need greater defensive or offensive capabilities.

There is a limit, however, to the expectations for power plants. The security regulations contain an exemption for security threats if they are deemed to arise from an 'enemy of the state'. Prior to 9/11, the term was generally understood to apply to attacks by nation states causing significant national harm. The 9/11 attacks demonstrated that non-nation states or terrorist groups, including those from within the state itself, could inflict significant national harm. Public and Congressional sentiment at that time, most clearly articulated in the Energy Policy Act amendments, dictated that some measures were necessary to address the new threat. The NRC could have justified no additional security measures by appealing to the enemy of the state exemption. This course of action would have been controversial but possible.

A simple terrorist adversary attempting to disrupt a single plant or a lone hacker disabling significant portions of the electrical grid could cause havoc. Exempting these threats from the DBT as an enemy of the state is not correct. Yet at some point, the counter measures against such threats exceed the capability of private companies in the US. That is the *de facto* definition of the enemy of the state exemption: actions by nuclear faculties that require government security capability. If the power plants or the electricity sector can use available tools to combat these threats, they will not likely fall under the enemy of the state exemption. But the debate over the limits of private sector response will continue.

In addition to the uncertainty over the level of security a private sector entity should provide, security also challenges the expectations for government transparency common within the US nuclear regulatory system. While the DBT is a public regulation, the details and elements of the adversary characteristic are restricted information. In addition, the security response elements are restricted. This challenges the government's ability to communicate security oversight and licensees' performance. Following the 9/11 attacks, the US government internally analysed the risks to nuclear power plants from the attacks. The details of the resulting NRC security orders were classified. Yet the public had strong interest in knowing what actions the NRC and other agencies were taking to address the attacks. This tension exists and will exist any time there is a significant change in the security threats facing nuclear power plants.

Information control also affects plant operation. The NRC and the government need to communicate sensitive information about threats to nuclear power plant licensees. However, only cleared individuals within the organizations can receive the most precise intelligence. Some information must be restricted to a subset of cleared individuals because of the sensitivity of the intelligence. This creates challenges within the companies operating the reactors, because they may need to act on information provided by the government but cannot share the rationale to all employees. This is contrary to many of the best safety practices encouraged for nuclear power plants.

This issue was heavily debated around the publication of results of force-on-force exercises after 9/11. Providing information about deficiencies in plant security is an important part of the oversight process. Unlike strictly safety-related errors, however, disclosing security vulnerabilities risks providing information to potential adversaries. The NRC struggled with the balance between assuring the public of its oversight and revealing information that could increase the likelihood of a successful attack.

For those security areas in which some element of risk analysis is possible, transparency concerns are exacerbated. Risk analysis methods suffer from transparency problems even for non-security analysis. Decisions made based on the results of complex computer models lack openness. Furthermore, even if they are accessible, the details of the models are understood only by the individuals who possess detailed knowledge of their construction and use. This challenge is further exacerbated when coupled with the information protection requirements of nuclear security. This is one reason why nuclear security risk analysis will continue to be a challenge. Just like interpreting the definition of enemy of the state, determining the appropriate level of transparency with security regulation is a persistent challenge.

Conclusion

The nuclear safety and security framework in the US has evolved significantly since the original adoption of the DBT in 1979. The DBT has become the global standard for nuclear security threat identification. Yet, in the US the details of its implementation continue to evolve and generate controversy. Following the 9/11 terrorist attacks, the NRC, partially in response to Congressional pressure, launched a major effort to update the security approaches for nuclear power plants and other regulated nuclear material. Reform of nuclear source security also followed from growing concern around dirty bombs. Many NGOs believed the NRC's updated DBT insufficiently reflected the capabilities of modern terrorist groups and the 9/11 attacker. The protection of sensitive security information contained in the DBT further complicates resolving these controversies. The NRC and other government agencies will continue to struggle with balancing the need to protect security information and the obligation to provide the public with information about its activities.

As the regulations for nuclear security changed, the analysis methods to address security risks also evolved. In particular, nuclear security methods attempted to follow broader trends in the nuclear safety arena of greater adoption of risk tools, such as probabilistic risk analysis. These approaches have unique challenges when applied to nuclear security, because security effects are primarily characterized by the initiation of the incident. Intentional acts of terror are more difficult to quantify in the context of a PRA than an earthquake, for example. Regulators must cautiously evaluate the results of risk tools in the security context, because of this inherent challenge for security. The ability to calculate a risk number does not guarantee the number itself has inherent value.

Security concerns will continue to be a major factor affecting the regulation, use, and operation of nuclear power plants and nuclear materials. Regulators, operators, and users of nuclear materials must continue to adapt to new information, inform the public of their actions, and resist the temptation to rely on potentially misleading information from risk tools. However, the pressure to address these concerns will continue to test their resolve.

Notes

1. George T. Mazuzan and J. Samuel Walker, *Controlling the Atom: The Beginnings of Nuclear Regulation, 1946–1962* (Berkeley, CA: University of California Press, 1984), 3.
2. US Nuclear Regulatory Commission, *2021–2022 Information Digest* (Washington, DC: US Nuclear Regulatory Commission, 2021), 3.

3. US Nuclear Regulatory Commission (NRC), 'Use of Probabilistic Risk Assessment Methods in Nuclear Regulatory Activities', *Federal Register*, 60/158 (1995), 42,622.
4. US Nuclear Regulatory Commission (NRC), Code of Federal Regulations, 2021, Title 10, 50.46(c).
5. For a summary, see Gregory Jaczko, *Confessions of a Rogue Nuclear Regulator* (New York: Simon and Schuster, 2019), Chapter 2.
6. Ibid., 23.
7. US Nuclear Regulatory Commission (NRC), 'Use of Probabilistic Risk Assessment Methods in Nuclear Regulatory Activities', 42,623.
8. Andrew C. Kadak and Toshihiro Matsuo, 'The Nuclear Industry's Transition to Risk-Informed Regulation and Operation in the United States', *Reliability Engineering and System Safety*, 92 (2007), 615.
9. Jaczko, *Confessions of a Rogue Nuclear Regulator*, Ch. 3.
10. Ibid., 33.
11. CFR is the code of federal regulations in the US.
12. US Nuclear Regulatory Commission, n.d. *Federal Register*, 44, 68,186.
13. US Nuclear Regulatory Commission (NRC), Code of Federal Regulations, Title 10. 73.1.
14. Ibid., 73.20.
15. President George W. Bush. 'State of the Union Address' (Washington, DC: The White House, 2002). https://georgewbush-whitehouse.archives.gov/news/releases/2002/01/20020129-11.html
16. U.S. Nuclear Regulatory Commission, 'Interim Compensatory Measures' (Washington, DC: 2002). https://www.nrc.gov/docs/ML0205/ML020510637.pdf
17. This provision was relied upon following the Fukushima nuclear accident to justify the ability of nuclear power plants to withstand a severe accident.
18. Nils J. Diaz, 'Letter to The Honorable Tom Ridge', Washington, DC: US Department of Homeland Security, 8 September 2004. https://www.nrc.gov/reading-rm/doc-collections/congress-docs/correspondence/2004/ridge-09082004.pdf.
19. See discussion in US Nuclear Regulatory Commission, Commission Order and Memorandum, CLI-2005-019, 2005, 29–30.
20. US Nuclear Regulatory Commission, 'Design Basis Threat', *Federal Register*, 19 March 2007, 12,725.
21. US Congress, Public Law 109-58, Washington, DC, 8 August 2005.
22. US Nuclear Regulatory Commission, 'Design Basis Threat'.
23. Risk-informed is used within the US commercial nuclear sector to refer to standards guided by risk information but not exclusively dependent on them. In practice, there is little difference between a risk-informed and risk-based approach, because even a risk-based approach uses risk information but does not rely exclusively on it.
24. Stanley Kaplan and B. John Garrick, 'On The Quantitative Definition of Risk', *Risk Analysis*, 1/1 (1981), 11–27.
25. US Nuclear Regulatory Commission, *The 2006 Radiation Source Protection and Security Task Force Report* (Washington, DC: Government Printing Office, 2006).
26. US Nuclear Regulatory Commission, 'National Source Tracking of Sealed Sources: Final Rule', *Federal Register*, 71 (8 November 2006), 65, 692.
27. US Government Accountability Office, *Nuclear Security: Actions Taken by NRC to Strengthen Its Licensing Process for Sealed Radioactive Sources Are Not Effective* (Collingdale, PA: Diane Publishing Co, 2007).

28. US Nuclear Regulatory Commission, *The 2018 Radiation Source Protection and Security Task Force Report* (Washington, DC: Government Printing Office, 2018).
29. US Nuclear Regulatory Commission and North American Electric Reliability Corporation, 2007, Memorandum of Understanding between the U.S. Nuclear Regulatory Commission and the North American Electric Reliability Corporation, 12 30.
30. US Congress, Public Law 109-58, Section 653.
31. US Nuclear Regulatory Commission, 'Design Basis Threat'.

Suggested Reading

Jaczko, Gregory, (2019), *Confessions of a Rogue Nuclear Regulator* (New York: Simon and Schuster).

Kadak, Andrew C. and Matsuo, Toshihiro, (2007), 'The Nuclear Industry's Transition to Risk-Informed Regulation and Operation in the United States', *Reliability Engineering and System Safety*, 92, 609–18.

Mazuzan, George T. and Mazuzan, George T. and Walker, J. Samuel, (1984), *Controlling the Atom: The Beginnings of Nuclear Regulation, 1946–1962* (Berkeley, CA: University of California Press).

President's Commission on the Accident at Three Mile Island, (1979), *The Need for Change: The Legacy of TMI*, Report on the President's Commission on the Accident at Three Mile Island.

US Nuclear Regulatory Commission (NRC), (1995), 'Use of Probabilistic Risk Assessment Methods in Nuclear Regulatory Activities', *Federal Register*, 60/158, 42, 622.

CHAPTER 24

THE UNITED KINGDOM'S TRANSITION FROM PRESCRIPTIVE TO OUTCOME-FOCUSED NUCLEAR SECURITY REGULATION

CHRISTOPHER HOBBS AND TOM PARKHOUSE

Introduction

The United Kingdom (UK) has a large and diverse civil nuclear programme, which dates back over 60 years and through which it generates, at the time of writing, 15 per cent of its electricity needs.[1] It is a sector in transition with half of its 6,500 MW(e) capacity set to be retired by 2025 and with construction ongoing on a new generation of nuclear plants, with the first expected to begin operations in 2027.[2] Several UK-based companies are also developing concepts for novel advanced reactors, including small modular reactors (SMRs), with designs currently under assessment by the UK Office for Nuclear Regulation (ONR).[3] The UK also has a large legacy programme, focused on decontaminating and dismantling its historical nuclear sites and is actively engaged in the international transport of both nuclear material and waste.

Nuclear security has long been an area of focus for the UK, given the strategic nature of nuclear technology and its historically 'mixed' programme, where both civil and military activities fell under the auspices of a single government body, the UK Atomic Energy Authority (UKAEA).[4] In the early years, like many countries, the UK adopted a classical 'guns, gates, and guards' approach to nuclear security. This focused on the physical protection of high-value targets from external adversaries, incorporating key security principles such as graded approach and defence-in-depth. From the 1950s to the early 2000s nuclear security was managed by a security directorate within the UKAEA, which functioned not only as a regulator, providing assurance to government, but also as a security consultant.[5] Under this arrangement nuclear operators were provided by the UKAEA with a series of prescribed security actions against which their adherence was checked.

This approach to regulating nuclear security was distinct from that adopted for nuclear safety regulation, the delivery of which was based on higher-level Safety Assessment Principles (SAPs).[6] These served as a guiding framework within which operators were able to develop their own bespoke safety cases for review and approval by the Nuclear Installations Inspectorate (NII), within the Health and Safety Executive (HSE). This more flexible approach was underpinned by a way of working that emphasized the minimization, but not elimination, of risks, where accidents and other risks to health and worker safety were reduced as low as reasonably practicable (ALARP). The approach recognized that once risks were reduced to a certain level, the cost of introducing additional measures becomes disproportionate to the further risk reduction achieved. In contrast, for nuclear security there was minimal discussion or understanding within government and the UKAEA about the limitation of security controls. Many senior officials operated under the misperception that security risks could be eliminated.

These parallel approaches to nuclear regulation co-existed for decades. However, the situation began to change in the early 2000s, with the establishment of the Office for Civil Nuclear Security (OCNS), within the UK Department of Trade and Industry.[7] This replaced the Security Directorate within the UKAEA and provided new independence to regulatory activities that had previously been conducted within an organization also responsible for the promotion of nuclear activities. This was followed in 2003 by the development of Nuclear Industries Security Regulations (NISR), which, rather than prescribe security measures, instead required nuclear organizations to develop security plans for assessment by OCNS. In 2007 OCNS was transferred to the NII in an effort to increase the integration of safety and security regulation. This was further cemented by the 2013 Nuclear Energy Act, which established the ONR with responsibility for both nuclear safety and security. However, despite these legislative and structural efforts to bring together the regulatory regimes, elements of the UK's prescriptive approach to nuclear security regulation have continued to endure. In large part this is due to the cultural shift and uplift in capability required to transition towards an outcome-focused regime.[8]

The chapter explores the UK's nuclear security regulatory shift from a prescriptive regime based on adherence to prescribed actions to one underpinned by security principles, focused on achieving higher-level security outcomes. It seeks to shed light on why this transition was initiated and the perceived benefits of an outcome-focused regulatory regime for nuclear security. The chapter also examines how this was implemented and the challenges encountered along the way, detailing the different initiatives launched in an effort to overcome these. It draws on official documents, academic and policy papers, and interviews with practitioners involved in the transition including regulators, operators, and security consultants.

An Overview of Regulatory Approaches

To provide broader context to the UK's regulatory transition for nuclear security, it is first necessary to discuss in a general sense different approaches to regulation, their core attributes, strengths, and weaknesses. Here, it is important to acknowledge that, while there exists a largely common understanding of what constitutes prescriptive or rule-based regulation, there is considerably more uncertainty around the definition and composition of alternatives. Examples of these include principles; standards; performance; outcomes; and goals-based approaches.[9] For the purpose of this discussion these alternative approaches are treated as synonymous and referred to collectively as outcome-based approaches, consistent with the terminology adopted by ONR, with focus placed on how they differ from rule-based regulation.

Rules-Based versus Outcome-Based Regulation

Prescriptive regulatory approaches are based on a set of rules, against which adherence is checked. These are typically detailed and unambiguous providing regulatees with advanced information on both allowed and prohibited actions and activities. Under such systems there is typically only limited flexibility in how these rules can be applied. Consequently, the application of rule-based regulation involves a largely mechanical decision-making process in which the regulatees follow a series of precisely formulated directives and the regulator collects and assesses information to determine compliance. In this system the regulator is ultimately accountable for failures within the regulatory regime.

In contrast, outcome-based regulation is focused on meeting regulatory objectives, with flexibility given to regulatees on how this is achieved. These objectives are typically formulated at a high level, through principles that must be satisfied. Under this system regulatees must exercise greater judgement in determining how to comply with a regulatory objective. This is also true for the regulator, which must undertake a careful assessment of whether the actions taken by the regulatee are sufficient to meet the set objectives. Under this approach responsibility for meeting regulatory objectives is shifted to a degree from the regulator to the regulatee.

Benefits and Challenges of Different Systems

The aforementioned approaches each have intrinsic strengths and weaknesses, with their effectiveness dependent on the manner and context in which they are applied. For example, rule-based regimes can provide high levels of predictability, with regulatees confident in their obligations and what constitutes compliance. They can also promote straightforward and cost-effective enforcement by the regulator, which is only required to make relatively simple judgements based on whether or not the prescribed rules have been followed. However, rule-based regimes can also suffer as a result of their rigidity and prescriptiveness, particularly if applied in environments undergoing significant change. To keep pace and avoid obsolescence, rules will continually need to be revised or added, and regulatees may struggle to keep aligned with the changing requirements. A rule-based approach can also serve to stifle innovation by the regulatee, resulting in a box-ticking mindset where the responsibility for achieving regulatory objectives lies almost completely with the regulator.[10]

An outcome-based regime shifts this focus and encourages regulatees not to simply follow the rules laid out by the regulator, but instead to consider how to meet these higher-level regulatory objectives most effectively. This approach offers potential efficiencies with regulatees able to select new methods that achieve the necessary goals, but at a reduced cost. It can also serve to facilitate a differentiated approach by the regulator, where resources are applied based on an organization's history and risk profile, rather than spread evenly across all regulatees. However, the flexibility afforded in outcome-based regimes can also create uncertainty, with regulatees unclear on how to meet high-level requirements. The effective application of outcome-based approaches also requires greater expertise and the increased application of professional judgement, within both the regulator and the regulatee, as a result of the wide range of possible solutions that may be advanced to meet regulatory objectives.

Pragmatic Hybrid Approaches

Given the advantages and disadvantages of rule-based and outcome-based regulation, it is not surprising that in practice regulatory regimes have tended to adopt a hybrid approach that attempts to combine the positive elements of the two approaches, in an effort to mitigate their potential shortcomings. For example, a rule-based regime may include higher-level regulatory goals and make provision for certain qualifications or exceptions to prescribed rules, if regulatees can justify why deviations are necessary. An outcome-focused regime may produce non-binding guidance and best practice for incorporation in regulatory solutions, reducing the uncertainty that may be felt by regulatees. However, there is also a risk that developing hybrid approaches may emphasize the more negative attributes of pure rule-based or outcome-based systems. For example, if guidance that illustrates how to comply with different goals is overused in a hybrid outcome-focused regime this can serve to reproduce the mechanistic box-ticking mentality of rule-based systems. Conversely, if a large number of exceptions are granted within a hybrid rule-focused regime, this may produce considerable uncertainty, removing one of the key advantages of this approach.

Drivers and Perceived Benefits behind a Regulatory Shift

The past few decades have seen outcome-based regulatory regimes replace more traditional prescriptive approaches across a wide range of industries, both in the UK and globally.[11] This has been driven in part by an increased government focus on 'better regulation' and reducing the burden on business, as well as a number of high-profile incidents, where rule-based regulatory systems were shown to have contributed to inefficiencies, poor performance, or failure.[12] In the nuclear sector, one of the most prominent examples is the 1979 nuclear accident that occurred at Three Mile Island in Pennsylvania, United States (US). The subsequent US Presidential Inquiry concluded that the prescriptive regulatory system in place had resulted in 'rule-following automatons' without sufficient understanding of the key 'principles of reactor safety'.[13] In the UK there exists a 'longstanding policy interest in alternatives to traditional regulation', with health and safety, financial services, and other sectors adopting outcome-based systems.[14] In was within this broader shifting regulatory landscape that the UK's decision to move from a prescriptive to an outcome-focused regime for nuclear security regulation was taken, precipitated by a number of interrelated drivers.

Arguably, the most prominent of these was a renewed interest in nuclear power within the UK in the mid-2000s that served to place a greater focus on the industry and how it was regulated. There was a perception in government at the time that the efficacy of existing nuclear security arrangements could be improved, with 'safety clearly better than security'.[15] Here, it was observed that while nuclear operators were active in the development of new safety systems and programmes, they remained relatively unengaged in security. Instead, security provisions depended largely on the expertise of the regulator to recommend appropriate security solutions. OCNS, the security regulator at this time, struggled to operate effectively due to both a small staffing footprint—approximately 10 times less than for nuclear safety—and high levels of staff turnover. Consequently, there was seen to be significant potential benefits in adopting a new regulatory approach for nuclear security, whose effectiveness had already been demonstrated for nuclear safety.

Secondary drivers included a realization that it would be intrinsically difficult to regulate certain new areas of nuclear security through a prescriptive framework. Chief among these was cyber security, for which it was recognized at the time that measures must be flexible, so they could be adapted in response to the rapidly changing nature of the threat.[16] Their success would also hinge significantly on the active engagement by industry, in particular, the ability of individuals working in nuclear organizations to identify and take appropriate actions when confronted by a cyber threat or potential vulnerability. In addition, an outcome-based regulatory system supported a more transparent approach to nuclear security. Such a system had the potential to inspire greater public confidence in support of a planned nuclear new build programme by sharing greater information on the higher-level principles and processes through which security was regulated.

Pursuing a New Regulatory Approach

It was recognized at the start of the process that it would take time and effort to shift away from what was a deeply embedded prescriptive regulatory system, and consequently a 'phased approach' was adopted.17 This involved a series of sequential actions aimed at encouraging the application of a more flexible outcome-based approach to nuclear security regulation. These actions are discussed below, within three key time periods linked to major developments, with a focus on the past 15 years.[17]

Preparing for Transformation (2007–2012)

As mentioned, post-2003 nuclear security regulation in the UK has been governed through NISR. This key legislative document is not intrinsically rule based, as it does not prescribe specific security measures. Instead, it requires that nuclear licensees develop and follow a security plan, detailing the design, implementation, evaluation, and maintenance of their security systems, for subsequent review and approval by the regulator. New security plans were required following the publication of NISR, in a move aimed at stimulating 'licensees to develop and justify their own arrangements'.[18] However, this initial effort to move away from prescription was stymied due to 'a lack of security capability and capacity within licensee organizations and a pervasive culture of supervision [within the regulator]'.[19] Unable to develop their own security plans, licensees reached out to OCNS for support, which in response produced a Technical Guidance Document (TGD) containing model security standards.

As discussed, the provision of non-binding guidance in support of higher-level objectives, as part of a hybrid outcome-based regulatory approach, can help reduce the uncertainty encountered by licensees when moving to a new system. However, in this case, many licensees incorporated key elements of the TGD largely verbatim into security plans, failing to consider its application in their own specific contexts.[20] This situation was further perpetuated by the subsequent decision by the OCNS to redraft the TGD as a Technical Requirements Document (TRD), with directive language and a 'prescriptive check list' of security measures against which licensee compliance was evaluated.[21] Consequently, when OCNS was transferred to the NII within the Health and Safety Executive in 2007, nuclear security was still regulated through a highly prescriptive rule-based approach.

Although nuclear security and safety initially operated out of separate offices within the NII, integration strengthened over time, resulting in the adoption of joint management practices and the increasing alignment of goals and regulatory approaches. These more common ways of working, combined with renewed impetus for a change in how nuclear security was regulated at the government level, paved the way for new initiatives aimed at moving the UK away from its highly prescriptive approach. Principal among these was the development of National Objectives, Requirements and Model Standards (NORMS), to replace the TRD. This was developed over several years by OCNS and contained a mix of high-level objectives, requirements, and standards.[22]

Transitioning Phase (2012–2017)

The publication of NORMS in 2012 can arguably be considered as the start of a transitioning period where the UK took its first major steps towards an outcome-based regulatory system for nuclear security. However, although the document was intended to encourage licensees to adopt an approach where security solutions were developed to meet higher-level objectives, in practice this suffered from a number of implementation challenges identified in a formal implementation review conducted in 2014. Firstly, while licensees were consulted prior to the publication of the review, this effort arguably did not reach key stakeholders, and consequently, the industry's ability to input into its formulation was relatively limited, reducing initial levels of buy-in and understanding.[23] Secondly, the language within NORMS was a mixture of directive and non-directive vocabulary, making it unclear what requirements should be followed and where flexibility existed for developing new approaches.[24] This was an artefact of how NORMS was created, having grown organically through the efforts of multiple, largely independent contributors, as opposed to having been authored to meet a clear requirement. It was also imbalanced in places and provided considerable amounts of information on certain areas of nuclear security, but relatively little on others.[25]

Thirdly, the inclusion of model standards within NORMS meant that these could be closely followed.[26] This effectively provided a risk-free route for nuclear operators to adopt without the need to develop their own in-house capability to devise and assess independent security solutions. These challenges were compounded by a lack of expertise and knowledge within the regulator to holistically assess whether security approaches developed by an operator met the high-level objectives within NORMS. The majority of inspectors at the time were still used to judging compliance through checking adherence to a set of prescribed rules.[27]

Given these issues, a decision was taken to revise NORMS in an effort to improve the effectiveness of nuclear security regulation in the UK. Initially titled NORMS 2, the framework was ultimately renamed Security Assessment Principles (SyAPs), in order to remove any potentially negative connotations associated with 'requirements' or 'model standards'.[28] The redevelopment drew on the findings of the 2014 NORMS implementation review and insights from other regulatory regimes. In particular, a deliberate decision was taken to mirror, in terms of language and approach, ONR's SAPs developed for nuclear facilities. SAPs were considered as relevant due to their perceived effectiveness and the clear benefits of a closely aligned approach to nuclear security and safety regulation, which would support greater engagement by key people outside of the security function.

The UK's Security Policy Framework (SPF), first published in April 2013, also served as a source of inspiration.[29] The SPF was deemed particularly relevant given its similar outcome-focused approach to protecting government assets (people, information, and infrastructure) and emphasis on elements increasingly recognized as crucial to the successful delivery of security, namely, leadership and governance; risk management; personnel security; and organizational culture.[30] The tiered structure of the SPF, outlining key information of relevance for different stakeholders, was also perceived to be beneficial and was consequently replicated in SyAPs. Some examples here include: the inclusion of a short 'Unifying Purpose Statement' to engage executive-level decision makers; the development of 'Fundamental and Key Security Principles' to inform the actions of senior nuclear security leaders; and the insertion of 'Security Delivery Principles' that could be utilized by security practitioners.[31]

Drawing lessons learned from NORMS on the importance of active stakeholder engagement, it was decided that SyAPs would be developed with much greater input from industry. This was facilitated through both informal discussions and dedicated workshops involving a wide cross-section of stakeholders responsible for protecting nuclear assets. A handful of these were held in 2014–2017, when ONR presented draft text and solicited feedback, with duty-holders also able to suggest new areas and concepts for inclusion.[32] Efforts were also made to align SyAPs with international guidance and good practice, in particular the Nuclear Security Series (NSS) of the International Atomic Energy Agency (IAEA), which forms the basis of the approach to nuclear security in many countries.[33] As part of this exercise, key security principles and concepts from the IAEA NSS were mapped onto the UK's regulatory framework. While helping flesh out the SyAPs, this process also served to identify areas within the NSS where further information and guidance of benefit to the international community could be provided. This was subsequently fed by the UK into the deliberations of the IAEA Nuclear Security Guidance Committee, which is tasked with developing and reviewing new publications within the series.[34]

Published in 2017, SyAPs was fundamentally different to the regulatory documents that had proceeded it. Firstly, the model standards of NORMS were removed and replaced with graded security outcomes, dependent on the potential consequence of a theft or sabotage attack. The new approach provided important flexibility for operators, allowing them to assess their specific risks and design and evaluate proportionate security systems that demonstrate an appropriate security outcome have been achieved. This represented a significant departure from previous approaches whereby adequacy could be demonstrated simply by having specified measures, rather than a goal for those measures to achieve.[35] Secondly, the list of requirements within NORMS was replaced by a series of principles, against which regulatory judgements by inspectors would be made.[36] These were formulated at both high (fundamental principles) and medium (delivery principles) levels in an effort to provide sufficient information against which security solutions could be designed but without advancing a specific approach. The lack of prescription in these is further reinforced by the acknowledgement within SyAPs that not all principles will apply to every situation or site.[37]

Thirdly, SyAPs expanded and emphasized regulatory expectations into new areas, including 'governance, leadership, competence management and supply chain assurance'.[38] Acknowledging these as vital not only to the delivery of security, but also in demonstrating its effectiveness, a necessary part of an outcome-focused regulatory approach—reliant on sampling rather than examining in detail every aspect of how security at a site—is implemented. In this new system, ONR confidence in an operators' submission is strongly 'influenced by the effectiveness and maturity of the dutyholder's governance and assurance processes'.[39] Fourthly, unlike TGD, TRD, and NORMS, SyAPs was openly published, albeit with its associated annexes redacted due to their sensitive nature.[40] This was a deliberate step taken to increase public understanding and awareness of how nuclear security is implemented in the UK and also to aid SyAPs' promulgation across the nuclear industry, including its easy integration into education and training courses for staff.

Developing Maturity (2017–2022)

Following the publication of SyAPs in March 2017, it was initially envisioned by ONR that operators' security plans would be developed, submitted, and approved by the end of 2019.[41] However, it soon became apparent that, for most operators, this time frame was overly ambitious; by September 2020, only 43 per cent of SyAPs security plans from across the UK civil nuclear estate were in force.[42] This elongated time frame was necessary for operators to increase their in-house security capability in order to effectively engage with SyAPs, as well as for the necessary back and forth with ONR to develop balanced security plans.[43] On the regulatory side the time frame was also hampered by a decision to disband the ONR team that had worked on SyAPs, on the assumption that the major work had been completed. This failed to fully recognize the implementation challenges that this major regulatory shift would present. Under NORMS, the assessment of security plans was relatively uncomplicated, with emphasis placed within this process on assessment and inspections to gather data for comparison with well-established security requirements. However, under SyAPs, a far more in-depth and holistic assessment of security outcomes was required, given the potentially far-greater range of security solutions.

Consequently, the focus needed to change from data gathering through inspections to the development and use of rigorous targeted assessment methodologies akin to those developed for nuclear safety. These assessments were fed with security information supplied by operators, and then confirmed by ONR through sampling. However, the enactment of this change also required a significant cultural shift within ONR, with staff historically used to directly checking all aspects of nuclear security. These challenges were recognized and a new ONR transition team was put in place, drawing in expertise from its safety function, which had garnered significant experience in aligning safety plans to outcome-focused regulation under SAPs.[44]

Within UK industry there has been considerable variation across nuclear operators, both with regards to the speed at which SyAPs was adopted and the extent to which new security solutions have been advanced under it. Such variation was further influenced by the background and approach of key individuals, as well as the fundamental make-up of different sites. For some, such as International Nuclear Services (INS), a specialist nuclear transport organization, the review and redevelopment of security under SyAPs was viewed by the executive as a way of realizing both security benefits and longer-term cost savings, as well as an opportunity to lead the industry in this transition.[45] Indeed, INS was the first organization to have an ONR-approved SyAPs plan and the associated internal assurance processes in place, which was as early as April 2017. By the end of 2017, INS had achieved under SyAPs a level of maturity that placed the organization at ONR's lowest level of nuclear security regulatory focus, reducing the frequency of regulatory inspections and their associated costs.[46]

However, for other nuclear organizations, the transition has been far more gradual, aligned with the development of suitably qualified and experienced personnel (SQEP) who are able to establish appropriate internal assurance systems and assess the effectiveness of existing security measures under SyAPs. Here, the involvement of safety specialists, familiar with SAPs, was perceived by ONR as an important enabler for speeding up this process.[47] However, even when established, the nature and history of certain nuclear sites may constrain the degree of security innovation possible under SyAPs. As mentioned, there exists a wide range of facilities in the UK regulated by ONR. These include well-established facilities with mature security systems, where it may not make sense to significantly change security measures or processes.[48]

There also exist industrial and research facilities that hold only very small amounts of nuclear materials, which in many cases do not present a serious hazard. For the operators of these facilities, it may not be economical to develop a significant security capability. However, this lack of expertise presents a challenge when it comes to developing security plans and assurance systems under SyAPs. ONR recognized this issue and noted that, for lower-hazard, smaller nuclear sites, the benefits of applying SyAPs may be outweighed by the costs.[49] For such sites, ONR therefore allowed for development of simplified security plans supported by a 'Guide for Smaller Dutyholders to the Application of the Security Assessment Principles'.[50] Formulated by ONR, this document functions as a tool, providing signposts and guidance for lower-hazard sites when developing security systems under SyAPs.

Key Lessons and Relevance to Others

While it is still too early to evaluate the relative performance of outcome-focused versus prescriptive nuclear security regulation in the UK, there are nevertheless a number of useful insights that can be extracted from this shift. Many of these are likely to be agnostic of the technology that is to be regulated and intrinsic to the general transitioning process, rather than UK specific. Consequently, these may serve to usefully inform similar future efforts in other sectors and countries. Here, there is already evidence of broader interest in the UK's transition, with the SyAPs document having been downloaded thousands of times by individuals from all around the world.[51]

Firstly, it is important to emphasize that a move away from prescriptive regulation is likely to take both considerable time and sustained effort. While the development of new regulatory documents that promote an outcome-focused approach may be a key part of this transition, this step alone is unlikely to be sufficient. Arguably far more important are associated efforts taken to change the mindset and culture of those both within the regulatory body and the licensees. Initiatives to this end can take many forms, from early engagement by the regulator with licensees—as a means of promoting buy-in and shaping the approach—to the development of dedicated training programmes aimed at upskilling existing staff and recruitment campaigns to create SQEP within both the regulator and regulatees. For SyAPs, in contrast to NORMS, there was a considerable period of consultation within industry, with operators able to provide suggestions that fed into the final document. This served to establish trust between ONR and its regulatees, who were also able to develop a clear picture of what a new outcome-focused approach to nuclear security aimed to achieve.[52]

Secondly, while a nuclear security system may be redeveloped to become outcome-focused if it is to remain effective and efficient, it will likely need to retain certain prescriptive aspects. For example, it is common for outcome-focused regulatory regimes to produce non-binding best practice guides or guidance documents that can support regulatees in the development of regulatory solutions. However, what these prescriptive elements are and how they are introduced needs to be carefully considered or they can serve to stymy efforts. For example, the directive language and model standards incorporated within NORMS served to both confuse nuclear operators as to where flexibility for new approaches existed, while providing them with a shortcut to meeting regulatory requirements, without considering the overarching objectives or developing their own expertise. In contrast under SyAPs, a deliberate decision was taken by ONR to remove the model standards of NORMS, so operators would have no choice but to develop and assess their own security solutions. While supplementary guidance was developed for small duty-holders under SyAPs, this was far more limited in scope and targeted to meet a specific need.

Thirdly, a shift to outcome-focused regulation will require significant changes at the working level in how both the regulator and licensees engage with security. Within the regulator this will require redistributing efforts from carrying out inspections to conducting assessments. Although inspections will always remain an important part of the regulator's toolkit, these should be deployed for the purpose of sampling the evidence provided by the regulatees, as opposed to the collection of a comprehensive set of data against which regulatory judgements on compliance against prescriptive standards can be made. On the part of the regulatees, focus will need to shift from implementing security measures prescribed by the regulator to developing their own solutions and establishing internal assurance systems that are able to demonstrate to the regulator that these are working effectively. These are major changes that will take significant effort and time to enact.

Conclusion

Considerable time and effort have been devoted by the UK nuclear industry to the pursuit of an outcome-focused regulatory regime for nuclear security. This has been applied in a phased manner, over more than a decade, through a series of interrelated initiatives. This long-term approach highlights the significant structural and organizational barriers within the industry that needed to be overcome for an outcome-focused approach for nuclear security to be effectively adopted. A similar level of sustained effort is likely to be needed by other countries or industries embarking on a similar journey, particularly if a culture of prescription is already deeply embedded within both the regulator and regulatees.

In the case of the UK this transition did not always follow a smooth linear process, with certain steps towards outcome-focused regulation having to be temporarily rolled back along the way, for example, ONR's decision in the early 2000s to redraft associated guidance for NISR as requirements when it became clear that nuclear operators where not yet ready to formulate their own security plans and solutions. These occurrences are likely to be common when undertaking a significant regulatory change, where the speed of innovation must be carefully balanced with the need to sustain a robust regulatory system throughout the process. It may not always be possible to get this right, although it is clearly key that lessons are captured and learned throughout. This can be seen at several stages within the UK's transition to outcome-focused nuclear security regulation. For example, the considerable efforts undertaken by ONR to garner industry buy-in and understanding of SyAPs, having recognized how the more limited promulgation of NORMS five years earlier had failed to generate sufficient levels of industry engagement.

The UK's experience also serves to demonstrate how certain elements of prescriptive and outcome-focused approaches do not have to be mutually exclusive and—if applied appropriately—can support the effective operation of a regulatory regime. As mentioned, while regulatory regimes may by positioned towards different ends of a rules- versus goals-based spectrum, in practice they may incorporate both elements. This hybrid approach to regulation can be observed in the UK's nuclear security efforts, where various requirements, standards, and guidance were introduced by ONR at key points throughout the journey to outcome-focused regulation. Here, it is clear that how these are formulated and the way in which they are presented matters, with these serving at different times to both undermine and support the transition.

Notes

1. World Nuclear Association, 'Nuclear Power in the United Kingdom', WNA [online], updated January 2023. https://world-nuclear.org/information-library/country-profiles/countries-t-z/united-kingdom.aspx
2. World Nuclear News, 'EDF Revises Hinkley Point C Schedule and Costs', *World Nuclear News* [online], 20 May 2022. https://www.world-nuclear-news.org/Articles/EDF-revises-Hinkley-Point-C-schedule-and-costs
3. UK Office for Nuclear Regulation (ONR), 'Generic Design Assessment (GDA) of New Nuclear Power Stations', ONR [online], 28 September 2028. https://www.onr.org.uk/new-reactors
4. William Walker, 'Entrapment in Large Technology Systems: Institutional Commitment and Power Relations', *Research Policy*, 29 (2000), 842.
5. Author interview with Adrian Freer, former Director of Civil Nuclear Security, ONR, 22 June 2021.
6. Geoffrey J. Vaughan, 'Safety Goals for Nuclear Materials and Radioactive Wastes—The UK Regulatory Approach', in *Proceedings of PSAM 7—ESREL '04 14–18 June 2004, Berlin, Germany* (London: Springer, 2004), 1648–53.
7. Select Committee on Trade and Industry, 'Memorandum by the Office for Civil Nuclear Security, Appendix 41', UK Parliament [online], 21 December 2006. https://publications.parliament.uk/pa/cm200506/cmselect/cmtrdind/1443/1443we50.htm
8. UK Office for Nuclear Regulation, 'Annual Report and Accounts 2020/21', Report HC 463, ONR [online], 20 July 2021. https://www.onr.org.uk/documents/2021/onr-annual-report-and-accounts-2020-21.pdf
9. Christopher Decker, 'Goals-Based and Rules-Based Approaches to Regulation', BEIS Research Paper Number 8 (London: Department for Business, Energy & Industrial Strategy), 15. https://assets.publishing.service.gov.uk/government/uploads/system/uploads/attachment_data/file/714185/regulation-goals-rules-based-approaches.pdf
10. Surendra Arjoon, 'Striking a Balance Between Rules and Principles-based Approaches for Effective Governance: A Risks-Based Approach', *Journal of Business Ethics*, 68 (2006), 69.
11. Peter J. May, 'Performance-Based Regulation', Jerusalem Papers in Regulation & Governance, Working Paper No. 2/April (2010), 2. http://regulation.huji.ac.il/papers/jp2.pdf
12. Philip Hampton, *Reducing Administrative Burdens: Effective Inspection and Enforcement* (London: HMSO, 2005), 2. https://www.regulation.org.uk/library/2005_hampton_report.pdf
13. John Braithwaite, 'Rules and Principles: A Theory of Legal Certainty', *Australian Journal of Legal Philosophy*, 27 (2002), 66; John G. Kemeny, et al., *Report of the President's Commission*

on the Accident at Three Mile Island: The Need for Change: The Legacy of TMI (Washington, DC: President's Commission on the Accident at Three Mile Island, 1979), 49.
14. Decker, 'Goal-Based and Rules-Based Approaches', 9.
15. Freer, 22 June 2021.
16. Ibid.
17. Matthew Sims, 'ONR's Experiences from Adopting an Outcome Focused Approach to Civil Nuclear Security Regulation', Paper #303, presented at the International Conference on Nuclear Security: Sustaining and Strengthening Efforts, 10–14 February 2020, Vienna, Austria, 3.
18. Ibid., 2.
19. Ibid.
20. Freer, 22 June 2021.
21. Sims, 'ONR's Experiences', 2.
22. Office for Nuclear Regulation, 'ONR Rolls Out New Security Guidance to Industry', ONR [online], 29 October 2012. https://news.onr.org.uk/2012/10/onr-rolls-out-new-security-guidance-to-industry
23. Author interview with Philip Hannah, Security Consultant to ONR, 22 June 2021.
24. Author interview with Matthew Sims, Professional Lead for Nuclear Security, ONR, 16 September 2021.
25. Ibid.
26. Hannah, 22 June 2021.
27. Ibid.
28. Sims, 'ONR's Experiences', 4.
29. UK Government, 'Security Policy Framework', Policy Paper, April 2014. https://www.gov.uk/government/publications/security-policy-framework/hmg-security-policy-framework
30. Sims, 16 September 2021.
31. Office for Nuclear Regulation, 2017, 'Security Assessment Principles for the Civil Nuclear Industry, 2017 Edition, Version 0', 12.
32. Hannah, 22 June 2021.
33. Wyn Q. Bowen, Matthew Cottee, and Christopher Hobbs, 'Multilateral Cooperation and the Prevention of Nuclear Terrorism: Pragmatism over Idealism', *International Affairs*, 88/2 (2012), 358.
34. Hannah, 22 June 2021.
35. Sims, 16 September 2021.
36. ONR, 'Security Assessment Principles', 12.
37. Ibid.
38. Sims, 'ONR's Experiences', 5.
39. ONR, 'Security Assessment Principles', 72.
40. These were published alongside associated Technical Assessment Guides (TAGs) and Technical Inspection Guides (TIGs).
41. Sims, 'ONR's Experiences', 6.
42. UK Office for Nuclear Regulation, 'Office for Nuclear Regulation Annual Report and Accounts 2019/20', Report HC 470, ONR [online], 29 September 2020, 17. https://www.onr.org.uk/documents/2020/onr-annual-report-and-accounts-2019-20.pdf

43. UK Office for Nuclear Regulation, 'Office for Nuclear Regulation Annual Report and Accounts 2018/19', Report HC 2271, ONR [online], 20 June 2019, 35. https://www.onr.org.uk/documents/2019/onr-annual-report-and-accounts-2018-19.pdf
44. Sims, 'ONR's Experiences', 6.
45. Author interview with Security Manager, International Nuclear Services (INS), 9 December 2019 (NB: in February 2021, INS was merged with Direct Rail Services to create a new business—Nuclear Transport Solutions (NTS)).
46. Ibid.
47. Sims, 16 September 2021.
48. Author interview with Security Director, UK Nuclear Operator, 8 August 2022.
49. Sims, 16 September 2021.
50. Office for Nuclear Regulation, 'Security Assessment Principles (SyAPs)', ONR [online], 16 May 2022. https://www.onr.org.uk/syaps/security-assessment-principles.pdf
51. Sims, 16 September 2021.
52. Security Manager, 9 December 2019.

Suggested Reading

Decker, Christopher, (2018), 'Goals-Based and Rules-Based Approaches to Regulation', BEIS Research Paper Number 8 (London: Department for Business, Energy & Industrial Strategy). https://assets.publishing.service.gov.uk/government/uploads/system/uploads/attachment_data/file/714185/regulation-goals-rules-based-approaches.pdf

Hobbs, Christopher, (2021) 'INSF Country Update: United Kingdom Nuclear Security—Exploring the UK's Recent Regulatory Transition, the Promotion of Nuclear Security Culture, and Response to the COVID-19 Pandemic', *Stimson Centre* [online], 27 October. https://www.stimson.org/2021/insf-country-update-united-kingdom-nuclear-security/

May, Peter J., (2010), 'Performance-Based Regulation', *Jerusalem Papers in Regulation & Governance*, Working Paper No. 2/April, 1–20. http://regulation.huji.ac.il/papers/jp2.pdf

Sims, Matthew, (2020), 'ONR's Experiences from Adopting an Outcome Focused Approach to Civil Nuclear Security Regulation', Paper #303, presented at the International Conference on Nuclear Security: Sustaining and Strengthening Efforts, 10–14 February 2020, Vienna, Austria.

Williams, Laurence G., (2019), 'Nuclear Safety and Nuclear Security Regulatory Challenges Facing a Country Embarking on a Nuclear Power Programme', *The Journal of World Energy Law & Business*, 12/1, 69–88.

CHAPTER 25

THE EVOLUTION OF NUCLEAR SECURITY IN RUSSIA

DMITRY KOVCHEGIN

Introduction

Nuclear security in Russia has been the focus of international attention since the collapse of the Soviet Union in 1991. In fact, nuclear security challenges created by the Soviet Union's dissolution were the first time that such issues took a high-profile position on the international security agenda. This was due to the vast amounts of nuclear materials attractive to potential bombmakers and nuclear facilities representing targets for sabotage with potentially catastrophic consequences. It was recognized that the poor security of these materials and sites, as well as changes in the international security environment, exposed Russia's nuclear complex to the growing threat of catastrophic terrorism. Russia, as one of the leading nuclear nations and exporter of nuclear technologies, serves as a role model for other countries. If Russia is seen to not be paying appropriate attention to nuclear security, other nations might potentially be tempted not to take the threat seriously. The evolution of nuclear security in Russia provides an example of creating modern nuclear security infrastructure almost from scratch to a mature and well-developed system and may serve as a source of lessons for newcomer countries starting to develop nuclear energy programmes.

Russian Nuclear Complex

Russia inherited most of the vast nuclear complex that the Soviet Union had developed before its collapse in 1991. This complex included nuclear power plants, nuclear fuel cycle sites, research and development institutions, a nuclear-powered icebreaker fleet, and a nuclear navy with supporting infrastructure, weapon assembly/disassembly enterprises, uranium mining organizations, and nuclear material storage sites. Most authoritative estimates of inventories of the nuclear materials in the Former Soviet Union (FSU) agree that, by 1991, the Soviet Union had accumulated around 130 tonnes of weapons-grade plutonium and around 1,200 tonnes of highly enriched uranium (HEU).[1] Russia today has the same nuclear complex and still possesses vast amounts of nuclear materials.

Organizations handling the largest amounts of nuclear materials and conducting the most sensitive activities are still located in 10 'closed cities' established during the Soviet period. These cities host major fuel cycle sites, such as Mayak, Siberian Chemical Combine, and Mining and Chemical Combine, nuclear weapon laboratories (VNIITF and VNIIEF), and nuclear weapon-production facilities. However, many other nuclear sites handling substantial amounts of weapons-usable nuclear materials are located in densely populated areas, including Kurchatov Institute and Bochvar Institute in Moscow, Luch and Electrostal in the Moscow region, Institute of Physics and Power Engineering in Obninsk, Research Institute of Atomic Reactors in Dimitrovgrad, and Novosibirsk Chemical Concentrates Plant.[2] In addition to these sites, substantial amounts of nuclear materials, typically in the form of fuel for research reactors and propulsion reactors of nuclear-powered surface ships and submarines, are handled at research and education institutes reporting to the Russian Academy of Sciences, in the Ministry of Science and Education or its predecessors, and at shipbuilding industry sites constructing, repairing, maintaining, and refuelling nuclear submarines and nuclear-powered surface ships. Russia's nuclear complex also requires substantial transportation, involving the movement of nuclear materials using railroad, road vehicles, marine and river ships, and occasionally aeroplanes.

Nuclear Security during the Soviet Union Years

The evolution of nuclear security over the past 30 years in Russia encompasses both a change in the threat environment and the approaches implemented to address these. Here, it should be emphasized that Russia had managed to accumulate substantial experience in ensuring the security of its nuclear complex before the collapse of the Soviet Union.[3] Considering the vast scale of the Soviet nuclear complex, its importance for the defence of the Soviet Union, and the general obsession of the Soviet authorities with security, this was a primary area of concern for the authorities.

The United States (US) was considered the key adversary to the Soviet Union, posing a threat to nuclear sites and materials, in the form of a military attack, an act of sabotage, and intelligence collection. The threat of an attack was addressed by military means, while sabotage against nuclear sites and intelligence collection were considered conventional threats that could be addressed by nuclear security systems to a large extent. Here, focus was placed on defending sites against military special operations forces, rather than terrorist groups or other non-state actors. Historically in the Soviet Union, the insider threat was considered a key security risk and security services considered all nuclear site employees as potential insiders. However, the focus was on detecting those who might be spying for foreign states or looking to conduct sabotage acts against nuclear sites, rather than stealing nuclear materials.

The approach to nuclear security in the Soviet Union included multiple layers that covered the whole country, not only nuclear sites and their vicinity.[4] Most critical nuclear sites were based in closed cities. These cities were located in remote locations in Russia, absent from public maps, and had no regular postal addresses (these being replaced by opaque postal office box designations).[5] Closed cities were surrounded by protected fences, with restricted and tightly controlled access, and people working and living there required special permission to leave. In addition, few foreigners could enter the Soviet Union and few of its citizens could travel abroad.

Every nuclear site in the Soviet Union had a division in its organizational structure dealing with the protection of state secrets. This was essentially a site office for the KGB (the Soviet Union's main security agency) that was staffed by KGB officers seconded to the site. The main goal of this division was the protection of information considered as state secrets or for official use only. The division served as an interface between the nuclear site and security services, and their responsibilities included a broad range of functions related to control over site personnel, for example, background checks, clearance, and routine surveillance. The division had substantial authority over the fate of each individual employee of the nuclear site and its work was an extremely important deterrence factor preventing insider threats to nuclear security.[6]

Protection and access control in closed cities and standalone nuclear sites were provided by paramilitary units of the Ministry of Internal Affairs assigned with responsibility to protect critical infrastructure facilities, including nuclear sites. Physical protection equipment like intrusion sensors was introduced at nuclear sites in the mid-1960s, when the approach to protection based on using guards became unsustainable as personnel number requirements achieved levels that could not be reasonably met due to the vast number of sites. In fact, the Soviet nuclear complex was the first critical infrastructure to introduce the widespread use of sensors and other technologies. However, physical protection systems at nuclear sites did not employ the graded approach (where protection system design is developed based on vulnerability assessment). Rather, these systems were based on requirements for equipment prescribed by the Ministry of Internal Affairs.[7]

Security measures at Soviet nuclear sites, including severe restrictions on personnel freedom, were counterbalanced by the privileged status afforded to nuclear complex employees. This included higher wages, welfare benefits, and living standards compared to the country in general.[8] The social status and economic benefits substantially reduced, if not totally eliminated, the temptation to steal nuclear material or commit acts of sabotage against a nuclear site. In addition, Soviet personnel selection and hiring processes targeting those perceived as the 'best' people created a distinct organizational culture that positively affected nuclear security.

The system of nuclear material accounting and control (NMAC) in the Soviet Union was not intended to detect diversion of nuclear materials for its use in an unauthorized nuclear explosive device. Rather, the primary concern was the monetary value of nuclear materials. Mass and isotopic composition of nuclear materials were measured for the purpose of technology and production processes control, not for NMAC. Inventory difference was determined by the difference between input and output of technology processes, and the only criteria for evaluation of the difference was the standard value of irretrievable losses assigned to each technology process. These criteria did not take into account proliferation-related considerations, not even a significant quantity of nuclear material going missing. If the difference between input and output was less than the sum of the standard levels of irretrievable losses for all technology steps involved in a process, the difference was not investigated.[9]

Control over nuclear materials was considered to be the personal responsibility of the employee handling nuclear materials. This personal responsibility was formally captured in logs documenting nuclear material movement through technology processes.[10] Nuclear security as a distinctive profession or area of activity did not exist in the Soviet Union. Instead, many stakeholders with official responsibilities often not related to nuclear security implemented individual activities pertinent to nuclear security.

Impact of the Collapse of the Soviet Union on Nuclear Security

The Soviet Union's dissolution in 1991—which created 15 constituent republics—and the political, social, and economic processes accompanying this breakdown presented new nuclear security challenges. Political changes brought more liberties and people obtained the right to freely move across the country. Closed cities, where nuclear sites were located, remained closed but people received more opportunities to move in and out of these cities. Borders between most of the constituent FSU republics, which essentially had not existed in the past, were now open. Russia faced a new threat environment with the collapse of old nuclear security systems that were inadequate to protect against evolving threats and lacking the necessary financial, technical, and organizational resources. The new border and customs control measures were also incapable of preventing the unauthorized movement of nuclear materials across the border.[11]

Government, which was the key customer of the nuclear industry and main source of its revenue, was less interested in—and much less capable of—providing financial support. This substantially weakened the second Soviet pillar of nuclear security: the wealth, living standards, social status, and morale of nuclear industry personnel. Nuclear personnel living in or close to major cities could leave their nuclear jobs to find better paying work. However, nuclear personnel living in closed cities did not have such opportunities, as nuclear sites were often the only employer and still inaccessible to outside commercial opportunities. Work in the nuclear complex also lost its prestige. This social and economic situation negatively affected the morale of nuclear industry personnel and created temptations to abuse access to nuclear materials to improve living conditions, for example, through stealing said materials to sell them on the black market.

The deteriorating conditions led to several confirmed cases of nuclear material thefts committed by insiders. One of the well-known cases was the theft of HEU from Luch, a research and production site close to Moscow that handled substantial amounts of nuclear materials in various forms. Leonid Smirnov, a site employee with daily access to nuclear material, committed protracted theft using his insider knowledge of weaknesses in the system of accounting and control of nuclear materials. Smirnov stole small amounts of nuclear materials on a regular basis between May and September 1992 without being detected.[12] He was sentenced to three years in prison; the relatively low severity of this punishment is another indicator of the lack of importance given to nuclear security and proliferation controls at the time.

The interview that Smirnov gave to US broadcaster PBS after his release from jail is the single most indicative document characterizing the status of nuclear security in Russia following the Soviet Union's dissolution.[13] In this interview, Smirnov notes that he had never thought about stealing nuclear material before 1992, but the dire economic conditions had prompted him. Luch, he says, 'was considered an elite plant, they paid well'; and 'everyone was satisfied'.[14] Smirnov had no specific buyer in mind when committing the theft. Instead, he learned from newspaper articles about others stealing 1,200 grams of uranium and imagined there was a market for stolen nuclear material. His plan was to look for the buyer later—he was going to 'wander around Moscow' knocking at the doors of 'all sorts of firms and offices there', hinting about the material in his possession.[15] Smirnov notes that there were very few chances he would be detected due to weaknesses of the NMAC system and lack of any detection equipment at the site exit. Notably, Smirnov was caught by accident, and not as a result of investigation or other planned effort of the law enforcement agencies.[16]

The Soviet Union's collapse gave rise to a new set of security threats, including terrorism. An open report of Russia's Foreign Intelligence Service (SVR) published in 1993 highlighted the growing threat from weapons of mass destruction (WMD).[17] In particular, the SVR report noted concerns regarding the threat of illicit trafficking in fissile materials, WMD technology documentation, and nuclear explosive device components. According to the report, the main goal of illicit trafficking was potential profits from selling material, technology, or equipment on the black market. However, the report did not exclude the use of illicit trafficking for acquiring WMD with the purpose of blackmail. It also highlighted the threat of sabotage against nuclear sites with potential catastrophic consequences. The report noted that this threat was not only limited to acts of terrorism specifically targeting nuclear sites, but also concerned any armed conflict in the region where a nuclear site was located. This threat was particularly valid during the 1990s for Russia, which was experiencing severe conflict in the North Caucasus and civil unrest in other parts of the country.

Chechen 'terrorists' were a major security threat in 1990s Russia, and had demonstrated their capability and interest in non-conventional terrorism, or at least using the threat of such terrorism, when they buried a container with highly radioactive cesium-137 in a park in Moscow in November 1995.[18] In October 2001, General Igor Valynkin, Head of Russia's Ministry of Defence directorate in charge of maintaining nuclear warheads, stated at a press conference that earlier that year Russian security services had prevented two attempts of terrorists conducting reconnaissance at nuclear warhead maintenance and storage sites, although, he said no actual intrusion attempt had taken place.[19]

International Cooperation as a Factor in the Evolution of Nuclear Security

The overwhelming share of international nuclear security assistance to Russia has been provided by the US, with the process starting shortly after the Soviet Union's collapse. This intervention was a major factor in the evolution of nuclear security in Russia. Potential nuclear security challenges were foreseen by the international community even before the Soviet Union formally disintegrated, when the huge implications of the collapse became obvious. US experts, who observed developments and had contacts with their Soviet counterparts and government officials, appealed to the US government, highlighting the threat and calling for a cooperative programme to be initiated that would help the Soviet Union—or its successor states—to address the challenges.[20]

Soviet officials shared these concerns. While being cautious about giving the US access to sensitive elements of the Soviet nuclear complex, they managed to come up with potential areas for cooperation to address nuclear security issues. In October 1991, Viktor Mikhaylov, then Deputy Minister of Atomic Energy and Industry of the Soviet Union soon to become first Russian Minister of Atomic Energy, visited the US and met with experts and government officials. He argued that the potential bottleneck in implementation of recently concluded US–Russian disarmament agreements was due to the absence of safe and secure storage for nuclear materials removed from dismantled warheads and urged Congress to provide funding necessary to build storage facilities.[21] In response to these calls, the Senate in a bipartisan effort led by Senators Sam Nunn and Richard Lugar passed the Soviet Nuclear Threat Reduction Act on 27 November 1991. The Act, later known as the Nunn–Lugar Act, was signed into law by President George H. W. Bush on 12 December 1991.

The US–Russian Agreement Concerning the Safe and Secure Transportation, Storage and Destruction of Weapons and the Prevention of Weapons Proliferation (known as Cooperative Threat Reduction or CTR) was the first agreement to facilitate cooperation in the area of nuclear security. Funded in accordance with the Nunn–Lugar Act, the CTR agreement was signed by US President Bush and Russian President Boris Yeltsin in Washington, DC in June 1992. Many other agreements that were later concluded to define specifics of cooperation in nuclear security-related areas referred to the CTR agreement, which had established a general framework for bilateral cooperation.

During the first stages of cooperation, the most important aspect was funding provided to Russia for the improvement of nuclear security systems at specific sites, as well as the development of national nuclear security infrastructure, since economic conditions in Russia did not allow any substantial investment in these areas. While this initial financial assistance was crucial, Russia had no past experience with modern nuclear security approaches; thus, the most valuable contribution over the whole period of cooperation was expertise shared by experts from the US and other countries. This expertise was used to develop critical elements of national nuclear security infrastructure. International cooperation served as a source of nuclear security best practices, shared with Russia via training, workshops, conferences, and a supply of modern equipment and technologies.

Nuclear Site Upgrades

After the Soviet Union's collapse, security systems had deteriorated due to a shortage of resources necessary to maintain them. Work to improve security at nuclear sites included rapid upgrades focused on closing gaping holes in nuclear site security, such as building secure fences and securing windows and doors. These upgrades characterized the early stages of cooperation, with simple, quick, and cost-effective solutions utilized to substantially improve security over a short period of time at a relatively low cost. Comprehensive upgrades focused on increasing the effectiveness of security systems and involved sophisticated technology solutions, such as electronic sensors, video surveillance systems, central alarms stations, and computerized NMAC systems. These improvements were implemented as longer higher-cost projects during more advanced stages of cooperation, typically following rapid upgrade work.

To provide security upgrades for nuclear sites, US project teams worked alongside their Russian equivalents. This work included defining the scope of required upgrades, developing the design of the security system, and implementing the designed system at the site. Typically, the Russian teams proposed the upgrades and implemented the agreed upgrades by themselves or involved Russian contractors, while the US side provided funding, expert review, and advice, and confirmed the intended use of the funding. During this period, the US provided upgrades to a total of 25 sites (excluding the Ministry of Defence), with work on weapon laboratories, nuclear fuel cycle facilities, research institutions, and shipbuilding and icebreaker fleet support facilities.[22]

A mandatory condition imposed by US Congress for the inclusion of a site into the upgrade programme was the availability of Category I–II (often referred to as weapons-usable) nuclear material. There were more sites with such material eligible for upgrades, but Russia refused cooperation at these sites due to information security concerns. Most of the upgrade efforts were completed in the 2000s, although some site upgrade activities continued until the collapse of cooperation in 2014. The US and Russia also cooperated in improving the security of nuclear materials transportation. This work included the provision of armoured vehicles, transportation containers, and railroad cars, as well as development of an Automated Transportation Security System (ATSS) that allows ongoing monitoring of nuclear material transportation and communication of alarms to response forces in case of accidents during transportation.[23]

In addition to improving security of nuclear materials and sites, the US and Russia worked to decrease the amount of sites with attractive weapons-usable nuclear materials. This work was implemented within the framework of the Materials Consolidation and Conversion (MCC) project,[24] which aimed to consolidate nuclear materials attractive to potential malicious non-state actors, such as HEU and plutonium, into fewer, more secure locations—either by consolidating the material into fewer buildings within one site or completely removing the material from the site. This approach helped to decrease the number of potential theft targets and the equipment and personnel costs associated with securing the material, while also working to convert weapons-usable HEU into low enriched uranium (LEU),[25] which is less attractive for potential adversaries.

Nuclear Security Sustainability

Nuclear security is not a state that can be achieved through a single upgrade of a security system. Ensuring the nuclear security of sites and materials over an indefinite period of time in a changing threat environment is often referred to as 'sustainability'. Sustainability as a concept is operationalized through a set of elements[26] that must be implemented at both the national and site levels. This became a focus of cooperation in the early 2000s, when the US realized that external large-scale support for Russian nuclear security programmes could not continue indefinitely and that Russia would have to maintain nuclear security using its own resources. The two sides initiated several projects aimed at developing sustainability elements at Russian nuclear sites as well as capabilities supporting them at the national level.

Nevertheless, there were differences in the US and Russian understandings of sustainability, which slowed down progress. In Russia, as a rule, sustainability referred to sustaining the equipment—keeping it operable, making repairs, keeping supplies of spare parts and consumables on hand, and replacing broken equipment—often accomplished with continuing funding support from the US. The US side interpreted 'sustainability' as high nuclear security system performance over time, going beyond equipment maintenance. This required a set of management practices adapted to the specific material protection, control, and accounting (MPC&A) arrangements at each site. If Russians would embrace this interpretation of the term, it could help eliminate existing gaps in sustainable nuclear security and facilitate effective use of available resources. However, the US and Russia failed to reconcile their differences before co-operation was suspended in 2014.

Sustainability of the Russian nuclear security programme was ultimately defined by a combination of legacy approaches to managing the nuclear industry and nuclear security in particular, Russia's own efforts to adjust them to a new reality, and best practices obtained through international nuclear security cooperation. Still, Russia arguably lags behind in implementing certain internationally accepted best practices, for example, in using performance testing as a source of data for evaluating systems effectiveness.

Legal Basis for Nuclear Security in Russia

Systemic efforts to improve nuclear security in Russia started in 1994 when Yeltsin issued a decree, 'On priority measures to improve system of accounting and physical security of nuclear materials'. This decree ordered the relevant government agencies to review the status of nuclear MPC&A in Russia and take actions necessary to improve them.[27] The federal law On Atomic Energy Use, the first Russian act establishing the legal basis for any civilian nuclear activity, was enacted in November 1995. This law established MPC&A as a mandatory condition for any activity involving nuclear materials and facilities. The law obliged operators of nuclear sites to ensure physical protection, control, and accounting at their sites, government agencies managing the use of atomic energy to implement MPC&A at the national level, and regulatory bodies to ensure oversight.

In 1996–2000, the Russian government issued several regulations that defined responsibilities of all stakeholders in the area of MPC&A, rules for their interactions, and fundamental requirements for national- and site-level systems of nuclear MPC&A. Work to develop nuclear security regulations started in the late 1990s within the framework of US–Russian cooperation. Before 2014, Russia, supported by the US, managed to develop and enact several dozen regulations that ensured comprehensive coverage of nuclear security in Russia.

The current regulatory development process in Russia is well established, and key nuclear security regulations are reviewed and revised approximately every five years. Revisions are intended to reflect feedback from implementation and new technology developments, and best practices into new regulations. Nuclear sites must develop instructions for their personnel based on requirements established in higher-level regulations and taking into account specifics of each site's operations. Regulations establish a mandatory set of MPC&A instructions that must be developed by each site.

Many nuclear security regulations have been revised several times to reflect feedback from the implementation of previous versions as well as the growing capability of nuclear sites to implement stronger requirements. For example, the Basic Rules of Accounting and Control of Nuclear Materials, a key regulation defining NMAC requirements to Russian nuclear sites was first issued in 2001 and then revised in 2005, 2012, and 2019.[28] While the first three versions were developed with the US financial and expert support, the 2019 revision used only Russian financial and expert resources. This practice of regular revision of nuclear security requirements indicates maturity in the regulatory development process. However, despite the continuous improvement process, at the time of writing, Russian regulations still have weaknesses and do not reflect all available international best practices. These weaknesses are due to either legacy practices or unpreparedness of many nuclear facilities in Russia to comply with certain mandatory requirements due to lack of personnel skills or financial and technical resources.

Defining Responsibilities for Nuclear Security

Russian legislation refers to several groups of stakeholders with varying responsibilities for nuclear security. A key group includes regulatory bodies, managing agencies, and operators, while a supporting group includes protective forces and government agencies with functions that support nuclear security operations. In addition, contractors providing various essential nuclear security goods and services support the various suites of stakeholders.

Regulatory Bodies

Russia has different regulatory bodies for nuclear security in civilian and defence nuclear applications. The Federal Service for Technology, Environment and Nuclear Oversight—Rostechnadzor—regulates nuclear security in the civilian domain, while nuclear security in defence applications is regulated jointly by the Ministry of Defence and Rosatom (the state atomic energy corporation). The separation of activities and sites subject to civilian and defence regulatory authorities is not always clear. While there are organizations conducting purely civilian or defence nuclear activities, there are many organizations that conduct both and have licences from the two regulatory bodies.

Regulation of nuclear security in the civilian nuclear complex has evolved substantially since the Russian independent regulatory body—Gosatomnadzor, predecessor of Rostechnadzor—was established by the President's Decree on 31 December 1991, five days after the Soviet Union's dissolution. Gosatomnadzor was well established with solid legal foundations. The Law on Atomic Energy Use that applies to civilian applications only provided the legal basis for the regulation of nuclear security, while subsidiary legal acts defined Rostechnadzor's authority and procedures for implementation. The regulation of nuclear security in Russia's defence nuclear complex is less developed. There is no fundamental law that provides a legal framework for defence nuclear activities.[29] The authority of responsible agencies and requirements for nuclear security is established in a set of loosely connected legal acts. Some regulations, for example the Rules of Physical Protection established by the Government Decree, as well as many agency-level nuclear security regulations, apply to both civilian and defence domains.[30]

Managing Agencies

Russia inherited the system of 'managing agencies' from the Soviet Union, where there was no private business and all organizations were owned by the state and reported to a specific ministry in accordance with their area of activity. Managing agencies are essentially government agencies with nuclear facilities under their control. Rosatom manages most of Russia's nuclear sites, although the Ministry of Defence has a substantial number of nuclear facilities under its control in addition to military units handling nuclear weapons. The Ministry of Defence's nuclear facilities comprise naval bases, where nuclear submarines and nuclear-powered surface ships are based, including those where nuclear fuel reloading is performed and fresh and spent nuclear fuel is stored. Another agency with a substantial nuclear complex under its control is the Ministry of Industry and Trade, which manages several shipbuilding sites that construct, maintain and repair nuclear-powered submarines and surface ships. The Russian Ministry of Science and Education manages two Russian universities that operate nuclear research reactors: Moscow Engineering Physics Institute (MEPhI) and Tomsk Polytechnical University (TPU).

Managing agencies implement state nuclear security policy and provide feedback for review and improvement, manage federal funding for nuclear security, coordinate nuclear security activities of subordinate nuclear sites, and reconcile certain nuclear security-related decisions on nuclear sites. While many nuclear security services can be provided by outside contractors, managing agencies with substantial resources, such as Rosatom and Minpromtorg, have in-house organizations providing these services, including vulnerability analysis, nuclear security system design and installation, equipment manufacturing, and many more. In addition to functions typical of any managing agency, Rosatom coordinates nationwide nuclear security activities of all agencies with nuclear security responsibilities, as well as maintaining NMAC at the national level, including through a national registry of nuclear materials. Rosatom also serves as Russia's point of contact for the purposes of implementation of international nuclear security agreements.[31]

Nuclear Site Operators

As in most countries with a nuclear sector, the ultimate responsibility for the security of a specific nuclear site or material belongs to the operator. Operating a nuclear facility or handling nuclear material without proper MPC&A measures is prohibited. The operator must obtain a licence from the regulatory body and one of the mandatory conditions for obtaining the licence is confirming capabilities to ensure nuclear security. While the operator is ultimately responsible for nuclear security, it can involve outside contractors to perform many nuclear security activities, such as vulnerability assessment and physical protection system effectiveness evaluation, design and construction of nuclear security systems, personnel training, and pro-force services. For some activities, contractors must obtain a licence from the regulatory body as well. While Russia's distribution of responsibility for nuclear security is similar to other countries, there are local specifics due to its legacy of centralized management without independent regulation and lack of clear borders between civilian and defence nuclear complexes.

Protective Forces

Nuclear sites in Russia are protected by armed protective forces. These forces can be provided either by the National Guard, which has responsibility for the protection of hazardous and critical infrastructure sites, or by agency pro-force organizations reporting to a managing agency. At the time of writing, Rosatom is the only managing agency that has substantial agency pro-forces that provide protection to Rosatom nuclear sites together with the National Guard. Nuclear sites of other agencies are protected either by federally funded National Guard troops or by the National Guard organization that provides protective services on a contractual basis. Ministry of Defence nuclear sites are protected by Ministry of Defence military units.

Other Government Agencies

There are several additional government agencies that have no nuclear sites under their control, but they implement a broad range of functions related to nuclear security, such as security and intelligence, pro-forces, nuclear materials transportation, nuclear materials measurements support, and cyber security. Cyber security requirements applicable to a broad range of computerized systems, including nuclear security systems, were introduced in Russia in the early 1990s while the first cyber security regulations specific to nuclear security were introduced in 1997.[32] Due to the range of technologies used and available information technology (IT) expertise, Rosatom was the leading Russian agency in terms of implementing advanced information protection approaches for its computerized systems. Cyber security became part of the US–Russian nuclear security cooperation in the early 2000s, when the parties worked together to ensure that nuclear security equipment installed at Russian nuclear sites complied with cyber security requirements.[33] As part of this cooperation a set of regulations establishing requirements for the cyber security of nuclear security systems and infrastructure necessary to implement these requirements were developed.

Personnel Education and Training in Nuclear Security

Efforts to develop nuclear security education and training capabilities in Russia started in the early 1990s. Nuclear security training included relatively short-term programmes (typically ranging from three days up to two weeks) aimed at developing specific nuclear security skills, for example, conducting vulnerability analysis and physical inventories. Today, training typically targets current employees of nuclear sites with nuclear security responsibilities or other organizations involved in nuclear security-related activities, such as nuclear materials transportation. Russian regulations require that nuclear security personnel pass training once every three to five years, depending on the category of personnel. Training is conducted either in specialized training centres or at nuclear sites.[34]

Two specialized training centres intended to provide training in MPC&A have been established in Obninsk, a small city 100 km from Moscow, which hosts several nuclear research institutions. The Interdepartmental Special Training Center (ISTC) was established in 1993 in accordance with a government order to provide physical protection training. The Russian Methodological and Training Center on Nuclear Materials Control and Accounting (RMTC) was established in 1994 by Rosatom and Rostechnadzor as a department within the Institute of Physics and Power Engineering (IPPE), one of the leading Russian nuclear energy research and development institutions. ISTC was merged into the newly established Rosatom Technical Academy in 2017.

Both training centres received substantial support from the US within the framework of US–Russian nuclear security cooperation. This support was used to develop training facilities, buy equipment to be used for training purposes, provide training to instructors (including through multiple visits to US national laboratories), develop dozens of training courses, and evaluate Russian nuclear industry training needs. US funding was also used to cover costs of nuclear sites to send employees for training at ISTC and RMTC, including tuition fees and travel costs. Although US support was terminated in 2014, these two centres have continued to provide training to personnel of nuclear sites, managing and regulatory agencies, and other organizations. Their curriculum includes several dozen training courses on a broad range of MPC&A topics. Individual nuclear sites took on responsibility to cover the costs of training their personnel.

Russian regulations require nuclear sites to establish training programmes for their employees and conduct regular evaluation of professional qualifications. Russian regulations provide model job descriptions and qualification requirements for nuclear security personnel. These qualification requirements can be used as a basis for developing training courses. In addition to training centres and training programmes at nuclear sites, there are two university-level programmes in Russia—at MEPhI and TPU—that focus on the security of nuclear materials and sites, again established within the framework of US–Russian nuclear security cooperation.[35] The programmes have substantially impacted the evolution of nuclear security in Russia, with previous graduates now holding management positions in Rostechnadzor's nuclear security department.

Funding for Nuclear Security in Russia

In the period following the Soviet Union's collapse, foreign assistance—with the lion's share of it provided by the US—was critical in terms of covering the costs of improving security at nuclear sites and developing national nuclear security infrastructure. Other initiatives, such as the International Science and Technology Center, have helped employ thousands of nuclear site staff who have critical knowledge of and access to nuclear materials.[36] In addition to this direct assistance, substantial US funding, including for nuclear security expenses and salaries, was provided through a HEU–LEU deal that facilitated the sale of HEU obtained from dismantled Russian nuclear weapons for down-blending and use as an LEU fuel for US nuclear power plants.[37] The total revenue from the HEU–LEU deal over the 20 years of its implementation, 1994–2013, could amount to as much as US$17 billion.[38] Improvements in the Russian economy from the 2000s to the 2010s led to higher domestic funding for nuclear security. Rosatom's export of nuclear power plants and associated technologies, goods, and services also contributed to a better financial situation for the Russian nuclear industry.

There are three key mechanisms for funding nuclear security in Russia. The federal budget funds national-level programmes (for example, regulations development and maintenance of national infrastructure), pays for pro-forces provided by the federally funded National Guard, and covers part of the nuclear industry's expenses for nuclear security. Nuclear sites must allocate a percentage of their revenue to a centralized nuclear security funding reserve. This centralized nuclear security funding reserve allows collection and redistribution of funding between multiple nuclear sites to cover capital expenses (for example, large-scale nuclear security upgrades at nuclear sites), without placing an extra burden on the budgets of individual sites. Sites must cover nuclear security expenses from their own budgets.

Russia's capacity to provide sufficient funding for nuclear security was a major international concern after the suspension in 2014 of US–Russian nuclear security cooperation. However, there is much evidence that current funding is sufficient. In particular, such evidence can be obtained through a nuclear industry procurement information system available to the public. According to Russian legislation, Rosatom and its subsidiary organizations must publish information related to procurements using federal budget funding as well as their own money, including procurements of goods and services related to nuclear security. This information is published at a dedicated website.[39] Analysis of this information shows that, since the suspension of US–Russian nuclear security cooperation in 2014, expenditure for nuclear security is comparable to funding support provided by foreign donors during the peak years of cooperation and amounts to hundreds of millions of US dollars annually.

Evaluation, Feedback, and Continuous Improvement

Over 30 years after the collapse of the Soviet Union, Russia has established a multilayered system for evaluating the status of nuclear security systems and the capability to defend against current threats. Elements of this system include vulnerability analysis and evaluation of the effectiveness of physical protection systems, site-level self-evaluation, agency monitoring by managing agencies, and oversight inspections conducted by regulatory bodies. Feedback from these activities is used to improve nuclear security. Russia's approach to these processes is generally consistent with international best practices, although there are some notable weaknesses. Vulnerability analysis and effectiveness in Russia only addresses physical protection systems and does not address NMAC—another critical element of overall nuclear security. There are also concerns regarding sources of data used as an input into system effectiveness evaluation.

From the Collapse of Cooperation to the Present

In late 2014, Russia declared its unwillingness to continue cooperation with the US. While this step was triggered by the general decline in US–Russian relations over multiple issues— the developments in Ukraine and Crimea in particular—there were also reasons related to cooperation itself that contributed to the termination. These reasons included dissatisfaction of both parties with the donor–recipient model of cooperation and general fatigue from the cooperation, where parties did not like the way cooperation was evolving but were unable to transform the cooperation framework in a mutually acceptable way.[40] Nevertheless, by that time cooperation had managed to achieve many of its goals. The security of multiple nuclear sites hosting hundreds of tonnes of weapons-usable nuclear material was substantially improved, Russia had built a national infrastructure supporting nuclear security, and nuclear security as a distinct area of activity had been established. Russia had also achieved substantial progress in developing elements of a sustainability programme to ensure it was capable of protecting nuclear sites over an indefinite period of time, without outside support, and in a changing threat environment.

Most of the concerns about Russia not being capable of maintaining its nuclear security programme independently have not been realized. Russia continues upgrades of nuclear security equipment at nuclear sites, trains nuclear security personnel, conducts inspections and follows up on findings, regularly reviews and revises nuclear security regulations, maintains other elements of national nuclear security infrastructure, and provides funding for these activities. However, there are issues stemming from the termination of cooperation that may decrease effectiveness and slow the progress of nuclear security in Russia.

First, international cooperation and associated interactions with foreign nuclear security experts and organizations provided excellent opportunities for peer reviews. These peer reviews were a source of fresh observations and independent insights and recommendations for further improvements. The need to prepare for such peer reviews and reveal results of nuclear security work to outside observers also motivated Russian organizations and nuclear security experts to ensure good performance and a favourable outcome from reviews. Obligations under cooperation contracts with foreign organizations, including multiple regular deliverables, also motivated day-to-day nuclear security work.

Second, international cooperation provided a venue for the exchange of nuclear security best practices—through joint projects, training courses, workshops, and conferences. When international cooperation ceased, opportunities for such exchanges became extremely limited, as by terminating joint projects, the Russian authorities also limited opportunities for Russian nuclear security experts to attend international nuclear security events. Third, while the US and Russian approaches to nuclear security have a lot in common, there are different priorities. Some activities that were supported within the framework of cooperation have lower priority for Russia, and therefore do not receive sufficient funding support.

While Russia has demonstrated its overall ability to ensure nuclear security without outside support, relying only on its own resources, there are several issues inherent to Russia in general, and to the Russian nuclear industry in particular, which can negatively affect nuclear security in Russia and increase the risk of nuclear material theft or successful act of sabotage against nuclear facilities. One of the most critical issues is pervasive corruption. The Russian nuclear complex is praised domestically and internationally for its anti-corruption efforts, but it cannot be isolated from the overall corruption environment in Russia. There are known cases when personnel of nuclear organizations in senior positions, including in positions related to nuclear security, abused their authority for personal gains. There is no evidence that these violations had the specific intent to harm nuclear security, steal nuclear material, or sabotage nuclear facilities. However, these crimes could result in severe nuclear security consequences.

The second issue is the lower priority given to nuclear security by nuclear sites and overall nuclear industry management compared to other aspects of operation, such as revenues and nuclear safety. This is evidenced by the extremely limited coverage of nuclear security in annual reports of Rosatom and nuclear sites, the absence of nuclear security performance from the list of key performance indicators against which a nuclear site and its management performance are evaluated, and lack of any substantial expert discussion of nuclear security issues.

Conclusion

Nuclear security in Russia has evolved substantially since the collapse of the Soviet Union. Developments in this area eventually led to the establishment of nuclear security as a distinctive profession that did not exist in the Soviet era. Russia has managed to transform its nuclear security programme from a system that worked well in the tightly controlled environment of the Soviet Union to a system appropriate to today's threat environment. International nuclear security cooperation, primarily with the US, was a major factor in this transformation. Such cooperation provided financial support for nuclear security improvements, when Russia could not afford them, and served as a source of nuclear security best practices. While international nuclear security cooperation collapsed in 2014, Russia has continued to improve nuclear security systems without outside support. Nevertheless, weaknesses remain and, consequently, renewed international cooperation would help further improve the effectiveness of Russia's nuclear security systems.

Notes

1. See, for example, International Panel on Fissile Materials, Global Fissile Material Report 2015: Nuclear Weapon and Fissile Material Stockpiles and Production, Eighth Annual Report. https://fissilematerials.org/library/gfmr15.pdf; David Albright, William Walker, and Frans Berkhout, *Plutonium and Highly Enriched Uranium 1996: World Inventories, Capabilities and Policies*, Stockholm International Peace Research Institute (SIPRI), SIPRI Monographs (Oxford: Oxford University Press, 1997).
2. Committee on International Security and Arms Control, *Proliferation Concerns: Assessing U.S. Efforts to Help Contain Nuclear and Other Dangerous Materials and Technologies in the Former Soviet Union* (Washington, DC National Research Council, 1997); Jon Brook Wolfsthal, Cristina-Astrid Chuen, and Emily Ewell Daughtry, eds., *Nuclear Status Report* (James Martin Center for Nonproliferation Studies (CNS), 2001).
3. Dmitry Kovchegin, 'International Nuclear Security Forum Country Update: Russian Nuclear Security', Stimson Center [online], 12 May 2021. https://www.stimson.org/2021/international-nuclear-security-forum-country-update-russian-nuclear-security

4. National Intelligence Council, *Annual Report to Congress on the Safety and Security of Russian Nuclear Facilities and Military Forces* (Washington, DC: NIC, 2004).
5. For a list and map of nuclear closed cities in the FSU, see Jon Brook Wolfsthal et al., *Nuclear Status Report.*
6. See V. N. Kuznetsov, *Regime of Secrecy at Factory #817*, State Security Agencies (Yekaterinburg: Sokrat, 2018), 244–53. [In Russian] http://elib.biblioatom.ru/text/kuznetsov_obespechenie-sekretnosti-zavoda-817_2018
7. Elena A. Kozlova, *Evgeny Trofimovich Mishin: Ideas Put into Effect* (Moscow: IzdAT, 2013) [In Russian]. elib.biblioatom.ru/text/kozlova_mishin_2013
8. Interview with Leonid Smirnov (*Frontline*, PBS, 1996). https://www.pbs.org/wgbh/pages/frontline/shows/nukes/interviews/smirnov.html
9. Dmitry Kovchegin, *Developing a Nuclear Material Control and Accounting System in Russia*, CISSM Working Paper, Center for International and Security Studies [online], 1 December 2013. https://cissm.umd.edu/research-impact/publications/developing-nuclear-material-control-and-accounting-system-russia
10. Ibid.
11. See Carnegie Endowment for International Peace and Monterey Institute of International Studies, *Nuclear Successor States of the Soviet Union: Nuclear Weapons, Fissile Materials, and Export Controls in the Former Soviet Union*, Nuclear Status Report, No. 5, Table II-B, 'Highlights of Significant Fissile Material Smuggling Cases' (Washington, DC: Carnegie Endowment for International Peace, 1998).
12. PBS Interview with Leonid Smirnov.
13. Ibid.
14. Ibid.
15. Ibid.
16. Ibid.
17. Foreign Intelligence Service of the Russian Federation, 'A New Kind of Terrorism', SVR [online], 2019 [In Russian]. http://svr.gov.ru/material/otkrytye-doklady-svr-rossii/novyy-vyzov-posle-kholodnoy-voyny-rasprostranenie-oruzhiya-massovogo-unichtozheniya-otkrytyy-doklad-/rasprostranenie-oruzhiya-massovogo-unichtozheniya-ugroza-vseobshchey-stabilnosti/novyy-vid-terrorizma.htm
18. Elena Masyuk, 'Main Enemy', *The New Times*, 22/250 [online], 25 June 2012 [In Russian]. https://newtimes.ru/articles/detail/53769
19. Pavel Koryashkin, 'Defence Officials Registered Two Cases of Terrorist Reconnaissance at Nuclear Warhead Storage Sites This Year—Ministry of Defence Representative', *ITAR-TASS* [online], 25 October 2001 [In Russian]. https://pircenter.org/wp-content/uploads/2022/10/YK_October-31-2001.pdf
20. See, for example, Ashton Carter, 'Origins of the Nunn–Lugar Program', paper presented at the Presidential Conference on William Jefferson Clinton: The 'New Democrat' from Hope, Hofstra University, 10–12 November 2005. https://www.belfercenter.org/sites/default/files/legacy/files/hofstra_presidential_conference_presentation_november2005.pdf; Graham Allison, 'What Happened To The Soviet Superpower's Nuclear Arsenal?', Discussion Paper No. 2012-04 (Cambridge, MA: Harvard Kennedy School: Belfer Center for Science and International Affairs, 2012). https://www.belfercenter.org/sites/default/files/files/publication/3%2014%2012%20Final%20What%20Happened%20to%20Soviet%20Arsenals.pdf

21. Mariana Budjeryn, Simon Saradzhyan, and William Tobey, '25 Years of Nuclear Security Cooperation by the US, Russia and Other Newly Independent States: A Timeline', Russia Matters [online], 16 June 2017. https://www.russiamatters.org/analysis/25-years-nuclear-security-cooperation-us-russia-and-other-newly-independent-states; V. N. Mikhaylov, 'Keys to the Nuclear Arsenal', in *I am a Hawk* (Moscow: Government Bulletin 1, 1992) [In Russian]. http://elib.biblioatom.ru/text/mihaylov_ya-yastreb_2008/go,100/?bookhl=
22. US Government Accountability Office (GAO), 'Nuclear Nonproliferation: Past U.S. Involvement Improved Russian Nuclear Material Security, but Little Is Known about Current Conditions', GAO-20-392 [online], February 2020. https://www.gao.gov/assets/gao-20-392.pdf
23. Nikolai N. Shemigon, Yuri L. Davydov, Vladimir M. Chernyshev, William J. Reich, Timothy D. Welch, and William C. Brunsdon, 'Main Aspects of the Automated Transportation Security System Development', in *Proceedings of the 51st Annual Meeting of the Institute for Nuclear Materials Management (INMM)*, 11–15 July 2010, Baltimore, Maryland (Red Hook, NY: Curran Associates, 2010).
24. Thomas Wander, 2003, 'The Material Consolidation and Conversion Project: Current Status and Future Prospects', in *Proceedings of the 44th Annual Meeting of the Institute for Nuclear Materials Management (INMM)*, 13–17 July 2003, Phoenix, Arizona (Red Hook, NY: Curran Associates, 2010).
25. Ibid.
26. Seven sustainability elements were used within the framework of US–Russian nuclear security cooperation. The IAEA also defines sustainability elements in its Implementing Guide 'Sustaining a Nuclear Security Regime', IAEA Nuclear Security Series, No. 30-G. These two sets of definitions differ in detail but are similar in general scope.
27. Government of the Russian Federation, 'On priority measures to improve system of accounting and physical security of nuclear materials', Decree of the President of Russian Federation, No. 1923, 15 September 1994.
28. For the latest revision of 'Basic Rules of Accounting and Control of Nuclear' (NP-030-19), see https://www.secnrs.ru/orders/ord-438-2019.pdf [In Russian.]
29. Matthew Bunn and Dmitry Kovchegin, 'Nuclear Security in Russia: Can Progress Be Sustained?', *The Nonproliferation Review*, 24/5–6 (2018), 527–51.
30. Ibid.
31. Authorities of managing agencies related to nuclear security are defined in 'Rules of Physical Protection of Nuclear Materials, Nuclear Facilities and Storage Sites of Nuclear Materials' approved by Government Decree #456 of 19 July 2007, and 'Regulation on the State System for Nuclear Material Accounting and Control' approved by Government Decree #352 of 6 May 2008.
32. See, for example, Guido Gluschke et al., 'Cyber Security at Nuclear Facilities: National Approaches', Institute for Security and Safety (Brandenburg: Brandenburg University of Applied Sciences, 2015). https://www.nti.org/analysis/articles/cyber-security-nuclear-facilities-national-approaches
33. Personal recollections of Dmitry Kovchegin, who worked as a consultant supporting relevant projects.
34. Bunn and Kovchegin, 'Nuclear Security in Russia'.
35. Rosatom, 'Performance of State Atomic Energy Corporation Rosatom in 2020', Moscow, 2021.

36. Nuclear Threat Initiative (NTI), 'International Science and Technology Center (ISTC)', NTI [online], 2022. https://www.nti.org/education-center/treaties-and-regimes/international-science-and-technology-center-istc
37. Aleksandr Pavlov and Vladimir Rybachenkov, 'HEU-LEU Project: A Success Story of Russian–US Nuclear Disarmament Cooperation', Center for Energy and Security Studies [online], 27 May 2013. http://ceness-russia.org/data/doc/HEU-LEU_Agreement_ENG.pdf
38. Ibid.
39. Rosatom, Official procurement portal. https://zakupki.rosatom.ru/en
40. Dmitry Kovchegin, 'Nuclear Material Protection, Control, and Accounting: New Realities', *Security Index: A Russian Journal on International Security*, 20/1 (2014), 79–86.

Suggested Reading

Bunn, Matthew and Kovchegin, Dmitry, (2018), 'Nuclear Security in Russia: Can Progress Be Sustained?', *The Nonproliferation Review*, 24/5–6, 527–51.

Hecker, Siegfried S., ed., (2016), *Doomed to Cooperate: How American and Russian Scientists Joined Forces to Avert Some of the Greatest Post-Cold War Nuclear Dangers* (Los Alamos, NM: Bathtub Row Press).

Nuclear Threat Initiative, (NTI), Securing the Bomb. [A series of reports issued in 2004, 2005, 2006, 2007, 2008, and 2010, focused on progress in locking down vulnerable nuclear materials.] https://www.nti.org/about/programs-projects/project/securing-bomb

Kovchegin, Dmitry, (2007), 'A Russian Perspective on Cooperative Threat Reduction', *BCSIA Discussion Paper* 2007-04 (Cambridge, MA: Harvard Kennedy School: Belfer Center for Science and International Affairs). https://www.belfercenter.org/sites/default/files/files/publication/kovchegin_2007_04.pdf

Kovchegin, Dmitry, *Developing a Nuclear Material Control and Accounting System in Russia*, CISSM Working Paper, Center for International and Security Studies [online], 1 December 2013. https://cissm.umd.edu/research-impact/publications/developing-nuclear-material-control-and-accounting-system-russia

CHAPTER 26

NUCLEAR SECURITY IN CHINA

XIAOBING GUO

Introduction

China[1] is a global actor worth special attention as far as nuclear security is concerned. On one hand, it has arguably the most ambitious nuclear power development plan in the world—about a quarter of the nuclear power reactors under construction today are in China.[2] On the other hand, it has maintained one of the best nuclear security records. According to the Chinese authorities, it 'has never ever lost either a single nuclear item or a gram of fissile material'.[3] How does China manage to do it? What kind of challenges does China face? What is the future of nuclear security in China? This chapter addresses these questions.

Overview of China's Nuclear Industry

Compared with other nuclear energy-producing countries, China has a relatively short history of nuclear power development. China issued its first nuclear power plan in 1970, although its first nuclear power plant was not commissioned until 1991. China's nuclear power sector has enjoyed rapid development in the past three decades. In 2020 there were 48 commercial nuclear power units in China with a total installed capacity of 49.88 gigawatts (GW), the second-highest nuclear power generator in the world.[4] The number of units under construction has consistently ranked first in the world for many years. According to the 'Fourteenth Five-Year Plan for National Economic and Social Development and Outline of Long-Term Goals for 2035', the installed operating capacity of nuclear power in China will reach 70 GW by 2025.[5] By 2030, that number is forecast to reach 120 GW, with nuclear power predicted to increase from 4.94 per cent to approximately 8 per cent of national power generation capacity.[6] China also operates 16 research nuclear reactors for the purpose of particle physics and nuclear physics, radiochemistry, activation analysis, materials science, isotope production, nuclear power, and nuclear medicine research.[7]

China has established a relatively complete nuclear power industry supply chain, which includes reactor design and construction and fuel cycle capabilities. It also explores emerging nuclear power technologies such as large pressurized water reactors, high-temperature gas-cooled reactors (HTGRs), and fast reactors. Although it is a relative newcomer to the nuclear energy sector, China now ranks at the forefront of the world in the field of nuclear power technology development, such as indigenous designs.

There are two main reasons behind the rapid nuclear power development in China. First, China's rapid economic growth has led to breath-taking developments in almost all energy sectors, including hydropower, solar power, and wind power, and nuclear power development is just a small part of that bigger picture. Second, concerns about environmental pollution and climate change are prompting more interest in nuclear power. As in other countries, nuclear power is regarded as an important alternative to fossil fuels. China also tries to export its nuclear technologies to overseas markets. Since the 1990s, China has built six reactors for Pakistan, all of which have a good operational record. China also signed an agreement with Argentina for the construction of the Hualong-1 reactor in 2020, and the project is scheduled to start by the end of 2022.[8]

Evolution of Nuclear security in China

International Engagement and Regulatory Development (1980s–1990s)

During this period, China evolved from a relative 'outsider' of the international nuclear security architecture to an active participant. As China opened up to the outside world in the late 1970s and early 1980s, its more positive attitude towards the nuclear non-proliferation regime and its needs for nuclear power development enabled it to engage with the international security architecture. Against that background, China joined the International Atomic Energy Agency (IAEA) in 1984, the Convention on the Physical Protection of Nuclear Material in 1989, the Treaty on the Non-Proliferation of Nuclear Weapons in 1992, and the Zangger Committee in 1997.

At the time when China started to develop its nuclear power industry, the 1986 Chernobyl nuclear accident occurred. This serious accident raised great concerns about nuclear safety and security in China. The Chinese government set the principle of nuclear power development as 'safety and security first, quality first' at this time.[9] Admittedly the priority was prevention and management of nuclear accidents, but the establishment and improvement of a nuclear emergency response system laid solid ground for nuclear security. For example, China paid close attention to the protection of nuclear facilities and developed a stringent confidentiality system. Against that background, China promulgated various domestic regulations related to nuclear security and introduced best practices for physical protection and materials.[10] China also established systems of nuclear material accounting and control (NMAC), nuclear emergency management, and nuclear export control, consistent with international practices.

Responding to an Increase in Terrorist Threats (2001–2009)

The 11 September 2001 ('9/11') terrorist attacks in the United States (US) alerted the world to the danger of potential nuclear terrorist attacks. The United Nations (UN) Security Council passed resolution 1540 to call all states to fight together against weapons of mass destruction (WMD) terrorism. China ratified the Amendment to the Convention on the Physical Protection of Nuclear Material (A/CPPNM) on 28 October 2008 and the International Convention for the Suppression of Nuclear Terrorism (ICSANT) on 28 August 2010. It also revised some domestic regulations and strengthened its nuclear security capabilities to reflect the new security landscape. In order to further restrict the behaviour of relevant domestic enterprises and units in exporting controlled items and technologies, and to make such organizations more self-disciplined, the Chinese government passed the amendment to the Criminal Law in December 2001 to criminalize the illegal manufacturing, trading, and transport of radioactive materials.[11] After joining the Nuclear Suppliers Group in May 2004, China revised the Regulations on Nuclear Export Control and Regulations on Export Control of Nuclear Dual-Use Products and Related Technologies by listing 'preventing acts of nuclear terrorism' for legislative purposes.[12]

Further Reform and Institutional Development (2010 to Present)

The US-initiated Nuclear Security Summits served to raise global awareness about nuclear security. China actively participated in that process and tried to integrate nuclear security into its general national security system. As a result, a series of laws and regulations related to nuclear security have been passed or updated in China. The State Security Law (2015) and Anti-Terrorism Law (2015) made it clear in legal terms that nuclear security is a vital aspect of national security and anti-terrorism issues.[13] China passed the Nuclear Safety Law on 1 September 2017. Although it is titled nuclear safety, it also addresses nuclear security. The Export Control Law was then passed on 17 October 2020, which enhanced China's export control system, including its nuclear export control system. The draft Atomic Energy Law and the Regulations on Nuclear Security are still in the process of being revised. In addition, some new departments or institutions related to nuclear security, such as the State Nuclear Security Technology

Center (SNSTC) and the Center of Excellence (CoE) on Nuclear Security, the Bureau of Industry, Security, Import and Export Control under the Ministry of Commerce, have all been established since 2011.

China's Nuclear Security Policy and Legal Framework

Nuclear Security Policy

In Mandarin, the term nuclear security has three different translations. Traditionally it was translated into HeBao An (核保安) or He An Bao (核安保), which are different from He An Quan (核安全), the Chinese translation of nuclear safety. But since the beginning of the Nuclear Security Summit process, He An Quan (核安全) has replaced He An Bao (核安保) as the official translation of nuclear security in the context of diplomatic communication and mass media coverage, while professional nuclear experts and arms control experts insist on using He An Bao (核安保) to discuss nuclear security.[14] Despite the confusion caused by different translations, it is clear that the Chinese government and nuclear sectors understand the technical meaning of nuclear security and its place at the top of the national security agenda. There are three reasons that explain why nuclear security is so important for China.

First, nuclear security is critical to China's energy security, general economic development, and environmental protection. From China's perspective, nuclear power can effectively help reduce carbon emissions and mitigate the impacts of climate change. Nuclear security, together with nuclear safety create the necessary preconditions for the implementation of China's large-scale nuclear power industry development plan. Second, nuclear security is critical to social stability and public safety. Nuclear terrorism is a real threat to the security of China and the world. History indicates that serious nuclear incidents not only pollute the environment and injure people in neighbouring areas, but also might create a systematic social and political crisis throughout a country, regardless of its size. Third, a nuclear security incident might affect world peace and thus threaten China's external environment of security and development. Indeed, at the 2016 Nuclear Security Summit held in Washington, DC Chinese President Xi Jinping recognized that 'incidents will have impacts that go beyond national borders' and expressed hope that all countries can work together to forge 'a community of shared future on nuclear security where we can enjoy the benefit of nuclear energy and fight against nuclear security threats together'.[15]

The Four Equal Emphases

President Xi defined China's approach to nuclear security as 'sensible, coordinated and balanced' in his statement at the 2014 Nuclear Security Summit held in The Hague.[16]

In his statement he explained that the essence of the Chinese approach is 'four equal emphases'.[17] The first emphasis is on development and security.[18] China holds that the nuclear industry must develop in a context of guaranteed security. Only by achieving development via these means can China truly control security risks, and only by guaranteeing security can China achieve sustainable nuclear energy development.

The second emphasis is on rights and obligations.[19] International nuclear security can only be promoted on the basis of respect for the rights and interests of all countries. Countries have the primary responsibility to fulfil their nuclear security obligations. But at the same time, China holds that the process of international nuclear security should take into consideration the specific situation of different countries and be advanced in a 'fair and pragmatic' way.[20] The right of all countries to adopt their own nuclear security policies and measures suited to their national conditions, and their right to protect sensitive nuclear security information should be respected.

The third emphasis is on independent efforts and coordination.[21] From China's perspective, universal nuclear security must be pursued with a mutually beneficial approach. Nuclear security is first and foremost a national goal, and the primary responsibility must be borne by national governments. But, at the same time, international cooperation is very important. China calls the world to work together to build a global nuclear security architecture featuring fairness and win–win cooperation.

The fourth emphasis is on treating symptoms and addressing root causes.[22] When China talks about efforts to enhance nuclear security, it not only means improvement of relevant policies and measures and development of related technologies. It also means broader international cooperation on efforts to counter the drivers of nuclear proliferation and nuclear terrorism.[23]

Legal Framework for Nuclear Security

Through decades of efforts, China has put in place a fairly complete legislative and regulatory framework for nuclear security. China's legal system for nuclear security includes laws at four levels: state laws, administrative regulations, department regulations, and normative documents. The law is formulated by the National People's Congress and the Standing Committee of the National People's Congress and promulgated by a presidential order, which has a more significant effect than administrative regulations, departmental rules, and normative documents. There is not an overarching state law governing nuclear security in China, but more than 20 state laws related to nuclear security.

As mentioned, the National Security Law has a separate article (Article 31) about security and safety in the nuclear area that identifies the role of nuclear security in the wider arena of national security, and sets the goals of nuclear security. The Counterterrorism Law requires production and import entities to add electronic track labels to nuclear and radioactive articles and requires a quick response to cases of stolen or otherwise missing nuclear and radioactive articles. The Law on Prevention and Control of Radioactive Pollution has provisions on the 'prevention and control of radioactive pollution that occurs in the process of the site selection, construction, operation, and retirement of nuclear facilities as well as the development and utilization of nuclear technologies, uranium (thorium) mines and associated radioactive mines'.[24] The Nuclear Safety Law passed in 2017 focuses on nuclear safety, but it also stipulates that the security of nuclear facilities and materials should be enhanced (Article 12) and encourages international exchange and cooperation related to nuclear security (Article 13).[25]

The most relevant legislation, however, is the long overdue Atomic Energy Law. The legislative process was delayed several times due to administrative reform and other reasons for several decades, but it has ramped up in recent years. In September 2018, the Ministry of Justice released the draft Atomic Energy Law for public comments. In June 2021, the draft Atomic Energy Law was submitted to the Standing Committee of the National People's Congress for deliberation. That draft has a separate article about nuclear security that requires the state to 'strengthen the security of atomic energy research, development and utilisation activities', and to 'prevent and combat relevant theft, destruction, unauthorized contact, illegal transfer or other malicious acts, and prevent nuclear terrorism'(Article 39). It includes another article about NMAC (Article 40).[26]

The second level of nuclear security legislation includes two dozen administrative regulations, promulgated by the State Council, that address issues about safety supervision and administration of civil nuclear facilities, control of nuclear materials, nuclear emergency management, cyber security, nuclear export control, internal security, safety and protection of radioisotopes and radiation devices, transportation safety of radioactive materials, safety management of radioactive waste, etc. The most relevant is the Regulations on Nuclear Security, which is still in the process of being drafted. The regulations drafted by the State Administration of Science, Technology and Industry for National Defense (SASTIND) aim to standardize and strengthen China's nuclear security work. As the first special regulation in the field of nuclear security in China, the regulations clarify the concept of nuclear security as well as define its scope and put forward its basic principles, technical measures, and management requirements; it also includes information regarding the punishment measures for the violation of relevant regulations. That draft was released for public comments in June 2016.[27]

The third level of nuclear security legislation includes about a dozen departmental regulations. These departmental regulations promulgated by relevant departments of the State Council within the scope of their authority are to address issues such as details about the Implementation of the Regulations on the Control of Nuclear Materials, Physical Protection of International Transportation of Nuclear Materials, and the transportation of radioactive materials, etc.

The fourth level of nuclear security legislation includes some normative documents promulgated by departments of the State Council or local governments to explain or supplement nuclear security regulations and recommend relevant methods and procedures. For example, SASTIND, the Ministry of Foreign Trade and Economic Cooperation, and the General Administration of Customs promulgated Administrative Measures for the Examination and Approval of Nuclear Product Transshipment and

Transit Transportation, and the State Environmental Protection Administration promulgated Methods for the Classification of Radioactive Sources. The National Nuclear Accident Emergency Coordination Committee and the National Nuclear Response Committee promulgated Measures for the Administration of Nuclear Emergency Reporting.[28]

Nuclear Security Regulation in China

There are more than a dozen departments involved in nuclear security governance in China, of which the China Atomic Energy Authority (CAEA) and the National Nuclear Safety Administration (NNSA) are the most important actors. The CAEA is under SASTIND and the Ministry of Industry and Information Technology. It is the authority for nuclear industry development in China and responsible for exchanges and cooperation with foreign governments and international organizations in the nuclear field. It takes a leading role in nuclear security and nuclear material control, national nuclear emergency response and management, and nuclear import/export review and management.

The CAEA established the SNSTC in 2011. This technology centre is the only institution in China that specializes in nuclear security management and technical support. It is mainly responsible for national nuclear security, nuclear material control, nuclear import and export management-related technical support work, as well as the construction and operation of the Sino–US Center of Excellence on Nuclear Security. The CAEA also established the National Nuclear Emergency Response Office, which is responsible for studying and formulating the national nuclear emergency plan and organizes its implementation.

As the nuclear and radiation safety regulator of China, the NNSA under the Ministry of Ecology and Environment undertakes overall regulation and management of national nuclear safety, radiation safety, and radiation environmental protection, and independently regulates national civil nuclear facilities and nuclear technologies in a centralized manner.

The NNSA was officially established under the leadership of the State Science and Technology Commission in July 1984. In March 1998, the NNSA was wholly incorporated into the State Environmental Protection Administration, which, in 2008, was upgraded to the Ministry of Environmental Protection, with the NNSA retained. In 2011, the NNSA expanded from one department into three, and underwent significant structural adjustment. It has issued some nuclear material accounting and physical protection guidelines. For example, the NNSA issued Nuclear Material Accounting Guidelines for Nuclear Power Plants in 2008 and Nuclear Facility Physical Protection Guidelines in 2018.

The NNSA is composed of a headquarters, six regional supervision stations, and two internal technical support units. Regional supervision stations and technical support units operate relatively independently, although they accept the leadership, management, and supervision of the headquarters in terms of administration and business. The NNSA has an important technical support unit named the Nuclear and Radiation Safety Centre. Its scope of business covers technical support for safety review and supervision of civil nuclear facilities, technical support for safety review and supervision of radiation environment, emergency response and evaluation of nuclear accidents and radiation environmental accidents, research on regulatory policies and regulations, scientific research on nuclear and radiation safety, and related technologies, consulting, and information services.

The Radiation Environment Monitoring Technology Centre is another technical support unit of the NNSA that specializes in the field of radiation environment monitoring. It is responsible for national nuclear and radiation environment supervision and technical support work, and organizes the construction of the national radiation environment monitoring network, technical guidance, and standard formulation, and emergency monitoring of major radiation environmental accidents, etc. Other departments also play important roles, such as the Ministry of Public Security, which is responsible for combatting nuclear terrorism and nuclear smuggling. Local governments are required to supervise and manage nuclear security work or provide necessary support within their respective responsibilities.

The Practice of Nuclear Security in China

By combining international and domestic experience, in recent years China has greatly improved its physical protection of nuclear materials and radioactive sources, nuclear export control, nuclear emergency management, and nuclear security culture. The following describes some key areas of nuclear security practice in China.

Since 9/11, China has paid more attention to the threat from terrorists and further enhanced the physical protection of nuclear facilities. All nuclear facilities in China are required to evaluate all potential security risks and necessary counter measures. China has invested billions of yuan in special funds to upgrade the security system of nuclear facilities. To date there have been no reported incidents of theft, loss, or illegal use of nuclear materials. China has also developed a nuclear material accounting system in accordance with international standards.

China has set strict security standards of radioactive sources management that cover all related aspects, including production, sales, transportation, use, and storage of the radioactive sources. In 1987, it published 'Security Requirements for Radioactive Waste Repositories in Cities' which necessitated upgrades to the security level of radioactive waste repositories in a number of cities. For example, China conducted security inspections of over 15,000 users of radioactive sources and properly disposed of disused radioactive sources from 2014 to 2016.[29]

China has installed over one thousand various types of radiation-detecting equipment nationwide at sea, air, highway, and railway ports. Almost all of China's large ports have radiation inspection and detection coverage. China has also worked to improve the capabilities of law enforcement personnel and held 45 training courses on radiation detection and commodity identification for over 1,280 officials from 2014 to 2016.[30] China has released the Nuclear and Nuclear Dual-Use Items Export Control List, and updates this regularly with reference to the latest control list of the Nuclear Suppliers Group (NSG).

China has established a fairly comprehensive system of legislation, regulations, and standards relating to nuclear emergency response, and has been constantly improving its nuclear emergency response plans and coordination mechanisms. At the time of writing, it has a nuclear emergency rescue force of approximately 320 people. In June 2015, China conducted the 'Shengdun-2015' exercise on nuclear emergency response. A nuclear power plant loading was the simulation scenario for the exercise. From 2016 to 2020, China conducted the 'Storm Series' nuclear security exercises, which were aimed at enhancing the security of nuclear facilities. In January 2016, China issued the white paper 'China's Nuclear Emergency Preparedness', further cementing its nuclear emergency response efforts.

China has put in place cyber security requirements for the management of industry control systems and is exploring the possibility of establishing a security risk notice mechanism that is supposed to share nuclear cyber security risk information among related stakeholders. Additionally, China has strengthened the protection of internet infrastructure and operation systems and conducted risk assessments regularly. Furthermore, it has enhanced emergency response capabilities concerning cyber security incidents and conducted a number of exercises in this regard.

China has established a three-tier radiation environment monitoring system at state, provincial, and municipal levels, as well as three networks—national radiation environment monitoring, surveillance monitoring of radiation environment in the vicinity of key nuclear facilities, and nuclear and radiation emergency monitoring—to monitor radiation environment around the clock in all areas. By the end of 2021, the state radiation environment monitoring network had 2,000 monitoring sites, with 500 automatic monitoring sites for atmospheric radiation.[31]

China has carried out publicity campaigns that promote nuclear security culture. In September 2014, the Ministry of Environmental Protection, the National Energy Administration, and the CAEA jointly published the Policy Statement on Nuclear Security Culture. Posters about countering nuclear terrorism are often seen on the streets and in the community. China has published a dedicated periodical titled *China Nuclear Security* and has also founded was a Nuclear Security Branch which operates within the Chinese Nuclear Society. This branch has experts from more than 100 units. It is attached to the National Nuclear Security Technology Center.

International Nuclear Security Cooperation

China maintains that nuclear security cannot be achieved without international cooperation and has sought to actively promote multilateral and bilateral cooperation. China has ratified all international legal instruments in the field of nuclear security, such as the CPPNM and its Amendment, and ICSANT. China not only fulfils its international legal obligations, but also actively promotes the universality and effectiveness of relevant conventions. Since the CPPNM Amendment came into effect, China has organized more than 20 national and regional training courses covering various topics, including physical protection of nuclear materials during transport, national physical protection system and measures, nuclear security culture, nuclear security risk identification, and threat assessment—thereby making practical efforts for the effective implementation and to demonstrate universal accession to the Amendment.[32]

China supports the central role of the IAEA in international cooperation on nuclear security and has provided all-round support to the IAEA in areas including nuclear materials security, radioactive sources monitoring, and nuclear and radiation emergency response. It has also recommended experts as participants in the formulation of IAEA nuclear security documents, as well as hosted nuclear security workshops jointly with the IAEA in China. Since 2012, China has contributed over US$3 million to the IAEA's Nuclear Security Fund.[33] The IAEA conducted its first International Physical Protection Advisory Service (IPPAS) mission in China in 2017 and concluded that China maintained 'strong and sustainable' nuclear security activities.[34] On 10 September 2019, the IAEA and the CAEA, through its SNSTC and the China Institute of Atomic Energy (CIAE), signed a Collaborating Centre Designation Agreement to improve the functioning of radiation detection equipment and physical protection systems.[35]

China has been closely involved in the work of the 1540 Committee of the UN Security Council. It received the first country visit by the 1540 Committee in October 2014 and hosted a training course on UN Security Council Resolution 1540's implementation for Points of Contacts in the Asia–Pacific region. China also actively participates in meetings and exercises on nuclear security organized under the frameworks of the Global Initiative to Combat Nuclear Terrorism (GICNT), International Criminal Police Organization (Interpol), the ASEAN Regional Forum, and Asia–Europe Meeting (ASEM), among others.

China attaches great importance to nuclear security policy exchange and cooperation between countries. It maintains close contact with France, the US, Russia, Japan, the Republic of Korea (ROK), and other countries, as well as emerging nuclear energy countries along the 'Belt and Road' region. China and the US have established a Center of Excellence on Nuclear Security, and the China Customs Radiation Detection

Training Center. Chinese and Russian customs have held joint exercises to prevent illicit trafficking of nuclear and other radioactive materials. China, Japan, and the ROK arrange meetings between regulators to share their experiences in nuclear safety and security regulation.

China also pays great attention to the safety and security of highly enriched uranium (HEU) and supports minimizing the use of HEU where technically and economically feasible. China collaborated with the US to convert an HEU reactor to run on low enriched uranium (LEU) in 2016. It also worked with the US and Russia to remove HEU from Ghana and Nigeria in 2017.[36]

Challenges and Prospects for Nuclear Security Improvements

Despite its successes, China faces some challenges in the area of nuclear security. First, emerging technologies might bring about new risks. For example, as nuclear power facilities increasingly rely on cyber technology, the vulnerabilities of nuclear cyber networks will increase. If the US and Israel were able to successfully launch cyber-attacks against Iran's nuclear facilities, the possibility to launch similar cyber-attacks also exists for non-state actors, for example, terrorist organizations.[37] The wide use of drones also poses threats to nuclear facilities and nuclear materials. Chinese nuclear security experts have paid special attention to the threats from so-called 'low altitude, small and slow' aircraft.[38]

Second, new facilities require new supervision capabilities. China's nuclear power industry is expanding rapidly and more professional nuclear security supervisors are needed. The Chinese Center of Excellence is expected to play an important role in this respect. In addition, as China has increasingly more new reactors and spent fuel reprocessing facilities, it needs to develop new nuclear security measures.[39]

Third, terrorist organizations pose real and evolving threats. For example, the Eastern Turkistan Islamic Movement (ETIM), an organization on the UN ISIL (Da'esh) and al-Qaida Sanctions List, remains very active, with more than a thousand fighters in Syria and hundreds of fighters in Afghanistan. It allegedly has contacts with al-Qaeda, organizes terrorist trainings, and often releases audio recordings and videos about terrorist attacks. Because the US and its allies withdrew their military force from Afghanistan, ETIM could potentially take advantage of the situation there, try to acquire nuclear materials or radiological sources from the international black market, and launch nuclear terrorist attacks against China.[40]

There is also room for further improving China's nuclear security systems. First, the legal framework could be strengthened. Given the important role of nuclear security in China's general security strategy and China's emphasis on the rule of law, milestone laws such as the Atomic Energy Law and Regulations of Nuclear Security are important and expected to be promulgated in the near future. They will provide a solid foundation for nuclear security efforts.

Second, China will seek to take leadership in nuclear security technology development. China has imported various nuclear reactors from France, Canada, Russia, and the US over several decades, and will build new types of nuclear reactors such as high-temperature, gas-cooled reactors and thorium-based molten salt reactors. China is developing nuclear material accounting and physical protection for these facilities. The significant nuclear security challenges posed by these technologies might transform into a significant opportunity for development.

Third, not only will China expand its nuclear security cooperation with IAEA and other countries, but also it will continue to support IAEA's nuclear security work and provide necessary assistance. Nuclear security is an area where major powers have more convergences than divergences. China will continue to cooperate with Russia, the US, and other countries to enhance multilateral nuclear security cooperation. The Center of Excellence on Nuclear Security can act as an important platform of regional nuclear security trainings and exchanges.

Conclusion

To date, the implementation of nuclear security in China has been a success. China has established comprehensive nuclear security systems and effectively implemented nuclear security regulations and standards. It has maintained an almost perfect nuclear security record. It has made great efforts to raise public awareness about the importance of nuclear security. This shows how a developing country can achieve rapid nuclear power expansion and nuclear security at the same time. As global interest in nuclear energy resumes out of concerns about climate change and energy supply chain security, China might lead by example for states new to nuclear power production to follow. Such success is the combined result of close attention at the top level, professional efforts, and international nuclear security cooperation. To effectively address challenges posed by new technologies, a fast-developing nuclear industry, and potential terrorist attacks in the future, China continues to improve its nuclear security systems and enhance its nuclear security capabilities. China does not think it can accomplish this tough task by itself—it needs to cooperate with the outside world and provide the necessary assistance to those countries in need.

Notes

1. This chapter does not cover the nuclear security situation in Hong Kong, Macao, or Taiwan.
2. For the number of under construction nuclear reactors worldwide, by country, see https://www.statista.com/statistics/513671/number-of-under-construction-nuclear-reactors-worldwide

3. Zhang Jianhua, Head of the Chinese Delegation, 63rd General Assembly of the IAEA, Vienna, 19 September 2019. 《中国代表团团长张建华在国际原子能机构第63届大会上的发言》,2019年9月19日。
4. Zhang Tingke, Li Minrong, and Yin Weiping, *China Nuclear Energy Development Report (2021)* (Beijing: Social Sciences Academic Press, 2021). (张廷克，李闽榕，尹卫平：《中国核能发展报告(2021),》社会科学文献出版社, 2021年11月。)
5. People's Republic of China, 'The Fourteenth Five-Year Plan for National Economic and Social Development of the People's Republic of China and the Outline of Long-Term Goals for 2035', 30 March 2021. (《中华人民共和国国民经济和社会发展第十四个五年规划和2035年远景目标纲要》, 2021年3月，第30页。)
6. Tingke et al., *China Nuclear Energy Development Report (2021)*.
7. World Nuclear Association, 'Research Reactors', WNA [online], June 2021. https://world-nuclear.org/information-library/non-power-nuclear-applications/radioisotopes-research/research-reactors.aspx
8. State-owned Assets Supervision and Administration Commission of the State Council (SASAC), 'China, Argentina Sign Agreement on Construction of Hualong One Nuclear Power Plant', 4 March 2022. http://en.sasac.gov.cn/2022/03/04/c_10896.htm
9. Pan Ziqiang, *Management of Nuclear and Radiological Terrorism Incidents* (Beijing: Science Press, 2005), 185. (潘自强：《核与辐射恐怖事件管理》, 科学出版社, 2005年4月第一版, 第185页。)
10. Regulations on the Safety Supervision and Administration of Civil Nuclear Facilities (29 October 1986); Regulations on the Control of Nuclear Materials (15 June 1987); Regulations on Radiation Protection of Radioisotopes and Radiation Devices (24 October 1989); Detailed Rules for the Implementation of the Regulations of Nuclear Materials (25 September 1990); Regulations on Nuclear Export Control (10 September 1997); Regulations on Export Control of Nuclear Dual-Use Products and Related Technologies (1 June 1998) are some of the most relevant ones.
11. People's Republic of China, Amendment to the Criminal Law of the People's Republic of China, 29 December 2001.
12. People's Republic of China, Interpretation of the decision to amend the export control regulations for nuclear dual-use items and related technologies, 16 February 2007.
13. People's Republic of China, 'National Progress Report on Nuclear Security of the People's Republic of China', 2016 Nuclear Security Summit, 1 April 2016, Washington, DC.
14. For a detailed discussion about the different translations of nuclear security in Chinese, please refer to China International and Strategic Studies, *Combating Nuclear Terrorism, Non-State Actors Nuclear Proliferation and Nuclear Security* (Beijing: Social Sciences Academic Press, 2012), 3. (中国国际战略研究基金会：《应对核恐怖主义：非国家行为体核扩散与核安全》，社会科学文献出版社, 2012年12月, 第3页。)
15. Xi Jinping, 'Strengthen Global Nuclear Security Architecture and Promote Global Nuclear Security Governance', 2016 Nuclear Security Summit, 1 April 2016, Washington, DC.
16. Xi Jinping, 2016 Nuclear Security Summit, 24 March 2014, The Hague. https://www.fmprc.gov.cn/mfa_eng/wjdt_665385/zyjh_665391/201403/t20140325_678144.html
17. Ibid.
18. Ibid.
19. Ibid.
20. Ibid.
21. Ibid.

22. Ibid.
23. Ibid.
24. People's Republic of China, The Law on Prevention and Control of Radioactive Pollution, 2003, Article 2,
25. People's Republic of China, The Nuclear Safety Law, 2017, Articles 12 and 13.
26. People's Republic of China, The Atomic Energy Law of the People's Republic of China (Draft), 20 September 2018. (《中华人民共和国原子能法(征求意见稿)》2018年9月20日。)
27. People's Republic of China, The Regulations on Nuclear Security (Draft), 3 June 2016.
28. Department II, State Administration of Science, Technology and Industry for National Defense, PRC, *Compilation of Laws and Regulations Related to Nuclear Security* (Beijing: China Legal Publishing House, 2016). (国家国防科技工业局系统工程二司：《核安保相关法律法规汇编》,法律出版社出版,2016年12月。)
29. People's Republic of China, 'National Progress Report on Nuclear Security of the People's Republic of China'.
30. Ibid.
31. China Environment News, 'Technological Innovation Facilitates the High-Quality Development of the National Radiation Environment Monitoring Network', 2 June 2022.
32. Sen Li, 'Statement by Minister Counsellor Li Sen at the Conference of the Parties to the A/CPPNM under Agenda Item 12.ii: Provisions on International Cooperation'. Conference of the Parties to the Amendment to the CPPNM, Vienna, 31 March 2022.
33. Ibid.
34. Ibid.
35. Inna Pletukhina, 'China's Atomic Energy Authority and IAEA to Collaborate to Improve Nuclear Security Equipment'. IAEA Press Centre [online], 26 September 2019. https://www.iaea.org/newscenter/news/chinas-atomic-energy-authority-and-iaea-to-collaborate-to-improve-nuclear-security-equipment
36. National Nuclear Security Administration, 'NNSA Removes All Highly Enriched Uranium from Ghana', NNSA [online], 29 August 2017. https://www.energy.gov/nnsa/articles/nnsa-removes-all-highly-enriched-uranium-ghana; Aaron Mehta, 'How the US and China collaborated to Get Nuclear Material out of Nigeria—and Away from Terrorist Groups', *Defense News* [online], 14 January 2019. https://www.defensenews.com/news/pentagon-congress/2019/01/14/how-the-us-and-china-collaborated-to-get-nuclear-material-out-of-nigeria-and-away-from-terrorist-groups
37. Ellen Nakashima and Joby Warrick, 'Stuxnet was Work of U.S. and Israeli Experts, Officials Say', *The Washington Post*, 2 June 2012.
38. Yang Zhimin, Liu Yongde, Deng Ge, and Han Yeliang, 'Challenges and Opportunities Facing China's Nuclear Security', *China Nuclear Power*, 12/5 (October 2019). (杨志民, 刘永德, 邓戈, 韩叶良：《中国核安保面临的挑战与机遇》, 《中国核电》, 第12卷, 第5期, 2019年10月。)
39. Ibid.
40. Hui Zhang, *It's Time for China to Turn Nuclear-Security Pledges into Reality* (Cambridge, MA: Harvard Kennedy School: Belfer Center for Science and International Affairs, 2016). https://www.belfercenter.org/publication/its-time-china-turn-nuclear-security-pledges-reality

Suggested Reading

China International and Strategic Studies, (2012), *Combating Nuclear Terrorism, Non-State Actors Nuclear Proliferation and Nuclear Security* (Beijing: Social Sciences Academic Press), 3. (中国国际战略研究基金会：《应对核恐怖主义：非国家行为体核扩散与核安全》，社会科学文献出版社，2012年12月，第3页。)

Department II, State Administration of Science, Technology and Industry for National Defense, PRC, December 2016, *Compilation of Laws and Regulations Related to Nuclear Security* (Beijing: China Legal Publishing House), (国家国防科技工业局系统工程二司：《核安保相关法律法规汇编》，法律出版社出版，2016年12月。)

Kutchesfahani, Sara Z., (2019), *Do As I Say, and As I Do: Chinese Leadership in Nuclear Security*. CISSM Working Paper (College Park, MD: Center for International and Security Studies (CISSM), University of Maryland).

Zhang, Hui and Zhang, Tuosheng, (2014), *Securing China's Nuclear Future* (Cambridge, MA: Harvard Kennedy School: Belfer Center for Science and International Affairs).

Ziqiang, Pan, (2005), *Management of Nuclear and Radiological Terrorism Incidents* (Beijing: Science Press). (潘自强：《核与辐射恐怖事件管理》，科学出版社，2005年4月。)

CHAPTER 27

INDIA'S APPROACH TO NUCLEAR SECURITY

RAJESWARI PILLAI RAJAGOPALAN

Introduction

India is not a stranger to the dangers of the misuse of nuclear and radiological materials. These dangers are not merely in terms of the safety of radioactive materials but also about the security of these materials. Located in a turbulent region, India is highly conscious of the security threats in its neighbourhood and is thereby mindful of the disastrous consequences of nuclear materials falling into the hands of terrorists or others who may have malign intentions. Global attention on nuclear security, which had gained traction after the collapse of the Soviet Union, became even more urgent after the terrorist attacks of 11 September 2001 ('9/11') in the United States (US). This increased global emphasis on nuclear security aligned with India's pre-existing concerns, making cooperation on such issues obvious for New Delhi. Meanwhile, incidents involving radioactive materials finding their way into scrapyards highlighted the dangers of such materials falling out of regulatory control.

Nuclear security is increasingly relevant for India, which is set to increase the nuclear component in its overall energy mix. According to government sources, the nuclear power component of the total electricity generation in India for the year 2021–2022 stood at 3.15 per cent.[1] With a total of 22 reactors, nuclear power is the fifth-largest source of electricity in India after coal, gas, wind power, and hydroelectric power.[2] In April 2022, the Indian government informed the parliament that it has given 'In-Principle' approval for five new sites to host future nuclear power plants. These projects will increase the current installed capacity of 6,780 megawatts (MW) to 22,480 MW by 2031.[3] To achieve this goal, the government is working on the construction of 10 domestic 700-MW pressurized heavy water reactors (PHWRs), for which administrative approval and financial sanction have already been cleared.[4] There are an additional 10 nuclear power reactors already in different stages of construction that are expected to add 8,000 MW combined generational capacity.[5]

India's nuclear journey began on 4 August 1956 when Apsara, India's first nuclear research reactor, attained operational status.[6] While the reactor was designed and built in India, the nuclear fuel was supplied by the United Kingdom (UK) under a lease agreement. India's second nuclear research reactor, CIRUS, was developed in partnership with Canada and reached operational status in the early 1960s. India was able to use nuclear power for its electricity production from October 1969, with two US-supplied reactors at Tarapur entering service. India's second nuclear power station was set up in Rajasthan, and the first unit from the station went operational in August 1972. India's first indigenously built nuclear power station was set up at Kalpakkam near Chennai.[7]

This chapter first examines India's approach to nuclear security, including its institutional and legal architecture, compliance aspects, and current practices. It then discusses India's nuclear security imperatives, namely, the country's expanding nuclear power sector and the changing nature of associated security challenges. Next, the chapter covers internal political and other official debates regarding the seriousness of nuclear security for India. It then details the domestic politics and governance regarding the policies, rules, and institutions that govern India's nuclear security activities. Next, the chapter deals with the operational aspects, including the design basis threat (DBT) and the defence-in-depth principle, which remain critical in ensuring the physical protection of nuclear materials. It then elaborates on India's nuclear and radiological material control and accounting system, which has become significant as a result of the 2010 Mayapuri incident—where the public was accidently exposed to a radioactive source following its loss from regulatory control.[8] The final sections deal with India's approach to security culture, the insider threat, and remaining challenges around nuclear regulation.

India's Nuclear Security Imperatives

The expansion of India's atomic energy programme provides an important imperative to streamline India's nuclear security approaches, particularly given how the threat landscape has evolved after the nuclear programme first began. Since independence, India has battled both internal and external security threats in the form of insurgencies in India's north-east and the disputed Jammu and Kashmir territory, as well as left-wing extremism, religious extremism, and cross-border terrorism. There are even some isolated and unconfirmed reports of Naxalites (Indian Maoist groups) harbouring intentions to target nuclear installations,[9] as well as speculation about possible collusion between Naxalites and Islamist jihadists, who might potentially combine their efforts under a broader anti-India coalition.[10] Furthermore, indigenous terrorist groups such as the Indian Mujahideen (IM) have reportedly expressed interest in using nuclear weapons sourced from Pakistan. In 2013, the IM co-founder Yasin Bhatkal allegedly wanted to bomb Surat, a city in western India, with nuclear weapons.[11]

However, these are not easy options for such groups to execute, given that India has deployed radiation detectors in border areas aimed at identifying smuggled nuclear materials.[12] And a decade on from these reports, there has been no demonstration of direct threats to nuclear facilities from terrorists and other non-state actors. Nevertheless, nuclear safety and security have remained an important concern of India's atomic energy agencies and security managers. Indeed, the catastrophic consequences of nuclear and radioactive incidents have made security managers calculate even the most improbable situations as they implement nuclear security measures. This has been translated into sustained efforts to improve both nuclear security and safety within India.

India's measures for nuclear security are captured in a number of legal and institutional measures, going back to the early 1960s. These include the foundational legislation, the Atomic Energy Act of 1962, upon which central government has established rules, regulations, and periodic notifications to execute its various provisions.[13] The Act has been periodically reviewed and updated based on India's changing security environment and threat perceptions. In more than six decades of operating an atomic energy programme, there has been no catastrophic event at India's nuclear power plants, but there have been significant incidents—most notably the Mayapuri radiological accident—that have focused global attention on India's nuclear and radiological policies and practices. In response, India has established a detailed and comprehensive countrywide architecture to deal with nuclear and radiological security threats. However, its national ranking in external global nuclear security assessments such as that of the Nuclear Threat Initiative (NTI), the Nuclear Security Index, remains relatively low.[14]

There is a prevalent sense—especially among the international civil society and think-tank community—that India does not have sound nuclear security practices. Although this is arguably far from reality, such assumptions in part stem from the fact that India has remained excessively cautious and secretive about its nuclear security policies and practices.[15] In recent years, India has made more progress in communicating its successes though. For example, prior to the 2014 Nuclear Security Summit, the Ministry of External Affairs published a brief booklet, *Nuclear Security in India*, which described India's nuclear security approach, with transparent information on components such as its governance, nuclear security practice and culture, institutions, technology, and international cooperation.[16] Nevertheless, the continuing absence of proactive Indian engagement with the global nuclear security community has led many countries to assume that either nuclear security is not a priority for India, or worse, that India does not have strong or comprehensive policy—which harms India's interests at a time when the country is making efforts to further integrate with the global non-proliferation architecture.[17]

India's Nuclear Security Approach

Indian nuclear facilities handle a wide of range of nuclear materials, including uranium ore concentrate, low and highly enriched uranium, plutonium used in power and research reactors, spent fuel from reactors, as well as radiological sources used for health, industrial and other purposes. Given India's security environment, non-state actors who could potentially seek to acquire and use these materials in an act of terrorism are at the top of India's nuclear security concerns.[18] Nevertheless, some have dismissed this as a hypothetical threat because the technological wherewithal and the processes involved in converting nuclear material into an actual weapon make it unlikely for any terrorist group to accomplish such a task.[19] Yet, while there exists significant barriers to weaponizing nuclear materials, the devastating consequences of such an eventuality means that contingencies need to be planned for. Furthermore, the profile of terrorism has changed over the decades, including in India.[20] Terrorists no longer come from underdeveloped, impoverished backgrounds but may be highly qualified and technically proficient individuals who join the ranks of terrorist organizations for a variety of reasons.[21]

In addition to the theft of nuclear materials, the Indian atomic energy establishment—the security managers as well as the political leadership, comprising ministers and members of parliament—has similarly considered the possibility of a sabotage attack on a nuclear facility.[22] Even if such an attack remains highly improbable, the consequences could potentially be disastrous. Here, other non-nuclear incidents serve as a reminder that these types of attacks could be launched at highly secured facilities. For instance, in the 2016 terrorist attack on the Indian Air Force (IAF) base in Pathankot, a heavily armed group of Islamic militants was able to kill or injure multiple security personnel.[23] Attacks also need not be limited to physical or conventional means. For example, cyber-attacks on critical infrastructure have been on the rise globally, including in India, with the Kudankulam Nuclear Power Plant, in the southern Indian state of Tamil Nadu, targeted in 2019.[24] Aerial threats such as through unmanned aerial vehicles (UAVs) also cannot be ignored,[25] particularly given how drone warfare in the border region between India and Pakistan has spiked in recent years.[26]

Mindful of such threats and their disastrous consequences, India has elevated these into its nuclear security calculus, despite the fact that traditionally the Indian establishment has tended to generally dismiss such threats as highly improbable.[27] At the highest decision-making level, this increased commitment and seriousness regarding nuclear security was demonstrated by the Indian prime minister's participation in three of the four Nuclear Security Summits.

Domestic Policy and Governance

India's nuclear security policy governance has been developed through rules, regulations, and institutions first established in the 1960s and 1970s. These have been reviewed and amended on a periodic basis depending on the changing security environment as well as technological advancements. The key legislation that addresses India's nuclear and radiological security is the Atomic Energy Act of 1962.[28] The Act is overarching, covering the entire spectrum of the industry's activities, including the development, control, and use of nuclear energy in India. The Act further directs central government to formulate rules and regulations, and to release notices implementing the provisions of the Act. The Act has been reviewed and modified several times over the years, providing a dynamic and strong legal foundation for India's nuclear security approach.

Other rules and laws also have been established that provide specific parameters for the location and operation of nuclear energy facilities. These include the 1986 Environment (Protection) Act, the 1996 Atomic Energy (Factories) Rules, and the 2003 Electricity Act. The Department of Atomic Energy (DAE) also framed the Guidelines for Nuclear Transfers (Exports) in 2006.[29] Some other significant legal instruments developed under the Atomic Energy Act include the 1971 Atomic Energy (Radiation Protection) Rules (further revised in 2004), the 1984 Atomic Energy (Working of the Mines, Minerals and Handling of Prescribed Substances) Rules, and the 1987 Atomic Energy (Safe Disposal of Radioactive Wastes) Rules.

The revised 2004 Atomic Energy (Radiation Protection) Rules authorize activities for nuclear fuel cycle facilities and radiation use in the areas of industry, medicine, and research.[30] These rules are comprehensive, laying out in clear terms the roles and responsibilities of the different stakeholders, including employers, radiation safety officers (RSOs), and others with regard to radiation protection. Most importantly, the rules clarify the role and powers of the competent authority—the Atomic Energy Regulatory Board (AERB), India's atomic energy regulator—on requirements concerning safety, health surveillance of workers, radiation surveillance, and records to be maintained, with directives, inspections, and enforcement actions being issued.[31] For instance, the Radiological Safety Division of the AERB is responsible for ensuring the compliance of the 2004 Radiation Protection Rules, as well as the 1987 Atomic Energy (Safe Disposal of Radioactive Wastes) Rules which prescribe the various factors for the decommissioning and disposal of radioactive wastes. Similarly, the Atomic Energy (Working of the Mines, Minerals and Handling of Prescribed Substances) Rules 1984 regulate activities in mining, milling, processing, and handling of prescribed substances.[32]

As part of the US–India Civil Nuclear Deal, the government passed the Weapons of Mass Destruction and their Delivery Systems (Prohibition of Unlawful Activities) Act in June 2005, which has further strengthened India's commitment to its nuclear security and non-proliferation goals.[33] With recent efforts to integrate with the global nuclear non-proliferation architecture, India has continued to review, revise, and update its technology export control rules and regulations, specifically in relation to the Nuclear Suppliers Group (NSG), the Australia Group (AG), and the Wassenaar Arrangement.[34] India has also established institutions such as the Atomic Energy Commission (AEC), the DAE, AERB, Nuclear Controls and Planning Wing (NCPW), and Bhabha Atomic Research Centre (BARC) to cater for any contingency in the nuclear and radiological arena. The AERB is responsible for regulating the atomic energy sector, although the DAE and BARC contribute to India's strategic nuclear programmes as well.

In addition to the aforementioned rules and regulations, in 2011 the Manmohan Singh government introduced a bill that would strengthen the independence and autonomous functioning of its current nuclear regulator, the AERB. The key goal of the Nuclear Safety Regulatory Authority (NSRA) Bill was to replace the current atomic energy regulator with a new body for nuclear safety issues.[35] However, more than a decade later, the NSRA Bill is yet to be debated or approved by India's parliament. Following a general election in 2014, the original bill had lapsed and had to be re-introduced in the parliament. At the time of writing, more than nine years have passed since the Narendra Modi government came to office, but still no progress has been made on this front.

Operational-Level Security

The DBT and defence-in-depth are two core principles employed in Indian nuclear facilities to ensure physical protection and security at nuclear facilities. Both are implemented from the time of building a new nuclear facility, through its entire life cycle, and to when it is decommissioned. India has been operating nuclear power and research reactors since the 1960s, but these measures have been adopted only in the last few decades since the relevant guidance was issued by the International Atomic Energy Agency (IAEA).[36] In the case of India, there is both a national DBT and a facility-level DBT. The latter is developed by the facility based on the national DBT but adapted to meet the local threat environment.[37] The facility-level DBT is used in designing the physical protection measures and must be approved by the nuclear regulator.

The different categories of threats that the Indian national DBT consider include external and internal adversaries and the possibility of collusion between them. The national DBT accounts for the various scenarios in which these threats can play out, such as inside a facility, during transportation of nuclear materials, or via cyber-attacks. Information is also included on the capabilities of potential adversaries against which security systems must be designed to defeat. The AERB as the nuclear regulator has an important role in assessing the threats and motivations for the three categories of threat actors—insiders, outsiders, and insider–outsider collusion—in carrying out an attack against a nuclear facility. Key questions debated when formalizing a DBT include: How many outsiders? How many insiders? What kind of vehicles? What motivation? What weaponry? Posing these questions at the time of designing a nuclear facility is important because appropriate security measures, processes, and procedures can be built into the design at the outset.

The DBTs are dynamic in nature and are reviewed and revised for several reasons, including the changing nature of threats—in terms of new adversaries, new tactics, and new vulnerabilities—or important changes in international or domestic rules, regulations, or policy guidance relevant to a DBT.[38] For instance, the 2008 Mumbai terrorist attacks were a significant driver for India to review and update the national

DBT. The attacks took place simultaneously at multiple locations in the city—an example of a meticulously planned and coordinated operation. There have been other sophisticated, coordinated terrorist attacks of various kinds in India, albeit, so far, not directed at nuclear establishments. India has also promoted adherence to the IAEA's INFCIRC/225/Rev.5, which contains specific recommendations on the effective physical protection of nuclear facilities.[39]

Other important factors that go into developing and designing a physical protection system for nuclear facilities in India include several interrelated functions: deterring an adversary; detecting an adversary if deterrence fails; delaying an adversary's access to the core of a facility and, in the case of an attack, defeating an adversary in a timely manner; and limiting the impact of the contingency that might arise. These together underpin India's defence-in-depth approach, which consists of a multi-layered system, with each level making the intrusion efforts more difficult. To this end, nuclear power plants in India incorporate a series of access control and denial measures, including spike strips, cement and steel barriers, and technology-aided measures such as biometric systems.[40] Each of these systems is useful in not only delaying an intruder's access to the facility but in providing additional time to the security agencies and first responders to be able to effectively respond to the evolving situation.

Within the vital areas of a facility, nuclear security measures are further increased, including efforts to protect key equipment and devices and materials whose tampering or misuse could be detrimental to nuclear operations (for example, where the triggering of a radiological release could endanger the broader public). Indian nuclear energy facilities also contain several important safety measures that have relevance for security, including fail-safe shutdown systems, active and passive cooling systems, and robust containment features that can limit the fallout from either a safety- or security-related incident.[41]

Nuclear Security Culture

Security culture is an increasing focus of nuclear regulatory inspections in India. However, this remains an intrinsically challenging aspect of nuclear security to assess, albeit one that had garnered significant attention in recent years—with recognition and appreciation that security culture is a critical element for any nuclear security policy and practice to be effective, including in the area of radiological materials.[42] For example, the AERB states that:

> Security culture should be inculcated among all employees. This should be done by means of training and awareness programmes, security drills, etc. It is not only the persons who are occupationally exposed to radiation in the facility but also the general employees working in the facility should be included in such programmes.[43]

The emphasis on responsibility for security across all staff working in the nuclear sector in India is reassuring.

India, like many other countries, has traditionally had a 'safety-first' approach as the overriding principle that directly influences operations at nuclear facilities. However, this 'safety-first' approach is in the process of transitioning to a 'safety and security first and foremost' attitude, where all staff members have an important role to play.[44] For instance, a decade ago, if one were to ask the facility managers as to who held responsibility for the security of that facility, they would have pointed to the guards protecting the property as the custodians of security. This mindset has changed to the extent where there is a sense that security is the responsibility of every single individual within a facility, from a guard to a manager to a scientist or technician.[45] This key concept has also been promoted by India in international engagement with other countries. For example, a track 1.5 workshop on nuclear security held in 2013, involving officials from the Indian atomic energy establishment and counterparts from the US, emphasized security culture as one that involves '(E)very person, from a custodian to a technician to a scientist to a guard in the protective force, [who] needs to believe in and support the nuclear security program for it to succeed.'[46]

Nuclear and Radiological Material Control and Accounting (MC&A)

India has established various regulations that provide for the accounting of nuclear and radiological materials. India's material inventory is maintained by the AERB, which has put in place strict protocols such as requirements for the possession of a valid licence and for the provision of a valid purpose for material requests by any institution requiring these materials. According to retired DAE officials, the procurement system established by the AERB is able to detect any discrepancies.[47] Nevertheless, the potential theft of nuclear and radiological materials remains an area of concern. As a further deterrent, facility procurement officers responsible for nuclear radiological materials can be charged with negligence if materials are unaccounted for under their watch. They are also provided with a direct contact within the AERB to report any material loss.

Despite these efforts, the Mayapuri incident in April 2010 involving the loss of a cobalt-60 radiological source brought to the fore the vulnerabilities of the system in place, as well as improper disposal practices. The incident exposed the violation of regulatory requirements pertaining to the decommissioning of a gamma unit on the part of Delhi University, which had enabled the gamma unit to end up in the hands of a metal scrap dealer in West Delhi.[48] It was later found that Delhi University had procured the gamma unit from abroad before the AERB was even set up—and therefore the AERB did not have any record of it.[49] Nevertheless, being in possession of the radioactive material, it was the responsibility of the university to inform the AERB. Specifically, the incident violated the 1987 Atomic Energy (Safe Disposal of Radioactive Waste) Rules and the 2004 Atomic Energy (Radiation Protection) Rules. The incident resulted in the death of one person and seven persons suffered from radiation sickness.[50]

Since the Mayapuri incident, the AERB and the University Grants Commission (the agency responsible for higher education institutions in India) have tightened the rules around the acquisition of nuclear and radiological materials. For instance, according to a new AERB directive, educational institutions must obtain a no-objection certificate from the AERB for all radioactive materials and related equipment, including X-ray machines. The regulations for decommissioning have also become stricter. To facilitate this, the AERB is empowered to undertake periodic inspections at relevant facilities with the goal of tightening processes in the handling and use of these materials.[51]

As a result of the Mayapuri incident, the Indian nuclear agencies are engaged in awareness and training programmes for Mayapuri scrap dealers on safety and security issues, as well as the legal and regulatory aspects that must be complied with in handling and disposing of nuclear and radiological materials.[52] Nevertheless, there remain questions around these follow-up activities, including the types of awareness and training programmes for different actors. In fact, some recent incidents are evidence of India's continuing challenges in the area of radioactive material control and accounting. The loss in 2019 of caesium-137 from a truck transporting machinery from an Oil and Natural Gas Corporation exploration site in the southern state of Andhra Pradesh is a case in point.[53] Even though the radioactive source was recovered a few days later, this was from a scrap shop 120 kms away from where it was lost, suggesting lingering weaknesses in India's material management system.

Given the vulnerability of nuclear and radiological materials during transport, this aspect of nuclear security has been given a high priority in India. For nuclear materials, a heavily armed security presence during transportation provided by India's security agencies is required under the rules prescribed by the AERB. The security agencies also plan the routes in a considered manner, altering routes periodically so as not to develop a pattern that can lead to easy predictability. India has also adopted a graded approach for nuclear transport, where the level of security depends on the type and quantity of material being transported. For example, in cases of materials commonly used by research institutions that are of low half-life and transported in small amounts, the health risk posed by those materials is very low and consequently the security measures applied are limited. In contrast, nuclear materials and highly radioactive sources are transported in specialized vehicles in order to ensure that there is no leakage or exposure to the material during transport, with real-time tracking of the materials so that a response can be swiftly launched should there be an incident.[54]

The Insider Threat

Like other countries, India has encountered challenges in mitigating the risk posed by insiders—individuals with authorized access to nuclear and radiological assets. A particular challenge concerns the issue of surveillance; for instance, to what extent should surveillance be placed on an employee's activities in the cyber domain—in order to determine with whom the person is interacting; whether these are terrorist groups, religious radical groups, or criminal groups; or if the person is leaking any information about nuclear plant operations or facility information? These are issues that are particularly problematic in democracies—like India—that value privacy and individual freedom. It is also possible for individuals to inadvertently release sensitive information, making it important to proactively engage employees through security awareness and training programmes. In Indian civilian nuclear facilities, such threats are addressed by the Computer Information and Security Advisory Group (CISAG), which is authorized to undertake audits of information systems, frame guidelines, and devise plans to mitigate cyber-attacks and their effects.[55]

One of the most effective ways to address insider threats is through a comprehensive personnel reliability programme (PRP). India's PRP measures are extensive, with security agencies assessing the personal background of individuals applying for access to nuclear facilities—including their social, economic, and political background, any criminal connections, their out of office social interactions, any changes in their religious and political orientations, and any drug and other substance abuse history. This vetting process occurs before and during the course of an individual's employment and is also triggered by a transfer to a more sensitive facility or a promotion that requires higher levels of security clearances. To be effective, PRPs need to include an element of behavioural observation, both during and beyond office hours, to see if there are significant changes in how an individual interacts. This may be done by colleagues and managers who need to be cognisant of what constitutes deviant behaviour and how this can be reported. India has a strong PRP process built into its nuclear security policy and practice; nevertheless, it must remain vigilant to changing threats and tactics in the neighbourhood and beyond, and continually update its PRP processes.[56]

Other challenges regarding PRP implementation in India include applying this to temporary workers who may only be attached to a nuclear facility for a short period of time. There may be practical difficulties due to the way temporary workers typically move from place to place looking for employment, making comprehensive background verification extremely problematic. Here, managers of nuclear facilities can decide to employ temporary workers whose credentials cannot be verified by restricting their access. For example, such workers may be limited to the periphery of facilities, without permission to enter the vital areas. However, even this level of limited access may put temporary workers within range of potential targets, and they may still be able to bypass several layers of security. Despite the ongoing challenges, the issue is being addressed through the digitalization of police data across different states and provinces, which has helped establish the credentials of many temporary workers.[57]

Regulatory Independence

Despite operating nuclear power plants for close to six decades, India claims a nearly perfect record in maintaining the safety and security of its programme without any significant event. Nevertheless, this should not lead to complacency, but instead efforts should be taken to improve security on a continuous basis. Here, nuclear regulatory aspects remain an area where improvements could be made, particularly given that the AERB is still not fully autonomous from the DAE. Even though India follows the standards and guidelines developed by the IAEA, the lack of a truly independent regulator has affected India's perceived degree of nuclear security compliance both within the country and outside. Since the Indian nuclear establishment architecture would not be required to make extensive adaptations to achieve regulatory independence, India could usefully come up with institutional modifications, which would bring obvious nuclear safety and security benefits as well as rectify the perception of the absence of an independent regulator. Such a regulator could report to the government, but one that is independent of the DAE. Here, India could learn lessons from how the Japanese nuclear industry was regulated prior to the 2011 Fukushima nuclear disaster and take appropriate steps over time to avoid similar outcomes.[58]

While Indian nuclear agencies may claim that the AERB is independent, others are cognisant of the need to establish a truly autonomous body. However, as mentioned, attempts by the previous Singh government to replace the current atomic energy regulator, the AERB, with the new NSRA failed, and the NSRA Bill is yet to be approved by parliament. The Bill will result in the creation of a Council of Nuclear Safety (CNS), under the leadership of the prime minister. However, there have been issues around the appointment of its members because, according to the Bill, the members will be 'appointed by the central government on the recommendations of the search committees'.[59] The problem with this arrangement is that the search committees will consist of members of the CNS, which include seven Indian central government ministers, the secretary of the DAE, and the Cabinet Secretary.

Due to the domain expertise of the DAE, there is a risk that it will have undue influence over the selection and appointment of the CNS members.[60] This is possibly due to a larger problem of there being an inadequate number of nuclear technology specialists in India outside of the country's nuclear establishment. Since there is no private nuclear industry in India, those who have studied nuclear science and technology are typically associated with the Indian atomic establishment at the DAE, AERB, or BARC. This problem is unlikely to go away until India opens up the current atomic establishment to allow for the private sector to run the nuclear industry, just as it has done in the case of other power plants based on thermal or hydroelectric energy. This is, however, unlikely to happen because nuclear power is kept entirely under government control for security reasons—and hence, the opening up of the sector is likely to be approached very cautiously, if ever.

In some respects, the Indian establishment's resistance to having an autonomous nuclear regulator is puzzling because this would not require much change in terms of policy measures and regulations, with India already complying with the IAEA's security and safety practices and guidelines. Moreover, such institutional tweaking could significantly improve the international perception of India when it comes to nuclear security, especially because this is an issue on which India has repeatedly been criticized, and global perceptions have become particularly important amid India's efforts to be aligned with global non-proliferation regimes.

Responding to a question in the Indian parliament in 2012, Minister of State for Personnel, Public Grievances and Pensions, and Prime Minister's Office, V. Narayanasamy highlighted that the NTI report had criticized India for not being transparent enough on its nuclear security measures, its lack of an independent nuclear regulator, and the need for more stringent physical protection measures.[61] Global perceptions about this subject are partly the result of internal Indian debates as well as official reports, such as that of the Comptroller and Auditor General, that have criticized India on the unhealthy relationship between the AERB and the DAE.[62] The NSRA legislation is an important measure in this regard, and as India strengthens nuclear cooperation with a number of international partners, it will be important to make progress in approving the legislation.

Other Areas for Improvements

In addition to its regulatory approach, there are other areas where India could further improve its nuclear security systems—most notably in strengthening nuclear security culture through creating a better appreciation for security threats across the board, particularly at the senior level. This is important as it has implications for capacity development, including training, availability of human and financial resources, and the development of appropriate institutional structures at state/province levels. Certain states and their leaders within India may be more attuned to nuclear security threats, whereas others are unaware of the serious consequences of nuclear security incidents, possibly because they are addressing what they consider to be more pressing threats. While some threats, such as left-wing extremism, have been more about political goals, such groups could potentially target nuclear facilities in an effort to attack the government. Therefore, a deeper understanding and appreciation of security threats and how they might apply to nuclear facilities is needed to ensure that security forces are more capable and effective at dealing with contingencies. The absence of this nuanced understanding can play out especially among first responders such as the fire service and local police force.

Another area that requires additional attention is India's ability to respond to emergency situations involving nuclear and radiological materials. While the security forces conduct annual exercises within their own services, exercises involving multiple security agencies are still not adequate. However, responding to a nuclear security incident involving the possibility of lost nuclear material or a radiological release would necessitate a coordinated multi-agency response. The challenge may not be unique to India, but the task of bringing multiple security agencies under a single unified leadership is problematic, going by past Indian experiences, such as the Mumbai terror attacks. In a real event, there could be problems of coordination, command, and control, and more importantly, lack of a common operating picture and common language so that commands are developed and transmitted for quick response. These problems could be reduced if different security agencies were to undertake joint exercises on a periodic basis. Given that India's population density, crowd control and managing the public are particular challenges, it might also be useful to explore using better equipped agencies such as the Indian Army which has a chemical, biological, radiological, and nuclear (CBRN) unit. Constables and retired army personnel could also be brought in to increase the strength of the National Disaster Response Force's (NDRF) CBRN battalions.[63]

Finally, as noted previously, India could also improve its global nuclear security outreach. While India has done a reasonably good job on nuclear safety issues, nuclear security also needs attention. This could be achieved by a more open and transparent approach, for example, by India publishing a high-level outline of its nuclear security framework. Fortunately, this is underway, for example, in the DAE's annual reports.[64] A decade ago, the annual reports did not mention nuclear security at all. India has a good track record and achievements of which it is proud, but it needs to engage in more conversations with its international partners—and others—about its nuclear security efforts and accomplishments, and more importantly, how these have been achieved. Here, its rigorous approach to PRP is potentially worth sharing as an example for others to emulate. A strengthened outreach effort could significantly benefit India's standing in the nuclear security realm.

Conclusion

India has instituted a series of institutional and legal measures to ensure that its nuclear security policies and practices are in line with IAEA guidelines and other global best practice guidance. India has so far maintained a good record in this realm, without any catastrophic incidents, but the changing nature of threats as well as the rapidly evolving tactics employed by adversaries call for constant vigilance in maintaining nuclear security. Despite the elaborate systems and checks in place, there have been issues confronting India's nuclear and radiological material control systems and ongoing concerns over the lack of independence of its nuclear regulator. The absence of an overarching approach to nuclear safety and security as two sides of the same coin is also an issue. India also needs to consider how to have more engaging conversations with the global nuclear community. Although India's excessively cautious approach has changed gradually for the better, there remains room for improvement.

Notes

1. Department of Atomic Energy, 'India's Nuclear/Atomic Energy', Unstarred Question No. 660 to be answered on 20.07.2022, Lok Sabha, Government of India, 20 July 2022. https://dae.gov.in/writereaddata/lsusq660.pdf
2. Department of Atomic Energy, 'Development of Nuclear Sector', Unstarred Question No. 5689 to be answered on 06.04.2022, Lok Sabha, Government of India, 6 April 2022. https://dae.gov.in/writereaddata/lsusq5689.pdf
3. Ibid.

4. Ibid.
5. Department of Atomic Energy, 'Share of Atomic Energy', Unstarred Question No. 2680 to be Answered on 04.08.2021, Lok Sabha, Government of India, 4 August 2021. https://dae.gov.in/writereaddata/lsusq%202680.pdf
6. M. R. Srinivasan, 'India's Atomic Energy Programme', Press Information Bureau, Government of India [online], 17 August 2017. https://pib.gov.in/newsite/printrelease.aspx?relid=170047; for an early institutional account of India's atomic energy programme, see H. N. Sethna, 'India's Atomic Energy Programme: Past and Future', *IAEA Bulletin* [online], 21/5 (1979), 2–11. https://www.iaea.org/sites/default/files/publications/magazines/bulletin/bull21-5/21505090211.pdf; for more recent information on India's nuclear programme, see Bhabha Atomic Research Centre, 'Indian Nuclear Power Program', n.d. http://www.barc.gov.in/randd/artnp.html
7. Srinivasan, 'India's Atomic Energy Programme'.
8. For details, see Atomic Energy Regulatory Board, *AERB Newsletter* [online], 23/1, January-June 2010. http://www.aerb.gov.in/AERBPortal/pages/English/t/NewsLetters/%202010/AERBNwsltrEnglVol23-1.pdf
9. Author interviews with intelligence and police officers as well as senior DAE officials in Delhi, Mumbai, Gujarat, and Rajasthan, October 2013.
10. Author interview with senior police officials in Jodhpur, 14 April 2014.
11. Neeraj Chauhan, 'Indian Mujahideen Wanted to Nuke Surat, Yasin Bhatkal Tells Cops', *Times of India* [online], 30 December 2013. https://timesofindia.indiatimes.com/india/indian-mujahideen-wanted-to-nuke-surat-yasin-bhatkal-tells-cops/articleshow/28116663.cms
12. Rita Guenther, Micah Lowenthal, Rajaram Nagappa, and Nabeel Mancheri, *India-United States Cooperation on Global Security: Summary of a Workshop on Technical Aspects of Civilian Nuclear Security* (Washington, DC: National Academy of Sciences, 2013). https://doi.org/10.17226/18412
13. Lok Sabha, 'The Atomic Energy Act, 1962', *The Gazette of India* [online], 13 August 1962. https://www.aerb.gov.in/images/PDF/Atomic-Energy-Act-1962.pdf
14. Nuclear Threat Initiative, 'The NTI Nuclear Security Index', NTI [online], 2020. https://www.ntiindex.org
15. Global nuclear security assessments, such as that of the NTI, which uses a quantitative approach to quantify nuclear security, have not been helpful. But that India has not pro-actively engaged the global nuclear community in raising awareness about India's nuclear security policies and practices has added to the problem. During the author's interactions with French nuclear security officials, it was found that even France, with which India cooperates on nuclear safety issues, was not aware of India's nuclear security policies and execution.
16. Ministry of External Affairs, 'Nuclear Security in India', MEA [online], 2014. https://www.mea.gov.in/Images/pdf/Brochure.pdf
17. Karthika Sasikumar, 'India's Emergence as a "Responsible" Nuclear Power', *International Journal*, 62/4 (2007), 825–44; Rajeswari Pillai Rajagopalan and Arka Biswas, 'Locating India within the Global Non-Proliferation Architecture: Prospects, Challenges and Opportunities', Observer Research Foundation [online], August 2016. https://www.orfonline.org/wp-content/uploads/2016/08/ORF_Monograph_NonProliferation.pdf; PNR Govind, 'India as a Responsible Nuclear Power: A Strategy for Stability', US Air War College, 2012.

18. Author interviews with intelligence and police officers in Delhi, Gujarat, and Rajasthan, as well as with senior officials of the DAE and the AERB during field visits between October 2013 and June 2014. There have been subsequent meetings in the following years, the latest of which took place in New Delhi in July 2022.
19. R. Rajaraman, 'Nuclear Security: An Overview', in Rajeswari Pillai Rajagopalan and Allard Wagemaker, eds., *Global Nuclear Security: Moving Beyond the NSS*, GP-ORF Series (New Delhi: Observer Research Foundation, 2018), 7–19. https://www.orfonline.org/wp-content/uploads/2018/07/GP-ORF_Nuclear-Security.pdf
20. Alexander Lee, 'Who Becomes a Terrorist? Poverty, Education, and the Origins of Political Violence', *World Politics*, 63/2 (2011), 203–45.
21. Ibid.
22. Department of Atomic Energy, 'Security of Indian Nuclear Installations from Cyber Attacks', Unstarred Question No. 963 to be answered on 10.02.2022, Rajya Sabha, Government of India, 10 February 2022. https://dae.gov.in/writereaddata/rsusq963.pdf; Department of Atomic Energy, 'Safety Measures in Nuclear Plants', Unstarred Question No. 561 to be answered on 20.07.2022, Lok Sabha, Government of India, 20 July 2022. https://dae.gov.in/writereaddata/lsusq561.pdf; Department of Atomic Energy, 'Safety at Nuclear Plants', Unstarred Question No. 2472 to be answered on 16.03.2022, Lok Sabha, Government of India, 16 March 2022. https://dae.gov.in/writereaddata/lsusq2472.pdf; Department of Atomic Energy, 'Nuclear and Radiation Safety', Unstarred Question No. 1547 to be answered on 28.07. 2021, Lok Sabha, Government of India, 28 July 2021. https://dae.gov.in/writereaddata/ls%20usq%201547.pdf. These views have been also corroborated during the author's interviews with security officials and atomic energy managers since 2014.
23. Agencies, 'Pathankot Attack: "All Terrorists Dead"', *The Hindu* [online], 3 January 2016. https://www.thehindu.com/news/national/Fresh-firing-at-Pathankot-air-base-Army-deploys-reinforcements-in-area/article60521192.ece
24. Department of Atomic Energy, 'Cyber Attack on KKNPP', Unstarred Question No. 659 to be answered on 20.11.2019, Lok Sabha, Government of India, 20 November 2019. http://164.100.24.220/loksabhaquestions/annex/172/AU659.pdf; Jay Mozoomdaar, 'Not only Kudankulam, ISRO, Too, Was Alerted of Cyber Security Breach', *Indian Express* [online], 6 November 2019. https://indianexpress.com/article/india/not-only-kudankulam-isro-too-was-alerted-of-cyber-security-breach-6105184
25. National Security Council Secretariat, 'National Counter Rogue Drone Guidelines', Government of India, 21 November 2019. https://www.civilaviation.gov.in/sites/default/files/Counter_rogue_drone_guidelnes_NSCS.pdf
26. Neeraj Chauhan, 'Drone Intrusions along India-Pak Border Have Risen in 2022: BSF Data', *Hindustan Times* [online], 31 August 2022. https://www.hindustantimes.com/india-news/drone-intrusions-along-india-pak-border-have-risen-in-2022-bsfdata-101661918247466.html; National Security Council Secretariat, 'National Counter Rogue Drone Guidelines'.
27. Author interviews with intelligence and police officers in Delhi, Gujarat, and Rajasthan, as well as with senior officials of the DAE and the AERB during field visits between October 2013 and June 2014. There have been subsequent meetings in the following years, the latest of which took place in New Delhi in July 2022.
28. Its predecessor was the Atomic Energy Act, 1948, which was replaced by the 1962 Act.
29. Ministry of External Affairs, 'Nuclear Security in India'.

30. Department of Atomic Energy, 'Atomic Energy (Radiation Protection) Rules, 2004', DAE [online], 25 August 2004. https://dae.gov.in/writereaddata/RPR2004.pdf
31. Ibid.
32. Department of Atomic Energy, 'Atomic Energy (Working of the Mines, Minerals and Handling of Prescribed Substances) Rules, 1984', DAE [online], April 1984. https://dae.gov.in/writereaddata/AE(WMMPS)%201984.pdf
33. Government of India, 'Weapons of Mass Destruction and their Delivery Systems (Prohibition of Unlawful Activities) Act' [online], 6 June 2005. http://www.mea.gov.in/Uploads/PublicationDocs/148_The-Weapons-Mass-destruction-And-Delivery-Systems-Act-2005.pdf
34. Rajagopalan and Biswas, 'Locating India within Global Non-Proliferation Architecture'.
35. PRS Legislative Service, 'The Nuclear Safety Regulatory Authority Bill 2011', PRS Legislative Service [online], 7 September 2011. http://www.prsindia.org/billtrack/the-nuclear-safety-regulatory-authority-bill-2011-1980
36. IAEA, 'Design Basis Threat (DBT)', IAEA [online], 2022. https://www.iaea.org/topics/security-of-nuclear-and-other-radioactive-material/design-basis-threat
37. Ministry of External Affairs, 'Nuclear Security in India'.
38. Department of Atomic Energy, 'Security Officer's Examination: Reading Material Paper-I & IV', Government Of India [online], 24 October 2019. https://dae.gov.in/writereaddata/BOSFW18112020.pdf
39. Ministry of External Affairs, 'Nuclear Security in India'; IAEA, *Nuclear Security Recommendations on Physical Protection of Nuclear Material and Nuclear Facilities (INFCIRC/225/Revision 5)*, Recommendations, IAEA Nuclear Security Series, No. 13. (Vienna: IAEA, 2010). https://www.iaea.org/publications/8629/nuclear-security-recommendations-on-physical-protection-of-nuclear-material-and-nuclear-facilities-infcirc/225/revision-5
40. K. Nityanandam IPS (Retd.), *Standards for Physical Security Management in Industry* (Hyderabad: Indian National Police Academy, 2015). https://www.svpnpa.gov.in/images/npa/pdfs/CompletedResearchProject/35_standardizationofphysicalsecurity.pdf; Department of Atomic Energy, 'Security Officer's Examination'.
41. Department of Atomic Energy, 'Protest of Nuclear Power Plants', Unstarred Question No. 6596 to be answered on 06.05.2015, Lok Sabha, Government of India, 6 May 2015. https://eparlib.nic.in/bitstream/123456789/662262/1/16697.pdf
42. Christopher Hobbs and Matthew Moran, 'Exploring the Human Dimension of Nuclear Security: The History, Theory, and Practice of Security Culture', *The Nonproliferation Review* [online], 9 June 2022. doi: 10.1080/10736700.2020.1811532
43. Atomic Energy Regulatory Board, *Security of Radioactive Sources in Radiation Facilities*, AERB Safety Guide No. AERB/RF-RS/SG-1 (Mumbai: AERB), 2011. https://www.aerb.gov.in/storage/uploads/documents/regdocFdOs8.pdf
44. Ibid.
45. Ibid., the changes in the mindset with personnel in the nuclear power plants was confirmed during an author's interview at the Department of Atomic Energy in October 2013.
46. Guenther et al., 'India-United States Cooperation on Global Security'.
47. Author's interviews with retired senior DAE officials, March, July, and September 2022.
48. NDTV Correspondent and Agencies, 'Six DU Professors Charged in Mayapuri Radiation Case', *NDTV*, 2 September 2011; Sandeep Sharma, Rakesh Kumar Sharma, and Rajesh Singh, 'Mayapuri Radiological Catastrophe: Good Practices and Lessons Learnt', *Current Radiopharmaceuticals*, 15/1 (2021), 21–31.

49. Author interviews with senior police officials at the Department of Atomic Energy, March 2022.
50. Atomic Energy Regulatory Board, 'AERB Newsletter'.
51. Manash Pratim Gohain, 'Atomic Board Nod Must for Colleges: UGC', *Times of India* [online], 17 January 2011. http://timesofindia.indiatimes.com/india/Atomic-board-nod-must-for-colleges-UGC/articleshow/7300519.cms
52. Atomic Energy Regulatory Board, *AERB Newsletter*. https://www.aerb.gov.in/images/PDF/Vol23-1.pdf
53. Sreenivas Janyala and Amitabh Sinha, 'Caesium-137: A Radioactive Isotope That Was Lost and Found in Andhra', *The Indian Express* [online], 24 January 2019. https://indianexpress.com/article/explained/this-word-means-caesium-137-5553013
54. Atomic Energy Regulatory Board, 'Safe Transport of Radioactive Material', AERB [online], March 2016. https://www.aerb.gov.in/images/PDF/CodesGuides/RadiationFacility/Transport/1.pdf; Atomic Energy Regulatory Board, 'Security of Radioactive Material During Transport', AERB [online], January 2008. https://www.aerb.gov.in/images/PDF/CodesGuides/RadiationFacility/Transport/2.pdf
55. Ministry of External Affairs, 'Nuclear Security in India'.
56. Author's interviews with senior police and DAE officials in October 2013. Subsequent interactions with these officials have helped validate the efficacy of India's PRP process.
57. Author's interviews with senior police and DAE officials in March 2022.
58. James M. Acton and Mark Hibbs, 'Why Fukushima Was Preventable', *Carnegie Endowment for International Peace* [online], 6 March 2012. https://carnegieendowment.org/2012/03/06/why-fukushima-was-preventable-pub-47361
59. PRS Legislative Service, 'The Nuclear Safety Regulatory Authority Bill 2011'.
60. Suvrat Raju and M. V. Ramana, 'It's Better to be Safe than Sorry', *The Hindustan Times* [online], 5 February 2014. https://www.hindustantimes.com/ht-view/it-s-better-to-be-safe-than-sorry/story-QpSS9S571iXtn6SaYaOlXK.html
61. Department of Atomic Energy, 'Security of Nuclear Materials', Unstarred Question No. 479 to be answered on 09.05.2012, Lok Sabha, Government of India, 9 May 2012. https://dae.gov.in/writereaddata/lssq479.pdf
62. See Comptroller and Auditor General of India, 'Report of the Comptroller and Auditor General of India on Activities of Atomic Energy Regulatory Board for the year ended March 2012', 22 August 2012. https://cag.gov.in/en/audit-report/details/2060
63. Author interviews with senior retired army officials in Delhi, March 2022.
64. For instance, see Department of Atomic Energy, 2020. Annual Report 2019–20. https://dae.gov.in/writereaddata/Annual2019-2020e.pdf

Suggested Reading

Guenther, Rita, Lowenthal, Micah, Nagappa, Rajaram, and Mancheri, Nabeel, (2013), *India–United States Cooperation on Global Security: Summary of a Workshop on Technical Aspects of Civilian Nuclear Security* (Washington, DC: National Academy of Sciences).

Pillai Rajagopalan, Rajeswari, 'Nuclear Security in India', Observer Research Foundation [online], February 2015. https://www.orfonline.org/wp-content/uploads/2015/02/NUCLEAR_SECURITY_IN_INDIA.pdf

Pillai Rajagopalan, Rajeswari, 'Radiological Security in India: Policies and Challenges', Observer Research Foundation [online], June 2020. https://www.orfonline.org/wp-content/uploads/2020/06/ORF-Occasional-paper-257-Radiological-Security.pdf

Pillai Rajagopalan, Rajeswari and Biswas, Arka, 'Locating India within Global Non-Proliferation Architecture: Prospects, Challenges and Opportunities', Observer Research Foundation [online], August 2016. https://www.orfonline.org/wp-content/uploads/2016/08/ORF_Monograph_NonProliferation.pdf

Sood, Rakesh, 'Charting the Course for Nuclear Security: An Indian Perspective', Carnegie India [online], 23 March 2006. https://carnegieindia.org/2016/03/23/charting-course-for-nuclear-security-indian-perspective-pub-63100

CHAPTER 28

NUCLEAR SECURITY IN THE REPUBLIC OF KOREA

JINHO CHUNG AND NA YOUNG LEE

Introduction

With an advanced civilian nuclear sector and as an international nuclear supplier, the Republic of Korea (ROK) shares responsibility with other countries to strengthen the international nuclear security regime. The implementation system for nuclear security was not fully systematized in the ROK until the mid-1990s. However, over the past two decades, the ROK has sought to demonstrate leadership in the nuclear security domain by actively participating in international dialogue and supporting efforts aimed at enhancing national and regional capacities in this area.

The ROK became a Member State of the International Atomic Energy Agency (IAEA) in 1957 when it began pursuing peaceful nuclear activities. The Atomic Energy Law was approved in 1958, and the Office of Atomic Energy was established in 1959. Since then, the ROK has actively supported the international nuclear non-proliferation regime and maintained a strong commitment to nuclear energy development. Beginning with commercial operations at the Kori Unit 1 reactor in 1978, at the time of writing there are 25 nuclear power plants in operation, which account for about 27 per cent of the total electricity mix.[1] A total gross capacity of 23.3 gigawatts electrical (Gwen) is installed across the operating plants, comprising 22 pressurized water reactors (PWRs) and three CANDU pressurized heavy water reactors (PHWRs).[2] There are plans to build four more units by 2030.[3]

This chapter explores the evolution of the nuclear security architecture in the ROK. Specifically, the chapter discusses the ROK's underpinning legal and regulatory framework and how this is implemented, with a focus on several key areas: physical protection, cyber security, research and development (R&D) activities, education and training programmes, and future vision and leadership in this field. The chapter also analyses several important drivers that influenced the development of nuclear security in the

ROK—such as its role in organizing the 2012 Nuclear Security Summit in the capital Seoul and the International Physical Protection Advisory Service (IPPAS) mission in 2014. In addition, it examines recent efforts by the ROK to effectively respond to emerging threats. Notably, the ROK has accelerated R&D activities for novel designs of nuclear power plants, such as small modular reactors (SMRs), and sought to combat new threats to nuclear facilities, such as unmanned aerial vehicles (UAVs). These efforts have served to minimize nuclear security risks through implementing effective measures to address theft, sabotage, unauthorized access, illegal transfer, and other malicious acts involving nuclear materials or their associated facilities.

Legal and Regulatory Framework

The ROK is a long-standing member of the international nuclear security regime, signing the Convention on the Physical Protection of Nuclear Materials (CPPNM) in 1982, its Amendment in 2014, and the International Convention for the Suppression of Acts of Nuclear Terrorism (ICSANT) in 2005. In parallel, the ROK has established and reinforced its domestic legal and institutional frameworks in the nuclear security domain. Although the Atomic Energy Act (AEA) was passed in 1957 as the legal basis to promote and regulate nuclear energy, a comprehensive nuclear security legal framework for regulating nuclear materials and associated facilities was not established until 1995 when the AEA was revised. In particular, the revision embraced IAEA guidelines for physical protection of nuclear materials and nuclear facilities (at the time contained in INFCIRC/225/Rev.4).

The terrorist attacks of 11 September 2001 ('9/11') emerged as a turning point for the ROK to strengthen the legal and institutional aspects of its nuclear security regulations. In response, the Act on Physical Protection and Radiological Emergency (APPRE) was developed, enacted in 2004, drawing on relevant international requirements and guidance.[4] The act is based on four fundamental elements. First, it expands the scope for nuclear security regulation. In contrast to existing laws, which prior to 2004 covered only the unauthorized removal of nuclear materials, the APPRE also covers the physical protection of nuclear facilities. Second, the act comprehensively specifies countermeasures against sabotage. Third, it embraces core principals from the CPPNM Amendment. Fourth, the act facilitates the establishment of an integrated national response system for addressing radiological emergencies.

The Ministry of Education, Science and Technology of Korea (MEST) was previously responsible for both promotional and regulatory functions in the ROK. However, it became increasingly apparent that the regulatory body should be separated from the promotional part to ensure independence of regulatory functions. This issue was brought to the fore in 2011 with the Fukushima nuclear disaster in neighbouring Japan. The Korean government reformed its nuclear sector, and the Nuclear Safety Act was enacted in

2011 to provide a legal ground for nuclear regulations.[5] In accordance with the act, the Nuclear Safety and Security Commission (NSSC) was created as an independent regulatory body responsible for physical protection and radiological emergency preparedness.[6] The NSSC is also responsible for inspecting, reviewing, and assessing the physical protection systems associated with nuclear facilities in the country.

The Korea Institute of Nuclear Non-proliferation and Control (KINAC) was established in 2006.[7] KINAC specializes in the implementation of nuclear security and non-proliferation measures. It is also responsible for implementing and managing physical systems for protecting nuclear facilities and materials against sabotage, theft, diversion, and other malicious acts. KINAC conducts regulatory reviews and inspections of physical protection systems for nuclear facilities and materials on behalf of the NSSC. To achieve this, it regularly reviews and revises the design basis threat (DBT) as well as 'vital areas' at facilities. Accordingly, KINAC performs periodic inspections to evaluate the robustness of physical protection systems, including resources and on-site management capabilities, against the DBT. KINAC has also established a regulatory framework for cyber security of nuclear facilities and conducts regulatory reviews and inspects these systems to prevent, detect, and respond to cyber-attacks that may threaten facilities.

Implementation of Nuclear Security

Physical Protection

As stipulated by the APPRE, licensees of nuclear power plants are required to regularly inspect the physical protection systems of their facilities and the transportation of their nuclear materials. The act stipulates that licensees must inspect their facilities at least every two years. In addition, nuclear materials must be inspected whenever transported. These inspections must be performed in accordance with the security and contingency plans prepared by the licensee. The integrity of the physical protection systems of nuclear facilities is evaluated through inspection and review processes led by the NSSC and KINAC.

KINAC is responsible for physical protection-related regulation under the APPRE. In accordance with Article 9 of the APPRE, the physical protection plans of the licensees are subject to approval by the regulatory body, with KINAC reviewing the physical protection plans submitted by licensees. The vital areas re-identified and approved for facilities are incorporated in the physical protection plans of these facilities. In accordance with Article 12 of the APPRE, licensees' facilities are subject to inspection by the regulatory body, with KINAC inspecting both the facilities and transportation of nuclear materials and specifying corrective or follow-up actions to be implemented by licensees. In accordance with Article 9-3 of the APPRE, the physical protection exercise plans of the licensees are subject to approval by the regulatory body, with KINAC reviewing these exercise plans and evaluating force-on-force exercises. Licensees are required to perform partial and force-on-force exercises semi-annually and annually, respectively. Evaluations may yield several follow-up actions to be taken by the licensees.

In accordance with Article 4 of the APPRE, the NSSC assesses the threats related to nuclear security every three years and establishes the DBT. A revised DBT is drafted based on the results of subject-matter expert meetings, and the new DBT is finalized upon the decision of the National Physical Protection Council. Subsequently, nuclear facilities must be evaluated, and physical protection systems must be updated to ensure protection against the revised DBT. KINAC is responsible for reviewing the threat and response scenarios developed by facility operators to assess the effectiveness of the physical protection systems against the DBT. Notably, KINAC provides guidelines for developing scenarios and hosts scenario-development workshops to enhance operators' understanding of these guidelines. KINAC reviews the attack and response scenarios developed by operators, emphasizing the timeliness in responses of the physical protection systems.

KINAC has sought to improve regulatory standards since its inception in 2006. For example, performance tests of external intrusion detection sensors and cameras for video surveillance have been conducted at the ROK's Security Research, Training and Test Facility. To test active infrared sensors, an indigenous mobile system has been constructed. In addition, KINAC has investigated nuisance alarms pertaining to bushes, slopes, and sunlight, evaluating the transmitter–receiver distance and the sensor height and angle. To test security cameras, surveillance zones have been identified based on the heights and angles of the installed equipment. The various test results have been used to update KINAC's documents on regulatory standards for intrusion detection and detection of prohibited items.

Vital Area Identification

Vital area identification is a first step in designing an effective physical protection system. By doing so, direct and indirect radiological effects from malicious acts such as sabotage can be minimized.[8] In the past, the ROK identified and established vital areas at nuclear power plants where nuclear materials were located or used. Seeking to improve effectiveness, however, the authorities decided to re-identify vital areas by utilizing probabilistic safety analysis (PSA) techniques to invoke radiological consequence. In 2015, the NSSC ordered vital areas of all the nuclear power plants to be re-established. In view of the varying characteristics of the different types of nuclear power plants in the ROK—including PWR and CANDU reactors—the operator Korea Hydro & Nuclear Power (KHNP) began reviewing the vital area identification for each type in 2017, with this process now completed.

Cyber Security

Since 2012, the DBT has been designed to include cyber threat attributes, as well as physical attributes, for nuclear facilities and materials. In accordance with Article 45 of the APPRE and Article 40 of the Enforcement Decree, KINAC is entrusted by the NSSC to perform activities related to cyber threat assessments.[9] In order to develop the cyber DBT, a threat assessment report is prepared by analysing new cyber threats. This is then used to evaluate, through a series of scenarios, appropriate security and response measures. Based on these threat assessment activities, as well as consultations with experts, the cyber DBT is regularly revised.

KINAC conducts both regular and special inspections of nuclear facilities, pursuant to Article 12 of the APPRE. One aspect of KINAC's regular inspections is verifying the implementation of cyber security plans (CSPs), prepared by each nuclear facility in accordance with Article 9 of the APPRE. In 2014, KINAC published KINAC/RS-015 as a regulatory standard to guide licensees in preparing CSPs.[10] Nuclear licensees subsequently developed and implemented their own CSPs, approved by the NSSC in 2015. KINAC's special inspections of nuclear facilities also cover CSP implementation. These special inspections have contributed to an overall improvement in cyber security competencies by refining existing procedures and strengthening policy. For example, an organization exclusively responsible for cyber security has been established at nuclear facilities.

Nuclear facilities must also perform full-scale cyber security exercises at least once a year and partial exercises at least twice a year, pursuant to Article 9 of the APPRE and Article 5 of the Enforcement Decree for the APPRE. These exercises help strengthen the cyber incident response system and response capability of nuclear facilities against cyber threats. After the licensees implement Step 4 of the CSP (defence-in-depth and incident response), each nuclear facility establishes a cyber security incident response procedure (CSIRP). Accordingly, the evaluation of the overall cyber security exercises is aimed at improving the CSIRP by verifying the adequacy and effectiveness of the procedures. A licensee's cyber security exercises are scenario based, with threat and response scenarios. Typically, a scenario involves seven steps: detection, initial response, establishment of the CSIRP, notification and reporting, collection of evidence, analysis, and elimination and restoration.

Inspections have been successful in examining key cyber security aspects, such as the adequacy of implementation of security regulations and technical measures. However, a challenge has been to implement these rules and procedures in practice. In particular, while roles and responsibilities are specified in regulations, taking accountability is another issue. In addition, developing regulations and conducting inspections are currently carried out by the same apparatus. For the sake of rigorousness, the independence of inspection functions needs to be taken into consideration.

International Nuclear Cooperation

The ROK has exported its new standard nuclear power plant, the APR-1400 (advanced power reactor for 1400 MW), to the United Arab Emirates (UAE), which is developing its own civilian programme at the Barakah Nuclear Power Plant.[11] In 2009, the two governments signed the Cooperation Agreement for the Peaceful Use of Nuclear Power; and in 2010, the MEST and the UAE's Federal Authority for Nuclear Regulation (FANR) signed a Cooperation Agreement. With the creation of the NSSC in 2011, an agreement subsequently transferred the signing authority from the MEST to the NSSC. In March 2011, KINAC and FANR signed the Implementing Arrangement for the Exchange of Technical Information and Cooperation in Nuclear Safeguards, Export Control of Nuclear Related Items and Physical Protection of Nuclear Material and Nuclear Facilities.[12]

In accordance with the Cooperation Agreement, the NSSC and FANR have held Annual Cooperation Meetings since 2011 to share progress on nuclear regulation-related work, including nuclear security. Additional bilateral meetings are held a few months before the Annual Cooperation Meeting to discuss issues at a working level. A ROK–UAE summit took place in March 2018 where the two countries also agreed to hold high-level consultations annually in recognition of the need to strengthen cooperation in the nuclear sector and develop comprehensive and strategic channels of communication.[13] These high-level consultations have three working groups, including one for the implementation of nuclear safety, nuclear security, and international cooperation.

Through bilateral consultations, the ROK and the UAE share information on Barakah operational technologies and nuclear security issues. They also discuss measures for practical cooperation, and conduct staff exchanges as and when needed. At the eighth Annual Cooperation Meeting in 2018, FANR requested the participation of Korean experts for Barakah's cyber security inspection, in order to share experiences and review the inspection process. During these inspections, Korean experts have assessed the status of cyber security in the UAE, including in areas such as digital systems, essential digital assets, protective digital asset identification procedures and documents, and the status of the facility's in-depth cyber security protection strategy.

The Nuclear Security Summits

In 2009, former US president Barack Obama claimed in a speech in Prague that 'nuclear terrorism is the most immediate and extreme threat to global security'.[14] To combat this threat, Obama proposed a 'Global Summit on Nuclear Security'. Supporting the cause, the ROK hosted the second Nuclear Security Summit in March 2012. Leaders from 53 countries and four international organizations gathered in Seoul. The second summit received a favourable evaluation for its specific and practical approach to strengthening global nuclear security.

The Seoul Summit took a different approach to the preceding Washington Summit in several ways. First, the Seoul Communiqué reiterated the central role of the IAEA in strengthening nuclear security and encouraged increased contributions to the Nuclear Security Fund (NSF) on a voluntary basis by states or industries.[15] Second, the Seoul Summit highlighted the participation of civilian sectors in nuclear security, such as academia, non-governmental organizations, and media. For example, a Nuclear Industry Summit and Nuclear Security Symposium were held as side events.[16] This approach was deemed necessary to sustain public attention and ensure cooperation with the private sector.[17] Third, while the previous summit had focused on the role of the government and nuclear industries in promoting and sustaining nuclear security culture, the Seoul Communiqué encouraged commitment from all stakeholders to enhancing security culture.[18] Fourth, the Seoul Summit paid more attention to transportation security. Although the Washington Work Plan briefly notes the transport of nuclear material, the Seoul Communiqué includes more detailed and practical measures to enhance the security of nuclear and other radioactive materials in transport.[19] Fifth, the nuclear security and safety interface was highlighted as a result of lessons learned from the Fukushima nuclear disaster in Japan the previous year.

Prior to hosting the Seoul Summit, the ROK comprehensively reviewed its legal and institutional system pertaining to nuclear security and identified key actions to strengthen its national nuclear security regime. First, a conversion project was launched to transform nuclear fuel for research reactors from highly enriched uranium (HEU) to high-density, low enriched uranium (LEU).[20] Second, the ROK provided Vietnam with a radiation source location tracking system in concert with the IAEA. Third, the ROK revised the APPRE by reflecting major elements from the Amendment to the CPPNM and the ICSANT. Fourth, the ROK pledged to establish a centre of excellence (CoE) to provide quality nuclear-security-related education and training programmes. To this end, the International Nuclear Non-proliferation and Security Academy (INSA) was opened in 2014. Fifth, the ROK decided to request an IPPAS mission of the IAEA.

IPPAS Mission to the ROK

The IPPAS missions are aimed at assisting IAEA Member States in strengthening their national nuclear security regimes. In response to the official request of the ROK government, the IAEA conducted a two-week IPPAS mission in the city of Daejeon from 24 February to 7 March 2014. The ROK requested a comprehensive review of the national physical protection regime and became the first Asian country to host an IPPAS mission of those operating nuclear power plants.[21]

The ROK mission covered all five IPPAS modules: national review, facility review, transport review, radioactive material security review, and computer security review. The IPPAS team consisted of eight international nuclear security experts from six countries and the IAEA. The team reviewed a number of areas, including: the ROK's nuclear-security-related legislative and regulatory frameworks for nuclear and other radioactive materials and associated facilities; the ROK's arrangements for the secure transportation of nuclear material and radioactive sources; the security of computer systems; and physical protection systems at the Hanbit Nuclear Power Plant, operated by the Korea Hydro and Nuclear Power Company (KEPCO), and the High-Flux

Advanced Neutron Application Reactor (HANARO), operated by the Korea Atomic Energy Research Institute.[22]

In preparation for the IPPAS mission, the ROK government and its technical support organizations conducted a thorough self-assessment of the national nuclear security system. These efforts and the subsequent recommendations from the mission served as a turning point for the country to enhance its physical protection regime. For example, the ROK had the opportunity to thoroughly examine how to set up vital areas at sites and determine a 'high radiological consequence/unacceptable radiological consequence' level, both critical to the graded approach for unauthorized removal or sabotage of nuclear material.[23] The IPPAS team concluded that a mature and well-established nuclear security regime was in place for protecting nuclear material and radioactive sources. Moreover, physical protection measures were adequately implemented at the visited facilities, with levels exceeding international guidelines in certain cases, including strong and well-coordinated response capabilities based on the graded approach and robust transport security implementation measures.[24]

Nuclear Security Research and Development

Nuclear Forensics

Nuclear forensics refers to the investigation of nuclear materials to identify the origin and history of radioactive materials out of regulatory control. The ROK has developed nuclear material analysis systems that focus on the technical aspects of nuclear forensics, including analytical techniques. It has also established a prototype national nuclear forensic library (NNFL). The ROK's NNFL includes signatures of radioactive material, algorithms to compare the material, and a pool of subject-matter experts. To efficiently implement nuclear forensics, there is a four-tier system in place: the legal basis; radiological crime scene management; a nuclear forensics library; and analysis. The legal basis pertains to law enforcement against nuclear terrorism. Radiological crime scene management includes the sampling of radiological nuclear materials and collecting of other traditional forensic evidence. The nuclear forensics library, the NNFL, has a database containing broad categories of information to identify and verify analysis results. Analysis is supported by technical capabilities, critical for identifying signatures.

The legal framework for nuclear forensics in the ROK was established in 2014 when the relevant provisions of the ICSANT and the CPPNM and its Amendment were reflected in domestic laws. To facilitate radiological crime scene management, the ROK established the necessary procedures for collecting evidence and developing equipment. The analysis programme was significantly advanced as a result of the Nuclear Security Summit process. Abiding to ICSANT guidelines, the ROK established a national law to set penalties for criminal offences.[25] However, the ROK needs to consider incorporating nuclear forensics activities into the national response plan to support investigations and also ensure that all relevant organizations are aware of the response procedures. If the investigation team does not react appropriately or contact the correct technical counterpart, technical capabilities may not be used in a timely manner.

A nuclear forensics database must be established by compiling data to support nuclear forensic interpretation, and the NNFL's database includes representative samples from nuclear fuel manufacturers. However, the ROK has not mandated the submission of nuclear materials data for the NNFL, and therefore the NNFL remains insufficient. Technical capabilities within the ROK include the analysis of the signatures of nuclear materials, essential for nuclear forensics. Nevertheless, traditional forensic evidence is required for this task. Hence, the ROK still needs to refine the relevant procedures and establish cooperative systems to analyse the evidence from both nuclear and traditional forensics domains.

Unmanned Aerial Vehicles

UAVs (i.e. drones) are among the significant threats to be considered in protecting nuclear power plants in the national threat analysis report used to update the national DBT. As a result, the Drone Law was enacted in April 2019, with the definition of a drone specified, and the Aviation Safety Law was also revised.[26] To establish a drone protection system for nuclear facilities to respond to threats from drones, operators need to introduce major changes to their nuclear security systems in accordance with the Drone Law and other applicable laws.

The Drone Law has introduced flight approval systems and pilot qualification systems as regulations, thereby impacting the physical protection systems of nuclear facilities. Nuclear facilities are classified as national critical infrastructure, and a radius of 18.3 km around sites is designated as a no-fly zone. The establishment of no-fly zones and flight approval systems prohibit drone operation near nuclear facilities in principle. Drone operation is permitted only when prior flight approval is granted. This system can help security personnel in performing anti-aircraft monitoring over nuclear facilities and in identifying malicious drones. By providing relevant information to pilot candidates, the pilot qualification system can also help prevent illegal flights near nuclear facilities. Other diverse measures, such as provision of device reports, operator certification, safety certification, operator compliance, insurance, and flight approval, have been implemented and revised to ensure the safe use of drones.

Despite the measures, drone flights have been intercepted near nuclear power plants, even though their purpose appears to have been leisure or filming. In response, further preventive measures have been implemented such as informing nearby residents of the no-fly zone and flight approval system. At the same time, the ROK has implemented policies to promote the drone industry, including the Drone Industry Development Basic Plan. With the ongoing development of the drone industry, however, regulations need to be regularly reviewed to effectively protect nuclear facilities against evolving threats associated with drones.

R&D for New Nuclear Facilities

The ROK has achieved considerable success in the nuclear industry with decades of experience of power plant construction since construction of its first nuclear plant in 1971. With steady investment in technical developments to improve safety and economic efficiency, the ROK has explored new type of reactors that enhance safety with simplified and reliable passive safety systems as well as easier operation and greater flexibility in their wider applications. The ROK has, however, experienced challenges in successfully deploying SMRs, such as the licensing of the novel technology, the economic disadvantages of small-scale technology, and public concerns about safety and spent fuel management.

With two decades of experience in this area, the ROK is the one of the pioneers of SMR technology. The system-integrated modular advanced reactor (SMART) was given standard design approval by the NSSC in 2012, while the modified version of SMART design has been submitted for approval, expected to be issued in 2023.[27] Competing in a global SMR market, the ROK has recently started to develop the next version of SMART called I-SMR, with an expectation that the development will gain standard design approval by 2028. The I-SMR is being developed with a focus on improved safety, enhanced economics, and increased flexibility.

There are different types of non-light-water SMRs in the ROK. Sodium fast reactor (SFR) technology-based SMRs are under development, called SALUS-100 (small, advanced, long-cycled and ultimate safe).[28] SALUS-100 is a spin-off reactor of PGSFR, the prototype Generation-IV SFR being developed for spent nuclear fuel reduction. In addition, the ROK is cooperating with Canada on a demonstration project of a high-temperature gas-cooled reactor (HTGR). This cooperation is expected to lead to the demonstration of the HTGR for supplying process heat and hydrogen production. The ROK also recently began developing the molten salt reactor (MSR), which is based on chloride salt. Shipbuilders are interested in developing MSRs to adopt zero-carbon engines for their containers, in order to meet regulations of the International Maritime Organization (IMO).

These new SMR types require new consideration from a regulatory perspective. At such facilities, nuclear security may be plagued with various issues, including the threat of stand-off attacks, inadequate staffing including security response personnel, insider threats, and cyber security issues at downsized remotely operating SMRs. Several measures have been considered in the ROK to address these challenges. First, in terms of defence-in-depth, the integrity of SMRs is influenced by gravity, pressure differences, and convection. Well-planned attacks might exploit these vulnerabilities, potentially leading to a loss-of-coolant accident. Safety weakness can be a vulnerability from the perspective of nuclear security. This is why the IAEA, the ROK, and other countries are focusing on 3S (safety, security, safeguards) harmonization.

Second, in terms of physical barriers, the containment structures of SMRs are small and inadequately robust. Therefore, the structures may be vulnerable to stand-off attacks, such as aircraft impacts and long-range cannon attacks. In such scenarios, underground sites might be used to enhance protection against attacks and to delay attacks, as well as protect against floods, thereby decreasing repair and maintenance costs.[29] Third, insider threats and cyber-attacks are significant issues facing SMRs. Remotely operating SMRs require remote unmanned autonomous security systems, and thus, effective cyber security controls are paramount. Other vulnerabilities include the limited staff, control rooms, and alarm stations, such as remote monitoring stations. In this context, more robust trustworthiness systems and cyber protection systems are required.[30]

The key advantage of SMRs is economic feasibility, although security regulation is likely to increase the associated cost. To facilitate the use of SMR technologies, the regulatory requirements must be updated to include substantial security assessments and changes in security concepts. Following the graded approach, the security requirements on SMRs might be relaxed proportional to the radiological consequence, which can ensure cost-effectiveness. Accordingly, specific security requirements must be identified and implemented for the various types of SMRs. For instance, more than 50 types of SMRs are currently being researched in the world. The regulatory body must develop prescriptive and scalable performance-based security requirements.

Nuclear Security Education and Training

Although the importance of training has always been emphasized in the ROK, systematic training programmes remained inadequate until recently. As a result, in 2014 INSA was established as an international education and training hub focused on nuclear non-proliferation and nuclear security in the Asia–Pacific region.[31] Because this organization was also to be engaged in R&D, an external facility for experiments was established. Experts on nuclear non-proliferation and nuclear security, from industry, academia, and research institutes, can all collaborate in R&D projects using this facility.

To satisfy demand from various users worldwide, the ROK has also developed customized training programmes, including e-learning programmes, in cooperation with the IAEA and the US. KINAC is responsible for designing and developing the curriculum and content of these programmes, which are available to both international and local attendees. To reduce duplication and facilitate practical cooperation, the ROK joined the International Network for Nuclear Security Training and Support Centres (NSSC Network) organized by the IAEA, and it closely cooperates with Japan and China, which have similar centres.[32] In 2012, the three countries set up the Asia Regional Network (ARN), as a subgroup within the NSSC Network, and hold annual meetings to facilitate information exchange and regional cooperation over human resource development in nuclear security.[33]

INSA has three objectives: to provide customized, high-quality nuclear security and non-proliferation training and educational programmes to satisfy the needs of domestic and regional Asia-Pacific personnel; to facilitate technical and scientific cooperation and assistance to emerging nuclear countries; and to promote R&D activities for physical protection systems.[34] INSA's programmes focus on providing theoretical concepts as well as practical knowledge. In addition to the nuclear security or non-proliferation programmes typically on offer, INSA provides training on import and export control, which is essential for individuals working in the business sectors related to the nuclear industry. A major distinguishing feature of INSA is the availability of large-scale testbeds for training and R&D. These testbeds facilitate efforts to acquire data regarding the vulnerabilities of a nuclear facility. For instance, the testbeds can be used to evaluate the performance of new equipment. Furthermore, INSA provides support to countries that wish to initiate their own nuclear industry by providing customized programmes, as well as technical and scientific assistance.

INSA's training programmes on nuclear security aim to enhance awareness of fundamental elements of the nuclear security regime and provide participants with essential information about nuclear security systems and international instruments and guidance. The courses, delivered in both the Korean and English languages, target domestic and international audiences. The domestic audience can be divided into two groups: individuals working in nuclear facilities, and the public. Individuals working in nuclear facilities have been mandated to take courses on nuclear security since 2010. All personnel must undergo nuclear security training for two to eight hours, depending on their roles and responsibilities.

As of 2021, the total number of personnel that have received training at INSA is 515. The mandatory training programme is expected to expand in the future, with the number of participants increasing gradually in proportion to the increasing number of nuclear facilities around the world. INSA plans to continue developing educational programmes to provide expert knowledge and experience and spread awareness regarding the importance of nuclear non-proliferation and nuclear security for Asian and Middle Eastern countries, thereby helping strengthen regional security systems.

Nuclear Security Culture

Unlike nuclear safety culture, which has been extensively promoted since the early 1990s, security culture has traditionally been regarded as a subordinate element of organizational culture. The importance of nuclear security culture has been emphasized only since the early 2000s.[35] In the ROK, national efforts to improve industrial security culture began around 2000 when the limited alignment of national nuclear security culture with international standards became apparent. These efforts included the formulation of a comprehensive national framework, introduction of a National Nuclear Security Culture Implementation Guide, and specification of high-level commitments. For instance, at the 2010 Washington Nuclear Security Summit, the ROK committed to the final Communiqué and articulated its willingness to 'work with the industry to prioritize physical protection, material accountancy, and security culture'.[36]

To address the challenges of low awareness and inconsistent efforts to enhance industrial security culture, the ROK has established a national framework supported by the National Nuclear Security Culture Implementation Guide, which was developed in 2013. This framework is directed at all those involved in the country's nuclear sector, including regulatory bodies and licensees. Prior to drafting the Implementation Guide, the relevant international literature was thoroughly reviewed. Notable influences included national commitments to the CPPNM, and international best practices outlined in the IAEA's Nuclear Security Series (NSS), particularly 'Nuclear Security Culture', NSS No. 7, which maintains that the state, rather than individual organizations, is responsible for creating a fundamental security environment, allocating responsibilities, and ensuring information security.[37]

Accordingly, KINAC and the NSSC have sought to adapt key principles within international guidance to reflect the country's indigenous nuclear security environment for both regulation and implementation. This task has required transforming the definition of nuclear security culture from its previous narrow focus on the unauthorized removal of nuclear materials to a broader concept. The ROK's Implementation Guide explicitly draws upon this expanded concept, noting the interrelationship of the human factor and leadership, as well as the interaction between safety and security culture pertaining to physical protection systems. Consequently, the Implementation Guide currently defines nuclear security culture as the 'mindset and behaviour of individuals and organizations that recognize nuclear security as the principal value in the peaceful use of nuclear energy'.[38]

To implement this new philosophy, a six-step framework has been designed to systematically foster and facilitate nuclear security culture. Based on the definition provided by the authorities, nuclear organizations are encouraged to reflect on the fundamental concepts of nuclear security culture and its core elements in their own internal principles and statements. To this end, organizations are encouraged to embed these principles into their own corporate cultures through the development of action plans. To ensure the sustainability of this process, staff are expected to engage in continuous education and training, and organizations are encouraged to assess and analyse their nuclear security culture as it develops.

In particular, organizations are expected to prepare their own security policy statements, establish management structures, provide resources for management, and conduct self-assessments. These practices can vary depending on the nature of each organization, although the core elements recommended by the authorities are expected to be incorporated in the principles. The Implementation Guide recognizes the key role that managers play in promoting nuclear security culture through their influence on employee attitudes. As such, the Implementation Guide extensively stipulates the manager's role, which covers—but is not limited to—definition of responsibilities, practice management, qualifications and training, motivation, and performance enhancement. Staff members are expected to recognize their current security environment and understand the potential consequences of their behaviour in relation to credible threats to nuclear security. Moreover, individuals must be conscious of how their roles and actions influence nuclear security culture.

However, these policies are only recommendations, and nuclear organizations in the ROK are not legally obliged to foster an effective nuclear security culture. Without the authority to compel organizations, the state focuses on encouraging all stakeholders to comply with the guidelines by specifying best practices and organizing education and training programmes. NSSC and KINAC regularly conduct surveys on workplace attitudes and assess the results. If the results indicate the need for improvements, NSSC and KINAC can choose to conduct further investigations in the necessary domains and organize additional training to enhance the standards, for example, through workshops.

Nuclear security culture is unlikely to be directly regulated for the foreseeable future, but it can be improved by various policy tools at the national and organizational levels. In addition to providing an implementation framework, the ROK emphasizes the provision of education and training programmes through its CoE, INSA, to nurture nuclear security culture domestically and abroad. Establishing and maintaining a nuclear security culture that is sustainable can improve the efficiency and effectiveness of the national nuclear security regime. This, in turn, helps to ensure the security of nuclear and other radioactive material, associated facilities, and related transport.

Conclusion

Responsibilities for establishing and operating an effective nuclear security regime for nuclear material and facilities rest with each country.[39] With a strong sense of international responsibility, the ROK has worked over decades to fulfil its obligations and duties of establishing, maintaining, and updating a robust national nuclear security system. The APPRE functions as a legal basis for implementing physical protection and cyber security measures in the country. NSSC and KINAC, as regulatory bodies, develop and implement necessary policies in a timely manner to strengthen the national nuclear security regime. The licensees actively cooperate with the government by developing and implementing their nuclear security systems and measures.

As a global nuclear supplier, the ROK emphasizes international cooperation and has sought to demonstrate leadership in the area of nuclear security. Recognizing the significance of the Nuclear Security Summit process, Seoul supported the global dialogue by hosting the 2012 Nuclear Security Summit. Furthermore, at the 2010 Nuclear Security Summit, the ROK pledged to establish a CoE, with INSA opened to the world in 2014. INSA has sought to provide quality education and training programmes to both domestic and international participants in the field of physical protection and cyber security.

To respond to emerging threats from the rapid changes in the nuclear security environment, the ROK continues to identify and analyse the prevailing situation at the national and regional level. The enhancement of R&D capacities is considered critical for appropriately reacting to these threats. To reinforce security measures at current nuclear facilities, KINAC is actively developing new processes aimed at identifying the vital areas of nuclear facilities. Meanwhile, nuclear forensics activities have been enhanced for the purpose of identifying and supporting the response to possible nuclear security events. Considerable efforts are also being devoted to security considerations for SMRs as well as combatting the threats posed by drones.

Although the ROK is committed to strengthening its national nuclear security regime, there is still room for improvement. For instance, the DBT reflects insider threats well, but the methodology of insider analysis needs to be upgraded to be more sophisticated. In addition, the ROK is well aware of the importance of the 'security-by-design' concept to reduce intrinsic vulnerabilities of nuclear facilities, but this concept has not been fully reflected in the designing process of a nuclear power plant.[40] The country understands the detrimental effect of complacency. Therefore, the ROK will continue to face these challenges and develop the best possible technical and policy solutions. As a global trustworthy partner, the ROK is committed to fostering a more peaceful and secure world.

Notes

1. World Nuclear Association, 'Nuclear Power in South Korea', [online], June 2022. https://world-nuclear.org/information-library/country-profiles/countries-o-s/south-korea.aspx
2. Ibid.
3. Heesu Lee, 'South Korea to Build Nuclear Plants to Meet Net-Zero Targets', *Bloomberg* [online], 5 July 2022. https://www.bloomberg.com/news/articles/2022-07-05/south-korea-to-build-nuclear-plants-to-meet-net-zero-targets
4. Nuclear Safety and Security Commission (NSSC), 'Act on Physical Protection and Radiological Emergency', Korea Legislation Research Institute [online], 2004. https://elaw.klri.re.kr/eng_mobile/viewer.do?hseq=46895&type=part&key=18
5. Ibid.
6. Nuclear Safety and Security Commission, 'Establishment of the Nuclear Safety and Security Commission', Korean Ministry of Foreign Affairs Press Release, Seoul, 26 October 2011.
7. Korea Institute of Nuclear Non-proliferation and Control (KINAC), 2022, 'History and Functions', 2022. https://www.kinac.re.kr/board?menuId=MN0000000410&siteId=00003
8. Minho Kang et al., 'The Analysis of Vital Area Identification Methodology', KINAC/TR-002/2018 (B8-7120), Daejeon, 2018.
9. Nuclear Safety and Security Commission, Act on Physical Protection and Radiological Emergency.
10. Korea Institute of Nuclear Non-proliferation and Control (KINAC), 'Computer and Information System Security of Nuclear Installations', Daejeon, 2016.

11. Amena Bakr and Cho Mee-young, 'South Korea Wins Landmark Gulf Nuclear Power Deal', *Reuters* [online], 27 December 2009. https://www.reuters.com/article/us-emirates-korea-nuclear-idUSLDE5BQ05O20091227
12. Korea Institute of Nuclear Non-proliferation and Control (KINAC), 'Non-proliferation Policy and Planning Division 2013 Annual Report', Daejeon, 2013.
13. Korean Ministry of Foreign Affairs, 'Establishment of the ROK-UAE High-level Consultation on Nuclear Cooperation', Korean Ministry of Foreign Affairs Press Release, Seoul, 14 November 2018.
14. Barack Obama, 'Remarks by President Barack Obama in Prague as Delivered, Hradcany Square, Prague, Czech Republic, 5 April' (Washington, DC: The White House, Office of the Press Secretary, 2009). https://obamawhitehouse.archives.gov/the-press-office/remarks-president-barack-obama-prague-delivered
15. Nuclear Security Summits, 'Seoul Communiqué', 2012 Nuclear Security Summit, Seoul, 28 March 2012.
16. Korean Ministry of Foreign Affairs, Press Release. Seoul, 23 March 2012.
17. Duyeon Kim, 'Fact Sheet: 2012 Seoul Nuclear Security Summit Results', Center for Arms Control and Non-Proliferation [online], 20 November 2021. https://armscontrolcenter.org/fact-sheet-2012-seoul-nuclear-security-summit-results
18. Chang Hoon Shin, *Harmonious Choice between Focus and Vision*, Issue Brief No. 23 (Seoul: The Asan Institute for Policy Studies, 2012).
19. Nuclear Security Summits, 'Seoul Communiqué'.
20. Korean Ministry of Foreign Affairs, 'ROK-US-France-Belgium Release a Joint Statement on Minimizing HEU', Korean Ministry of Foreign Affairs Press Release, Seoul, 27 March 2012.
21. Jung Soo Kim, Myung-Tak Jung, Kookheui Kwon, and YoungWook Lee, 'Effort and Prospective on Nuclear Security in ROK', *International Journal of Nuclear Security*, 7/1 (2022), Article 14.
22. IAEA, 'IAEA Completes Nuclear Security Review Mission in Republic of Korea', IAEA Press Centre [online], 7 March 2014. https://www.iaea.org/newscenter/pressreleases/iaea-completes-nuclear-security-review-mission-republic-korea
23. Kim et al., 'Effort and Prospective on Nuclear Security in ROK'.
24. Myung-tak Jung, Jungsoo Kim, and Youngwook Lee, '2014 International Physical Protection Advisory Service (IPPAS) Mission to the Republic of Korea (ROK)', in *Transactions of the Korean Nuclear Society Autumn Meeting*, Gyungju, Korea, 29–30 October (Daejeon: KNS, 2015), 1–3.
25. Korean Ministry of Foreign Affairs, 'ROK Deposits Instruments of Ratification of Two International Nuclear Conventions', Korean Ministry of Foreign Affairs Press Release, Seoul, 30 May 2014.
26. Korean Ministry of Land, Infrastructure and Transport, Act on Promotion of Utilization of Drones and Creation of Infrastructure Therefore, 2019.
27. Nuclear Safety and Security Commission, The 145th Nuclear Safety and Security Committee, Seoul, 13 August 2021.
28. The Korean Physical Society, 'Sodium-Cooled Small Modular Reactors', *Physics and High Technology* [online], 26 May 2022. https://webzine.kps.or.kr/?p=5_view&idx=16716
29. Ross Peel, George Foster, and Professor Sukesh Aghara, *Nuclear Security and Safeguards Considerations for Novel Advanced Reactors* (London: Centre for Science and Security

Studies, 2022), 22–5. https://www.kcl.ac.uk/csss/assets/nuclear-security-and-safeguards-considerations-for-novel-advanced-reactors.pdf
30. Ibid., 18–20.
31. Hosik Yoo, *History of International Regime of Nuclear Security and Its Implementation in the ROK*, KINAC Technical Report (Daejeon: KINAC, 2022), 44–5.
32. Hyun Chul Kim, 'KINAC/INSA International Training Activities and Lessons Learned', in *Transactions of the Korean Nuclear Society Spring Meeting*, 12–13 May 2016, Jeju, Korea (Daejeon: KNS, 2016).
33. IAEA, IAEA NSSC Network Newsletter, 4, Vienna, 2019. https://www.iaea.org/sites/default/files/19/09/iaea_nssc_network_newsletter_issue4_v1_welcome_remarks.pdf
34. Jongsook Kim, Hyung-kyung Lee, and Jin-young Lee, 'Challenges and Responses for Ensuring Sustainability of INSA Training Programs', *International Journal of Nuclear Security*, 2/1 (2016), Article 16.
35. Christopher Hobbs and Matthew Moran, 'Exploring the Human Dimension of Nuclear Security; The History, Theory, and Practice of Security Culture', *The Nonproliferation Review* [online], 9 June 2022. doi: 10.1080/10736700.2020.1811532
36. Nuclear Security Summit, 'Communiqué', 2010 Nuclear Security Summit, 13 April 2010, Washington, DC.
37. IAEA, 'Nuclear Security Culture', Implementing Guide, IAEA Nuclear Security Series, No. 7, Vienna, 2008.
38. Nuclear Safety and Security Commission, *National Nuclear Security Culture Implementation Guide* (Seoul: NSSC, 2013).
39. IAEA, 'Amendment to the Convention on the Physical Protection of Nuclear Material', INFCIRC/274/Rev.1/Mod.1 (Corrected), Vienna, 18 October 2021.
40. Kim et al., 'Effort and Prospective on Nuclear Security in ROK'.

Suggested Reading

Choe, Kwan Kyoo, (2015), 'International Relations and HRD Activities of the International Nuclear Non-proliferation and Security Academy of the ROK', *International Journal of Nuclear Security*, 1/1, Article 9.

Government of the Republic of Korea, (2016), 'National Progress Report', 2012 Nuclear Security Summit, 31 March 2016, Washington, DC. http://www.nss2016.org/document-center-docs/2016/3/31/national-progress-report-republic-of-korea

Kim, Jung Soo, Jung, Myung-Tak, Kwon, Kookheui, and Lee, YoungWook, (2022), 'Effort and Prospective on Nuclear Security in ROK', *International Journal of Nuclear Security*, 7/1, Article 14.

Kim, Seungmin, et al., (2019), 'Cyber Security Strategy for Nuclear Power Plant through Vital Digital Assets', in *Proceedings of the 6th Annual Conference on Computational Science and Computational Intelligence (CSCI 2019)*, 5–7 December 2019, Las Vegas, NV (Piscataway, NJ: IEEE, 2019), 224–6.

Yoo, Hosik, et al., (2018), 'An Analysis of the Survey Results on Nuclear Security Culture for Personnel at Nuclear Facilities', *Progress in Nuclear Energy*, 112, 75–9.

PART VI

COMBATTING NUCLEAR SMUGGLING

CHAPTER 29

ILLICIT NETWORKS AND NUCLEAR MATERIAL TRAFFICKING

EGLE E. MURAUSKAITE

INTRODUCTION

OVER the past 20 years, 'nuclear smuggling', 'nuclear contraband', 'nuclear black market', and similar phrases have been used to describe a whole spectrum of activities: from rogue individuals stealing a few hundred grams of radiological materials, hoping to sell it for a quick profit, to state-run nuclear weapons programmes systematically sourcing controlled dual-use components for the construction of nuclear centrifuges. Aside from creating a degree of confusion, the liberal use of these terms—associated with the shocking prospect of a nuclear explosion should efforts to stop such proliferation activities fail—has also been used to maintain political and public urgency of the trafficking issue even as its nature continued to change.

From the start of the nuclear age, states and individuals have managed to illicitly obtain and transfer nuclear and radiological materials; indeed, a small number of states actually succeeded in also procuring the necessary know-how and technologies, and thus developed the nuclear bomb against prevailing international norms. However, as international controls on these materials were tightened (and, arguably, as the norms against nuclear proliferation strengthened), the majority of non-proliferation and counter-proliferation efforts came to focus on dual-use technologies and financing used to advance state-level nuclear programmes—all the while keeping the aforementioned imagery very much alive in the public psyche. In line with the scope of this volume, this chapter narrowly focuses on historical trends concerning the illicit trafficking of radiological and nuclear materials, while acknowledging that this is a lesser component of the overall proliferation problem.

Current State of Affairs

One of the main sources of public information concerning nuclear and radiological materials trafficking is the Incident and Trafficking Database (ITDB) of the International Atomic Energy Agency (IAEA) which reports annual incident statistics. It is worth noting that one of the reasons for the ITDB's establishment was the degradation of the nuclear security environment following the collapse of the Soviet Union, and a subsequent dramatic increase in trafficking of nuclear materials from East to West Europe. The ITDB distinguishes between three groups of incidents: (1) those related to trafficking or malicious use; (2) those of undetermined intent; and (3) those unrelated to trafficking or malicious use. For the purposes of this chapter, the focus is on the first type of incidents. Based on voluntary self-reporting by IAEA member states, the ITDB provides a decent general overview of the situation, while necessitating caution about the data it provides. On the one hand, the historical tendencies of under-reporting the incidents, particularly among non-Western countries, are widely recognized.[1] On the other hand, certain years show increases in the number of reported incidents due to the improvement and spread of detection technologies, rather than due to an increase in trafficking activity.

In addition to this official source of statistics, non-governmental organizations such as the James Martin Center for Nonproliferation Studies (CNS) and the Nuclear Threat Initiative (NTI) have been regularly publicizing incident statistics, based on publicly available information. Their case descriptions provide considerably more detail, but have shorter temporal coverage, and can suffer from double counting where cases get recorded multiple times (the year an incident took place, as well as the year it was uncovered). Furthermore, in the growing number of cases where equipment containing radioactive materials (medical, logging, etc.) is stolen with malicious intent, it is difficult to ascertain whether the intended target of the perpetrators was the radioactive material or, instead, the often-expensive equipment that contained it.[2] Figure 29.1 summarizes the latest ITDB data available at the time of writing.

Since 2015, fewer than 10 incidents have been reported internationally each year, suggesting an overall decline in radiological and nuclear trafficking. The sharp dip in 2020 activity was most likely related to the Covid-19 global travel restrictions. Over the past two decades, the extremely low number of reported malicious incidents precludes any statistically significant trend analysis, necessitating contextual extrapolation—but also giving rise to speculation. In particular, the question concerning the existence and extent of a nuclear black market has remained largely a matter of contested expert opinions based on such extrapolations. While some[3] believe that the low number of recorded incidents points to the absence of such a market, and instead characterize the situation as a declining series of scattered isolated events, others[4] suggest the existence of dark networks whose sophistication allows them to avoid detection—with only the unsuccessful amateurs getting caught and reported. This chapter identifies and discusses

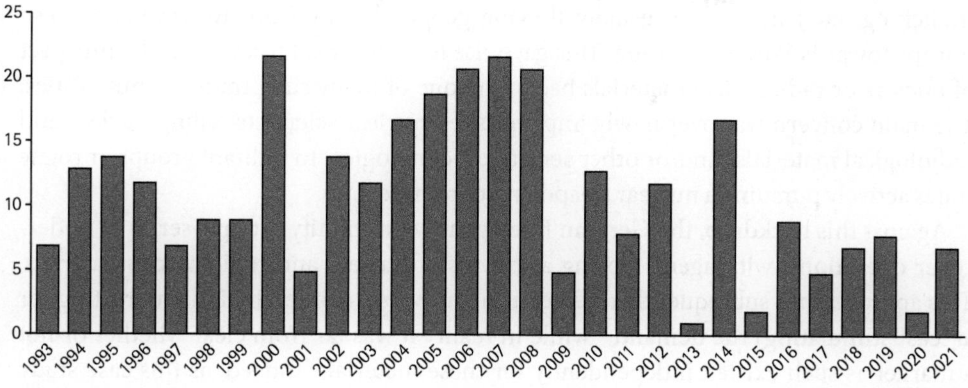

FIGURE 29.1. Incidents reported to the ITDB that are confirmed, or likely relate to trafficking or malicious use, 1993–2021.

IAEA, '2022 Factsheet', IAEA Incident and Trafficking Database (ITDB). https://www.iaea.org/sites/default/files/22/01/itdb-factsheet.pdf

certain discernible patterns based on information concerning some of the best-known nuclear and radiological materials trafficking networks and invites readers to make the judgement call for themselves.

Trafficking Networks through the Decades

This section discusses three case studies, selected to represent what can be considered a typical *modus operandi* for a particular period, based on publicly available information. These case studies also offer some insight on how the network structures and operational decisions have evolved and adapted over time. The dates in the subsection titles indicate the year when a network was uncovered, although, as will be apparent in the discussion, the temporal span of their activities differed considerably. The goal here is to characterize and compare these nuclear and radiological trafficking networks using a fixed set of typological features. It is nevertheless important to note that the vast majority of incidents involving nuclear and radiological materials smuggling are carried out by individuals acting alone—so any networked activity is already an atypical occurrence.

Vagner Network (1994)

The collapse of the Soviet Union in 1991 marked a period of considerable turmoil in the territories that emerged from it, as well as some European countries bordering them. Smuggling and illicit trade of all types (drugs, guns, art, stolen vehicles, human trafficking, etc.) flourished, usually flowing geographically from Russia via Eastern Europe towards Western Europe. This gave rise to widespread fears about the prospect of nuclear or radiological materials becoming one of many such traded commodities. The main concern was over newly impoverished nuclear scientists selling nuclear and radiological materials (and/or other sensitive technologies) to militant groups or rogue states actively pursuing a nuclear weapons programme.[5]

Against this backdrop, the German intelligence community set up a series of undercover operations, with agents posing as buyers of nuclear and radiological materials. This approach has subsequently received near-universal condemnation for creating (or at least stimulating) the demand,[6] while in reality it was far from clear whether or not a market as such existed independently for these materials. Indeed, as this case study illustrates, rather than identifying criminal networks at work, the perceived demand led amateur commodity traders to try and act as facilitators for nuclear and radiological material transfers. Rather than being a typical example of trafficking activities in the 1990s, the case constitutes a self-fulfilling prophecy and, as such, illustrates the key concerns of the period—some of which were subsequently justified, but through illicit activities by individuals rather than networks.

In early 1994, a physicist from Russia's Obninsk Institute for Physics and Power Engineering (IPPE) allegedly managed to divert 2.73 kilograms (kg) of U-235 enriched to 87.7 per cent; this represents the largest-known quantity of highly enriched uranium (HEU) to be illegally diverted.[7] Through several intermediaries, the physicist got in touch with Jaroslav Vagner, a nuclear physicist turned commodity trader from the Czech Republic. Vagner, after whom the network was named, acted as the main facilitator and merchandise broker for this loosely connected group of profit-driven traders.[8] The network initially involved eight Russian, Belarusian, Polish, and Czech commodity traders, many of whom had a background in physics or technologies, and reportedly knew one another from either work or their neighbourhood. None of these individuals had previous criminal records or connections. Poor and/or indebted, they were enticed by the same prospective buyer in Germany—the undercover agent Walter Boden.[9]

Through a series of documented meetings and telephone conversations, the traders, travelling between respective countries in planes, trains, and cars, arranged for the stolen HEU to be tested in an official laboratory in the Czech Republic, and subsequently distributed smaller samples to interested buyers—without being intercepted in transit or at any of the national borders.[10] The traders met with the undercover agent in small groups several times, for him to establish their possession of genuine HEU, and each group was subsequently arrested—but due to the very loose nature of their connections, they remained unaware of each other's fates. By mid-1994, word was spreading about the HEU in the traders' possession, with interested parties allegedly reaching out to them from North Korea, Nigeria, Austria, and Italy.[11]

Subsequently, one Columbian and two Spanish traders were documented getting involved in negotiations to source the materials from the Vagner network, and (inadvertently) sell them to an undercover buyer from the Spanish police.[12] The network was finally thwarted by a Czech sting operation in December 1994, when Vagner and two accomplices met with the undercover agent in a bank vault in Prague, bringing the main HEU batch.[13] Two more samples from that batch were later recovered in June 1995—both of which are recorded in the ITDB. In 1997, Czech courts finished trying the case, sentencing Vagner to eight years' imprisonment; the others got away with lighter sentences due to their lack of demonstrable connections to criminal organizations.

A. Q. Khan Network (2004)

The fears of 'loose nukes' during the 1990s and growing awareness of such low-probability high-impact incidents led European and American authorities to boost the development of tracking mechanisms and legal instruments for nuclear and radioactive materials. The terrorist attacks of 11 September 2011 (9/11) heightened the fears of a terrorist organization acquiring a nuclear or radiological bomb. In addition, the United States (US) was concerned with 'rogue regimes' in the Middle East potentially seeking nuclear weapons. Thus, arguably, the focus of concern was gradually shifting away from the nuclear and radioactive materials and individuals on the supply side towards designs and technologies and their prospective demand, mainly constituted by state clients. Indeed, states looking to gate-crash the nuclear club were looking to acquire the capabilities to enrich the required uranium or plutonium themselves, as in the case of Iran. Thus, focusing on the interdiction of technologies such states were seeking was chosen as an approach that could help nip the problem in the bud.

Against this backdrop, the network led by Pakistani scientist Abdul Khadir Khan is perhaps the most elaborate, geographically global, and widely known case of the trafficking of nuclear and radiological materials. Throughout its existence from the early 1970s to 2004, the A. Q. Khan network traded in the full spectrum of commodities (nuclear materials, bomb schemes, components for enriching uranium, etc.). Unlike most other (individual) traffickers caught mid-act, this network managed to successfully deliver the trafficked goods to more than one buyer. It is also crucial to note that the network was originally designed to assist a state-run nuclear programme (in Pakistan) and served other state-level clients (Iran, Libya, North Korea, possibly Syria, and others); indeed, state-level connections and protection were responsible in no small part for the longevity and success of the operation. A. Q. Khan publicly came forward in 2004 to reveal the network, after the Pakistani government came under increasing pressure from the US intelligence community about its activities. Nevertheless, A. Q. Khan was swiftly pardoned and remained a national hero in Pakistan, living under house arrest but revered until his death in 2021.

In the early 1970s, A. Q. Khan had graduated from studies in metallurgy in Europe and was subsequently employed by a Dutch company subcontracting to Urenco (a national consortium that, among other commodities, produced centrifuges). There, he was able to view Dutch and German centrifuge design blueprints several times, as well as compile a large roster of contacts of firms producing components for those centrifuges—the basis for his later network.[14] Interestingly, A. Q. Khan had been monitored by Dutch and

American intelligence from the start and was removed from his position in 1975 under suspicion of espionage.[15]

Shortly thereafter, A. Q. Khan left for Pakistan to head a national research laboratory, where he successfully replicated the centrifuge designs. By 1978, this first generation of centrifuges was producing enriched uranium for Pakistan's national nuclear weapons programme.[16] A. Q. Khan's subsequently recounted how decades of friendship between Pakistani and Chinese governments eventually led to technology and nuclear material transfers involving his network. Namely, over the course of 1982, dozens of C-130 planes were flown from Pakistan to China carrying centrifuge components produced by companies connected to the A. Q. Khan network; in return, using the same planes, China sent 15 tonnes of uranium hexafluoride gas for Pakistan's centrifuges.[17] Indeed, A. Q. Khan reported flying to Urumqi in China on one occasion, and enjoying a barbecue while 50 kg of 'weapon-grade enriched uranium' were loaded onto a C-130 plane destined for Pakistan, along with blueprints for a basic nuclear bomb design.[18]

During its first decade of functioning, the A. Q. Khan network was mainly importing sensitive dual-use materials, components, and know-how for a state-run programme, rather than dealing in nuclear or radiological materials directly. It is also worth noting that in 1983, A. Q. Khan was tried and sentenced in absentia in a Dutch court, but the verdict was later overturned, and the case dismissed due to the difficulty of forcibly summoning him from Pakistan.[19]

By the mid-1980s, the network had turned towards sensitive technology exports as well. As Pakistan's centrifuge programme progressed to use the more advanced P-2 model, Khan had started to sell the surplus P-1 model components to other state clients (and order extras to satisfy the demand).[20] Iran was among the first of such clients, receiving uranium enrichment centrifuge components and blueprints, as well as conducting training to build and operate this technology. North Korea became another major state client, seemingly on barter basis. Pakistan was made privy to North Korea's missile technology (and soon developed its own, based on the No-dong missile design), while sharing P-1 centrifuge drawings and components with North Korea.[21] Both sides exchanged dozens of visits by scientists, technicians, and military officials during this period. Allegedly, Pakistan had also shared depleted uranium hexafluoride, hence marking the transfer of actual radioactive materials by the A. Q. Khan network.[22]

In the mid-1990s, A. Q. Khan reached out to Syria and Iraq, offering similar 'nuclear weapons' starter packs'.[23] While Syria may have ultimately received technology transfers via North Korea, the Iraqi president Saddam Hussein dismissed the offer, suspecting it was an undercover Western intelligence operation. Nevertheless, towards the late 1990s, Libya became one of the most profitable clients for the network, purchasing 20 assembled P-1 centrifuges, as well as parts for constructing a pilot enrichment facility of 200 P-1 centrifuges. A front business was set up in Dubai to route these transfers, and provisions continued to flow into Libya until 2003, eventually involving P-2 centrifuges, 1.87 tonnes of uranium hexafluoride (sourced via North Korea), and Chinese blueprints for a nuclear bomb.[24]

The network's international operations were intercepted by the US authorities in 2003; large shipments of aluminium tubes used in centrifuges were seized en route to North Korea and Libya, and A. Q. Khan came forward shortly thereafter. Starting in 2004, various members of A. Q. Khan's international network of suppliers were arrested and put to trial—albeit the proceedings concerned transfers of dual-use technology, rather than nuclear materials or bomb schemes. Buhary Seyed Abu Tahir, a Sri Lankan businessman who acted as the key facilitator and middleman in the network, was arrested in Malaysia in 2004 and sentenced to four years' imprisonment.[25] Drawing on Tahir's testimony, the following year Swiss authorities arrested a German engineer Gotthard Lerch in 2004 and extradited him to Germany for trial. Lerch had been supplying vacuum technology for uranium enrichment, earning an estimated 25 million DM (approximately US$13.7 million) from the network. In 2008, he was sentenced to five-and-a-half years' imprisonment and fined US$4.7 million.[26]

The same year, German authorities arrested a family of German engineers operating businesses in Switzerland—Urs, Marco, and Frederic Tinner—and extradited them to Switzerland to be tried for violations of export control laws. They had been producing centrifuge parts for the network in Dubai and Malaysia—and later Urs Tinner had tipped off the US Central Intelligence Agency (CIA) about A. Q. Khan's deal with Libya.[27] In 2004, South African authorities arrested Gerhard Wisser, a German citizen, and Daniel Geiges, a Swiss citizen, both of whom had been supplying the network with dual-use parts out of South Africa. In 2007 Wisser was sentenced to 18 years' imprisonment, and in 2008 Geiges was sentenced to 13 years.

In terms of motives, arguably the initial driving force was A. Q. Khan's ideological sentiments, looking to ensure Pakistan could stand up to a nuclear armed India. However, once the network turned towards exports and worked to expand them, profits seemed to have played a central role. Many of the businesses from which the network was sourcing the sensitive components were middle-sized and not necessarily aware of the scheme they were being drawn into, operating as regular profit-making entities with insufficient due diligence. However, the radiological materials were rarely shifted through this network; rather, they were exclusively traded between state enterprises (albeit allegedly as part of a package rather than individual items), so distinguishing between ideological and profit motives is difficult. Another notable characteristic was the network's 'professionalism' (with the exception of the unwittingly involved business suppliers). In contrast to most known instances of nuclear or radiological trafficking, the parties were not only experts in their field, but also specifically involved in pitching and selling the means to starting state-run nuclear weapons programmes to willing clients.

Dadayan Network (2010)

The revelations concerning the international activities of the A. Q. Khan network spurred significant advances in efforts by the US, as well as its Western partners, to curb nuclear proliferation, with a focus on dual-use goods. Historically, export controls and proliferation financing regulations have been particularly geared towards preventing any more states from joining the nuclear weapons club. It is worth noting that for years some of the existing nuclear weapons states (for example, China and Israel) also continued to procure dual-use goods for their programmes.[28] Following the revelations about the A. Q. Khan network, efforts were redoubled to curb such practices. While the trafficking of radiological and nuclear materials has not received comparable renewal of interest or countermeasures, it certainly remained on the radar as being of likely interest to extremist organizations.

Trafficking pathways have also shifted geographically in the past two decades. Although there had been a considerable outflow from Russia, Europe no longer figured so prominently as a route or destination. Instead, publicly available data from the year 2000 onwards consistently points to the Black Sea region as an area with the greatest incidence of thefts and losses of nuclear and radiological material. Indeed, links to drug traffickers and other criminal elements identified in nuclear and radiological trafficking cases in the Black Sea region[29] fuelled concerns about the emergence of a terror–crime nexus setting up an operation to acquire such materials. This case illustrates some of the logistical and typological trends in Black Sea trafficking issues. However, compared to the two networks discussed previously, publicly available information on this case remains relatively sparse.[30]

Garik Dadayan, after which this small but resourceful network is named, was an Armenian and a decorated and disabled veteran of the first Nagorno Karabakh war, living in Moscow. He managed to obtain an unknown quantity of HEU from one of the local research facilities in the Siberian city of Novosibirsk (Russian intelligence services noted him travelling there at least twice).[31] In 2002, Dadayan approached an Armenian acquaintance, retired physicist Hrant Ohanian, to test the material's quality. Dadayan brought a sample of HEU in an envelope to Ohanian's apartment in the Armenian capital Yerevan. Ohanian confirmed that it was HEU, and the two men proceeded to try and sell it.[32]

Dadayan subsequently connected with a Turkish middleman, Teimur Sadik, who said he was acting on behalf of 'a Muslim client'.[33] In June 2003, Sadik and Dadayan agreed to meet in Armenia, and Dadayan set out to drive there from Russia via Georgia, carrying 170 g of HEU in a metal box alongside tea. Meanwhile, Georgian police had received an anonymous tip and were following his car.[34] Dadayan was detained when his cargo set off radiation detection monitors at the Armenian border. He was subsequently put to trial in Armenia and sentenced to two-and-a-half years' imprisonment but released early in 2005.[35]

In the meantime, the physicist Ohanian had been approached by another acquaintance—heavily indebted Armenian businessman Sumbat Tonoyan—who was looking to source some HEU to sell for a quick profit (specifically to Turkish radical groups).[36] Subsequently, Ohanian reached out to Dadayan, who offered a sample of 100 g for sale. In February 2010, Tonoyan travelled to the Georgian city of Batumi seeking out the Turkish connections but, instead, one of the people he contacted turned out to be a Georgian police informant. A sting operation was set up, with a Georgian operative posing as a Turkish buyer.[37] In March 2010, Ohanian wired US$1,000 to Dadayan, who managed to travel unimpeded from Moscow to Armenia that same day carrying 18 g of HEU (seemingly from the same batch that he had been detained with in 2003). With the HEU in hand, Tonoyan and Ohanian set out to meet the Turkish buyer in the Georgian capital Tbilisi.

Their logistical operation is worth noting in some detail, as it reflects not only a level of professionalism rarely seen among other (individual) smugglers, but also hints at the considerable increase in international detection capabilities for radioactive cargo. Awareness of these capabilities forced the Dadayan network to take extra precautions. The 18 g of HEU was placed in a lead-lined packet of cigarettes and hidden between compartments of a train travelling from Yerevan to Tbilisi. At the Armenian–Georgian border, the smugglers disembarked the train, leaving the hidden HEU unattended, and proceeded by taxi to Tbilisi, while the package arrived undetected by radiation monitors.[38]

In Tbilisi, Tonoyan met with another Georgian smuggler to collect the HEU and meet the undercover buyer; upon doing so they were both arrested, and Ohanian was picked up by the authorities while sleeping at another hotel. In April 2010, Georgian authorities tried Ohanian and Tonoyan for the smuggling of nuclear materials, and sentenced them to 13 and 14 years' imprisonment, respectively. Dadayan was put to trial in Armenia for assisting the smugglers and was sentenced to seven years' imprisonment and his property confiscated. Interestingly, in 2015, Dadayan was released by presidential pardon, and in 2018, the European Court of Human Rights found that Armenia had not ensured Dadayan a fair trial, overturning the verdict and ordering a new trial,[39] which at the time of writing has not concluded.

Evolving Understanding of Trafficking Issues

In a growing body of research, the trafficking of nuclear and radiological materials has been compared to the trafficking of drugs, arms, fine art, and a series of other, more common illicit goods that established networks of criminal elements tend to traffic.[40] However, the actual functioning of these different networks in practice shows very few indications of smugglers in one trade learning in a systematic manner from other types of smuggling. This is largely due to the individualized and amateurish nature of the vast majority of traffickers dealing in radioactive materials. Nevertheless, one of the most studied—and elusive—aspects has been the search for indications of an emerging nexus between organized crime groups and extremist movements with the purpose of sourcing, moving, and acquiring radioactive materials.

While most cases of trafficking by networks and individuals—including those discussed in this chapter—were supply driven, there have been several (publicly undisclosed) incidents where investigators had identified buyers (i.e. the demand side) from extremist organizations.[41] In addition, there is historical evidence of interest in radioactive materials from extremist organizations. For instance, al-Qaeda's consistent (yet unsuccessful) pursuit of nuclear or radiological materials throughout the 1990s is well documented.[42]

One of the best-known cases of an extremist organization acquiring radiological materials is that of Chechen separatists.[43] In 1995, one of their leaders Shamil Basayev notified the media that he had buried a container with radiological materials in a park in Moscow—as a demonstration of capabilities.[44] Another separatist movement in Nagorno-Karabakh has also been accused of stocking up on spent nuclear fuel with the view of using it in a radiological bomb, although no caches were identified or seized.[45] More recent examples include the detention in 2006 of two Columbians attempting to sell a bar of depleted uranium to undercover authorities posing as the Colombia guerrilla organization Revolutionary Armed Forces of Colombia (FARC), and a raid in 2008 on FARC's[46] cache containing nearly 30 kg of depleted uranium.[47]

Despite these indications of demand from violent non-state actors, there has been seemingly limited interest among international/transnational organized crime networks in partnering for such a trafficking operation due to the significant risk–reward mismatch. Namely, well-established criminal networks that would possess the technical and financial capabilities to carry out such a supply operation (especially more than once) deem the risks too high; this includes the risk of jeopardizing their institutional links, often vital for other trades. For instance, the most lucrative known nuclear proliferation deal (involving both nuclear and dual-use materials) was the one struck between Libya and the A. Q. Khan network in 2003, estimated at around US$100 million in total, whereas most of the drug-smuggling submarines intercepted by US authorities have tended to be valued at around US$200 million each.[48]

Nevertheless, research on criminal involvement in nuclear and radiological trafficking suggests that smaller local crime groups, or even individuals with a previous criminal record, have indeed become involved in such incidents. For instance, in 2001, several members of Russia's Balashikha smuggling gang were detained with 1 kg of low enriched uranium (LEU) and discovered to have a cache of 200 g of caesium-137.[49] In another relevant incident, in 2005 Turkey detained two 'criminals' trying to sell 173 g of LEU to an undercover agent posing as a buyer; the Turkish authorities also alleged that the suppliers were trying to sell to a 'terrorist group'.[50]

Most recently, in 2019, Austrian authorities arrested three men suspected of smuggling nuclear materials, one of whom had a previous conviction for a similar offence and the two others had criminal records.[51] Reflecting on the issue of a terror–crime nexus, such small local criminal enterprises would have lesser capabilities to deliver to terrorist organizations, even if their risk aversion may be mitigated. However, as discussed in the final section of this chapter, in regions plagued by turmoil and/or characterized by low state capacity, such risks may be lowered considerably even for more established criminal groups, as new pathways and networks are being formed in the turmoil.

Finally, it is worth noting a relatively new methodological approach to studying nuclear and radiological trafficking issues, including terrorist or criminal links: that of 'hypothetical studies'. For instance, although Steve Sin and Marcus Boyd acknowledge the current absence of any identifiable terror–crime nexus, their study explores and models the hypothetical routes and modes of trafficking that could be enabled by such a partnership.[52] When considering a hypothetical generalized smuggler engaged in an operation, Sin and Boyd point to the Caribbean basin as the most likely end-point, while a hypothetical terrorist organization would mostly likely use the African Great Lakes region. Similarly, Justin Hastings expounds on the potential for trafficking to become a much more active and dangerous enterprise as the civilian nuclear energy industry continues to expand to states that have not dealt with nuclear technology before.[53]

Future Concerns

Non-proliferation experts[54] agree that weak/failing states and ungoverned/under-governed territories will continue to pose the greatest challenges of all along the trafficking chain (with radioactive sources potentially diverted from their facilities, moved through their territories, and their citizens involved in trafficking). Matching the conceptual profiles with empirical reality over the short- to medium-term, this puts the focus on (1) unstable great powers—Russia in particular (and to some extent, China), and (2) war-torn Ukraine—being at the heart of an already problematic Black Sea region.

Over the past 40 years, Russia has been the origin source for around two-thirds of radioactive materials seized in trafficking, pointing to a relatively high availability of poorly managed resources.[55] Combined with considerable human expertise available and limited (and, arguably, currently declining) law enforcement capabilities, trafficking risks emanating from Russia are likely to increase. Indeed, whereas historically major concerns focused on Russia's impoverished scientists getting involved in trafficking networks, presently a much greater concern has become the evolving links between Russia's corrupt officials with access to nuclear materials and local organized crime groups—due to their potential to enable ever-more-sophisticated operations.[56] Furthermore, combining the unstable great power and under-governed territory characteristics, suspicions have been raised about an active nuclear trafficking route leading from Russia via Armenia towards Iran.[57]

In terms of Ukraine, the NTI and CNS trafficking databases suggest that during 1992–2014, that is, even before the start of the conflict with Russia, the number of nuclear and radiological smuggling incidents involving Ukrainian citizens or territory was among the highest in the Black Sea region—exceeding the number of cases linked to Georgia or Moldova.[58] As Ukraine has continued to slide deeper into conflict since 2014 and, most significantly, since the Russian invasion began in February 2022, its capacity to identify and address the issues related to the smuggling of nuclear or radiological materials (as well as dual-use goods) has been declining. For instance, the security of non-strategic objects and sites (e.g. cancer treatment facilities or construction sites) has been impacted, whereas all types of criminal activity have been on the rise.[59] Combined with the overall bureaucratic turmoil and the difficulties of ensuring continuous border security during conflict, this puts into question Ukraine's ability to curb radioactive trafficking. Moreover, external visibility on the issue, and thus the ability to effectively account for any developments, has also been reduced. This does not bode well in the context of considerable under-reporting of nuclear and radiological trafficking incidents in the Black Sea region even before this war. As discussed previously, since 2010 this has been an area where the majority of such incidents have unfolded.

Over the past decade, the CNS nuclear trafficking database (last updated in 2021) has recorded three to four incidents each year unfolding in Ukrainian territory or involving Ukrainian citizens. While some of these involved radiation-contaminated scrap metal or other goods (e.g. watches and guns), around 20 per cent of these cases involved thefts and attempted sales of radiological materials removed from industrial or medical equipment, and in eight cases perpetrators attempted to sell spent nuclear fuel stolen from Ukrainian power plants.

Since 2014, the incidence and quality of reporting seems to have declined considerably, due to the ongoing geopolitical tension with Russia, but the serious nature of the incidents that were publicized suggests a deepening problem. For instance, according to the CNS trafficking database, in August 2015, Ukraine detained four members of a criminal group attempting to sell uranium-238 of unspecified level of enrichment. In June 2018, six individuals described by Ukrainian authorities as 'members of an international nuclear smuggling ring' were detained in a sting operation attempting to sell radium-226. In July 2017, individuals were apprehended in an attempt to excavate hundreds of barrels contaminated with caesium-137 from a radioactive waste facility. Incidents like these illustrate not just the risks associated with poorly protected sources, but also point to the potential development of systematic trafficking network formations involving criminal elements.

Furthermore, the disregard of Russian military forces for the potential risks associated with nuclear sites became apparent when most of the calibration instruments containing radioactive materials were stolen from a laboratory near Chernobyl, with pieces mailed home as souvenirs.[60] This surprisingly limited awareness of the radiation risks was further highlighted by the decision of Russian forces to dig trenches in an extremely radiation-contaminated forest area behind the Chernobyl nuclear power plant,[61] not to mention the targeted shelling of Zaporizhzhia nuclear power plant.[62] Given the documented efforts by prominent Russian criminal groups (such as the Solntsevo or Mogilevitch gangs) to immediately forge networks with Ukrainian counterparts and bureaucratic circles from the very start of the conflict in 2014,[63] the risk of an emergence of a trafficking enterprise dealing in radioactive materials in this region is high.

Conclusion

This chapter has offered a review of the current state of affairs, in terms of trends and patterns, in nuclear and radioactive materials smuggling. Pointing out the potential confusion due to conflicting definitions, as well as the types of publicly available reporting, the chapter has nevertheless presented the latest available data, including incidents of particular concern in the currently war-torn Ukraine. Three case studies illustrated the typical nature of the smuggling of radioactive materials, from the early 1990s following the collapse of the Soviet Union, to the next decade of rising concerns about states attempting to jump-start nuclear weapons programmes, to the post-9/11 decade plagued by fears of non-state actors wielding a radiological weapon. Despite continued speculation about a looming unknowable black market or a nexus between terrorist and criminal networks, publicly available information discussed hardly supports such prospects. The evidence presented suggests that nuclear and radiological materials smuggling remains largely an activity of profit-driven individual amateurs, with failed or weak states presenting particularly fertile territorial and institutional conditions.

Notes

1. Lyudmila Zaitseva, 'Nuclear Trafficking in Ungoverned Spaces and Failed States', in Anne L. Clunan and Harold A. Trinkunas, eds., *Ungoverned Spaces: Alternatives to State Authority in an Era of Softened Sovereignty* (Stanford, CA: Stanford University Press, 2010), 193–212.
2. Steve Sin and Marcus Boyd, 'Searching for the Nuclear Silk Road: Geospatial Analysis of Potential Illicit Radiological and Nuclear Material Trafficking Pathways', in Brecht Volders and Tom Sauer, eds., *Nuclear Terrorism: Countering the Threat* (New York: Routledge, 2016), 159–81.
3. See, for example, Rens Lee, *Smuggling Armageddon: The Nuclear Black Market in the Former Soviet Union and Europe* (New York: St Martin's Press, 1998); Rens Lee, 'Why Nuclear Smuggling Matters', *Orbis*, 52/3 (2008), 434–44.
4. See, for example, Lyudmila Zaitseva and Kevin Hand, 'Nuclear Smuggling Chains: Suppliers, Intermediaries, and End-Users', *American Behavioral Scientist*, 46/6 (2003), 822–44.
5. Zaitseva and Hand, 'Nuclear Smuggling Chains'; William C. Potter, 'Nuclear Smuggling from the Former Soviet Union', in David R. Marples and Marilyn J. Young, eds., *Nuclear Energy and Security in the Former Soviet Union* (New York: Routledge, 1997), 139–66.
6. Galya I. Balatsky and William R. Severe, 'Illicit Trafficking of Radioactive and Nuclear Materials', in James E. Doyle, ed., *Nuclear Safeguards, Security, and Nonproliferation*, 2nd edn (Oxford: Elsevier, 2019), 357–87; Potter, 'Nuclear Smuggling from the Former Soviet Union'; Lee, 'Why Nuclear Smuggling Matters, Wyn Q. Bowen, 'Indicators and Warnings: A Framework for Assessing the Threat', in *International conference on illicit nuclear trafficking: Collective experience and the way forward, 19–22* November 2007 *Edinburgh* (Vienna: IAEA, 2008). https://www-pub.iaea.org/MTCD/publications/PDF/Pub1316_web.pdf

7. Jeffrey Fleishman, 'Bumbling Gang Took Gray for Gold', *Philadelphia Inquirer* [online], 13 January 1999. http://articles.philly.com/1999-01-13/news/25492337_1_sasha-uranium-nuclear-warheads; Ondrej Benda and Ross Larsen, 'Uranium Smuggling Case Botched, Judge Tells Cops', *The Prague Post*, 23 September 1997; Alex P. Schmid and Charlotte Spencer-Smith. 'Illicit Radiological and Nuclear Trafficking, Smuggling and Security Incidents in the Black Sea Region since the Fall of the Iron Curtain–An Open Source Inventory', *Perspectives on Terrorism*, 6/2 (2012), 117–57. https://www.pbs.org/wgbh/pages/frontline/shows/nukes/readings/potterarticle.html
8. Egle E. Murauskaite, *Failure Points in Smuggling Networks: A Case Study of Two Nuclear Smuggling Networks* (College Park, MD: START, 2014).
9. Ibid.
10. Ibid.
11. Lee, *Smuggling Armageddon*.
12. Der Spiegel, 'Panic Made in Pullach', *Der Spiegel* [online], 10 April 1995. https://www.spiegel.de/politik/panik-made-in-pullach-a-4e5a73eb-0002-0001-0000-000009181696?context=issue; Mary Williams Walsh, 'Stung by a Nuclear Sting', *Los Angeles Times* [online], 26 January 1996. http://articles.latimes.com/print/1996-01-26/news/mn-29024_1_nuclear-smuggling
13. Schmid and Spencer-Smith, 'Illicit Radiological and Nuclear Trafficking'.
14. Mark Fitzpatrick, *Nuclear Black Markets: Pakistan, A.Q. Khan and the Rise of Proliferation Networks* (London: International Institute for Strategic Studies (IISS), 2007).
15. Michael Laufer, 'A. Q. Khan Nuclear Chronology', Carnegie Endowment for International Peace [online], 7 September 2005. https://carnegieendowment.org/2005/09/07/a.-q.-khan-nuclear-chronology-pub-17420
16. Fitzpatrick, *Nuclear Black Markets*; Laufer, 'A. Q. Khan Nuclear Chronology'.
17. Jeffrey R. Smith and Joby Warrick, 'A Nuclear Power's Act of Proliferation', *Washington Post*, 13 November 2009.
18. Ibid.
19. Fitzpatrick, *Nuclear Black Markets*; Laufer, 'A. Q. Khan Nuclear Chronology'.
20. Ibid.
21. Ibid.
22. Mike Chinoy, 'How Pakistan's A.Q. Khan Helped North Korea Get the Bomb.' *Foreign Policy* [online], 11 October 2021. https://foreignpolicy.com/2021/10/11/aq-khan-pakistan-north-korea-nuclear
23. Fitzpatrick, *Nuclear Black Markets*; Laufer, 'A. Q. Khan Nuclear Chronology'.
24. Ibid.
25. Peter Crail, 'Germany Convicts Khan Associate', *Arms Control Today* [online], November 2008. https://www.armscontrol.org/act/2008-11/germany-convicts-khan-associate
26. Ibid.
27. Catherine Bosley, 'Swiss Family Involved in Nuclear Smuggling Ring Avoids Jail', *Reuters* [online], 25 September 2012. https://www.reuters.com/article/us-swiss-judicial-nuclear-idUSBRE88O14Z20120925
28. David Albright and Andrea Stricker, 'The World of Illicit Nuclear Trade: Present and Future', in Matthew Bunn, Martin B. Malin, William C. Potter, and Leonard Spector, eds., *Preventing Black-Market Trade in Nuclear Technology* (Cambridge: Cambridge University Press, 2018), 23–47.
29. Zaitseva, Nuclear Trafficking in Ungoverned Spaces and Failed States.'

30. This is largely due to the reluctance of Russia to share intelligence concerning many of the key details.
31. Lawrence Scott Sheets, 'A Smuggler's Story', *The Atlantic* [online], 1 April 2008. http://www.theatlantic.com/magazine/archive/2008/04/a-smuggler-s-story/306736
32. Murauskaite, *Failure Points in Smuggling Networks*.
33. Elena Sokova, William C. Potter, and Cristina Chuen, *Recent Weapons Grade Uranium Smuggling Case: Nuclear Materials are Still on the Loose* (Monterey, CA: James Martin Center for Nonproliferation Studies, 2007.
34. Alexander Kupatadze, 'Organized Crime and the Trafficking of Radiological Materials: The Case of Georgia', *Nonproliferation Review*, 17 (2010), 219–34.
35. Ibid.
36. Julian Borger, 'Tiny Box Causes Explosive Trouble', *Mail & Guardian* [online], 12 November 2010. http://mg.co.za/article/2010-11-12-tiny-box-causes-explosive-trouble
37. Ibid.
38. Ibid.
39. Araks Mamulyan, 'Armenia's Uranium Smuggling Case: Back in the Courts Ten Years Later,' *Hetq* [online], 6 February 2021. https://hetq.am/en/article/127126
40. See Joshua D. Foss, 'Plutonium and Picasso – A Typology of Nuclear and Fine Art Smuggling'. *Interagency Journal*, 8/2 (2017), 34–43; Sin and Boyd, 'Searching for the Nuclear Silk Road'; Brandon Behlendorf, 2015, 'Failure Points in Smuggling Networks'. https://www.start.umd.edu/research-projects/failure-points-smuggling-networks;
41. Balatsky and Severe, 'Illicit Trafficking of Radioactive and Nuclear Materials'; IAEA, *Combating Illicit Trafficking in Nuclear and other Radioactive Material*, Technical Guidance. IAEA Nuclear Security Series, No. 6 (Vienna: IAEA, 2007). https://www-pub.iaea.org/mtcd/publications/pdf/pub1309_web.pdf
42. See, for example, Sammy Salama and Lydia Hansell, 'Does Intent Equal Capability? Al-Qaeda and Weapons of Mass Destruction', *Nonproliferation Review*, 12/3 (2005), 615–53; David Albright, *Peddling Peril: How the Secret Nuclear Trade Arms America's Enemies* (New York: Free Press, 2010); *United States of America v Usama bin Laden et al.*, 126 F. Supp. 2d 290 (S.D.N.Y. 2001), 6 February 2001. Digital files from the Court Reporters Office. https://cryptome.org/usa-v-ubl-dt.htm
43. Separatist organizations, even militant ones, are not considered terrorist organizations by all states.
44. Zaitseva, Nuclear Trafficking in Ungoverned Spaces and Failed States'; Balatsky and Severe, 'Illicit Trafficking of Radioactive and Nuclear Materials'.
45. N. Aliev and I. Yusifoglu, 'One More Threat: Armenian NPP Supplies Karabakh Separatists with "Material" for Terrorist Acts', *Ekho (Baku)*, 17 July 2002, 133.
46. While it remains a militant political movement, in 2021 the US removed FARC from its list of terrorist organizations.
47. Lawrence Livermore National Laboratory (LLNL), *Open Source Reporting: Illicit Trafficking of Nuclear and Radiological Material* (Livermore, CA: Lawrence Livermore National Laboratory, 2006).
48. Justin V. Hastings, 'Consequences of the Nuclear Renaissance for Nonstate Nuclear Trafficking', in Adam N. Stulberg and Matthew Fuhrmann, eds., *The Nuclear Renaissance and International Security* (Stanford, CA: Stanford University Press, 2013), 223–49.
49. Zaitseva and Hand, 'Nuclear Smuggling Chains: Suppliers, Intermediaries, and End-Users'
50. LLNL, *Open Source Reporting*.

51. Europol, 'Crime Group Suspected of Smuggling Nuclear Materials Arrested in Vienna', Europol [online], 6 December 2019. https://www.europol.europa.eu/media-press/newsroom/news/crime-group-suspected-of-smuggling-nuclear-materials-arrested-in-vienna
52. Sin and Boyd, 'Searching for the Nuclear Silk Road'.
53. Hastings, 'Consequences of the Nuclear Renaissance for Nonstate Nuclear Trafficking'.
54. For example, Zaitseva, Nuclear Trafficking in Ungoverned Spaces and Failed States'; Hastings, 'Consequences of the Nuclear Renaissance for Nonstate Nuclear Trafficking'; LLNL, *Open Source Reporting*.
55. Matthew R. Kershner, *Trafficking Nuclear and Radiological Materials And the Risk Analysis of Transnational Criminal Organization Involvement*, Future Warfare Series, 57 (Maxwell Air Force Base, AL: USAF Center for Unconventional Weapons Studies, 2014). https://media.defense.gov/2019/Apr/11/2002115521/-1/-1/0/57TRAFFICKING.PDF
56. Louise Shelley, 'The Links Between Organized Crime and Terrorism in Eurasian Nuclear Smuggling', Talk at the Kennan Institute, Wilson Center, Washington, DC, 12 December 2005. https://www.wilsoncenter.org/event/the-links-between-organized-crime-and-terrorism-eurasian-nuclear-smuggling
57. Zaitseva, Nuclear Trafficking in Ungoverned Spaces and Failed States.'
58. Nuclear Threat Initiative (NTI), 'NIS Nuclear Trafficking, 1991–2012', NTI [online], 2012. http://www.nti.org/analysis/articles/2012-nis-nuclear-trafficking; Nuclear Threat Initiative (NTI), 'The CNS Global Incidents and Trafficking Database', NTI [online], 20 October 2021. http://www.nti.org/analysis/reports/cns-global-incidents-and-trafficking-database
59. Egle E. Murauskaite, 'Nuclear Smuggling and Threats to Lithuanian Security', *Lithuanian Annual Strategic Review*, 2015-2016, 14/1 (2016), 181–96. https://journals.lka.lt/journal/lasr/article/133/info.
60. For example, Michael Lipin and Igor Tsikhanenka, 'Photos Indicate Russian Looting of Radioactive Materials from Ukraine's Chernobyl', *Voice of America (VOA)* [online], 12 April 2022. https://www.voanews.com/a/photos-indicate-russian-looting-of-radioactive-materials-from-ukraine-s-chernoby/6526620.html
61. Yogita Limaye, 'Inside Chernobyl: We Stole Russian Fuel to Prevent Catastrophe,' *BBC News* [online], 9 April 2022. https://www.bbc.com/news/world-europe-61048256
62. Geoff Brumfiel et al., 'Video Analysis Reveals Russian Attack on Ukrainian Nuclear Plant Veered near Disaster'. *NPR* [online], 11 March 2022. https://www.npr.org/2022/03/11/1085427380/ukraine-nuclear-power-plant-zaporizhzhia?t=1653468732389
63. Mark Galeotti, 'Moscow's Spy Game: Why Russia Is Winning the Intelligence War in Ukraine', *Foreign Affairs* [online], 30 October 2014. http://www.foreignaffairs.com/articles/142321/mark-galeotti/moscows-spy-game

Suggested Reading

Albright, David, (2010), *Peddling Peril: How the Secret Nuclear Trade Arms America's Enemies* (New York: Free Press).

Corera, Gordon, (2006), *Shopping for Bombs: Nuclear Proliferation, Global Insecurity, and the Rise and Fall of the A. Q. Khan Network* (Oxford: Oxford University Press).

Lee, Rens, (1998), *Smuggling Armageddon: The Nuclear Black Market in the Former Soviet Union and Europe* (New York: St Martin's Press).

Zaitseva, Lyudmila, (2010), 'Nuclear Trafficking in Ungoverned Spaces and Failed States', in Anne L. Clunan and Harold A. Trinkunas, eds., *Ungoverned Spaces: Alternatives to State Authority in an Era of Softened Sovereignty* (Stanford, CA: Stanford University Press), 193–212.

Zaitseva, Lyudmila and Hand, Kevin, (2003), 'Nuclear Smuggling Chains: Suppliers, Intermediaries, and End-Users', *American Behavioral Scientist*, 46/6, 822–44.

CHAPTER 30

OPEN SOURCE INTELLIGENCE AND INVESTIGATIVE TECHNIQUES FOR LOCATING RADIOACTIVE MATERIALS

MICHAEL DUITSMAN AND
MARGARITA KALININA-POHL

INTRODUCTION

THIS chapter examines the problem of radioactive sources outside of regulatory control, traditional methods for locating and recovering them, and how open source intelligence (OSINT) can contribute to these efforts. It describes a variety of OSINT tools, illustrated with real-world examples. It examines the benefits and challenges of using OSINT, as well as the future prospects of OSINT for the recovery of radioactive sources. The OSINT applications described in this chapter are fairly simple applications of tradecraft. Some of the suggested readings at the end of the chapter describe more complex tools, such as automated data collection and the use of software to process text data.

The International Atomic Energy Agency (IAEA) defines radioactive sources as:

> radioactive material that is permanently sealed in a capsule or closely bonded, in a solid form and which is not exempt from regulatory control. It also means any radioactive material released if the radioactive source is leaking or broken, but does not mean material encapsulated for disposal, or nuclear material within the nuclear fuel cycles of research and power reactors.[1]

This chapter focuses on radioactive sources that have a wide range of applications in industry, medicine, agriculture, and other practical uses. For example, radioactive sources are used for weld inspection, radiotherapy, sterilization, crop breeding, and smoke detection. They contain such isotopes as cobalt-60 (Co-60), caesium-137 (Cs-137), iridium-192 (Ir-192), or americium-241 (Am-241), all of which pose security concerns because of their high levels of ionizing radioactivity and easy public access (such as at hospitals and construction sites). These items are typically kept under regulatory control through registration and inventory management systems.

Loss of Control of Radioactive Sources

A radioactive source that has left regulatory control, whether through being abandoned, misplaced, stolen, or transferred without authorization, is considered an orphan source.[2] The James Martin Center for Nonproliferation Studies (CNS) and the Nuclear Threat Initiative (NTI) found that more than 150 radioactive sources left regulatory control each year between 2013 and 2019.[3] Orphan sources pose a risk to public safety and security and can be used for the production of radiological dispersal devices (RDD), which are often referred to as 'weapons of mass disruption'.[4] National regulatory authorities invest significant time and resource in tracking down and locating orphan sources.

Incidents involving orphan sources have raised significant public health, safety, and environmental concerns. For example, in 1987 scrap dealers in Goiânia, Brazil cut open a high-activity caesium-137 source from an abandoned teletherapy machine, resulting in substantial contamination, the deaths of four people, and significant economic losses.[5] In 2001, an abandoned building in Webster, Texas, was found to contain over 300 radioactive sources; one room was so contaminated that workers would reach their maximum yearly dose in only 13 minutes.[6] In 2015, several people were arrested in Moldova for attempting to sell radioactive material to an undercover law enforcement officer posing as a representative of the so-called Islamic State.[7]

The IAEA maintains the Incident and Trafficking Database (ITDB), which catalogues reports, voluntarily provided by member states, on lost or stolen nuclear and other radioactive material. Over 3,500 incidents have been reported to the ITDB since 1993, with over 250 incidents of theft of radioactive sources reported over the past decade alone. About one-third of these sources were reported as unrecovered and their whereabouts unknown.[8] The ITDB is not publicly accessible, although it produces an annual factsheet with aggregated data. The only publicly available database, the Global Incidents and Trafficking Database, is produced by CNS for the NTI in the United States (US). It reports that 72 per cent of trafficking incidents involve non-nuclear radioactive sources or material, some suitable for radiological terrorism.[9]

Traditional Methods

The location and recovery of orphan sources are important functions of a national regulator. Several IAEA documents provide guidance for conducting search and recovery operations.[10] These documents describe two main types of searches: administrative and physical.[11] Administrative searches look for information about the sources, rather than for the sources themselves. A variety of resources are suggested for these searches. These include national inventories of radioactive sources and licenses, information provided by the public, and interviews with people who have worked with radioactive sources.[12] These methods can involve significant time expenditure. If government databases are not online, personnel must travel to various locations and may need to sift through large quantities of paperwork.[13] Appeals to the public for information take time to generate replies, and interviews require scheduling. For example, a Spanish government orphan source search and recovery programme lasted nearly two years.[14]

Physical searches are field operations to locate sources visually or by using radiation detection equipment. Although these operations can be triggered by information from administrative searches, they still require significant planning and substantial financial resources unavailable in many countries. Search plans must be drafted, transportation arranged, equipment prepositioned, local authorities contacted, and actions coordinated.[15] These expenses mean that physical searches are almost always the last step in efforts to locate orphan sources. However, they are still a necessary step for finding sources. For example, Mexican law enforcement used physical search methods in 2013 to locate a stolen cobalt-60 source.[16] Similarly, during inspection of the previously mentioned site in Webster, Texas, radiation detection equipment was used to locate radioactive sources hidden behind false walls.[17]

OSINT and the Identification of Orphan Sources

One type of information that receives comparatively little attention in existing guidance documents and academic literature is open source data. The rise of the internet has created new freely available data streams useful to analysts and experts at research institutions and regulatory bodies, resulting in a new generation of OSINT practitioners. Open source data includes, among others, geospatial data, social media, news media, trade publications, advertisements, videos, academic literature, and grey literature.[18] For the purposes of this chapter, the authors use this definition of OSINT: 'the process of identifying, harvesting, processing, analysing, and reporting data obtained from publicly available sources for intelligence purposes'.[19]

Government inventories of radioactive sources and materials are not publicly available and are considered sensitive due to safety and security concerns. Open source research to help responsible parties locate radioactive sources must therefore focus on indirect methods. One potential focus point is the facilities that are suspected to possess radioactive sources. The concept of the facility as an organizational method is worth explaining. Using little more than a name and address, OSINT techniques can build a picture of each facility's past and present, the work performed there, and the people who have worked there. Since the facility, as a potential host of radioactive sources, is the object of concern, all this information can be presented in a facility-centric manner, as shown in Figure 30.1.

This OSINT facility-centric method is especially useful in states that have experienced significant upheaval or are concerned about the completeness of existing source registries. For example, in the former Soviet Union radioactive sources were widely distributed across the constituent republics. Radioactive source inventories, however, were centrally maintained and were not passed on to successor states. In the economic chaos following the Soviet Union's collapse in 1991, many military and industrial sources were abandoned, and the related documentation lost.[20] While most of the post-Soviet states have utilized international and domestic funding to create radioactive source inventories, there are still concerns about their completeness. Orphan sources are still being located at Soviet-era facilities and this provides an opportunity to utilize OSINT to help track down radioactive sources that remain outside of regulatory control.

OSINT is increasingly used by non-governmental entities for research and analysis of a wide range of safety and security threats. This contests governments' monopoly on information and allows for access to information that was previously off limits. In turn, governments are increasingly relying on OSINT. One estimate shows that 90 per cent of useful information acquired by intelligence comes from open sources.[21] The use of OSINT to help governments locate orphan radioactive sources complements traditional methods that are still widely used.

OSINT can be viewed as both a toolbox and a process. As a toolbox, each type of data and the methods used to analyse it form a stand-alone tool available whenever the situation requires it. OSINT practitioners can, for example, use geospatial data such as

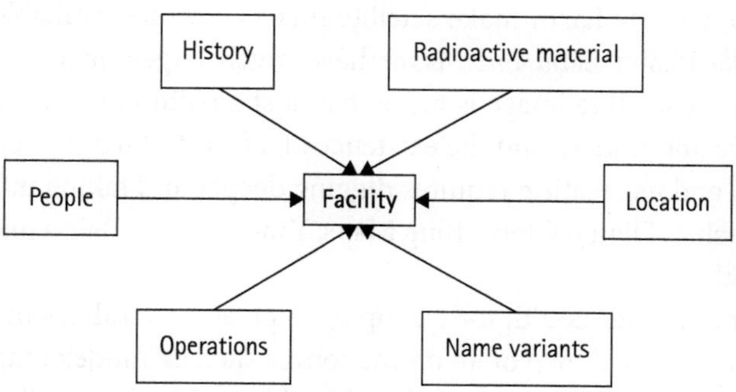

FIGURE 30.1. Facility-centric organization of information.

Figure courtesy of the authors.

satellite imagery and mapping resources to determine the perimeter of an abandoned facility. However, even when applying an OSINT technique as a stand-alone tool, an OSINT practitioner is unlikely to rely on a single data type.[22] Using the previous example, although an OSINT practitioner might rely primarily on satellite images and maps to determine a facility's perimeter, other resources, such as blog posts and news reports, could confirm the location and indicate if the facility was known by multiple names.[23]

As a process, the outputs of one OSINT tool are used to focus the collection or analysis of data for the benefit of another tool. For example, an OSINT practitioner can use information about the location and names of a facility to guide a search of other data types to learn about the facility's history or function, or to find social media accounts of current and former employees.[24] This is not a purely linear process, as information found during this research can further refine the geospatial analysis by indicating how a facility changed over time, or if it had branch locations.

In the context of radioactive sources, OSINT can also be used to build a history of individual radioactive sources. This is important if there is a period of time during which a regulator is missing large quantities of registration data.[25] Radioactive sources containing long-lived isotopes like caesium-137 can have useful lives measured in decades. Even if it is no longer usable for its original purpose, the entire source can be re-encapsulated in a different form useful for other purposes.[26] Thus, determining when a radioactive source was acquired, what it was used for, and when it passed into regulator custody for disposal helps the regulator understand what the radioactive materials situation was within their territory over time.

Characterizing Radioactive Source Facilities Using Geospatial Data

Geospatial data, and satellite imagery in particular, is perhaps the most visually striking of all the open source data forms. Formerly the preserve of government agencies, programmes like Google Earth make satellite imagery readily available to the public. Corporations like Planet Labs and Maxar have much larger imagery libraries available but for a fee. A satellite image is more than just a picture. It is a window through which to perceive and understand the existence of a facility. Such understanding is the OSINT product, and its creation requires digging deeper and taking information from other sources, such as Digital Globe, Bing Maps, Yandex maps, Baidu maps, Timescape, Polymaps, and others.[27]

An OSINT practitioner could, for example, be given the address of a factory. Map data, taken from a physical map or an online source such as Yandex maps, Baidu maps, or Google maps, can provide the location of the address. If only a facility name is given, the address can usually be found using the internet. Company websites are generally the most reliable source of information for this, followed by maps and business listings. However, an address is just a point on a street—a door, or a gate in a fence, or perhaps an office or storage unit in a larger building. The address does not indicate the boundaries of the factory.

For large, multi-building facilities, there are two ways to find this boundary data: visual inspection and information from other sources. Visually, there are several indicators of facility boundaries. Fences are a common method to separate a facility from its neighbours or to secure areas within a facility. Internal roads also control movement; while fences control where people and vehicles cannot go, roads control where they can go.[28] The roads within a facility are unlikely to connect with an external road or another facility without an additional form of access control, like a gate or guardhouse.

A variety of other sources can also provide data for geospatial analysis. The website Wikimapia.org allows users to enter information about locations and annotate a detailed world map with street names, building and facility outlines, and other features. However, Wikimapia is dependent on the users for data, so the quality of data is not universal across all regions.[29] For example, the information for former Soviet Union states and much of Europe is very thorough, but parts of China are quite bare.

Another form of data useful for geospatial analysis is 'ground truth' imagery. This refers to pictures taken at a known or knowable location and containing features identifying where the picture was taken and what features are in the picture itself.[30] Street view images, such as those offered by Google, Yandex, and Baidu, are an example of ground truth imagery. Street-level images are especially useful when mapping services are unable to parse an address and can only provide a general location. Most facilities have identifying signs visible from the street, and these can indicate a facility's precise location.

Ground truth imagery can also be found in press releases, videos, blog posts, and social media. The pictures included with North Korean press releases, for example, are crucial for understanding that country's defence industry.[31] Perhaps the most curious sources of ground truth images are so-called 'urban explorers'—people who trespass on abandoned buildings and properties for the purpose of taking pictures and videos. There are many websites dedicated to urban exploration, several of which include organized systems for users to post pictures with notes for the facility explored. For the OSINT practitioner, these can be useful for learning about defunct facilities.

The text contained in press releases and other resources is also important. This might describe different buildings at a site, the work performed, dates of construction, work conditions, and other useful information. The variations of names used to refer to a facility should also be noted. These name variations can be the result of changes in government, changes in ownership, or corporate structure, abbreviation, or truncation. For example, the Moldovan factory RAUT SA was known as 'The Factory named after Vladimir Ilyich Lenin' during the Soviet era. However, in most texts, Lenin's first name and patronymic are truncated to initials. In some cases, other words in the name—even the word factory itself—are also truncated.[32] This results in many variations of the same name, all of which are valuable search terms for finding data.

A further use for satellite images is detecting changes at a facility. Over time, buildings at a site can fall into disuse, undergo renovation, or be destroyed. New buildings may be constructed, either to replace older buildings, add new capabilities, or as part of a change in use. The Chisinau Tractor Factory in Moldova, for example, transitioned to being a business park after 2011, and has experienced significant rebuilding since then.[33] All of these processes can result in the displacement, sale, discarding, or destruction of equipment that may contain radioactive sources.

Satellite imagery can also be used to locate large pieces of equipment that may contain radioactive materials. Some Soviet aircraft and many Soviet helicopters were equipped with the RIO-3 ice detection system, which used sensors containing a small strontium-90 source.[34] In the Republic of Moldova, these radioactive sources are subject to regulatory control. Moldovan regulators applied satellite imagery and drones to find aircraft and helicopters equipped with the system on open-air display and elsewhere in the country.[35] This included non-operational aircraft and helicopters in an open-air museum, and others at airports that were operational.

OSINT in Preventing Radioactive Sources Loss: Malaysia Case Study

Satellite imagery has applications in searching for lost radioactive sources or preventing such losses. A significant portion of radioactive source losses occur during transportation, and such losses happen in many countries. For example, between 2015 and 2020, there were two source loss incidents in Malaysia. In one case, the source was never recovered.[36] The authors used geospatial analysis to evaluate the country's transportation routes and highways infrastructure. The purpose of this particular analysis was to examine which routes are most likely to be used for transporting non-destructive testing (NDT) equipment containing radioactive sources.

Using geospatial analysis, OSINT practitioners can examine the relative locations of industrial radiography providers and their clients and analyse the routes that radiography teams are most likely to use when driving to worksites. This can reveal obstacles drivers are likely to encounter, such as poor-quality roads, construction, traffic, and circuitous or confusing routes. While these features may at first seem minor, large potholes, sudden lane changes, swerving, and hard brake application have the potential to dislodge equipment in the vehicle, which could subsequently fall off, resulting in a radioactive source loss.

The routine transportation of sealed radioactive sources over long distances makes route planning especially important. This is given further salience by two additional factors. First, in order to avoid delaying construction, much of Malaysia's industrial radiography is performed at night. While this means that radiography teams avoid daytime traffic, it also results in decreased situational awareness on the road. This exacerbates

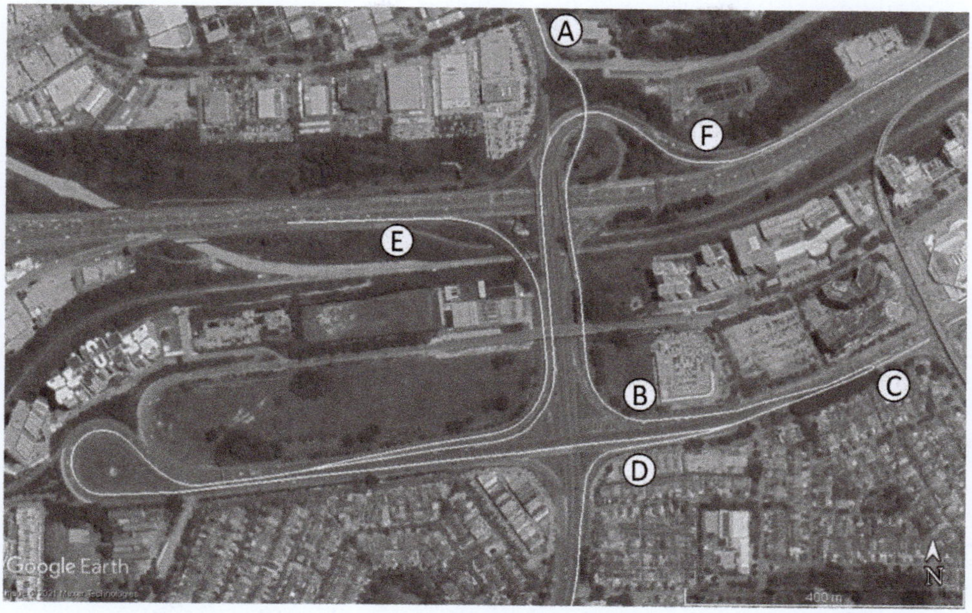

FIGURE 30.2. Interchange between Jalan Pengaturcara U1/51, Federal Route 2, and the E10 highway in Kuala Lumpur, Malaysia.

Image courtesy of Google, 2021. Maxar Technologies.

the second potential problem: navigation. Although Malaysia's road infrastructure is well maintained, its highway system has many complex interchanges. As a result, drivers must remain vigilant when entering, exiting, and transferring highways.

A notable example is shown in Figure 30.2, the primary highway access point for one of Malaysia's industrial radiography providers (note that cars in Malaysia drive on the left-hand side of the road). The local, north–south road at point A does not have a direct connection to the highway. Instead, drivers must use an intermediate road to the south. As the local road approaches the intermediate road, it merges with the highway off-ramps, and then joins the intermediate road at point B, where six lanes rapidly narrow to three lanes. The driver would then move across those three lanes, and pull a tight U-turn in a designated lane at point C. From this point, the driver can enter the southbound highway at point D or proceed around the loop and join the east–west highway at points E and F.

Open Source Datasets

There are also complete open source datasets available to OSINT practitioners. These have several applications. For example, the aforementioned Global Incidents and Trafficking Database maintained by CNS and the NTI in the US tracks nuclear and radioactive materials outside of regulatory control using news reports, press releases, and other open source data.[37] While the information in this dataset may not be directly applicable to a given country's regulatory agency, it is illustrative of how and why radioactive material leaves regulatory control. Knowing how and why sources were lost in the past can provide guidance on preventing future losses. This could be accomplished through eliminating trafficking pathways, improving regulatory oversight, or requiring that sensitive items be secured before transport, among other measures. The information in the dataset can also reveal trends about which types of sources with which isotopes go missing most frequently and from which sectors (medical facility, commercial site, scrap metal, etc.).

Another example is publicly shared radiation monitoring data obtained by volunteer groups and concerned citizens. The value and credibility of data generated by these groups have been internationally recognized. The NGO summit 'Solutions for a Secure Nuclear Future' on the side lines of the 2016 Nuclear Security Summit featured one such group called Safecast. The group employs innovative technologies to create useful, accessible, and granular environmental data.[38] Initially formed as a response to the lack of public information about radiation levels after the 2011 Fukushima Daiichi nuclear accident in Japan, Safecast continues to monitor and share data about radiation levels and air quality. Another group known to the authors is a three-person cycling group in Moldova that monitors radioactive levels in and around the capital Chisinau with Geiger counters and recreational bicycles. Such groups have the potential to detect elevated levels of radiation emitted from a lost or abandoned radioactive source.

Academic and Grey Literature

Academic articles, government reports, and internal corporate reports can provide valuable historical context for a facility, including what activities were performed, who the important people were, and, in some cases, when and how radioactive sources were used. They also include some of the most reliable and authoritative forms of open source data,[39] as the authors tend to be the same people and organizations responsible for studying, storing, and using the radioactive source in question.

For academic literature, an OSINT practitioner should consider both active academic journals and older conference proceedings and inactive journals. The IAEA and other organizations maintain scanned copies of old conference proceedings and academic publications.[40] These resources are invaluable for learning about previous activities at research institutes and universities. For example, when researching one of the leading Agrarian universities in Moldova, the authors found several abstracts in Soviet radiobiology conference proceedings from the 1980s. Some of these articles mentioned the use of radioactive sources for generating mutations in crop breeding research.[41] One of the researchers cited on those abstracts was listed as a researcher in genetics and crop breeding on the university's current website.[42]

Although the Moldovan regulatory agency had located and removed the radioactive sources from the university's laboratory in the 2000s, at the authors' recommendation the agency contacted the university seeking a meeting with the researcher. The university informed the regulator that the researcher was deceased, but a meeting was arranged with an electrical engineer formerly employed in her laboratory. While the engineer indicated that his work had never involved the handling of radioactive sources, he directed regulators to other researchers closely involved in the laboratory, several of whom were subsequently interviewed. This helped the Moldovan regulatory body build a history of the radioactive sources held in the laboratory's possession.

A general trait of grey literature[43] is that it is not published via traditional methods. While press releases, legislation, and news reports are published in a deliberate manner and are easy to find and identify, grey literature is often released as part of a process or following an event. Put another way, grey literature is not intended for general consumption in the same way as other forms of media.[44] Although such documents may lack the refinement of a proper publication, they may contain much greater technical detail and candour.

In terms of government documents, grey literature includes items like policy papers, drafts, and declarations. These will occasionally impact suspect facilities; for example, the aforementioned researcher from the agriculture institute was cited in several Moldovan government documents as the director of the National Center for Radioecology and the head of the university's radiobiology laboratory.[45] Other examples of government grey literature include committee and agency documents, such as annual reports, and technical documents, such as environmental impact assessments and pollution surveys. These technical reports sometimes document radiation in the environment, discussing known and new sources of contamination and the facilities responsible. Annual reports to shareholders are one example of corporate grey literature.[46] These reports might contain information on expenditures related to the use and storage of radioactive sources. Other forms of corporate grey literature include internal reports on topics like safety, research, and construction.[47] The documents could include descriptions or pictures of building interiors or mitigation efforts to secure radioactive sources located on a site.

A challenge with grey literature is that, since it is not formally published, it can be harder to find. For information related to radioactive materials, especially from government and academic sources, the IAEA's International Nuclear Information System (INIS) is an excellent resource, while the archive of the Russian Academy of Sciences (arran.ru) and Rosatom's online library (elib.biblioatom.ru) have resources related to Russia and the Soviet Union. Some government agencies, such as the US Department of Information and Office of Scientific and Technical Information also provide documents, although these are best searched using an external engine like Google. Corporate grey literature is much more ephemeral. Documents might be removed after several years, and entire websites might come down if corporations change ownership or go bankrupt. It is therefore important to always save copies of documents locally, and back up websites using the Internet Archive (archive.org).

Business Advertisements, News Reports, and Videos

Other forms of media published online also have application for OSINT. Business advertisements, company product catalogues, and press releases show how companies want to present themselves. Even if the contents are less informative than an internal report, these documents can still provide useful details. For example, if equipment at a defunct facility is going to be auctioned off, the company responsible will likely advertise the auction to attract more buyers. In such advertising materials, the company may include images or a list of the items for sale. Even when this information is not provided, knowing that an auction took place tells an OSINT practitioner when equipment may have left a facility.

Press releases and other documents on company websites are also valuable sources of information. For example, to attract customers, an industrial radiography company might provide a list of past projects. The authors have also encountered a case where a radiography provider listed the training and certifications held by each of its employees. In another case, the website of a factory that formerly manufactured non-destructive testing equipment indicated that it had transitioned to making wiring harnesses for automobiles and was renting out part of its buildings to third parties. This suggests that some equipment at the factory could now be disused or removed.

News reports can provide important updates about the status of a facility. From a local perspective, bankruptcies, acquisitions, and investments are often worth reporting. These can aid the OSINT practitioner by providing information about a facility. For example, an announcement that a facility believed to possess radioactive sources has auctioned some of its equipment may indicate that the radioactive sources have passed on to a new organization.[48] News reports can also provide information about events at a facility, including fires, accidents, and criminal prosecutions. Although damage from events such as fire or flooding is often visible in satellite images, news reports can provide a much more precise date and better information on the extent of damage. Pictures included with news reports, whether historical or current, can provide ground truth imagery. Similarly, videos, such as those found on YouTube or on news websites, are very valuable as a source of ground truth imagery, especially building interiors.

In factories, wide angle and panning shots are useful for learning about process flow (the path an item takes through a production line during the course of manufacture). For example, while reading a news article about a local brewery in Moldova, a regulatory inspector noticed a piece of equipment with the radiation trefoil in one of the pictures.[49] A YouTube video confirmed the presence of this equipment. Reviewing the agency's registration data, the inspector discovered that no radioactive sources or radiation-generating equipment were registered to the brewery. The inspector reached out to the brewery and learned that the device was a fill level detector that uses radiation from americium-241 to determine the fill level of liquid in a container. In the Republic of Moldova this type of radioactive source is subject to regulatory control, so the inspector scheduled a site visit and worked with the brewery to bring it into regulatory compliance.

Social Media

Social media data are, generally, a very promising source of data for recovering radioactive sources, although achieving that potential can be challenging. While other forms of data provide information about a facility, the value of social media comes from the fact it contains information about the employees.[50] The IAEA's Specific Safety Guide 'National Strategy for Regaining Control over Orphan Sources and Improving Control over Vulnerable Sources' notes that people who work at facilities of interest may have key information regarding the possible presence of radioactive sources there.[51] Although personal memories are fallible, in cases where no other records exist, they may be the only option.[52]

Social media data could potentially aid the search for radioactive sources due to how many social media users list their current and former employers on their profiles. Through their social media searches, the authors generated lists of almost 1,800 employees from former Soviet facilities in Moldova and Georgia. To narrow down these large groups to key individuals who may be of potential interest, the authors conducted social network analysis to identify people with more connections to each other and facilities of interest.[53] Social network analysis identifies, differentiates, and quantifies the relationships between people and organizations.[54] Theoretically, it should be possible to interact with these people. These interactions, however, are elusive.

When creating a social media profile on most websites, a user is prompted to enter a wide variety of information, such as current and former employers. This applies for most social media platforms, including Facebook, LinkedIn, the Russian platforms VKontakte, Odnoklassniki, and others. Unless the user activates certain privacy options, this information can be viewed by all other users of that platform. Finding these people is quite easy, either using the social media platform's built-in search function or with a constrained query with a search engine. Quite often, however, the same facility will be listed several times with different name variations.

Interacting with people over social media is a challenge on its own. Social media profiles rarely list phone numbers or email addresses. Although it is possible to converse with other members using the chat functions on social media platforms, this is not conducive to collecting information in a systematic fashion. In order to ensure data are collected in a consistent manner, both in online and offline interactions, regulators may find it useful to develop a structured interview or a survey. Many survey hosting websites offer in-depth creation tools and the ability to collect both multiple choice and text data, although a paid account is often required for advanced features. The survey can also be produced in a printable format for use during field inspections.

For drafting the survey, the IAEA provides guidance on what information interviewers should seek and suggests using standardized questions.[55] This guidance, however, is not comprehensive, and regulators might find further assistance in other publications, such as the US General Accounting Office's 'Using Structured Interview Techniques'.[56] Generally speaking, most social media users associated with a facility will not have worked with radioactive sources, and many will not even be aware of the regulatory agency's existence. The survey should therefore include a short description of the regulator and its role and provide assurance that the survey-taker will not face punitive action for providing information. The survey should also include pictures of the radiation trefoil and radioactive sources, materials, and devices commonly found in the regulator's jurisdiction. Questions should be simple and straightforward, focusing on if, when, and where the objects were seen. If desired, the final questions might ask the survey-taker to attach their name and contact information.

Incentivizing people to actually take the survey is another matter entirely. In Moldova, the regulatory body created an official Odnoklassniki profile and populated it with press releases and other information. This profile was used to send survey invitations to over 200 Odnoklassniki users. None of the users responded to the survey; most ignored the message, several people blocked the regulator's account, and a few, curiously, attempted to add the account as a friend. Odnoklassniki administrators also locked the account on two occasions for 'suspicious activity'. On one hand, this is a positive result, as it shows users are unlikely to provide information to accounts they consider suspicious. On the other hand, it points to the difficulty government agencies encounter in trying to build trust on social media platforms. Nonetheless, the survey still proved useful as an online tool. Inspectors would provide facility managers with a link to the survey. The managers would then distribute the link to employees via email, thus establishing the authenticity of the survey to the employees. The survey was also useful as an offline tool for in-person interviews.

Social media also has other challenges. Although individuals should have no expectation of privacy for information published in a publicly visible profile, laws governing online and social media privacy are constantly evolving. Regulators must be aware of and understand local laws and how they may affect the gathering and use of social media data. Regional laws, such as the European Union's General Data Protection Regulation (GDPR), can also affect data collection and analysis operations. Additionally, social media platforms themselves are constantly evolving, with new features being added and others removed. It is never safe to assume that information present today will still be there tomorrow. For example, during the mid-2010s, it was possible to search Odnoklassniki profiles using external search engines such as Yandex or Google. Since that time, Odnoklassniki's policies have changed, and this is no longer possible. Now, only registered users can see members' workplaces. OSINT practitioners should also pay attention to terms of service. Most social media platforms prohibit the use of automated scraping tools to collect data, although manual collection exists in a sort of grey zone, as website algorithms sometimes have trouble detecting it.

The Future of OSINT in Locating Radioactive Sources

Digitization of data has led to the democratization of knowledge and increased the accessibility of data to wider audiences. Commercially available satellite images and other data used for research in nuclear security and other fields have enabled academia, think tanks, and practitioners to produce top-quality analyses that were unthinkable two decades ago. It is also imperative to create awareness among various governmental stakeholders about OSINT and change negative perceptions that link OSINT to the breach of a nation's classified or sensitive information.

OSINT tools provide regulators with access to information that may otherwise be unavailable. As society becomes increasingly digitalized and the realm of open source data continues to grow, OSINT capabilities will only increase in value. The bottleneck in their application is likely to be in training. Outside of Europe and North America, very few institutions offer skills-based training in OSINT techniques that meets the needs and schedules of government employees. However, it is only through increased application that regulators will see the full potential of OSINT.

Increased use of social media by regulatory agencies, for example, has potential side benefits. When a radioactive source is lost or stolen, such as in the theft of an industrial radiography source carrier, rapid response is crucial to prevent it being passed on to a third party and disappearing. In these cases, public outreach through social media is potentially valuable, and the regulator may actually benefit from the public's general wariness of anything related to radiation. When informed that a device containing radioactive material has been stolen, people are likely to exhibit greater vigilance. Such events may also build trust and rapport between the regulatory agency and the public and create future opportunities for education and outreach.

Additionally, the challenge of incentivizing responses to surveys distributed via social media is theoretically solvable. This is an especially important goal for regulators seeking information on defunct facilities. One potential solution could be to work with the social media platform to attach some form of marker to the agency's account to denote its status as a government entity. Nonetheless, the ability to pull names of current and former employees of a suspect facility is still useful. This information could potentially be used to find contact information via other means, such as phone books and government databases.

OSINT techniques cannot replace other methods of locating orphan sources. However, they have an important role supplementing administrative searches, providing information that is otherwise unavailable. This information can fill the holes in a regulator's data, answer questions that the regulator's data cannot, and help offset costs and other resources associated with traditional searches. The concepts and techniques covered in this chapter are merely a sampling of the world of OSINT. The field will continue to grow and change as new data and tools become available.

Conclusion

OSINT can form a valuable part of efforts to locate radioactive sources. However, it must be used as part of a broader strategy. While OSINT can guide and inform the planning of inspections and physical searches, it can never replace them. Nor can it replace administrative searches of internal data held by the regulatory agency and partner agencies. OSINT complements, rather than replaces, these methods. Thus, when considering the use of OSINT tools, it is important to identify both the advantages and disadvantages.

One of OSINT's greatest strengths in the context of locating orphan sources is that it considers an entirely different set of data than that normally present in regulatory agencies' databases. The outcome of open source research is therefore not tied to any preconceptions built into regulatory documents or regulator expectations. Analysis of open source data can lead to unexpected results, revealing new facilities and connections and prompting new types of interactions. Therefore, cases where the regulatory agency has little to no pre-existing knowledge of a facility are some of the best situations to apply OSINT techniques.

This feeds into another strength of OSINT: the constant growth and ubiquity of open source data. Hundreds of news stories are published every day. Governments and corporations publish reports and press releases on a regular basis. Social media platforms are always abuzz with activity. The information available to the open source researcher is constantly evolving. Even if a question cannot be answered today, the information necessary to answer it may appear at some point in the future, or a researcher can seek an answer through crowdsourcing, by appealing to the public.

On the other hand, the information retained by regulatory agencies is usually selected because it relates specifically to the role and function of the agency. The opposite is true for open source data. Outside of academic literature and IAEA publications, very little of the open source data encountered by an OSINT practitioner will be directly related to radioactive sources. Instead, an OSINT practitioner's job is to sift through large quantities of tangentially related information and pull out the relevant data points. The practitioner's skill is in their ability to select the correct data, organize and analyse it, and present the results in an understandable format. This process can, in some situations, require more time than traditional administrative searches. The time requirement may, however, be somewhat counter-balanced by another benefit of open source research and analysis: cost. Although OSINT can require a significant time investment to collect, sort, and analyse data, all of this work can be conducted from a single location, such as the home or office. By contrast, even administrative searches can involve expenditures for transportation to archives and time spent organizing and preparing for interviews.

Lastly, a further advantage of OSINT tools is that they have value beyond the tools themselves. Developing and applying strategies for the exploitation of open source data as part of a given project requires an open-minded approach to new problems that may not be covered in regulations or guidance documents. For example, IAEA guidance for interviewing employees at facilities using radioactive sources focuses on those employees who are directly involved with the materials. This advice is less applicable for interacting with people who may have only a basic understanding of radiation. In order to resolve this problem, the authors worked with partner agencies to examine different interview methods and determine what information to present to interviewees and what questions to ask. As such, training in OSINT techniques gives regulators the tools to develop new and innovative solutions.

Notes

1. IAEA, *Code of Conduct on the Safety and Security of Radioactive Sources*, Non-serial Publications, IAEA/CODEOC/2004 (Vienna: IAEA, 2004), 3. https://www.iaea.org/publications/6956/code-of-conduct-on-the-safety-and-security-of-radioactive-sources
2. IAEA, *IAEA Safety Glossary*, 2018 Edition, Non-serial Publications, STI/PUB/1830 (Vienna: IAEA, 2019), 220. https://www.iaea.org/publications/11098/iaea-safety-glossary-2018-edition
3. Sam Meyer and Jakob Lengacher, *The CNS Global Incidents and Trafficking Database* (Nuclear Threat Initiative (NTI) [online], 23 September 2021. https://nti.org/analysis/resource-collections/the-cns-global-incidents-and-trafficking-database
4. Nuclear Regulatory Commission, 'Backgrounder on Dirty Bombs', NRC [online], 23 February 2022. https://nrc.gov/reading-rm/doc-collections/fact-sheets/fs-dirty-bombs.html
5. IAEA, *The Radiological Accident in Goiânia*, Non-serial Publications, STI/PUB/815 (Vienna: IAEA, 1988).
6. Interstate Technology and Regulatory Council, *Decontamination and Decommissioning of Radiologically Contaminated Facilities* (Washington, DC: ITRC, 2008), 137. https://connect.itrcweb.org/HigherLogic/System/DownloadDocumentFile.ashx?DocumentFileKey=3bf91bf2-bcc2-4947-a928-27c3635400c2
7. Desmond Butler and Vadim Ghirda, 'AP INVESTIGATION: Nuclear Smugglers Sought Extremist Buyers', *Associated Press* [online], 6 October 2015. https://apnews.com/article/688d82738c6f4e89b9636edfbc868de6
8. Charlotte East and Kendall Siewert, 'Incident and Trafficking Database: Combatting Illicit Trafficking of Radioactive Material for 25 Years', *IAEA Bulletin* [online], February 2020. https://www.iaea.org/bulletin/incident-and-trafficking-database
9. Nuclear Threat Initiative (NTI), 'CNS Global Incidents and Trafficking Database Archived Report and Graphics', Fact Sheet [online], 24 July 2018. https://www.nti.org/analysis/articles/cns-global-incidents-and-trafficking-database-archived-reports-and-graphics
10. See, for example, IAEA, *National Strategy for Regaining Control over Orphan Sources and Improving Control over Vulnerable Sources*, Specific Safety Guide, IAEA Safety Standards Series, No. SSG-19 (Vienna: IAEA, 2011); IAEA, *Methods to Identify and Locate Spent Radiation Sources*, IAEA-TECDOC-804 (Vienna: IAEA, 1995); IAEA, *Strengthening Control Over Radioactive Sources in Authorized Use and Regaining Control Over*

Orphan Sources, IAEA-TECDOC-1388 (Vienna: IAEA, 2004); IAEA, *Nuclear Security Recommendations on Nuclear and Other Radioactive Material out of Regulatory Control*, Recommendations, IAEA Nuclear Security Series, No. 15 (Vienna: IAEA, 2011).

11. They also describe other methods to learn about orphan sources and radioactive material, including reports of loss or theft, radiation monitoring, and radiation poisoning incidents. See IAEA, *National Strategy for Regaining Control over Orphan Sources and Improving Control over Vulnerable Sources*, 36.
12. Ibid., 37–41.
13. Ibid., 43.
14. Pedro Carboneras, Maria Teresa Ortiz, and Carmen Rueda, 'Spanish National Campaign for the Search and Recovery of Orphan Radioactive Sources', in *Proceedings of the Third European IRPA Congress*, 14–18 June 2010, Helsinki, Finland (Vantaa: STUK–Radiation and Nuclear Safety Authority, 2010), 2, 190.
15. IAEA, *National Strategy for Regaining Control over Orphan Sources and Improving Control over Vulnerable Sources*, 44–9.
16. Rafael Romo, Nick Parker, and Mariano Castillo, 'Mexico: Stolen Radioactive Material Found', *CNN* [online], 4 December 2013. https://www.cnn.com/2013/12/04/world/americas/mexico-radioactive-theft/index.html
17. Interstate Technology and Regulatory Council, *Decontamination and Decommissioning of Radiologically Contaminated Facilities*, 137.
18. Nihad A. Hassan and Rami Hijazi, *Open Source Intelligence Methods and Tools: A Practical Guide to Online Intelligence* (New York: Apress, 2018), 5.
19. Catherine Chipeta, 'Open Source Intelligence (OSINT): Top Tools and Techniques', *UpGuard* [blog], 12 May 2022. https://www.upguard.com/blog/open-source-intelligence
20. Committee on Opportunities for U.S.—Russian Collaboration in Combating Radiological Terrorism, *U.S.—Russian Collaboration in Combating Radiological Terrorism* (Washington, DC: National Academies Press, 2007), 44–5.
21. Hassan and Hijazi, *Open Source Intelligence Methods and Tools*, xix.
22. Economist Staff, 'Trainspotting, But with Nukes: Open-Source Intelligence Challenges State Monopolies on Information', *The Economist* [online], 7 August 2021. https://www.economist.com/briefing/2021/08/07/open-source-intelligence-challenges-state-monopolies-on-information
23. Jeffrey Lewis, 'Commercial Satellite Imagery—Ch. 2: Applying New Tools to Nonproliferation', in 'Applying New Tools to Nonproliferation: A Nuclear Detective Story', NTI [online], 29 April 2016. https://nti.org/analysis/articles/commercial-satellite-imagery
24. Michael Duitsman, Bryan L. Lee, Margarita Kalinina-Pohl, and William Potter, 'Using Targeted Social Media to Locate Orphaned Radiological Sources: Results of a Pilot Study in Moldova', in *Proceedings of the 58th Annual Meeting of the Institute of Nuclear Materials Management (INMM 2017)*, 16–20 July 2017, Indian Wells, CA (Red Hook, NY: Curran Associates, 2017), 1956–63.
25. IAEA, *Strengthening Control Over Radioactive Sources in Authorized Use and Regaining Control Over Orphan Sources*, 11–12, 24.
26. See, for example, US Nuclear Regulatory Commission, 'Registry of Radioactive Sealed Sources and Devices, No. NR-1235-S-103-S, NRC [online], 4 October 2006. https://www.nrc.gov/docs/ML0627/ML062770491.pdf
27. Hassan and Hijazi, *Open Source Intelligence Methods and Tools*, 288.

28. US Department of the Army, 'Terrain Analysis: US Army Field Manual 5-33' (Washington, DC: Government Printing Office, 1990), 1–15.
29. Wikimapia, 'About Wikimapia', 2012. https://wikimapia.org/about
30. Jesse Boyce, Susan Hassol, John Katzenberger, Martha Kirpes, Anthony Leiserowitz, and William Stapp, *Ground Truth Studies Teacher Handbook*, 2nd edn (Aspen, CO: Aspen Global Change Institute, 2003), 34.
31. Jeffrey Lewis, 'Identifying DPRK Machine Plants', *Armscontrolwonk* [online], 17 January 2019. https://www.armscontrolwonk.com/archive/1206662/identifying-dprk-machine-plants
32. Raut Sa, 'History', n.d. http://raut.md/istoric/?lang=en archived at https://web.archive.org/web/20210612114942/http://raut.md/istoric/
33. TRACOM, 'Our History', n.d. https://tracom.md/en/pi-tracom/history
34. I. Balan, B. Lee, M. Kalinina-Pohl, and W. Potter, 'The Use of Innovative Approaches in the Search for and Regaining Control over Orphan Radioactive Sources in Republic of Moldova', in *International Conference on the Security of Radioactive Material: The Way Forward for Prevention and Detection*, 3–7 December 2018, Vienna, Austria (Vienna: IAEA, 2018).
35. Ibid.
36. Francisco Parada, Margarita Kalinina-Pohl, and Miles Pomper, 'Radioactive Material Is Still Missing in Malaysia: Cause for Concern?' *The Bulletin of Atomic Scientists* [online], 14 September 2018. https://thebulletin.org/2018/09/radioactive-material-is-still-missing-in-malaysia-cause-for-concern
37. Nuclear Threat Initiative (NTI), 'The CNS Global Incidents and Trafficking Database', NTI [online], 23 September 2021. https://www.nti.org/analysis/resource-collections/the-cns-global-incidents-and-trafficking-database
38. Safecast, 'About', n.d. https://safecast.org/about
39. Hassan and Hijazi, *Open Source Intelligence Methods and Tools*, 9.
40. IAEA, 'Publications', n.d. https://iaea.org/publications
41. See, for example, Olga Blyandur, 'Сельскохозяйственная радиобиология и задачи в селекции и генетике кукурузы' [Agricultural radiobiology and problems in the breeding and genetics of maize]', in *I Всесоюзный радиобиологический съезд* [I All-union radiobiological congress], Vol. 5 (Pushchino: Scientific Center for Biological Research of the USSR Academy of Sciences in Pushchino, 1989), 1175. https://inis.iaea.org/collection/NCLCollectionStore/_Public/22/059/22059786.pdf
42. 'Cercetare stiinţifică' [Scientific research], State Agrarian University of Moldova, 2016. https://web.archive.org/web/20161103181507/http://uasm.md/ro/cercetare-stiintifica
43. Grey literature is information produced outside of traditional publishing and commercial channels. Examples include reports, working papers, newsletters, government and company documents, speeches, white papers, etc.
44. GreyNet, 'About GreyNet', n.d. https://greynet.org/home/aboutgreynet.html
45. See, for example, Republic of Moldova, 'О создании Национального комитета радиационной защиты' [On the establishment of the National Committee for Radiation Protection], Decree 748, Government of Moldova [online], 3 November 1995. https://www.legis.md/cautare/getResults?doc_id=76226&lang=ru
46. Welch Medical Library, 'What is Grey Literature', Johns Hopkins University, 28 July 2022. https://browse.welch.jhmi.edu/c.php?g=1195274
47. Ibid.

48. TRACOM, 'Licitația "cu strigare" privind comercializarea activelor neutilizate a întreprinderii' [The "shout" auction regarding the sale of the unused assets of the enterprise], 28 December 2017. https://tracom.md/2017/12/28/28-12-2017
49. Names are not included in order to preserve the privacy of the entities involved.
50. Hassan and Hijazi, *Open Source Intelligence Methods and Tools*, 205.
51. IAEA, *National Strategy for Regaining Control over Orphan Sources and Improving Control over Vulnerable Sources*, 40.
52. Ibid., 40.
53. Duitsman et al., 'Using Targeted Social Media to Locate Orphaned Radiological Sources'.
54. Christina Prell, *Social Network Analysis* (Thousand Oaks: Sage, 2012), 7–9.
55. IAEA, *National Strategy for Regaining Control over Orphan Sources and Improving Control over Vulnerable Sources*, 43–4; IAEA, *Strengthening Control Over Radioactive Sources in Authorized Use and Regaining Control Over Orphan Sources*, 45.
56. US General Accounting Office, *Using Structured Interviewing Techniques*, GAO/PEMD-10.1.5 (Washington, DC: Government Printing Office, 1991).

Suggested Reading

Carboneras, Pedro, Ortiz, Maria Teresa, and Rueda, Carmen, (2010), 'Spanish National Campaign for the Search and Recovery of Orphan Radioactive Sources', in *Proceedings of the Third European IRPA Congress*, 14–18 June 2010, Helsinki, Finland (Vantaa: STUK–Radiation and Nuclear Safety Authority).

Duitsman, Michael, Lee, Bryan L., Kalinina-Pohl, Margarita, and Potter, William, (2017), 'Using Targeted Social Media to Locate Orphaned Radiological Sources: Results of a Pilot Study in Moldova', in *Proceedings of the 58th Annual Meeting of the Institute of Nuclear Materials Management (INMM 2017)*, 16–20 July 2017, Indian Wells, CA (Red Hook, NY: Curran Associates).

Fiorella, Giancarlo, 'First Steps to Getting Started in Open Source Research', Bellingcat [online], 9 November 2021. https://www.bellingcat.com/resources/2021/11/09/first-steps-to-getting-started-in-open-source-research

Hassan, Nihad A. and Hijazi, Rami, (2018), *Open Source Intelligence Methods and Tools: A Practical Guide to Online Intelligence* (New York: Apress).

Hobbs, Christopher, Moran, Matthew, and Salisbury, Daniel, (2014), *Open Source Intelligence in the Twenty-First Century: New Approaches and Opportunities* (London: Palgrave Macmillan).

IAEA, (2011), *National Strategy for Regaining Control over Orphan Sources and Improving Control over Vulnerable Sources*, Specific Safety Guide. IAEA Safety Standards Series, No. SSG-19 (Vienna: IAEA).

CHAPTER 31

RADIATION DETECTION FOR BORDER MONITORING

CRAIG MARIANNO

Introduction

Detecting illicit radioactive material at border crossings is challenging. Authorities at these entry points must be vigilant and monitor vehicle traffic, pedestrian traffic, and cargo. This monitoring must be completed efficiently and quickly so as to not impede normal commerce. Unfortunately, depending on the type and amount of radiation emitted, detecting certain material can be extremely difficult. This is further complicated by people who have been injected with medical isotopes or the legal movement of naturally occurring materials, both of which can trigger radiation alarms. With border radiation monitoring it is imperative for authorities to understand both what they are trying to detect and the equipment necessary to detect it. Detection instruments can range in size from hand-held instruments to fixed portal monitors that are several metres high. To be effective these systems must all work in conjunction with one another as part of a defence-in-depth protection strategy.

The development of radiation detection programmes at borders has been driven by incidences where dangerous radioactive materials have crossed international boundaries, either through being smuggled or by mistake. In 1995 the International Atomic Energy Agency (IAEA) started a programme aimed at assisting member states in combatting the trafficking of illicit nuclear material.[1] This programme, known as the Incident and Trafficking Database (ITDB), was originally established to only track the incidents of illicit trafficking of nuclear and other radioactive materials. However, it was later expanded to include all incidents of these materials out of regulatory control. As of 2021, 142 IAEA member states contribute to this database and have reported 3,928 incidents since 1993.[2] In that time there have been 320 recorded cases related to trafficking or malicious use of radiological material. Of that, 38 per cent have been from radioactive sources like caesium and cobalt and 45 per cent have included special nuclear material such as uranium and plutonium. The remaining 17 per cent included mostly non-radioactive material used in scams and/or frauds.

The United States (US) has significant radiation detection capability at its borders and has played an important role in strengthening not just its own but the ability of other countries to detect radioactive materials at border crossings. This programme known as the Second Line of Defense (SLD) was meant to 'strengthen the capabilities of partner countries to deter, detect and interdict illicit trafficking of special nuclear and other radioactive materials at international border crossings including airports, seaports and other points of entry'.[3] Currently, this programme is conducted from the Office of Nuclear Smuggling Detection and Deterrence (NSDD) and has installed over 3,500 radiation portal monitors (RPMs) and supplied hand-held equipment to over 600 sites in over 50 countries.[4] Billions of dollars have been spent on developing and installing these systems, with a dramatic upsurge in funding and increased internationalization of these efforts in the period following the 11 September 2001 ('9/11') terrorist attacks, driven by a perceived increase in the threat of nuclear terrorism.

Types of Radiation Relevant for Border Monitoring

It is important to review the types of radiation that a border security agent may encounter and the challenges of detecting them. Alpha radiation is the most difficult type to detect. Alpha particles are released from heavy, unstable nuclei such as plutonium or uranium. While these types of materials must be interdicted for security reasons, in most situations alpha radiation is not detectable. These particles are shielded by only a few centimetres of air and are completely blocked by a sheet of paper. This makes them impossible to detect from a distance or if they are contained in any type of material.

The next most difficult type to register is beta radiation. These particles are essentially electrons and, like alphas, take little effort to shield. However, depending on their energy, they may be detected from 1 to 2 metres away in air. Betas are harder to shield than alphas, but nevertheless can be blocked by materials only a few centimetres thick. The only way that alpha and beta radiation will be detected in a trafficking scenario is when the radioactive sources are completely exposed, and a border security agent is using the proper instrumentation. Fortunately for the purposes of border security, most alpha and beta emitters also release gamma rays.

Gamma radiation is like X-rays and therefore is more easily detected than alpha and beta particles. Gamma radiation travels tens of metres in air and penetrates many materials. Because of these properties, gamma radiation is the primary radiation type being monitored at border locations. Gammas can penetrate people, containers, and many types of materials registered by many types of radiation detectors. If the proper detector is being used, the material emitting the gamma radiation can be identified.

Gamma detectors with identification capability are the only passive, non-destructive way to identify radioactive material if it is within an object such as a vehicle or cargo container. Identification is important because it provides the ability to distinguish radioactive sources that may be a threat from naturally occurring materials and medical isotopes.

The last radiation type of interest at border crossings are neutrons. Like gamma rays, neutrons are highly penetrating and can be detected from several metres away. They can be shielded, but usually from different types of materials that are effective against gamma rays.[5] Unlike gamma rays, neutrons cannot easily be used to identify a material. However, no medical sources emit neutrons, and it is only man-made radiation sources that emit them to any significant levels.[6] Consequently, the detection of neutrons can provide important information to a border security agent or customs official that may lead them to trigger a further investigation. For the purposes of this chapter, discussion is focused on gamma and neutron detection, given their primary importance in border monitoring.

Detection Media and the Radiation they Detect

For border monitoring, the materials used for radiation detection purposes are solid and gas. A well-known type of radiation detector is Geiger–Müller instruments. These are gas-filled devices that detect alphas, betas, and photons (gamma rays and X-rays), but they do not register neutrons. Geiger–Müller detectors are primarily used for safety as opposed for security. They can register radiation, but they cannot distinguish between the radiation type (alpha, beta, gamma) and therefore cannot be used to identify the radioactive substance. These types of detectors are very portable, easy to use, and can detect and localize radioactive substances.

Another common type of detector used in border monitoring is scintillation-type detectors. The materials used to construct these detectors can either be a salt or plastic and are optically transparent. When radiation interacts with the detection material, it produces visible light or scintillates. This light is collected and converted into an amplified voltage pulse that is registered by the detection system electronics. These types of detectors can be used in hand-held instruments or made very large for vehicle-mounted systems and RPMs. The most widely used scintillation detector material is sodium iodide (NaI). This is followed by lanthanum bromide (LaBr) and caesium iodide (CsI). Detection systems using scintillators will detect and identify gamma-emitting materials. Another common type of scintillation material primarily employed in RPMs is a plastic known as polyvinyl toluene (PVT). It is also used to detect gamma and X-rays but has limited ability to identify the materials emitting the radiation.

A less-common detector found at border crossings utilizes semiconductors as the detection media. These are limited in availability usually due to the cost of the systems. Hand-held gamma detectors made with scintillators can range in cost from a few thousand US dollars to almost US$20,000. Similar semiconductor systems that utilize high purity germanium (HPGe) can exceed US$100,000. Their cost differences are due to the sophistication of these instruments and the difficulty to produce the HPGe material. Their advantage though is that they have a much better capability to identify radioactive sources and distinguish illicit sources from medical or natural gamma-emitting nuclides. In addition to the cost, the other disadvantage of these types of detectors is that they must be kept near liquid nitrogen temperature (~100°K) to operate.

Like photons, neutrons are detected by gas, scintillators, and semiconductors. The most common type of neutron detector used in border security is metal tubes filled with helium-3 gas. This isotope of helium is ideal for neutron detection because of its absorption properties. Unfortunately, its availability has decreased in the past two decades. The main source of helium-3 gas was from nuclear weapons programmes. This gas is a decay product of tritium, which is used in these devices. Starting in the 1990s, weapons production drastically decreased, which corresponded to a decrease in tritium production. Without new tritium the rate of helium-3 production also decreased.

Following the 9/11 terrorist attacks, the potential for the illicit trafficking of radioactive materials to support terrorist activities became even more of a concern. As a result, the international need for neutron detection systems increased. More systems were being developed and produced at a faster rate than previously experienced for border security, customs enforcement, and law enforcement. For instance, between 2002 and 2011 the US Department of Homeland Security deployed nearly 1,400 RPMs to be utilized both domestically and internationally.[7] This increase in demand paired with a decrease in supply caused almost an order of magnitude increase in cost.[8] As a result, alternative forms of neutron detection have been investigated and deployed. These include boron trifluoride gas (BF_3), scintillation detectors such as CLYC, and lithium- or boron-loaded semiconductors.[9] All of these can detect neutrons, but at a lower efficiency than helium-3.

Radiation Alarms at Borders and the Sources that Trigger Them

Radiation alarms at borders are triggered when radioactive sources get close enough to a detection system for the radiation they emit to be registered. 'Real' alarms occur when sources out of regulatory control are detected. Alarms caused from harmless sources like medical and naturally occurring radioactive material (NORM) are classified as 'nuisance' alarms. Nevertheless, every radiation alarm, either nuisance or real, must be investigated. This can include pulling individuals aside or moving vehicles to secondary inspection areas. Sometimes gamma identification can be used immediately and can separate the nuisance alarm from a real alarm, without having to do time-consuming secondary activities to determine the cause of the alarm.

Some hand-held and fixed-location detection systems will classify the detected source into four categories that are based on where the radioactive material is used. Radiation instruments will alert law enforcement with an Industrial (IND) alarm if the radiation that is being detected appears to be from an isotope commonly used in industry. The most common isotopes of concern here include caesium-137 (^{137}Cs), strontium-90 (^{90}Sr), and Co-60 (^{60}Co).[10] These isotopes are commonly found in industrial settings but could be used to create radiological dispersal devices (RDDs) or radiological exposure devices (REDs) to expose individuals to high radiation fields.

The Medical (MED) alarm classification is for radioactive sources used in medicine. These are typically injected into patients for diagnostic reasons and sometimes for therapeutic. The isotopes can exist in the human body and be detected for several hours or days after treatment. The most common medical isotope is technetium-99m (^{99m}Tc) followed by isotopes used for positron-electron tomography (PET) scans such as fluorine-18 (^{18}F). The most serious alarm category is special nuclear material (SNM) and includes radionuclides such as uranium and plutonium. These are the most concerning because isotopes of these elements can be used to build nuclear explosives. Lastly, there are alarms in the Natural (NAT) category. These are caused by NORMs such as potassium-40 (^{40}K) or isotopes in the uranium decay chain (bismuth-212/214, radium, etc.). NORMs can be found in normal foodstuffs, fertilizer, granite, and ceramics.

A potential source of radiation that has caused alarms at borders comes from metallurgical scrap.[11] Metal scrap is transported all over the world because of recycling activities. Occasionally, radioactive material or sealed sources are inadvertently included in the scrap shipments.[12] Sometimes these sources are even melted along with the scrap producing radioactive metal. Shipments such as these have crossed borders and been detected by existing radiation detection systems used at borders. The types of sources involved in these shipments would be identified in the IND alarm category.

The IAEA has developed a tool that helps border security agents and customs officials to determine the potential threat of a detected source of radiation alarms. This application is called the Tool for Radiation Alarm and Commodity Evaluation (TRACE) and can be loaded onto a smart phone.[13] This app was developed through an IAEA-led research project and uses information on commodities and isotopes that may be scanned by RPMs. TRACE can be used in two ways. A customs official could review the manifests of incoming cargo shipments. After reviewing the incoming commodities, TRACE can be used to ascertain the potential NORM that may be detected. Similarly, TRACE can be used after an alarm is triggered. While investigating the alarm the customs agent can enter the information about the commodity into the application. In both cases, the application will reveal the potential isotopes that might be identified, as well as some physical and chemical information regarding the material being scanned.

Operational Environment

There are many different types of entry points into a country. Regardless of their operation, each presents its own challenges and therefore must use a variety of radiation detection systems to efficiently scan for radiation, whether the material is being carried by a person, a vehicle, or a cargo container being removed from a ship.[14]

Border radiation monitoring must operate effectively in all types of environments. Most border crossings, either for vehicular and/or pedestrian crossings, can be loud and extremely busy areas. Detection systems must have both audible and vibration alarm indicators so they can be acknowledged by operators working in such conditions. These systems also operate in especially harsh working environments. Weather conditions can include rain, high humidity, desert, extreme heat, and extreme cold. Operationally, these systems can be dropped or hit by vehicles (in the case of RPMs). Therefore, these systems must be ruggedized to not only operate in extreme conditions, but to withstand them for long periods of time.

Official Border Crossing Points

Manned locations where people and commerce cross borders are difficult radiation detection environments due to the variety of ways that illicit radiological material can be smuggled. It is also important to note that if anything is detected it is vital to cause as little disruption as possible so as to not disturb traffic or commerce flow. These points of entry can be extremely busy. Figure 31.1 is an image of the San Ysidro border crossing between the US and Mexico. In 2019, the year before the Covid-19 pandemic began, over 40,000 personal vehicles and buses, in addition to over 29,000 pedestrians, crossed this location on a daily basis into the US.[15] Other border crossing points can include rail and sea terminals. The detection regime at all border crossings must be flexible enough to scan everything that crosses. At each of these entry points it is important to have plans in place of what to do if something is detected. Even small disruptions in border traffic can cause significant delays.

Seaports

Seaports are vital for international commerce as 90 per cent of the world's trade is transported through cargo containers.[16] They are also great areas of concern for radioactive material smuggling. For example, over 500,000 cargo containers came into the Port of Rotterdam in 2021, equating to over 436 million tonnes of goods and materials.[17,18] In 2006, the US established the SAFE Port Act, which requires all containerized cargo entering the country to be screened for radiation. Due to the importance of these

FIGURE 31.1. San Ysidro border crossing from Mexico to the US.

Image courtesy of Matt Wade, 'San Ysidro Border Crossing', Wikimedia Commons, 12 August 2007. https://commons.wikimedia.org/wiki/File:SanYsidroBorderCrossing.JPG

facilities around the world, great care has gone into developing radiation detection systems and security plans for this unique environment so as not to disrupt commerce.

Radiation Detection Instrumentation and Their Use

Radiation detection instrumentation employed at borders takes many forms and sizes. It can range from large, fixed installations to small detectors that fit in the palm of a hand. Regardless of the number and type, they all work together to provide a defence-in-depth strategy that spreads detection capability while providing the best coverage possible.

Radiation Portal Monitors

The primary radiation detection instrument in border security is the RPM. These instruments are placed at choke points where vehicles and pedestrians must pass.

The base component of every RPM is a large-size scintillator either made of NaI or PVT paired with large helium-3 tubes. The scintillators are several litres in volume with the largest face ranging in width from 10 cm (for NaI) to almost 30 cm (for PVT). They can range in length from 40 cm (NaI) to greater than 1 m (PVT) (Figure 31.2). The large size provides a large surface area through which emitted photons can pass. If radiation can make it to the face of the detector, it at least has a chance to be detected. RPMs allow for

FIGURE 31.2. Different types and sizes of scintillators. The detector on the left is an NaI detector almost 40 cm high. The other two detectors are PVT with the largest being almost 1.5 m high.

Image courtesy of Texas A&M University, June 2022.

FIGURE 31.3. Vehicles queueing at the border crossing in Calexico, California. The large yellow structures are RPMs.

Image courtesy of General Services Administration, 'Border Infrastructure: Actions Needed to Improve Information on Facilities and Capital Planning at Land Border Crossings', GAO-19-534, US Government Accountability Office [online], 11 July 2019. https://www.gao.gov/products/gao-19-534.

pedestrian or vehicle traffic to pass through checkpoints without seriously hampering movement at borders. It should be noted that RPMs can also be made using arrays of HPGe systems. However, these can be ten times more expensive than scintillator-based systems.

Figure 31.3 shows a series of RPMS that are placed at the San Ysidro border crossing between the US and Mexico. The large yellow boxes on the sides of each lane are the RPMs. The RPM is on both sides of the vehicles to ensure the whole vehicle is scanned. Similar RPMs with similar placement strategies can be found at shipping ports (Figure 31.4) and at rail crossings. RPMs can also be designed for pedestrian entry points at border crossings.

There are other types of large radiation detection systems that can enhance fixed-location RPMs. Mobile RPMs (MRPMs) are used extensively near border crossings and in shipping ports. One disadvantage of most RPMs is that they are stationary, and as such, adversaries know where they are and may be aware of their capabilities. On the other hand, MRPMs can be placed along any roadway or choke point at any time. Research has indicated that MPRMs may greatly reduce a smuggler's ability to avoid detection by adding variability to a detection system's location that an adversary cannot predict.[19]

Another type of detection system being investigated is one that would be placed on spreader-bar cranes. These are the cranes that load and unload large cargo containers from ships. Such a detection system could assure that every cargo container is scanned before it is set on shore. It will also have the advantage of being subject to

FIGURE 31.4. Example of RPMs at sea ports or large amounts of commercial shipping points at borders.

Image courtesy of General Services Administration, 'Border Infrastructure: Actions Needed to Improve Information on Facilities and Capital Planning at Land Border Crossings', GAO-19-534. US Government Accountability Office [online], 11 July 2019. https://www.gao.gov/products/gao-19-534.

less background radiation interference from NORM found on the dock and inside the ports. This is because as the container is transferred from ship to shore it is tens of metres off the ground, placing it much further from the surface of the dock. In addition, because the container is being transited in close proximity to the detectors on the spreaders bar, it allows the container to be scanned longer than when moved through an RPM. Lower background noise and longer scan times are advantageous and indicate that these type of systems could detect lower amounts of radiation than current fixed-location RPMs.[20]

Personal Radiation Detectors

A common radiation detection system used by law enforcement and customs inspectors is the personal radiation detector (PRD) (Figure 31.5).[21] These small electronic devices can fit in the palm of a hand, weigh less than 450 g, and are designed to clip on the belt of the user. These radiation-sensitive instruments can be inexpensive, costing less than US$5,000. Little training is required to operate them. PRDs for untrained users can provide a simple 1–10 scale to indicate the magnitude of the radiation field. More advanced PRDs can provide digital displays indicating the radiation exposure rate or type of radiation being detected (gamma or neutron). All PRDs have visual, audible, or vibratory alarms to indicate when elevated radiation levels are detected. Some PRDs have the ability to record

FIGURE 31.5. Example of a PRD.

Image courtesy of Texas A&M University, June 2022.

and analyse gamma radiation signals to identify the material. These are known as spectral PRDs (SPRDs).

PRDs hold several advantages in border radiation detection. Being relatively inexpensive, most—if not all—personnel at a border crossing will carry these unobtrusive instruments. This allows for wide area monitoring of vehicles, pedestrians, and cargo. They are sensitive instruments capable of alarming in less than two seconds due to radiation fields ranging from 5 μR/hr to 2 mR/hr above background levels (typical background levels are near 10 μR/hr).[22] If elevated levels of radiation are detected, the instrument can be used to scan suspected items or persons to localize the radiation source.

Radioisotope Identification Devices

Radioisotope identification devices (RIIDs) are other important tools for border security agents. These devices are sophisticated radiation detection and isotope identification instruments that require little to no training to operate. If a device such as an RPM or PRD indicates a radiation source is present, a border security agent will use an RIID to better localize the source and identify it to determine its threat level. These are handheld instruments that can vary in size, shape, and mass. They range from approximately 2–8 kg in weight and can range in cost of approximately US$20,000 to over US$100,000 (Figure 31.6).[23] Each has easy-to-follow user interfaces that allow the user to search for, and identify, concealed radioactive material. Lastly, all RIIDs have the ability to save and transfer collected data in case additional analysis is required.

The main advantage of the RIID is its ability to automatically identify gamma-emitting nuclides. Gamma radiation is unique for each nuclide. What the RIID does is collect the emitted gamma signals and uses sophisticated software to analyse the data and provide the identity to the user. RIIDs can use either scintillator materials like NaI or semiconductor materials like HPGe. Scintillators are much less expensive than their semiconductor counterparts. However, semiconductor-type RIIDs can interpret more complicated gamma signals and give the user higher confidence in their results. The disadvantages of these types of RIIDs include their high cost; they usually must operate at liquid nitrogen temperatures and are much bigger (i.e. heavier) than the scintillation type. RIIDs also can incorporate other types of detectors to enhance their capabilities. Many have neutron detectors that aid in the ability of the instruments to

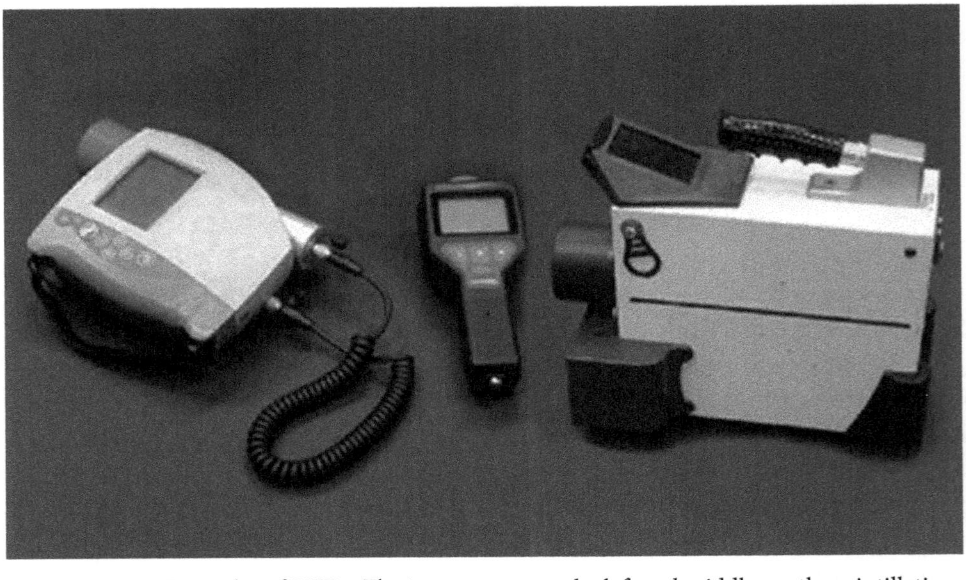

FIGURE 31.6. Examples of RIIDs. The instruments on the left and middle are the scintillation-type while the one on the right is a mechanically cooled HPGe.

Image courtesy of US Department of Homeland Security, 'Radioisotope Identification Devices (RIIDs)', TechNote, October 2009.

detect SNM. Dose rate meters can also be included in the devices to warn users if they are approaching radiation fields that may be harmful.

Training

Not only is the right technology important for border radiation monitoring, but the individuals who use the equipment must be trained. RPMs, PRDs, and RIIDs have all been designed so that non-radiation experts can use them. Many of the hand-held instruments have few buttons and most can be used with a single hand. The instruments display simplified menus so that non-experts do not get intimidated by the instruments and can easily follow prompts during critical situations. There are no standards for the type or length of training individuals must have to use these instruments, but all who do must should be required to undertake at least one to two hours of classroom work and hands-on exercises. The training should provide information on what radiation is, how it is detected, and cover radiation safety topics. It should also include hands-on training with radioactive sources so that individuals can see how the instruments work, how they react when radiation is present, and how to use the instrument's functions. Hands-on training should at least be conducted annually so that the trainees maintain familiarity with the radiation detection equipment.

Strategies for Border Monitoring

All the radiation detection equipment mentioned in this chapter must work in a collaborative system to prevent the trafficking of illicit radioactive material. In most cases radiation monitoring at borders begins with the RPM. Vehicles and cargo will move through the portals and be scanned. If an RPM registers increased radiation levels it is important that further investigation takes place. Therefore, in order to not impede traffic, the vehicle (or cargo) is directed to a secondary inspection point. Sometimes, the vehicle is directed to a second RPM for a confirmation scan. This is because normal background radiation fluctuations may cause a false reading. If a second scan does not reveal increased levels, the vehicle may be allowed to continue its journey or be directed to a secondary inspection point out of the flow of traffic. Once at a secondary inspection point border agents will scan the item that caused the alarm (person, vehicle, or cargo). In the same way as vehicles, pedestrians may pass through an RPM. If elevated levels are registered, the same procedures are used for pedestrians: confirmation scans, secondary screening, and removal from the flow of traffic.

The most efficient detector for secondary scanning is the RIID. Some RPMs will indicate the approximate location of a potential source, which helps the secondary inspection. In either case, the RIID slowly passes around the item being inspected until increased radiation levels are noted. At the location where the highest reading is registered the user will activate the identification capability of the RIID. If the identification results are not conclusive, the data can be uploaded and examined by a trained scientist through 'technical reachback'.[24] These scientists are radiation experts and will have dedicated tools to further analyse the RIID results. After a detailed analysis they will communicate with the responders regarding their conclusions and recommendations. If the results can be explained the vehicle will be allowed to continue on its journey. If not, more invasive inspections may take place.

PRDs also play a vital role in monitoring for radiation at borders. As stated previously, because they are relatively inexpensive and small, nearly all border security personnel will have a PRD. While these officials carry out their normal duties, the PRDs will continuously scan the individuals, vehicles, and cargo, and will either vibrate and/or produce an audible alarm when exposed to elevated levels of radiation. At this point, a border agent may remove the PRD from their belt, silence the alarm, and then use the instrument's display to isolate the source of the alarm. If the alarm is caused by a person, the individual may be asked to accompany the border agent to a secondary location to be interviewed. Once in this location, an RIID would be used to identify the radionuclide. Similar procedures would be executed for vehicles or cargo.

Conclusion

The need for radiation monitoring at borders is essential because the threat of illicit radioactive material trafficking is real. The IAEA established the ITDB to monitor these types of incidents; between 1993 and 2021, 320 incidents of malicious trafficking were reported. The highest percentage of interdicted material was special nuclear material—such as uranium and plutonium—that can be used to create a nuclear explosive; and the second highest percentage was industrial sources like caesium and cobalt, which can be used to make RDDs. If materials like these cross borders they could be used in criminal or terrorist activities that might threaten lives and cause severe economic consequences.

Monitoring for this illicit radioactive material is challenging. Of the four major types of radiations, only gamma rays and neutrons can be effectively detected by border security instruments. Both of these radiations travel metres in air and pass through different types of materials. However, they can also be shielded. Special nuclear material emits both of these types of radiation, while industrial sources primarily emit gamma rays. Using the proper detection system, border security agents can localize and identify gamma-emitting material.

Unfortunately, gamma rays are also emitted by NORM and medical sources. These can both be present at border crossings and interfere with gamma monitoring. Neutrons, unlike gamma rays, cannot be used to identify a material but are only emitted from man-made materials. Therefore, if they are detected at border crossings it is a clear indication that man-made radioactive material is present and nearby. Knowing these challenges of radiation detection, it is important that border security and custom officials have the proper types of radiation detection systems in place.

Radiation monitoring at borders is composed of several layers that work in conjunction with each other. The foundation of border radiation detection is the radiation portal monitor. Vehicles, pedestrians, and cargo are directed through choke points so they move near these large detection systems and low levels of radiation may be registered. Monitoring for radiation with portals does not impede the busy traffic typically found at borders or cargo ports. When radiation is detected, secondary inspection occurs with hand-held detection systems that both localize and can identify the source of radiation. This separates illicit source trafficking from the natural radiation or medical sources that can also be detected.

Notes

1. G. Anzelon, W. Hammond, and M. Nicholas, 'The IAEA's Illicit Trafficking Database Programme', in *Measures to Prevent, Intercept and Respond to Illicit Uses of Nuclear Material and Radioactive Sources*, C&S Papers Series. No. 12/P (Vienna: IAEA, 2002), 595–6.
2. IAEA, *2022 Factsheet*, IAEA Incident and Trafficking Database (Vienna: IAEA, 2021).
3. Stephanie P. Clarke, 'Radiation Detection, Response and Recovery', presentation at Control and Management of Radioactive Material Inadvertently Incorporated into Scrap Metal: Proceedings of an International Conference Held in Tarragona, Spain, 23–27 February 2009. https://www-ns.iaea.org/downloads/rw/meetings/tarragona2009/presentations-wednesday/s-clarke.pdf
4. Oak Ridge National Laboratory (ORNL). 'Nuclear Smuggling Detection and Deterrence: FY 2016: Data Analysis Annual Report', ORNL/SR-2017/91, ORNL [online], January 2017. https://info.ornl.gov/sites/publications/files/Pub72426.pdf
5. G. Knoll, *Radiation Detection and Measurement*, 4th edn (Hoboken: John Wiley & Sons Inc., 2010), 53–4.
6. Ibid., 19–26.
7. Government Accountability Office, *Managing Critical Isotopes: Weaknesses in DOE's Management of Helium-3 Delayed the Federal Response to a Critical Supply Shortage.* Washington: US Government Printing Office, 2011).
8. D. Kramer, 'DOE Begins Rationing Helium-3', *Physics Today*, 63/6 (2010), 22–5. https://doi.org/10.1063/1.3455244
9. US Government Accountability Office (GAO), *Technology Assessment Neutron Detectors, Alternatives to Using helium-3*, GAO 11-753 (Washington, DC: Government Accountability Office, 2011).
10. National Council on Radiation Protection and Measurements (NRCP), 'Report No. 165, Responding to a Radiological or Nuclear Terrorism Incident: A Guide for Decision Makers', January 2011.
11. IAEA, *Detection of Radioactive Materials at Border*, IAEA-TECDOC-1312 (Vienna: IAEA, 2002).

12. J. O. Lubenau and J. G. Yusko, 'Radioactive Materials in Recycled Metals: An Update', *Health Physics*, 74/3 (1998), 293–9.
13. IAEA, 'About TRACE', n.d. https://nucleus.iaea.org/sites/FLONetwork/SitePages/About%20TRACE.aspx
14. Please note that the discussion in the chapter focuses on official border entry points. There are of course many unofficial routes into most countries, many of which are also monitored for the presence of radiation.
15. US Department of Transportation, 'Border Crossing/Entry Data', 1 March 2020. https://www.bts.gov/browse-statistical-products-and-data/border-crossing-data/border-crossingentry-data
16. R. Gil Kerlikowske, 'Tenth Anniversary of SAFE Port Act', U.S. Customs and Border Protection [online], 6 October 2016. https://www.cbp.gov/newsroom/blogs/tenth-anniversary-safe-port-act
17. Port of Rotterdam, 'Throughput Port of Rotterdam', February 2022. https://www.portofrotterdam.com/sites/default/files/2022-02/port-of-rotterdam-throughput-2021.pdf
18. Port of Rotterdam, 'Facts and Figures', June 2021. https://www.portofrotterdam.com/sites/default/files/2021-06/facts-and-figures-port-of-rotterdam.pdf
19. Ryan Coogan, Craig Marianno, and William Charlton, 'A Strategic Analysis of Stationary Radiation Portal Monitors and Mobile Detection Systems in Border Monitoring', *Nuclear Engineering and Technology*, 52/3 (March 2020), 626–32.
20. Matthew D. Grypp, Craig M. Marianno, John W. Poston, and Gentry C. Hearn 'Design of a Spreader Bar Crane-Mounted Gamma Ray Radiation Detection System', *Nuclear Instruments and Methods in Physics Research*, Section A. 743 (April 2014), 1–4.
21. US Department of Homeland Security, 'Personal Radiation Detectors (PRDs)', *TechNote*, April 2014.
22. National Committee on Radiation Instrumentation, 'American National Standard for Evaluation and Performance of Radiation Detection Portal Monitors for Use in Homeland Security', N42, ANSI N42.35-2016, 17 March 2016.
23. US Department of Homeland Security, 'Radioisotope Identification Devices (RIIDs)', *TechNote*, October 2009.
24. W. M. Buckley and R. W. Allen, 'The Importance of Technical Reachback in the Adjudication of Radiation Alarms', LLNL-CONF-411428 (Livermore, CA: Lawrence Livermore National Laboratory, 2009).

Suggested Reading

Anzelon, G., Hammond, W., and Nicholas, M., (2002), 'The IAEA's Illicit Trafficking Database Programme', in *Measures to Prevent, Intercept and Respond to Illicit Uses of Nuclear Material and Radioactive Sources*, C&S Papers Series, No. 12/P (Vienna: IAEA), 595–6.

Coogan, Ryan, Marianno, Craig, and Charlton, William, (2020), 'A Strategic Analysis of Stationary Radiation Portal Monitors and Mobile Detection Systems in Border Monitoring'. *Nuclear Engineering and Technology*, 52/3, 626–32.

IAEA, (2021), *IAEA Incident and Trafficking Database: 2022 Factsheet* (Vienna: IAEA).

IAEA, (2002), *Detection of Radioactive Materials at Border*, IAEA-TECDOC-1312 (Vienna: IAEA).

US Department of Homeland Security, (2009), *Radioisotope Identification Devices (RIDDs), TechNote* (Washington, DC: DHS).

US Department of Homeland Security, (2014), *Personal Radiation Detectors (PRDs), TechNote* (Washington, DC: DHS).

CHAPTER 32

THE EVOLUTION OF THE LEGAL AND TECHNICAL ASPECTS OF NUCLEAR FORENSICS

VITALY FEDCHENKO

Introduction

Forensic science is commonly referred to as the application of science to law, or the utilization of the sciences in legal proceedings.[1] A subset of this is nuclear forensic science, also referred to as 'nuclear forensics'. The International Atomic Energy Agency (IAEA) defines nuclear forensics as 'a discipline of forensic science involving the examination of nuclear and other radioactive material, or of other evidence that is contaminated with radionuclides, in the context of legal proceedings'.[2]

The legal context in which nuclear forensics is applied defines its scope and function. The most common application of nuclear forensics is to combat the illicit trafficking of nuclear or other radioactive materials. Illicit trafficking of such materials, also known colloquially as 'nuclear smuggling', is defined by the IAEA as criminal or unauthorized acts involving nuclear or other radioactive materials outside of regulatory control, such as their 'import, export, possession, sale, delivery, movement, use, storage, disposal or transfer'.[3] As such, nuclear forensics is seen today as one of the 'tools' of nuclear security, which seeks to prevent, detect, and respond to a nuclear security event initiated by non-state actors.

Nuclear forensics is most often employed during the response phase, for example, by contributing to investigations and prosecutions through confirming that an intercepted smuggled material is indeed nuclear or that it is consistent with specific material found elsewhere. Nuclear forensics also can contribute to prevention, for example, by identifying a specific facility where the control over an illicitly trafficked material was lost, and therefore pinpointing where in the nuclear fuel cycle the material protection, control, and accounting (MPC&A) measures should be strengthened.[4]

From a Crime Scene to Court

A scene of a nuclear security event, such as an interception of nuclear material being smuggled, or a radiological dispersal device (RDD) explosion site, will likely be designated as a 'radiological crime scene'. This is, in essence, a regular crime scene but one that contains some nuclear or other radioactive material in intact or dispersed form. Effective response to nuclear security events requires a dedicated national response system to be developed in advance and documented in a national response plan. Such a plan should contain, among many other things, a description of procedures on the conduct of radiological crime scene management (RCSM), as well as on the conduct of three types of forensic analysis: traditional evidence, radioactive evidence, and traditional evidence contaminated with radionuclides (see Figure 32.1). The IAEA's guidance documents refer to the totality and sequence of these procedures as the 'nuclear forensics model action plan'.[5]

The point of a model action plan is that the sequence of RCSM and the three kinds of forensics must produce findings, combined in forensic conclusions, that are both useful for, and admissible in, legal proceedings. For the findings to be useful to investigative authorities, the RCSM and forensics processes must be guided by specific questions formulated by those authorities. For example, a prosecutor may wish to know if radioactive material found in a suspect's car matches material stolen from a nuclear facility. The whole nuclear forensics process will then be focused on answering that question, avoiding unnecessary types of analysis.

The Process of RCSM and Nuclear Forensics

According to the model action plan (see Figure 32.1), once the response personnel arrive at the scene of a nuclear security event, an incident commander conducts reconnaissance (crime scene assessment), which includes: establishing scene control; ensuring the safety of the public; protecting the environment and personnel from all hazards; locating radioactive areas within the crime scene; and finding radioactive materials and other evidence. By that time, the investigative authority is expected to have formulated questions about the event, and so the event action plan is developed on the basis of those questions and the crime scene assessment.[6]

The event action plan describes the content and sequence of investigative actions at a crime scene. This includes: safety and security measures; communication procedures; rules of evidence and the chain of custody procedures; analytical measurements that have to be taken on site; an evidence collection plan; the categorization of materials; and the packaging and transportation to a lab or a holding facility.[7] Categorization is the quick assignment of the material of interest to a pre-defined group, thus determining its further handling, storage, and transport. An evidence collection plan ensures the safe and effective collection of evidence and maintenance of the chain of custody.

Once evidence has been seized at the crime scene, and possibly transported to a temporary holding site, the investigative authorities consult with laboratories, which have been previously designated in the national response planning as capable of conducting nuclear and other forensic examinations. Together, they formulate a forensic examination plan that ensures that the needs of the investigation are met (i.e. that the investigative questions are answered). This plan defines the sequence of analyses (for example, traditional forensic techniques such as the lifting of fingerprints is done first, and nuclear forensics afterwards, or vice versa), the laboratories that conduct those analyses, and the expected time required. In particular, the plan defines at what point the nuclear or other radioactive materials are delivered to a nuclear forensics laboratory for analysis.

In a nutshell, a nuclear forensics laboratory is expected to conduct the characterization of the material, interpret the results of that characterization, and formulate its nuclear forensic findings on the basis of those two processes (see the 'Nuclear forensics' section in Figure 32.1).[8] This is done on the basis of a nuclear forensics analytical plan, which is, in essence, a translation of the forensic examination plan into a correct sequence of specific analytical techniques applied to the sample in order to answer the questions of the investigative authority. A nuclear forensics analytical plan is written by the nuclear forensics laboratory in cooperation with the investigative authority. It is possible that the characterization and interpretation stages, as described in the following sections, will yield unexpected results that would justify changing a nuclear forensics analytical plan, for example, by necessitating adding more measurements at the characterization stage. In that event, the nuclear forensic process becomes iterative, that is, requiring the repetition of all or some of its steps (see Figure 32.1).

Characterization is the determination (i.e. measurement) of a sample's characteristics, of which there are four categories: physical characteristics (for example, the diameter of a nuclear fuel pellet); chemical composition; elemental composition (for example, the fuel pellet is made of uranium dioxide with trace amounts of gadolinium); and isotopic composition of the chemical elements present (for example, uranium in the recovered fuel pellet contains 4 per cent of an isotope uranium-235). Characterization provides a description of the sample in the form of raw data.

The process then moves to interpretation—through correlating the characteristics of the sample with information on known methods of material production, handling, and use. At this point, the raw, meaningless data collected at the characterization stage is transformed into meaningful information. For example, characterization might reveal that a uranium sample contains 0.37 per cent of the isotope uranium-236, as well as traces of plutonium. Interpretation of this data would lead to the conclusion that this material was most likely irradiated in a nuclear reactor.

Finally, once the characterization and interpretation processes are concluded, the nuclear forensic laboratory formulates the nuclear forensic findings—in an output document that is defensible in court. This must demonstrate the implementation of concepts and procedures, such as the rules of evidence, quality assurance, and quality control. An investigative authority collates nuclear forensic findings

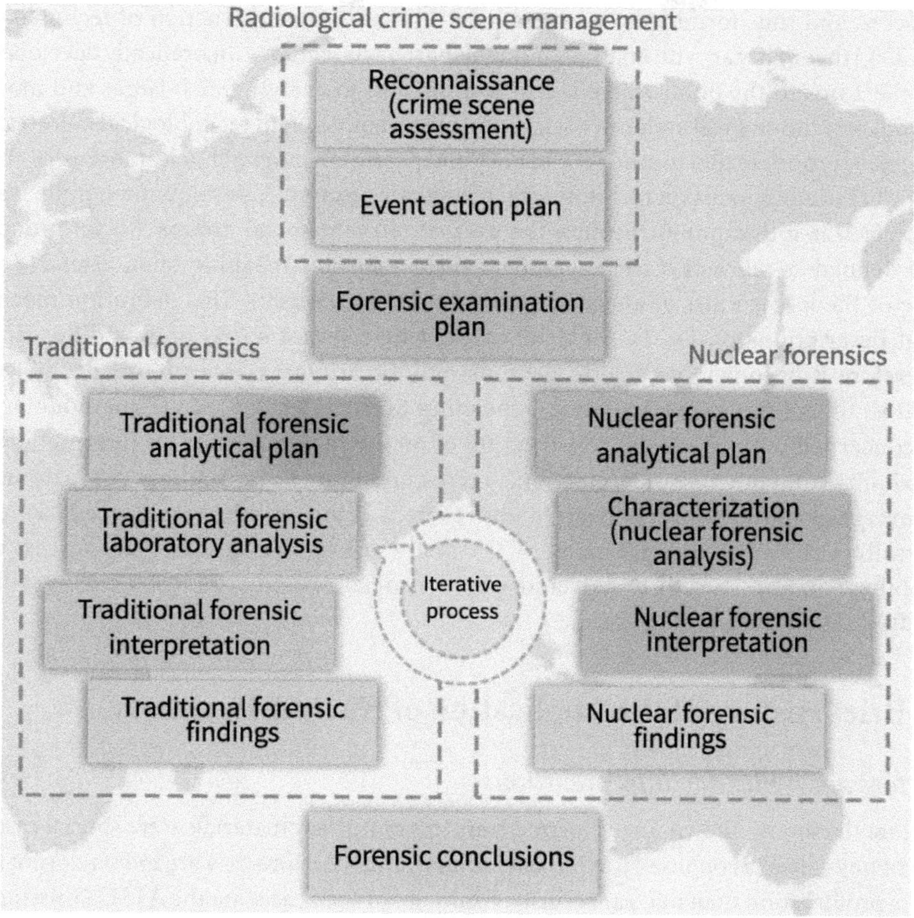

FIGURE 32.1. Nuclear forensics model action plan.

Figure courtesy of the IAEA, *Introduction to Nuclear Forensics*, IAEA Nuclear Security E-Learning Course (Vienna: IAEA, 2022). https://elearning.iaea.org/nsns/training

with those of the traditional forensics laboratory, as well as the analysis of evidence contaminated with radionuclides, in order to produce forensic conclusions. These can then be combined with other information obtained by the investigative authorities or received from other entities in order to develop associations between material, production processes, people, places, and events of interest to the investigation.[9]

The History and Social Construction of Nuclear Forensics

Sociology of technology, a relatively new discipline that emerged in the 1980s, studies technology and society as 'tied together, co-constructed in a symbiotic, inseparable process, and thus forming a "seamless web".[10] The social construction of technology (SCOT) theory began with analysis of singular artefacts, and was theoretically developed by the 2010s to the point where it can be employed to analyse much larger and more complex technological systems, sociotechnical ensembles, and technological cultures.[11] It is an appropriate tool to study the history and potential future of nuclear forensics.

SCOT studies focus on technological change (for example, development of nuclear forensics as a discipline) through the eyes of relevant social groups. Social groups are defined as relevant if their members share a specific meaning of an artefact (or one of the larger units of analysis, as mentioned previously). This definition means that descriptions of, for example, an artefact by relevant social groups differ, and therefore it becomes possible to demonstrate the artefact's 'interpretive flexibility', that is, different ways of assigning a meaning to the artefact.[12] SCOT methodology is concerned with the social construction of an artefact, following its interpretative flexibility from the time of its creation, and documenting how meanings compete and converge in a process of stabilization until there is a closure—the emergence of a universally accepted meaning. Finally, SCOT methodology is focused on the process of stabilization and tries to understand why the social construction took one specific course, rather than another.

A Brief History of the Application of Nuclear Forensics

Before 2001: Flexible Interpretations

Before the 1990s, known unauthorized transfers of nuclear materials were special cases involving states (as opposed to non-state actors), and therefore they are more relevant to non-proliferation than nuclear security. Examples of such cases are the ALSOS missions (the removal of metallic uranium and uranium oxides from Germany and Austria in 1945 by the Soviet Union and the United States; US) and 'the NUMEC affair' (an alleged theft of highly enriched uranium (HEU) from a Nuclear Materials and Equipment Corporation (NUMEC) plant in the US and its transportation to Israel in the 1960s).[13] Despite the application of analytical techniques to identify the characteristics and origins of these and other materials, the term 'nuclear forensics' was not notably utilized before the 1990s. Before this time, nuclear forensics activities tended to fall within the related fields of technical intelligence and treaty verification.[14]

By the time of its collapse in December 1991, the Soviet Union had developed an enormous nuclear fuel cycle at its territory, with very large stocks of nuclear and other radioactive materials. Dissolution led to rapid and profound economic decline and poverty for most of the population, including people working at nuclear fuel cycle facilities. This process also resulted in the degradation of Soviet political and security controls, with hundreds of reports of nuclear smuggling happening in the countries of the former Soviet Union.[15] While some of these reports turned out to be hoaxes, a significant number of incidents did occur, including at least 18 involving HEU and plutonium.[16]

Over the subsequent decade, the European Union (EU), the US, and the Russian Federation worked to combat this threat through investigating illicit trafficking cases and tracing the origin of intercepted nuclear materials. However, such efforts were driven largely by non-proliferation concerns, rather than the possibility of nuclear terrorism. The 1996 Moscow Nuclear Safety and Security Summit, attended by the G7 heads of state and government and co-chaired by the presidents of Russia and France, issued a declaration stating that 'illicit trafficking of nuclear material is a public safety and non-proliferation concern'.[17] Notably, the threat of terrorism was mentioned in the subsidiary summit documents, but not in the main declaration. Contemporary scholars also saw nuclear smuggling as mainly a proliferation threat.[18] By 1996, nuclear forensics existed as a discipline, but it was not yet being constructed by policy makers and diplomats as a counter-terrorism or a nuclear security tool.

The beginnings of nuclear forensics in the US, the EU, and Russia were largely influenced by two relevant social groups that can be termed 'scientists' and 'policy makers and diplomats'. A third social group, 'law enforcement officials', was present from the beginning, but had less influence at that time. According to one first-hand account, contemporary nuclear forensics in the US began from an interaction between a prominent policy maker and scientist, Hans Mark, and staff at the Lawrence Livermore National Laboratory (LLNL) in 1991.[19] Mark posed a question that was novel at the time: what can be found out about the origin and history of the nuclear material (or nuclear explosive device) from the analysis of its characteristics? Already in 1991 it was understood that nuclear forensics may have to deal with three separate scenarios: nuclear and other radioactive materials; nuclear explosive devices (in exploded or unexploded form); or radioactive dispersal devices.

During the early 1990s, the LLNL scientists 'gave numerous briefings to US Government officials about the need to develop a nuclear forensics capability to address nuclear smuggling'.[20] These briefings were done explicitly for the two relevant social groups outlined previously: 'scientists' and 'policy makers and diplomats'. In parallel, international discussions concerning the combatting of nuclear smuggling were happening in the framework of the G7 summits in Naples, Italy in 1994, and Halifax, Canada, in 1995.

Due to proximity to the former Soviet territory, most illicit trafficking cases were happening in Europe. The first known nuclear smuggling cases were reported in 1991 in Switzerland and Italy, with subsequent incidents in Germany, Czech Republic, Hungary, and other European states.[21] Unlike in the US, where reliance on experience of the national nuclear weapons laboratories was habitual, in the European context that question could be best answered by a well-established laboratory with experience in nuclear material analysis and data interpretation from the nuclear safeguards and material science areas, namely the European Commission's Joint Research Centre's Institute for Transuranium Elements (ITU) in Karlsruhe, Germany.

Nuclear material seized from nuclear smuggling was analysed at the ITU for the first time in March 1992. The so-called 'Find-1' materials were intercepted in Augsburg, Germany on 5 March 1992, and determined by the ITU to be fuel pellets for a Russian-type graphite moderated reactor (RBMK), with the possible origin being either the Ulba Metallurgy Plant in Kazakhstan or Elektrostal in Russia. Since then, the ITU (later renamed JRC Karlsruhe) has become the foremost European nuclear forensic science institution.

In 1995, the US Department of Energy (DOE) sponsored the International Conference on Nuclear Smuggling Forensic Analysis, attracting 'an unusual mix of experts in law enforcement, laboratory science and governmental policy.'[22] As a result of that deliberation, the Nuclear Smuggling International Technical Working Group, which later became the Nuclear Forensics International Technical Working Group (ITWG), was created. The first meeting of the group was convened in Karlsruhe in 1996, and the meetings have since continued annually. The ITWG has now become the focal point and mechanism for the development of nuclear forensics.

In the 1990s, both nuclear forensics and the ITWG were developed by 'scientists' with a mandate from 'policy makers and diplomats' in the presence of 'law enforcement officials'. Applying SCOT theory, it is possible to show that each of the three groups attributed their own meaning and purpose to nuclear forensics (and the ITWG). The scientists' view of nuclear forensics is captured in the earliest known definition of nuclear forensics put forward in 1996:

> Nuclear forensic analysis is the process by which intercepted illicit nuclear materials and their associated materials […] are analyzed to provide clues to attribution. The goal of nuclear forensic analysis is to identify attribution indicators in an interdicted nuclear materials sample or its surrounding environment […]. These indicators arise from known relationships between material characteristics and illicit activity.[23]

For scientists at that time, nuclear forensics was a branch of science, characterizing intercepted materials, interpreting data, and providing an 'essential input to attribution assessment'.[24] In 1997, the ITWG scientists began to develop an 'action plan'—a general guidance and a step-by-step description of a nuclear forensics process. At a meeting in Vienna in June 2000, the ITWG adopted a reference model action plan for the first time.[25] The ITWG received political support when nuclear forensics was first recognized (by the 'diplomats and policy makers' social group) as an element of the response to nuclear smuggling events at the 1995 G7 Summit in Halifax and the 1996 Moscow Nuclear Safety and Security Summit. Later the ITWG was endorsed by the Non-Proliferation Experts Group (NPEG) of the Group of Eight (G8) in May 1999, and by the G8 Nuclear Safety and Security Group in 2005.[26]

Analysis of early ITWG documents shows that this group saw nuclear forensics at that time as a credible tool providing timely answers to policy makers' questions about a smuggling incident, specifically the questions concerning the source attribution (for example, 'where did the material originate?') and route attribution (for example, 'is this an isolated event or series of shipments?').[27] As discussed previously, nuclear smuggling at that time was mostly seen as a phenomenon representing a risk of nuclear proliferation, although risks associated with terrorism and organized crime were also considered. The third group, law enforcement officials, have regarded nuclear forensics as a sub-discipline of forensic science, and therefore their concern has been with incorporating into the work of the ITWG dimensions such as the rules of evidence and chain of custody. As a result, nuclear forensics had 'interpretive flexibility' at that time; it could be described differently depending on which of the three relevant social groups one would belong to.

Post-9/11: Concept Stabilization and Involvement of International Organizations

Even though the attacks of 11 September 2001 ('9/11') did not involve radioactive materials, they demonstrated the risk of mass-casualty terrorism, including the possibility of nuclear terrorism, and precipitated the development and formalization of nuclear security as a discipline.[28] Nuclear terrorism was also formally described in the 2005 International Convention for the Suppression of Acts of Nuclear Terrorism (ICSANT). Another international initiative was the 2006 Global Initiative to Combat Nuclear Terrorism (GICNT).[29] GICNT partner nations endorsed a Statement of Principles, one of which was essentially a voluntary commitment to develop nuclear forensic capabilities.[30] These initiatives served to conceptualize nuclear security as a defence against nuclear terrorism and placed nuclear forensics firmly within that context. The IAEA subsequently played a major role in diminishing the interpretive flexibility of nuclear forensics and stabilizing its definition.

The results of almost a decade of work by the ITWG on defining nuclear forensics and developing the model action plan were published in 2004.[31] The IAEA used that publication, as well as the expertise of many ITWG participants, as a major input into the preparation of its first Nuclear Security Series (NSS) guidance document on nuclear forensics. Published in 2006, the document codifies the discipline internationally and defines nuclear forensics as 'the analysis of intercepted illicit nuclear or radioactive material and any associated material to provide evidence for nuclear attribution', where attribution refers to 'the process of identifying the source of nuclear or radioactive material used in illegal activities, to determine the point of origin and routes of transit involving such material, and ultimately to contribute to the prosecution of those responsible'.[32] Unlike the 1996 definition, this one explicitly refers to prosecution—a sign of the influence of law enforcement officials.

Further stabilization of nuclear forensics occurred as part of the four Nuclear Security Summits, which took place from 2010 to 2016. By the time of the first summit in 2010, nuclear forensics had developed into a mature scientific discipline routinely used to provide nuclear forensic science findings for the process of attribution. Nevertheless, a comparison of documents from the 2010 summit with documents from later summits and from the work of GICNT shows that nuclear forensics was initially perceived as laboratory-driven, and that perceptions shifted towards placing nuclear forensics into the context of investigations and prosecutions by the final 2016 summit.[33]

In terms of the three relevant social groups, unlike in the case of the ITWG, where most of the work is done by nuclear forensics scientists, GICNT is run by diplomats and policy makers, with participation of law enforcement officials. In September 2010, GICNT established the Nuclear Forensics Working Group (NFWG). The NFWG educates policy makers and decision makers about nuclear forensic science; conducts workshops and tabletop exercises to increase understanding of the benefits of implementing nuclear forensic capabilities; and promotes the exchange of best practices. In 2012, the NFWG published a guidance document, 'Nuclear Forensics Fundamentals for Policy Makers and Decision Makers', which, for the first time, explicitly defined nuclear forensics as a subdiscipline of forensic science in general. The document takes the definition of 'traditional' forensics ('comprehensive scientific analysis of physical, biological, behavioural, and documentary evidence in the context of civil, criminal or international law') and specifies the kind of evidence, resulting in the following language: 'Nuclear forensics is the comprehensive scientific analysis of nuclear or other radioactive materials or of evidence contaminated with radioactive material in the context of civil, criminal, or international law.'[34]

In line with these developments at GICNT, the treatment of nuclear forensics changed at the 2012 Nuclear Security Summit. The paragraph of the summit's Communiqué dedicated to nuclear forensics encourages combining 'the skills of both traditional and nuclear forensics through the development of a common set of definitions and standards'.[35] This part of the Communiqué is also in line with previous work done by the Netherlands, which, in preparation for the 2012 summit, tabled a White Paper on nuclear forensics, aimed at strengthening 'the links between traditional and nuclear forensics through the development of a common set of definitions and standards'.[36]

The 2014 Nuclear Security Summit further stabilized the concept of nuclear forensics through the integration of nuclear and traditional forensics. The 2014 summit Communiqué also contained language on nuclear forensics that welcomed the development of traditional forensic science methods and encouraged 'connecting and enhancing traditional and nuclear forensic capabilities'.[37] In addition, the Netherlands put forward a Joint Statement, supported by 23 other states, that proposed a 'Nuclear Forensics Lexicon'—a glossary of pertinent terms publicly available as a mobile telephone app. Nuclear forensic science is defined there as '[a] discipline of forensic science ... the examination of nuclear or other radioactive material, or of other evidence that is contaminated with radioactive material, in the context of legal proceedings, including national or international law or nuclear security'.[38] The Dutch definition is very similar to the GICNT NFWG definition in that they both classify nuclear forensics as a chapter of wider forensic science traditionally applied in criminal law.

Around the same time, the IAEA was finalizing the revision of its Implementing Guide on nuclear forensics, previously published in 2006. The revision, published in 2015, is titled 'Nuclear Forensics in Support of Investigations' (as opposed to 'Nuclear Forensics Support' in 2006), and contains a definition of the discipline that is very similar to the one in the Dutch lexicon: 'the examination of nuclear or other radioactive materials, or of evidence that is contaminated with radionuclides, in the context of legal proceedings under international or national law related to nuclear security'.[39]

A comparison of the IAEA's 2015 definition with the one from 2006 shows that the concept of attribution is not used any more, and the emphasis is placed on the needs of law enforcement and criminal prosecution. Influenced by the same process of concept stabilization that produced the GICNT and the Dutch definitions, the IAEA's 2015 definition firmly placed nuclear forensic science as an integral part of legal proceedings. The concept of nuclear forensics was stabilized by the final 2016 Nuclear Security Summit. The summit documents focused on strengthening political support for the discipline in general and on its practical implementation and sustainability.[40]

Techniques and Methods of Nuclear Forensics

Destructive and Non-Destructive Analytical Techniques for Material Characterization

When initiating a nuclear forensic investigation, the list of material characteristics subject to measurement at the characterization stage is not necessarily known in advance. In some cases, an obvious set of basic measurements may be sufficient but, in others, the characterization process will continue after it has been informed by the interpretation of initially obtained results, and new, perhaps unexpected, measurements will have to be performed. Unlike other scientific disciplines dealing with predefined sample types, nuclear forensic investigation may demand an analysis of almost anything—from gases and dust particles to wines and human bodily excretions. Regardless of the form of the material in the sample, there are four categories of characteristics that can be obtained at the characterization stage: physical, chemical, elemental, and isotopic. Various analytical techniques (i.e. measurement methods) and the appropriate equipment are used to extract characteristics from a sample.[41] The analytical techniques can be non-destructive or destructive.

Non-destructive analysis (non-destructive assay; NDA) is the measurement of sample characteristics without causing significant physical or chemical changes to the material. It is carried out by measuring the radiation or heat emitted from the sample and comparing that emission to a calibration based on essentially similar material whose contents are known. NDA can be passive or active. Passive NDA measures spontaneously emitted radiation or heat. For example, gamma spectrometers register gamma radiation emitted from a sample and measure its energy, which allows identification of the isotope whose decay has led to the emission of the photon with this particular energy. In active NDA the emission to be measured is stimulated by the measurement instrument, for example by a beam of neutrons.

Destructive analysis (destructive assay, DA) is the measurement of the content and elemental or isotopic concentration of the nuclear material in a sample that involves a change in the material's physical and chemical form (i.e. destruction). It is normally conducted after the sample has been processed in some way into a state suitable for the specific analysis (for example, by dissolution in acid). The prepared material is then used in, and consumed by, the analysis. DA is usually much more sensitive and precise, but also more expensive and time-consuming, than NDA.[42]

Tools and Methods of Nuclear Forensic Interpretation

The purpose of the nuclear forensic interpretation stage is to convert the raw data produced at the characterization stage into meaningful information. This is done using the essential concept of a nuclear forensic signature, which is defined as 'a set or sets of data characteristics of a given sample of nuclear or other radioactive material that may enable the sample to be identified as being consistent with, or as not being consistent with, particular nuclear or other radioactive material'.[43] That is, a signature is a combination of characteristics that enable the material to be identified. The choice of signatures depends on the questions that the investigative authority has put forward, and the nuclear forensic analytical plan. A simple measurement of the size and shape of a fuel pellet might be enough to determine its producer, if it is known that pellets of the measured shape are only manufactured at a certain facility. Additional signatures must be sought if the investigative authority is interested in, for instance, the manufacturing date.[44]

There are two types of signature: comparative (also known as empirical) and predictive. Predictive signatures are produced from modelling chemical and physical processes happening in the nuclear or other radioactive materials or involved in the nuclear fuel cycle. The two most important examples of predictive signatures are radioactive material age determination (i.e. calculation of time passed since the latest chemical purification from ratios of 'parent' and 'daughter' isotopes) and reconstruction of the history of material irradiation in a nuclear reactor (using reactor modelling software).[45]

Comparative signatures are discovered by comparison with reference materials and already existing data, for example, in scientific literature or in a national nuclear forensics library (NNFL). An NNFL is a national system for the identification of nuclear and other radioactive materials found out of regulatory control, as recommended by the IAEA.[46] An NNFL is essentially an organized collection of subject matter expertise and reference information on nuclear and other radioactive materials produced, used, or stored within a state that may be used to identify these materials.

Good Practices in Implementing Nuclear Forensics

The IAEA's Implementing Guide 'Nuclear Forensics in Support of Investigations' contains best practice guidance for the development and implementation of a national nuclear forensic capability to support prevention and response to a nuclear security event.[47] The document recommends that national planning for response to nuclear security events should incorporate the model action plan, and that states should define roles and responsibilities and ensure that expertise, instrumentation, and procedures are in place, including procedures for the safe and secure storage and transportation of nuclear or other radioactive materials involved in the nuclear security event.

The IAEA makes a number of further recommendations. First, a significant part of internationally recognized best practice in nuclear forensics is focused on international cooperation in terms of requesting, receiving, or providing assistance, either in capability development or in joint investigation of a case. Nuclear forensic capability development assistance can include, among other elements: awareness raising at the national level of the benefits of nuclear forensics; the development of education and expertise; building of infrastructure; development of legal and regulatory frameworks; operational support; and assistance with procurement of specialized equipment. Development assistance is provided either bilaterally by donors such as the US and EU, or through multiple programmes coordinated by the IAEA.

Second, assistance during all stages of investigating a nuclear security event may be set up through international organizations or bilateral or multilateral arrangements. Some states may choose not to develop and maintain national capabilities to perform full characterization of nuclear or other radioactive material or of evidence contaminated with radionuclides. This is because they may judge that the specialized tools necessary may only be required in a handful of rare cases, and that it is logistically, financially, and politically easier to establish arrangements with foreign laboratories (national or international, for example, JRC Karlsruhe), which can be activated in a rare case of a nuclear security event.

Multilateral cooperation in nuclear forensics happens through four international organizations that promote awareness of the importance of nuclear forensics and provide, on request, various forms of support:

1. GICNT focuses on political leadership in its partner states and their capability to develop and maintain national nuclear forensic capabilities, through raising awareness, fostering inter-governmental relationships, conducting joint exercises, and promoting best practices.
2. The IAEA develops nuclear security guidance documents, including on nuclear forensics, and assists states in applying that guidance through training or joint research.
3. The International Criminal Police Organization (Interpol) supports national police organizations, including on matters of combatting nuclear terrorism (nuclear security), facilitates the exchange of information, conducts intelligence analysis, delivers training, and is able to provide operational support during a nuclear security event.
4. The ITWG works to develop a preferred approach to nuclear forensic investigations that is widely understood and accepted as credible.

Third, for the forensic findings and conclusions to be admissible, forensics must follow clearly defined procedures in compliance with the rules of evidence (including the maintenance of the chain of custody), as well as maintain confidence in analytical results. Rules of evidence may differ depending on the country where the investigation is being conducted, taking into consideration that country's legal system and national legislation. Chain of custody is defined as the procedures and documents that account for the integrity of physical evidence by tracking its handling and storage, from its point of collection to its final disposition.[48]

Synergies between National Nuclear Forensic Capabilities and other Capabilities Available

The 'Nuclear Forensics in Support of Investigations' document points out that the creation of nuclear forensic capabilities

> should begin by identifying existing capabilities, including facilities that are already established and relevant expertise that is already used for other purposes, and creating mechanisms for their use in an investigation. Relevant capabilities may exist, for example, at radiation protection institutions, [. . .]. Some states may be able to use experience or infrastructure created to help to verify compliance with, or adherence to, international treaties[. . .].[49]

This suggestion is based on the fact that there are no measurement techniques or even methods of interpretation of their output that are unique to nuclear forensic science. Measurement techniques used for nuclear forensics are widely used in other applications in areas 'adjacent' to nuclear security, for example, IAEA safeguards, international and national technical means of verification of multilateral and bilateral treaties, and nuclear intelligence.

In many countries, a significant part of the nuclear forensic capabilities available derive from science and measurement techniques developed historically. Some of the earliest known analyses of radioactive material for security purposes happened in the early 1940s in the US. In order to monitor the potential operation of Germany's prospective plutonium-producing reactors, in 1944 the US searched for radioactive isotopes in environmental samples (air, water, and wine).[50] In 1945, the US conducted the first analysis of microparticles in the atmosphere to verify the results of nuclear explosions, and this procedure was successfully used to detect the first Soviet nuclear test in 1949.[51] These techniques continued to be developed in many countries during the Cold War and later 'spilled over' from nuclear intelligence to non-proliferation, disarmament and, later still, nuclear security and nuclear forensics. Table 32.1 gives examples of the radioactive material analysis use in a range of applications.

Table 32.1. Material analysis methods for nuclear security and treaty application.[a]

Framework[b]	Typical sample form, content (and source)	Information to be inferred on the material or item's history
Non-proliferation and disarmament		
NPT (environmental sampling for safeguards)	Particles (on swipes from material handling areas)	Age, production process (consistency with declaration)
	Air, water, sediments, vegetation, soil, biota	Age, production process (consistency with declaration)
Partial Test Ban Treaty	Particles and gases (weapon debris)	Nuclear explosive origin of debris, their age and location, especially if leaked from an underground test
CTBT (IMS and on-site inspections)	Particles and gases (weapon debris)	Nuclear explosive origin of debris
Fissile Material (Cut-Off) Treaty (if and when negotiated)	Noble gases (reactors, isotope production, or reprocessing facilities)	Origin and age of nuclear materials and fuel cycle effluents
	Bulk graphite samples (shut-down plutonium-producing reactors)	Reactor's lifetime plutonium output
Nuclear security		
Prosecution in an illicit trafficking case	Nuclear or other radioactive materials, items, or bulk form (nuclear fuel cycle facilities)	Age, production process, manufacturer
Nuclear terrorism event attribution	Particles and gases (weapon debris), particles, and contaminated objects (RDD)	Design features of explosive device, material used, explosion yield, device, and material origin
	Human body, excretions and remains (RDD and poisoning)	Identity and origin of radioactive material
Nuclear intelligence or radiation protection		
Monitoring of foreign explosions	Particles and gases (weapon debris)	Explosive device's characteristics Total radioactivity of fallout
Monitoring of foreign facilities and materials	Noble gases (reactors, isotope production, or reprocessing facilities)	Nuclear material production Early warning of nuclear and radiation accidents or nuclear terrorism events
	Particles in man-made media (e.g. wine or clothing), in air, water, sediments, vegetation, soil, and biota	

[a] Table courtesy of the author. See V. Fedchenko, 'Nuclear Material Analysis for Forensic and Other Security Purposes', presented at 53rd Annual Meeting of the Institute of Nuclear Materials Management (INMM 53), Orlando, 10–14 July 2012 (2012), 4.

[b] CTBT = Comprehensive Nuclear- Test-Ban Treaty; IMS = International Monitoring System; NPT = Non-Proliferation Treaty; RDD = Radiological Dispersal Device.

Conclusion

The chapter has discussed the social construction of nuclear forensics, interpretive flexibility of the discipline, and the process of stabilization of its meaning and purpose over the years. At the time of writing, nuclear forensics is understood at an international level as a discipline pertinent to nuclear security and intended to support investigations of nuclear smuggling or nuclear terrorism cases. The future of the discipline may depend on two kinds of potential developments.

First, the three relevant social groups identified—scientists, policy makers and diplomats, and law enforcement officials—are by far the most relevant at an international level, as is reflected in the list of four international organizations dealing with nuclear forensics issues: GICNT, Interpol, IAEA, and ITWG. However, current practice in at least some countries shows that implementation of nuclear forensics at a national level will involve a wider set of relevant social groups including, for example, national nuclear regulatory authorities and national emergency response teams. National experience, as well as an outcome of the exercises conducted by GICNT, shows that those additional social groups may still have diverging interpretations of nuclear forensics and its place in response to nuclear security events. One reported approach sometimes utilized at a national level, without necessarily a direct link to existing national legislation or international recommendations, is the prioritization of radiation safety regarding the discovered source of radiation by transportation of it to a safe location; such an approach may not facilitate the development of criminal investigation or even result in no such investigation.

Even though the initial circumstances of finding material out of regulatory control (MORC) may not appear to require the involvement of a prosecutor or law enforcement, a subsequent investigation can often reveal the criminal intent of perpetrators and, in some cases, the involvement of organized crime.[52] This means that there may in fact be much more MORC trafficked with a malicious intent than is known to competent authorities. The clarification to additional relevant social groups of the internationally stabilized meaning of nuclear forensics as a sub-discipline of forensic science supporting investigations will likely lead to a larger number of properly investigated and correctly reported cases.

Second, nuclear forensics received a powerful political impetus through the Nuclear Security Summits in 2010–2016. However, political support could not be sustained at such a high political level with the end of the summit process. The level of funding earmarked for nuclear security programmes is likely to follow accordingly. Notably, the number of publications per year using 'nuclear forensic' as a keyword in the Science Direct database has declined every year since 2018.[53]

Methods and capabilities relevant to nuclear forensics are applicable across a range of fields and applications. An argument can be made that, in the situation of diminishing political attention and funding, nuclear forensics would benefit from taking stock of its roots and improving awareness. It might also collaborate with other frameworks, such as the verification of non-proliferation and disarmament treaties, including IAEA safeguards. The consolidation of the various dimensions of nuclear forensics—or of nuclear material analysis for security purposes—would cut costs and provide new and possibly unexpected synergies between different applications.

Another way of dealing with a problem of diminished political attention is for individual states to not only increase the share of correctly investigated cases of illicit trafficking of nuclear or other radioactive materials, but also to publish the results and circumstances of those that are finalized. This would serve to draw attention to the existing problem, help stabilize the meaning of nuclear forensics, and provide a means to keep the data on investigated cases for future use.

Notes

1. F. Crispino and M. M. Houck, 'Principles of Forensic Science', in Jay A Siegel, Pekka J Saukko, and Max M B T Houck, eds., *Encyclopedia of Forensic Sciences*, 2nd edn (Waltham: Academic Press, 2013), 278–81.
2. IAEA, *Nuclear Forensics in Support of Investigations*, Implementing Guide. IAEA Nuclear Security Series, No. 2-G (Vienna: IAEA, 2015), 62.
3. IAEA, *Combating Illicit Trafficking in Nuclear and Other Radioactive Material*, Technical Guidance. IAEA Nuclear Security Series, No. 6 (Vienna: IAEA, 2007), 2. http://www-pub.iaea.org/MTCD/Publications/PDF/pub1309_web.pdf
4. IAEA, *Combating Illicit Trafficking in Nuclear and Other Radioactive Material*, 6.
5. Ibid., 6–7.
6. IAEA, *Radiological Crime Scene Management*, Implementing Guide. IAEA Nuclear Security Series, No. 22-G (Vienna: IAEA, 2014), 9, 31. https://www-pub.iaea.org/MTCD/Publications/PDF/Pu1672web-85447671.pdf
7. For more information, see IAEA, e-Learning Course on Nuclear Forensics, n.d. https://elearning.iaea.org/nsns/training
8. Characterization may also be referred to as 'nuclear forensic laboratory analysis'. See IAEA, 'Nuclear Forensics in Support of Investigations', 18–20.
9. J. A. Tilden and D. Boyd, 'Nuclear Forensics and Deterrence', *National Interest*, 174/May/June (2021), 72–82.
10. Tarja Cronberg, *Renegotiating the Nuclear Order: A Sociological Approach* (New York: Routledge, 2021), 12. https://doi.org/10.4324/9781003104667
11. Wiebe E. Bijker, 'How Is Technology Made?—That Is the Question!', *Cambridge Journal of Economics*, 34/1 (2010), 66. https://doi.org/10.1093/cje/bep068
12. W. E. Bijker, 'Technology, Social Construction of', in James D. Wright, ed., *International Encyclopedia of the Social & Behavioral Sciences*, 2nd edn (Elsevier, 2015), 137. https://doi.org/10.1016/B978-0-08-097086-8.85038-2
13. R. Rhodes, *The Making of the Atomic Bomb* (New York: Simon and Schuster, 1986), 605–13; P. Oleynikov, 'German Scientists in the Soviet Atomic Project', *The Nonproliferation Review*, 7/2 (2000), 1–30; V. Gilinsky and R. J. Mattson. 'Did Israel Steal Bomb-Grade Uranium from the United States?', *Bulletin of the Atomic Scientists* [online], 17 April 2014. https://thebulletin.org/2014/04/did-israel-steal-bomb-grade-uranium-from-the-united-states/#post-heading; V. Gilinsky and R. J. Mattson. 'Revisiting the NUMEC Affair',

Bulletin of the Atomic Scientists [online], 1 March 2010. https://thebulletin.org/2010/03/revisiting-the-numec-affair-2/.
14. Vitaly Fedchenko, *The New Nuclear Forensics: Analysis of Nuclear Materials for Security Purposes* (Oxford: Oxford University Press, 2015), 4–7, 159–267.
15. Graham T. Allison, Owen R. Coté, Richard A. Falkenrath, and Steven E. Miller, 'Avoiding Nuclear Anarchy', *Washington Quarterly*, 20/3 (1996), 190.
16. L. Zaitseva and F. Steinhäusler, 'Nuclear Trafficking Issues in the Black Sea Region', EU Non-Proliferation Consortium. Non-Proliferation Paper, No. 39 (April 2014), 5. https://www.sipri.org/sites/default/files/EUNPC_no-39.pdf
17. IAEA, 'Moscow Nuclear Safety and Security Summit Declaration', INFCIRC/509. IAEA [online], 4 June 1996. https://www.iaea.org/sites/default/files/infcirc509.pdf
18. Allison et al., 'Avoiding Nuclear Anarchy', 50–3.
19. Sidney Niemeyer, 'The Early Development of International Nuclear Forensics', Nuclear Forensics International Technical Working Group Virtual Meeting, 15 June 2021.
20. Sidney Niemeyer, 'The Origin Story of the Nuclear Forensics International Technical Work Group', *ITWG Nuclear Forensics Update* [online] 13, 1, December 2019. https://www.nf-itwg.org/newsletters/ITWG_Update_no_13.pdf
21. K. Mayer, M. Wallenius, and T. Fanghänel, 'Nuclear Forensic Science—From Cradle to Maturity', *Journal of Alloys and Compounds*, 444–5 (2007), 50–6. https://doi.org/10.1016/j.jallcom.2007.01.164
22. L. Koch and S. Niemeyer, 'Status of International Cooperation on Nuclear Smuggling Forensic Analysis', Nuclear Forensics International Technical Working Group [online], 20 March 1996. https://www.nf-itwg.org/newsletters/ITWG_Update_no_24.pdf.
23. Ibid., 19.
24. Ibid.
25. Vitaly Fedchenko, 'Nuclear Forensics as Nuclear Material Analysis for Security Purposes', *EU Non-Proliferation Consortium Non-Proliferation Papers*, 60, 4, June 2017. https://www.sipri.org/sites/default/files/2017-07/eunpc_no_60.pdf
26. IAEA, Nuclear Forensics Support: Reference Manual, Technical Guidance. IAEA Nuclear Security Series, No. 2 (Vienna: IAEA, 2006), 1; P. Thompson, 2010. '15 Years of ITWG: A Personal View', Paper presented to the ITWG-15 Meeting, Oxford, June 2010.
27. Koch and Niemeyer, 'Status of International Cooperation on Nuclear Smuggling Forensic Analysis', 19–20.
28. Fedchenko, 'Nuclear Forensics as Nuclear Material Analysis for Security Purposes', 5.
29. George Bush and Vladimir Putin, 'Joint Statement by US President George Bush and Russian Federation President V. V. Putin Announcing the Global Initiative to Combat Nuclear Terrorism', G8 Summit, St Petersburg, 15 July 2006. http://en.kremlin.ru/supplement/4721
30. GICNT, 'Statement of Principles', October 2006. http://www.gicnt.org
31. M. J. Kristo, D. K. Smith, S. Neimeyer, and G. B. Dudder, *Model Action Plan for Nuclear Forensics and Nuclear Attribution*. UCRL-TR-202675, US Department of Energy (Livermore, CA: Lawrence Livermore National Laboratory, 2004).
32. IAEA, 'Nuclear Forensics Support'.
33. Fedchenko, 'Nuclear Forensics as Nuclear Material Analysis for Security Purposes', 7.
34. GICNT, 'Nuclear Forensics Fundamentals for Policy Makers and Decision Makers', February 2012, 2.
35. Nuclear Security Summits, 'Seoul Communiqué'. 2012 Nuclear Security Summit, Seoul, 27 March 2012. http://en.kremlin.ru/supplement/1185

36. US Department of State, 'Forensics in Nuclear Security'. Joint Statement in the context of the Nuclear Security Summit, The Hague, 2014. https://2009-2017.state.gov/documents/organization/235503.pdf
37. Ibid.
38. Netherlands Forensic Institute, 'Coming to Terms with Nuclear Terms', Netherlands Forensic Institute [online], 12 March 2014. https://www.forensicinstitute.nl/news/news/2014/03/12/coming-to-terms-with-nuclear-terms
39. IAEA, *Nuclear Forensics in Support of Investigations*, 62.
40. Nuclear Security Summits. 'Joint Statement on Forensics in Nuclear Security: Joint Statement in the Context of the 2016 Nuclear Security Summit'. 2016 Nuclear Security Summit, Washington, DC, 5 April 2016. http://www.nss2016.org/document-center-docs/2016/4/1/joint-statement-on-forensics-in-nuclear-security
41. For more information on applicable measurement techniques, see IAEA, *Nuclear Forensics in Support of Investigations*, 51–8.
42. For more information, see Fedchenko, *The New Nuclear Forensics*, 19–22.
43. IAEA, *Nuclear Forensics in Support of Investigations*, 26.
44. Fedchenko, *The New Nuclear Forensics*, 23–4.
45. Ibid., 24, 93–127.
46. IAEA, 'Development of a National Nuclear Forensics Library: A System for the Identification of Nuclear or Other Radioactive Material Out of Regulatory Control', IAEA-TDL-009, 2018. https://www-pub.iaea.org/MTCD/Publications/PDF/TDL009web.pdf
47. IAEA, *Nuclear Forensics in Support of Investigations*, 7.
48. Ibid., 13–14, 19–20, 32, 61; IAEA, *Radiological Crime Scene Management*, Implementing Guide. IAEA Nuclear Security Series, No. 22-G (Vienna: IAEA, 2014), 6.
49. IAEA, *Nuclear Forensics in Support of Investigations*, 8.
50. Fedchenko,ss *The New Nuclear Forensics*, 160–3.
51. Ibid., pp. 164–65, 169–71.
52. A. Apostol, 'Nuclear forensics: 10 years a full-time job'. *ITWG Nuclear Forensics Update* [online] 19, 2, June 2021. https://www.nf-itwg.org/newsletters/ITWG_Update_no_19.pdf.
53. Z. Varga, et al., 'Trends and Perspectives in Nuclear Forensic Science', *Trends in Analytical Chemistry* [online], 146 (2022), 116503. https://doi.org/10.1016/j.trac.2021.116503.

Suggested Reading

Dudder, G. B., Niemeyer, S., Smith, D. K., and Kristo, M. J., (2004), Model Action Plan for Nuclear Forensics and Nuclear Attribution (Livermore, CA: Lawrence Livermore National Laboratories). https://doi.org/10.2172/15009803. https://www.osti.gov/servlets/purl/15009803

Fedchenko, Vitaly, (2015), *The New Nuclear Forensics: Analysis of Nuclear Materials for Security Purposes* (Oxford: Oxford University Press). https://global.oup.com/academic/product/the-newnuclear-forensics-9780198736646

Fedchenko, Vitaly, (June 2017), 'Nuclear Forensics as Nuclear Material Analysis for Security Purposes', EU Non-Proliferation Consortium Non-Proliferation Papers, No. 60, 4. https://www.sipri.org/sites/default/files/2017-07/eunpc_no_60.pdf

IAEA, (2015), *Nuclear Forensics in Support of Investigations*, Implementing Guide. IAEA Nuclear Security Series, No. 2-G (Vienna: IAEA).

Mayer, K., Wallenius, M., and Fanghänel, T., (2007), 'Nuclear Forensic Science—From Cradle to Maturity', *Journal of Alloys and Compounds*, 444–5, 50–6. https://doi.org/10.1016/j.jallcom.2007.01.164

CHAPTER 33

CASE STUDIES OF A NUCLEAR FORENSIC EXAMINATION

DAVID K. SMITH

INTRODUCTION

NUCLEAR forensics, as a subdiscipline of forensic science, involves the examination of nuclear or other radioactive material, including evidence contaminated by radionuclides, in the context of national or international legal instruments related to nuclear security.[1] Inherent to a nuclear forensics examination are measurements of the physical characteristics (for example, density, colour, and grain size) and the chemical properties (for example, isotope composition, major elements, and trace elements) of materials, as these can provide information on its geologic and manufacturing history (for example, isotopes separated for enrichment from natural uranium and in fissionable nuclear fuels).[2]

Nuclear forensics examinations that support potential criminal prosecutions under national laws are a priority. These require that evidence returned from the examination must be admissible within a court of law and include a plan for the examination that is acceptable to law enforcement and judicial officials. For this reason, the collection and management of evidence at a crime scene including nuclear and other radioactive material out of regulatory control (MORC) must encompass hazard mitigation and documentation. This requires the establishment of a 'chain of custody' to track the location and movement of evidence from the crime scene to its presentation in court.

Case studies can inform improvements to nuclear forensics practices and examinations and are important because they offer objective lessons that illuminate the nature of the threat as well as procedures taken to identify and secure the materials.[3] As such, they are illustrative of the many resources, infrastructure, and expertise required to implement a nuclear security investigation. The purpose of this chapter is not to exhaustively review individual nuclear forensics case studies in detail, but rather to use diverse published case studies to extract lessons learned and assess the strength of measures designed to prevent and respond to the diversion of nuclear and other radioactive MORC. An additional objective is to illustrate the effectiveness of nuclear forensics in support of investigations, including the initial response to a nuclear or radiological security event, as a tool for states to better meet their nuclear security responsibilities.

Model Action Plan for Nuclear Forensics

Incidents in central and eastern Europe involving the interdiction of up to kilogram (kg) quantities of fissionable uranium and plutonium began to be reported after the former Soviet Union disbanded at the end of 1991. Concerns regarding high-consequence terrorist acts involving nuclear material were further heightened after the 11 September 2001 attacks in the United States (US).[4] In parallel, international concerns mounted about the security of large inventories of fissionable materials in the newly formed Commonwealth of Independent States, and elsewhere, generating urgency over how to best secure these materials in production, use, and storage. Current armed military conflicts amplify this threat.

Scientists and law enforcement officials came together in the mid-1990s to address technical questions surrounding seized, but unattributed, nuclear material.[5] These experts recognized that nuclear science, radiochemical, and analytical skills maintained during the Cold War, and utilized in nuclear materials production as well as in underground weapons testing, could also be adapted to provide insight into the origin and history of smuggled or inadequately secured nuclear materials.

To address international concerns overs reports of illicit nuclear trafficking, the Nuclear Smuggling International Technical Working Group was founded in 1995. Later renamed the Nuclear Forensics International Technical Working Group (ITWG), this informal, unaffiliated association of nuclear scientists, law enforcement officials, nuclear regulators, forensic science specialists, and nuclear security professionals has focused on sharing best approaches. This has included generalized guidance for the conduct of a nuclear forensics examination, nuclear forensics evidence collection, implementation of interpretative and analytical collaborative material exercises, and outreach, as well as more recent scientific and technical advancements by international nuclear forensics laboratories.[6]

One of most significant contributions since the founding of the ITWG has been the formulation of the 'nuclear forensics model action plan' as a guide to the conduct of a nuclear forensics examination. This plan was embraced by the International Atomic Energy Agency (IAEA). Originally published in 2006 and revised in 2015, the nuclear forensics model action plan is set out in detail in the IAEA's Nuclear Security Series Implementing Guide, 'Nuclear Forensics in Support of Investigations'.[7]

The evaluation of the nuclear forensics case studies in this chapter is facilitated through the common lens of the nuclear forensics model action plan. The evaluation compares the elements of the model action plan and the actual conduct of evidence recovery, analysis, and interpretation in the case studies. While nuclear forensics involves the analytical aspects of an investigation, the model action plan further incorporates the initial safe and secure response to incidents of MORC. This includes the management of a radiological crime scene as well as the recovery of evidence and its documentation through a chain of custody (Figure 33.1).

Incident response

Crime scene analysis

Forensics examination plan
including chain of custody

Traditional forensics analysis

FIGURE 33.1. The principal elements of the nuclear forensics model action plan.

Photographs courtesy of the author. Also see IAEA, *Nuclear Forensics in Support of Investigations.*

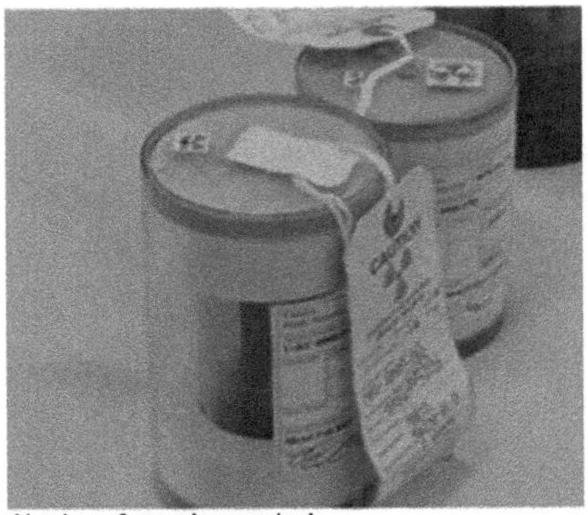

Nuclear forensics analysis

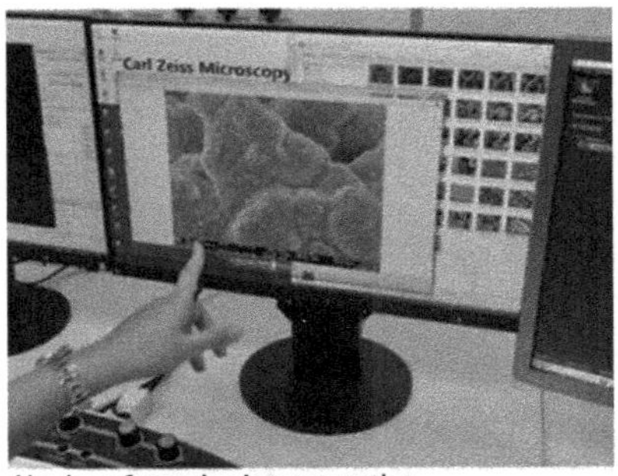

Nuclear forensics interpretation

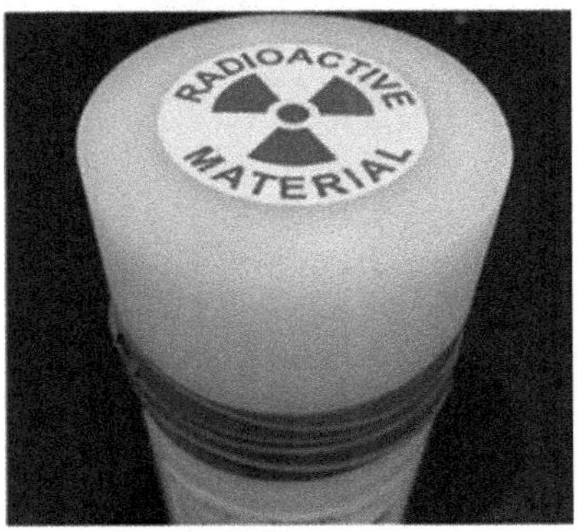

Nuclear forensics findings

FIGURE 33.1. Continued

The model action plan specifies a generalized approach to an examination of the traditional evidence recovered at a radiological crime scene (including fingerprints, DNA, digital evidence, fibres, and explosive residues) as well as nuclear forensics evidence (nuclear and other radioactive material, as well as traditional evidence contaminated by radionuclides). Often traditional forensics and nuclear forensics analyses are done in parallel by separate specialized laboratories. As each case is unique, planning and the resulting implementation of forensics examinations will each be different. However, each and every examination must return scientifically defensible findings.

Nuclear Forensics Categorization and Characterization

The initial stages of response to a radiological crime scene involve protection of the public and the integrity of potential evidence. Initial categorization at the radiological crime scene requires determination of the nature and extent of radioactivity above background, namely, the specific radionuclides present (for example, plutonium-239, plutonium-238, uranium-235, irdium-192, caesium-137, strontium-90, cobalt-60, and others) and their dose rate (for example, micro-sievert or milli-sievert/hour).[8] This information will inform access to the radiation area, health physics implications, and considerations for the safe handling, packing, and transport of the evidence to the appropriate laboratory or laboratories for further analysis. The forensics examination plan involves collection of evidence contaminated by radionuclides as well as evidence that is not contaminated and might be encountered outside the radiation cordon.[9] It is crucial to establish an unbreakable chain of custody that documents the control over the sample, including transfer of the samples sequentially from the point of collection through to their ultimate disposition.

Nuclear forensics analysis and interpretation involves measurements of material characteristics, or signatures—comprising isotopes, trace elements, and physical properties—that are indicative of their geologic and industrial origins.[10] Analyses are tailored to meet requirements set by the lead investigator. Because the laboratory results may be used in legal proceedings or to provide information on the effectiveness of the nuclear regulatory regime, careful handling of evidence as well as a clear understanding of the confidence levels for the findings resulting from the analyses is paramount. Non-destructive analyses are typically prioritized first before destructive analyses involving sample digestion.

Categorization including gamma spectrometry of radioactive material (for example, caesium-137 and cobalt-60), coupled with physical determination of sample weight, density, morphology, and molecular composition using X-ray techniques, can provide an abundance of information in the initial stage of a nuclear forensics examination. Subsequent characterization involving dissolution or destruction of a solid sample is typically required for the mass spectrometric measurements of the isotope ratios. This helps to sequester potential nuclear forensics signatures, including enrichments in isotopes of uranium, plutonium, and transuranic elements in metals and oxides; the decay of radiogenic 'parent' isotopes to their 'daughters' to determine the 'age of materials' (i.e. their date of last radiochemical purification); and measurement of trace elements concentrations (at dilute levels of parts per billion—nanogram per gram (g)—or lower) to differentiate natural from industrial samples.

As findings are returned from the examination and are interpreted, a national nuclear forensics library, or a network of laboratories, maintained by the state may be used to identify the origin and history of MORC.[11] The library contains reference information and access to expertise in nuclear or other radioactive material used, produced, or stored within the state aiding in the comparison of samples. In particular, such a library is helpful to indicate whether diverted materials originated within the state or elsewhere.

Nuclear Forensics Case Studies

Case studies can be used to provide broader insights into the effectiveness of a nuclear forensics examination in response to incidents involving MORC. These may originate from the civilian and military nuclear fuel cycle or encompass radioactive sources used in science, industry, and medicine. Such examinations are as varied as the individual incidents and materials involved.

Nuclear forensics case studies are organized in the frame of categories recognized by the IAEA's Incident and Trafficking Database (ITDB). The ITDB compiles confirmed incidents reported by Member States.[12] These incidents involve:

- malicious use or illicit trafficking involving nuclear or other radioactive materials
- incidents with insufficient information on whether the incident is connected or not to malicious use or illicit trafficking
- incidents where there is sufficient information to determine incidents are not connected to malicious use or illicit trafficking.

All case studies discussed in this chapter were reported by IAEA Member States to the ITDB.

HEU Seizures in Bulgaria, France, and Moldova

The most prominent examples of nuclear forensics case studies involve the seizures of highly enriched uranium (HEU) in Ruse, Bulgaria, in 1999, in Paris, France, in 2001, and in Chisinau, Moldova, in 2011. These cases are unique because the investigations were informed by the nuclear forensics model action plan and a significant amount of information on the ensuing nuclear forensics examinations was subsequently published.[13] Nuclear forensics findings strongly suggest that not only was the isotopically enriched uranium oxide in the three cases similar nuclear material, but also that other evidence—including packaging and shielding used to smuggle this material—was comparable. Of further significance is the more than 10-year time span between the cases, suggesting illicit trafficking of these materials involved organized and lasting networks of sellers and buyers, extending from western Europe to the Black Sea region. The outcomes from these cases emphasize the crucial role of the law enforcement partner as well as the necessity of a robust response plan to MORC.

The Ruse, Bulgaria, Seizure in 1999

The 29 May 1999 seizure of 4 g of 72.2 atom % enriched uranium-235 in the form of uranium oxide particles from a driver in Ruse, Bulgaria, has been well documented, as has the ensuing nuclear forensics examination of the HEU both in Bulgaria and the US.[14] Once recovered, the HEU was only one component of a diverse inventory of evidence that was subsequently exploited by investigators in the case. The response to the Bulgaria seizure highlights the many opportunities available to forensics examiners. Evidentiary and investigative streams can be divided into three general categories: the vehicle and the driver transporting the HEU, the packaging and related evidence not involving the HEU, and the nuclear material itself. The following description of the Bulgaria seizure is drawn from published accounts of Kenton Moody and colleagues, and of Sidney Niemeyer and Ian Hutcheon.[15]

A sealed transparent glass ampoule containing the HEU was recovered by vigilant Bulgarian border officials from a vehicle driven through a checkpoint at the Danube River. After the vehicle was stopped, questioning of the driver indicated that the material had previously transited south along the western edge of the Black Sea where the driver then headed to Turkey to potentially sell the material to an interested buyer. When the sale fell through, the driver had travelled back along the same western Black Sea route before being stopped during the night at the Ruse checkpoint. The individual reportedly displayed a suspicious demeanour during the questioning. A subsequent search of the vehicle returned an abundance of clues, including the discovery of a cast lead cylinder filled with a bright yellow paraffin wax that was shielding a glass ampoule containing approximately 4 g of extremely fine-grained black uranium oxide powder, as well as written paper documentation relevant to the smuggled material.

Using gamma spectroscopy and optical measures, initial categorization by the Bulgarian authorities confirmed the material consisted of uranium oxide particles with an enrichment in excess of 20 atom % uranium-235.[16] These measurements confirmed this was a high-priority nuclear smuggling case. In 2000, the Bulgarian authorities partnered with the US government to characterize the uranium using technical assets and expertise available within the US national laboratory complex. These analyses found the uranium oxide was also enriched with uranium-236 to 12.1 atom %, as well as with trace quantities of plutonium.[17] Furthermore, residual fission products of antimony-125, caesium-134, and caesium-137 were detected; and concentrations of sulphur, bromine, and chlorine were also elevated. Radiochronometry provides the time since the last chemical purification of the sample, where radioactive parents are separated from their decay daughters (for example, natural decay of uranium-235 to protactinium-231, decay of plutonium-241 to americium-241, and others); in this instance the time of last separation was determined to be October 1993 (with a measurement uncertainty of 50 days). Taken together, the findings indicated the seized material was derived from a reprocessing enterprise involving nuclear fuels.[18]

The examination of evidence returned from the Bulgaria seizure also included non-nuclear evidence: the paper documents, lead shielding, and the yellow wax and a glass ampoule containing the HEU. A 'billing of lading' collected from the vehicle was written in Cyrillic. The lead container used for shielding was hand-fashioned and lined with a yellow paraffin wax containing high concentrations of barium chromate colorant, a carcinogen not common to North America. The crafting and production of the glass ampoule containing the uranium oxide was indicative of industrial-scale production.

Taken together, these findings suggested the material was diverted from an industrial processing facility exclusive of a North American origin and likely to be from former Soviet Union states.[19] The smuggler was tried and convicted in a Bulgarian court.[20] However, due to the absence of national laws in Bulgaria that prohibited the unauthorized possession, use, and transport of nuclear materials at the time of the seizure and trial, the judiciary only sentenced the perpetrator to violating customs formalities and the individual was released from jail for time already served with a reduced fine.[21]

The Paris, France, Seizure in 2001

In July 2001, French authorities reported the seizure of 468 mg of HEU in Paris following an inspection of a transit van.[22] In this case, an arrest was made for the unauthorized possession of gram-sized quantities of HEU oxide powder being offered for clandestine sale. The uranium sample was contained in a glass ampoule secured within a shielded lead container lined with bright yellow paraffin wax. Both the uranium oxide sample and the packaging materials were analysed by France's Commissariat à l'Energie Atomique[23] for forensics evidence.[24] The isotopic abundance of the material seized was 72.6 atom % uranium-235 and 12.2 atom % uranium-236. Trace concentrations of plutonium were also detected. The French examiners reported the radiochemical date of last purification was November 1994 (with a measurement uncertainty of 100 days) and noted similarities between their results and those reported after the Bulgarian analysis, with the suggestion that the two analysed materials had a similar origin.[25]

Analyses of the packaging material also had remarkable similarities to the material seized in Bulgaria. The lead container was fashioned almost identically to that recovered in the Bulgaria case, being enriched in antimony, while the protective yellow wax encompassing the sample vial was enriched in barium chromate and the shape of the glass ampoule mimicked that seen in Bulgaria.[26] The comparative evidence returned by examiners in these two cases implicated the transboundary nature of the illicit trafficking.[27] Investigations suggested that the nuclear material was smuggled through the Black Sea region on its way to Paris and, like the Bulgaria case, involved reprocessed spent nuclear fuel originating in former Soviet Union states.[28] Prosecutors tried these smugglers with an attempt to link them to organized criminal networks.[29]

The Chisinau, Moldova, Seizure in 2011

Of significance for investigations of nuclear trafficking, and nuclear forensics as a response to these incidents, was the June 2011 seizure of 4.4 g of HEU oxide powder in Chisinau,

Moldova. Several arrests were made in the sting operation, aimed at countering the illicit trafficking of nuclear material. Evidence collected by law enforcement officials included a lead container, paraffin shielding, a glass ampoule, and fine-grained uranium oxide powder, all remarkably similar to those recovered in Bulgaria and France.[30]

Subsequent nuclear forensics analysis of the HEU by the European Commission's Joint Research Centre in Karlsruhe, Germany indicated the uranium particulates intercepted in Chisinau had a uranium-235 enrichment of 72.7 atom per cent and a uranium-236 enrichment of 12.1 atom per cent.[31] Age dating of swipe samples of uranium oxide indicated a purification date of December 1992 (with a measurement uncertainty of 1.0 year).[32] Based on the many analytical parallels between the traditional and nuclear forensics evidence in the Bulgaria, France, and Moldova cases, taken together, the investigations concluded the materials involved a similar origin.[33]

Lessons Learned from the Bulgaria, France, and Moldova Case Studies

The examinations performed on both traditional and nuclear forensics evidence in the Bulgaria, France, and Moldova cases highlight the crucial role that nuclear forensics techniques can play in investigations of nuclear trafficking cases. Through the publication of the nuclear forensics findings, it was possible to determine the close similarity of the materials involved in each case. The isotopic abundance measurements of the HEU, determinations of the time (i.e. age) of the last radiochemical purification; the documentation of the uniform and nanometre grain size of the uranium oxide powder; as well as the traditional forensics examination of the confiscated lead containers, the bright yellow paraffin wax within owing to the presence of barium chromate, and the manufacture of the sealed glass ampoule containing the uranium powder, were all consistent sample-to-sample—and suggested a common origin from a nuclear fuel fabrication or reprocessing facility.

For the analyses of smuggled uranium in all three cases, access to mass spectrometry was important to provide the required accuracy and precision for comparing measurements of isotopes of uranium-234, uranium-235, uranium-236, and uranium-238. Variations in the radiochemical age of purification were interpreted as differences between production batches. Forensics studies involving the origin of the material analysed in the Bulgaria seizure, with links to the France and Moldova cases, were unable to identify the exact source of this material but focused on likely origins. The presence of uranium-236 and fission products in addition to the uniformly fine grain size of the uranium oxide suggested this material had been returned from chemical reprocessing of previously irradiated nuclear fuels. The reprocessing steps, together with the barium chromate in the wax and accompanying documents written in Cyrillic, discounted a North American origin for this material.[34]

The role of law enforcement in these cases cannot be underestimated. In particular, the Bulgaria case commenced due to the vigilance of an alert Bulgarian border guard who became suspicious of the driver transiting north to Romania and who took appropriate action, including detaining the driver for additional questioning, as well as initiating a search of the driver and his vehicle. The role and preparedness of law enforcement was also crucial in the France and Moldova cases to disrupt the attempted sale of these materials.[35] The response by authorities to each of these cases was predicated on prior knowledge of national forensics capabilities to fully exploit both the nuclear as well as the traditional evidence, consistent with requirements of the lead investigative authority. Additionally in the Bulgaria and Moldova cases, bilateral arrangements facilitated the sharing of the confiscated samples and the full characterization of the uranium oxide powder in nuclear forensics laboratories in the US and European Union.

The seizures of gram-sized quantities of nuclear material were far less than the published significant quantity of 25 kg of uranium-235, the HEU required to construct a fissionable nuclear device.[36] However, these smaller quantities do provide insight into the smuggling enterprise and, in particular, emphasize vulnerabilities associated with the physical protection and detection of smaller samples and the response to interdictions of these materials. In all three cases, the seized gram-sized nuclear materials were examples of much larger inventories of nuclear materials that might be transferred later. Nevertheless, these cases indicate that similar nuclear material was misappropriated as a result of deficiencies in the nuclear security regime for more than a decade (1999–2011), or perhaps even longer. Together, these case studies teach investigators not to treat individual cases in isolation but to share results, where possible, and compare the findings between individual cases. The comparative aspects between the three cases are an additional endorsement for states to collaborate internationally and to establish a national nuclear forensics library to help in the identification of nuclear material under their jurisdiction.[37]

The traditional forensics evidence returned from the three cases was also extremely useful to the goals of the investigation. In each, those responsible for the trafficking of the material used a lead container to shield the material from detectable radiation emissions, a bright yellow paraffin wax to cushion the sample, and crafted sealed glass ampoule to contain the material. This evidence suggests similar, careful, and extensive preparations went into the diversion and smuggling of these samples. For the purposes of the investigation, responders and examiners must look broadly at the entirety of evidence available. Exploiting traditional forensics (for example, fingerprints and digital evidence) can also be powerful in linking a suspect to MORC.

That the samples from the three cases were encountered out of regulatory control remains worrisome, however. The samples discovered had disrupted the potential transfer of these materials between sellers and buyers who recognized these materials were rare, of strategic importance, and could be used for malicious purposes. The smugglers were organized in their approach and methods. The similarity and timing of the three cases suggests the presence of a network of sellers, middlemen, and buyers, possibly working together over a timeline that extended over many years. The presence of buyers for this uranium has potential implications for the threat of nuclear terrorism.[38] Less certain is whether the HEU in the three cases was diverted out of regulatory control during a single episode, or over a longer period of time.

Loss of a Cobalt-60 Source in Mexico

In December 2013, Mexico reported that a 60g, 95-terabecquerel cobalt-60 source located within the head of a medical teletherapy unit[39] had gone missing. The teletherapy source was being transported from a hospital in Tijuana to a radioactive waste storage site in central Mexico state, when the truck carrying it was stolen by armed men. The truck and medical radiotherapy unit were then immediately sold to a scrap metal dealer, with the source extracted from its shielding.[40] Elevated radiation levels (in excess of 100 μSv/hour) were detected in the vicinity using vehicle-borne radiation detectors.[41]

The entire site was subsequently secured by the Mexican authorities and a search, using a robot, ensued for the exact location of the cobalt-60 source. Two days after the theft the Mexican authorities informed the public about the radiation hazards associated with proximity to or handling of the source. The source was subsequently identified in uninhabited farmland outside the town of Hueypoxtla approximately a week after the theft, and was secured in a shielded container provided by the authorities using the robotic technology.[42] It was thought that 10 individuals with proximity to the source had been exposed to gamma radiation, necessitating assessment for acute radiation exposure.[43] Biodosimetry indicated that one individual, the person who had assisted the authorities in locating the source, exceeded the Mexican regulation for radiation exposure (500 mSv annual whole body effective dose). The individual required medical intervention. The police later opened a criminal investigation into the responsibility for the theft.[44] In 2014, charges were filed by prosecutors in the case that included vehicular theft, theft of the cobalt-60 source, and abandonment of a radioactive substance.[45]

Lessons Learned from the Loss of a Cobalt-60 Source in Mexico

The Mexican case study underscores the importance of a national response to criminal acts involving MORC, as well as the management of a radiological crime scene. Other than nuclear forensics categorization, the response involved recovery and re-establishing regulatory control over the source without accompanying nuclear forensics laboratory analysis. The radiation levels encountered were extremely high; this case involved a Category I source, defined by the IAEA as most dangerous to human health.[46] Responders must be prepared to determine the type and levels of radiation present, applying nuclear forensics categorization, and use this information to plan and implement a strategy to re-secure the stolen source.

The case further illustrates how a criminal act involving the theft of a truck has grave implications for the security and safe handling of radioactive materials. The exact motives of the assailant were unclear but likely involved their intention to steal the truck without a broader plot to weaponize the cobalt-60. However, organized and sophisticated criminal activity involving gangs and cartels throughout the region remains a concern and the outcomes thereof could have implications for potential terrorist acts.[47] The case study also points to the particular vulnerabilities associated with the transport of highly radioactive material; while the cobalt-60 source was secured within the confines of the approved facility, once it left by truck, the source became susceptible to criminal acts that resulted in the radioactive source being diverted out of regulatory control.[48] For this reason, the case emphasizes the necessity of maintaining the integrity of nuclear security measures throughout the lifecycle of radioactive material.

The case study also underscores the need for responders to address evolving nuclear security events that may occur at different locations and over weeks of time. The case was notable as it involved many different authorities, including the national nuclear regulatory body, civil protection agencies, federal police, military, nuclear plant operators, and a national institute for nuclear research.[49] Due to the extraordinarily high levels of radiation encountered, field categorization involved the use of robots and remote area surveys to locate the source. This case also required intervention by medical practitioners to survey for acute radiation exposure and provide bioassay. Although extensive nuclear forensics laboratory analysis was not required, national preparedness by the Mexican authorities as part of a criminal investigation to respond to the hazards of a highly radioactive source out of regulatory control was essential, including the collection of traditional forensic evidence encompassing fingerprints and genetic samples. Lessons learned have implications for the development of a national response plan to a nuclear security event.[50]

Nuclear Forensics Examinations of Contaminated Metal Scrap

Nuclear and other radioactive material is often inadvertently incorporated in the salvaging streams of metal recyclers.[51] Sealed cobalt-60, iridium-192, and caesium-137 gamma-emitting sources are commonly used in the oil refining and manufacturing industries and can be lost or improperly disposed of along with benign excess or waste metal.[52] Scrap metal is a global business that continues to be a lucrative market driven by higher prices for raw materials sourced in recycled materials and reflective of demands by global supply chains during the Covid-19 pandemic beginning in 2020, and beyond.

While cases involving contaminated scrap metal do not involve malicious intent, they do underscore deficiencies in the nuclear security regime that allows these materials to be diverted out of regulatory control and be transported to scrap metal recycling yards. The principal concern is removing radioactive contamination from the waste stream before it is smelted; the cost to a recycler to decontaminate and clean a metal smelter is prohibitive. Despite these precautions, there are reported incidents of radionuclides (including cobalt-60) that have made their way into recycled household goods.[53] For this reason, radiation portal monitors are outfitted at many scrapyards to screen entering haul trucks; scrapyard handling claws affixed to crane manipulators are also fitted with gamma detectors. Nuclear forensics examinations have been conducted to categorize and characterize the radioactive materials encountered in scrap, in order to determine the inherent radiation profile and any potential threat, decide how these sources can be handled and secured, and provide insight into their history.

Beside sealed gamma-emitting sources, HEU has been discovered within contaminated steel scrap sent to European recyclers. The HEU is affixed as powders, solutions, or gases to the surfaces of the host metal.[54] Nuclear forensics analyses of contaminated samples using both non-destructive and destructive bulk and micro-analytical techniques has indicated variable enrichments that were the result of the inhomogeneity of the distribution of uranium in the samples.[55] In one instance, the highest uranium-235 enrichment in this metal scrap was 28.3 weight % measured in bulk samples by gamma spectroscopy, with the corresponding highest uranium-235 enrichment of 25.9 weight %, measured in bulk samples by thermal ionization mass spectrometry. Individual uranium particles had a maximum enrichment of 94.4 weight % uranium-235 using spatially-resolved laser ablation inductively coupled plasma mass spectrometry.[56] The detection of uranium-236 in these samples is indicative that the samples were reprocessed.

Lessons Learned from Nuclear Forensics Examinations of Contaminated Scrap Metal

Nuclear forensics examination of contaminated scrap metals highlights the role of industry working in concert with national regulators to ensure that nuclear and other radioactive materials are detected and secured. Although nuclear security is a national responsibility, industry has a financial incentive to utilize nuclear forensic science to categorize radioactive materials and keep them out of recycled metal product streams in order to alleviate contamination of the manufacturing process and final products. The recycling yards serve as 'choke points' where larger quantities of metals can be aggregated and routinely surveyed and nuclear and other radioactive materials removed and secured. Increased awareness and preparedness of recyclers and regulators to the threat has resulted in the removal of dangerous materials from metal scrap and heightened measures to prevent future incidents.

Nuclear forensics also plays a role. The outcome of these examinations highlights lapses in the originator's regulatory regime where radioactive sources and other radionuclides have become inadvertently incorporated with metals bound for recycling yards as well as shortcomings in transport during which time these offending radioactive materials are shipped without detection to a recycler. In the case of the nuclear materials encountered by European recyclers, scrap streams blend lots of metal from many different sources, and forensics findings represent this extreme heterogeneity and mix of materials. Enrichments in uranium-235 and uranium-236 in metal scrap provide indications that there has been a compromise in the physical security measures at a source nuclear facility; isotope ratios and physical information on metals and particles provide information on the processing history of the uranium out of regulatory control, and together with other nuclear forensics signatures, may be used to identify the facility from where the material was diverted.

Conclusion

Nuclear forensics case studies yield information on those individuals, groups, or networks who would possess, use, or store nuclear and other radioactive materials for malicious purposes, as well as the nature of the materials that are stolen or trafficked. By providing information on the origin and history of MORC, nuclear forensics provides insights on the effectiveness of the nuclear security regime, including preventive measures and regulations. Findings from these studies offer confidence to examiners, law enforcement officials, and the judiciary that nuclear forensics is an effective tool that supports the criminalization of unauthorized acts.

The case studies contained in this chapter focus on both the challenges and opportunities in conducting a nuclear forensics examination and highlight essential lessons learned that can improve the state of practice. Most importantly, results from nuclear forensics examinations provide demonstrated confidence in the conduct of nuclear security investigations to include criminal prosecutions. While information concerning incidents of MORC, and the associated results of nuclear forensics examinations, is typically managed and controlled by the state, the results directly inform broader nuclear forensics methodologies as well as contributions to a national response plan. When this information can be shared, outcomes from case studies further benefit regional and international partners who may need to respond to similar threats using the tools of radiological crime scene management and nuclear forensic science as capabilities to meet their nuclear security responsibilities.

Some case studies emphasize the crucial role of law enforcement officials who, in the case of the Bulgaria seizure in 1999, thoroughly questioned and searched an individual who was behaving suspiciously. Follow-on cases in Paris in 2001 and Chisinau in 2011 involved intelligence-driven investigations where sting operations and information alerts allowed nuclear materials to be intercepted. The threat of nuclear and radioactive material is ever present with over 100 incidents being registered each year by the IAEA's ITDB.[57]

Case studies also emphasize the crucial role of a national response plan to a nuclear security event. Movement of MORC occurs internationally and often involves different jurisdictions. In the Mexican example, the cobalt source was recovered in a rural agricultural setting that necessitated remote recovery operations, including specialized robotics, as a result of high levels of gamma radiation. Additionally, medical intervention was required. The ability of first responders, regulators, law enforcement officials, health and environmental specialists, and transport experts working in a coordinated and comprehensive manner was essential to the response.

These case studies also point to the need for responders to request and receive technical assistance that allows high confidence in the material identification and the nature of the threat, including use of instrumentation that can detect radionuclides, protect the health and safety of responders, and securely collect nuclear forensics or traditional forensics evidence. The ability to categorize evidence in the field and develop a forensics examination plan is a necessity and must include the receiving laboratory where the ensuing examination and necessary characterization will be conducted.

Case work is the ultimate test of nuclear forensics as a preventive response to the serious threat of MORC. These studies inform the international community that nuclear and other radiological material continues to be diverted outside of regulatory control. The studies provide information of what the material is, what threat it may pose, as well as its origin and history. When shared, the cases allow direct comparison of materials to one to another; this was the situation involving the Bulgarian, French, and Moldovan nuclear forensics examinations that collectively spanned more than a decade.[58]

Without compromising sources and methods that may need to be protected as part of an investigation, dissemination of conclusions from case studies demonstrates the successful conduct of the model action plan encompassing a nuclear forensics investigation leading to successful prosecutions of smugglers. For this reason, more nuclear forensics cases need to be publicized internationally. This outreach also demonstrates the availability of nuclear forensics as a key tool to assist states in meeting their nuclear security obligations. By linking materials to people, places, and events, nuclear forensics can further serve as a preventive measure to deter those who might contemplate unauthorized acts, as well as remedy deficiencies in physical protection of detection measures that allowed this material to be diverted. Findings from case studies continue to strengthen the nuclear security regime and are only more relevant given the vulnerabilities of nuclear and radioactive materials located in conflict zones of eastern Europe.

Notes

1. IAEA, *Nuclear Forensics in Support of Investigations*, Implementing Guide. IAEA Nuclear Security Series, No. 2-G (Rev. 1) (Vienna: IAEA, 2015).
2. Michael J. Kristo, ' Nuclear Forensics', in Michael L'Annunziata, ed., *Handbook of Radioactivity Analysis*, 4th edn, Vol. 2 (London: Elsevier, 2020), 921–51; IAEA, 'Nuclear Forensics in Support of Investigations'; Klaus Mayer, Maria Wallenius, Zsolt Varga, Magnus Hedberg, and Nicole Erdmann, 'Inorganic Mass Spectrometry as a Tool of Destructive Nuclear Forensic Analysis', in Vitaly Fedchenko, ed., *The New Nuclear Forensics* (Oxford: Oxford University Press, 2015), 47–73.
3. Christopher Hobbs and Matthew Moran, *Insider Threats: An Educational Handbook of Nuclear and Non-Nuclear Case Studies* (London: King's College London, 2015); Wyn Bowen, Matthew Cottee, Christopher Hobbs, Luca Lentini, Matthew Moran, and Sarah Tzinieris, *Nuclear Security Briefing Book*, 2020 edn (London: King's College London, 2020).
4. Graham T. Allison, *Nuclear Terrorism: The Ultimate Preventable Catastrophe* (New York: Times Books, 2004).
5. Sidney Niemeyer and Lothar Koch, 'The Historical Evolution of Nuclear Forensics: A Technical Viewpoint', in *Proceedings of the International Conference on Advances in Nuclear Forensics: Countering the Evolving Threat of Nuclear and Other Radioactive Material out of Regulatory Control., 7–10 July 2014 Vienna, Austria*. (Vienna: IAEA, 2015), 1–19. https://www-pub.iaea.org/MTCD/Publications/PDF/SupplementaryMaterials/P1706/Plenary_Session_1A.pdf

6. Benjamin Garrett, Klaus Mayer, Paul Thompson, Boon Kin Pong, and Gauthier Lasou, 'The Nuclear Forensics International Technical Working Group (ITWG)' in *Proceedings of the International Conference on Advances in Nuclear Forensics: Countering the Evolving Threat of Nuclear and Other Radioactive Material out of Regulatory Control, 7–10 July 2014 Vienna, Austria*. (Vienna: IAEA, 2015), 6; David Smith, 'The ITWG: 20 Years of Science Supporting Law Enforcement and Nuclear Security Investigations', ITWG Nuclear Forensics, 1 (2016), 1. International Technical Working Group. No. 1; Ruth Kips, Sidney Niemeyer, Michael Kristo, Naomi Marks, David Smith, Stephen LaMont, James Borgardt, Gary Eppich, Michael Curry, and Erica Wolfe, *25 Years of ITWG: Current Needs and Future Challenges in Nuclear Forensics*. LLNL-PROC-837394 (Livermore, CA: Lawrence Livermore National Laboratory, 2022).
7. IAEA, *Nuclear Forensics in Support of Investigations*.
8. Vitaly Fedchenko, 'The Process of Nuclear Forensic Analysis', in Vitaly Fedchenko, ed., *The New Nuclear Forensics* (Oxford: Oxford University Press, 2015), 11–33.
9. IAEA, Radiological Crime Scene Management, Implementing Guide. IAEA Nuclear Security Series, No. 22-G (Vienna, IAEA, 2014).
10. Kristo, 'Nuclear Forensics'; IAEA, *Nuclear Forensics in Support of Investigations*; and Mayer et al., 'Inorganic Mass Spectrometry as a Tool of Destructive Nuclear Forensic Analysis'.
11. IAEA, *Development of a National Nuclear Forensics Library: A System for the Identification of Nuclear or Other Radioactive Material out of Regulatory Control*, IAEA Non-Serial Publication, IAEA-TDL-009 (Vienna: IAEA, 2018).
12. IAEA, 'Incident and Trafficking Database (ITDB), 2022 Factsheet', IAEA [online], n.d. https://www.iaea.org/resources/databases/itdb
13. Lyudmila Zaitseva and Friedrich Steinhäusler, 'Nuclear Trafficking Issues in the Black Sea Region', *Non-Proliferation Papers*, EU Non-Proliferation Consortium, No. 39, 2014; Elizabeth Keegan, Michael J. Kristo, Kaitlyn Toole, Ruth Kips, and Emma Young, 'Nuclear Forensics: Scientific Analysis Supporting Law Enforcement and Nuclear Security Investigations', *Analytical Chemistry*, 88/3 (2016), 1, 496; Michael J. Kristo, Amy M. Gaffney, Naomi Marks, Kim Knight, William S. Cassata, and Ian D. Hutcheon, 'Nuclear Forensic Science: Analysis of Nuclear Material Out of Regulatory Control', *Annual Reviews of Earth and Planetary Sciences*, 44 (2016), 555.
14. Sidney Niemeyer and Ian D. Hutcheon, 'Forensic Analysis of a Smuggled HEU Sample Interdicted in Bulgaria', in *Proceedings of the International Conference on Advances in Destructive and Non-Destructive Analysis for Environmental Monitoring and Nuclear Forensics, 21–23 October 2002, Karlsruhe* (Vienna: IAEA, 2002), 17–19; Kenton J. Moody, Patrick M. Grant, and Ian D. Hutcheon, *Nuclear Forensic Analysis*, 2nd edn (Boca Raton: CRC Press, 2015), 417.
15. Moody et al., *Nuclear Forensic Analysis*; Niemeyer and Hutcheon, 'Forensic Analysis of a Smuggled HEU Sample Interdicted in Bulgaria'.
16. Moody et al., *Nuclear Forensic Analysis*.
17. Ibid.; Niemeyer and Hutcheon, 'Forensic Analysis of a Smuggled HEU Sample Interdicted in Bulgaria'.
18. Moody et al., *Nuclear Forensic Analysis*.
19. Zaitseva and Steinhäusler, 'Nuclear Trafficking Issues in the Black Sea Region'.
20. Moody et al., *Nuclear Forensic Analysis*.
21. Ibid.

22. S. Baude, 'HEU Seized in July 2001 in Paris Analytical Investigations Performed on the Material', in *Illicit Nuclear Trafficking: Collective Experience and the Way Forward* 397; S. Baude, B. Chartier, D. Kimmel, F. Mariotte, D. Masse, H. Peron, and D. Tilly, 2008. 'The French Response in Cases of Illicit Nuclear Trafficking Lessons Learned from a Real Case', in *Illicit Nuclear Trafficking: Collective Experience and the Way Forward* (Vienna: IAEA, 2008), 363.
23. Now named Commissariat à l'Energie Atomique et aux Energies Alternatives (CEA).
24. Baude, 'HEU Seized in July 2001 in Paris Analytical Investigations Performed on the Material'; Baude et al., 'The French Response in Cases of Illicit Nuclear Trafficking Lessons Learned from a Real Case', 363.
25. Baude, 'HEU Seized in July 2001 in Paris Analytical Investigations Performed on the Material'.
26. Ibid.
27. Zaitseva and Steinhäusler, 'Nuclear Trafficking Issues in the Black Sea Region'; Baude, 'HEU Seized in July 2001 in Paris Analytical Investigations Performed on the Material'.
28. Zaitseva and Steinhäusler, 'Nuclear Trafficking Issues in the Black Sea Region'; William Potter and Elena Sokova, 'Illicit Trafficking in the NIS: What's New, What's True?', *The Nonproliferation Review*, Summer (2002), 112.
29. James Russell and James Wirtz, *Globalization and WMD Proliferation: Terrorism, Transnational Networks and International Security* (New York: Routledge, 2009).
30. Zaitseva and Steinhäusler, 'Nuclear Trafficking Issues in the Black Sea Region'; US Senate, 'Enhancing Non-Proliferation Partnerships in the Black Sea Region', Committee on Foreign Relations (Washington, DC: US Government Printing Office, 2011).
31. Keegan et al., 'Nuclear Forensics'; Zsolt Varga, Maria Wallenius, Adrian Nicholl, Klaus Mayer, Ionel Balan, and Vasile Benea, 'Measurement of Production Date (Age) of Nanogram Amount of Uranium', *Journal of Radioanalytical and Nuclear Chemistry*, 322 (2019), 1585–91.
32. Varga et al., 'Measurement of Production Date (Age) of Nanogram Amount of Uranium'.
33. Zaitseva and Steinhäusler, 'Nuclear Trafficking Issues in the Black Sea Region'.
34. Moody et al., *Nuclear Forensic Analysis*.
35. Ibid.; Baude, 'HEU Seized in July 2001 in Paris Analytical Investigations Performed on the Material'.
36. IAEA, *IAEA Safeguards Glossary*, 2001 Edition. IAEA International Nuclear Verification Series, No. 3 (Vienna: IAEA, 2002).
37. IAEA, *Development of a National Nuclear Forensics Library: A System for the Identification of Nuclear or Other Radioactive Material out of Regulatory Control*. IAEA Non-Serial Publication, IAEA-TDL-009 (Vienna: IAEA, 2018).
38. Zaitseva and Steinhäusler, 'Nuclear Trafficking Issues in the Black Sea Region'.
39. Now often known as 'external beam radiotherapy'.
40. Timothy Farnsworth, 'Cobalt 60 Stolen, Recovered in Mexico', *Arms Control Today* [online], January 2014, 44. https://www.armscontrol.org/act/2014-01/cobalt-60-stolen-recovered-mexico; IAEA, 'Arrangements for the Termination of a Nuclear or Radiological Emergency, General Safety Guide'. IAEA Safety Standards, No. GSC-11 (Vienna: IAEA, 2018); IAEA, 'The Radiological Incident in Hueypoxtla', IAEA/RAD/INC (Vienna: IAEA, 2022).
41. IAEA, 'Arrangements for the Termination of a Nuclear or Radiological Emergency'.
42. Ibid.; IAEA, 'The Radiological Incident in Hueypoxtla'.

43. Ibid.
44. Tom Bielefeld, 'Mexico's Stolen Radioactive Source: It Could Happen Here', *Bulletin of the Atomic Scientists* [online], 23 January 2014. https://thebulletin.org/2014/01/mexicos-stolen-radiation-source-it-could-happen-here/.
45. IAEA, 'The Radiological Incident in Hueypoxtla'.
46. IAEA, *Categorization of Radioactive Sources*. IAEA Safety Standards Series, No. RS-G-1.9 (Vienna: IAEA, 2005).
47. Bielefeld. 'Mexico's Stolen Radioactive Source'.
48. IAEA, *Security of Radioactive Material in Transport*, Implementing Guide. IAEA Nuclear Security Series, No. 9-G (Rev.1) (Vienna: IAEA, 2020).
49. IAEA, 'Arrangements for the Termination of a Nuclear or Radiological Emergency'.
50. IAEA, 'The Radiological Incident in Hueypoxtla'.
51. IAEA, *Control of Orphan Sources and Other Radioactive Material in the Metal Recycling and Production Industries*, Specific Safety Guide. IAEA Safety Standards, No. SSG-17 (Vienna: IAEA, 2012); Klaus Mayer, Maria Wallenius, and Zsolt Varga, 'Sample Characteristics and Nuclear Forensic Signatures', in Vitaly Fedchenko, ed., *The New Nuclear Forensics* (Oxford: Oxford University Press, 2015), 93–127.
52. IAEA, 'Sealed Radioactive Sources', IAEA/PI/A.98/13-32051 (Vienna: IAEA, 2013).
53. Ibid.
54. Mark Elert, *Import of Metal Scrap—Risks Associated with Radioactivity*. KEMAKTA AR 92-22 (Stockholm: Kemakta Konsult AB, 1992).
55. Michael Krachler, Maria Wallenius, Adrian Nicholl, and Klaus Mayer, 'Spatially-Resolved Uranium Isotopic Analysis of Contaminated Scrap Metals Using Laser Ablation Multi-Collector ICP-MS', *Royal Society of Chemistry Advances*, 10 (2020), 16.
56. Ibid.
57. IAEA, *2022 Factsheet*.
58. Kristo et al., 'Nuclear Forensic Science'.

Suggested Reading

Apostol, Andrei I., (2021), 'ITWG Nuclear Forensics Update: 10 Years a Full Time Job', in Vitaly Fedchenko, ed., *ITWG Nuclear Forensics Update*. International Technical Working Group. No. 19.

Dinu, Elena, (2019) 'Contaminated Playing Cards: An Illustration of the Prosecutor's Role in Investigations Concerning Nuclear and Radioactive Materials', in Vitaly Fedchenko, ed., *ITWG Nuclear Forensics Update*. International Technical Working Group. No. 12.

Fedchenko, Vitaly, (2015), *The New Nuclear Forensics: Analysis of Nuclear Materials for Security Purposes* (Oxford: Oxford University Press).

Kristo, Michael J., Amy M. Gaffney, Naomi Marks, Kim Knight, William S. Cassata, and Ian D. Hutcheon, (2016), 'Nuclear Forensic Science: Analysis of Nuclear Material Out of Regulatory Control', *Annual Reviews of Earth and Planetary Sciences*, 44, 555–79.

Moody, Kenton J., Grant, Patrick M., and Hutcheon, Ian D., (2015), *Nuclear Forensic Analysis*, 2nd edn (Boca Raton: CRC Press).

PART VII

PERMANENT THREAT REDUCTION

CHAPTER 34

INTERNATIONAL EFFORTS TO MINIMIZE THE USE OF HEU

FERENC DALNOKI-VERESS AND MILES A. POMPER

Introduction

Minimizing the global trade in highly enriched uranium (HEU) has long been a goal for the international community, as HEU represents a highly attractive target for terrorists and state proliferators. HEU can be used to construct the simplest nuclear explosive device, referred to as a 'gun-type' weapon, which was the type of weapon that destroyed the Japanese city of Hiroshima during the Second World War. Such a device explosively collides one subcritical piece of HEU with another to form the supercritical mass required for a nuclear detonation. This process is well publicized, and most experts agree that an improvised nuclear bomb based on this design is technically feasible for a financially and organizationally powerful terrorist group. The fissile material of choice for such a scenario would be HEU with more than 80 per cent uranium-235 (U-235).

To make matters worse, HEU is generally safe to handle and difficult to detect due to its low radioactivity and high density (small volume) and is therefore easy to hide. Even HEU waste is far less radioactive than one would hope from a security perspective. Within months, HEU waste loses its 'self-protection', in that it would not deliver a would-be thief an immediately incapacitating radiation dose.[1] It is important to emphasize that not all HEU is the same: as the uranium-235 enrichment decreases from weapons-grade to lower grades it becomes more and more challenging to use for a gun-type weapon as more HEU is needed to form a supercritical mass. It is also utilized in ship propulsion reactors, such as those aboard submarines, aircraft carriers, and icebreakers. And, of course, most state-owned nuclear bombs (but not all) include HEU.

As part of President Dwight D. Eisenhower's 1953 'Atoms for Peace' programme, the United States (US) shipped research reactors and low enriched uranium (LEU) fuel to many countries. However, the performance of LEU fuel technology used in the 1950s was limited and consequently research reactor designers sought to utilize HEU fuel to develop more powerful reactors and enhance the performance of existing reactors. As a result, the US, the Soviet Union, and other nuclear weapon nations developed and exported HEU-fuelled research reactors and HEU fuel in the decades that followed—spreading HEU around the world.

HEU's primary civilian use has been to create a regulated high flux of neutrons in research reactors. Research reactors carry out a range of functions, from education and basic scientific research to producing medical isotopes and 'doping' silicon for semiconductors. According to a 2016 National Academy of Sciences report, other techniques for creating neutrons with equivalent spectra and flux levels to accomplish these tasks do not exist.[1] However, India's 'Smiling Buddha' nuclear test in 1974 raised concerns about the potential misuse of exports of nuclear technology and material and led the international community to reconsider additional transfers of HEU.[2]

Following this test, the US and the Soviet Union initiated measures to reduce the risk of misuse of fissile materials, particularly HEU, at civilian facilities.[3] For example, they initiated the establishment of the Nuclear Suppliers Group (NSG; states that agree to coordinate the transfer of civilian nuclear material, equipment, and technology) as well as guidelines to guard HEU at civilian facilities issued by the International Atomic Energy Agency (IAEA). Despite this, HEU security remains a worry at many civilian installations that remain active but were not initially constructed with security in mind.

Even today, many research reactors are on university campuses with little security, leaving nuclear materials susceptible to theft. According to US official assessments, a well-funded terrorist group might be able to acquire the materials to build a bomb.[4] As a result, in the late 1970s both the US and the Soviet Union began programmes to convert research reactors from the use of HEU to LEU where technically and economically feasible. In the intervening years, dozens of reactors around the world have been converted or shut down. Altogether, over 100 reactors have been shut down or converted since the US Reduced Enrichment for Research and Test Reactors (RERTR) programme began.[5]

Global Efforts to Minimize Use of HEU

Converting research reactors from HEU fuel to LEU fuel is an important step towards reducing the possibility that HEU could end up in the wrong hands. Conversion principles dictate that minimum influence be made on reactor operations, which means that the reactor's infrastructure and footprint should not be increased, or its productivity diminished. Numerous research reactor designs were unique and already designed for maximum productivity utilizing compact cores, making it difficult to adapt them to utilize LEU fuel without breaching conversion principles. Therefore, the reactor conversion projects throughout the world needed meticulous calculations and testing, as well as the collaboration of international scientists and diplomatic efforts. For example, an international partnership of scientists from the US, China, and Nigeria successfully converted a Chinese origin miniature neutron source reactor (MNSR) in Nigeria from

HEU to LEU and repatriated a kilogram (kg) of weapons-grade HEU back to China. This case is particularly noteworthy as it occurred during a period of high tension between the US and China in late 2018 when the Donald Trump Administration announced extensive sanctions on China and Canada arrested a Huawei executive.[6]

In the decades following the initiation of the US and Soviet conversion programmes, progress was slow, although US efforts were enhanced by the passage of the 1992 Schumer Amendment, legislation that makes US exports of HEU to existing reactors conditional on pledges to convert. These efforts accelerated following the 11 September 2001 terrorist attacks in the US that pushed the spectre of nuclear terrorism to the forefront of public awareness. Following the attacks, the George W. Bush Administration consolidated existing efforts, boosted funding, and widened their scope to incorporate new types of infrastructure and materials. President Bush and his Russian counterpart Vladimir Putin agreed in 2005 to supply LEU for any research reactor constructed by the US or Russia in third countries that was already operating with HEU and was able to be converted.[7] The expended or unused HEU would then be returned to its native country. Russia was reimbursed by the US National Nuclear Security Administration (NNSA) for aiding in the repatriation of HEU from countries such as Poland, Serbia, and Ukraine.

Altogether, more than 33 nations and Taiwan were cleared of HEU. Furthermore, almost seven tons of HEU were removed from broader civilian usage.[8] It is difficult to quantify the equivalent number of nuclear weapons because the HEU varies in enrichment and, the lower the enrichment level, the more HEU needed for a given yield, and vice versa; but, if the material was weapons-grade, only a few kgs of material would be required for a nuclear weapon. These steps have strengthened non-proliferation and nuclear security regimes while lowering the worldwide threat presented by HEU usage. However, at the time of writing there are still 80 research reactors in 14 nations that utilize HEU yet to be shut down or converted.[9]

Technical Challenges with High-Performance Research Reactors

A significant technological impediment to further progress has been the inability of certain high-performance research reactors (HPRRs) in the US and Europe to be converted with existing LEU fuels, necessitating the development of novel fuels. As such, converting reactors is a 'time consuming and technically demanding process akin to using a new kind of fuel in a car engine while seeking to maintain the car's performance and safety and not altering its basic dimensions or operating costs'.[10] The starting point for this task is finding ways to enhance the fuel's uranium density. To retain the same performance characteristics for research reactors using LEU fuel, fuel designers must raise uranium density proportionate to the drop in enrichment and then add a small density increase to compensate for the 'neutron intensity penalty'. Additionally, one of the conditions for a successful conversion is that no significant structural modifications to the reactor occur, in order to maintain ongoing experiments and beamlines.

To qualify, a new LEU fuel must preserve the same mechanical and geometric integrity as the existing HEU fuel. The cladding must offer the same barrier, and the cooling routes must be unaffected. Scientists in the US and Europe have been working for decades to produce high-density LEU fuels that may be utilized in place of the existing HEU fuels used in HPRRs. Unfortunately this has been a difficult process, and the time frame for converting the most difficult reactors, which was originally anticipated for 2013, has been pushed back to 2035.[11]

The main engineering difficulty has been developing the high-density LEU fuels necessary for conversion. The issue is that uranium occurs in a variety of crystallographic forms with varied dimensions and structures, resulting in varying degrees of irradiation resistance. The only state of uranium that is stable under irradiation is the gamma phase, which is stable only above 776°C. The gamma phase breaks down to other crystallographic forms at lower temperatures.[12] Under irradiation, this instability leads to swelling of the fuel and localized deformations ('pillowing') of the fuel cladding's dimensions. If the swelling and displacement are excessive, it is possible for the cladding to be damaged and radioactive fission products to be released. However, when uranium is alloyed with specific metals, the breakdown mechanism is reduced, making the fuel more resistant to heat cycling and irradiation. Numerous metals were considered as potential options, with researchers selecting molybdenum because of its superior mechanical qualities and corrosion resistance, as well as its stability in the gamma phase.

The US and Europe have concentrated their research efforts on two different primary fuel types. US researchers have concentrated on monolithic uranium–molybdenum (U–Mo) fuels, which are composed entirely of uranium molybdenum alloys. European scientists have concentrated their efforts on producing U–Mo dispersion fuels, which include globules of the U–Mo alloy suspended in an aluminium matrix. Figure 34.1 illustrates the differences between the two forms of fuel. Dispersion fuels typically employ U–6Mo or U–7Mo (uranium alloys containing 6 or 7 per cent molybdenum by weight), while monolithic fuels under development often use a greater molybdenum proportion of U–10Mo.[13]

Each HPRR has performance requirements for safe and anticipated operation. These requirements are measured using two parameters ranging from zero to specified

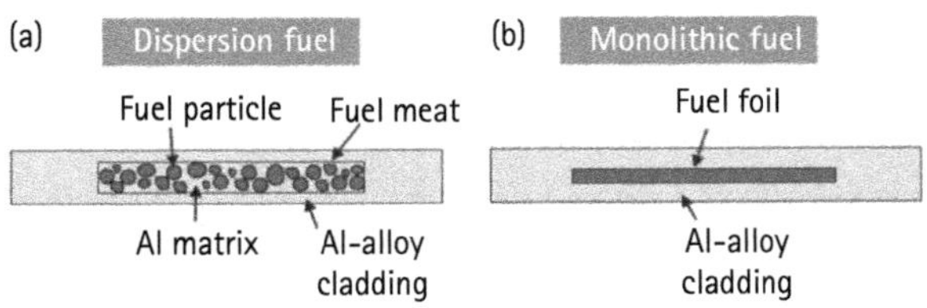

FIGURE 34.1. Schematic cross sections of (a) plate-type dispersion fuel and (b) plate-type monolithic fuel.

Figure courtesy of National Academies of Sciences, Engineering, and Medicine, *Reducing the Use of Highly Enriched Uranium in Civilian Research Reactors*, 62.

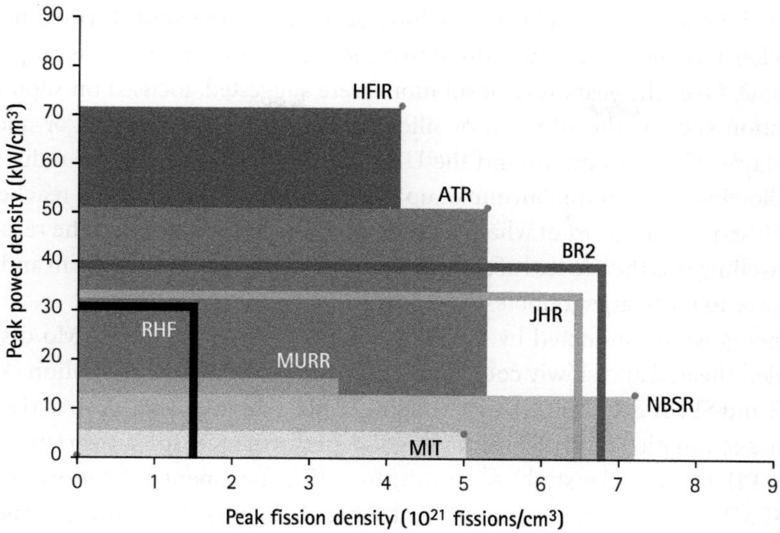

FIGURE 34.2. Graph depicting peak power density and fission density for several high-performance reactors in Europe and the US. A new LEU high-density fuel must be able to achieve 6.8×10^{21} fissions/cm^3 and a peak power density of 40 kW/cm^3 to meet safety criteria.

Figure courtesy of National Academies of Sciences, Engineering, and Medicine, *Reducing the Use of Highly Enriched Uranium in Civilian Research Reactors*, 63–4. [Figures 4.3 and 4.4 merged.]

maximum values (referred to as the 'performance envelope') that should not be exceeded. These are the local *power density* of the fuel and the *fission density*. The density of fission is related to the amount of uranium-235 consumed (burnup) and 'provides a measure of the accumulated fission products and fission gases and radiation damage that can lead to failure of the fuel'.[14] The power density is proportional to the neutron flux, which is a measure of the reactor's usefulness or capabilities. As a result, the two requirements vary for different reactors and impact the performance that fuels must achieve to satisfy reactor operator goals (see Figure 34.2 for the performance envelopes of various high-performance reactors).

European HPRRs

Developing high-density fuel for any high-performance reactor is an ambitious undertaking. Throughout the years, European investigators have formed several consortia to tackle the task (ALPS in 2008–2013, LEONIDAS in 2010–2014, HERACLES in 2013–2020, and now LEU-FOREVER [an EU initiative to study uranium silicide, U_3Si_2, as a fuel]). HERACLES is a collaboration of German, French, and Belgian researchers to develop high-density HPRR fuels. European efforts in the words of one account 'experienced a number of unexpected setbacks'.[15]

Early experiments in 2003 irradiated two flat plates ('FUTURE' test) with U-7Mo (7 per cent U by weight) atomized powder and found that as the burnup increased to about 2×10^{21} fissions/cm^3, fuel plate swelling occurred. Large voids formed between the U–Mo kernels and the aluminium matrix, accounting for as much as 70 per cent of the volume. Over the years several solutions were suggested, focused on suppressing the interaction such as the addition of silicon to the aluminium matrix or adding a protective layer of zirconium around the U–Mo kernels. These solutions reduced the swelling, allowing a maximum burnup of up to 4.7×10^{21} fissions/cm^3 to be reached, but this was still 30 per cent short of what was needed. It was recognized that the reason for the rapid swelling was the process of recrystallization, where gas bubbles form and eventually coalesce to form larger bubbles.

Experiments were conducted by the HERACLES group, with the U–Mo dispersal fuel annealed (heated and slowly cooled to remove stresses) before irradiation (known as EMPIrE and SEMPER FIDELIS experiments).[16] Results from non-destructive post-irradiation examinations (PIE) indicated 'good performance' for all variants, while destructive PIE is still under study. According to a US Department of Energy (DOE) report, the SEMPER FIDELIS experiment had some challenges. Nevertheless, one plate did reach a burnup of 80 per cent of the power levels that envelope the European BR2 and RHF reactors and the PIE indicated the performance of that plate was 'favorable'.[17]

Other recent irradiation experiments conducted on U_3Si_2 dispersion fuel are either in the process of being carried out or the results are being analysed. One experiment, COBRA–FUTURE, tested a full-size LEU element manufactured in the US at BWXT, while another, HiPROSIT, tested various high-density fuels fabricated by CERCA. The HiPROSIT experiment, part of the LEU–FOREVER consortium, is investigating the utility of U_3Si_2 fuel for European HPRRs rather than U–Mo monolithic fuels. An early result was reported at the 2021 RERTR conference in late April showing encouraging results, indicating a 'credible alternative pathway for LEU conversion of high-performance reactors still using HEU'.[18]

The full regulatory qualification process requires irradiating several plates in the reactor core along with a mixed fuel element containing standard HEU fuel and LEU fuel. In addition, a further complication is that the licensing of the fuel will be different for different reactors and regulatory bodies. Therefore, it is expected that the full testing programme will extend at least until 2027 for U–Mo dispersion fuel and the fuel will not be available until 2031–2036. At the April 2021 RERTR meeting, the schedule presented a more optimistic schedule, suggesting that all testing, PIE, and non-destructive testing for all fuels will be completed by 2024.[19]

European High-Performance Research Reactors

The US stance on supplying fuel for European reactors is governed by the 1992 Schumer Amendment, which prohibits the export of HEU unless nations have committed to or are in the process of converting.[20] This amendment offers critical support for NNSA's efforts to convert the reactors to LEU. Four European high-performance research reactors are incapable of conversion using commercially accessible fuels.

The first reactor is the French 100 MW JHR reactor scheduled to start up in 2025–2027. The proposed LEU fuel for the JHR is U–7Mo, but an independent evaluation published in 2015 recommended that JHR operators investigate uranium silicide dispersion fuel as a 'backup' since previous high flux reactors (with peak fluxes as high as 10^{14} n·cm^{-2}·s^{-1}) have been converted using this fuel and the fabrication is well understood and less costly, but the drawback is that a higher enrichment will be required to reach the design neutron flux (exceeding 5×10^{14} n·cm^{-2}·s^{-1}) in the core and in the reflector. According to NNSA officials, their French CEA counterparts have pledged to convert to use high-density LEU dispersion fuel when such fuel is available.

The second reactor is the BR2 reactor at the Belgian nuclear research centre known as SCK.CEN.[21] The reactor uses about 29 kg of weapons-grade, 93 per cent enriched, fuel per year, and the operator has agreed to convert to LEU as soon as an appropriate LEU fuel is qualified. NNSA obtained a licence in 2006 to export 85.5 kg of 93.4 per cent enriched HEU to France for BR-2 fuel production. At the time the US Nuclear Regulatory Commission (NRC) complied with NNSA's request because it was expected that the reactor would be converted to LEU fuel within three years, although this did not happen. Even though the reactor was not converted as expected, further transfers of 93.5 kg and 144 kg of weapons-grade HEU occurred in 2010 and 2014, respectively. Both U–7Mo Dispersal Fuel or U_3Si_2 Dispersal Fuels are being considered for the BR-2 reactor.

The third reactor is the French RHF reactor based in Grenoble, managed and operated by the Institut Laue-Langevin (ILL). In 1998, operators committed to converting the RHF to LEU; however 'by that time France began receiving HEU from Russia, so the conversion efforts were suspended'.[22] In the late 1990s there were several exports of HEU from Russia to France for the RHF reactor in Grenoble and the ORPHEE research reactor in Saclay.[23] The NNSA Y-12 National Security Complex applied for a NRC licence in 2010 to transfer 160 kg of 93.4 per cent enriched HEU to France to fabricate fuel for the RHF in two shipments of 80 kg. ILL predicts that under the current fuel inventory it can operate until 2020–2025.[24] According to the 2016 National Academy report, the RHF can meet its operating envelope and safety margin requirements with currently developed U_3Si_2 dispersion fuels.

The fourth reactor is the German Forschungs-Neutronenquelle Heinz Maier-Leibnitz-II Reactor (FRM-II) operated and managed by the Technische Universität München (TUM). FRM-II is one of the few research reactors in Europe that still uses HEU fuel. It was constructed after a *de facto* global moratorium on building new reactors using HEU. The reactor is fuelled with 93 per cent HEU, even though eight years before first criticality Argonne National Labs had proposed an alternative design. The Bavarian government set the bar for conversion so high (a performance reduction of less than 5 per cent and no modifications to the core structure) that it has required decades of research to find fuel to satisfy these objectives.

In response to worldwide criticism, the German government committed to converting the FRM-II to 50 per cent HEU by 2010. However, when the deadline drew near, Berlin chose to postpone any action and further extend the deadline until

2018. Russia acknowledged in July 2016 that it is providing HEU for the FRM-II, undermining US attempts to use its control over HEU shipments to compel Germany towards conversion.[25] In October 2019, the Green Party's parliamentary faction petitioned the Bavarian Ministry of Environment and Consumer Protection to halt the operation of the plant until it changed to fuel that contained less than 50 per cent U-235.[26]

A recent press release stated that the FRM-II operators are pursuing all three fuel types but have agreed to have made the decision to select 'which of the three fuel variants' will be considered for conversion.[27] According to NNSA, TUM is required to submit a report outlining its plans to convert the reactor to operate on fuel with an enrichment of less than 40 per cent which is an additional step from the government-mandated 50 per cent enrichment fuel. While 40 per cent enriched fuel does not qualify as LEU, it is a positive step and aligns with Recommendation 4a of the 2016 National Academies study on reactor conversion.[28]

US Reactors

The US has six high-performance research reactors (USHPRR) that all use HEU fuel cores. There are plans to convert all of them when appropriate fuel is available. In collaboration with European colleagues, NNSA has successfully developed a high-density U–10Mo alloy LEU fuel that can be used to convert all USHPRRs. In 2009, researchers completed feasibility studies demonstrating the ability to convert the National Institute of Standards Reactor (NBSR), the University of Missouri Research Reactor (MURR), and the Massachusetts Institute of Technology Reactor (MITR). The other three reactors, the Advanced Test Reactor (ATR), and its associated critical assembly (ATRC), and the High Flux Isotope Reactor (HFIR), have all completed preliminary designs for U–10Mo fuel. However, unlike the other reactors, the HFIR has selected a 'silicide (U3Si2) fuel design with a fuel meat density of 4.8 gU/cm^3. The HFIR fuel choice will 'require additional fuel qualification due to the complex fuel design containing boron and since HFIR exceeds existing silicide qualification limits'.[29]

A significant milestone was NNSA's submission of a report on U–Mo Monolithic fuel to the NRC in 2017. According to NNSA, the report contains data about the performance of the new fuel and how it holds up under a variety of conditions.[30] Each of the USHPRRs is unique in design and utilization, and each reactor's fuel element needed to be redesigned to use LEU fuel. BWXT and Y-12 struggled for several years to achieve consistency in fabrication, but BWXT now has the capability to fabricate fuel for all NRC-licensed reactors. However, experiments still need to be conducted to understand the limits on bounding performance specifications (performance limits for safe use of the fuel). LEU fuel for the MIT reactor is set to be qualified in 2024 and the reactor converted to using the new fuel a few years after that. Other conversions are then expected to follow. The successful conversion of US reactors requires meeting a series of milestones, which operators seem well on the way to achieving.[31]

Russian Reactors and HEU Production: Steps Forward but One Step Back

Conversion of Russian Reactors

An investigation by the James Martin Center for Nonproliferation Studies (CNS) in 2021 found that Russia still has 43 HEU-fuelled facilities.[32] Converting most of the remaining Russian research reactors is possible with existing or soon-to-be-qualified LEU fuel, but it is not a 'high priority for Russia'.[33] The 2016 National Academies study highlighted Russia's enormous number of reactors and its unique hazards and concerns. Many of these reactors contain critical and subcritical assemblies that can pose a particular risk because the fuel is only lightly irradiated and hence vulnerable to theft.

Earlier Russia–US feasibility studies on converting six Russian HEU-fuelled research reactors have not led to action amidst the demise of US–Russian relations and nuclear cooperation.[34] The six reactors are the IR-8, OR, and ARGUS reactors at the Kurchatov Institute, IRT-M at the Moscow Engineering Physics Institute, IRT-T at the Tomsk Polytechnic University, and MIR.M1 at the Scientific Research Institute of Atomic Reactors.[35] Conversion to LEU fuel was judged to be viable for all six reactors, and one of the six reactors (Argus) was successfully converted in 2014. Conversion of the IRT-M reactor at the Moscow Engineering Physics Institute (MEPhI) was also envisaged. Nonetheless, a decade later, the other five research reactors continue to operate on HEU, with apparently no intention to convert.

HEU Production

Russia continues to produce HEU both for internal use at research reactors and for export. In 2011, Russia's Rosatom announced plans to boost the capacity of an enrichment plant in Zelenogorsk by approximately 30 per cent.[36] Much of this work was done to make fuel for research and fast neutron reactors domestically.[37] In terms of exports, the Zelenogorsk facility generated HEU in 2015 for Germany's FRM-II research reactor.[38] Rosatom supplied a shipment of HEU fuel to China's CEFR fast neutron reactor in July 2019.[39] It is unclear why Russia continues to produce HEU while having a great deal of HEU stocks. However, the CNS report suggests that 'it may be considered easier to produce fresh HEU rather than retrieve it from the stocks' and that HEU from the weapons programme may have a 'chemical composition unsuitable for the needs of foreign customers opt for freshly enriched HEU'.[40]

Other Countries

Belarus

It is difficult to assess the current stockpile of HEU in Belarus since the country does not report this information voluntarily. However, 2014 estimates compiled by CNS showed 80–280 kg of HEU, with approximately 40 kg of it enriched to 90 per cent or higher levels.[41] The 2020 estimates of the International Panel on Fissile Materials provide a far larger range of 100–1,000 kg of HEU.[42] In a joint Belarus–US statement made on the side-lines of the 2010 Astana summit of the Organization for Security and Co-operation in Europe (OSCE), Belarus committed to eliminating its HEU stockpile by 2012.[43] In 2011, after several batches of HEU had been successfully transported to Russia, Belarus halted the HEU-removal process in retaliation for EU and US sanctions.[44] Nevertheless, there is an ongoing project on developing high-density LEU fuel for the Hyacinth critical assembly facility, involving the US and Russia. Moreover, the two critical facilities (Hyacinth and Crystal) and one subcritical assembly (Yalina-B) already could run with either HEU or LEU for fuel and prefer using LEU in the fast critical assembly, although have not yet parted with their HEU. Belarus is also planning to build a new storage facility for non-irradiated nuclear material, Yavor-1.[45]

South Africa

South Africa's HEU stockpile was developed at its pilot enrichment plant in the late 1970s as part of its nuclear weapons programme, which was dismantled in the early 1990s as apartheid came to an end.[46] South Africa's government considers information regarding its HEU stockpile confidential. The IAEA safeguards the material and is aware of the amount because of South Africa's first declaration to the agency in 1991. The Washington-based Institute of Science and International Security estimated that South Africa possessed 700–750 kg of HEU at the end of 2014.[47]

South African Nuclear Energy Corporation's NTP Radioisotopes SOC Ltd, in co-operation with the US NNSA, completed the conversion of its molybdenum-99 medical isotope production process from HEU to LEU in 2008,[48] and NTP has used only LEU targets since August 2017.[49] Hence, the last issue related to the HEU minimization effort in South Africa is its current HEU stock (both fresh and in spent fuel). After 6.5 kg of US-origin, South African HEU spent fuel was shipped back to the US in 2011,[50] South Africa's HEU stock is said to only contain material produced indigenously.

South Africa refuses to give up its indigenously produced stock of HEU. It considers that HEU might be useful for socio-economic development in the future and also that the possession of HEU remains a symbol of the country's international status and sovereignty in nuclear matters and insistence that non-nuclear-weapon states' rights under the NPT not be curbed.[51] Finally, South Africa might use HEU possession as a bargaining chip for global nuclear risk reduction through nuclear weapons disarmament.[52]

Kazakhstan

For the past five years, Kazakhstan has been steadily working with international partners, including the US, the IAEA, and Russia, through bilateral and multilateral initiatives to eliminate HEU from its Soviet-built research reactors. These efforts include the completion of the conversion of the VVR-K (or WWR-K) reactor at the Institute of Nuclear Physics (INPh) in Alatau and ongoing conversions of two research reactors IGR and IVG.1M at the National Nuclear Center (NNC) in Kurchatov, as well as down-blending or removing fresh and spent HEU fuel. The complete elimination of all HEU from the VVR-K reactor, including the down-blending of 49.3 kg of fresh HEU fuel and the removal of 158.3 kg of irradiated HEU, was reported by NNSA in September 2020.[53]

At the 2016 Nuclear Security Summit, Kazakhstan pledged to convert two additional reactors, the IGR reactor and the IVG.1M reactor at the NNC on the former Semipalatinsk Nuclear Test Site.[54] In 2019, a decision was made to down-blend fresh HEU graphite fuel from the IGR reactor at the Ulba Metallurgy Plant (UMP). Because the HEU in the reactor was mixed with graphite, it required 25 trucks to move only 2.9 kg of unirradiated HEU, accomplished in late 2020.[55] In 2017, two experimental channels with LEU fuel, manufactured by Russia's Federal State Unitary Enterprise Scientific-Research Institute 'LUCH', were loaded into the IVG.1M reactor for testing and were found to be successful.[56]

A significant milestone in the IVG.1M conversion process was accomplished in February 2021, when LUCH-produced LEU fuel was transported to the reactor site. Then, in March 2022 the reactor was fuelled with LEU fuel and May 2022 the reactor went on-line fully converted to LEU.[57] Despite considerable progress with research reactors conversions, Kazakhstan remains the only country in the Central Asian region with significant HEU holdings. Around 10 tons of HEU are expected to be included in spent fuel from the deactivated BN-350 fast reactor.[58] While Kazakhstan has announced that spent fuel would be housed in a secure long-term storage facility at the Semipalatinsk Test Site and will be subject to IAEA safeguards, Kazakhstan has made no preparations for permanent storage or disposal.

Japan

The Kyoto University Critical Assemblies (KUCA) have been working with Argonne National Laboratories of the US. For the second LEU fuel used in KUCA, Kyoto University has identified a solution using fuel fabricated by Framatome CERCA. The U-7Mo powder has been atomized by KAERI and has formed into aluminium-clad coupons by Framatome-CERCA and others. When the conversion takes place, it will be the first critical assembly to be fully converted.[59] In August 2022 it was announced that 45 kg of HEU was removed and repatriated to the US, fulfilling a 2016 Nuclear Security Summit commitment.

Remaining Challenges

Molybdenum-99

The Belgian firm IRE, the last major producer to convert from HEU to LEU targets, has begun processing irradiated LEU targets to produce the vital medical isotope molybdenum-99.[60] Mo-99 (often referred to as 'Moly-99') is used in around 30 million medical operations each year to identify a large range of medical disorders. It is produced by irradiating LEU 'target' plates with a half-life of less than a week, and then separating the Mo-99, which decays into the even shorter lifetime Tc-99 (technicium-99). Mo-99 is the most widespread used diagnostic tracer, being utilized in around 30 million medical operations each year to identify a wide variety of illnesses affecting several bodily systems, including cancer, heart disease, and neurological disorders such as dementia.[61]

IRE's total conversion to LEU is now underway but will take some time to complete, as its extensive network of global clients must first acquire local regulatory permission for the new LEU-based Mo-99 and resultant Tc-99M. Nonetheless, according to one US official: 'With IRE there is a finality there. They have made finalizing this conversion a true priority and they are communicating it to their customers.'[62] Likewise, the Dutch High Flux Reactor's operators declared in March 2021 that they were irradiating only LEU targets for Mo-99 processing (at IRE and Curium).[63] As a result, it was announced on 20 December 2021 that there was a sufficient US domestic supply of Mo-99 and therefore the Joe Biden Administration could meet a Congressional mandate to end exports of HEU for Mo-99 production.[64]

To reach this stage, a succession of technological, economic, and political obstacles had to be overcome. The OECD/NEA High-Level Group on the Security of Supply of Medical Radioisotopes (HLG-MR) concluded in a series of reports that previous Mo-99 supply shortages had occurred because older HEU-fuelled reactors had their capital costs effectively subsidized, making it difficult for new LEU-based competitors to enter the market and compete successfully.[65] Furthermore, numerous European host nations for large isotope manufacturers vowed at the 2012 Nuclear Security Summit to press their companies to transition to LEU, pending regulatory permission.[66] In addition, the American Medical Isotopes Production Act of 2013 increased pressure on international reactors to convert. The bill increased support for LEU-based Mo-99 production via financial incentives, provided financing for the development of domestic non-HEU Mo-99 production, and called for an end to HEU export for isotope production by 2020, although it authorized extensions if there was insufficient non-HEU Mo-99 production for the US market.[67,68]

Reactors for Outer Space Missions

There has been a revival of interest in deep space missions using fission reactors. The first use is to generate energy for a moon or Mars base, referred to as a Surface Power Reactor. Radioisotope Thermoelectric Generators (RTGs) that depend on the heat generated by the natural decay of radioactive isotopes without a fission chain reaction is not practicable for delivering the required 50 kilowatts (kW) of power.[69] The US National Aeronautics and Space Administration (NASA) conducted a test of a solution in 2018 with the KRUSTY reactor experiment, generating heat using a fission reactor.[70] This heat was then transformed into mechanical energy in a Stirling Engine, which utilizes heat to move a piston up and down. Finally, the mechanical energy is converted to electricity, which could then be used to power the outpost.

However, the KRUSTY reactor takes 30 kg of weapons-grade HEU to create 1 kW.[71] This amount of HEU would be sufficient to provide the nuclear fuel for a half-dozen nuclear warheads. Nonetheless, it would only generate enough electricity to heat an electric kettle. According to NASA, a 10-kW version would need 50 kg of weapons-grade uranium and a LEU version of this reactor would be much heavier—a critical factor to consider. According to recent research, a LEU-based reactor would be around twice as heavy.[72] In December 2020, the Trump administration issued a directive to restrict (but not ban) HEU use in future space missions. HEU use 'should be limited to applications for which the mission would not be viable with other nuclear fuels or non-nuclear power sources', the new regulation states. This high bar all but bans the use of HEU for future power missions.[73]

A fission reactor might also be used to assist with rocket propulsion. The reactor generates enough heat to 'burn hydrogen in rocket engines'. Although the thrust is equivalent to that of conventional liquid-fuel rockets, the efficiency is likely to be far better. The advantage of this application is that astronauts could return to Earth for up to three months of their scheduled seven-month journey if anything goes wrong on the way. This window is just a few days if conventional rockets are used. A call for proposals has been issued for designs of a reactor that can power hundreds of megawatts. Four companies are anticipated to compete on the award (BWX Technologies, General Atomics, X Energy, and Ultra Safe Nuclear Corp.); all the designs employ LEU.

Conclusion

A great deal of effort has been made over the past 50 years to minimize the use of HEU around the world, reducing the ability of terrorists or rogue states to gain access to the fissile material easiest to fashion into a basic nuclear weapon. In fact, well over 100 reactors have thus far either been converted or shut down since the US RERTR programme began. In addition, the major producers of the critical isotope molybdenum-99 are all now using LEU for their neutron targets.

Progress has slowed, however, as the 'low-hanging fruit' has been picked and efforts have hit technical and political obstacles. Many of the remaining HEU-fuelled reactors are in Russia, and the war in Ukraine as well as Russia's ensuing isolation and pinched finances are likely to continue holding up progress for the foreseeable future. Meanwhile, efforts to convert HPRRs in the US and Europe are moving ahead but very slowly given the challenges of developing highly unique fuel as well as the ongoing need for these reactors. Finally, there is a new interest in HEU-based space reactors which could complicate worldwide efforts to reduce HEU use. Absent terrorist acquisition and use of such material, maintaining the necessary political and financial commitment for at least another decade will remain a constant challenge given competition with other challenges that tend to be viewed as more urgent.

Notes

1. National Academies of Sciences, Engineering, and Medicine, *Reducing the Use of Highly Enriched Uranium in Civilian Research Reactors* (Washington, DC: The National Academies Press, 2016), 2. www.nap.edu/catalog/21818/reducing-the-use-of-highly-enriched-uranium-in-civilian-research-reactors
2. The explosion involved a fission device using plutonium derived from natural uranium metal fuel (not HEU) produced in the CIRUS (Canada India Research Utility Service) reactor. In a bilateral agreement, Canada and India had agreed the reactor would be employed for 'peaceful purposes only', but India separated the plutonium from spent nuclear fuel in a domestic reprocessing facility, then used it for what was effectively a nuclear weapon. See Comprehensive Nuclear-Test-Ban Treaty Organization, '18 May 1974–Smiling Buddha', n.d. https://www.ctbto.org/specials/testing-times/18-may-1974-smiling-buddah/#:~:text=On%2018%20May%201974%2C%20India,a%20fission%20device%20using%20plutonium
3. Anya Loukianova and Cristina Hansell, 'Leveraging U.S. Policy for a Global Commitment to HEU Elimination', in William C. Potter and Cristina Hansell, eds., *The Global Politics of Combating Nuclear Terrorism: A Supply-Side Approach* (Abingdon: Routledge, 2010), 33–4.
4. Charles D. Ferguson and William C. Potter, *The Four Faces of Nuclear Terrorism* (New York: Routledge, 2005), 134.
5. Miles Pomper, Ferenc Dalnoki-Veress, Margarita Kalinina-Pohl, and Artem Lazarev, *The Final Stretch: Tackling Remaining HEU Challenges*, Occasional Paper 51 (Monterey, CA: James Martin Center for Nonproliferation Studies (CNS), 2021), Appendix 2. https://nonproliferation.org/the-final-stretch-tackling-remaining-heu-challenges
6. Council on Foreign Relations, 'U.S.–China Relations', 2022. https://www.cfr.org/timeline/us-china-relations
7. Matthew Bunn and Eben Harrell, *Consolidation: Thwarting Nuclear Theft* (Cambridge, MA: Project on Managing the Atom, Belfer Center, 2012), 22; The White House, 'U.S.–Russia Joint Fact Sheet: Bratislava Initiatives', 2005. https://georgewbush-whitehouse.archives.gov/news/releases/2005/02/20050224-7.html

8. Author communication with National Nuclear Security Administration (NNSA) personnel, 25 March 2021.
9. Pomper et al., 'The Final Stretch', Appendix 3.
10. Miles A. Pomper, 'The 2012 Seoul Nuclear Security Summit and HEU Minimization'. Working Paper, US–Korea Institute at SAIS, January 2012. https://www.jstor.org/stable/resrep11124?seq=1
11. National Academies of Sciences, Engineering, and Medicine, *Reducing the Use of Highly Enriched Uranium in Civilian Research Reactors*; author communication with National Nuclear Security Administration (NNSA) personnel.
12. M. K., J. Meyer, J. F. Gan, D. D. Jue, E. Keiser, A. Perez, D. M. Robinson, N. Wachs, N. Woolstenhulme, G. L. Hofman, and Y. S. Kim, 'Irradiation performance of U-Mo monolithic fuel', *Nuclear Engineering and Technology*, 46/2 (2014), 169–82.
13. Jian Gan, D. Dennis, D. M. Keiser, Adam Robinson Wachs, Brandon Miller, and Todd R. Allen, 'Transmission Electron Microscopy Characterization of Irradiated u-7mo/al-2si dispersion fuel', *Journal of Nuclear Materials*, 2/3 (2010),396, 234–239. https://doi.org/10.1016/j.jnucmat.2009.11.015; Jian Gan, Brandon Miller, Dennis D. Keiser, Adam Robinson, James W. Madden, Pavel Medvedev, and D. M. Wachs, 'Microstructural Characterization of Irradiated u-7mo/al-5si Dispersion Fuel to High Fission Density', *Journal of Nuclear Materials*, 1/3 (2014), 454, 434–45. https://doi.org/10.1016/j.jnucmat.2014.08.052
14. National Academies of Sciences, Engineering, and Medicine, *Reducing the Use of Highly Enriched Uranium in Civilian Research Reactors*.
15. Sven Van den Berghe and Patrick Lemoine, 'Review of 15 Years of High-Density Low-Enriched U–Mo Dispersion Fuel Development for Research Reactors in Europe', *Nuclear Engineering and Technology*, 46/2 (2014), 125–46.
16. National Academies of Sciences, Engineering, and Medicine, *Reducing the Use of Highly Enriched Uranium in Civilian Research Reactors*.
17. Robert Birgeneau et al., *The Scientific Justification for a US Domestic High-Performance Reactor-Based Research Facility* (Washington, DC: Office of Science, US Department of Energy, 2020).
18. Sven van den Berghe, 'Heracles / LEU Forever Campaign Progress and Plans', Virtual Presentation at RERTR 2021, 22 April 2021; Update on European progress at; Wight, Jared, Stéphane Valance, and Stefan Holmström, 'Innovation and Qualification of LEU Research Reactor Fuels and Materials', *EPJ Nuclear Sciences & Technologies*, 9 (2023), 3.
19. Ibid.
20. Edwin Lyman, 'The Congressional Assault on RERTR', Presentation at the 25th Annual RERTR Meeting, Chicago, 10 October 2003. https://inis.iaea.org/collection/NCLCollectionStore/_Public/35/066/35066249.pdf
21. 'Infrastructure', SCK CEN, 2022. https://www.sckcen.be/en/about-sck-cen/corporate-information/infrastructure
22. International Panel on Fissile Materials, 'Russia to Supply HEU Fuel for French Research Reactor', *IPFM Blog* [online], 18 September 2013. http://fissilematerials.org/blog/2013/09/russia_to_supply_heu_fuel.html
23. International Panel on Fissile Materials, 'US to Supply HEU for the High Flux Reactor in France', *IPFM Blog* [online], 11 March 2010. https://fissilematerials.org/blog/2010/03/us_to_supply_heu_for_the_.html
24. Ibid.

25. International Panel on Fissile Materials, 'Russia Confirmed Supplying HEU to the FRM-II Reactor in Germany', *IPFM Blog* [online], 18 July 2016. http://fissilematerials.org/blog/2016/07/russia_confirmed_supplyin.html
26. Research Neutron Source, 'Report on Legality of the Operation of FRM II with Highly Enriched Uranium (HEU)', Technical University of Munich, 16 January 2020. https://www.frm2.tum.de/en/news-media/press/news/news/article/gutachten-zur-rechtmaessigkeit-des-betriebs-des-frm-ii-mit-hochangereichertem-uran-heuo
27. Research Neutron Source, 'New Lower Enrichment Fuels Designed for the FRM II', Technical University of Munich, 23 September 2021. https://www.frm2.tum.de/en/frm2/news-media/press/news/article/neue-brennstoffe-zur-umruestung-des-frm-ii-auf-niedrigere-anreicherung
28. Author communication with National Nuclear Security Administration (NNSA) personnel; National Academies of Sciences, Engineering, and Medicine, *Reducing the Use of Highly Enriched Uranium in Civilian Research Reactors*.
29. E. H. Wilson, et al., *US High Performance Research Reactor Preliminary Design Milestone for Conversion to Low Enriched Uranium Fuel* (Washington, DC: Office of Scientific and Technical Information, US Department of Energy, 2019). https://www.euronuclear.org/download/rrfm-2019-part-3/?wpdmdl=4557&refresh=60197569b98b51612281193
30. NNSA, 'NNSA Reaches Milestone in Developing New Nuclear Fuel for U.S. High-performance Research Reactors', *Energy.gov* [online], 20 December 2017. https://www.energy.gov/nnsa/articles/nnsa-reaches-milestone-developing-new-nuclear-fuel-us-high-performance-research
31. E. H. Wilson, et al., 'Preliminary Fuel Development and Reactor Design Milestones for LEU Conversion of US High Performance Research Reactors', Presentation at the International Conference on Nuclear Security: Sustaining and Strengthening Efforts, 10–14 January 2020, Vienna, Austria.
32. Miles Pomper, et al., *The Final Stretch*.
33. National Academies of Sciences, Engineering, and Medicine, *Reducing the Use of Highly Enriched Uranium in Civilian Research Reactors*, 5.
34. US Department of Energy, 'US, Russian Federation Sign Joint Statement on Reactor Conversion', *Energy.gov* [online], 26 June 2012. https://www.energy.gov/articles/us-russian-federation-sign-joint-statement-reactor-conversion
35. Jordi Roglans, 'NNSA's Russian Reactor Conversion Program: Historical Overview, Major Accomplishments, Current Status', Presentation, National Academy of Sciences, Status and Progress on Eliminating HEU Use in Fuel for Civilian Research and Test Reactors, 16 April 2015. https://nas-sites.org.
36. World Nuclear Association, 'Russia's Nuclear Fuel Cycle', WNA [online], December 2021. https://www.world-nuclear.org/information-library/country-profiles/countries-o-s/russia-nuclear-fuel-cycle.aspx
37. ECP, Annual Report of Electrochemical Plant, 2012, 13. https://www.ecp.ru/sites/default/files/download/go_ecp_12.pdf
38. TVEL, 'Report 2015', 79. https://web.archive.org/web/20161108121546/http://www.tvel2015.ru/static/pdf/tvel_report_2015_ru.pdf
39. World Nuclear News, 'TVEL Delivers Fuel for China's Experimental Fast Reactor: Uranium & Fuel', *WNN* [online], 10 July 2019. https://www.world-nuclear-news.org/Articles/TVEL-delivers-fuel-for-China%E2%80%99s-experimental-fast-r
40. Pomper et al., *The Final Stretch*.

41. CNS Researchers, 'Civilian HEU: Who Has What?', Nuclear Threat Initiative (NTI) [online], November 2014. https://media.nti.org/pdfs/heu_who_has_what_2014_11.pdf
42. International Panel on Fissile Materials, 'Highly Enriched Uranium', IPFM [online], 4 September 2021. https://fissilematerials.org/materials/heu.html
43. Robert Golan-Vilella, 'Belarus Agrees to Give Up HEU Stockpile', *Arms Control Today* [online], January 2011. https://www.armscontrol.org/act/2011-01/belarus-agrees-give-up-heu-stockpile
44. Miles A. Pomper and Philippe Mauger, 'Crossing the Finish Line: Ending the Civilian Use of Highly Enriched Uranium', Policy Analysis Brief, The Stanley Center [online], 14 May 2014. http://www.stanleyfoundation.org/publications/PomperPAB514.pdf
45. World Bank, *Strengthening Hydromet and Early Warning Services in Belarus: A Road Map* (Washington, DC: The World Bank Group, 2020), 23. https://openknowledge.worldbank.org/handle/10986/34336
46. Thomas B. Cochran, 'Highly Enriched Uranium Production for South African Nuclear Weapons', *Science & Global Security*, 4/2 (1994), 161–76; Minister of Energy, 'Reply to Question 1807 (NW2199E)', Parliamentary Monitoring Group, August 2012. https://pmg.org.za/question_reply/342
47. David Albright, 'Highly Enriched Uranium Inventories in South Africa Status as of End of 2014', Institute for Science and International Security [online], 16 November 2015, 8. http://large.stanford.edu/courses/2019/ph241/greene1/docs/isis-2015.pdf#:~:text=South%20Africa%20produced%20highly%20enriched%20uranium%20%28HEU%29%20for,early%201990%2C%20was%20designed%20to%20make%20weapon-grade%20uranium
48. NNSA, 'NNSA Collaborates with South African Firm on Groundbreaking Conversion to Low-enriched Uranium-based Molybdenum-99 Production', *Energy.gov* [online], 22 September 2017. https://www.energy.gov/nnsa/articles/nnsa-collaborates-south-african-firm-groundbreaking-conversion-low-enriched-uranium
49. National Academies of Sciences, 'Current Molybdenum-99 Supply', in *Opportunities and Approaches for Supplying Molybdenum-99 and Associated Medical Isotopes to Global Markets: Proceedings of a Symposium* (Washington, DC: The National Academies Press, 2018), 13–16. https://www.nap.edu/read/24909/chapter/5
50. World Nuclear News, 'US-Origin HEU Returned from South Africa', *WNN* [online], 19 August 2011. https://www.world-nuclear-news.org/RS-US_origin_HEU_returned_from_South_Africa-1908114.html
51. Robin Moser, '"The Major Prize": Apartheid South Africa's Accession to the Treaty on the Non-Proliferation of Nuclear Weapons, 1988–91', *The Nonproliferation Review* [online], 6 January 2020, 26/5–6. https://www.tandfonline.com/doi/full/10.1080/10736700.2019.1696543
52. Jacob Zuma, 'Statement by President of the Republic of South Africa at Leaders' Working Dinner hosted by President of the Republic of Korea', Nuclear Security Summit, Seoul, 26 March 2012. http://www.thepresidency.gov.za/speeches/statement-president-republic-south-africa-he-jacob-zuma%2C-leaders%E2%80%99-working-dinner-hosted
53. NNSA, 'Kazakhstan and U.S. Cooperate to Eliminate Highly Enriched Uranium in Kazakhstan', *Energy.gov* [online], 22 September 2020. https://www.energy.gov/nnsa/articles/kazakhstan-and-us-cooperate-eliminate-highly-enriched-uranium-kazakhstan
54. Government of Kazakhstan, 'National Progress Report: Kazakhstan', 2016 Nuclear Security Summit, Washington, DC, 31 March 2016. https://www.nss2016.org/document-center-docs/2016/3/31/national-progress-report-kazakhstan

55. Office of the Federal Register, National Archives and Records Administration, 21 December 2020. https://www.dateas.com/en-us/docs/federal-register/2020/12/2; author communication with National Nuclear Security Administration (NNSA) personnel, December 2020.
56. Christina Nunez, 'The Ongoing Effort to Convert the World's Research Reactors', *ANS Newswire* [online], 10 July 2020. https://www.ans.org/news/article-211/the-ongoing-effort-to-convert-the-worlds-research-reactors
57. World Nuclear News, 'Kazakh Research Reactor Restarts after Conversion to LEU', *WNN* [online], 11 May 2022. https://www.world-nuclear-news.org/Articles/Kazakh-research-reactor-restarts-after-conversion
58. World Nuclear News, 'Kazakhstan's Unique Contribution', *WNN* [online], 7 July 2014. https://world-nuclear-news.org/Articles/Kazakhstan-s-unique-contribution
59. H. Unesaki, et al., 'KUCA Support of Japan's Nuclear Engineering and Science', RERTR Virtual Conference, 20 April 2021.
60. International Panel on Fissile Materials, 'Belgium Produced First Commercial Batch of Mo-99 with LEU', *IPFM Blog* [online], 11 May 2020. https://fissilematerials.org/blog/2020/05/belgium_produced_firs
61. OECD/NEA, *The Supply of Medical Isotopes: An Economic Diagnosis and Possible Solutions* (Paris: OECD Publishing, 2019), 19–47. https://www.oecd-nea.org/jcms/pl_15144/the-supply-of-medical-radioisotopes; OECD/NEA, *The Supply of Medical Radioisotopes 2018 Medical Isotope Demand and Capacity Projection for the 2018–2023 Period* (Paris: OECD Publishing, 2018). https://www.oecdnea.org/cen/docs/2018/sen-hlgmr2018-3.pdf
62. Author interview with US officials, December 2020.
63. World Nuclear News, 'Dutch Research Reactor Ends Use of HEU Targets', *WNN* [online], 18 March 2021. https://www.world-nuclear-news.org/Articles/Dutch-research-reactor-ends-use-of-HEU-targets
64. US Department of Energy, 'U.S. Secretaries of Energy and Health and Human Services Jointly Certify Sufficient Worldwide Supply of Critical Medical Isotope', *Energy.gov* [online], 20 December 2021. https://www.energy.gov/articles/us-secretaries-energy-and-health-and-human-services-jointly-certify-sufficient-worldwide
65. High-Level Group on the Security of Supply of Medical Radioisotopes, 'Final Report of the Fourth Mandate of the High-level Group on the Security of Supply of Medical Radioisotopes (HLG-MR)', OECD/NEA [online], February 2021. https://www.oecd-nea.org/jcms/pl_55496/the-supply-of-medical-radioisotopes-final-report-of-the-fourth-mandate-of-the-high-level-group-on-the-security-of-supply-of-medical-radioisotopes-hlg-mr?details=true
66. Nuclear Security Summits, 'Belgium–France–Netherlands–United States Joint Statement: Minimization of HEU and the Reliable Supply of Medical Radioisotopes', 2012 Nuclear Security Summit, Seoul, 2012.
67. The act was passed as part of the fiscal year 2013 Defense Authorization Act, National Defense Authorization Act for Fiscal Year 2013 (2010), PL 112-239. https://www.congress.gov/112/plaws/publ239/PLAW-112publ239.pdf
68. Pomper et al., *The Final Stretch*.
69. David Kramer, 'No Bomb-Grade Uranium in Space, Says White House', *Physics Today* [online], 23 December 2020.
70. David I. Poston, Marc A. Gibson, Thomas Godfroy, and Patrick R. McClure, 'KRUSTY Reactor Design', *Nuclear Technology*, 206/Supp 1 (2020), S13–S30.
71. Kramer, 'No Bomb-Grade Uranium in Space, Says White House'.

72. Leonardo de Holanda Mencarini and Jeffrey C. King, 'Fuel Geometry Options for a Moderated Low-Enriched Uranium Kilowatt-Class Space Nuclear Reactor', *Nuclear Engineering and Design*, 340 (2018), 122–32.
73. Office of the Federal Register, National Archives and Records Administration.

Suggested Reading

Kuperman, Alan J., ed., (2013), *Nuclear Terrorism and Global Security: The Challenge of Phasing Out Highly Enriched Uranium* (New York: Routledge).

National Academies of Sciences, Engineering, and Medicine, (2016), *Reducing the Use of Highly Enriched Uranium in Civilian Research Reactors* (Washington, DC: The National Academies Press). www.nap.edu/catalog/21818/reducing-the-use-of-highly-enriched-uranium-in-civil ian-research-reactors

Pomper, Miles, Dalnoki-Veress, Ferenc, Kalinina-Pohl, Margarita, and Lazarev, Artem, *The Final Stretch: Tackling Remaining HEU Challenges*, James Martin Center for Nonproliferation Studies (CNS), Occasional Paper 51, 17 August 2021. https://nonproliferation.org/the-final-stretch-tackling-remaining-heu-challenges

Pomper, Miles A. and Mauger, Philippe, 'Crossing the Finish Line: Ending the Civilian Use of Highly Enriched Uranium', Policy Analysis Brief, The Stanley Foundation, May 2014. http://www.stanleyfoundation.org/publications/PomperPAB514.pdf

von Hippel, Frank, 'Banning the Production of Highly Enriched Uranium', International Panel on Fissile Materials, Research Report, No. 15, March 2016.

CHAPTER 35

CONVERSION OF A RESEARCH REACTOR FROM HEU TO LEU IN NIGERIA

HAMISU A. ADAMU AND SUNDAY A. JONAH

INTRODUCTION

IN 1976, the Federal Government of Nigeria selected Ahmadu Bello University (ABU), Zaria as a 'Centre of Excellence' in nuclear technology, giving impetus to the establishment of a nuclear programme in the country. The Centre of Excellence was initiated at ABU a year later, funded by a grant from the Nigerian government and managed by the Nuclear Science Programme Steering Committee (NSPSC).[1] The Committee established a Centre for Nuclear Science and Technology at ABU, now known as the Centre for Energy Research and Training (CERT). In 1986, the University Governing Council approved the establishment of CERT through Statute 29 of university law.[2] At about the same time, the Nigerian government enacted Act No. 46 of 1976, which established the Nigeria Atomic Energy Commission (NAEC), although it was not until 2006 that the NAEC was fully operational. In 1995, the Nigeria Nuclear Regulatory Authority (NNRA) was established by Act No. 19 as an independent nuclear regulatory body in line with global best practice, becoming operational in 2001.

Since its inception, CERT has undergone three remarkable transitions. The first transition was its move from the temporary site at the university's main campus to the present site near Jama'a village close to the Industrial Development Centre (IDC). Starting in March 1993, this involved the relocation of major equipment such as a neutron generator (NG), a transmission electrons microscope (TEM), X-ray fluorescence (XRF), X-ray diffractometer (XRD), and a liquid nitrogen plant (LNP). Final occupation of the laboratory was completed in 1996.[3]

The next transition was a technological one. This involved technical cooperation efforts with international organizations such as the International Programme in the Physical Sciences (IPPS) at Uppsala, Sweden,[4] the German Academic Exchange Programme (DAAD), and the International Atomic Energy Agency (IAEA). This transition included training (and re-training) of available personnel and acquisition of various equipment including the Nigeria Research Reactor 1 (NIRR-1) and related facilities. The third transition, which was related to the second, involved building confidence and developing analytical capabilities relevant to the organization. In this transition, capabilities such as analytical theoretical and technical expertise in CERT were developed and contacts made with stakeholders.

The development of indigenous capability in nuclear science and technology through these transitions was key to the successful conversion in 2018 of NIRR-1 from highly enriched uranium (HEU) to low enriched uranium (LEU), with the assistance from the IAEA, the United States Department of Energy (DOE), and other international partners. Nigeria had joined the IAEA in 1964 and the conversion project was in line with its aspiration to advance the peaceful applications of nuclear science and technology for the socio-economic development of the country. This chapter describes the historical background of CERT, the operation of NIRR-1, the motivations behind its conversion to LEU, the steps in the conversion programme in 2006–2018, and the challenges and lessons learned in the process.

Motivations Behind the Conversion of NIRR-1

Political Motivations (Insecurity in Nigeria: Nuclear Security Summits)

Converting a research or test reactor from the use of HEU fuel to LEU fuel is a hugely complicated process, and may be hindered by political disagreements, technical complications in the reactor, and the large financial outlay. In the case of Nigeria, the decision to convert encompassed a combination of political and technical drivers. Regarding the former, research reactors such as NIRR-1 are often located in areas with minimal security controls, such as in university campuses. This makes them especially vulnerable to non-state actors, including terrorist groups, which could potentially seek to acquire nuclear and radiological materials to use in an improvised nuclear device (IND).[5]

Since the terrorist attacks of 11 September 2001 ('9/11'), the international community has become increasingly concerned about non-state actors wanting to acquire nuclear materials in any form as a means of facilitating acts of nuclear terrorism.[6] In the Nigerian context this possibility has been further amplified in recent decades through the rise of homegrown violent extremists and jihadist militant groups.[7] Such groups, most notably Boko Haram, have often emerged on the basis of ethno-religious sentiments to perpetrate violent activities against the state and citizens of Nigeria. Nevertheless, there is no evidence to date that terrorist groups in Nigeria have specifically sought to acquire nuclear and radiological weapons.[8]

The rise in domestic insecurity has coincided with a growth in the nuclear sector in Nigeria, with facilities now scattered all over the country. These include the NIRR-1 research reactor and other smaller facilities that house radiological materials used in the country's medical and petroleum sectors.[9] The violent activities perpetrated by radical groups in Nigeria include suicide bombings, kidnapping for ransom, bombings of schools, among other attacks, although until now these have not involved the use of nuclear and radiological materials. However, there is widespread recognition that these groups pose a significant threat to nuclear facilities and must be protected.[10]

For these reasons, the Federal Government of Nigeria actively participated in the Nuclear Security Summits convened by then-US president Barak Obama. At the second summit held in 2012 at Seoul, South Korea, Nigeria put forward a pledge to convert the NIRR-1 reactor from HEU to LEU. In his summit speech, then-Nigerian president Goodluck Jonathan declared that Nigeria would accelerate 'efforts in partnership with the United States, China and the IAEA to convert the highly enriched uranium (HEU) research reactor in Nigeria to low enriched uranium (LEU) as a means of reducing threats for theft of fissile materials by potential rogue groups'.[11]

Technical Motivations (Global HEU Minimization: Reactor Aging/Modernization)

For decades, numerous international activities have been underway to convert reactors from the use of HEU fuel to LEU fuel, or to shut down and decommission HEU-fuelled research reactors as they reach the end of their service life. These activities support the long-standing non-proliferation objective of reducing and eventually eliminating the civilian use of HEU. To this end, in 1978 the US developed the Reactor Conversion Program—previously known as the Reduced Enrichment for Research and Test Reactors (RERTR) Program—which is operated by the Material Management and Minimization (M3) Conversion Program of the National Nuclear Security Administration (NNSA), under the US DOE. Efforts in this area have been supported by the IAEA, which in 2006 launched a Coordinated Research Project (CRP) aimed at supporting states that operated miniature neutron source reactors (MNSRs), run on HEU, to convert them to run on LEU.[12]

MNSRs are low power (nominal ~30 kilowatt; kW) research reactors, designed and manufactured by the China Institute of Atomic Energy (CIAE) in Beijing, which operate with particularly high-enrichment HEU reactor cores, over 90 per cent uranium-235, and consequently present potential risk from the perspective of nuclear terrorism. MNSRs are mainly used for neutron activation analysis, training, and testing of nuclear instrumentation.[13] The first MNSR was the Prototype MNSR sited at the CIAE, which went critical in 1984 and has been operational since.[14] There were eight additional commercial MNSRs developed in China, five of which were supplied to other countries, including Ghana, Iran, Pakistan and Syria, as well as Nigeria—all of which were originally fuelled by HEU enriched to 90 per cent or higher.[15]

NIRR-1 is one of the MNSR facilities outside China. Having been in operation since first criticality on 3 February 2004 with an HEU core, by the mid-2010s the NIRR-1 facility was confronted with failure of components due to ageing, with challenges encountered in obtaining replacement parts. From a technical perspective, conversion to LEU provided an opportunity to obtain new equipment including upgrading some of the reactor modules that were obsolete. Fortunately for NIRR-1 and facilities with similar core configurations, conversion to LEU would have little or no impact on performance, and therefore on utilization.[16]

Pre-Conversion Activities

Overview of the NIRR-1 HEU Core

Research on the development of MNSR reactors formally started in the early 1980s at the CIAE in Beijing. It was decided that the conceptual design should use uranium aluminium alloy (26.18 wt.per cent) of 90.3 per cent enriched uranium-235 as fuel meat and metal beryllium as reflectors. The preparation group of the strong neutron source reactor was established in February 1982, and at the same time the design and fabrication of fuel elements and beryllium reflectors were carried out. Some of these included full-scale zero-power reactor physics mock-up experiments,[17] thermal hydraulic simulations,[18] and chemical corrosion tests. The construction work of the Prototype MNSR building started in December 1981 and was completed in early 1983. In late 1983, the final core lattice design was set up to be lattices in the core arranged in 11 concentric circles. The Prototype has a total number of 376 fuel rods and 35 equally distributed dummy elements at the outermost circle.[19]

The commissioning of the Prototype was completed in March 1984, with the reactor reaching criticality shortly after and achieving the rated power of 27 kW. In view of the fact that the MNSR was to be installed in communities with dense populations and most of the staff of the operating organization might not have broad technical knowledge of nuclear reactors, it was decided that more safety margins in the commercial design would be provided to enhance control over release of radioactivity. The major differences between the commercial and the Prototype MNSR are as follows:

1. Kinetic calculation and experiments done using the Prototype indicated that the main factor in controlling the reactor is the moderator temperature coefficient which affects the peak power when a positive reactivity is quickly inserted. The Doppler coefficient of this reactor is relatively less important. The designed ratio

of hydrogen to uranium-235 (N_H/N_{U235}) was therefore reduced from 244 to 197. This meant the core was more under-moderated, leading to better suppression of power excursion. As a result, the commercial design had better inherent safety features and a longer core lifetime than the Prototype.

2. The main factors that affect the maximum steady state operable time are the inlet flow orifice that in turn determines the temperature effect. The height of the inlet flow orifice for the commercial design was increased up to 6 mm from the original value of 4.5 mm, while the cold state excess reactivity was increased to 4 metre Kelvin (mK), up from 3.0–3.5 mK. The change in orifice provided the commercial MNSR with better safety features.
3. A new protection system and reactivity regulators were employed in the commercial design, making it safer for installation in densely populated environments like university campuses.

The aforementioned features were incorporated into the design of the commercial MNSR (C-MNSR). In November 1988, the first MNSR was built at Shenzhen University, Guandon Province in China. A second was installed at the Shandong Experimental Centre of the Ministry of Mining Industry in May 1989, and a third in Shanghai in December 1991. The first MNSR commissioned outside China was at the Pakistan Institute of Nuclear Science and Technology (PINSTECH) in November 1989. Other MNSRs were commissioned in Pakistan in November 1989, Iran in May 1994, Ghana in January 1995, Syria in April 1996, and Nigeria in February 2004.[20] In 2019, the CIAE initiated a move to construct another MNSR in Thailand designed to operate at a maximum power level of 45 kWth, with use as both a neutron source and for medical application via the boron neutron capture therapy (BNCT) technique.[21] This facility was designed to start up with LEU unlike the other facilities as, by this time, all MNSRs were undergoing conversion to HEU.

Since 1978, the US NNSA Reactor Conversion Program has worked to convert research and test reactors from the use of HEU fuel to LEU fuel and to verify the shutdown and decommissioning of HEU-fuelled research reactors. But it was not until the IAEA's 2006 CRP that a viable fuel design was found to convert the MNSRs from their original HEU fuel to LEU fuel with no degradation in overall reactor performance.[22] Following this development, the Prototype MNSR in China was converted in March 2016 and the Ghana MNSR (GHARR-1) in August 2017. NIRR-1 was also converted and on 2 November 2018 went critical with LEU fuel. Other MNSR facilities sited in Iran, Pakistan, and Syria are yet to be converted, but discussions on this are underway as part of global HEU minimization efforts.

Technical Specifications of NIRR-1

The Nigeria MNSR is a tank-in-pool type reactor. The reactor complex contains five major components: reactor assembly, control console, auxiliary systems, irradiation system, and the pool. Figure 35.1 shows a vertical cross-section of the reactor. The reactor assembly consists of the reactor core, beryllium reflector, small fission chambers for detecting neutron fluxes, one central control rod and its drive mechanism, and

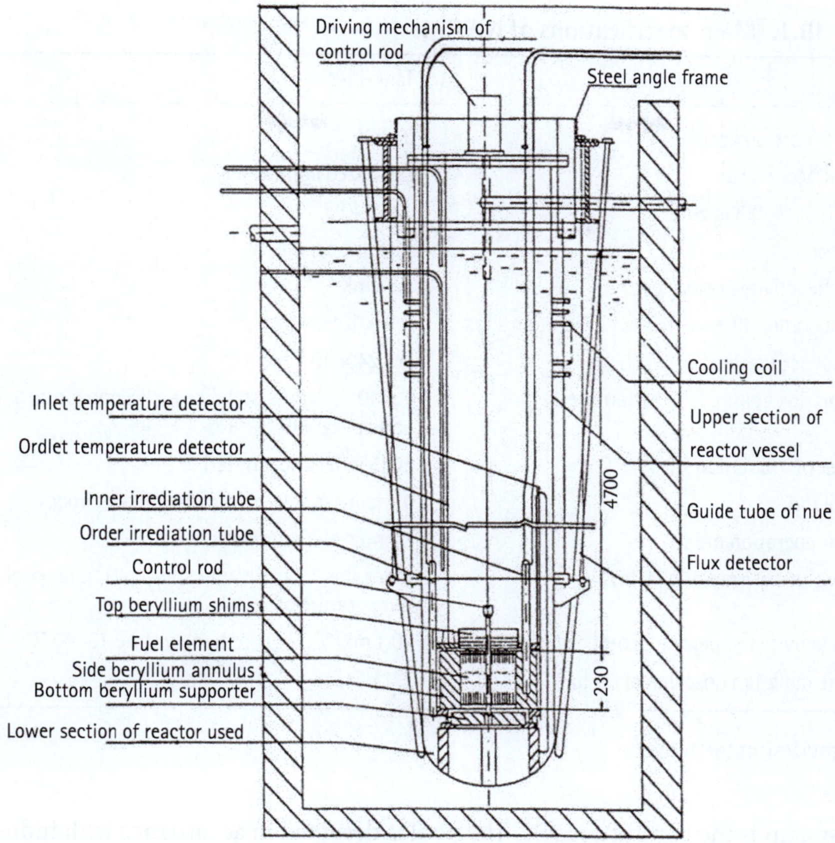

FIGURE 35.1. Vertical section of MNSR (NIRR-1).

Figure courtesy of the Centre for Energy Research and Training (CERT), 'Safety Analysis Report for NIRR- 1 LEU core', Technical Document. Ahmadu Bello University, Zaria, 2019.

thermocouples for measuring inlet and outlet temperatures of the coolant. The container, which is built in two sections, is suspended in a stainless steel-lined water pool made of reinforced concrete. Table 35.1 lists the main specifications of NIRR-1.

The reactor core is cooled by natural convection and there are a number of safety features built into its design. An aluminium tray holds the upper reflector, which is designed to hold semi-circular beryllium shims. The beryllium shims can be added when needed to compensate for fuel burn-up and other fission product poisoning. The reactor is designed to have a self-limiting power excursion characteristic. Only one control rod is at the centre of the reactor core. A fail-safe principle is adopted in the design of the reactor control system. A single cadmium rod is used for regulating the power level, compensating for fuel consumption, and starting up and shutting down the reactor. There are two control systems for the reactor, each of which can be operated independently of the other. In addition, each of the systems can be operated in either automatic or manual mode. One of the systems is a microcomputer closed-loop control system, developed for the reactor with a microcomputer. In addition to controlling the reactor, the system acts as a data acquisition system and for monitoring reactivity. The

Table 35.1. Main specifications of NIRR-1.

Type	Tank-in-pool
Nominal core power	34 kW
Coolant/Moderator	Deionized light water
Loading of U-235 in core	1350.40 g
Reflector	Metallic beryllium
Excess Reactivity—cold, clean	3.94 mK
Daily operation fluence in inner irradiation sites	$< 9 \times 10^{15}$ n/cm^2
Core fluence lifetime	$> 3.24 \times 10^{19}$ n/cm^2
Neutron flux at inner irradiation sites	1×10^{12} n/cm^2.s, stability $\pm$ 1 %, horizontal and vertical variation < 3%
Number of irradiation sites	10 (5 inner and 5 outer)
Control rod	1, stainless steel clad, cadmium absorber
Reactor operation mode	Manual and automatic
Temperature in irradiation sites	Inner site < 54 °C; outer sites < 40 °C (at pool temperature of 25 °C)
Core reactivity temperature coefficient	-0.1 mK/°C; for core temperature 15–40 °C
Average radiation dose in reactor hall	< 1 μSv/h

Table courtesy of the authors.

other system is the control console. The pool is designed in accordance with industrial building standards.

The reactor incorporates several auxiliary systems which, again, incorporate important safety features. Two purification systems for the reactor vessel water and the pool water are used for controlling the water quality.[23] A reactor gas purge system is employed to pump out the gas accumulated in the top space of the reactor vessel. There is a vessel water cooling system, monitoring systems for water temperature and level, a radiation detecting system for measuring the dose level at the top of the reactor vessel, and the reactor water deionizer column.

The HEU to LEU Conversion Programme of Activities

Role of the IAEA (Coordinated Research Project: LEU Feasibility Search)

Work to convert or decommission HEU-fuelled reactors had begun decades earlier but the turning point came in 2006 with the launch of the IAEA's CRP.[24] Representatives from each country currently operating a MNSR facility participated in the CRP, along with representatives from the DOE and US national laboratories.[25] The CRP entailed the use of neutronics, thermal hydraulics, and other related computer codes to search for qualified fuels that could be used to convert MNSRs to LEU with little or no modifications to the HEU core configuration.[26] The LEU search was focused mainly on high-density fuels used to fuel power and research reactors, with uranium dioxide (UO_2) fuel qualified for applications in power reactors being converted. When the CRP concluded in 2010, a MNSR Working Group was established—comprising MNSR operators, the designer, and stakeholders—to continue coordinating activities related to the conversion of MNSRs to LEU, as well as the repatriation of the HEU to China.[27]

In contrast to the previous conversion projects managed bilaterally under the US RERTR programme, the MNSR conversion projects in China, Nigeria, and Ghana were notable for their involvement of various international stakeholders.[28] The iterative process, with clarifications provided at each development stage, has facilitated efficient progress, with completion of the three conversions in three years. Multilateral cooperation has centred in particular on conversion feasibility analysis and early definition of regulatory and bureaucratic procedures, both facilitated by regular and clear communication among stakeholders.[29]

Removal Activities and Spent Fuel Repatriation

The NIRR-1 removal activities involved a multinational group that included Idaho National Laboratory (INL), UJV of the Czech Republic, SOSNY of the Russian Federation, the Nigerian Armed Forces under the aegis of the Office of National Security Adviser (ONSA), and the NNRA.[30] The NIRR-1 HEU core-removal process involved upgrades to the reactor facility to facilitate the shipment process. CERT staff at ABU underwent training to attain proficiency at core removal and package loading operations, including in licensing and packaging for the handling of the MNSR nuclear materials, which consisted of the irradiated HEU core and three fresh HEU MNSR fuel pins.[31] These efforts were conducted with the support of the IAEA, CIAE, and M3 under the US DOE. The upgrade process also covered reactor instrumentation and radiation monitoring instruments, loading pad and road resurfacing capable of supporting the 30-ton TUK 145/C cask, a 5-ton dual-axes crane, and a power back-up provision via a 50-kVA generator.[32] The licensing process comprised of two categories: authorizations/approvals/licences in Nigeria, and those processed in Russia and China.

Core Discharge and Loading

In order for technicians and engineers from the respective MNSR facilities to participate actively in the core removal activities, the International MNSR Training Centre (IMTC) was built at the Ghana Atomic Energy Commission (GAEC) in the Ghanaian capital Accra.[33] A MNSR mock-up facility was designed and installed at the site to enable technicians and engineers to become proficient at undertaking removal activities performed at the respective facilities with little or no supervision. Another reason for the IMTC was to enable a 'one-light day' operation, owing to insecurity at some MNSR sites. Consequently, a team of CERT personnel received several training sessions at the IMTC in Accra, so as to be proficient at core removal and package loading operations.

The team comprised three groups, named: Technical Engineering System (TES), Skoda/MNSR, and Leak Test. The removal and packaging of the NIRR-1 irradiated HEU core and three fuel pins was carried out in one light-day on 23 November 2018. The reactor was prepared for the core discharge by a team of CIAE personnel prior to the activity.[34] The entire activity was carried out by staff of CERT with a technical observation team from SOSNY R&D of Russia and UJV-Rez of the Czech Republic in attendance.[35] The TES team was responsible for the discharge of the irradiated HEU core from the reactor vessel into the interim transfer cask (ITC), which was used to transport the core to the loading pad outside the reactor hall.[36] Figure 35.2 shows the TES team inserting a guide tube in preparation for the core discharge from the reactor vessel. The yellow frame provided support for the 3-ton ITC to protect the vessel against physical damage.

The irradiated HEU core was discharged from the NIRR-1 vessel into the ITC using the electronic control module. Figure 35.3 shows the electronic module and the ITC, containing the discharged core, being lifted from the reactor top.

FIGURE 35.2. The TES team inserting the guide tube into the NIRR-1 vessel.
Courtesy of Ahmadu Bello University, 'NIRR-1HEU Core Discharge Photo Album', 8.

FIGURE 35.3. (a) Electronic control module; (b) Lifting of the ITC after core discharge.

Courtesy of Ahmadu Bello University, 'NIRR-1HEU Core Discharge Photo Album', 12.

FIGURE 35.4. Loading of the HEU core into the transport cask.
Courtesy of Ahmadu Bello University, 'NIRR-1HEU Core Discharge Photo Album', 17.

While the TES was carrying out the core removal from NIRR-1, the Skoda/MNSR team prepared the transport cask in readiness for the loading and packaging of the nuclear materials. Aligned to the IAEA's classification system, the TUK 145/C was certified as a Type C package for the transportation of the research reactor spent fuel assemblies, by all transport modes including the aircraft.[37] The Nigeria military provided security for the transportation of the HEU within Nigeria. Armed guards secured the NIRR-1 site from the day the HEU was discharged from the reactor and packaged for shipment, until its departure from the site.

Figure 35.4 shows the HEU core being loaded into the transport cask. The Leak Test team conducted their checks after the Skoda cask was fully closed and tightened. The top half of the TUK 145/C transport cask was finally closed and laced with the IAEA safeguard seal. Figure 35.5 shows the CERT technical team who conducted the core removal and loading operations.

The initial plan was to ship the nuclear materials the same day, but due to unfinished processes to receive the irradiated fuel in China, the loaded cask was kept at the NIRR-1 site until 4 December 2018 when the Chinese authorities completed the necessary arrangements for its reception in China. Under military escort the TUK 145/C transport cask departed the NIRR-1 site in Zaria by road to Kaduna International Airport, where it was loaded into the waiting air cargo plane, as shown in Figure 35.6. Kaduna Airport was secured 24 hours before the arrival of the cargo plane used to transport the HEU to China.

FIGURE 35.5. The CERT technical team.

Courtesy of Ahmadu Bello University, 'NIRR-1HEU Core Discharge Photo Album', 27.

FIGURE 35.6. Loading of the TUK 145/C cast at Kaduna International Airport.

Courtesy of Ahmadu Bello University, 'NIRR-1HEU Core Discharge Photo Album', 30.

FIGURE 35.7. Arrival of the NIRR-1 HEU materials at the CIAE facility in Beijing, China.
Courtesy of Ahmadu Bello University, 'NIRR-1HEU Core Discharge Photo Album', 31.

The nuclear material arrived in China on 6 December 2018. The transfer of ownership between the Nigerian and Chinese authorities was signed at Shijiazhuang Zhengding International Airport in China. Figure 35.7 shows the arrival of the NIRR-1 HEU materials at the CIAE facility in Beijing. The shipment of the NIRR-1 irradiated HEU core together with the three spare HEU fuel pins added Nigeria to the list of countries that are free from weapons-grade nuclear materials.

Conversion Activities (Determining the Make-up of the LEU Core)

In order to perform an enrichment search, the Monte Carlo N-Particle transport code (MCNP) was used for neutronics analysis of the HEU core. On this basis an accurate model of the HEU core was established and used to search for LEU fuels for the conversion of the NIRR-1 facility from HEU to LEU. A computational method was also developed to assess the impact of conversion to LEU on utilization. In order to address the risk of single point failure posed by the use of a single control rod in both commercial MNSRs and the Prototype MNSR, the inclusion of two additional control rods in the design of MNSR facilities was checked using the NIRR-1 model.[38] The analysis demonstrated that it would be feasible to use additional safety rods in the unconnected inner irradiation channels of NIRR-1 for enhanced safety. Detailed versions of the results have been published by scholars, including Sunday A. Jonah.[39] These include results of neutronics data obtained for the HEU core and the proposed LEU core for the flux performance.

Post-Conversion Activities

NIRR-1 has been successfully converted from HEU to LEU, making it the third MNSR in the world to be fuelled with UO_2 pellets enriched to 13 per cent. Like the HEU cores of MNSR facilities, a major safety characteristic in the converted facility is the self-power regulating capability, even with the single control completely withdrawn and the reactor allowed to operate unattended. This functionality was tested in the NIRR-1 LEU core during the on-site commissioning. NIRR-1 was operated in the manual mode with all safety and protection features disabled so as to monitor the transient behaviour of the facility with the insertion of total cold core excess reactivity of 3.94 mK. Results obtained indicate that the reactor power reaches a peak value of 70 kW and power gradually reduces over the entire transient time of 3,500 seconds. Results of measured and simulated reactivity insertion transients of 3.94 and 3.0 mK for NIRR-1 LEU core compare well with data obtained for the HEU core and have been used to develop a PARET LEU model for the analysis of the reactivity insertion transient in MNSR facilities and low power research reactors with similar thermal hydraulics characteristics.

Following the successful conversion of the NIRR-1 to LEU, a Safety Analysis Report (SAR) was updated to conform with relevant IAEA guidelines[40] and their domesticated versions as issued by the NNRA. In preparing the final SAR, each chapter of the HEU Safety Analysis Report[41] was revised and updated to reflect the extensive work carried out in producing the new LEU core. The SAR outlined the safety analysis of NIRR-1 with LEU fuel of 13 per cent enrichment, having been converted from 90.2 per cent. The reactor went critical on 2 November 2018 and was commissioned for full operation on 27 November 2018, incorporating the results of the design, safety, and accident analyses performed on the LEU. The successful commissioning of the converted reactor has opened a new chapter in the life cycle of NIRR-1.

There was an active physical protection system (PPS) at the facility prior to the conversion project. Nevertheless, following the completion of the conversion the system was upgraded. This involved, among a number of security improvements, replacing the access door to the facility with a fortified security door, as well as installing additional closed circuit television (CCTV) and motion detectors to cover any blind spots.

Challenges Encountered and Lessons Learned

Contracting and Regulatory Review

Contracts were required for the reactor-specific conversion analysis, conversion implementation work, and transportation of the LEU and HEU cores. Since the LEU fuel was fabricated by China and supplied through the IAEA, along with technical assistance for reviews of research reactor safety, a project and supply agreement (PSA) between the IAEA, China, and the receiving country was needed.[42] Similarly, several contracts were needed for replacements and upgrades of some modules of NIRR-1. Specifically, contracts were initiated for spent fuel repatriation from the NIRR-1 site in Nigeria to the CIAE in China. Funding for all contracts for the LEU fuel procurement, core removal, and transportation of fresh and irradiated fuels was provided by the IAEA and two US organizations: INL and the Argonne National Laboratory (ANL).

In conversion projects, the conversion SAR is required to undergo regulatory review. Depending on the country, approvals by one or more bodies are also often needed for the transportation of cask licenses, operator permits, material storage, among other aspects. The need to meet regulatory requirements is often responsible for delays in the implementation of this type of project. In Nigeria, the regulatory organization, the NNRA was engaged in the project immediately after the LEU feasibility search under the aegis of the IAEA's CRP. This enabled a smooth relationship with the regulatory organization during the licensing process and eventual removal of the HEU as well as the LEU on-site loading. A noteworthy example is the participation of a regulatory officer as part of the Nigerian team in the off-site zero-power test that took place at the CIAE facility in China in 2018, prior to the LEU shipment to the NIRR-1 site later that year.

Communication

It was important to keep such a large, diverse group of stakeholders informed and coordinated, but it was difficult to do so across organizational, geographic, and cultural boundaries. The teams experimented with several different methods to ensure that information sharing and coordination were clear, efficient, and effective. The methods included telephone calls with all stakeholders on a weekly and monthly basis, emails as frequently as desirable, and physical meetings organized in the owner country (i.e. Nigeria), at the IAEA Headquarters in Vienna, and at the sites of contractors engaged in the project, in the Czech Republic as well as SOSNY in Russia. The in-person meetings helped keep communication lines open and ensured quick decision making on electronic platforms.

Lessons Learned

For reactors acquired using a PSA arrangement, as in the case of NIRR-1, there is a long lead time to conclude all processes involved in the conversion to LEU. Some of the protocols involved preparing a 'full power contract' to enable the National Liaison Officer (NLO) of the owner country to approve a tripartite contract, comprising the owner country, the IAEA, and the CIAE. In the case of Nigeria, the Nigerian NLO was empowered by the Nigerian Foreign Minister to endorse the contract. The Nigerian experience has shown that sufficient time should be provided for approvals of the three parties. The Full Powers letters for PSA signatories must be received in original hardcopy and contracts signed as early as possible. In the same vein, all related documents should be prepared and submitted as early as possible to avoid unnecessary bottlenecks. Furthermore, timely contract review is recommended to avoid unnecessary delay and, if possible, a reduction in the number of contracting parties helps.

Similarly, the use of an off-site core removal training for MNSR facilities is a valuable tool—completed at the IMTC in the NIRR-1 case—and saves deployment time at the site of the facility. Because of the diversity of project management philosophies across international teams, there is a need for patience. During the on-site conversion, a video feed can be useful to accommodate those wishing to see the conversion process but who have no need to be physically present in the reactor hall. There is also a need for early involvement of national/local security organizations in the development of the Transport Security Plan (TSP) and its submission to the regulator for approval.

Conclusion

In order to reduce proliferation risks and make the country HEU-free, Nigeria joined the global effort to convert its only research reactor to LEU under the aegis of the IAEA's CRP. The work performed under the CRP indicated that the NIRR-1 HEU core could be replaced by LEU fuel consisting of UO_2 with nominal enrichment of 13 per cent. Based on the results obtained under the CRP, the conversion of the NIRR-1 to LEU fuel did not present any risks of new accidents nor did the conversion increase the consequences of any of the postulated design basis accidents (DBAs) identified in the current approved SAR. At the same time, the conversion of NIRR-1 from HEU to LEU enhanced nuclear security in the country by depriving terrorist organizations such as Boko Haram. Consequently, NIRR-1 was converted to LEU in 2018 with the cooperation of the IAEA, US, China, and other countries.

The success of the conversion project was demonstrated by multilateral cooperation on conversion feasibility analysis, early definition of regulatory and bureaucratic procedures, and frequent, regular communication among all stakeholders. With the successful 2018 MNSR conversion in Nigeria, and the equivalent conversion in Ghana, all research reactors on the continent of Africa are now using LEU fuel. This makes Africa the third continent to have completed all HEU to LEU reactor conversions, following Australia and South America. The conversion project was also enabled because suppliers of the reactors are expected to take back spent fuel. The project was a first of its kind experience in core repatriation from Nigeria, and has provided lessons for the Nigerian nuclear industry, amid aspirations to add nuclear power to diversify Nigeria's energy options.

Notes

1. The Committee was established in November 1977 by Vice-Chancellor Professor Iya Abubakar, who also appointed Head of Department of Physics, Professor Deborah Ajakaiye, as the Chairman of the Committee.
2. Ahmadu Bello University, Ahmadu Bello University Law, Zaria, 1986/88.
3. Centre for Energy Research and Training, 'CERT, 2012. CERT in ABU Zaria @ 50 (Thirty-Six Years of Nuclear Science and Technology Studies at ABU)' (Ahmadu Bello University. Zaria, 2012).
4. The IPPS is one of the three main activities of the International Science Programme (ISP) at Uppsala University.
5. Christoph Wirz and Emmanuel Egger, 'Use of Nuclear and Radiological Weapons by Terrorists?', *International Review of the Red Cross*, 87/859 (2005), 497–510.
6. Graham Allison, *Nuclear Terrorism: The Ultimate Preventable Catastrophe* (New York: Henry Holt, 2004).
7. ReliefWeb, 'Global Terrorism Index', UN Office for the Coordination of Humanitarian Affairs (OCHA), 1 March 2022. https://reliefweb.int/report/world/global-terrorism-index-2022
8. T. Akintunde Orunmuyi, M. Mike Sathekge, and R. Joh Buscombe, 'Ensuring Effective and Sustainable Radionuclide Delivery and Its Impact on the Development of Nuclear Medicine in the Developing World with Special Mention to Nigeria', *World Journal of Nuclear Medicine*, 18/1 (2019), 2–7.
9. A full list of these facilities is available on the website of the Nigerian Nuclear Regulatory Authority (NNRA): NNRA, 'Regulated Industries', 2022. https://www.nnra.gov.ng/nnra/page-regulated_industries
10. Goodluck Jonathan, 'National Statement', 2012 Nuclear Security Summit, Seoul, 2012; Aaron Mehta, 'How the US and China Collaborated to Get Nuclear Material out of Nigeria—and Away from Terrorist Groups', *Defense News* [online], 14 January 2019. https://www.defensenews.com/news/pentagon-congress/2019/01/14/how-the-us-and-china-collaborated-to-get-nuclear-material-out-of-nigeria-and-away-from-terrorist-groups
11. Goodluck Jonathan, 'National Statement'.
12. IAEA, *Analyses Supporting Conversion of Research Reactors from High Enriched Uranium Fuel to Low Enriched Uranium Fuel: The Case of the Miniature Neutron Source Reactor*, IAEA-TECDOC-1844 (Vienna: IAEA, 2018).
13. Ibid.
14. Richard Stone, 'U.S.–China Mission Rushes Bomb-Grade Nuclear Fuel out of Africa', *science.org* [online], 31 August 2017. https://www.science.org/content/article/us-china-mission-rushes-bomb-grade-nuclear-fuel-out-africa

15. IAEA, 'Conversion of Miniature Neutron Source Research Reactors (MNSR) to Low Enriched Uranium (LEU)', 2020. https://www.iaea.org/projects/crp/t12020
16. S. A. Jonah, Y. V. Ibrahim, A. S. Ajuji, and M. Y. Onimisi, 'The Impact of HEU to LEU Conversion of Commercial MNSR: Determination of Neutron Spectrum Parameters in Irradiation Channels of NIRR-1 Using MCNP Code', *Annals of Nuclear Energy*, 39 (2012), 15–17.
17. Gao Jijin, 'MNSR Commissioning Programme', CIAE Technical Report, Beijing, 1998.
18. Guo Chenzhan, Xu Hanming, Hu Zhiyi, and Shi Shangkai, 'Prototype MNSR Dynamic Feedback Experiments and Calculations', CIAE Technical Report, Beijing, 1985.
19. Zhou Yongmao, 'Design and Construction of the Prototype MNSR', CIAE Technical Report, Beijing, 1984.
20. S. Tozser, I. Bolshinsky, J. Dewes, S. Moses, K. Gaines, I. J. K. Aboh, H. C. Odoi, F. Ofosu, and P. Chakrov, 'International MNSR Training Center in Operator Training for MNSR HEU Core Discharge and Packaging', paper presented at the 60th Annual Meeting of the Institute of Nuclear Materials Management (INMM), Palm Desert, 14–18 July 2019.
21. Thailand Office of Atoms of Peace, 'National Report of Thailand Convention on Nuclear Safety', The 8th Review Meeting, Ministry of Higher Education, Science, Research and Innovation, August 2019.
22. Jaci Dickerson, 'Conversion of Miniature Neutron Source Reactors from High Enriched Uranium to Low Enriched Uranium', Contribution ID 394, Proceedings of the International Conference on Nuclear Security (ICONS), 10–14 February 2020, Vienna, Austria.
23. S. A. Jonah, J. R. Liaw, and J. E. Matos, 'Monte Carlo Simulation of Core Physics Parameters of Nigerian Research Reactor-1 (NIRR-1)', *Annals of Nuclear Energy*, 34 (2007), 953–7.
24. Dickerson, 'Conversion of Miniature Neutron Source Reactors'.
25. Ibid.
26. Prince Amoah, 'Thermal Hydraulic Analysis of Low Enriched Uranium (LEU) Core of the Ghana Research Reactor-1 (GHARR-1)', University of Ghana, July 2019. https://inis.iaea.org/search/search.aspx?orig_q=RN:51084079
27. Dickerson, 'Conversion of Miniature Neutron Source Reactors'.
28. Ibid.
29. Ibid.
30. J. Dewes, I. Bolshinky, S. Tozser, Y. A. Ahmed, and H. Adamu, 2019, 'Shipment of Irradiated HEU Nuclear Fuel from Nigeria to China', Proceedings of the 19th International Symposium on the Packaging and Transportation of Radioactive Materials, 4–9 August 2019, New Orleans, LA, USA.
31. Ibid.
32. Ibid.
33. Igor Bolshinsky, John Dewes, Kris Gaines, and Debbie Browning, 'International MNSR Training Center in Ghana Provides HEU Removal Expertise', *DNN Sentinel*, 4/2 (2018), 4–5.
34. Ibid.
35. Ibid.
36. Ibid.
37. Dewes et al., 'Shipment of Irradiated HEU Nuclear Fuel from Nigeria to China'.
38. Y. V. Ibrahim, H. C. Odoi, R. L. Njinga, M. O. Adeleye, and S. A. Jonah, 'Monte Carlo Simulation of Additional Safety Control Rod for Commercial MNSR to Enhance Safety', *Annals of Nuclear Energy*, 44 (2012), 71–5.

39. S. A. Jonah, K. Ibikunle, and Y. Li, 'A Feasibility Study of LEU Enrichment Uranium Fuels for MNSR Conversion using MCNP', *Annals of Nuclear Energy*, 36/8 (2009), 1285–8; Jonah et al., 'The Impact of HEU to LEU Conversion of Commercial MNSR', 15–17; Jonah et al., 'Monte Carlo Simulation'.
40. IAEA, *Safety Assessment for Research Reactors and Preparation of the Safety Analysis Report*. IAEA Safety Standards Series, No. SSG-20 (Vienna: IAEA, 2016); IAEA, *Use of a Graded Approach in the Application of the Safety Requirements for Research Reactors*. IAEA Safety Standards Series, No. SSG-22 (Vienna: IEA, 2012); IAEA, *Safety of Research Reactors*. IAEA Safety Standards Series, No. SSR-3 (Vienna: IAEA, 2016).
41. Centre for Energy Research and Training (CERT), 'NIRR-1 Safety Analysis Report (SAR)', Technical Document, Ahmadu Bello University, Zaria, 2012.
42. F. E. Osaisai, E. D. Mshelia, L. A. Dim, and S. A. Jonah, 'Planning and Implementation of a TC Project for Acquisition of a New Research Reactor for Multipurpose Applications in Nigeria', Conference Paper, International Group on Research Reactors, January 2013.

Suggested Reading

Centre for Energy Research and Training (CERT), (2019), 'Safety Analysis Report for NIRR-1 LEU Core', Technical Document. Ahmadu Bello University, Zaria.

China Institute of Atomic Energy (CIAE), (1993), *Training Materials for Nigeria*, MNSR Books 1–4 (Beijing: CIAE).

Evans, Gareth and Kawaguchi Yoriko, (2009), Eliminating Nuclear Threats: A Practical Agenda for Global Policymakers, Report of the International Commission on Nuclear Non-proliferation and Disarmament (Canberra: ICND).

Mehta, Aaron, 'How the US and China Collaborated to Get Nuclear Material out of Nigeria—and away from Terrorist Groups', *Defense News* [online], 14 January 2019. https://www.defensenews.com/news/pentagon-congress/2019/01/14/how-the-us-and-china-collaborated-to-get-nuclear-material-out-of-nigeria-and-away-from-terrorist-groups

Nuclear Security Summits, (2012), 'Communiqué', 2012 Nuclear Security Summit, Seoul, Korea.

CHAPTER 36

ALTERNATIVE TECHNOLOGIES FOR RADIOACTIVE SEALED SOURCES

CHRIS ENGLEFIELD

Introduction

There are many important beneficial uses of radioactive materials in medicine, industry, research, and education. The processes in which they are used are manifold but include medical therapies and imaging, materials testing, scientific analytical techniques, and process control. Increasing international threat perceptions regarding the possibility of nuclear terrorism over the past two decades has led—in some parts of the world—to the adoption of so-called 'alternative technologies' to radioactive sources. The underlying premise is that the use of alternative technologies will reduce inventories of radioactive sources that could be used in terrorist acts, resulting in permanent threat reduction. There are also supplementary safety arguments that stem from loss of control incidents like the notorious 1987 Goiânia accident in Brazil,[1] which would not have occurred if radioactive sources had not been involved.

The chapter reviews the range of radioisotopic and alternative technologies that are currently on the market and provides examples of each. These include the various medical applications of teletherapy, brachytherapy, and blood irradiation, as well as industrial applications such as non-destructive testing and process control for road building. It considers the scope and limitations of each as replacements for radioactive sealed sources and also briefly reviews the extent to which alternative technologies can be effectively implemented in both developed and developing countries, and across a range of sectors. In support of this analysis, the chapter begins with a brief technical overview of the radioactive materials and alternative technologies.

The Science of Radioisotopic and Alternative Technologies

Radioactive Materials

Radioactive material contains unstable atoms that emit ionizing radiation as they decay, a process known as radioactivity. The rate of radioactive decay is dependent on the specific radioisotope and is independent of its chemical and physical form.[2] The materials that are referred to in this chapter are mainly metallic or soluble crystalline salts that do not have the fissionable properties of nuclear materials such as uranium, plutonium, or thorium and, as such, cannot be used to make a nuclear weapon. To enable safe use, these materials are usually incorporated into robust containers that seal them off from the environment, while allowing, when required, energetic radiations to escape. Strict regulatory controls govern the use of these materials to help ensure that users stay within suitable operational limits to minimize the potential risk to workers, the public, or the environment. These controls also aim to protect radioactive sources from an adversary intent on weaponizing radioactive materials by the creation of an improvised radiological device (IRD).[3] An example is a 'dirty bomb' that combines explosives with radioactive materials which, if successfully detonated, could cause serious harm, while also contaminating the local environment and causing significant disruptive effects.[4]

Alternative Technologies

Physical protection measures and effective security management and culture are commonly employed to protect radioactive inventories from an adversary. However, there is another approach to risk reduction that may accomplish the same goal: the replacement of radioactive materials with alternative technologies.[5] While common radioactive source applications utilize alpha, beta, gamma, and neutron sources (plus a few rare and more exotic emitters), alternative technologies most commonly utilize photons emitted by X-ray generators or, in rarer cases, electrons or neutrons.

Certain alternative technologies may not use radiation of any kind. For example, some metallurgical testing methods use penetrating dyes to detect superficial flaws in materials. This non-destructive testing technique, also known as liquid penetrant inspection (LPI), is a cost-effective method used to locate surface-breaking flaws such as cracks, porosity, and other surface discontinuities in welds, as well as stress cracking within in-service components.[6] LPI can replace the use of radiographic instruments that have been traditionally used to detect industrial defects. However, alternatives may not be able to fully replicate the functions of radioactive materials. For example, LPI does not enable the detection of enclosures that might weaken a structure, or voids or cracks within welds. To detect these, the use of radiographic methods is still required.

Traditionally, radiographic methods have utilized radioactive materials as a source of radiation. In some cases, a similar function is provided by X-ray generators that only emit radiation when they are powered by electricity. In a basic X-ray generator, a voltage is applied to a coil (known as the cathode), which causes the emission of electrons from its surface into the vacuum of the glass envelope in which it is contained. Another voltage is applied across the cathode and a target (known as the anode); this accelerates the electrons towards the target. The kinetic energy of the electrons is largely converted into heat energy during the extreme braking that they are subject to when they hit the target. This also results in the generation of X-rays, which can be directed out through a suitable window. As the target is bombarded, it heats up; excess heat must be removed from the system by an air-cooled heatsink or a cooling water circuit.

Although radioactive gamma sources and X-ray generators rely on different types of physics, the types of radiation that they produce are essentially similar. Generally, radioactive gamma emitting sources produce higher energy photons than X-ray machines and they do so without the complexities of power and cooling systems. The outcome of these similar physical properties is that X-rays and gamma rays interact with matter in a similar fashion and so produce broadly similar biological effects for a given energy. To a first approximation, X-rays are a substitute for gamma rays. Unlike gamma rays, X-rays from radiation generators can be 'turned off'.

Security Arguments for the Use of Alternative Technologies

Until the start of the twenty-first century, the drive for minimization of radioactive source inventories was primarily based on one of the safety principles of radioactive waste management. This is recognized by the Safety Standards Series of the International Atomic Energy Agency (IAEA), with the principle that 'the generation of radioactive waste must be kept to the minimum practicable level'.[7] Regulators have sought to implement this fundamental principle and ensure that radioactive source inventories held by operators are comprised only of sources in current use. However, following the attacks of 11 September 2001 ('9/11'), threat perceptions broadened to include the potential terrorist use of nuclear or radiological materials, leading to the international community pursuing efforts to reduce this possibility by looking to secure and decrease the inventories of such materials.

As a result of concerns about the potential for nuclear terrorism, a security-driven argument for the reduction in radioactive source inventories using alternative technologies has developed. States that opt for this approach have a range of options available. They may, for example, focus on a single radioisotope of concern or they may focus on a particular practice such as the use of radioisotopes in blood irradiators. The former approach was taken by Norway, which has worked to replace a range of applications that utilized caesium-137.[8] The latter approach has been pursued under the US Global Threat Reduction Initiative (GTRI) programme, which has worked with countries around the world to replace blood irradiators with alternative technologies.[9]

Applications of Alternative Technologies and their Benefits

The Medical Sector

Alternative Methods of Teletherapy

High doses of radiation kill biological cells. This is especially effective on rapidly dividing cells but, with a high enough dose, even very long-lived cells will be adversely affected. Cancer cells very rapidly divide and as such are vulnerable to the application of external beams of radiation. Traditionally, cancer treatment has been performed through 'cobalt therapy', where small but strong sources of radioactive cobalt-60 are manipulated and focused by remotely operated machines. The source is safely enclosed in a heavily shielded 'treatment head' with an opening shutter to allow the release of radiation for therapy, when required.

The aim of this 'teletherapy' is to focus a thin beam of radiation onto a solid tumour within a patient. In the early days, the problem was that healthy tissue in front of the tumour received equally high doses. Later redesigning of the treatment head enabled it to rotate in an arc around the patient with the tumour at its centre of rotation. At the focal point, the tumour receives multiple doses. Surrounding healthy tissue is somewhat spared from the effects of irradiation by the spread of dose across a range of paths of entry into the patient's body. The expectation is that the tumour cells will have been killed while healthy tissue will eventually recover. Although cobalt therapy can be used to treat cancer, studies have also demonstrated how, as a result of its properties as a strong gamma, potentially it could be readily utilized in a 'dirty bomb'.[10]

Alternative technologies to cobalt therapy all rely on the acceleration of particles to either directly or indirectly irradiate the tumour. These are large entities like ions, neutrons, and protons. The most common form of generator of such species for teletherapy in the medical context is the linear accelerator (or 'linac'). This device uses high-frequency microwaves to accelerate charged particles towards a target. It does not include any radioactive material and so only emits its radiation when energized by a power supply. The gross mechanical engineering of linacs is very similar in principle to that of cobalt machines: the beam is focused on an isocentre by rotating the device around the patient in a very precisely controlled manner.

The beam can be modulated electronically to enhance the therapeutic benefits on both tumours and healthy tissue. Variables such as beam intensity and shape can be tailored to the patient's individual clinical need and as a result the clinical outcome can often be improved compared to even modern cobalt therapy. These developments drove a move away from radioactive source-based machines on the basis of clinical need even before the events of 9/11 that created new security grounds for a shift from cobalt therapy. The result has been a widescale adoption of linac methods in developed countries, with the disused cobalt treatment machines often being redeployed to developing countries that otherwise lacked any form of radiotherapy.[11]

However, things are less straightforward in low- and middle-income countries. Wroe and colleagues demonstrated that the United Kingdom (UK) as a high-income country has much higher day-to-day availability of medical linacs compared to Botswana as an upper middle-income country and Nigeria as a lower middle-income country.[12] A range of factors was found to have affected downtimes. These included some physical factors, for example, the variation in climate and stability of power supply, which were clearly key influences on linac performance. In short, a linac that works well in a high-income country may not work as well in a low-income country. The result can be increased running costs, and more importantly, reduced throughput of patients.

A similar effect has been found for other medical devices, including X-ray sets.[13] Film-based and digital imaging equipment, and particularly computers and network equipment, require a reliable and stable supply of power. Extreme temperatures and very high humidity levels can permanently damage electrical and electronic equipment in radiography departments. Most computer and imaging equipment requires certain temperature and humidity ranges for safe and reliable operation. This may necessitate specific building requirements to provide for air cooling/heating and humidity control, with appropriate housing and maintenance. Weatherproof and climate-controlled housing of imaging equipment is thus not a luxury but should be considered essential for the safe and reliable operation of such equipment over any significant length of time.[14] In summary, high-specification alternative technologies may not be appropriate for sustainable operation in low-income countries that cannot yet provide suitable infrastructure.

Alternative Technologies for Brachytherapy

Sometimes tumours occur superficially or within body cavities. In certain treatments, radiotherapy is administered by applying the source directly to the affected tissue. This is called brachytherapy, which is mostly used to treat prostate cancer, but it also can be used for gynaecological cancers, such as cervical cancer and uterine (endometrial) cancer, as well as cancers of the breast, rectum, eye, and skin. As an example, brachytherapy of the prostate gland involves the permanent injection of a number of radioactive iodine 'seeds' into the prostate. Each is the size and shape of a grain of rice. The seeds stay in the prostate forever and give a steady dose of radiation over a few months; this damages the prostate cells and stops them dividing and growing. The cancer cells cannot recover from this damage and die, but healthy cells repair themselves. For certain cancers, these low dose-rate methods have clinical benefits over teletherapy and in many cases, no alternative technology has been developed or approved for clinical use.

The level of radioactivity/specific activity of the individual sources is generally low compared to cobalt teletherapy sources, and most brachytherapy sources are Category IV or V on the IAEA's categorization scale.[15] High dose-rate brachytherapy (HDR) is also undertaken, for example with cobalt-60 and iridium-192, which are categorized as Category II sources. It is noteworthy that some HDR treatments are now being made with electronic alternative sources by the generation of X-rays, such as the 'Papillon technique' for early-stage rectal cancer.[16] Mostly though, radioisotopic methods are still used.

Alternative Technologies for Blood and Cell Irradiators

Radiation can also be applied to tissues that have been removed from the body, through the use of irradiators to deliver a defined dose of gamma radiation. Blood is the most commonly treated tissue, immediately prior to transfusion into patients. Transfusion-associated graft-versus-host disease (TA-GvHD) is a rare and usually fatal complication of blood transfusion in which some white cells from the transfused blood attack the recipient's tissues. This only occurs if the patient is immune-deficient or immunosuppressed. Treatment with radiation prevents the occurrence of this condition in patients who are susceptible. In several countries, including the UK, many blood transfusion centres have blood irradiators for this purpose.

As medical technologies improve, additional clinical treatments (such as stem cell transplants) have been shown to have improved outcomes with appropriate irradiation of transplanted cells. In the research arena, other tissues or even whole organisms may be irradiated, for example, to examine cancer induction rates and radiation dose susceptibility. By far the most used radioisotope for these devices is caesium-137, sources of which emit a very useful medium-energy gamma photon. The radioactive material exists as the chloride salt and is contained within a double-walled, welded stainless steel container. Most chlorides are water soluble, enabling them to be easily manipulated chemically into highly dispersible forms. For these reasons caesium-137 sources have attracted special attention in the assessment of security threats. Approaches to this vary: some states have specifically eradicated the use of caesium-137;[17] others have promoted the use of non-isotopic technologies instead, for example, with blood irradiators that use X-ray generators.[18] Different approaches are influenced by regulatory requirements, especially with regard to clinical effectiveness, as well as value-for-money arguments.

The Industrial Sector

Alternative Technologies for Industrial Radiography

Modern construction methods often use very high standards of quality control to ensure public safety. Load-bearing steel structures, pressure vessels, and pipelines are usually subject to radiography of welds to ensure the absence of flaws such as cracks, porosity, lack of fusion of a weld, or lack of penetration through the thickness of the substrate. This involves the juxtaposition of a source and a detector of some kind (usually film) in a process known as plane radiography. In this way, incomplete welds or other flaws may be visualized and appropriately remediated. In the case of pressure vessels, for example, the steels involved may be very thick (~10 cm) and highly penetrating radiation is required to achieve useful results.

Most often, this means using gamma rays; a suitably high-energy gamma ray source will be selected to achieve the penetration required with a reasonably short exposure time. Two of the more common industrial gamma ray sources for industrial radiography are iridium-192 and cobalt-60.[19] These isotopes emit radiation as a few discreet energies across a useful range. While these energies can be achieved by suitable X-ray generators, radioisotopes do this without the need for a power source or a cooling system. As a result, radioisotope methods are sometimes preferred due to limits on accessibility to the item being radiographed; the radiography equipment is usually smaller and more flexible, and this is sometimes an issue. In remote locations, the logistical support required to ensure power supplies and sometimes water as a coolant for X-ray sets make radioactive source-based methods much more practicable. However, it is important to note that in some cases, the ability of an X-ray set to provide low energy emissions means that X-radiography may be preferable to gamma radiography methods, especially for low atomic number materials.

Alternative Technologies for Road Building

There are several parameters that must be controlled during road building and tunnelling operations.[20] These include the density of the materials, which is a function of the amount of compaction they are subject to, and the moisture content. Techniques vary according to the geotechnical need but, typically, the devices use the interaction of gamma radiation with matter to measure density, either through direct transmission or the 'backscatter' method. The device determines the density of material by counting the number of photons emitted by a radioactive source (caesium-137), read by the detector tubes in the base of the gauge. Radioisotopic gauges are very widely used in routine field measurements.

Non-radioisotopic gauges that can deliver similar measurements are now available.[21] An electrical density gauge measures pavement density indirectly by measuring its dielectric constant. A small current passes through the pavement and creates an electrical sensing field. Instead of measuring radiation scattering, density is measured by the response of the electrical sensing field to changes in the pavement's impedance (consisting of the pavement's resistivity and dielectric constant). One advantage of using the electrical density gauge is that the readings can be obtained more rapidly than radioisotopic methods.

Decision Making around the Adoption of Alternative Technologies

As mentioned, decisions on whether or not to move from the use of radioactive materials to alternative technologies can be far from straightforward, dependent on a mix of safety, security, operational, and economic factors, and influenced by the broader political climate. In the early to mid-2000s, following the events of 9/11, arguably security concerns trumped other drivers, resulting in decisions in certain national contexts that, while reducing the risk of radiological terrorism, served to adversely impact operations, resulting, for example, in negative patient outcomes in the healthcare sector.

Over time consensus has developed on the need to consider risks in a much more holistic way. This is reflected in a report published by the World Institute for Nuclear Security (WINS) on considerations for the adoption of alternative technologies.[22] The report provides a nine-step method for evaluating end-user operational needs of alternative technologies compared to radioactive source devices. This methodology is recommended for its systematic and logical approach to these issues. A range of stakeholders is identified, including end users, national government, regulators, professional associations, and device manufacturers. The WINS report recognizes that the choice of technology in the medical field is heavily influenced by the professionals who use the equipment to perform medical procedures and associated tasks. Decision making is also influenced by management, which approves the investment. The report acknowledges that each application is subject to its own dynamics in terms of risk, viability, and cost, and that the opportunity to switch technology may vary depending on the application.

The WINS report also recognizes the role of other government agencies beyond the nuclear regulatory body. An example is the UK Medicines and Healthcare products Regulatory Agency (MHRA), which is responsible for regulating the UK medical devices market. At the time of writing, two companies are registered with the MHRA for blood irradiators; both of these provide *in vitro* UV-based systems. No X-ray-based irradiator provider is currently registered with the MHRA. However, in other countries (for example, Norway and Japan) X-ray-based blood irradiators are now in exclusive use. In the context of the security of radioactive sources used in hospitals, the question to be answered is: What is the net benefit of installing or requiring security measures (for example, an alternative technology) compared to the potential negative consequences? A multi-disciplinary, risk-based assessment of the potential impacts of adopting security measures is essential today. It should be emphasized that the discussion here does not preclude the use of security measures to protect radioactive material from an adversary. Instead, it is an argument for a well-considered, risk-informed, flexible, and optimized approach to the selection of security measures, including alternative technologies.

Broader Considerations When Adopting Alternative Technologies

To further illustrate the importance of taking a holistic approach to risk assessment when considering the adoption of alternative technologies, this section briefly outlines a hypothetical example, where decision making is driven primarily by security concerns. In this case a hospital is required by the regulator to install security measures that results in a radiotherapy department reducing its patient throughput by 15 per cent. This can be considered equivalent to the replacement of a teletherapy unit with a linac system on security grounds, in a developing country, where throughput is similarly reduced by 15 per cent due to power outages.

The throughput of a radiotherapy department will vary according to the hours that it is in operation per day. One study concluded that 10,000 radiotherapy courses per annum for a particular facility is not untypical (this equates to around 5,000 patients).[23] A 15 per cent reduction in throughput equates to 750 patients per annum who could be disadvantaged by a switch to a restrictive and time-consuming security measure or the use of an alternative technology. Not all those disadvantaged would necessarily die, but their quality of life would be affected. Clearly this is a complex decision that should not rely solely on judgements by security specialists; rather, regulators should recognize the need to consult other specialisms outside their direct functions.

Here, key questions that need to be considered include: Is there a net gain in the security of radioactive materials as a result of the alternative technology? If there is not a balanced security system, for example, that also includes measures to ensure the security of waste radioactive sources, then vulnerabilities may be increased instead of decreased. What are the capital costs of disposal of disused radioactive sources released by adopting alternative technologies? Is there sufficient funding to enable this or will they simply be kept in vulnerable storage? What are the potential clinical impacts on loss of quality of life or excess mortalities and how would this affect assessments of long-term treatment planning at the societal level? Could this impact on public health targets and policy? Governments are sensitive to under-achievement in healthcare and the effects of reputational harm to treatment facilities or a national oncology programme. This latter point may have political implications that policy makers may have to manage, even though the regulator should be independent of these considerations.

A number of other factors need to be considered for alternative technologies. In the design specification of a brand-new treatment facility with adequate funding in a developed country, the choice of non-radioactive source-based treatment may be automatic simply for clinical reasons. For example, intensity-modulated radiation therapy (IMRT) is an advanced mode of high-precision radiotherapy that uses computer-controlled linear accelerators to deliver precise radiation doses to a malignant tumour. IMRT allows for the radiation dose to conform more precisely to the three-dimensional (3D) shape of the tumour by modulating—or controlling—the intensity of the radiation beam in multiple small volumes.[24] In short, it is a leading-edge clinical technique that is likely to be affordable and supportable in the context of a developing country. Concerns about the use of radioactive sources do not apply to this treatment mode and the building will be designed accordingly, without extensive physical protection measures for security.

However, developing countries may not be able to afford or be able to effectively operate an advanced and expensive alternative technology such as IMRT. Here, equally effective treatments for the head, principally the brain, can be delivered through the use of radioactive sources, including a technology known as a 'Gamma Knife'.[25] This uses approximately 200 precision-aimed, converging beams of cobalt-60 gamma sources and advanced robotic technology to move the patient in sub-millimetre increments during treatment to focus radiation successfully on all parts of the target (the tumour). A special rigid head frame incorporating a 3D coordinate system is attached to the patient's skull to accurately target the tumour. Using 3D coordinates determined in the planning process, the frame is then precisely attached to the Gamma Knife unit to guarantee that when the unit is activated, the target is placed exactly in the isocentre of the beams.

A Gamma Knife unit is a collection of IAEA Category I sources and so security measures will need to provide a 'high level of protection of radioactive material against unauthorized removal'.[26] As an essential clinical intervention, however, such a device should not be refused solely on security grounds, particularly in countries where alternatives are not viable. That said, it is important to recognize that the adoption of alternative technologies can be very successful.[27] A paper by Kamen and colleagues summarizes around nine years of effort by Mount Sinai—one of the largest healthcare institutions in New York with about 50,000 employees—to migrate completely from radioactive irradiators to X-ray irradiators without compromising patient care or research studies. Mount Sinai Hospital collaborated with the National Nuclear Security Administration (NNSA) at the US Department of Energy (DOE) to enhance the security of irradiators through the US GTRI programme, now known as the Office of Radiological Security (ORS). The hospital used state-of-the-art security equipment such as biometric access control and 24/7 video monitoring. In addition, a remote monitoring system with alarms was installed and connected to local law enforcement agencies for constant monitoring and intervention, if necessary, in a timely manner.

By the end of the programme, Mount Sinai had successfully disposed of all its Cs-137 irradiators. The hospital has now migrated completely to alternative technology for blood irradiation. Mount Sinai is a multi-billion dollar business and finances of that scale, particularly when applied to flagship NNSA projects, enable achievements that would struggle to be funded in the UK, not to mention in a developing country. Unfortunately, the paper does not provide any information about the total costs of this work, but it is clear that purchasing costs were heavily subsidized. This is an unfortunate omission as a key consideration is running costs; maintenance costs of X-ray irradiators are significantly higher if annual service contracts are used. Performance of the two irradiator types appears to be equivalent, but in some cases X-ray irradiations may need to be performed in multiple configurations to achieve adequate uniformity in dose.[28]

Consider now two other hypothetical cases. First, a national blood transfusion service has several tens of blood irradiators distributed around a country. The operator comes under pressure from an overseas single interest pressure group to change these devices for X-ray-based irradiation machines. The current radioisotope-based machines are all subject to the national regulatory regime, which closely follows IAEA best practice. The manager of the regulation of radioactive sources within the regulatory body has consulted field inspectors and concluded that current regulatory standards are being met. The regulator is content with the status quo. An X-ray irradiator costs about US$270,000[29] which is similar to the capital cost of a Cs-137 irradiator.[30] The national blood transfusion service requires 50 X-ray irradiators, which means the total cost of new equipment is US$13.5 million. The decision to be made is whether spending US$13.5 million on replacement irradiators is the best use of the operator's funds. It might be but, what is the wider national situation with regard to other types of radiological sources that may need protection?

Second, industrial operators within the same country have several thousand nucleonic gauges for measuring level, density, thickness, and moisture content. These too are protected to current standards, but they are very numerous and are often deployed in relatively remote locations. Their absence from a process control system should be promptly apparent, but during outages up to 100 sources are stored together pending recommissioning. They may be 'only' Category IV sources, but they are still radioactive enough to cause significant problems and costs, especially if used in multiples. Could more value be obtained from spending some (or maybe all) of the money on protecting these instead?

There is no generalized correct answer to this question, except of course that the IAEA methodology prioritizes Category I sources over Category III and IV sources in industrial gauges. But an agency, or a multi-disciplinary team at the state level, could assess the national situation (the 'bigger picture') and make such a decision. Their challenge is to decide if spending money on alternative technology for blood irradiation is a 'better deal' than protecting other sources that are 100 times more numerous. Implementation of this may be achieved by national adjustments to the IAEA categorization of sources.

Conclusion

Radioactive materials have been used beneficially in many sectors for decades, but since 9/11 there have been concerns about their implications for the security of society, should they fall into the wrong hands. A range of alternative technologies are now available to replace radioactive sources where this is technically feasible and appropriate. However, this is often not a straightforward 'like-for-like' exchange. Here, there should be a holistic assessment by regulators, operators, and security specialists of how risks are altered, taking into account security considerations but also broader factors relating to the local conditions, cost, and operational effectiveness. This is essential to avoid focusing on one particular aspect in isolation, which could produce an outcome that reduces risk in one area but serves to increase it in others.

Thankfully, there now exists international guidance that can support this process, for example, the WINS guide on the replacement of high-activity radioactive sources. Insights should also be drawn from real-life cases where radioactive sources have been replaced by alternatives. There exist examples where states and operators have undertaken substantial multi-year programmes, in some cases eliminating an entire class of radioactive source. However, there are currently only limited details available on these efforts and in particular what drove the decision-making calculus. As such, greater priority should be given to extracting and sharing lessons from these cases.

Notes

1. IAEA, *The Radiological Accident in Goiânia*, IAEA Non-serial Publications (Vienna: IAEA, 1998).
2. Alan Martin, Sam Harbison, Karen Beach, and Peter Cole, *An Introduction to Radiation Protection*, 7th edn (Boca Raton: Taylor & Francis Group, 2018).
3. World Health Organization (WHO), 'Development of Stockpiles for Radiation and Chemical Emergencies', Report of the Radio-Nuclear Working Group (Geneva: WHO, 2007).
4. US Nuclear Regulatory Commission (NRC), 'Fact Sheet on Dirty Bombs', n.d. https://www.mass.gov/service-details/nuclear-regulatory-commission-nrc-fact-sheet-on-dirty-bombs
5. Miles Pomper, 2020, 'Permanent Risk Reduction: A Roadmap for Replacing High-Risk Radioactive Sources and Materials', paper presented at The National Academies of Sciences Radioactive Sources: Applications and Alternative Technologies Meeting, 31 January 2020.
6. Piping Engineering, 'Inspection and Testing: Dye Penetrant Inspection', 14 December 2017. https://www.pipingengineer.org/dye-penetrant-inspection
7. IAEA, *Fundamental Safety Principles*, IAEA Safety Standards, Safety Fundamentals, No. SF-1, para. 3.29 (Vienna: IAEA, 2006).
8. Andrew J. Bieniawski, Ioana Iliopulos, and Michelle Nalabandian, 'Case Study: An Analysis of Several National Approaches to Alternative Technologies for Radioactive Sources', Nuclear Threat Initiative (NTI) [online], July 2017. https://www.nti.org/wp-content/uploads/2021/09/NTI_INMM_Alternative_Technology_Paper_Jul2017_oDGeMET.pdf
9. National Nuclear Security Administration (NNSA), 'Office of Radiological Security (ORS)', US Department of Energy, 2021. https://www.energy.gov/nnsa/office-radiological-security-ors
10. Frank Barnaby, 'Dirty Bombs and Primitive Nuclear Weapons', Oxford Research Group [online], June 2005. https://www.files.ethz.ch/isn/22204/dirtybombs.pdf
11. Vic Levin, Ahmed Meghzifene, Joanna Izewska, and Hideo Tatsuzaki, 'Improving Cancer Care: Increased Need for Radiotherapy in Developing Countries', *IAEA Bulletin*, No. 43/02 (2001), 25–32.
12. L. M. Wroe, T. A. Ige, O. C. Asogwa, S. C. Aruah, S. Grover, R. Makufa, M. Fitz-Gibbon, and S. L. Sheehy, 'Comparative Analysis of Radiotherapy Linear Accelerator Downtime and Failure Modes in the UK, Nigeria and Botswana', *Clinical Oncology*, 32/4 (2020), 111–18. https://doi.org/10.1016/j.clon.2019.10.010
13. C. Nwobi, C. C. Nzotta, A. D. Obotib, and G. Luntsi, 'Failure Mode Analysis of Radiologic Equipment in a Tertiary Institution in South-Eastern Nigeria', *IOSR Journal of Dental and Medical Sciences*, 16/10 (2017), 82–90.
14. IAEA, *Worldwide Implementation of Digital Imaging In Radiology: A Resource Guide In Cooperation With The World Health Organization*, IAEA Human Health Series, No. 28 (Vienna: IAEA, 2015).
15. IAEA, *Categorization of Radioactive Sources*, IAEA Safety Standards, Safety Guide No. RS-G-1.9 (Vienna: IAEA, 2005).

16. National Institute for Health and Care Excellence (NICE), 'Low Energy Contact X-ray Brachytherapy (the Papillon Technique) for Early Stage Rectal Cancer', Interventional procedures guidance [IPG532], 23 September 2015. https://www.nice.org.uk/guidance/ipg532
17. Bieniawski et al., 'Case Study: An Analysis of Several National Approaches to Alternative Technologies for Radioactive Sources'.
18. C. Bergman and B. G. Pettersson, 'Radiation Applications and Waste Management: Taking the Final Steps', *IAEA Bulletin*, 36 (1994), 36–41. https://www.iaea.org/sites/default/files/publications/magazines/bulletin/bull36-1/36104683640.pdf
19. Ibid.
20. IAEA, *Identification of Radioactive Sources and Devices*, Technical Guidance, IAEA Nuclear Security Series, No. 5 (Vienna: IAEA, 2007).
21. See for example, Troxler, n.d., 'PaveTracker™ Model 2701-B Plus'. http://www.troxler.ca/field_equipment/2701_b.html
22. WINS, 'Considerations for the Adoption of Alternative Technologies to Replace High Activity Radioactive Sources', WINS Special Report Series [online], 22 January 2021. https://www.wins.org/document/considerations-for-the-adoption-of-alternative-technologies-to-replace-high-activity-radioactive-sources-2
23. Eduardo Rosenblatt, 'Planning National Radiotherapy Services', *Frontiers in Oncology* [online], 4/315 (2014). https://www.frontiersin.org/articles/10.3389/fonc.2014.00315/full
24. RadiologyInfo.org, 'Intensity-Modulated Radiation Therapy (IMRT)', 20 July 2021. https://www.radiologyinfo.org/en/info/imrt
25. RadiologyInfo.org, 'Gamma Knife', 20 January 2018. https://www.radiologyinfo.org/en/info/gamma_knife
26. IAEA, *Security of Radioactive Material in Use and Storage and of Associated Facilities*, Implementing Guide, IAEA Nuclear Security Series, No. 11-G (Rev.1) (Vienna: IAEA, 2019).
27. J. Kamen, W. Y. Hsu, B. Boswell, and C. Hill, 'Successful Migration from Radioactive Irradiators to X-ray Irradiators in One of the Largest Medical Centers in the US', *Health Physics*, 117/5 (2019), 558–70.
28. Brian Dodd and Richard J. Vetter, 'Replacement of 137Cs Irradiators with X-ray Irradiators', *Health Physics*, 96/2 Suppl (2009), 27–30.
29. E. Bakken, K. Cary, A. Derrick, E. Hildebrand, K. Schroeckenthaler, and M. Taalbi, *Cost-Benefit Analysis of Switching from Cesium-Chloride to X-ray Blood Irradiators* (Madison, WI: University of Wisconsin-Madison, 2013).
30. Ibid.

Suggested Reading

Bieniawski, Andrew J., Iliopulos, Ioana, and Nalabandian, Michelle, 'Case Study: An Analysis of Several National Approaches to Alternative Technologies for Radioactive Sources', Nuclear Threat Initiative (NTI) [online], July 2017. https://www.nti.org/wp-content/uploads/2021/09/NTI_INMM_Alternative_Technology_Paper_Jul2017_oDGeMET.pdf

Danish Health Authority, (2020), 'Radiation Generators: For Service Companies (Guide)'. https://www.sst.dk/-/media/Viden/Straaling/Vejledninger/Straalingsgeneratorer/Radiation-generators-For-service-companies.ashx?la=en&hash=6EE51CE24FD859FBEAF921EEA311C4B9F48F3CCA

Englefield, C., (2014), 'Radioactive Source Security: Why Do We Not Yet Have a Global Protection System?', *Nuclear Engineering and Technology*, 46/4, 461–6.

Englefield, C., (2014), 'Radioactive Source Security: The Cultural Challenges'. *Radiation Protection Dosimetry*, 164/1–2, 13–17.

Ferguson, Charles D. and Potter, William C., (2005), *The Four Faces of Nuclear Terrorism* (New York: Routledge).

Gale, Robert Peter and Lax, Eric, (2013), *Radiation: What It Is, What You Need to Know* (New York: Knopf).

Martin, Alan, Harbison, Sam, Beach, Karen, and Cole, Peter, (2018), *An Introduction to Radiation Protection*, 7th edn (Boca Raton: Taylor & Francis Group).

CHAPTER 37

REMOVAL OF GAMMA BLOOD IRRADIATORS IN NORWAY

STYRKAAR HUSTVEIT, SINDRE ØVERGAARD, TONE-METTE SJØMOEN, AND HÅVAR A. SOLLUND

Introduction

A rare complication of blood transfusion known as transfusion-associated graft-versus-host disease (TA-GvHD) occurs when viable donor T-lymphocytes in a transfused blood product react against the tissues of the recipient. Normally, the donor T-lymphocytes will usually be recognized as foreign and destroyed by the recipient's immune system. However, this will not always happen if the recipient is severely immunocompromised or if the donor and recipient share tissue type antigens (human leukocyte antigens; HLAs) that are used by the immune system to differentiate 'self'-cells and 'non-self'-cells. The risk of TA-GvHD increases with the degree of HLA homogeneity between donor and recipient, with a higher degree of homogeneity found in blood relatives, as well as in populations in some parts of the world, for example, Japan and Israel.[1] TA-GvHD can affect the skin, gastrointestinal tract, liver, and bone marrow of the patient, and may lead to pronounced and irreversible pancytopenia (i.e. too few red and white blood cells and platelets in the blood). There is no effective treatment for TA-GvHD, and the mortality has been estimated to be above 90 per cent.[2,3] Hence, effective preventive measures are highly important.

Irradiation is the best-established method for prevention of TA-GvHD.[4,5,6] High doses of ionizing radiation abrogate the ability of T-lymphocytes in donor blood to proliferate. There has been some controversy over the appropriate radiation dose to ensure prevention of TA-GvHD,[7] but delivered doses typically range from 25–50 gray (Gy).[8,9] Doses above 50 Gy could harm the function or viability of red blood cells and platelets.[10,11] Although irradiation is effective for prophylaxis against TA-GvHD, drawbacks of the method include operational costs, time delays, and logistics problems if irradiation is not done in-house, and a shortened shelf-life of irradiated red blood cells.[12,13,14]

A major drawback of using conventional gamma-based blood irradiators is the security concerns associated with the irradiators' high-activity caesium-137 (Cs-137) radioactive sources. The International Atomic Energy Agency (IAEA) has developed a categorization system for sealed radioactive sources, which comprises five categories ranking radioactive sources in terms of their potential to cause harm to human health.[15] Radioactive sources in Category I, to which caesium irradiator sources belong, have the highest activity, and are deemed to be the most dangerous. An exposure of only a few minutes to an unshielded Category I source might cause permanent injury or death.

In addition to their high activity, Cs-137 sources used in blood irradiators have physical properties that significantly increase the security risks associated with their use. The sources have conventionally been produced in the form of a steel-encapsulated compressed powder. The powder consists of caesium chloride (CsCl) salt, which is water soluble and highly dispersible, making the material easy to handle and suitable for use in a radiological dispersal device (RDD).[16,17] Dispersed CsCl binds easily to surfaces, may migrate into concrete, and is very mobile in the environment. Moreover, Cs-137 has a long half-life of about 30 years, and it emits penetrating gamma radiation (the main gamma emission has an energy of 662 keV, with a half-value thickness of 8 mm lead).[18] These characteristics make decontamination of an area affected by accidental or malicious dispersal of CsCl very complicated and costly. The security threat posed by CsCl irradiators is exacerbated by the widespread presence around the world of blood irradiators in hospitals and blood banks located in major population centres.

Following the terrorist attacks of 11 September 2001 in the United States (US), the threat posed by malicious acts involving high-activity radioactive sources has received increased international attention. The IAEA's Code of Conduct on the Safety and Security of Radioactive Sources, first issued in March 2001, was revised and strengthened following the attacks.[19] At the time of writing, 140 states have made a political commitment with regard to the Code of Conduct, and endorsed the objectives and principles set out in the Code.[20] To complement the Code of Conduct, the IAEA has issued a number of guidance documents, including recommendations for security of radioactive sources, suggesting a graded approach based on the IAEA's categorization system.[21] Additional steps towards international awareness of radioactive source security were taken at the 2014 and 2016 Nuclear Security Summits through the Joint Statements on 'Enhancing Radiological Security' (in 2014) and on 'Strengthening the Security of High Activity Sealed Radioactive Sources' (in 2016), which were endorsed by 23 and 28 states, respectively.[22,23]

Along with the increased international focus on security, it has been recognized by states and other stakeholders that technological advances can create opportunities for replacement of high-activity radioactive sources with alternative non-isotopic technologies. In an influential report from 2008, the US National Research Council conducted a comprehensive review of the feasibility of replacing high-activity radioactive sources with alternative technologies.[24] One of the main recommendations of the report was to eliminate the use of Category I and II CsCl sources and replace CsCl irradiators with X-ray-based irradiators.

Alternatives to CsCl irradiators include the use of less dispersible radioactive sources, such as cobalt-60 (Co-60) or vitrified Cs-137 in pencil-shaped form,[25] the use of medical linear accelerators (linacs),[26] and the use of pathogen reduction by means of UV irradiation combined with photosensitizers.[27,28] The latter method has the advantage of also eliminating harmful bacterial and viral pathogens from the blood components. Pathogen reduction methods are promising alternatives for the future but are currently restricted to treatment of platelets and plasma and cannot yet replace gamma- or X-ray irradiation of other blood components. Modern X-ray irradiators may on the other hand be adopted as direct replacements for conventional CsCl irradiators for all relevant blood products.

While X-ray irradiators two decades ago were prone to breakdowns from X-ray tube and power source failures, modern second-generation X-ray irradiators are broadly regarded as reliable and cost-effective alternatives to CsCl irradiators.[29] The throughputs of X-ray and gamma-based irradiators are similar, both with regard to exposure times and with regard to the number of blood bags that can be processed during each operation cycle.[30,31] In a study prepared for the Global Threat Reduction Initiative in the US, the lifecycle costs of switching from CsCl to X-ray irradiators were examined, including costs from installation, operation, and termination.[32] The results indicated a cost-benefit of switching to X-ray irradiators, with the main reasons for this being the large costs of securing and disposing of CsCl irradiators.

The security threats associated with CsCl irradiators and benefits of replacing them with X-ray irradiators have led to increased adoption of X-ray irradiators in several states, such as Japan, which has replaced more than 80 per cent of its gamma-based irradiators, and France, which completed replacing all its 30 gamma-based irradiators in 2016.[33] In the US, the transition to X-ray-based blood irradiators progressed slowly until the Cesium Irradiator Replacement Project (CIRP) was launched in 2014. CIRP provides financial incentives to facilitate replacement of CsCl irradiators, and as of April 2021 CIRP had contributed to removing more than 165 of the approximately 750 CsCl irradiators in the US, with another 150 irradiators awaiting removal.[34] While the success of CIRP in the US demonstrates the power of financial incentives, other states have had success with different strategies, for example, through regulatory approaches, such as a strict implementation of the justification principle in licensing processes.[35] States successfully adopting the latter strategy include France and Switzerland.

All of Norway's 13 CsCl blood irradiators were replaced with X-ray irradiators within the three-year period of 2012–2015. This chapter outlines the factors contributing to the swift and successful removal of gamma-based blood irradiators in Norway and highlights key elements in the Norwegian strategy.

Legal Basis

Norway implemented a new Act on Radiation Protection and Use of Radiation in 2000.[36] Supplementary to the Radiation Protection Act, new Regulations on Radiation

Protection and Use of Radiation were passed in 2003.[37,38] The updated Radiation Protection Act and Regulations harmonized Norwegian legislation with international guidelines from the IAEA and the International Radiation Protection Commission (ICRP). Importantly, both the act and the regulations incorporated the three main principles of radiation protection—justification, optimization, and dose limitation—and stated that all use of radiation sources shall be justified. In the assessment of justification, it shall be considered whether the benefits of the activity outweigh the risks.

Furthermore, the new regulations contained a requirement of substitution, which complements the justification requirement. It stated that any undertaking using ionizing radiation sources must assess alternatives and choose methods that do not involve the use of ionizing radiation if feasible. Moreover, the requirement explicitly stated that, for non-medical use of radiation, X-ray devices shall be used, rather than radioactive sources when practically achievable. Note that any use of radiation sources not involving direct exposure of a patient is considered 'non-medical' use of radiation in Norwegian regulatory practice. Blood irradiators are therefore treated as 'non-medical' radiation sources, and the imperative to choose X-ray devices when practically achievable thus applied for blood irradiation. The substitution requirement in the Radiation Protection Regulations proved to be instrumental for initiating the phase out of gamma-based blood irradiators in Norway.

In 2010 the regulations were revised, adding to the substitution requirement that a reassessment shall be performed whenever new information emerges relating to the justification of existing methods. This further enabled the regulator, the Norwegian Radiation Protection Authority (NRPA), to reject renewal of licences for gamma-based blood irradiators when improved X-ray-based alternatives became available.

Supplementary to the Security Act of 1998, the new Regulations on Object Security that came into force in 2011 provided another important legal basis for Norway's removal of gamma-based irradiators.[39] The regulations required every Norwegian ministry to designate 'Critical National Objects and Infrastructure' within their area of responsibility and report these to the Norwegian National Security Authority (NSM). Critical National Objects—which included buildings, areas, and property of high importance to national security—were subject to strict security provisions, reminiscent of the IAEA's security recommendations for Category I radioactive sources.[40] When blood irradiators, due to the security threat they posed, were designated as Critical National Objects, it provided an implicit financial incentive to replace the gamma-based irradiators in order to avoid inflated security costs. It also facilitated cooperation between the NRPA and NSM in regulating and inspecting the security of the irradiators.

Stricter Security Requirements for Radioactive Sources

The first Radiation Protection Act in Norway had come into force in 1938, but traditionally radiation protection has been mostly about safety, which has also been reflected in the legislation. Stricter security requirements for radioactive sources in Norway are the result of a more recent process that started about 20 years ago and is still ongoing at the time of writing. The development of stricter security requirements for radioactive sources in Norway has coincided with—and to a large extent been driven by—the development of international guidelines and directives in this area. Indeed, the NRPA has benefitted greatly from international cooperation and information exchange on the subject. The NRPA followed the IAEA Technical Meetings on implementation of the Code of Conduct closely in the 2000s, and in the following years the NRPA has aimed to harmonize Norwegian regulatory practice with the IAEA's Code of Conduct and associated guidance documents.[41]

Regional cooperation between the Nordic countries has also been important in this context. The Nordic regulators on radiation protection established the Nordic Group on Use of Sealed Sources (NORGUSS) in 2003. The group was established on basis of the increased international focus on security of radioactive sources. Incorporation of the objectives and principles in the Code of Conduct and the European Union's 'Council Directive 2003/122/Euratom of 22 December 2003 on the control of high-activity sealed radioactive sources and orphan sources' in national legislation was a common challenge that the Nordic countries faced at that time.[42] Even though Norway is not part of the Euratom Treaty, the Council Directive 2003/122/Euratom was an important guideline in the preparation of national regulations. In 2010, the Nordic working group was expanded to work with all types of ionizing radiation sources in non-medical applications. The group has provided an important platform for sharing information on subjects such as new emerging technologies, security of high-activity sealed sources, experiences from inspections, and development of new regulations.

During 2004 the NRPA conducted inspections at eight of the 10 hospitals that had gamma blood irradiators at that time. The inspections were carried out due to the new Radiation Protection Regulations from 2003 and focused on both safety and security of gamma blood irradiators. Information was provided to all hospitals about the requirements in the new regulations, which included a new requirement to perform risk assessments related to the use of radiation sources. The requirement stated that new activities involving radiation sources may not start before a risk assessment had been completed and necessary preventive measures implemented. If the assessment showed an unacceptable risk to employees, patients or other persons, or that radiation sources may come out of the licensee's control, the undertaking should take preventive measures to reduce the risk, including to secure the radiation sources against theft, sabotage, and other damage. At that time, it was only a recommendation, and not yet a requirement, for high-activity sources to be separated from unauthorized personnel with at least two technical barriers and with access control. The pictures in Figure 37.1 show gamma blood irradiators at two different hospitals during inspections in Norway in 2004: one irradiator located near an open office landscape and another in a corridor. As observed from the pictures, basic security measures, like access control, were lacking.

A classified Norwegian RDD study was performed in 2006 where potential consequences of dispersal of radioactive material from a gamma blood irradiator were assessed. This proved an important impetus for the regulator in Norway to increase ongoing efforts to improve security for radioactive sources. The majority of Category

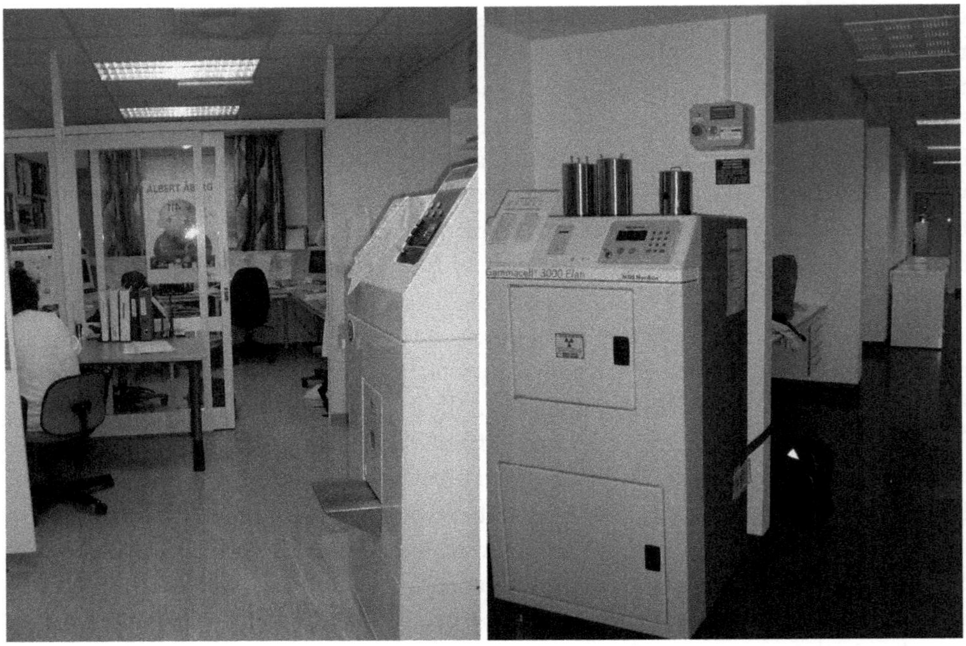

FIGURE 37.1. Pictures from inspections at two different hospitals in Norway in 2004. One gamma blood irradiator located near an open office landscape (left) and another in a corridor (right).
Image courtesy of the Norwegian Radiation Protection Authority.

I radioactive sources in Norway at that time were gamma blood irradiators, and the use of such irradiators was expanding to more hospitals. By 2006, two technical barriers and access control had already become standard security requirements for high-activity sources in Norway. However, at the end of 2006 the hospitals were instructed to implement even stricter security measures for gamma blood irradiators. Within a short timeframe (two months), the security level had to be increased with a third independent technical or administrative security measure against sabotage. This could be, for example, locking the irradiator in a separate room with access only for legitimate users or the establishment of video surveillance of the facility. In the longer term (six months), it was required to locate the blood irradiator in a separate room with access only for legitimate users and with individual access registration upon entry and exit. This implied that the room had to have its own key or code that only was available to staff who needed access to the gamma blood irradiator. In addition, hospitals were required to carry out an assessment of the staff with access to the room, to reduce any possible insider threat.

In the following years, the stricter security requirements were followed up both administratively and with new inspections by the NRPA. The following security requirements were issued in licences:

- access control to the department where the irradiator was located
- irradiator to be located in a separate room with access only for legitimate users, that is, the room had to have its own key or code that only was available to legitimate users

- system for individual access registration upon entry and exit
- possibility to retrospectively find out who had been in the room
- carrying out of assessment of staff with access to the room, to reduce any possible insider threat, and
- use of code or key to operate the irradiator.

Follow-up inspections showed that some hospitals met the security requirements, while others struggled.

Decision to Phase Out

In May 2010 inspections at the University Hospital of North Norway in Tromsø found that regulatory requirements for access control to the irradiator were not being met. The main issue was to restrict access to the area where the irradiator was located to only those required to operate the irradiator during working hours. The existing building layout put this requirement into conflict with other logistical demands at the hospital. After assessing the practicalities and costs of remedying the situation, the hospital decided later that year to replace their caesium-based irradiator with an X-ray-based irradiator. This meant the hospital would be exempt from the security requirements for high-activity sources and could avoid the costly and cumbersome modification that would otherwise be required at the site.

To the NRPA, as the regulator, this was a pivotal decision. There had been internal discussions, in part based on the US National Research Council study from 2008, on whether X-ray irradiators had become mature enough as a technology to be a direct replacement for most, or all, gamma-based blood irradiators.[43] The hospital operators had up to this point been quite determined that they required the (presumed) higher reliability of gamma irradiators. When one of the operators at this point had assessed for themselves that an X-ray irradiator would meet their needs, it was an important confirmation of the maturity of the then-available X-ray equipment.

In 2012, a hospital in the Norwegian capital Oslo submitted an application to replace their aging gamma blood irradiator with a new gamma irradiator. With reference to the requirement of substitution in the Radiation Protection Regulations, the NRPA requested that the application for replacement included an assessment of alternatives to gamma irradiation. The hospital's assessment was that X-ray was not a suitable alternative to gamma due to the desired levels of availability and reliability, and that a switch to X-ray would represent infrastructural challenges because of the required power and water supply. X-ray irradiators require substantially more electricity than gamma irradiators since the radiation has to be generated, and they also require water for cooling, which is not an issue for gamma irradiators. The hospital also argued that they had already put physical security measures in place for their existing irradiator that would also provide adequate protection for a new gamma irradiator.

In assessing the application, the NRPA concluded that the arguments for continuing with gamma irradiation were not sufficient to outweigh the continued risk of housing substantial quantities of radioactive material at the site and that the use of a gamma-based irradiator would not be justified. With regard to the question of availability, the NRPA's position was that the hospital was in possession of other blood irradiators, and also had access to additional irradiation capabilities on other sites, including nearby blood irradiators and radiotherapy linacs. Consequently, it should be possible for them to set up contingency planning in the case that the irradiator in question became unavailable. According to Norwegian law, the applicant is entitled to appeal decisions by the regulator to the Ministry of Health and Care Services. The hospital appealed the decision, but it was upheld by the Ministry in April 2013; it then became clear that the hospital would not be allowed to install a new gamma-based blood irradiator.

During the time that the application was being processed, the NRPA in 2013 also took advantage of its international connections through the European Radiation Protection Authorities Network (ERPAN) to send out a questionnaire about the practice in other European countries. The study revealed that Sweden and Germany had some experience with X-ray-based blood irradiation, and that France already had a high percentage of X-ray-based blood irradiators.[44] This information proved that also other countries had assessed the X-ray technology as sufficiently mature and was an important confirmation of the NRPA's assessments.

Since the application process from the hospital in Oslo had involved the blood irradiator in Norway with the highest throughput in terms of annual irradiated blood volume, this case made the NRPA conclude that it would be highly unlikely that any other hospital could come up with a strong enough justification for installing new gamma-based irradiators. The conclusion from this case was that the NRPA would probably not grant licences to install new gamma-based blood irradiators in Norway, and that gamma irradiators would be gradually phased out and replaced with X-ray irradiators as existing irradiators aged and eventually were decommissioned. The NRPA informed the hospitals about this decision and published a bulletin explaining the new regulatory policy in 2013.[45]

Decision to Substitute

During this time period Norway was also struck by a serious domestic terrorist attack. The 22 July 2011 terrorist attack was carried out by the right-wing extremist Anders Behring Breivik against the state administration, random civilians in Oslo, and the Workers' Youth League (AUF) summer camp on Utøya. A total of 77 people lost their lives in the deadliest attack in Norway since the Second World War.[46] Prior to the attack, Breivik released a 1,518-page manifesto titled '2083–A European Declaration of Independence' by distributing it to more than 1,000 email addresses. The manifesto partly described his motives for committing the terrorist attacks and partly reviewed the planning phase. Tens of pages were devoted to detailed descriptions of nuclear and radiological terrorism. The attack firmly and devastatingly demonstrated that Norway was not shielded from the international trend of a heightened threat from terrorism.

This led to an increased national focus on security and reports recommending enhanced security of important objects. In the August 2012 investigation report from the July 22 Commission was released, and one of the main conclusions was that the attack on the government quarter in Oslo could have been prevented through effective implementation of already-adopted security measures.[47] The commission recommended that leaders at all levels in the public administration should work to strengthen their own organizations' culture for risk recognition. The Regulations on Object Security had come into force just a few months prior to the attack (1 January 2011), and it was stated that implementation of security requirements for Critical National Objects should be expedited. The NSM was encouraged to follow up by performing inspections.

According to the Security Act and the Regulations on Object Security, the NSM was the supervisory authority, with the main responsibility for assuring compliance with the provisions of the Security Act and its supplementary Regulations. This included supervisory responsibility for the *process* of designating Critical National Objects or infrastructure in society. Both the Ministry of Health and Care Services and NRPA also had responsibilities related to these regulations. The Ministry of Health and Care Services was responsible for deciding which undertakings and objects within the health and care service sector the Security Act and associated regulations should apply to. As the national authority and expert body in matters concerning radiation protection, the NRPA had authority to propose for the Ministry of Health and Care Services objects to be placed under the Security Act. The NSM could also on its own initiative propose to a ministry which objects they considered to be Critical National Objects or infrastructure.

The classified RDD study from 2006 had already highlighted the security threat and potential economic impact of a malicious act involving gamma blood irradiator sources—a finding later supported by other studies, such as the US National Research Council report from 2008.[48] Moreover, national and international threat assessments and signals from intelligence experts indicated that hospitals were attractive targets for cybercrime, industrial espionage, and state intelligence. Based on a proposal from the NRPA, the Ministry of Health and Care Services decided that all gamma-based blood irradiators in Norway should be designated as Critical National Objects and classified as worthy of protection.

The fact that the blood irradiator facilities became designated as Critical National Objects meant a full review of existing security measures and a mapping of the need for additional measures. This process required both the NRPA's expertise on radiation safety and radiological threats and the NSM's expertise on security. The NRPA and NSM therefore cooperated by conducting joint inspections of blood irradiation facilities. The findings revealed that the existing security measures were not in line with the provisions of the Security Act. Among other things, the inspections led to an increased focus on the demands for resistance in wall materials and doors, as well as the ability to detect and respond to malicious acts. Implementation of new, comprehensive, and costly security measures had to be expected in order to provide a sufficiently high level of protection of the radioactive sources and meet the requirements of the Security Act. More or less all the hospitals in Norway struggled to meet the new security requirements, especially when it came to demands for resistance in doors and wall materials in rooms where irradiators were located.

Improved security measures had to be implemented as soon as possible and no later than three years after the regulations came into force. Upon application to the NSM, the implementation could be postponed for a period of up to one year, leaving 1 January 2015 as a deadline for the implementation of the required security measures. As the responsible and decision-making authority, the Ministry of Health and Care Services, in collaboration with the NRPA and NSM, now initiated and carried out a process with the regional health authorities to assess what relevant security measures had to be implemented, to consider if available alternative technology was satisfactory, and to make a qualitative cost–benefit assessment of the different alternatives.

As described, the NRPA had already concluded that the X-ray alternative was satisfactory and had decided to gradually phase out gamma-based blood irradiators by means of not granting authorization for new acquisitions. One Norwegian hospital had already replaced its gamma-based irradiator with an X-ray irradiator, and through the ERPAN study, it had also been documented that X-ray blood irradiators were allowed and in use in several other European countries. There were few signals from the Norwegian health sector indicating that there would be major problems associated with a replacement, except from the two blood irradiators used for research. It was argued that new experimental data could not be compared with the old data. Whether these irradiators should be either replaced or secured was therefore carefully considered before the decision was taken.

Disregarding the cost associated with the security measures, the conclusion was that the two technologies (X-rays and gamma) were considered comparable in terms of costs. Expenses related to new security measures come in addition to other costs related to purchase, maintenance, and disposal. Given that implementation of new comprehensive security measures was required for gamma-based blood irradiators, retaining these irradiators turned out to be the most expensive option.

Based on available knowledge, it was concluded that the X-ray alternative was a satisfying alternative to gamma-based blood irradiators for the hospitals. It also turned out that replacing was the less expensive alternative. On this basis, the Ministry of Health and Care Services, NSM, NRPA, and regional health authorities agreed to replace all the gamma-based blood irradiators in Norway with X-ray based blood irradiators, including the gamma-based blood irradiators used for research. An ambitious scheme was drawn up to complete the replacement process within the 1 January 2015 deadline in the Regulations on Object Security. Due to logistical challenges related to lack of transport capacity and difficult weather and driving conditions during the winter in Norway, the process was delayed by a few months.

After the decision to substitute was taken, the regional health authorities in cooperation with the NRPA acquired new X-ray based irradiators and disposed of the old gamma-based irradiators. In accordance with IAEA recommendations, all the gamma-based blood irradiators were sent back to the manufacturer in the country of origin for reuse or recycling. All but one of the irradiators were returned to Best Theratronics Ltd., Canada. The manufacturer provided approved transport containers and specialized personnel to assist in the transports. The last irradiator was of a different type and was shipped to Gamma-Services Recycling GmbH in Germany. The communication between the NRPA and the importing state followed the recommendations of the IAEA's 'Guidance on Import and Export of Radioactive Sources'.[49]

Conclusion

There are some key lessons that can be learned from the Norwegian decision to replace gamma-based blood irradiators with X-ray irradiators. Security requirements for high-activity radioactive sources have gradually, but consistently, become stricter, thereby providing an implicit incentive for replacement. This has been achieved by various means, such as by revision of acts and regulations; by changes in regulatory practice through stricter enforcement or new interpretation of existing regulatory requirements; or by introducing new conditions when licences for blood irradiators have been issued or renewed. Stricter requirements have always been communicated to all licensees and often followed up by inspections, resulting in reduced risk of malicious acts and providing a strong incentive for replacement due to the costs of implementing new security measures.

The development of international guidelines and participation in international forums for information exchange have also been key drivers for the adoption of stricter security requirements. When faced with stricter security requirements, one of the operators, the University Hospital of North Norway, was able to reassess their needs and to take a look at the situation as part of a larger picture. Rather than being locked into implementing costly and impractical physical security measures, the operator decided that a better overall result could be achieved by using an alternative technology. Through the requirement of substitution, the regulator had an important legislative tool to deny new gamma-based irradiators. However, it was also crucial that the regulator's assessment of the maturity of the alternative technology was sufficiently well founded that an appeal of the decision did not change the outcome.

The phase-out policy could have been implemented by the NRPA without any dedicated funding. Costs would be incurred when the irradiators had to be replaced in any case, and there were compelling arguments that the overall lifetime cost of the irradiator could be lower when using X-ray technology, since costs for additional security and disposal were not applicable for X-ray irradiators. The downside of a gradual phase out would be that it would take many years before all irradiators were replaced.

The decision to substitute in Norway was a result of mutual collaboration between the Ministry of Health and Care Services, the regional health authorities, the NRPA, the NSM, and other relevant qualified expertise. Knowledge-based advice and recommendations to the Ministry of Health and Care Services were essential during the process. The overall conclusion was that the X-ray alternative was a cost-effective, risk-eliminating, and technically satisfactory alternative for the hospitals.

Notes

1. Burak Bahar and Christopher A. Tormey, 'Prevention of Transfusion-Associated Graft-Versus-Host Disease With Blood Product Irradiation: The Past, Present, and Future', *Archives of Pathology & Laboratory Medicine*, 142/5 (2018), 662–7.
2. Sadaf N. Bhatti, Alice Gustavsen, Ingvild S. Hausberg, et al., 'Irradiation to Prevent a Fatal Transfusion Complication', *Tidsskr Nor Legeforen* [online], 141/9 (2021). https://tidsskriftet.no/en/2021/06/kronikk/irradiation-prevent-fatal-transfusion-complication
3. Ilana Kopolovic, Jackie Ostro, Hideki Tsubota et al., 'A Systematic Review of Transfusion-Associated Graft-versus-Host Disease', *Blood*, 126 (2015), 406–14.
4. Bahar and Tormey, 'Prevention of Transfusion-Associated Graft-Versus-Host Disease'.
5. Bhatti et al., 'Irradiation to Prevent a Fatal Transfusion Complication'.
6. Kopolovic et al., 'A Systematic Review of Transfusion-Associated Graft-versus-Host Disease'.
7. IAEA, *Effects of Ionizing Radiation on Blood and Blood Components: A Survey*. IAEA-TEDOC-934 (Vienna: IAEA, 1997).
8. Helsedirektoratet, *Veileder for transfusjonstjenesten i Norge* (Oslo: Utgave, 2017). https://helsedirektoratet.no/retningslinjer/veileder-for-transfusjonstjenesten-i-norge [In Norwegian].
9. Office of Blood Research and Review, *Recommendations Regarding License Amendments and Procedures for Gamma Irradiation of Blood Products*, Technical Report (Bethesda, MD: Food and Drug Administration (FDA), 1993).
10. Paola Pinnarò, Antonella Soriani, Daniela D'Alessio et al., 'Implementation of a New Cost Efficacy Method for Blood Irradiation Using a Non-Dedicated Device', *Journal of Experimental Clinical Cancer Research*, 30/1 (2011), 7.
11. J. Chapman, R. D. Finney, K. Forman et al., 'Guidelines on Gamma Irradiation of Blood Components for the Prevention of Transfusion-Associated Graft-versus-Host Disease', *Transfusion Medicine*, 6/3 (1996), 261–71.
12. Kopolovic et al., 'A Systematic Review of Transfusion-Associated Graft-versus-Host Disease'.
13. Pinnarò et al., 'Implementation of a New Cost Efficacy Method for Blood Irradiation Using a Non-Dedicated Device'.
14. Chapman et al., 'Guidelines on Gamma Irradiation of Blood Components for the Prevention of Transfusion-Associated Graft-versus-Host Disease'.
15. IAEA, *Categorization of Radioactive Sources*, IAEA Safety Guide, No. RS-G-1.9 (Vienna: IAEA, 2005).
16. Committee on Radiation Source Use and Replacement, *Radiation Source Use and Replacement*, abbr. edn (Washington, DC: The National Academies Press, 2008). https://www.nap.edu/catalog/11976
17. Miles Pomper, Egle Murauskaite, and Tom Coppen. 'Promoting Alternatives to High-Risk Radiological Sources: The Case of Cesium Chloride in Blood Irradiation', *CNS Occasional Paper*. No. 19 (Washington, DC: James Martin Center for Nonproliferation Studies, 2014).

18. D. Delacroix, J. P. Guerre, P. Leblanc, and C. Hickman, 'Radionuclide and Radiation Protection Data Handbook 2002, Second Edition', *Radiation Protection Dosimetry*, 98/1 (2002), 9–168.
19. IAEA, *Code of Conduct on the Safety and Security of Radioactive Sources*, Non-serial Publications (Vienna: IAEA, 2004).
20. IAEA, 'Status List of Countries', IAEA [online], 15 September 2001. https://nucleus-new.iaea.org/sites/ns/code-of-conduct-radioactive-sources/Documents/Status_list%20 15%20September%20%202021.pdf
21. IAEA, *Security of Radioactive Material in Use and Storage and of Associated Facilities*, IAEA Nuclear Security Series, No. 11-G (Rev. 1) (Vienna: IAEA, 2019).
22. Nuclear Security Summit, 'Statement on Enhancing Radiological Security', 2014 Nuclear Security Summit, The Hague, 24 March 2014. https://2009-2017.state.gov/documents/organization/235507.pdf
23. Nuclear Security Summit, 'Joint Statement on Strengthening the Security of High Activity Sealed Radioactive Sources', 2016 Nuclear Security Summit, Washington, DC, 11 March 2016. http://www.nss2016.org/document-center-docs/2016/4/6/joint-statement-on-the-security-of-high-activity-radioactive-sources
24. Committee on Radiation Source Use and Replacement, *Radiation Source Use and Replacement*.
25. Committee on Radioactive Sources, *Radioactive Sources: Applications and Alternative Technologies* (Washington, DC: The National Academies Press, 2021). https://doi.org/10.17226/26121
26. Pinnarò et al., 'Implementation of a New Cost Efficacy Method for Blood Irradiation Using a Non-Dedicated Device'.
27. Bahar and Tormey, 'Prevention of Transfusion-Associated Graft-versus-Host Disease With Blood Product Irradiation'.
28. Committee on Radioactive Sources, *Radioactive Sources: Applications and Alternative Technologies*.
29. Ibid.
30. Ibid.
31. Pomper et al., 'Promoting Alternatives to High-Risk Radiological Sources'.
32. Erik Bakken, Katie Cary, Allison Derrick, Ellen Hildebrand, Kyle Schroeckenthaler, and Malika Taalbi, *Cost-Benefit Analysis of Switching from Cesium-Chloride to X-ray Blood Irradiators*, Prepared for the Global Threat Reduction Initiative, National Nuclear Security Administration (Madison, WI: University of Wisconsin-Madison, 2013).
33. Andrew J. Bieniawski, Ioana Iliopulos, and Michelle Nalabandian, *Case Study: An Analysis of Several National Approaches to Alternative Technologies for Radioactive Sources* (Washington, DC: Nuclear Threat Initiative (NTI), 2017). https://www.nti.org/wp-content/uploads/2021/09/NTI_INMM_Alternative_Technology_Paper_Jul2017_oDGeMET.pdf
34. Committee on Radioactive Sources, *Radioactive Sources: Applications and Alternative Technologies*.
35. Bieniawski et al., *Case Study*.
36. Government of Norway, Act on Radiation Protection and Use of Radiation, Norwegian Act No. 36, Oslo, 12 May 2000. https://dsa.no/regelverk/Act%20and%20Regulations%20on%20Radiation%20Protection%20and%20Use%20of%20Radiation.pdf

37. Government of Norway, Regulations on Radiation Protection and Use of Radiation (Radiation Protection Regulations), Norwegian Regulations No. 1362, Oslo, 21 November 2003. https://lovdata.no/dokument/SFO/forskrift/2003-11-21-1362?q=str%C3%A5levernforskriften [In Norwegian].
38. Government of Norway, Regulations on Radiation Protection and Use of Radiation (Radiation Protection Regulations), Norwegian Regulations No. 1659, Oslo, 16 December 2016. https://dsa.no/regelverk/Regulations%20on%20Radiation%20Protection%20and%20Use%20of%20Radiation%20-%202020.pdf
39. Government of Norway, Regulations on Object Security. Norwegian Regulations No. 1362, Oslo, 22 October 2010. https://lovdata.no/dokument/SFO/forskrift/2010-10-22-1362 [In Norwegian].
40. IAEA, *Security of Radioactive Material in Use and Storage and of Associated Facilities*.
41. Norwegian Radiation Protection Authority, 'Implementation of the IAEA Code of Conduct on the Safety and Security of Radioactive Sources and its Supplementary Import/Export Guidance', presented at the IAEA International Conference on the Safety and Security of Radioactive Sources: Maintaining the Continuous Global Control of Sources throughout their Life Cycle, 27–31 October 2013, Abu Dhabi; Strålevern Rapport 6:2013 Implementation of the IAEA Code Of Conduct on the Safety and Security of Radioactive Sources and its Supplementary Import/Export Guidance. https://dsa.no/en/search/_/attachment/inline/235a32da-7ba5-44ec-a86c-075e0eb3d49a:8a44e875d4ed25eb874424ad899557e10412118e/StralevernRapport_06-2013-.pdf
42. Council of the European Union, 'Council Directive 2003/122/Euratom of 22 December 2003 on the Control of High-Activity Sealed Radioactive Sources and Orphan Sources', *Official Journal of the European Union*, 346 (2003), 57–64. https://eur-lex.europa.eu/legal-content/EN/TXT/PDF/?uri=CELEX:32003L0122&from=EN
43. Committee on Radiation Source Use and Replacement, *Radiation Source Use and Replacement*.
44. Sindre Øvergaard and Gunnar Saxebøl, 'Norway is Phasing Out Gamma Based Blood Irradiators', *European ALARA Newsletter*, 33 (2013), 6–7. https://www.eu-alara.net/images/stories/Newsletters/Newsletter33/alara-newsletter%2033-september%202013.pdf
45. Norwegian Radiation Protection Authority, 'Traditional Blood Irradiators Based on Radioactive Sources Are Phased Out', *NRPA Bulletin*, 2 (2013) [In Norwegian].
46. Norwegian Parliament, 'Rapport fra 22. juli-kommisjonen', NOU 2012:14, 2012. https://www.regjeringen.no/contentassets/bb3dc76229c64735b4f6eb4dbfcdbfe8/no/pdfs/nou201220120014000dddpdfs.pdf [In Norwegian].
47. Ibid.
48. Committee on Radiation Source Use and Replacement, *Radiation Source Use and Replacement*.
49. IAEA, Guidance on the Import and Export of Radioactive Sources, 2012. https://www-pub.iaea.org/MTCD/Publications/PDF/8901_web.pdf. IAEA, *The Radiological Accident in Goiânia*. Non-serial Publications, 1998. https://www.iaea.org/publications/3684/the-radiological-accident-in-goiania

Suggested Reading

Bieniawski, Andrew J., Iliopulos, Ioana, and Nalabandian, Michelle, (2017), *Case Study: An Analysis of Several National Approaches to Alternative Technologies for Radioactive Sources* (Washington, DC: Nuclear Threat Initiative (NTI)). https://www.nti.org/wp-content/uploads/2021/09/NTI_INMM_Alternative_Technology_Paper_Jul2017_0DGeMET.pdf

Committee on Radiation Source Use and Replacement, (2008), *Radiation Source Use and Replacement*, abbr. edn (Washington, DC: The National Academies Press). https://www.nap.edu/catalog/11976

Committee on Radioactive Sources, (2021), *Radioactive Sources: Applications and Alternative Technologies* (Washington, DC: The National Academies Press). https://doi.org/10.17226/26121

IAEA, (1998), *The Radiological Accident in Goiânia*, Non-serial Publications (Vienna: IAEA). https://www.iaea.org/publications/3684/the-radiological-accident-in-goiania

IAEA, (2019), *Security of Radioactive Material in Use and Storage and of Associated Facilities*, IAEA Nuclear Security Series, No. 11-G (Rev. 1) (Vienna: IAEA).

Pomper, Miles, Murauskaite, Egle, and Coppen, Tom, (2014), 'Promoting Alternatives to High-Risk Radiological Sources: The Case of Cesium Chloride in Blood Irradiation', *CNS Occasional Paper*, No. 19 (Washington, DC: James Martin Center for Nonproliferation Studies).

PART VIII

NUCLEAR SECURITY DURING TIMES OF CRISIS

CHAPTER 38

NUCLEAR EMERGENCY PREPAREDNESS AND RESPONSE

ANGELA E. LEEK AND JEFFREY D. SEMANCIK

INTRODUCTION

ANY incident that results in the release of radioactive material requires prompt assessment of health and safety risks. This assessment supports effective decision making to protect facility staff, the public, and the responders. When an incident involves a nuclear power facility, coordination between the utility organization and the surrounding off-site jurisdictions is vital. Radiological emergency response plans must incorporate both the on-site considerations, as well as the anticipated activities and authorities of the many jurisdictions and organizations off-site throughout all stages of the assessment and decision-making process. Often, these plans are practised in the scope of an incident, natural disaster, or equipment malfunction, but over the past two decades, there has been an increased need to focus on incidents that involve security issues and intentional breaches of facility safety systems.

When considering nuclear power facility contingency plans, it is important to recognize that three characteristics make nuclear energy special and unique in the field of emergency preparedness:

- **Energy**: There is a tremendous amount of stored energy in nuclear fuel. The nuclear fuel burned by a 1,000 megawatt reactor every 18 months is the energy equivalent to over 2,000,000 metric tonnes of coal.[1]
- **Heat**: Nuclear fuel continues to generate a significant amount of heat through radioactive decay of fission products.[2] Heat continues to be emitted even after the reactor is shut down, known as decay heat. Decay heat can be as much as 7 per cent of full power heat. Decay heat is generated for years after shutdown. As a result, this heat must be continuously removed to prevent additional damage to the systems.

- **Radioactivity**: The by-products of nuclear fission are radioactive. Some of these fission products decay after only a few seconds; others remain radioactive for hundreds of years. If these fission products are released into the environment as a result of an incident, then action must be taken to protect the public from radiation exposures and to limit the contamination of surrounding ecosystems.

Safety systems at nuclear power facilities are specifically designed to account for these special characteristics by controlling the energy, removing decay heat, and containing the radioactive fission products. If these safety systems fail, the tremendous amount of stored energy could disperse significant quantities of radioactive material over large areas. It is also important to note that the nuclear fuel exists not only in the reactor vessel, but also may be stored in other locations on-site in areas, such as wet spent fuel storage pools or dry storage vaults.

Decay heat requires cooling mitigation for long periods to ensure that the nuclear fuel and its protective metal cladding remain intact, and the fission products remain contained. Since decay heat is produced even when a reactor is shut down, nuclear power plants remain vulnerable to failures that can cause an incident during planned refuelling and maintenance outages. Such outages may represent an elevated risk because safety systems may be intentionally out-of-service for maintenance, and barriers to fission product release (such as the containment building[3]) may be purposely bypassed to facilitate access. Additionally, refuelling and maintenance outages require many contract staff who are not typically employed at the nuclear power facility to be on site to perform these specific outage-related tasks.

In all conditions, security at a nuclear facility must be robust and effective in order to maintain safe operations by (1) controlling access to the facility, (2) preventing the release of security-related information, and (3) responding to hostile threats.

Radiological Incident Response Priorities

Radiation Emergency Response Priorities

There are common priorities for all types of radiological and nuclear incidents, and they begin with the primary goal of (1) *protecting facility personnel, emergency responders, and the public*. Assessing the potential radiation dose and associated risks guides appropriate protective action recommendations, such as evacuation or sheltering in place for the public. For facility personnel and emergency responders, understanding the radiation hazard can assist in determining where work efforts should be focused and the permissible time for working in hazardous areas. A second goal is (2) *to secure the scene to limit the impacts of all hazards*, including radiation, by controlling movement in and around the area and by assessing for additional hazards/incidents. The third goal is (3) *to mitigate the radiation hazards*, including efforts to stop any on-going release of radioactive material and reduce other hazards to facilitate recovery and support the eventual resumption of normal activities in that area. Depending on the radionuclides[4] involved in the incident, it may be months or years before the radiation hazard is completely mitigated, and in some cases, areas may remain uninhabitable indefinitely.

These three priorities are important for all entities responding to a radiological or nuclear incident to consider, so agencies and responders across local and national borders must coordinate to ensure a consistent response and effective protection from radiation exposure. The International Atomic Energy Agency (IAEA) fosters coordination through the 'Joint Radiation Emergency Management Plan'.[5] This Joint Plan encourages response entities to collaborate and reinforces the concept of one focused response effort shared across partner agencies. The United States (US) emphasizes this concept by including the 'whole community' in the Nuclear Radiological Incident Annex of its plan.[6]

The health and safety risk from the radiation exposure hazard is just one of the many considerations for protection of facility staff, responders, and the public during a nuclear power facility incident. Appropriate actions are determined by weighing the radiation risks against many other factors which include restoring and maintaining critical infrastructure; weather or other environmental conditions; political views, risk perception, and other socio-economic factors that influence the amount of risk that affected communities are willing to accept; and other impediments or competing risk factors that may require compensatory adjustments to planned radiological response actions.

Radiological Emergency Preparedness (REP)

Significant efforts have been made to establish a foundation of basic elements of radiological emergency preparedness in an international context. These plans promote the integration of on-site and off-site planning and response that focuses on protecting the public and responders throughout the incident. Established protective and emergency action levels are outlined in various guidance documents both in the US[7] and internationally.[8]

The IAEA has defined specific safety goals for nuclear and radiological contingency[9] planning, including where to focus initial priorities to control the situation and save lives.[10] The IAEA guidance recommends that all available resources and assistance from off-site responders be used when necessary and that contingency plans recognize the potential for concurrent emergencies and the need to take the most effective actions considering the radiological situation in addition to other hazards.[11] Radiation reference values guide protective actions, at pre-defined 'operational intervention limits',[12] that are implemented to prevent acute radiation syndrome and minimize the risk of cancer development in later years.[13]

IAEA guidance also provides recommendations for other important preparedness concepts. Medical support personnel should be trained to recognize and treat radiation injuries and to understand the risks in the context of other injuries, as well as long-term monitoring and psychological care. The public must be kept informed through frequent and honest communications that explain the situation and what people can do to protect themselves, their families, and their property. Preventing the use of food and water that is radioactively contaminated is a key intervention to minimize adverse long-term public health effects. In the later stages of an incident, contingency plans must address societal and economic impacts, including when and how to restore access to and use of affected areas.[14] When implementing actions to save lives, decision makers must ensure that the proposed protective actions provide more benefit than harm. For example, sheltering may be preferred over evacuation during severe weather events.

Plans specific to nuclear power facilities often have predesignated emergency classification levels that help to quantify the radiological risk based on plant conditions and radiation calculations or measurements (Table 38.1). Examples of these levels include the emergency classification levels required by the US Nuclear Regulatory Commission (NRC) for nuclear power facilities in the US that range across the lowest to highest risk levels with standard terminology: Unusual Event, Alert, Site Area Emergency, and General Emergency.[15] Internationally, IAEA emergency classification levels correspond to similar common risk levels.[16] In both cases, these predetermined levels and scales are used to allow for ease and consistency of communication across all members of the response community regarding complex facility parameters and radiation dose units. These emergency classification levels can be socialized before an incident and used during the initial phases of the incident response, but a transition to actual incident-specific details based on data regarding current and projected radiation measurements will be necessary to make an appropriate assessment of risk during each incident.

Table 38.1. Emergency Classification Levels.

Emergency Classification Level			Managing Risk to:		
International	Increasing Severity	United States	Facility Staff	Responders	Public
GENERAL EMERGENCY		GENERAL EMERGENCY	Protective actions	Protective actions	Protective actions
SITE AREA EMERGENCY		SITE AREA EMERGENCY	Protective actions	Protective actions (<u>if</u> on-site)	Precautionary Actions[vii]
FACILITY EMERGENCY			Protective actions	Protective actions (<u>if</u> on-site)	Use Normal Operational Limits
ALERT		ALERT	Use Normal Operational Limits	Use Normal Operational Limits	Use Normal Operational Limits
		UNUSUAL EVENT	Use Normal Operational Limits	Use Normal Operational Limits	Use Normal Operational Limits

Precautionary actions are taken to prepare to take Protective Actions, for example, dismissing school children in anticipation of evacuation. Table courtesy of authors.

Planning for Radiation Incidents

Planning is an important part of ensuring effective coordination between the nuclear power facility and off-site response partners during a response. Planning supports a common understanding of the terminology used in nuclear incidents and establishes how each entity intends to apply the commonly referenced protective action guidance levels. Additionally, plans can help response partners understand the overall assumptions and anticipated actions that each entity will be likely to make during the earliest phases of the response when communication and coordination will be most difficult.

The pre-identification of priorities and needs based on the scope of the incident is a critical outcome of coordinated planning. However, because each incident has its unique issues and considerations, the most important component of plans and the planning process is the identification of how the nuclear power facility, the various response partners, and supporting jurisdictions will communicate critical information throughout the early, intermediate, and later phases of the incident. Actively communicating, training, and exercising established plans during non-emergency times allows partners to remain familiar with one another and build proficiency with interoperability. This will help to ensure that all partners are comfortable with, and know how to effectively share, information to guide assessment and actions in a coordinated fashion.

The integration of specialized technical resource partners to gather data, assess radiation risks, and put those risks in context with the other incident considerations is critical to quantify the overall incident radiation risk and guide protective action decision making. During severe radiological emergencies that have the potential to extend to areas beyond the facility boundaries, off-site authorities are responsible for the safety of the public. As a result, many response organizations are activated to take appropriate actions off-site and support the on-site response efforts. Specially trained teams take independent radiation measurements off-site to identify or confirm the release of radioactive material and quantify the amount.

Other specialists run environmental dispersion models to predict the ultimate extent of any radioactive contaminant. These measurements and models are compared to pre-established intervention levels that inform decision makers on appropriate actions to protect the health and safety of the public and to limit the impact of radioactive contamination and exposure. Data must be collected from the field to determine the make-up of the radioactive material deposition, measure resulting exposure rates, and gather samples to assess contamination. Some of these facility and radiation release-related parameters may be considered by the facility security personnel to be sensitive information during a radiation incident, particularly during an incident with security- or threat-related considerations. Planning must incorporate considerations on how to navigate any procedural restrictions that may prevent access to or sharing of this data.

Beyond the radiological assessment, other response organizations must prepare for operational implementation of the protective measures. For example, evacuation requires effective traffic management, provisions for mass transportation, establishment and activation of community reception centres, and conducting epidemiological tracking. Logistics and planning sections prepare to support these missions and any specific needs beyond the capacity of the facility to terminate the release of radioactive material or to safely mitigate the event. Each of these supporting organizations will need plant and radiological information to put the risks related to their aspect of the mission into context for their responders.

If the incident is related to security- or threat-related issues, there may be additional partners responding with a critical role of assessing, investigating, or mitigating the threat, but who are not typically a part of a facility or off-site jurisdiction's radiological response planning. These security-related response agencies may bring their own command structure that may require alteration of and coordination with the typical response command structure that is outlined in the radiological response plans. It is important to understand how partners, roles, and information sharing can change response plans when both radiation and security aspects are a part of an incident.

Due to their knowledge of and expertise in the technology and risks at the site, facility operators are well prepared to develop contingency plans (Table 38.2) and must be a significant contributor in coordinating these plans and supporting the effective overall implementation of the contingency plan both on-site and off-site. They should instruct and assist off-site responders in developing plans that support the facility because many of the response needs can only reasonably be provided by off-site authorities, such as critical infrastructure (electricity and fuel oil); emergency medical, fire, and security support; and the ability to surge various resources to the facility site. Off-site authorities bring an existing understanding of the best approach to coordinate off-site response actions, address public communications, and have the authority to issue broad public protective actions. The contingency plan should establish clear roles and responsibilities for both on-site and off-site response organizations. Clear protocols allow for better allocation of scarce resources and a more manageable span of control for all response personnel.

Table 38.2. Contingency Plan Responsibilities.

On-Site Organization	Off-Site Authorities
Mitigate the incident	Make appropriate responder and public protective action decisions
Protect on-site personnel	Prevent consumption of contaminated food and water
Communicate to off-site officials the status of the incident	Inform the public
Identify key resources necessary to bring the incident under control	Provide additional services and resources as requested by the facility

Table adapted from IAEA, *Arrangements for Preparedness for a Nuclear or Radiological Emergency*, IAEA General Safety Guides, No. GSG- 2.1 (Vienna: IAEA, 2011). https://www.iaea.org/publications/7503/arrangements-for-preparedness-for-a-nuclear-or-radiological-emergency

Security Considerations Related to Nuclear Emergency Preparedness

Safety systems at nuclear power facilities are designed, maintained, and tested to ensure high reliability and robustness. Designs employ multiple independent and fully diverse systems to safely control the nuclear reactor and to limit the release of radiation into the environment and its exposure to the public. These redundant and diverse safety systems generally enhance physical security by requiring hostile actors[17] to disable multiple targets to damage the nuclear fuel or cause a release of radioactive material.

When such robust facilities with multiple layers of defence do fail, contingency plans should question how that occurred including consideration of terrorism or an insider hostile actor. To prevent hostile actors from intentionally disabling multiple layers of protection, contingency plans should compartmentalize the protection of diverse safety systems. In a hostile incident situation, security must work with plant operations personnel to protect critical systems, ensuring at least one of multiple response strategies is successful. Particular attention should be given to protecting vulnerable off-site interfaces, such as connection to the electrical grid and cooling water intakes.

Response to hostile actions will usually require the response of off-site security, military, or law enforcement. Contingency planning and training exercises are critical to ensure that both on-site and off-site forces are familiar with methods to verify friend from foe to effectively terminate the threat and avoid friendly fire engagements. Frequent coordination is required to establish trust between the facility and off-site organizations. Such coordination involves training, site familiarization, and radiation safety briefings. These activities are important to ensure protocols exist that verify off-site personnel are adequately trained, credentialed, and prepared to come on-site during an incident. It is also important that such personnel are properly vetted to minimize the potential for hostile actors to gain access to critical areas during emergencies. On-site and off-site responders training and coordinated cross-organizational training exercises help to identify interoperability gaps and develop proficiency with contingency plans.

Emergent incidents unavoidably complicate security. In many cases, security actions that are initiated in response to an incident may complicate or even contradict prudent operator actions to mitigate the issues causing the incident and gain control of the facility operations. For example, an active security threat may require the sheltering of personnel on-site even though operators need to access vital plant equipment. In such hostile incident events, security risks must be carefully balanced against risks to nuclear and radiological safety. Clear communication is central when taking an action based on security or safety, and priority should be given to protecting trained plant operators who will be required to manage the challenges to the facility. Contingency plans should also address sequestration of plant personnel following facility evacuation to ensure a proper law enforcement investigation is accomplished. Active engagement between security and emergency response personnel on areas of potential tension can enhance contingency plans by leading to the development of innovative strategies that effectively meet both security and safety response needs.[18]

To effectively implement any of these strategies, both security and emergency response personnel must understand the driving reason to take an action. The most effective strategy is ultimately to quickly terminate the active threat and gain physical control of the facility. This enables site personnel to respond using their normal emergency procedures and protocols. Planning and communication between security, emergency response, and radiation protection personnel before an incident will allow the time needed for the consideration of various situations where security and radiological risks will need rapid assessment and balance.[19]

Case Studies

A key element of an effective emergency preparedness culture is learning from past events. Event assessment and evaluation offer insights into potential gaps in existing emergency plans. By critically evaluating emergency plans against the parameters of these events, organizations can continuously improve their preparedness. The scope of incidents from which to learn involves both prior radiological events as well as other significant, non-radiological emergency incidents that have required large-scale public protective action considerations. These incidents offer critical insights into all-hazards preparedness and process improvement. Over the past two decades, the potential for complex terrorist events, the impact of rare but severe environmental disasters, and the realities of a global pandemic have resulted in a re-evaluation of nuclear preparedness planning, protocols, and strategies.

Case 1: Integrating Security with Emergency Preparedness and Response: The Impact of the 9/11 Terrorist Attacks

Historically, security plans were developed independently from emergency response plans for nuclear facilities and were exercised separately. The rise of international terrorism, including the attacks on 11 September 2001 ('9/11') in the US, and the subsequent investigations of these incidents identified the very real possibility for a nuclear or radiological emergency initiated by terrorist acts.[20] These reviews have revealed that security forces and emergency responders must integrate their actions and operations to ensure an effective response; they have led not only to integrated training but also to collaboration between security and emergency preparedness personnel in developing plans and protocols.

Emergency plans became informed by security considerations and security plans also prioritized the protection of assets vital to emergency response needs. Testing conducted at Sandia National Laboratories in 1988 had previously verified the ability of reinforced concrete construction used in nuclear power plant containment buildings to withstand direct aircraft impacts.[21] However, the 9/11 terrorist attacks demonstrated that the risk from the subsequent wide-area conflagration that resulted from jet fuel could have a significant impact on mitigating the event.[22] Facility designs and emergency fire and response plans were modified to account for this hazard.

Case 2: Preparing for Extreme Events: Fukushima Daiichi Following the 2011 Tohoku Earthquake and Tsunami

Nuclear power plants are robust facilities designed to withstand severe incidents and a range of external events. Engineers establish the facility design constraints from parameters that are extrapolated from historical occurrences of various natural phenomena. The engineers define the design basis events that the facilities can safely withstand based on reasonably probable scenarios. The probability of an event, the design of the facility, and the amount of radioactive material that may be released are considered in a graded approach to risk that is used to determine the level of detail and the extent of off-site coordination required in contingency planning. When employing a graded approach, it is important to account for the fact that some low-risk events, while improbable, have unacceptably high consequences. For these events, specific strategies should be built into contingency plans to mitigate these 'cliff-edge' (low-probability high-consequence) events.[23]

The tsunami that struck the Fukushima Daiichi nuclear plant in Japan was a cliff-edge event. The earthquake and tsunami exceeded those hypothesized in the design of the plant. As a consequence, contingency plans were not adequate for the resulting scenario. In the aftermath, nuclear facility contingency plans and equipment have been upgraded to provide strategies to mitigate these beyond design basis events. By installing general purpose connections to key systems and incorporating multi-purpose portable equipment (pumps, generators, and batteries), response strategies and principles are developed to provide operators and emergency personnel multiple methods to meet safety functions in response to unanticipated failures, so they can successfully limit damage and subsequent releases of radiation.[24] These protocols are designed with maximum flexibility to provide safety functions given multiple failures through diverse strategies, including ad hoc systems, should all on-site equipment fail. These protocols and flexible systems also improve response options to security events in which unpredictable damage has been caused by hostile action.

Case 3: Flexible Response to Concurrent Incidents: Preparedness during the 2020 Public Health Emergency

During the Covid-19 pandemic, decision makers needed to understand the risks of disease transmission in public gathering spaces in order to compare the risk of the protective action recommendation of congregating in shelters versus the risk of the radiation exposure from remaining in homes or other buildings during a radiation release. While it is generally considered to be most protective, from a radiation protection standpoint alone, to evacuate the public before any release of radiation, the high risk of Covid-19 transmission and mortality introduced a new risk consideration to the balance. The key protective action strategy implemented to prevent rapid transmission and spread of the Covid-19 virus was to maintain social distancing and minimize the movement and gathering of large populations. Throughout 2020, the earliest stages of the pandemic, there were many uncertainties regarding the morbidity and mortality risks, and there were no established mitigation measures in place (such as masking and vaccination). The risk of transmission of the virus from the movement of populations and large gatherings of people greatly outweighed a postulated radiation exposure risk from most nuclear power facility releases, particularly for elderly populations; so, sheltering in place was the preferred protective strategy for radiation incidents that might occur during the pandemic.[25]

Over time, as information about the Covid-19 virus transmission characteristics improved and people became vaccinated, the risk balance was continually reassessed, and evacuation was again considered practicable in appropriate circumstances. If decision makers are not provided with all of the relevant information (both radiological and non-radiological), they may choose to take an action that is inadvertently more detrimental to health and safety than the radiological risk alone. Decision makers must continuously balance risks in each incident, which will be unique to each aspect of the situation or jurisdictional priorities and will continually change over the timeline of the incident. The lessons learned throughout the Covid-19 pandemic could be applicable for any future pandemic or other concurrent situations requiring wide-scale public protections.

International Best Practice

The IAEA provides detailed guidance for developing effective contingency plans at nuclear power plants and other radiological facilities.[26] As with any contingency plan, those for nuclear facilities should be designed with maximum flexibility. Priority should be given to recognizing radiation hazards and providing guidance to minimize exposure.

Sharing Technical Response Aspects and Resources

There are important technical considerations for nuclear incidents, which are compounded when there are security- or threat-related aspects. Coordination is critical. Partner agencies should be aware of each respective organization's plans and resources.

A unified approach to data collection ensures that all agencies understand priorities, effectively allocate resources, and do not duplicate efforts. Reducing duplication of efforts saves time and resources, but agreements and discussions must happen ahead of the incidents for these coordinated efforts to be successful. If there is a security aspect to the incident and areas of the incident site are restricted from access, this may limit the amount of radiological data available to make risk assessments. Security authorities may need to allow for access by technical radiological teams in a timeframe that is earlier than their security plans or procedures account for. If access is not possible, the technical assessment groups may need to complete their radiological assessments with less information than their technical procedures require. By facilitating partnerships ahead of time, responders will be able to adapt and find appropriate solutions that will enable an effective response.

Nuclear incidents will almost immediately require coordination across regional, national, and international boundaries. Every affected jurisdiction will want the same limited assets and information, so plans should include agreements on how to coordinate across boundaries. The IAEA facilitates planning and coordination across international partners through the Joint Radiation Emergency Management plan and the IAEA's 'Operations Manual for Incident and Emergency Communication'.[27] The IAEA guidance documents outline the coordination necessary across local governments, national governments, international partners, and all responding entities. The coordination of decisions, assessment of risk, and sharing of resources may not follow the typical flow of communication between agencies in daily operations. Understanding differences and coordination opportunities are very important to establish before an incident occurs.

Stakeholder Coordination and Information Sharing

Security-related threats to radiological and nuclear incidents introduce specific concerns.[28] The need to share information, especially in nuclear power facility incidents, contradicts typical nuclear power facility security protocols. Plans that outline the designation of and communications with trusted entities and organizations will foster familiarization of their responders and support the ability to establish a framework where important information can be shared. However, the practicality of sharing data during an incident is typically much more difficult than it appears in plans.

If an organization defaults to routine strict sharing restrictions during an emergency without consideration of the overall incident response, it may not be aware of the impact of this data restriction on other response organizations actions. Establishing a common designation of sharing protocols ahead of time and practising the sharing of this information through exercises and drills can expose these gaps and allow jurisdictions to find solutions to share this potentially sensitive or restricted data. Additionally, it allows each organization to have realistic expectations about what data will and will not be available, so they can plan their respective actions appropriately.

During incidents, challenges may arise in both communication and coordination, such as:

- **Data Sharing and Message Coordination**: Timeliness of data sharing and coordination of messages is a challenge. Critical responders and technical assets can be overwhelmed. For instance, a brief telephone call outlined as a step in the plans and procedures could inadvertently affect other aspects of the incident response if it is delayed or made to the incorrect organization. Consequently, during planning, considerations should be made to establish approvals in advance to allow for immediate sharing with certain personnel to limit potential delays. Among the US off-site partners, for example, security/law enforcement personnel often coordinate briefing meetings several times per year with nuclear power facility personnel, emergency management, and technical resources to foster ongoing communication during non-emergency times.
- **Communication Channels**: Normal communication channels can also be physically disrupted and inhibit communications. Coordination in planning establishes common assumptions and guidance levels for radiation risk. Planning allows entities who are unable to talk to one another to manage the incident using the same assumptions because they aligned them ahead of time.
- **Interorganizational Exercises**: Plans also often reference sharing of information. Consequently, a concerted effort should be made to practise communication between partners who do not often communicate. These exercises can identify planning gaps and make it easier to adapt to the inevitable and unforeseen gaps that will occur during incidents.

Integrating Radiological Expertise in Off-site Response Organizations

Many off-site response organizations have limited or no radiological expertise. The knowledge and skills to properly assess radiological hazards and to inform responder and public protective actions during a radiological or nuclear incident is vital. Training can help to close this gap, but it can require a significant investment in resources and requires ongoing maintenance. Subject matter experts in radiation from government, hospitals, academic institutions, and industrial facilities, with proper training and support, can serve as an augmentation to existing technical expertise to provide radiation risk perspectives and interpretation for responders who are not as familiar with radiation.

In the US, there is an effort to build and incorporate these technical experts into response plans through the Radiological Operations Support Specialist (ROSS) programme,[29] which outlines qualifications to designate individuals as a ROSS, so response agencies can incorporate this support into their nuclear emergency plans. ROSSs are specially trained and credentialed radiation experts that can be deployed to provide technical support at any level of a response organization to assist with radiological considerations.[30] ROSSs complete rigorous training and participate in exercises to ensure they are prepared to effectively integrate into the incident command structure. They can provide practical guidance on the consequences and response strategies applicable to a variety of radiological emergencies, including nuclear power plant incidents, radiological dispersal devices, and nuclear detonations. By training a core cadre of ROSSs across the US, local ROSSs are available to mobilize to support responders within their jurisdictions as well as supplement other organizations across the nation.

Leveraging Network Effects to Develop a Common Operating Picture

Mobile cellular technology provides a tool for geographically diverse responders to share data in real time. Common situational awareness during a nuclear incident is critical to coordinate and align protective action recommendations. Rapid access to information and data measurements from the response site are important to adequately assess radiation risks. Often, plans include the use of electronic systems to share this information and data, such as the Unified System for Information Exchange in Incidents and Emergencies (USIE) through the IAEA.

In the US, the RadResponder network,[31] a part of the CBRNResponder network, is designated as the system for sharing radiological information and data across the US federal, state, and local response agencies.[32] The use of these integrated data management tools enables responders to enter radiological data and share this information across the network in real time, leveraging network effects to efficiently build a common operating picture. By providing a common environment to collect, assess, and share radiological data, these data management tools ensure that responders from different organizations can rapidly share and assess field data across a wide geographic area so they can synthesize this information into a common tactical picture. It is important to practise and exercise using these systems to maintain access, proficiency, and sharing agreements. Most of all, the use of these systems can only be maintained and improved through training, drills, assessment, and continuous review and adjustment.

Horizon Scanning and Emerging Challenges

As the international nuclear power industry shifts and grows, security-related threats must be monitored. Communicating with new and differing partner groups must also be a focus. A strong foundation in communication that includes nuclear power facility efforts to include non-traditional stakeholders in emergency preparedness discussions ensures an effective network that is prepared to protect the public and responders during any potential incident. Several emerging concepts offer considerations for continued proactive coordination efforts.

Advanced Reactor Technologies

There is a growing interest in deploying the next generation of advanced reactors, particularly as they provide reliable power without carbon emissions. These include advanced light water, sodium-cooled, high-temperature gas, molten salt, and fusion reactors. These reactors improve on past safety and security designs. In addition, they are designed to limit the size of the emergency planning zone to the property line of the facility by limiting the radiation source and mitigating severe incident progression to reduce the probability of releasing radioactive fission products.[33]

While eliminating any off-site emergency zone simplifies siting, reduces costs, and streamlines regulatory approvals for the nuclear power facility, it can leave off-site response organizations reliant on all-hazards contingency plans that may not fully support response to the unique aspects associated with nuclear power facilities. Without a formal off-site emergency planning zone, fewer opportunities for training, coordination, and proficiency between on-site personnel and off-site responders exist, and the result may cause a delay in information flow and an increased reliance on the facility to provide emergency response resources. Since formal off-site response contingency planning may not be required for these reactor facilities, proposed facilities and the surrounding off-site response organizations must coordinate and agree on the critical aspects necessary to not only ensure the safety of the public off-site, but also to effectively respond to and share information during a wide range of incidents, including security-related threats.

Drones

The emergence of highly capable unmanned aerial vehicles (UAVs; or drones) introduces both opportunities and threats to radiological contingency planning. UAVs offer the ability to efficiently conduct wide-area surveillance and data acquisition. UAVs can be deployed into radiological and security hazard areas without threatening the safety of responders. Data links have been developed to enable the rapid transmission of radiological, visual, and other information.[34] However, UAVs also represent a potential security threat to nuclear facilities. For example, UAVs could potentially be used to deliver explosive threats or conduct surveillance of security forces. In some jurisdictions, UAVs may have statutory protections similar to any aircraft challenging the ability of security forces to employ countermeasures.[35] Security and emergency contingency plans must include UAV protocols consistent with local laws.

Digital Technologies and Cybersecurity

Digital devices are used throughout operations, security, and emergency response organizations. Uses include plant control and protective devices, security systems (such as video surveillance and intruder detection systems), communications, and information sharing systems. These systems are vulnerable to a coordinated cyber-attack and must be adequately protected from these threats. In addition, contingency plans must include the use of diverse and redundant systems and protocols to ensure critical functions are maintained if primary systems are affected. A common concern during emergency response is cellular traffic congestion affecting both communications and data sharing. Emergency response organizations should ensure functions relying on cellular data use a robust public safety data network to ensure priority is given to emergency response personnel and data.

Conclusion

Efforts continue in both the security and emergency preparedness aspects of nuclear power facilities to maintain and to improve the protection of the health and safety of the public, emergency responders, and the personnel working within the facility. The emergence of new technologies and continually changing threat portfolios requires ongoing efforts to maintain robust communication and coordination between on-site personnel and critical off-site authorities for ensuring both security and emergency preparedness. While security and emergency preparedness efforts may be well established independently of one another, facilities and surrounding jurisdictions are often not adequately prepared for an efficient response to an incident at a nuclear power facility when preparedness includes a combined security and emergency response focus. Fixed nuclear power facilities benefit from the general requirement to continuously conduct significant planning and procedure development for a predetermined area based on the anticipated scope of potential impact during a radiological or security incident.

Intentional coordination with off-site partners, including both the technical and security assets, can ensure that these shared plans, along with coordinated efforts to train and exercise these plans, provide a strong foundation of communication and coordination to ensure rapid and effective situational assessment and protective actions. An important aspect to solidify this coordination is pre-established arrangements made between partner organizations to facilitate sharing of critical sensitive security information and to highlight the priorities of both the security and technical response aspects of an incident before it occurs. Continued leveraging of these principles and partnerships, before an incident occurs, will improve the effectiveness of plans and responses. These efforts also ensure that facility personnel and response agencies are prepared to protect the facility staff, the public, and the responders.

Notes

1. Calculations by authors using US Energy Information Administration energy conversion calculator, estimating approximately 12,000,000 MWh produced by 1000MW reactor over an 18-month period. US Energy Information Administration (EIA), 'Energy conversion calculators', EIA [online], n.d. https://www.eia.gov/energyexplained/units-and-calculators/energy-conversion-calculators.php
2. When a uranium or plutonium atom fissions, it splits into two smaller atoms, called fission products, as well as releasing heat and subatomic particles (neutrons).
3. Commercial nuclear reactors are enclosed within a containment building—a robust concrete structure designed to retain radioactive material and prevent its release to the environment under extreme accident conditions.
4. An unstable form of a chemical element that releases energy in the form of ionizing radiation as it moves towards a stable state.
5. IAEA, *Joint Radiation Emergency Management Plan of The International Organisations* (Vienna: IAEA, 2019). https://www.iaea.org/publications/11163/joint-radiation-emergency-management-plan-of-the-international-organizations
6. Office of Response and Recovery, *Nuclear/Radiological Incident Annex to the Response and Recovery Federal Interagency Operational Plans* (Washington, DC: Federal Emergency Management Agency, US Department of Homeland Security, 2016). https://www.fema.gov/sites/default/files/2020-07/fema_incident-annex_nuclear-radiological.pdf
7. Office of Radiation and Indoor Air, PAG Manual, EPA-400/R-17/001, Radiation Protection Division, US Environmental Protection Agency, 2017. https://www.epa.gov/sites/default/files/2017-01/documents/epa_pag_manual_final_revisions_01-11-2017_cover_disclaimer_8.pdf
8. IAEA, *Preparedness and Response for a Nuclear or Radiological Emergency*, IAEA Safety Standards Series, No. GSR-7 (Vienna: IAEA, 2016). https://www.iaea.org/publications/10905/preparedness-and-response-for-a-nuclear-or-radiological-emergency
9. The US uses the term radiological emergency plan as synonymous to IAEA contingency plan terminology.
10. IAEA, *Developing a Nuclear Security Contingency Plan for Nuclear Facilities*, IAEA Nuclear Security Series, No. 39-T (Vienna: IAEA, 2019). https://www.iaea.org/publications/13433/developing-a-nuclear-security-contingency-plan-for-nuclear-facilities
11. IAEA, *Considerations in the Development of a Protection Strategy for a Nuclear or Radiological Emergency*, IAEA Emergency Preparedness and Response Series (Vienna: IAEA, 2021). https://www.iaea.org/publications/14801/considerations-in-the-development-of-a-protection-strategy-for-a-nuclear-or-radiological-emergency
12. The US uses the term protective action guidelines, and this is synonymous to IAEA terminology of Operational Intervention Limits.
13. IAEA, *Preparedness and Response for a Nuclear or Radiological Emergency*.
14. Ibid.
15. US Nuclear Regulatory Commission, 'Appendix E to Part 50–Emergency Planning and Preparedness for Production and Utilization Facilities', Regulations (NRC, 10 CFR), NRC Library, 2021. https://www.nrc.gov/reading-rm/doc-collections/cfr/part050/part050-appe.html
16. IAEA, *Criteria for Use in Preparedness and Response for a Nuclear or Radiological Emergency*, IAEA General Safety Guides, No. GSG-2 (Vienna: IAEA, 2011). https://www.

iaea.org/publications/8506/criteria-for-use-in-preparedness-and-response-for-a-nuclear-or-radiological-emergency

17. The term 'actors' is used in this context to define individuals with mal intent towards the operation of the nuclear power facility and the health and safety of the populations surrounding the facility. In contrast, the terms 'stakeholder' or 'partner' are used to describe individuals with a shared priority of ensuring safe operations of the nuclear power facility and maintaining the health and safety of the surrounding populations.
18. Colin Glesner, Robbe Geysmans, and Catrinel Turcanu, 'Two Sides of the Same Coin? Exploring the Relation between Safety and Security in High-Risk Organizations', *Journal of Safety Research*, 82 (2021), 184–93. https://doi.org/10.1016/j/jsr/2022.05.010
19. Sonal Gandhi and Jungmin Kang, 'Nuclear Safety and Nuclear Security Synergy', *Annals of Nuclear Energy*, 60 (2013), 357–61.
20. National Commission on Terrorist Attacks Upon the United States, *The 9/11 Commission Report: Final Report of the National Commission on Terrorist Attacks Upon the United States* (Washington, DC: US Government Printing Office, 2004). https://www.govinfo.gov/content/pkg/GPO-911REPORT/pdf/GPO-911REPORT.pdf
21. W. A. von Riesemann, R. L. Parrish, D. C. Bickel, S. R. Heffelfinger, K. Muto, T. Sugano, H. Tsubota, N. Koshika, M. Suzuki, and S. Ohrui, 'Full-Scale Aircraft Impact Test for Evaluation of Impact Forces. Part 1: Test Plan, Test Method, and Test Results', in Proceedings of the International Conference on Structural Mechanics in Reactor Technology (SMIRT), 14–18 August 1989, Anaheim, CA, USA.
22. National Commission, *The 9/11 Commission Report*.
23. US Nuclear Regulatory Commission, 'Appendix E to Part 50'.
24. Charles Miller, Amy Cubbage, Daniel Dorman, Jack Grobe, Gary Holahan, and Nathan Sanfilippo, 'Recommendations for Enhancing Reactor Safety in the 21st Century–The Near-Term Task Force Review of Insights from the Fukushima Dai-ichi Accident', US Nuclear Regulatory Commission [online], July 2011. https://www.nrc.gov/docs/ML1125/ML112510271.pdf
25. Angela E. Leek and Jeffrey D. Semancik, 'Compensatory Considerations for Radiological Emergency Response and Public Protective Actions during the COVID-19 Pandemic', *Health Physics*, 221/2 (2022), 333–40.
26. IAEA, *Joint Radiation Emergency Management Plan of The International Organisations*.
27. IAEA, *Operations Manual for IAEA Assessment and Prognosis during a Nuclear or Radiological Emergency*, IAEA Emergency Preparedness and Response Series (Vienna: IAEA, 2020). https://www.iaea.org/publications/12362/operations-manual-for-iaea-assessment-and-prognosis-during-a-nuclear-or-radiological-emergency
28. As noted previously, the term 'stakeholder' is used here to outline an individual with interest in supporting safe operations of the nuclear power facility and the protection of public health and safety.
29. CBRN Office, *Radiological Operations Support Specialist Fact Sheet* (Washington, DC: Federal Emergency Management Agency, US Department of Homeland Security, 2020). https://www.fema.gov/sites/default/files/2020-07/fema_cbrn-ross.pdf
30. Federal Emergency Management Agency, 'Radiological Operations Support Specialist Position Qualification', US Department of Homeland Security, Description ID: 17-509-1415, FEMA [online], October 2019. https://rtlt.preptoolkit.fema.gov/Public/Position/ViewFile/17-509-1415?type=Pdf&s=FemaId&p=8

31. The RadResponder network is the national standard and whole community solution for the management of radiological data in the United States. It is a product of collaboration between the US Federal Emergency Management Agency, the US Department of Energy/ National Nuclear Security Administration, the US Environmental Protection Agency, and the US Defense Threat Reduction Agency. www.radresponder.net
32. Office of Response and Recovery, *Nuclear/ Radiological Incident Annex*.
33. US Department of Energy Nuclear Energy Research Advisory Committee and the Generation IV International Forum, 'A Technology Roadmap for Generation IV Nuclear Energy Systems', GIF-002-00, December 2002. https://www.gen-4.org/gif/jcms/c_40481/technology-roadmap
34. Interagency Security Committee, *Protecting Against the Threat of Unmanned Aircraft Systems (UAS)* (Washington, DC: Cybersecurity & Infrastructure Security Agency, 2020). https://www.cisa.gov/sites/default/files/publications/Protecting%20Against%20the%20Threat%20of%20Unmanned%20Aircraft%20Systems%20November%202020_508c.pdf
35. Ibid.

Suggested Reading

Glesner, Colin, Geysmans, Robbe, and Turcanu, Catrinel, (2021), 'Two Sides of the Same Coin? Exploring the Relation between Safety and Security in High-Risk Organizations', *Journal of Safety Research*, 82, 184–93. https://doi.org/10.1016/j/jsr/2022.05.010

IAEA, (2016), *Preparedness and Response for a Nuclear or Radiological Emergency*. IAEA Safety Standards Series, No. GSR-7 (Vienna: IAEA). https://www.iaea.org/publications/10905/preparedness-and-response-for-a-nuclear-or-radiological-emergency

IAEA, (2019), *Developing a Nuclear Security Contingency Plan for Nuclear Facilities*. IAEA Nuclear Security Series, No. 39-T (Vienna: IAEA). https://www.iaea.org/publications/13433/developing-a-nuclear-security-contingency-plan-for-nuclear-facilities

Leek, Angela E. and Semancik, Jeffrey D., (2022), 'Compensatory Considerations for Radiological Emergency Response and Public Protective Actions during the COVID-19 Pandemic', *Health Physics*, 221/2, 333–40.

Office of Response and Recovery, (2016), *Nuclear/Radiological Incident Annex to the Response and Recovery Federal Interagency Operational Plans* (Washington, DC: Federal Emergency Management Agency, US Department of Homeland Security). https://www.fema.gov/sites/default/files/2020-07/fema_incident-annex_nuclear-radiological.pdf

CHAPTER 39

RESPONDING TO RADIOLOGICAL TERRORISM

JULIA M. PEARCE

Introduction

In October 2001, in the wake of the 9/11 terrorist attacks, the US National Council on Radiation Protection and Measurements (NCRP) released a report entitled 'Management of Terrorist Incidents Involving Radioactive Material'.[1] This report provided a ground-breaking analysis of the crisis consequence management challenges presented by radiological and nuclear terrorism, identifying social, psychological, behavioural, and communication issues as 'absolutely *central* to the success or failure of preparedness and response efforts'.[2] Since that time, radiological terrorism has been characterized as a growing threat.[3] This has been attributed to the increasingly widespread use of radiological materials in a variety of industrial and healthcare settings,[4] as well as lack of effective regulation and security.[5] For example, in 2021 the International Atomic Energy Agency (IAEA) reported 120 unauthorized activities and events that year involving nuclear and other radioactive material.[6] Furthermore, attempts by terrorist groups to acquire and develop radiological capability have been reported[7] and there is evidence of increased interest in the use of radiological agents among far-right extremists.[8]

It is important to distinguish between nuclear and radiological terrorism. Nuclear terrorism would likely involve the detonation of an improvised nuclear device (IND) constructed from components of a stolen state-built nuclear weapon or built using stolen nuclear material such as uranium to produce a massive explosion. The expected impacts of an IND have been compared to the atomic bombing of Hiroshima in 1945.[9] This is very different from radiological terrorism, which would likely involve a radiological dispersal device (RDD) or radiological exposure device (RED). An RDD (or 'dirty bomb') combines radioactive material with conventional explosive. The most severe health impacts from this type of incident would come from blast injuries as it is unlikely that an RDD would contain sufficient radioactive material to produce acute radiation sickness.[10] Furthermore, there is an inverse relationship between the impact of blast injuries and the radiation threat, as a larger blast will increase the spread of radioactive material thus diluting the radioactivity.[11] The primary health impacts from an RDD are therefore more comparable to the immediate deaths and elevated long-term cancer risk associated with the 2011 tsunami and subsequent radioactivity release from the Fukushima Daiichi nuclear power plant in Japan.[12] A RED is a source of radiation that has been deliberately hidden, with the purpose of exposing people in the local area to ionizing radiation without their knowledge. Unlike RDDs, these devices do not incorporate any explosive material and once a RED has been found and removed, it no longer presents a risk as it does not result in any long-lasting contamination.[13]

Radiological terrorism is generally considered more probable than nuclear terrorism, due to higher levels of security surrounding nuclear weapons materials and there being fewer technical obstacles to making an RDD in comparison with an IND.[14] Consequently, there tends to be greater focus on prevention than mitigation in the context of nuclear terrorism,[15] although the benefits of preparing the public to increase adherence with protective instructions in the event of a nuclear attack have been established.[16] Radiological terrorism is of particular interest from a social, psychological, behavioural, and communication perspective, as radiological incidents tend to score highly on psychometric measures of fear and dread[17] and even incidents that result in limited or no actual exposure can cause widespread anxiety and behaviour change.[18] Psychological and behavioural responses can therefore pose greater long-term challenges for radiological incident management than the structural damage caused by direct physical impacts[19] and it has been posited that the fear of radiation poses a greater threat to public health than radiation itself.[20]

Common response priorities for all types of radiological incident include securing the scene, preventing ongoing release, assessing the potential radiation dose and associated risks to guide protective health recommendations, as well as restoring and maintaining critical infrastructure. An effective response will depend on the capacity for monitoring and medical response[21] and the ability and willingness of first responders to attend the scene.[22] If the radiological incident is malicious and targeted to provoke fear in the wider public, it will also involve a security response and the management of intense reactions from the public relating to perceived risk of repetition and moral violations.[23] As recognized in the NCRP report, public reactions therefore play a major role in determining the overall health, economic, social, and psychological impacts of terrorist attacks involving radioactive materials.[24]

Initial health impacts will be influenced by whether individuals take appropriate action in the immediate aftermath of an attack, such as sheltering in place rather than spontaneous evacuation.[25] Economic impacts will be exacerbated if large numbers of low-risk patients seek healthcare[26] or if long-term concerns about safety influence willingness to purchase food or other products manufactured in affected regions. For example, following a relatively minor radioactive contamination incident that took place in Goiânia, Brazil in 1987, crops and other products from the region were refused or could only be sold at a substantial discount.[27] Social impacts may include unwillingness to return to previously contaminated areas and stigmatization of communities, both of which were also seen following the Goiânia incident.[28] Psychological impacts include the extent to which high levels of distress persist among those not exposed to radiation as well as those directly affected,[29] and have been associated with a range of negative health impacts following radiological incidents, including alcoholism, unsafe sexual practices, and insomnia.[30]

Factors That Influence Risk Perceptions

As described previously, an effective response to radiological terrorism will likely require the management of heightened risk perceptions to reduce unnecessary care-seeking from low-risk patients and spontaneous avoidance behaviours that negatively impact health, social, and economic recovery. However, low perceptions of risk that are based on inaccurate assessments of relative harm can also pose a response challenge, as insufficient recognition of risk severity is associated with reduced public compliance with protective health advice and behavioural responses that increase individual risk exposure and delay recovery.[31] It is increasingly recognized that there will be variation in risk perceptions among different population subgroups. For example, in what has been termed the 'white male' effect there is a tendency for white men to have lower risk perceptions than women or ethnic minorities, reflecting uneven risk distribution and hazard impacts.[32] An effective response to radiological incidents therefore requires an understanding of how the associated risks are perceived and experienced by different sections of the public.

There is a well-established literature regarding factors that are likely to influence public perceptions of and attitudes towards risk. The extent to which people are familiar with and understand the risk, feel that they have choice and control over risk exposure, the perceived fairness of exposure, and whether the hazard is natural or technological have all been shown to influence risk perceptions.[33] As radiological terrorism is a novel, unnatural threat that is involuntary and indiscriminate, risk theory would predict low risk tolerance and heightened risk perceptions. Although radiological terrorist attacks are unlikely to cause major physical health impacts, as discussed, there is evidence to suggest that the public associate them with nuclear accidents and attacks.[34] Radiological terrorism therefore falls into the category of 'dread risks'.[35] Dread risks are low-probability, high-impact events that are perceived to have catastrophic potential. These types of risk are particularly frightening for members of the public and are associated with a sense of fatalism, which can be a barrier to people taking action to protect themselves and others.[36]

Heightened risk perceptions are also driven by lack of certainty. Radiological incidents involve many uncertainties around magnitude and range of contamination, risk of exposure, and hidden health effects.[37] During the initial response to an RDD, for example, it is likely to take some time to establish reliable dosimetry data. Furthermore, interpretation of the health impacts of low doses of ionizing radiation is challenging, which means there will likely be differences between assessments by experts of what constitutes a safe level of radiation.[38] Uncertainty around exposure may be further exacerbated by similarities in the physical symptoms associated with psychological stress (for example, nausea and skin rashes) and radiation sickness.[39] The focus of uncertainty will vary for different groups. For example, scientists will need to manage uncertainty relating to methodological and data-driven issues (such as accuracy of measurements), whereas decision makers face uncertainty around consequences of and public reactions to different response options, and the public may be uncertain about risk exposure and the trustworthiness of experts and decision makers.[40]

Trust in experts and decision makers is crucial for an effective response and directly influences risk perceptions.[41] For example, surveys conducted after the Fukushima nuclear accident identified distrust in the government as a primary driver for public anxiety concerning the impacts of low-dose radiation and for ongoing food safety concerns two years after the incident.[42] Trust is influenced by factors including perceived fairness, competency, and objectivity.[43] It can be undermined by lack of consensus among experts, but also by differences between expert and public understandings of risk. Experts tend to discuss risk in terms of probability and quantifiable measures of harm, whereas the public may be more influenced by the emotional significance of events.[44] Consequently, the public tends to have lower risk tolerance than experts and may perceive a cancer risk to be higher. For example, following the detonation of an RDD, the authorities will need to set a mitigation goal for radiation levels that will differ from the previously established acceptable level of excess radiation exposure in that area. Public questions about why this new level is now considered safe and what impact this has on cancer risk will need to be satisfactorily addressed to maintain trust.[45]

In summary, radiological terrorism scores highly on psychometric measures of fear and dread. As such, it is likely to result in heightened risk perceptions and public anxiety. Furthermore, it may require interventions such as mass decontamination that are also unfamiliar and scary.[46] Consequently, counter-terrorism planning that takes into consideration public responses has traditionally focused on managing mass panic. However, more than 50 years of disaster research has consistently demonstrated that panic is rare and tends only to be exhibited in a narrow range of circumstances, for example, in situations where people are trapped by fire with no possible escape route.[47] This is important, as the assumption of panic can lead to response management that in itself produces a heightened fear response.[48] Recognition that panic is not an inevitable response to dread risks raises the question of how the public will respond to a radiological attack.

How Will the Public Respond?

Although radiological terrorism is recognized as a global threat, occurrences remain rare and of 446 chemical, biological, or radiological (CBR) terrorist incidents that occurred between 1970 and 2017, only 13 involved a radiological agent.[49] Consequently, expectations regarding likely public reactions tend to be based on research that employs hypothetical scenarios to examine behavioural intentions.[50] This research provides valuable insight into public beliefs and concerns, but even the most immersive scenarios cannot fully replicate the threat or stress that would be experienced during a real incident. To understand likely reactions, public responses to other types of radiation incidents are therefore examined here through three case studies: the aforementioned 1987 event in Goiânia, Brazil, the deliberate 2006 radiation poisoning of Alexander Litvinenko in London, United Kingdom (UK), and the 2011 Fukushima nuclear accident in Japan.

Goiânia

The Goiânia incident is often cited in the risk communication literature as it was a relatively small-scale event that had large-scale impacts due to public behavioural responses linked to heightened risk perceptions. It occurred when scavengers discovered a source of caesium-137 in an abandoned teletherapy unit; without understanding that the material was dangerous, they were attracted by its blue glow, leading them to distribute the source among family and friends. This resulted in four deaths and 249 contaminations. However, despite the relatively low numbers of fatalities and injuries, more than 112,000 individuals sought monitoring to confirm that they were not contaminated.[51] The scale of the public response in Goiânia has been attributed to poor risk communication in the immediate aftermath of the incident.[52] Local officials initially presented the evacuation of contaminated houses as a response to a gas leak to avoid disruption to an international motorcycling event that was taking place in city.[53] Trust in official communications was further undermined by attempts to control information through spokespeople who lacked credibility with the local community, and lack of consensus between authorities and independent scientists.[54]

Alexander Litvinenko

In contrast, there was a relatively muted response from the public when it was reported in November 2006 that former Russian spy, Alexander Litvinenko, had been deliberately poisoned by a radioactive substance in London.[55] Prior to being admitted to hospital with vomiting and stomach pains, Litvinenko met with two Russian ex-KGB officers in a central London hotel bar and had lunch in a sushi restaurant. He was initially treated for gastroenteritis, but as his condition deteriorated it was established that he had been poisoned and subsequent investigations found traces of polonium-210 at numerous locations, including the bar and restaurant that he had visited the day he became ill.

Even among those with most cause for concern—that is, members of the public who had visited sites that were confirmed to have been contaminated—there is little evidence of persistent anxiety.[56] This has been attributed in part to the perception that it was a targeted attack rather than terrorism.[57] This message was reinforced by hoardings that were placed outside the restaurant Litvinenko visited while it was checked for contamination. These hoardings characterized the poisoning as 'an international espionage incident' and used imagery that directly referenced a widespread advertising campaign for a new James Bond film. The lack of public over-reaction lack also been attributed to an effective communication strategy employed by the UK Health Protection Agency, which successfully conveyed that the risk to public health was geographically restricted.[58] While the peculiarities of this case limit its use as an example of what might happen in a radiological terrorist attack, it does demonstrate that heightened risk perceptions are not inevitable, and that effective risk communication can help reduce unnecessary healthcare-seeking behaviour.

Fukushima

A third example that is more comparable to the type of impacts that would need to be managed following an RDD attack is the Fukushima nuclear accident. On 11 March 2011, a powerful tsunami triggered by the Tōhoku earthquake severely damaged the Fukushima Daiichi nuclear power plant, resulting in a large-scale release of radioactivity into the environment. It has been calculated that there were 15,770 confirmed deaths due to direct trauma from the earthquake and tsunami,[59] but to date there have been no deaths due to deterministic radiation injury and experts believe that long-term radiation injury deaths will be too rare to be statistically detectable.[60]

Following the incident, a 3-km mandatory evacuation zone was immediately set up and this was extended to 10 km the following day. Approximately 150,000 people evacuated, including voluntary evacuations from neighbouring prefectures.[61] Almost a month later, the evacuation area was extended, based on data which indicated that residents in a wider area would be exposed to unacceptable levels of excess radiation over the coming year. This new evacuation zone included areas to which evacuees had initially relocated, causing some people to have to evacuate for a second time, creating mistrust and confusion. Trust was further eroded by perceived inconsistencies in the distribution of iodine prophylaxis, a medical intervention used to protect the thyroid gland from the effects of radiation.[62] By the end of 2016, the government started reopening some evacuation zones, but to date there has been reluctance among evacuees to return. This has been attributed not only to ongoing anxiety about health risks, but also to lack of employment opportunities, poor housing, and educational and health infrastructure.[63]

As with the radiation incident response in Goiânia, multiple issues have been identified in Japan's response to the Fukushima nuclear accident, with official risk communications described as 'often not transparent, timely, clear, nor factually correct'.[64] This has been attributed to a one-way, top-down communicative approach that prioritizes the provision of reassurance, leading to lack of clarity around radiation exposure levels, changes to acceptable dose levels, and the scale of food contamination.[65] Multiple, often contradictory messages have been linked both to risk-averse and risk-taking behavioural responses, including unnecessary limiting of outdoor activities and unintended consumption of potentially contaminated food.[66] A recent review found that although there have been some improvements in risk communications regarding the impacts of the incident on food safety 10 years on, official messages continue to focus primarily on disputing consumer concerns rather than supporting informed decision making.[67]

These three case studies demonstrate that public psychological and behavioural reactions to radiological incidents are not only influenced by *a priori* perceptions of the risk posed by radiation but are also determined by the way they are managed and communicated, and by material as well as psychological constraints. These cases also provide further evidence that while radiation release may create anxiety, the public response is likely to be proportional to the incident and people will not panic. However, if the risk is poorly understood or the official response is considered unfair or incompetent, public reactions may create both acute and longer-term demand that can overwhelm healthcare and other systems. The primary intervention that is used to mitigate this is risk and crisis communication.

What Is the Role of Risk and Crisis Communication?

There are multiple factors likely to influence variation in public responses to radiological terrorism, including cultural context, prior knowledge and experience, and unequal distribution of risk, as well as specific features of the incident, such as scale and perceived motivation. However, as the discussed case studies illustrate, the extent and quality of official communications also play a central role in determining how the public will react. It has been shown that large-scale radiological incidents have a range of societal impacts beyond public health consequences relating to the environment, economy, and social cohesion, all of which require risk and crisis communication interventions.[68] Consequently, as has been seen during the Covid-19 response, official recommendations need to balance multiple impacts and will reflect decision makers' values as well as scientific facts.

Risk communication and crisis communication can be differentiated in relation to focus and timing. Crisis communication is more event focused and short term, with broader aims that take in reputation management as well as the provision of information about risk.[69] However, both communication types share many basic principles and in the context of major incident response, both are used to enhance knowledge and understanding to influence behavioural responses to risks.[70] In practice, health communicators therefore tend to employ risk and crisis communication in combination, as exemplified by the US Centers for Disease Control (CDC) Crisis and Emergency Risk Communication (CERC) programme.[71]

The role of risk and crisis communication is well established in relation to reducing anxiety and confusion, providing a sense of control, promoting trust, increasing compliance with public health guidance, and enabling appropriate behaviours.[72] A successful response to radiological terrorism therefore requires a coordinated communication strategy that provides consistent, timely messages to inform public risk perceptions and behavioural responses.[73] Radiation communication is complex, so it is crucial that information is provided in clear language that can be understood by the public.[74] This is particularly important in the aftermath of an attack when substantial fear may impede people's information processing capacity.[75] Risk communication is most straightforward in situations when the risk is well known, and mitigations are clear. However, it plays a particularly important role in situations of uncertainty and when people may not fully understand data limitations.[76] As discussed, radiological incidents involve a great deal of uncertainty, as well as differences in expert and public acceptance and understanding of risk. Effective risk and crisis communication is therefore crucial in this context.

Accurate risk information will need to be communicated quickly, consistently, and with clarity to inform behavioural responses and reduce rumours in the immediate aftermath of an attack. The swift coordination of accurate information can be a challenge for any incident response, but especially so if radiation is involved, when lack of data and scientific certainty can impede clear communication.[77] It should be recognized that people routinely manage uncertainty in their everyday lives, but in situations of heightened concern, it can be very stressful.[78] Research has shown that it is beneficial be clear about what is and is not known when communicating about radiological emergencies.[79] However, a false assumption of public panic can lead risk and crisis communicators to withhold information or to overly focus on providing reassurance. While reassurance can play an important role, it must be proportionate to the risk to avoid unintended consequences, such as increasing inappropriate risk taking. Furthermore, reassurances that later turn out to be false will reduce trust in responding organizations and decrease the likelihood that future guidance will be followed.[80]

Differences in the ways that experts and the public conceptualize and accept risk also need to be managed. Scientific conventions regarding the expression of uncertainty mean that experts are likely to be cautious in assessing evidence. They may also focus on aspects of uncertainty in the data that may not be particularly relevant for decisions the public need to take. As observed with discussions of climate change, this can create the impression of greater scientific uncertainty than the data actually suggest.[81] Furthermore, if the risk assessments underpinning official guidance are not clearly explained, recommended actions may not be considered acceptable by the public. For example, there was public outrage following the 1986 Chernobyl nuclear accident when the Norwegian government recommended that it was safe for children to play outside but also advised that cows should be kept indoors for six weeks. This was due to the perception that greater protections were being afforded to livestock than children. Had the guidance explained the risks associated with ingesting grass (relative to walking on it) and the aims of the guidance relating to protecting children from potential harms from contaminated milk supply, this reaction could have been avoided.[82]

Risk and crisis communication therefore has a key role to play in radiological terrorism response but can be hindered by uncertainties relating to risk of exposure and differences in public and expert risk perceptions. Lack of clarity in conveying scientific uncertainty is associated with a wide range of interpretations of guidance intended to support public understanding of risk.[83] This may be compounded when risk communicators employ commonly used terms without explaining their precise meaning in a scientific context, for example, the use of 'normal level' in public communications following Fukushima.[84] Risk communicators should therefore avoid withholding information or minimizing risk and instead be transparent about any uncertainties underlying recommendations to encourage appropriate behaviours and maintain confidence in the ability of responding organizations. To do so effectively, they need to understand public knowledge and concerns, and use clear, mutually understood terms to convey risk. To this end it is important to draft and test messages ahead of an event.[85]

Developing and Testing Communication Interventions

Multiple research projects have been undertaken to develop and test messages with the public to inform response plans for a range of CBR incidents.[86] Projects that have developed and tested messages for RDD and RED scenarios demonstrate the value of pre-preparing messages for release following a radiological attack and the need to tailor messages for different incident types.

Rogers and colleagues employed a two-phase UK-based focus group study to assess the impact of public health messages on levels of knowledge, understanding, and intended compliance with official advice in response to a hypothetical RDD scenario.[87] The scenario was conveyed using a series of mock TV news injects. The final inject included an 'independent scientist' who used inaccurate information to challenge official advice. Phase 1 focus groups explored baseline risk perceptions and information needs. Issues identified during this phase were then used to inform the design of a leaflet intervention that provided information about radiation health impacts and treatments, as well as likely actions that would be taken by authorities in the event of an RDD. This leaflet was presented to Phase 2 participants before they were shown the final TV news inject.

The research found low baseline levels of knowledge about radiation and its effects, with elevated levels of confusion and fear in response to the scenario. Participants' concerns about health impacts focused on long-term increased cancer risk and possible adverse effects on future generations. Heightened risk perceptions were associated with intention to collect children from school despite official warnings; intention to leave the area, even if outside of the affected zone; unwillingness to attend a treatment centre for fear of contamination; and unwillingness to return home or resume normal activities despite being informed that it was safe to do so. These findings suggest that official recommendations need to go beyond the simple provision of factual information to explain how and why recommended actions are protective. For example, participants questioned how showering and disrobing would be helpful if radioactive material had been inhaled and why it was considered safe to stop sheltering after the plume had dissipated.

Phase 2 results indicated that the additional information provided in the leaflet increased trust in and intention to comply with official advice, although it should be noted that intended compliance was already high. The additional information had most influence on intention to return to an area following decontamination. It also improved knowledge and understanding about radiation and RDDs and addressed some short-term concerns. Participants reported that this reduced their anxiety and made them feel more in control. However, they still wanted additional information relating to water and food contamination, radiation transmission (person-to-person and to/from pets), and radiation measurement and monitoring. Furthermore, they remained concerned about inconsistency in messages provided by officials and the independent scientist, and wanted long-term repeated measurement of radiation levels to feel reassured that the affected area was now safe.

Pearce and colleagues employed a similar two-phase study to assess public intentions and information needs in response to a RED scenario in the UK and Germany.[88] Again, the first phase was used to identify baseline information needs and behavioural intentions, and results informed the design of a communication intervention that was evaluated in a second round of focus groups. As with the original study, the additional information was provided to Phase 2 participants prior to showing them a news item that included an interview with an 'independent scientist' who challenged official guidance. This study confirmed low baseline levels of public knowledge about different types of radiation, as identified in the original research. It also found participants in both countries primarily associated radiological terrorism with nuclear bombs (Hiroshima) and disasters (Chernobyl). Participants were not aware that radiological material could be distributed using a covert device and therefore initially assumed that an explosive device had failed to 'go off'. Consequently, initial risk perceptions were lower than predicted, although some concerns were reported regarding the potential severity, contagion, and pervasiveness of the attack.

Most participants indicated that they would intend to continue with their daily routine in the days and weeks following an attack of this type, but a sizeable minority indicated that they would unnecessarily attend a monitoring centre even though they had been outside of the immediate impact zone, in order to obtain reassurance they had not been exposed to radiation. Primary information needs identified at Phase 1 focused on additional information about health and security. Participants responded positively to the independent scientist who challenged official guidance due to his perceived independence and when his comments resonated with existing concerns. However, others indicated that they found his appearance and style of presentation unconvincing.

Phase 2 results demonstrated positive benefits of the communication interventions in relation to increasing intention to follow official advice. Additional information about the way that REDs work reduced concern about cordon size, incident severity, and lack of quarantine, while better understanding of personal risk reduced intention to unnecessarily attend monitoring centres. Furthermore, there was more critical engagement with the challenge posed by the independent scientist, which drew directly on the additional information that was provided. Participants were particularly positive about the credibility of information provided in the form of leaflets—an important point to note in the context of increasing reliance on social media to communicate messages during and after emergencies. Leaflets were considered more credible due to the costs involved in producing them and that they cannot be easily updated. However, some participants expressed concern that the provision of leaflets—particularly in advance of an event—might cause alarm.

These studies demonstrate the potential for risk and crisis communication to inform risk perceptions and intended behavioural responses to radiological terrorism by increasing knowledge, managing expectations, and building trust. Consistent with the evidence from real-world case studies, these results suggest that heightened risk perceptions can lead to feelings of anxiety and some unnecessary care-seeking. However, the RED scenario, where the mode of dispersal was inconsistent with existing preconceptions about radiological terrorism, produced a more muted response. This suggests that generic radiation communication designed to reduce anxiety may not be appropriate for this type of scenario and that targeted communication that explains the mechanism and impacts of REDs should be prepared and tested in advance to allow rapid deployment should the need arise. This study also highlighted that audience reception is influenced by the perceived credibility of the communicator, as well as the message that is being conveyed.

Enhancing Risk and Crisis Communication

This chapter has established that risk and crisis communication is an essential part of responding to radiological terrorism. It has also highlighted communication challenges relating to low baseline knowledge about radiation, the impact of fear on people's ability to comprehend information in the immediate aftermath of an incident, and the speed at which information needs to be disseminated while the scale of impacts may still be unclear. Effective communication is crucial at this time, as the low perceived likelihood of an attack occurring means that the public is unlikely to have engaged with information in advance.[89] There is also evidence to suggest similar levels of fear and knowledge gaps among frontline responders. For example, a survey of emergency responders in smaller cities in the US found low knowledge of radiation health risks, with limited preparedness for radiological terrorism attributed to low perceived risk to their cities, even though these locations housed materials that could be used for attack planning.[90] Risk communications should therefore also target the responder community.

To enhance risk and crisis communication, it is necessary to recognize that increased information seeking and anxiety is a normative response to radiological terrorism. Rather than characterizing those seeking reassurance as 'the worried well', provision should be made for low-risk patients reasonably seeking information or reassurance in a situation of uncertainty.[91] Attempts to control information via top-down, one-way communications have been shown to have the opposite to the intended effect and will likely erode trust in responding organizations. Rather, risk and crisis communication is most effective when it involves two-way information sharing, considering public risk perceptions and enabling decision making while respecting individual autonomy.[92] This can be achieved by involving the public in designing and testing messages to ensure appropriate content, language, and framing.

Effective risk and crisis communication needs to provide actionable advice that is targeted at encouraging or discouraging specific behaviours. Messages that briefly explain the reason for advice are more likely to be successful than those that simply provide instructions. This is particularly important when guidance might be counter-intuitive or runs contrary to preferred actions. For example, parents may be disinclined to comply with 'shelter-in-place' advice if it prevents them from collecting their children from school.[93] Psychological models of health behaviour, such as Protection Motivation Theory,[94] the Extended Parallel Process Model,[95] and COM-B,[96] highlight the need not only to take into consideration public appraisals of the threat (relating to perceived personal risk of and consequences of exposure), but also perceptions about the recommended response (relating to the perceived efficacy of recommended actions, perceived ease of carrying out these recommendations, and the costs associated with this, such as the aforementioned separation from loved ones).

To maximize the effectiveness of risk and crisis communication it is also important to consider the suitability of the messenger and most effective modes of message delivery. The success or failure of risk communication is strongly and consistently mediated by levels of trust in the communicators of the message.[97] It is therefore important that messengers and the organizations that they represent are trusted. Scientific experts who are perceived to be independent of political influence are established trusted sources of information,[98] although there is evidence to suggest increased anti-science sentiment during Covid, which means this cannot necessarily be assumed for future incidents.[99]

For messages to be effective, it is also important that they can be successfully conveyed via the mass media.[100] This requires the use of engaging communicators who can provide information and recommendations in language that is understood by and resonates with the public. It also involves effective use of social media to facilitate dialogue and disseminate messages at pace.[101] Preferred social media channels need to be identified to maximize reach. For example, younger generations may not use the platforms typically employed by responding organizations.[102]

Conclusion

This chapter has argued that public reactions play a key role in determining the overall economic, social, physical, and psychological impacts of radiological terrorism. Understanding and informing public responses is therefore a crucial part of a successful response. A review of the risk perception literature and public responses to real-world case studies indicates this will likely require the management of widespread fear and heightened anxiety to reduce the burden on healthcare systems and support social and economic recovery. However, it also demonstrates that impacts are unlikely to be evenly distributed and it is therefore important for responders to understand variation in public risk perceptions and reactions. Furthermore, psychological and behavioural reactions to radiological terrorism are influenced not only by a priori risk perceptions, but also by the way that the response is managed and communicated, and by material considerations, such as the provision of adequate infrastructural support to follow guidance.

Effective risk and crisis communication can encourage protective health behaviours and reduce unnecessary care-seeking. It can also reduce anxiety, provide a sense of control, and help build and maintain trust in responding organizations. To communicate effectively following a radiological attack requires understanding of public risk perceptions and likely behaviours. It also requires responders to demonstrate trustworthiness in the fairness and competency of recommendations. To this end, it is important to communicate uncertainty and to be transparent about any trade-offs underpinning recommendations. Messages also need to be developed and tested with multiple audiences to ensure that they are clear and mutually understood, and to establish preferred communication channels and credible communicators. Finally, it is important to recognize that generic principles of risk communication may need adaption for different types of incidents, as well as for different audiences and specific national and cultural contexts.

Notes

1. John W. Poston et al., 'Management of Terrorist Events Involving Radioactive Material (2001)', Report No. 138 (Bethesda, MD: National Council on Radiation Protection and Measurements (NCRP), 2001).

2. Steven M. Becker, 'Addressing the Psychosocial and Communication Challenges Posed by Radiological/Nuclear Terrorism: Key Developments Since NCRP Report No. 138', *Health Physics*, 89/5 (2005), 521.
3. Holly Carter and Richard Amlôt, 'Mass Casualty Decontamination Guidance and Psychosocial Aspects of CBRN Incident Management: A Review and Synthesis', *PLoS Currents: Disasters* [online], 27 September 2016, 8.
4. Robert J. Ingram, 'Emergency Response to Radiological Releases: Have We Communicated Effectively to the First Responder Communities to Prepare Them to Safely Manage These Incidents?', *Health Physics*, 114/2 (2018), 208–13.
5. BreAnne K. Fleer, 'Radiological-Weapons Threats: Case Studies from the Extreme Right', *The Nonproliferation Review*, 27, 1–3 (2020), 225–42.
6. IAEA Incident and Trafficking Database (ITDB), 2022 Factsheet, 2021.
7. Louis Gauntlett, Richard Amlôt, and G. James Rubin, 'How to Inform the Public About Protective Actions in a Nuclear or Radiological Incident: A Systematic Review', *Lancet Psychiatry*, 6/1 (2019), 72–80.
8. Fleer, 'Radiological-Weapons Threats: Case Studies from the Extreme Right'.
9. Timothy J. Jorgensen, 'Predicting the Public Health Consequences of a Nuclear Terrorism Attack: Drawing on The Experiences of Hiroshima and Fukushima', *Health Physics*, 115/1 (2018), 121–5.
10. Cristiane P. S. Andrade, Claudio J. Souza, Eduardo S. N. Camerini, Isabela S. Alves, Helio C. Vital, Matthew J. F. Healy, and Edson Ramos de Andrade, 'Support to Triage and Public Risk Perception Considering Long-Term Response to a Cs-137 Radiological Dispersive Device Scenario', *Toxicology and Industrial Health*, 34/6 (2018), 433–8.
11. Jorgensen, 'Predicting the Public Health Consequences of a Nuclear Terrorism Attack'.
12. Ibid.
13. Julia M. Pearce, G. James Rubin, Piet Selke, Richard Amlôt, Fiona Mowbray, and M. Brooke Rogers, 'Communicating with the Public Following Radiological Terrorism: Results from a Series of Focus Groups and National Surveys in Britain and Germany', *Prehospital and Disaster Medicine*, 28/2 (2013), 110–19.
14. Andrade et al., 'Support to Triage and Public Risk Perception Considering Long-Term Response to a Cs-137 Radiological Dispersive Device Scenario'.
15. Jorgensen, 'Predicting the Public Health Consequences of a Nuclear Terrorism Attack'.
16. Jaspa Pandza, 'Managing the Consequences of Nuclear Terrorism', *Survival*, 53/3 (2011), 129–42.
17. Steven M. Becker, 'Emergency Communication and Information Issues in Terrorist Events Involving Radioactive Material', *Biosecurity and Bioterrorism: Biodefense Strategy, Practice, and Science*, 2/3 (2004), 113–37.
18. Louise Lemyre, Colleen Johnson, and Wayne Corneil, 'Psychosocial Considerations for Mass Decontamination', *Radiation Protection Dosimetry*, 142/1 (2010), 17–23.
19. Akiko Sato and Yuliya Lyamzina, 'Diversity of Concerns in Recovery after a Nuclear Accident: A Perspective from Fukushima', *International Journal of Environmental Research and Public Health*, 15/2 (2018), 350.
20. David Ropeik, 'Critical Areas for Improvement in Communications Regarding Radiological Terrorism', *Health Physics*, 114/2 (2018), 214–17.
21. Robert C. Whitcomb, Armin J. Ansari, Jennifer J. Buzzell, M. Carol McCurley, Charles W. Miller, James M. Smith, and D. Lynn Evans, 'A Public Health Perspective on the U.S. Response to the Fukushima Radiological Emergency', *Health Physics*, 108/3 (2015), 357–63.

22. Ingram, 'Emergency Response to Radiological Releases'.
23. Aino Ruggiero and Marita Vos, 'Communication Challenges in CBRN Terrorism Crises: Expert Perceptions', *Journal of Contingencies and Crisis Management*, 23/3 (2015), 138–48.
24. Becker, 'Emergency Communication and Information Issues in Terrorist Events Involving Radioactive Material'.
25. James M. Acton, M. Brooke Rogers, and Peter D. Zimmerman, 'Beyond the Dirty Bomb: Rethinking Radiological Terror', *Survival*, 49/3 (2007), 151–68.
26. G. J. Rubin and P. Dickmann, 'How to Reduce the Impact of "Low-Risk Patients" Following a Bioterrorist Incident: Lessons from SARS, Anthrax and Pneumonic Plague', *Biosecurity and Bioterrorism: Biodefense Strategy, Practice, and Science*, 8/1 (2010), 37–43.
27. Erik R. Svendsen, Ichiro Yamaguchi, Toshihide Tsuda, Jean Remy Davee Guimaraes, and Martin Tondel, 'Risk Communication Strategies: Lessons Learned from Previous Disasters with a Focus on the Fukushima Radiation Accident', *Current Environmental Health Reports*, 3/4 (2016), 348–59.
28. M. P. Curado, 'The Communication of Radiological Risk to Populations Exposed to a Radiological Accident: Considerations Concerning the Accident in Goiânia', *Radiation Protection Dosimetry*, 68/3/4 (1996), 283–86.
29. Pandza, 'Managing the Consequences of Nuclear Terrorism'.
30. Meritxell Martell, Tanja Perko, Nadja Zeleznik, and Susan Molyneux-Hodgson, 'Lessons being Learned from the Covid-19 Pandemic for Radiological Emergencies and Vice Versa: Report from Expert Discussions', *Journal of Radiological Protection* [online], 42/1 (2022), 011518.
31. M. Brooke Rogers and Julia M. Pearce, 'Risk Communication, Risk Perception and Behaviour as Foundations of Effective National Security Practices', in Babak Akhgar and Simeon Yates, eds., *Strategic Intelligence Management* (Amsterdam: Butterworth-Heinemann, 2013), 66–74.
32. Christina Palmer, 'Risk Perception: Another Look at the "White Male" Effect', *Health, Risk and Society*, 5/1 (2003), 71–83.
33. Paul Slovic, *The Feeling of Risk: New Perspectives on Risk Perception* (Abingdon: Earthscan, 2010).
34. Pearce et al., 'Communicating with the Public Following Radiological Terrorism'.
35. Slovic, *The Feeling of Risk*.
36. Gauntlett et al., 'How to Inform the Public About Protective Actions in a Nuclear or Radiological Incident'.
37. Svendsen et al., 'Risk Communication Strategies'.
38. Ibid.
39. Lemyre et al., 'Psychosocial Considerations for Mass Decontamination'.
40. Ferdiana Hoti, Tanja Perko, Peter Thijssen, and Ortwin Renn, 'Radiation Risks and Uncertainties: A Scoping Review to Support Communication and Informed Decision Making', *Journal of Radiological Protection*, 40/2 (2020), 612–32.
41. Bjoern Hagen, Adenike Opejin, and K. David Pijawka, 'Risk Perceptions and Amplification Effects over Time: Evaluating Fukushima Longitudinal Surveys', *Sustainability*, 14/13 (2022), 7896.
42. Tine Walravens, Paul O'Shea, and Nicolai Ahrenkiel, '"Let's Eat Fukushima": Communicating Risk and Restoring "Safe Food" after the Fukushima Disaster (2011–2020)', *Japan Forum*, 34/1 (2022), 79–102.
43. Hagen et al., 'Risk Perceptions and Amplification Effects over Time'.

44. M. Brooke Rogers, Richard Amlôt, G. James Rubin, Simon Wessely, and Kristian Krieger, 'Mediating the Social and Psychological Impacts of Terrorist Attacks', *International Review of Psychiatry*, 19/3 (2007), 279–88.
45. Jorgensen, 'Predicting the Public Health Consequences of a Nuclear Terrorism Attack'.
46. Carter and Amlôt, 'Mass Casualty Decontamination Guidance and Psychosocial Aspects of CBRN Incident Management'.
47. Rogers and Pearce, 'Risk Communication, Risk Perception and Behaviour as Foundations of Effective National Security Practices'.
48. Carter and Amlôt, 'Mass Casualty Decontamination Guidance and Psychosocial Aspects of CBRN Incident Management'.
49. Burc Aydin, 'Global Characteristics of Chemical, Biological, and Radiological Poison Use in Terrorist Attacks', *Prehospital Disaster Medicine*, 35/3 (2020), 260–6. Aydin identified 446 incidents fulfilling terrorism and poisoning definitions using data from the Global Terrorism Database (GTD) and RAND Database of Worldwide Terrorism Incidents (RDWTI). This figure includes 401 enacted events and 45 attempted events. The majority of incidents (63.5 per cent) involved chemical agents.
50. For example, Becker, 'Emergency Communication and Information Issues in Terrorist Events Involving Radioactive Material'; Pearce et al., 'Communicating with the Public Following Radiological Terrorism'.
51. IAEA, *The Radiological Accident in Goiania*, Non-Serial Publications (Vienna: IAEA, 1988).
52. Pandza, 'Managing the Consequences of Nuclear Terrorism'.
53. Svendsen et al., 'Risk Communication Strategies'.
54. Curado, 'The Communication of Radiological Risk to Populations Exposed to a Radiological Accident'.
55. G. James Rubin, Lisa Page, Oliver Morgan, Richard J. Pinder, Paul Riley, Stephani Hatch, Helen Maguire, Mike Catchpole, John Simpson, and Simon Wessely, 'Public Information Needs after the Poisoning of Alexander Litvinenko with Polonium-210 in London: Cross-Sectional Telephone Survey and Qualitative Analysis', *British Medical Journal*, 335 (2007), 1143–56.
56. Acton et al., 'Beyond the Dirty Bomb'.
57. G. James Rubin, Richard Amlôt, and Lisa Page, 'The London Polonium Incident: Lessons in Risk Communications', *Health Physics*, 101/5 (2011), 545–50.
58. Rubin et al., 'Public Information Needs after the Poisoning of Alexander Litvinenko with Polonium-210 in London'.
59. Shinji Nakahara and Masao Ichikawa, 'Mortality in the 2011 Tsunami in Japan', *Journal of Epidemiology*, 23/1 (2013), 70–3.
60. Ropeik, 'Critical Areas for Improvement in Communications Regarding Radiological Terrorism', 214–17.
61. Tsutomu Shimura, Ichiro Yamaguchi, Hiroshi Terada, Erik Robert Svendsen, and Naoki Kunugita, 'Public Health Activities for Mitigation of Radiation Exposures and Risk Communication Challenges after the Fukushima Nuclear Accident', *Journal of Radiation Research*, 56/3 (2015), 422–9.
62. Ibid.
63. Sato and Lyamzina, 'Diversity of Concerns in Recovery after a Nuclear Accident', 350.
64. Tanja Perko, 'Risk Communication in the Case of the Fukushima Accident: Impact of Communication and Lessons to Be Learned', *Integrated Environmental Assessment and Management*, 12/4 (2016), 683.
65. Shimura et al., 'Public Health Activities for Mitigation of Radiation Exposures and Risk Communication Challenges after the Fukushima Nuclear Accident'.

66. Svendsen et al., 'Risk Communication Strategies'.
67. Walravens et al., '"Let's Eat Fukushima"'.
68. Martell et al., 'Lessons Being Learned from the Covid-19 Pandemic for Radiological Emergencies and Vice Versa'.
69. M. Brooke Rogers and Julia M. Pearce, 'The Psychology of Crisis Communication', in Andreas Schwarz, Matthew W. Seeger, and Claudia Auer, eds., *The Handbook of International Crisis Communication Research* (Hoboken, NJ: John Wiley & Sons, 2016), 34–44.
70. Deborah C. Glik, 'Risk Communication for Public Health Emergencies', *Annual Review of Public Health*, 28 (2007), 33–54.
71. Centers for Disease Control and Prevention (CDC), 'Crisis & Emergency Risk Communication (CERC)', 2018. https://emergency.cdc.gov/cerc
72. Carter and Amlôt, 'Mass Casualty Decontamination Guidance and Psychosocial Aspects of CBRN Incident Management'.
73. Whitcomb et al., 'A Public Health Perspective on the U.S. Response to the Fukushima Radiological Emergency'.
74. Svendsen et al., 'Risk Communication Strategies'.
75. S. B. Bass, J. R. Greener, D. Ruggieri, C. Parvanta, G. Mora, C. Wolak, R. Normile, and T. F. Gordon, 'Attitudes and Perceptions of Urban African Americans of a "Dirty Bomb" Radiological Terror Event: Results of a Qualitative Study and Implications for Effective Risk Communication', *Disaster Medicine and Public Health Preparedness*, 9/1 (2015), 9–18.
76. Baruch Fischhoff and Alex L. Davis, 'Communicating Scientific Uncertainty', *Proceedings of the National Academy of Sciences of the United States of America*, 111/Suppl. 4 (2014), 13664–71.
77. Walravens et al., "Let's Eat Fukushima"'.
78. Rogers et al., 'Mediating the Social and Psychological Impacts of Terrorist Attacks'.
79. Martell et al., 'Lessons being Learned from the Covid-19 Pandemic for Radiological Emergencies and Vice Versa'.
80. Ibid.
81. Fischhoff and Davis, 'Communicating Scientific Uncertainty'.
82. Lemyre et al., 'Psychosocial Considerations for Mass Decontamination'.
83. Fischhoff and Davis, 'Communicating Scientific Uncertainty'.
84. Perko, 'Risk Communication in the Case of the Fukushima Accident'.
85. Fischhoff and Davis, 'Communicating Scientific Uncertainty'.
86. For example, Becker, 'Emergency Communication and Information Issues in Terrorist Events Involving Radioactive Material'; Ricardo J. Wray, Steven M. Becker, Neil Henderson, Deborah Glik, Keri Jupka, Sarah Middleton, Carson Henderson, Allison Drury, and Elizabeth W. Mitchell, 'Communicating with the Public about Emerging Health Threats: Lessons from the Pre-Event Message Development Project', *American Journal of Public Health*, 12 (2008), 2214–22.
87. M. Brooke Rogers, Richard Amlôt, and G. James Rubin, 'The Impact of Communication Materials on Public Responses to a Radiological Dispersal Device (RDD) Attack', *Biosecurity and Bioterrorism: Biodefense Strategy, Practice, and Science*, 11/1 (2013), 1–10.
88. Pearce et al., 'Communicating with the Public Following Radiological Terrorism'.
89. G. James Rubin, Alexander K. Chowdhury, and Richard Amlôt, 'How to Communicate with the Public about Chemical, Biological, Radiological, or Nuclear Terrorism: A Systematic Review of the Literature', *Biosecurity and Bioterrorism: Biodefense Strategy, Practice, and Science*, 10/4 (2012), 383–95.
90. Ingram, 'Emergency Response to Radiological Releases'.

91. Rubin and Dickmann, 'How to Reduce the Impact of "Low-Risk Patients" Following a Bioterrorist Incident'.
92. Svendsen et al., 'Risk Communication Strategies'.
93. Julia M. Pearce, G. James Rubin, Richard Amlôt, Simon Wessely, and M. Brooke Rogers, 'Communicating Public Health Advice Following a Chemical Spill: Results from National Surveys in the UK and Poland', *Disaster Medicine and Public Health Preparedness*, 7/1 (2013), 65–74.
94. Ronald W. Rogers, 'A Protection Motivation Theory of Fear Appeals and Attitude Change', *The Journal of Psychology*, 91 (1975) 93–114.
95. Kim Witte, 'Putting the Fear Back into Fear Appeals: The Extended Parallel Process Model', *Communication Monographs*, 59 (1992), 329–49.
96. Susan Michie, Lou Atkins, and Robert West, *The Behaviour Change Wheel: A Guide to Designing Interventions* (London: Silverback, 2014).
97. Thomas A. Glass and Monica Schoch-Spana, 'Bioterrorism and the People: How to Vaccinate a City against Panic', *Clinical Infectious Diseases*, 34 (2002), 217–23.
98. F. P. Stone, *The 'Worried Well' Response to CBRN Events: Analysis and Solutions* (Montgomery, AL: USAF Counterproliferation Center, 2007).
99. Danielle M. McLaughlin, Jack Mewhirter, and Rebecca Sanders, 'The Belief that Politics Drive Scientific Research & Its Impact on COVID-19 Risk Assessment', *PLoS ONE* [online], 21 April 2021, 16/4.
100. Stone, *The 'Worried Well' Response to CBRN Events.*
101. T. Perko and M. Martell, 'Communicating Nuclear and Radiological Emergencies to the Public: How and to What Extent Are European Countries Prepared?', *International Journal of Disaster Risk Reduction*, 50 (2020), 101722.
102. Martell et al., 'Lessons Being Learned from the Covid-19 Pandemic for Radiological Emergencies and Vice Versa'.

Suggested Reading

Becker, Steven M., (2005). 'Addressing the Psychosocial and Communication Challenges Posed by Radiological/Nuclear Terrorism: Key Developments since NCRP Report No. 138', *Health Physics*, 89/5, 521–30.

Pearce, Julia M., Rubin, G. James, Selke, Piet, Amlôt, Richard, Mowbray, Fiona, and Rogers, M. Brooke, (2013), 'Communicating with the Public Following Radiological Terrorism: Results from a Series of Focus Groups and National Surveys in Britain and Germany', *Prehospital and Disaster Medicine*, 28/2, 110–19.

Rogers, M. Brooke, Amlôt, Richard, and Rubin, G. James, (2013), 'The Impact of Communication Materials on Public Responses to a Radiological Dispersal Device (RDD) Attack', *Biosecurity and Bioterrorism: Biodefense Strategy, Practice, and Science*, 11/1, 1–10.

Rogers, M. Brooke and Pearce, Julia M., (2013), 'Risk Communication, Risk Perception and Behaviour as Foundations of Effective National Security Practices', in Babak Akhgar and Simeon Yates, eds., *Strategic Intelligence Management* (Amsterdam: Butterworth-Heinemann), 66–74.

Slovic, Paul, (2010), *The Feeling of Risk: New Perspectives on Risk Perception* (Abingdon: Earthscan).

CHAPTER 40

NUCLEAR SECURITY IN THE WAKE OF THE 2011 EARTHQUAKE AND TSUNAMI IN JAPAN

NICKOLAS ROTH AND DANIEL SALISBURY

Introduction

On 11 March 2011, an earthquake and resultant tsunami destroyed large swathes of the Japanese coastline, killing thousands of people. The nuclear power plant at Fukushima Daiichi was badly damaged by the tsunami, leading to a radioactive release. This chapter considers the evolution of nuclear security in Japan before and after this safety incident and is structured in four main sections. The first section provides background on the nuclear sector in Japan and other contextual information for the following sections. The second section considers the Fukushima disaster and its immediate and longer-term impacts for nuclear security. Rather than focusing on the potential for a non-state actor to maliciously cause a Fukushima-style accident, discussion centres around the challenges of maintaining nuclear security during times of crisis. The third considers issues in Japanese nuclear security since the disaster, particularly focusing on materials and minimization, insider threats, and security culture.

The final section of the chapter looks forward and draws conclusions about the future of nuclear security in Japan. The chapter argues that Japan has made significant progress in its enhancement of nuclear security since the 11 September 2001 ('9/11') attacks in the United States (US) and especially since the 2011 Fukushima disaster. While Fukushima was a safety, rather than security, event, the disaster and crisis heightened interest in the safe, secure, and effective management of nuclear assets in Japan, and led to significant developments in Japan's regulation of nuclear technology with knock-on effects for security. At the time of writing, the future of nuclear security—like the future of nuclear power—in Japan remains unclear, with fewer than one-third of Japan's 33 nuclear reactors having been cleared to restart, largely as a result of safety and security concerns following the disaster a decade ago.

Background Context

As a result of historical events and extensive nuclear assets and material holdings, Japan has a strong interest in securing its nuclear materials. Japan has extensive experience of nuclear technologies and, until the Fukushima disaster, relied upon nuclear power for a significant proportion of its nuclear energy mix. In 2011, Japan relied on nuclear energy for approximately 30 per cent of its total energy requirements, a figure that was planned to rise to 40 per cent by 2017.[1] As well as a large number of nuclear reactors for power generation, Japan has a range of other sites hosting nuclear materials and assets that require nuclear security measures. This includes a sizable stockpile of separated plutonium and highly enriched uranium (HEU) for a state without a military nuclear programme.

The only country in the world to suffer the effects of the use of nuclear weapons in war, Japan is a non-nuclear weapons state under the nuclear Non-Proliferation Treaty (NPT), having acceded to the treaty in 1976. The rise in nuclear weapons-related security threats in East Asia—namely from China, and also regional threats like North Korea—as well as Japan's large plutonium stockpile and advanced defence and space industries has led some policymakers and analysts to consider the risk that Japan may have a future interest in nuclear weapons.[2] However, the Japanese government has always assured of its peaceful intentions and the country has never breached international safeguards.

While the country had a very limited nuclear weapons research effort during the Second World War, Japan's history with nuclear technology dates back to the 1960s, with the nuclear power industry taking off in the 1980s. Prior to the 2011 disaster—over the 45 years since Japan's first power reactor went live in 1966—Japan hosted over 50 operational nuclear reactors alongside many of the elements of a civil nuclear fuel cycle. The country has a small uranium enrichment capability and several fuel fabrication facilities that produce fuel for Japan's nuclear reactors.[3] At the back end of the fuel cycle, Japan also has reprocessing and waste storage facilities.[4] This complex and capable fuel cycle provides a wide range of sites, assets, and materials which require nuclear security.

While all sites hosting nuclear material require some level of protection, it is Japan's plutonium stockpile that arguably presents the greatest risk. A 2021 report by the Japanese government records the quantity of separated material owned by Japan as 46.1 tonnes, 8.9 of which are stored in Japan.[5] The country has a further 149 tonnes of plutonium contained in spent fuel at civil reactor sites, as well as a further 27 tonnes in spent fuel at reprocessing plants.[6] Prior to the 2011 disaster, Japan had 4.4 tonnes of separated unirradiated plutonium at its reprocessing facilities, alongside 4.9 tonnes mostly held in other forms elsewhere in Japan.[7] Reprocessed plutonium—while highly radioactive and likely 'self-protecting' against theft—poses one of the highest risks due to its utility in nuclear weapons. Much of the plutonium in Japan is located at the Rokkasho Reprocessing Plant site, but some, in the form of mixed-oxide (MOX) fuel, is located in at least four other reactor sites.

Japan has also hosted one of the largest stockpiles of HEU outside of the NPT *de jure* nuclear weapons states. HEU enriched to 93 per cent—an enrichment typically used in nuclear weapons—has been repatriated to the US. Japan worked with the US to convert the Kyoto research reactor from HEU to low-enriched uranium (LEU) fuel in 2010, with the HEU later removed to the US for downblending. Japan also removed HEU and plutonium from the Japan Atomic Energy Agency (JAEA) Fast Critical Assembly and Kyoto University Critical Assembly.[8] In March 2022 more than 30 kilograms (kg) of HEU were removed from three sites in Japan in a joint effort with the US.[9] According to international guidance, just 2 kg of unirradiated plutonium or 5 kg of HEU on a particular site merits the highest Category I security measures.[10]

Japan's nuclear security threat perceptions have developed as a result of both international and more local events. In 1995, the Aum Shinrikyo religious cult used homemade sarin—a chemical nerve agent—killing 13 and wounding many more in a terror attack on the Tokyo subway. Like many states around the world, Japan took new measures after the 9/11 attacks to strengthen its approach. For example, it was only after 9/11 that armed guards were deployed at Japanese nuclear facilities, albeit seemingly in limited numbers.[11] In 2005, Japan overhauled its legal structure for nuclear security to bring it in line with international standards—notably INFCIRC/225. This included the adoption and implementation of a 'design basis threat' (DBT) for the first time.[12] Other steps were taken in the latter half of the decade when Japan signed the 2005 International Convention for the Suppression of Acts of Nuclear Terrorism (ICSANT) and codified it into national law, and also revised its licensing system for radiological sources.

In 2010, less than a year before the Fukushima disaster, Japan participated in the 2010 Nuclear Security Summit in Washington, DC. At the summit, the Japanese national statement noted progress in the area of nuclear security since the 9/11 attacks. Around this time Japan also announced it would be establishing a nuclear security centre of excellence, the Integrated Support Center for Nuclear Nonproliferation and Nuclear Security (ISCN). ISCN would be established under the JAEA and would focus on strengthening nuclear security in East Asia and contributing to regional capacity building.[13]

Nuclear Security in the Wake of the Fukushima Daiichi Crisis

On 11 March 2011, at approximately 14:45, a 9.0-magnitude earthquake struck off the coast of Honshu, Japan.[14] The quake, one of the largest to strike Japan in decades, caused the seafloor to shift, triggering a tsunami that engulfed large swathes of Japan's northeastern coast. The shock caused by the combined earthquake and tsunami devastated Honshu, destroying towns and killing tens of thousands. Damage was discovered as far as 4 km from the shore.[15] The natural disaster also triggered the most significant nuclear disaster since the 1986 Chernobyl accident in the former Soviet Union.

Five nuclear power plants along the coast of Japan were directly impacted by the earthquake and tsunami: Higashidori, Onagawa, Fukushima Daiichi, Fukushima Daini, and Tokai Daini Nuclear Power Stations.[16] All but Higashidori had operating reactors at the time of the tsunami. All the facilities were impacted by the earthquake and tsunami in some way, but it was the Fukushima Daiichi plant that was damaged the most. The first wave of the tsunami struck Japan's shore 20 minutes after the earthquake began.[17] The first wave hit the Fukushima Daiichi Nuclear Power Station about 11 minutes later, with the second wave hitting Fukushima Daiichi about 10 minutes later.

There were six reactors at Fukushima Daiichi. On the day of the earthquake, Units 1, 2, and 3 were operating and Units 4, 5, and 6 were in outage for maintenance. Units 1, 2, and 3 automatically shut down when the earthquake occurred. The earthquake severed the plant's connection to off-site power and the backup diesel generators successfully kicked in to provide cooling for the reactor and power the site. The tsunami that struck Daichi was 13 m high, flooding the areas around Units 1–4 over 5 m. It destroyed a concrete seawall designed to protect the facility from a much smaller wave. Water rushed into the reactor and turbine buildings causing significant damage to equipment. The water destroyed the backup generators and most cooling systems, causing a station blackout.

In the 72 hours following the power loss, Units 1, 2, and 3 reactors melted down, releasing enormous amounts of radiation. Hydrogen explosions in Units 1, 3, and 4 caused significant damage to those buildings and chaos across the site. Off-site emergency responders were overwhelmed by the earthquake and tsunami. Transportation and communication to and from the site was extremely difficult as the tsunami had destroyed much of the infrastructure surrounding the site, and deposited debris across the terrain. On 12 March, the Japanese government imposed a 10-km evacuation zone around the site, eventually increasing it to 20 km.

The aftermath of the accident led to significant efforts to study it and extract lessons learned. This included an extensive report published in 2012 by the Fukushima Nuclear Accident Independent Investigation Commission of the Japanese National Diet (Japan's parliament), which built on hundreds of hours of testimony and interviews with over 1,000 witnesses. In the executive summary, the Chairman noted that the disaster was 'made in Japan', elaborating that 'its fundamental causes are to be found in the ingrained conventions of Japanese culture: our reflexive obedience; our reluctance to question authority; our devotion to "sticking with the program"; our groupism; and our insularity'.[18] The report also acknowledged that the accident was a 'manmade' disaster and 'the root causes were the organizational and regulatory systems that supported faulty rationales for decisions and actions'.[19]

Since no major nuclear security incident occurred, discussion of the accident has focused, for good reason, on nuclear safety, radiation release, and health physics. Yet, the Fukushima crisis can provide important lessons for nuclear security. It has also shaped nuclear security in Japan, in large part because the disaster has—and continues to—shaped the trajectory of the nuclear industry in the country. While there was widespread catastrophic destruction at the Fukushima Daiichi site and it seems unlikely an adversary would or could have attacked during that period, there have been numerous crises at nuclear facilities that resulted in significant disruption without meltdowns or explosions.[20] It is not difficult to imagine, in such scenarios, a malevolent actor taking advantage of the turmoil by carrying out an act of sabotage or theft.

A strong nuclear security system is one that protects against all plausible threats, both insiders and outsiders. It must remain resilient and adaptable even in the face of crises. This requires an organizational culture focused on security, where staff are constantly looking for ways to improve security and have a well-developed understanding of threats.[21] It also requires adequate training so that staff are prepared for a crisis, effective communication is in place between key facilities and staff, and that hardened security equipment and systems are capable of withstanding a major event.

Numerous accounts provide a vivid picture of the events that unfolded in the minutes, hours, and days following the earthquake and tsunami.[22] While many of these reports are dedicated to safety, they provide important insights about the state of the facility's security system and lessons for the future. Reports written about security during the crisis have largely focused on how to strengthen security at nuclear sites in case a terrorist seeks to trigger a Fukushima-type catastrophe.[23] Part of the increased concern about such a threat is that the Fukushima accident revealed vulnerabilities at nuclear reactors that could potentially be exploited by adversaries. Rather than addressing future scenarios, this analysis focuses on security during the crisis itself. Many of the lessons from the crisis related to safety also apply to security. This crisis demonstrated the inextricable relationship between safety and security—a theme frequently alluded to in discussion of nuclear security. The security-related impacts of the disaster are explored below in three categories: security systems, staffing and access control, and communications.

Security Systems

A 2014 US National Academies of Science report noted that Fukushima Daiichi was particularly vulnerable to major catastrophes. The facility was not equipped with 'survivable' on-site power and communications equipment, both of which would impact security operations in the event of an emergency.[24] Additionally, the facility's emergency procedures did not include scenarios where the facility would completely lose on-site power and operators had not been trained for such scenarios.[25] Partly because of these deficiencies, the 11 March 2011 earthquake and tsunami had an enormous impact on the plant's security system. It is unclear how much damage initially occurred as a result of the earthquake. Unfortunately, according to reports, 'plant workers were not able to complete an inspection of the plant for earthquake-related damage before the tsunami struck'.[26]

What is known is that there was significant damage from the quake itself before the tsunami hit. Operations buildings on the Daiichi site were functionally destroyed. The combination of physical damage to the facility, loss of electric power, and the evacuation of personnel after the tsunami, however, made it very likely that 'security systems at the plant were substantially degraded by the earthquake and tsunami and the subsequent accident'.[27] Lights, physical barriers, intrusion detection systems, assessment equipment, and communications were all impacted. All security equipment requiring electrical power was inoperable for more than a week.

Staffing and Access Control

The catastrophe also had a significant impact on staffing and access control. At the time of the earthquake, approximately 6,400 personnel were on site at Fukushima Daiichi.[28] Staff monitoring protected areas were moved to higher ground before the tsunami hit, and the entire security guidance team, which was responsible for access control, was temporarily evacuated in the days after the disaster.[29] Security patrols in some areas were reduced to avoid exposing guards to radiation. Most personnel were evacuated during the first days of the emergency. By 15 March 2011—just four days after the initial earthquake—only approximately 700 personnel remained at the site.[30] Those who stayed and continued to work for days and weeks throughout the catastrophe deserve to be commended for their efforts, putting their lives on the line, and many without knowing the fate of their families following the tsunami.

At the same time, additional staff began arriving at the site almost immediately after the crisis began. On 11 March, and for the next few days, about 400 additional personnel arrived each day to help restore power, cool the reactor, address issues related to radiation, and provide general support.[31] This influx of new personnel created security challenges, with limited screening as personnel accessed the site. Emergency workers entering Fukushima Daiichi were offered access passes without having to provide proper documentation like their address, phone numbers, and other data that would allow them to be later identified. This led to individuals being able to access the site in the early days of the crisis who were later never identified.[32] Indeed, at the time, background checks were not required for personnel at Japanese nuclear facilities. As will be discussed, this changed in the decade following the crisis.

Communications

Communications were challenging throughout the crisis—with implications for nuclear security as well as safety. Communication between reactor control rooms and the on-site emergency response centre (ERC) was limited. At the ERC, section chiefs, including those responsible for the guard force, provided guidance to those in the control room. The teams backing up the section chiefs, including those responsible for security, were in booths.[33] When a team received information that needed to be shared with the entire group, it would broadcast this throughout the ERC using a microphone. The system did not work well. The ERC did not transmit accurate information and instructions in a timely manner. There was tremendous confusion about roles and responsibilities within the ERC. Staffing was not adequate for addressing the 'scope and long duration' of the crisis.[34]

There was also an off-site ERC at Fukushima Daiichi—a separate location kept ready for contingencies—but the damage from the tsunami and earthquake rendered it inaccessible and mostly inoperable.[35] All of its telecommunications equipment was damaged except for that enabling a satellite connection. Those who were able to access the ERC needed to be evacuated on 14 March 2011 because of concerns about radiation. As a result of the poor communications, staff at the headquarters of Tokyo Electric Power Company (TEPCO)—the Fukushima Daiichi operator—in Tokyo and within government facilities took on some of the roles usually undertaken by those based in the ERC, leading to further confusion.[36]

Communication at TEPCO headquarters was equally chaotic and problematic. TEPCO's chairman and president were unable to get to the headquarters for more than 17 hours after the crisis began and after the meltdowns occurred.[37] Communication from the prime minister's office was equally chaotic. The prime minister was not fully aware of the situation at TEPCO, and apparently lacked confidence in the organization's ability to manage the accident.[38] Additionally, the prime minister's emergency operations centre did not have adequate communications arrangements, forcing the prime minister to switch locations to his office, which also was not equipped with necessary communications equipment, requiring his staff to run information up and downstairs.

Other nuclear sites not as hard hit by the earthquake and tsunami managed to respond to the emergency more effectively than Fukushima Daiichi. Despite suffering damage and facing significant challenges, the Fukushima Daini and Onagawa Nuclear Power Stations largely avoided catastrophe. At Daini, site superintendent Naohiro Masuda was celebrated after the crisis for his part in saving the facility. According to one account, Masuda concentrated on persuading his team to act despite the chaos around them: 'with each problem they encountered, they recalibrated, iteratively creating continuity and restoring order … they acted their way into a better understanding of the challenges they faced'.[39] While Onagawa's superintendent was not on site when the earthquake occurred, an acting superintendent took charge who was well prepared because of the frequent emergency response exercises that were regularly conducted at the site.[40]

Effectively establishing and maintaining communications was also a critical element of the response at both facilities. After the initial tsunami warning period was over, patrols were sent to assess damage at Onagawa while also securing lines of communication in anticipation of the greater crisis.[41] The results of the patrols were consolidated by the main control room and transmitted to the ERC that had been established in a new office building designed to withstand earthquakes. The responses at different sites clearly helped to mitigate the effects of the disaster, with important implications for maintaining nuclear security.

Natural disasters, and resulting safety incidents, can significantly disrupt and degrade nuclear security systems. Fukushima—one of the most serious nuclear disasters of the atomic age—showcases such impacts on security systems and provides insights into the types of challenges that mounting an emergency response can pose for security. However, it is important to recognize the severity of the earthquake and tsunami. The earthquake was among the strongest in recorded history, and the strongest ever recorded in Japan. It is perhaps unrealistic to expect security equipment, systems, and processes to fully withstand the effects of a disaster of such severity. Most crises will not be as severe, but security will still need to be maintained. The case of the Fukushima disaster shows the importance of planning ahead for nuclear security in times of crisis.

Nuclear Security in Japan Since the 2011 Disaster

The nuclear catastrophe that occurred on 11 March 2011 at the Fukushima Daiichi nuclear plant had an immediate impact on site security, but it was also the catalyst for a complete re-evaluation of nuclear security oversight and implementation in Japan. According to a report published by Japan's Advisory Committee on Nuclear Security after the crisis, 'The accident revealed the possibility that terrorism at a nuclear facility may have the same serious effects on society.'[42] Based on what occurred at Fukushima, the Committee made other important recommendations. The report recommended improving the durability of and 'perfectly' maintaining protection equipment; developing plans for the availability of extra personnel and equipment in the event of a terrorist attack; security training and assessment; and protecting against insider threats.

The Fukushima disaster formed the backdrop to the 2012 Nuclear Security Summit in Seoul, attended by then-prime minister Noda Yoshihiko. Noda spoke of progress in responding to the disaster, and announced a number of security-related measures, including an increase of redundancy in site security protection systems, drills to enhance coordination of response, an increase in armed security personnel and patrolling, and a strengthening of computer security.[43] Noda also noted that 'The cause of the accident at TEPCO's Fukushima Daiichi Nuclear Power Station may have been a natural disaster, but there are lessons learned from that experience that can be applied to countermeasures for human threats such as terrorist attacks against nuclear facilities.'[44] He highlighted three lessons for nuclear security: the importance of preparing for unanticipated risks; the necessity of having a common response plan; and the need to keep in mind worst-case scenarios.

One of the key findings from investigations into the Fukushima disaster was that Japan's lack of an independent nuclear regulatory body was a major factor in Daiichi's lack of preparation for the accident. As the Diet report noted, 'The Commission has concluded that the safety of nuclear energy in Japan and the public cannot be assured unless the regulators go through an essential transformation process ... Japan's regulators need to shed the insular attitude of ignoring international safety standards and transform themselves into a globally trusted entity.'[45] In addition to highlighting the issue of 'regulatory capture', the report emphasized the lack of independence, expertise, and transparency.

Just over a year after the accident, in September 2012, Japan founded the Nuclear Regulatory Authority (NRA). The NRA provides independent oversight of both nuclear safety and security in Japan. Since the NRA was established, Japan has made significant progress in addressing some of the security flaws revealed during the Fukushima disaster, including conducting force-on-force exercises, enhancing cyber security, and upgrading equipment to function better during severe conditions.[46]

Since the accident, Japan has developed new rules to protect against insider threats. In 2014, the NRA established a working group to consider how to determine trustworthiness of nuclear facility staff. In 2016, new regulations were adopted that require private operators of nuclear power facilities to apply new background checks before employees were able to access weapons-useable nuclear material or sensitive information.[47] The checks, however, are carried out by individual companies with little experience in this area, not by the government. Additionally, employees are not actually required to submit information for the checks.[48] Strikingly, much more extensive background checks are in place for gun ownership in Japan.[49] Japan's new regulations also require a two-person rule for those handling nuclear materials and in sensitive areas, and the installation of security cameras in sensitive areas of facilities.

In addition, the NRA has made strengthening nuclear security culture a priority in the decade since Fukushima. Nuclear security culture and the importance of the human factor have increasingly emerged as key issues in international nuclear security over the past decade. Japanese regulations require operators to establish and implement programmes dedicated to strengthening security culture within nuclear organizations. In January 2015, the NRA developed a Code of Conduct on Nuclear Security Culture for NRA staff. It emphasized, among other things, awareness of threats, senior management involvement, education, and self-improvement. As one initiative, NRA commissioners themselves walk through facilities with senior plant executives to discuss opportunities for strengthening security culture. The NRA has also distributed nuclear security culture self-assessment questionnaires to all operators, who then develop their own self-evaluation procedures.[50]

These recent developments in nuclear security in Japan reflect developments in many countries during the post-9/11 period, and with heightened attention paid to nuclear security as a result of the Nuclear Security Summit process. In Japan, however, developments have also been heavily shaped by the Fukushima disaster and its aftermath. Perhaps the most consequential impact of Fukushima for nuclear security in the country is the establishment of the NRA, and the implications of this new body that regulates security as well as safety.

Ongoing and Future Nuclear Security Challenges

Two major factors affect the future of nuclear security in Japan. The first is the future of nuclear power in Japan, and the second relates to the country's nuclear material stockpiles. While Japan's nuclear security arrangements have been significantly strengthened since the Fukushima crisis, most of its nuclear power plants are still not operational. Only 10 of the country's 33 operable nuclear power reactors have been approved to restart since the Fukushima accident.[51] Japan is under pressure to restart nuclear power plants in order to meet its planned reduction in carbon dioxide (CO_2) emissions, which had been a political aim long before 2011 and has remained a priority as climate change becomes more evident. Japan also faces pressure to begin removing the growing stocks of spent fuel at reactor sites around the country.

Several of these delays to plant restarts have been related to nuclear security issues. On 14 April 2021, the NRA issued an order that effectively prevented the Kashiwazaki-Kariwa Nuclear Power Station from restarting due to serious nuclear security issues at the site. The NRA discovered malfunctioning of the intruder detection equipment, violations of security protocols by the company's employees, and negligence related to construction.[52] In 2020, several Japanese utilities shut down their reactors after they had failed to complete construction of anti-terrorism facilities at their plants.[53]

Japan also has ongoing challenges related to managing its nuclear material stocks. Every facility that eliminates its weapons-useable nuclear materials is one less target that needs to be protected against theft or sabotage, and therefore material minimization is a key tenet of nuclear security. While Japan eliminated some nuclear material when it shipped 331 kg of plutonium and 214.5 kg of weapons-grade HEU stored at the Fast Critical Assembly at Tokai-mura to the US in 2016, and a further 30 kg in 2022, large quantities, as already noted, remain in the country.[54]

Japan's strategy for addressing its spent fuel—separating out the plutonium at its reprocessing plant in Rokkasho—has faced continuous setbacks and cost overruns.[55] The project has reportedly been pushed back 24 times, with the reprocessing facility originally intended to open in 1997. While Japan Nuclear Fuel Limited's current plan is to open the facility in early 2022, Rokkasho's MOX facility that would fabricate the separated plutonium into fuel is not scheduled to open until 2024.[56] Japan's stated policy is to reduce its plutonium stocks. This timeline, however, raises serious questions about the extent to which Japan is committed to this goal.[57] Moreover, it raises the possibility that, at least in the short term, Japan could increase its stocks of weapons-useable plutonium. This could heighten risks and burdens for nuclear security, or at least not reduce them as Japan moves in the opposite direction to minimization.

Beyond these core factors shaping future nuclear security, Japan—like most countries around the world—is having to deal with the challenges and opportunities posed by emerging technologies. This wide-ranging and fairly ill-defined set of technologies includes those as diverse as unmanned aerial systems (UASs, or drones), artificial intelligence, 3D printing, and the 'internet of things'. Some of these technologies could be harnessed by malicious actors to pose threats to nuclear facilities in new and innovative ways. They could also be harnessed by security teams and guard forces for the purpose of enhancing nuclear security.

Drones represent a good example. In 2015 a Japanese citizen was arrested after landing a drone containing a small amount of mildly radioactive sand on the roof of the Japanese prime minister's office in order to protest energy policy.[58] This and other incidents around the world have led the Japanese authorities to take the threat posed by UASs to nuclear facilities seriously. Two years later, likely in part as a result of the 2015 incident, the Ikata Nuclear Power Plant in Ehime Prefecture conducted its first counter-terrorism drill to simulate a UAV attack on a nuclear facility.[59] Japan, as a high-tech economy globally known for its innovation, is also well placed to benefit from emerging technologies in enhancing nuclear security.

Furthermore, the shocking assassination of former prime minister Shinzo Abe in July 2022, using a homemade firearm, highlights new concerns over the potential misuse of 3D printing technology in a country that has very strict gun control regulations.[60] While there is no public evidence that 3D printed parts were used in the weapon, such technology could be appealing to a malicious actor with limited options. The medium- and long-term implications of the murder of one of Japan's most well-known political figures—for nuclear security and more broadly—are yet to be seen.

Conclusion

The March 2011 earthquake and tsunami in Japan, and the nuclear disaster at Fukushima Daiichi, have dominated Japan's experience with nuclear power over the past decade. This chapter has considered nuclear security in Japan before and after this safety incident. It explored the evolution of Japanese nuclear security, considering key issues in Japanese nuclear security since the disaster. Rather than focusing on the potential for a non-state actor to maliciously cause a Fukushima-style accident, the chapter has focused on maintaining nuclear security during times of crisis—particularly with regard to security systems, staffing and access control, and communications. The chapter has argued that Japan made significant progress in the area of nuclear security since the 9/11 attacks and especially since the 2011 Fukushima disaster. At the time of writing, however, the future of nuclear security—like the future of nuclear power—in Japan remains unclear. This is largely a result of the disaster a decade ago.

While Fukushima was a safety, rather than security, event, the disaster and crisis heightened interest in the safe, secure, and effective management of nuclear assets in Japan, and led to significant developments in Japan's regulation of nuclear technology with knock-on effects for security. Notably, the conclusions of various inquiries into the disaster highlighted the role of 'regulatory capture', as well as a lack of independence, expertise, and transparency in the regulator. The creation of the NRA in 2012 had significant implications for nuclear security as well as preventing safety lapses which enabled the Fukushima accident. The accident has also framed the discussion and development of nuclear security more generally in Japan.

Fukushima Daiichi was not prepared for the crisis it faced, but it would likely not be alone in being unprepared for such a significant disaster. Natural disasters—whether the 2011 Japanese earthquake and tsunami, the 2000 Cerro Grande fire surrounding Los Alamos National Laboratory in the US, or the 2020 wildfires around the Chernobyl site in Ukraine—clearly have the potential to create serious problems for nuclear security. Learning how to maintain security in times of crisis is of great importance and will become more so over the coming decades as the impacts of climate change magnify.[61] In this respect, Fukushima holds wide-ranging lessons for nuclear security as well as nuclear safety.

Notes

1. For a summary, see World Nuclear Association, 'Nuclear Power in Japan', WNA [online], September 2021. https://world-nuclear.org/focus/fukushima-daiichi-accident/japan-nuclear-power.aspx
2. See for example, Mark Fitzpatrick, 'How Japan Could Go Nuclear', *Foreign Affairs* [online], 3 October 2019. https://www.foreignaffairs.com/articles/asia/2019-10-03/how-japan-could-go-nuclear
3. World Nuclear Association, 'Nuclear Power in Japan'.
4. Ibid.
5. IAEA, 'Communication Received from Japan Concerning its Policies Regarding the Management of Plutonium', INFCIRC/549/Add.1/24, 16 August 2021. The other 37.2 tonnes are currently stored abroad, including around 21 tonnes in the UK and 15 tonnes in France.
6. Ibid.
7. Around 35 tonnes were held outside of Japan in the UK and France. See International Panel on Fissile Materials, 'Japan's 2011 Civilian Plutonium Declaration', *IPFM Blog* [online], 3 October 2012. https://fissilematerials.org/blog/2012/10/japans_2011_civilian_plut.html
8. Office of the Press Secretary, 'FACT SHEET: Removal of All Highly Enriched Uranium and Plutonium from Japan's Fast Critical Assembly', The White House [online], 1 April 2016. https://obamawhitehouse.archives.gov/the-press-office/2016/04/01/fact-sheet-removal-all-highly-enriched-uranium-and-plutonium-japans-fast
9. National Nuclear Security Administration, 'US Removes over 30 kilograms of Highly Enriched Uranium from Japan', US Department of Energy [online], 23 May 2022. https://www.energy.gov/nnsa/articles/us-removes-over-30-kilograms-highly-enriched-uranium-japan
10. HEU is defined as U-235 enriched to 20 per cent or higher, according to INFCIRC225/Rev.5 and the 2005 Amendment to the Convention on the Physical Protection of Nuclear Material (A/CPPNM).
11. Martin Fackler, 'Japan's Nuclear Sites Raised U.S. Concerns, Cables Show', *New York Times* [online], 8 May 2011. https://www.nytimes.com/2011/05/09/us/politics/09tokyo.html

12. Ministry of Foreign Affairs of Japan, 'Japan's National Statement at the Washington Nuclear Security Summit', 2010 Nuclear Security Summit, Washington, DC, 12 April 2010. https://www.mofa.go.jp/policy/un/disarmament/arms/nuclear_security/2010/national_statement.html
13. Ibid.
14. For an excellent overview of the disaster, see Scott Sagan and Edward Blandford, eds., *Learning from a Disaster: Improving Nuclear Safety and Security after Fukushima* (Stanford: Stanford University Press, 2016).
15. International Tsunami Information Center, '5th Anniversary, 11 March 2016, Great East Japan Earthquake and Tsunami', ITIC [online], 11 March 2011. http://itic.ioc-unesco.org/index.php?option=com_content&view=article&id=2081&Itemid=2982
16. Account of the disaster taken from National Diet of Japan, 2012, *The Official Report of the Fukushima Nuclear Accident Independent Investigation Commission*. https://warp.da.ndl.go.jp/info:ndljp/pid/3856371/naiic.go.jp/wp-content/uploads/2012/09/NAIIC_report_lo_res10.pdf
17. For a detailed timeline, see Ibid., 24.
18. National Diet of Japan, *The Official Report of the Fukushima Nuclear Accident Independent Investigation Commission*, Executive Summary, 9.
19. Ibid., 16.
20. See Geoffrey Chapman, Rebecca Earnhardt, Christopher Hobbs, Nickolas Roth, Daniel Salisbury, Amelie Stoetzel, and Sarah Tzinieris, *Nuclear Security in Times of Crisis*, Occasional Paper (London: Centre for Science and Security Studies/Stimson Center, 2021). https://www.stimson.org/wp-content/uploads/2021/07/nuclear-security-in-times-of-crisis-handbook.pdf
21. For more on developing a strong nuclear security culture, see Matthew Bunn, 'Incentives for Nuclear Security', in Proceedings of the Institute for Nuclear Materials Management 46th Annual Meeting, July 16–20, 2006, Nashville, TN. https://scholar.harvard.edu/matthew_bunn/publications/incentives-nuclear-security
22. Beyond the Diet's official study, others were undertaken, including by the IAEA. See IAEA Director General, *The Fukushima Daiichi Accident: Report by the Director General*, 2012; Other valuable studies, including those from Japan's Advisory Committee on Nuclear Safety and the US National Academy of Sciences' Committee on Lessons Learned, are cited in subsequent references.
23. For example, see Graham Allison and William H. Tobey, 'Could There Be a Terrorist Fukushima?', *The New York Times*, 4 April 2016.
24. Committee on Lessons Learned, *Lessons Learned from the Fukushima Nuclear Accident for Improving Safety and Security of U.S. Nuclear Plants* (Washington, DC: The National Academies Press, 2014), 109. https://doi.org/10.17226/18294
25. Ibid.
26. Ibid., 86.
27. Ibid., 76.
28. Ibid., 108.
29. Ibid.
30. Ibid.
31. Ibid.
32. Ibid.
33. Ibid., 322.

34. Ibid., 6.
35. Ibid., 111.
36. Ibid.
37. National Diet of Japan, *The Official Report of the Fukushima Nuclear Accident Independent Investigation Commission*, Ch. 3, 2.
38. Committee on Lessons Learned 2014. *Lessons Learned from the Fukushima Nuclear Accident for Improving Safety and Security of U.S. Nuclear Plants*, 129.
39. Ranjay Gulati, Charles Casto, and Charlotte Krontiris, 'How the Other Fukushima Plant Survived', *Harvard Business Review* [online], July–August 2014. https://hbr.org/2014/07/how-the-other-fukushima-plant-survived
40. Japan Nuclear Safety Institute, 'Responses to the Tohoku-Pacific Ocean Earthquake and Tsunami at the Onagawa Nuclear Power Station and Tokai No.2 Power Station', August 2013. http://www.genanshin.jp/english/archive/disastersitereaction/data/report_OnTk.pdf
41. Ibid., 20.
42. Advisory Committee on Nuclear Security, 'Strengthening of Japan's Nuclear Security Measures', Japan Atomic Energy Commission, 9 March 2012, 22. http://www.aec.go.jp/jicst/NC/senmon/bougo/kettei120309.pdf
43. Ministry of Foreign Affairs of Japan, 'Attendance by Prime Minister Noda at the Seoul Nuclear Security Summit (Outline)', MOFA Japan [online], 27 March 2012. https://www.mofa.go.jp/policy/un/disarmament/arms/nuclear_security/2012/outline.html
44. Ibid.
45. National Diet of Japan, The Official Report of the Fukushima Nuclear Accident Independent Investigation Commission, Executive Summary, 20.
46. See Sara Z. Kutchesfahani, Kelsey Davenport, and Erin Connolly, *The Nuclear Security Summits: An Overview of State Actions to Curb Nuclear Terrorism 2010–2016* (Washington, DC: Arms Control Association and Fissile Materials Working Group, 2018), 53. https://www.armscontrol.org/sites/default/files/files/Reports/NSS_Report2018_digital.pdf
47. See Yuzo Yamaguchi, 'Japan's NRA to Introduce Nuclear Security Regulations in August', *Platts Inside NRC* [online], 25 July 2016; Masakatsu Ota, 'Potential Insider Threat Against Japanese Nuclear Facilities: Challenge and Progress Since the Fukushima Nuclear Disaster', *NAPSNet Special Reports* [online], 16 November 2017.
48. See Yamaguchi, 'Japan's NRA to Introduce Nuclear Security Regulations in August'; Editorial, 'State Secrets Law Still Deeply Flawed', *The Japan Times*, 8 December 2015.
49. In order to buy a gun in Japan, a person needs to pass a written test, take classes, and pass a psychological test at a hospital. The police conduct a background check, interview relatives, and check to see if the person is a member of certain political or activist groups. Gun owners are required to keep their gun in a locker and provide a map to the police showing its location. The police conduct an annual gun inspection and licences must be renewed every three years. See David Kopel, 'Japanese Gun Control', *Asia-Pacific Law Review*, 2/2 (1993), 26–52.
50. Nobuaki Eguchi, 'Efforts to Enhance Nuclear Security Culture in Japan', Presentation to the Second International Regulators Conference on Nuclear Security, 11–13 May 2016, Madrid, Spain; Satoru Tanaka, 'A Nuclear Security Regime in Japan: Enhancement Efforts and Global Contributions', Presentation to the IAEA International Conference on Nuclear Security: Commitments and Actions, 5–9 December 2016, Vienna, Austria.
51. Stephen Stapczynski, 'Japan Restarts First Nuclear Reactor Since 2018 Amid Hurdles', *Bloomberg* [online], 23 June 2021. https://www.bloomberg.com/news/articles/2021-06-23/japan-restarts-first-nuclear-reactor-since-2018-amid-hurdles

52. Jiji Press, 'Japan's Nuclear Regulator Inspects Tepco's Kashiwazaki-Kariwa N-Plant', *Nippon.com* [online], 26 October 2021. https://www.nippon.com/en/news/yjj2021102600188/
53. Nuclear Engineering International, 'Only One Power Reactor Remains in Operation in Japan', *NEI Magazine* [online], 10 November 2020. https://www.neimagazine.com/news/newsonly-one-power-reactor-remains-in-operation-in-japan-8354484
54. For more on the 2016 removal, see Nickolas Roth, 'Japanese Nuclear Material Arrives in the United States', *Nuclear Security Matters* [blog], 25 August 2016. https://www.belfercenter.org/publication/japanese-nuclear-material-arrives-united-states
55. For an extensive analysis of Japan's plutonium management challenges, see Alan J. Kuperman and Hina Acharya, 'Japan's Misguided Plutonium Policy', *Arms Control Today* [online], October 2018. https://www.armscontrol.org/act/2018-10/features/japan%E2%80%99s-misguided-plutonium-policy; James M. Acton, 'Wagging the Plutonium Dog: Japanese Domestic Politics and Its International Security Implications: Brief', Carnegie Endowment for International Peace [online], 29 September 2015. https://carnegieendowment.org/2015/09/29/wagging-plutonium-dog-japanese-domestic-politics-and-its-international-security-implications-pub-61427
56. World Nuclear News, 'Further Two-Year Delay to Japanese MOX Plant', *WNN* [online], 17 December 2020. https://world-nuclear-news.org/Articles/Further-two-year-delay-to-Japanese-MOX-plant
57. Tatsujiro Suzuki and Masa Takubo, 'Japan's New Policy on Its Plutonium Stockpile', International Panel on Fissile Materials, 20 August 2018. https://fissilematerials.org/blog/2018/08/japans_new_policy_on_its_.html
58. BBC, 'Japan Radioactive Drone: Tokyo Police Arrest Man', *BBC News* [online], 25 April 2015. https://www.bbc.co.uk/news/world-asia-32465624
59. Kyodo, 'First Counterterror Drill for Drone Attack Held at Nuclear Plant in Ehime', *Japan Times* [online], 19 June 2017. https://www.japantimes.co.jp/news/2017/06/19/national/first-counterterror-drill-drone-attack-held-nuclear-plant-ehime
60. Giulia Carbonaro, 'Homemade Gun Used in Abe Shooting Fuels Concerns over DIY Weapons and 3D-Printed "Ghost Guns"', *Euronews* [online], 11 July 2022. https://www.euronews.com/next/2022/07/11/what-are-3d-printed-ghost-guns-and-are-they-a-threat-in-europe-too
61. Christopher Hobbs, Nickolas Roth, and Daniel Salisbury, 'Security Under Strain? Protecting Nuclear Materials During the Coronavirus Pandemic', *RUSI Journal*, 166/2 (2021), 40–50.

Suggested Reading

Advisory Committee on Nuclear Security, (2012), *Strengthening of Japan's Nuclear Security Measures* (Tokyo: Japan Atomic Energy Commission).

Chapman, Geoffrey, Earnhardt, Rebecca, Hobbs, Christopher, Roth, Nickolas, Salisbury, Daniel, Stoetzel, Amelie, and Tzinieris, Sarah, (2021), *Nuclear Security in Times of Crisis*, Occasional Paper (London: Centre for Science and Security Studies/Stimson Center).

Gulati, Ranjay, Casto, Charles, and Krontiris, Charlotte, 'How the Other Fukushima Plant Survived', *Harvard Business Review* [online], July–August 2014. https://hbr.org/2014/07/how-the-other-fukushima-plant-survived

Hobbs, Christopher, Roth, Nickolas, and Salisbury, Daniel, (2021), 'Security Under Strain? Protecting Nuclear Materials During the Coronavirus Pandemic', *RUSI Journal*, 166/2, 40–50.

Committee on Lessons Learned, (2016), *Lessons Learned from the Fukushima Nuclear Accident for Improving Safety and Security of U.S. Nuclear Plants: Phase 2* (Washington, DC: The National Academies Press).

Sagan, Scott and Blandford, Edward, eds., (2016), *Learning from a Disaster: Improving Nuclear Safety and Security after Fukushima* (Stanford: Stanford University Press).

CHAPTER 41

NUCLEAR SECURITY DURING THE COVID-19 PANDEMIC

CHRISTOPHER HOBBS AND AMELIE STOETZEL

Introduction

The Covid-19 pandemic has had a considerable impact on a wide range of industry sectors, serving to disrupt and complicate operations and reduce financial turnover. Most affected have been 'high-contact' industries, such as wholesale and retail, transportation, accommodation, and entertainment, which are heavily dependent on physical interaction.[1] However, 'low-contact' industries, such as manufacturing, agriculture, construction, and the energy sector, have been still significantly impacted. As such, the emergence of Covid-19 in late 2019 presented a crisis for both governments and industry due to a combination of the largely unforeseen nature of a global pandemic, the fundamental challenges it posed to normal ways of working, the threat it posed to fundamental values of entire industries, and the rapid speed at which solutions had to be developed, based on what was initially a very incomplete understanding of the virus and its effects.[2]

This chapter explores how the pandemic affected the global nuclear sector, with a focus on the delivery of security. It starts by considering the pandemic's initial impact, the major challenges encountered, and how pre-existing emergency preparedness plans and systems shaped crisis responses. Attention then turns to the security vulnerabilities created because of the pandemic and how adversaries have sought to exploit these. For example, it will be demonstrated how the need for remote online-based work has opened new windows of opportunity for those seeking to exploit potential cyber vulnerabilities at nuclear facilities. Building on this, the chapter discusses the continuing impact of Covid-19, and whether newly advanced security solutions, developed in response to the pandemic, are likely to be adopted in the longer term. In exploring these issues, the chapter examines changes that have taken place at the international, national, and organization levels and the links between them. Here it should be emphasized that there is no single approach to achieving effective nuclear security, although there exist commonalities in how adversaries have sought to take advantage of the pandemic and the measures developed to mitigate them.

The Initial Response: Unpreparedness, Supply Chain Challenges, and Activity Slowdown

Many governments conduct regular risk assessments aimed at identifying and classifying future threats, informing the development of emergency preparedness plans and broader security and resilience capabilities. Typically, these encompass a wide range of risks, including, for example, environmental, health, societal, accidents, and malicious attacks, which may emanate both domestically and from abroad. It has also been relatively common in recent decades for countries to experience sudden outbreaks of infectious disease, for example, SARS in China (2002–2004), Swine Flu in Mexico and Europe (2009–2010), and Ebola in West Africa (2013–2016).[3] Consequently, the possibility of a global pandemic and the resultant crisis potential was both well known within the international scientific community and featured prominently in many governments' risk assessments. As an example, in the National Risk Register of the United Kingdom (UK), an influenza pandemic has been the number one non-malicious risk since 2008 and was assessed in 2017 to have a significant likelihood of occurring within the next five years.[4] However, despite pandemics being at the forefront of national-level risk discussions, the emergence of Covid-19 in late 2019 found many governments and organizations initially ill-prepared, including in terms of medical equipment and supply chain management.

In the UK, preparedness efforts had focused largely on two scenarios: the emergence of an influenza pandemic; and a very infectious disease with a high fatality rate, with similar characteristics to Ebola. According to the UK's National Audit Office, this narrow focus meant that the development of pandemic plans, and the resultant emergency planning and preparedness training, had not been conducted for a disease with a relatively low mortality and asymptomatic transmission, such as Covid-19.[5] Consequently, the government initially struggled to respond to the wide-ranging 'whole-system' impact that the pandemic had on society, the economy, and public services.[6] In the UK nuclear sector, it is common for organizations to maintain their own risk registers, which, akin to those at the national level, aim to capture key future threats and associated mitigation measures. However, despite its presence at the top of the 2017 national risk register, a pandemic did not feature within UK nuclear organizations' top 10 threats.[7] Consequently, while some high-level nuclear sector plans existed, these were both focused on influenza scenarios and lacked operational level detail.

Despite the disconnect between the threat posed by Covid-19 and the focus of national and organizational risk registers and their associated emergency preparedness plans, existing risk management systems nevertheless had utility in shaping the response to Covid-19. For example, the United States (US) drew on expertise, routines, and medical equipment stockpiles created as a result of its Pandemic and All-Hazards Preparedness and Advancing Innovation Act,[8] revised in 2019, which laid out preparations necessary to mitigate threats of pandemics, such as the Strategic National Stockpiles and Hospital Preparedness Program.[9] While a 2022 government assessment shows that the implementation of most requirements outlined in the Act have been limited,[10] the US Nuclear Regulatory Commission (NRC) was able to draw upon and adapt components of existing influenza pandemic preparedness plans, when considering, for example, how to mitigate the potential of large-scale absenteeism as a result of illness at nuclear facilities.[11] US nuclear power plant operators adapted their security trainings, after approval by the NRC, in accordance with steps outlined in a pandemic licensing plan from 2007.[12] Similarly in the UK, existing pandemic plans developed by nuclear operators, although imperfect and formulated at a high level, served as a useful framework within which a more detailed response could be developed.[13]

To combat the lack of information that is typical at the onset of a crisis such as the Covid-19 pandemic, countries made use, where available, of pre-existing networks and mechanisms for information sharing between key stakeholders. In the UK, the government was able to rapidly expand its information gathering for the nuclear industry, collating and updating on a weekly basis data of relevance to operations, security, and safety. This data collection effort enabled the identification of potentially concerning trends, feeding into a national approach where resources could be pooled and focused if necessary. For example, if high Covid-19 infection rates in certain regions led to outbreaks among the security and guarding forces at a particular nuclear facility, personnel stationed at other facilities could—in theory—be brought in to avoid any potential degradation of security.[14]

Thankfully, reported Covid-19 infections have been relatively low among nuclear workers, because of steps taken by governments and industry to reduce the risk of staff exposure. In Russia, for example, key nuclear power plant workers were isolated from the general public, and provided with separate on-site accommodation, meals, and healthcare.[15] Although outbreaks of Covid-19 have nevertheless been reported at certain nuclear sites, staff reductions have largely been caused by close contact isolation requirements.[16] There have also been cases of staff absences as a result of dissatisfaction with the personal protection equipment (PPE) provided against Covid-19. For example, France reported walk-outs of more than a dozen employees at three nuclear power plants (Fessenheim, Civaux, and Chooz).[17] At its highest point, the UK nuclear sector reported a total of 18–20 per cent absence of staff during the onset of the country's second Covid-19 wave in September 2020.[18] While these numbers might seem concerning, EDF Energy, a subsidiary of Électricité de France that operates eight nuclear power plants in the UK, estimated that it could maintain operations for up to three months with only 75 per cent staffing levels and for three weeks with only 60 per cent of its usual staff.[19]

Despite relatively high levels of resilience to low staffing levels in the nuclear sector of most countries, Covid-19 still posed significant challenges for particular parts of the nuclear sector. One of the most visible impacts of the pandemic was the unfolding of supply chain disruptions caused by the re-introduction of border checkpoints in economic and trading blocs such as the European Union (EU), resulting in long queues,[20] and the cancellation of flights due to the quarantine obligations of crew and passengers.[21] These transport and supply chain issues also had dramatic effects on the availability of radioisotopes, with serious disruption in the medical nuclear sector. This was a particular issue for the supply of molybdenum-99 (Mo-99), which is used to diagnose heart disease and treat cancer and must be regularly restocked because of its short half-life.[22] Since Mo-99 is commonly transported by air, disruption in the aviation industry made obtaining these isotopes difficult. For example, in March 2022 the South African radioisotope producer NTP Radioisotopes, a major global supplier of Mo-99, reported a loss of goods worth US$912,000 due to delays and re-routing of flights.[23] This led to the delay and cancellation of nuclear medical procedures, which according to the International Atomic Energy Agency (IAEA) were down globally 73.3 per cent in June 2020 and 56.9 per cent in October 2020, in comparison to pre-pandemic levels, with the biggest impact seen in low- and middle-income countries.[24]

In contrast, the nuclear power sector fared better, with operators in some cases choosing to lessen the impact of supply chain interruptions by delaying refuelling and maintenance outages. As a result, 'no country has reported the enforced shutdown of a nuclear power reactor due to the effects of Covid-19 on the workforce or supply chains', according to the IAEA.[25] While nuclear facility shutdowns contributed to the overall global nuclear energy output reduction of 4 per cent from 2019 to 2020, the lower energy output was also a result of reduced energy demand. By 2021, nuclear energy output had bounced back to pre-pandemic levels.[26]

Supply chain problems and centralized production also caused shortages in Covid-19-related PPE, such as masks, gloves, and gowns. In some cases, this generated disgruntlement within nuclear workforces and delayed certain activities. For example, at the Hanford nuclear decommissioning site in the US, operations were largely suspended in March 2020 for over a month following the onset of Covid-19.[27] When these resumed, an inability to social distance while working and a lack of PPE meant that employees were reportedly terrified about returning to work.[28] Similar stories can be found at other nuclear facilities in the US, such as the Millstone Nuclear Power Plant, where it was reported in May 2022, that staff had to 'fight' for PPE, generating in the words of one security officer, 'a lot of frustration, there's a lot of concern, and I would say there's anger'.[29]

It was not only the global supply of goods that was impacted by Covid-19 but also services and human capital. In the nuclear sector this was particularly acutely felt when it came to nuclear new-build efforts, with several projects largely reliant on foreign contractors. For example, the IAEA reported in June 2020 that Covid-19 had served to delay construction at sites in Belarus, the United Arab Emirates, Turkey, and Bangladesh.[30] In May 2022, the managing director of Hinkley Point C in the UK claimed that the project had 'lost in excess of half a million individual days of critical work in

2020 and 2021' as a result of Covid-19, pushing back the planned start date of the nuclear power station to June 2027, with the delay causing an extra cost of GB£3 billion.[31]

Evolving Threats and Vulnerabilities

The pandemic precipitated a significant move to remote working across the world, with organizations largely embracing this change, including in the nuclear sector. While necessary to keep operations running, this created other risks, most notably an increased vulnerability to cyber-attacks, particularly during its early stages when organizations rushed to link remote workers to the office, with many taking a 'connectivity first' approach.[32] According to Gallup, a global analytics company in the US, the percentage of full-time employees working from home jumped from 33 per cent to 61 per cent in just two weeks, from the middle to the end of March 2020.[33] This rapid transition meant that, for many organizations, there was simply not time to adapt cyber security policies, adopt secure collaboration tools, and upskill staff on cyber risks, such as the use of personal devices to connect to work-related online services. Recognizing this vulnerability, cyber criminals and other actors sought to target organizations through a number of scams and phishing attempts, with the creation of fake websites and weblinks pertaining to pandemic-related information, embedded in phishing emails and more than 2,000 malicious domains by March 2020.[34] The target of attacks was also redirected 'from individuals and small businesses to major corporations, governments and critical infrastructure'.[35] Multinational corporation IBM reported 20 times more cyber-attacks on critical national infrastructure in the first year of the pandemic compared to previous levels.[36]

Despite the sudden emergence of this threat, at the time of writing, cyber-attacks on nuclear facilities and governance bodies are limited and appear to be unconnected to the Covid-related shifts in working procedures.[37] However, this should be caveated by the fact that it is not common practice for nuclear organizations to openly disclose cyber- or other security-related incidents. Yet, it is clear that the nuclear sector is cognisant of this changing risk landscape, with organizations seeking to strengthen their resilience in the cyber area. For example, in the UK, nuclear companies have placed significant emphasis, following the onset of Covid-19, on awareness raising and training for staff on changing cyber threats and new security protocols for working remotely.[38] Nevertheless, this will remain an important area of focus for the global nuclear community, with the need for continued efforts in this area highlighted by the Nuclear Threat Initiative (NTI) in its Nuclear Security Index, which seeks to benchmark countries' progress on nuclear security. In its 2020 edition only 34 per cent of countries with weapons-usable material on their territory were awarded a high score in the cyber security category.[39]

Changes in working practices due to Covid-19 also have the potential to increase other risks, most notably those posed by 'insiders'—individuals with malicious intent employed at nuclear facilities with authorized access to materials and systems.

Historically, insiders have been responsible for some of the most serious nuclear security breaches, having successfully stolen nuclear materials and sabotaged facilities.[40] In the early days of the pandemic, numerous reports documented the frustration felt by on-site nuclear workers unable to get access to PPE and fearing workplace exposure to Covid-19.[41] This is highly relevant because studies have shown that disgruntlement can be a primary motivator of revenge by an employee against their organization.[42] Thankfully, PPE shortages have eased since the start of the crisis, reducing this potential driver.

However, there exist other triggers of insider action, including financial and psychological issues, with historical cases demonstrating how financial pressure has triggered individuals to steal or attempt to steal nuclear materials.[43] Here, the nuclear sector has done better than most in preventing a deterioration of wages and maintaining jobs following the onset of Covid-19.[44] For example, in June 2020, the UK National Skills Academy Nuclear (NSAN) announced a 'Nuclear Bursary' offering financial support and training for those furloughed to ensure their successful return to work after the pandemic.[45] Simultaneously, nuclear power advocates have taken the opportunity to push for a 'nuclear renaissance' as a means of restarting economies—a path that might also tackle other pressing challenges such as climate change and energy security.

Despite these positive developments, the move to remote working has nevertheless served to disrupt existing employee monitoring systems, thus complicating the identification of insider threats. These includes basic checks, for example, bag searches and security pat downs, which have at times been limited due to the need for social distancing.[46] Many countries also conduct staff observation programmes in their nuclear and other critical sectors, which involve monitoring employees for concerning changes in their behaviour. As many staff are now spending a significant proportion of their time working from home, the utility of in-person behavioural observation programmes has been somewhat negated. To counter this, some nuclear organizations have begun exploring remote activity monitoring software, commonly referred to as User and Entity Behaviour Analytics, or UEBA,[47] which can flag anomalies from system users that might signal malicious or otherwise improper activity. The insider threat will remain an area of concern and focus for the nuclear and other sectors given the ongoing strain felt by many individuals because of the pandemic, including increased self-isolation, health concerns, employment insecurity, or other issues.

Although the use of national or regional lockdowns as a means of containing the pandemic has become less common over time, travel restrictions and constraints on number and proximity of staff and contractors at nuclear and other organizations have continued to be deployed.[48] These interventions have potential negative impacts on nuclear security, due to the limitations they impose on security system maintenance, testing, training, exercising, and assurance activities. It is common for security technologies to be installed and serviced by specialist third-party suppliers, whose activities have had to be delayed to minimize the infection risk. National and organizational training has also been affected by the pandemic. For example, Slovenia suspended in-person reactor training for its nuclear engineers and shifted to an online simulator setting in September 2020.[49] Following the onset of Covid-19, the NRC in the US has also issued exemptions from the usual training and qualification plans for security personnel and suspended in-person scenario-based security personnel training that cannot accommodate social distancing rules.[50] Here, it was deemed that '[…] it is reasonable to conclude that security personnel will continue to maintain their proficiency even though the requalification periodicity is temporarily exceeded'.[51] This was part of a broader series of exemptions to pre-pandemic requirements within the US, which allowed operators to lengthen the time between staff training in order to increase 'flexibility in using personnel resources to most effectively manage the impacts of the Covid-19'.[52]

In terms of broader security performance testing, many countries have delayed planned large-scale security exercises due to the pandemic. For example, in the US force-on-force exercising, which is seen as a vital means of evaluating the proficiency of a security at a facility, was suspended by the NRC in various US states including Virginia, North Carolina, South Carolina, Georgia, Florida, and others.[53] Sustaining robust nuclear security systems also requires regular and rigorous external assurance, which has traditionally relied heavily on on-site inspections by an independent national nuclear regulator. However, in many countries the frequency of these have been impacted by the pandemic, due to restrictions placed on internal movement.[54] For example, in March 2022, the pandemic completely disrupted nuclear regulatory activities in Mauritius, resulting in a period of 'no regulatory oversight' where the Radiation Safety and Nuclear Security Authority was unable to travel or even access their own offices.[55]

Exploring the Impact of Covid-19 on International Nuclear Security Initiatives

While it is widely acknowledged that nuclear security is primarily a state responsibility, over the past 40 years a broad global nuclear security architecture has been developed in support of national and organizational efforts in this area. This is composed of a complex web of treaties, guidance, international training and education programmes, awareness raising, and other activities aimed at ensuring the development and implementation of robust nuclear security systems. Covid-19 has clearly affected international nuclear security cooperation, with travel restrictions leading to the postponement of certain activities, altering the format of others, and potentially diverting political attention.

Arguably, one of the most important elements of the global nuclear security regime are regular high-level international meetings, which help to maintain political salience in nuclear security, while also serving as a focal point for finalizing and reporting national achievements. As a result of the pandemic, the Conference of the Parties to the Amendment to the Convention on the Physical Protection of Nuclear Material

(A/CPPNM), was delayed a year to March 2022. As the first review conference of the only legally binding treaty that requires countries to protect nuclear materials and nuclear facilities on their territory and in international transport, the A/CPPNM is a cornerstone of international nuclear security efforts.[56] To help ensure the conference's value, the IAEA, the depository organization for A/CPPNM, utilized remote means to promote adherence to the amendment, furthering its universalization, while also holding virtual meetings with key countries helping them prepare for the review conference.[57] Despite the significant challenges—including Russia's seizure of a nuclear power station in Ukraine, just a month before the conference—the A/CPPNM conference was widely considered a success. It benefited from broad ministerial representation with states using the opportunity to present on steps taken individually and collectively to strengthen nuclear security. The conference also achieved its major goals, concluding in its outcome document that the 'Convention as Amended is adequate' and requesting the IAEA to hold a subsequent review conference within five years.[58]

In terms of directed international support for nuclear security, travel restrictions due to Covid-19 led to the postponement of a range of activities, such as workshops, training courses, and security reviews. The disruption also affected the International Physical Protection Advisory Service (IPPAS), coordinated by the IAEA upon the request of states, which provides peer advice on the implementation of national nuclear security systems, helps benchmark them against international instruments and guidance, and facilitates the exchange of experience and good practices. As a result of the pandemic, no IPPAS missions were conducted in 2020, but they eventually resumed in May 2021.[59] Nevertheless, preparatory work continued in the interim with the IAEA holding multiple virtual workshops with countries in advance of planned in-person visits.[60] The IAEA also took the opportunity to complete over this period a significant update of its IPPAS Good Practices Database, which contains anonymized information on positive security practices identified during IPPAS missions and can be accessed remotely.[61]

A shift to online engagement due to the restrictions imposed by the pandemic was also reflected across other IAEA nuclear security activities. Here, the IAEA sought to encourage Member States to make use of existing resources, such as their Cyber Learning Platform for Network Education and Training (CLP4NET), while also looking to develop new ones.[62] From July 2020 to June 2021 over 1,000 users from 126 states completed over 3,000 IAEA nuclear security e-learning modules, while materials from in-person workshops were adapted and delivered virtually and a new nuclear security webinar series was launched.[63]

The IAEA also played a role in facilitating the sharing of experiences on new practices developed by the global nuclear industry as a result of the pandemic on its new information platform, the Covid-19 Nuclear Power Plant Operating Experience (OPEX), which produced its first reports by April 2020, less than one month after the World Health Organization (WHO) declared Covid-19 a pandemic.[64] This platform supported peer-to-peer engagement on how to maintain safe and secure operations during the pandemic. Separately the IAEA sought to gauge the impact of the pandemic on the radioactive source sector, surveying national regulatory activities and uncovering the collective challenges encountered and lessons learned.[65] However, while these initiatives have provided useful opportunities for engagement across the global nuclear sector, the majority of this sharing has been on operations and safety-related issues, with relatively little information available on how the pandemic has affected security.

Similar moves were also seen by other organizations involved in international nuclear security capacity building work. For example, the World Institute for Nuclear Security (WINS) moved most of its workshops online, while continuing to promote training and certification through its WINS Academy Programme, with courses and exams delivered virtually.[66] The Institute for Nuclear Material Management (INMM) took the decision to move its 2020, 2021, and 2022 Annual Meetings online, in order to negate travel restrictions and facilitate international participation. Today, most meetings have returned to an in-person setting. However, a hybrid component has usually been retained. This trend has enabled wider access and reduced the travel burden, thereby reaching a wider range of stakeholders than the pre-pandemic in-person only format.

Adaptation of National and Organizational Nuclear Security Approaches

Following the onset of the pandemic, emphasis at the operational level was understandably placed on keeping nuclear systems running while protecting the health of personnel. Over time this has broadened to include an increased focus on the maintenance of nuclear safety and security while limiting in-person interactions. Although fundamental security principles such as graded approaches and defence-in-depth continue to be applied, changes have been made to the delivery of security in relation to the use of certain technologies, procedures, and oversight mechanisms.

In terms of adversary detection, some organizations have sought to deploy alternative methods that minimize human contact, for example, by setting up additional closed circuit television (CCTV) cameras and increasing cooperation between facilities' security forces and the local police.[67] As for access control, advanced touch-based fingerprint biometric devices at some nuclear facilities have been replaced by former technology, such as radio frequency identity cards (RFID) cards, which do not require physical contact.[68] Particular emphasis was also placed on the risks posed by possible insiders, both online and in person. To monitor psychological stress levels among staff, some operators offered increased psychological support, including resilience training via counselling and mobile apps, electronic mood trackers, and relaxation rooms, with questionnaires introduced to assess staff health.[69]

As for licensing and regulatory inspections, countries have explored the greater use of remote mechanisms for ensuring that nuclear security systems are being appropriately designed and maintained, with increased emphasis on information collection and assurance by the operator. For example, the Federal Authority for Nuclear Regulation (FANR) in Abu Dhabi utilized surveillance camera-based electronic licensing services during the pandemic.[70] In the US, the NRC has shifted some of its inspections to an online setting by having on-site employees replace the presence of inspectors by conducting live video-walkdowns in facilities under the virtual direction of inspectors.[71] Similarly, in the UK, the onset of Covid-19 resulted in a reduction in physical inspections by the Office for Nuclear Regulation (ONR), with greater emphasis placed on the operator to provide evidence on the efficacy of security through frequent digital reporting.

Although these actions may have been driven by necessity, there are potentially significant longer-term benefits to be had by placing greater reliance on the use of alternative regulatory mechanisms. In the UK, for example, even though the disruptive effects of the pandemic have decreased, the ONR has chosen to scale-up the remote proportion of its security assessment work, with resource-intensive onsite inspections focused on sites that require the greatest attention.[72] Here, it should be noted that such changes in the UK have been facilitated through its outcome-focused approach to regulating nuclear security, where onus is placed on operators to devise appropriate security systems and associated assurance mechanisms, for sampling by the regulator.

While it is difficult to judge the impact of these changes on nuclear security, some have hailed the benefits. For example, Director General Agnieszka Kaźmierczak of the Euratom Supply Agency said that the pandemic provided an 'enforced opportunity to accelerate the adoption of new technologies' to cut down on 'red tape'.[73] Others have emphasized that online settings allow for more frequent, more inclusive, and even more productive meetings, trainings, and webinars.[74] Furthermore, the renewed attention on cyber issues and technology-supported security has been considered positive because of its increasing relevance as nuclear facilities become ever more digitized.[75] Organizations have also applied emergency preparedness plans developed in response to Covid-19 to other crises. For example, the national uranium operator in Kazakhstan, Kazatomprom, was able to avoid delays or disruptions to uranium production during the large-scale political unrest in Kazakhstan in January 2022, driven by a sharp increase in gas prices, through utilizing risk assessment and business continuity plans established during the pandemic.[76]

Conclusion

The Covid-19 pandemic has had a significant impact on the global nuclear industry, disrupting operations and complicating nuclear security. However, more than two years since its onset, it is clear that the sector has fared better than most, thanks in part to the way that governments, regulators, and operators have embraced virtual means of engagement and alternative means of delivering nuclear security. This has been reflected within all levels of the international nuclear security regime. International organizations—most obviously the IAEA—have supported national governments remotely, continuing dialogue, keeping nuclear security high on the political agenda, and developing products and services. While key actions have been cemented through in-person activities, these would arguably not have been as successful without this important preparatory work.

At the operational level, the pandemic provided industry with an (albeit forced) opportunity to trial and implement new security approaches. Although some of these were short-term temporary fixes, others where benefits were realized will continue into the longer term. For example, many nuclear organizations will continue to support flexible working by incorporating remote elements due to their perceived efficiencies. This move will serve to accelerate the widespread adoption and use of secure information management and communication systems, and the upskilling of staff on cyber risks. Similarly, although the importance of in-person security training is widely recognized, many organizations are looking to also continue with virtual elements. In this context, hybrid offers are able to cement key lessons, while also bring in diverse perspectives and reaching a greater number of stakeholders. Furthermore, regulators have recognized the benefits of alternative approaches to providing external assurance that nuclear security systems are functioning adequately. The related increase in desk-based activities, online engagement, and commensurate reduction in time-intensive, in-person site visits is likely to continue. While physical inspections will remain a key part of this process, this new approach will allow an increased focus on sites or activities where it can provide the most insights.

Notes

1. UK Office for National Statistics (ONS), 'Effects of the Coronavirus (COVID-19) Pandemic on "High-Contact" Industries', Census 2021, UK ONS [online], 2021. https://www.ons.gov.uk/economy/grossvalueaddedgva/articles/effectsofthecoronaviruscovid19pandemiconhighcontactindustries/2022-05-06
2. Charles F. Hermann, 'Some Consequences of Crisis Which Limit the Viability of Organizations', *Administrative Science Quarterly*, 8/1 (1963), 61–82.
3. Pooja Bhadoria, Gaurisha Gupta, and Anubha Agarwal, 'Viral Pandemics in the Past Two Decades: An Overview', *Journal of Family Medicine and Primary Care*, 10/8(2021), 2745–50. https://www.ncbi.nlm.nih.gov/pmc/articles/PMC8483091
4. Comptroller and Auditor General, 'The Government's Preparedness for the COVID-19 Pandemic: Lessons for Government on Risk Management', National Audit Office [online], 19 November 2021. https://www.nao.org.uk/wp-content/uploads/2021/11/The-governments-preparedness-for-the-COVID-19-pandemic-lessons-for-government-on-risk-management.pdf; Cabinet Office, 'National Risk Register Of Civil Emergencies', 2017 edition, Cabinet Office, September 2017, 9. https://assets.publishing.service.gov.uk/government/uploads/system/uploads/attachment_data/file/644968/UK_National_Risk_Register_2017.pdf
5. Comptroller and Auditor General, 'The Government's Preparedness for the COVID-19 Pandemic: Lessons for Government on Risk Management', 8.
6. Ibid.

7. Findings from author interviews with various UK nuclear industry stakeholders.
8. Homeland Security Council, 'National Strategy for Pandemic Influenza–Implementation Plan', Washington, DC, May 2006. https://www.cdc.gov/flu/pandemic-resources/pdf/pandemic-influenza-implementation.pdf
9. Kim Rodgers, 'Pandemic and All-Hazards Preparedness and Advancing Innovation Act Signed into Law', National Association of County Health Officials (NACHO), 16 July 2019. https://www.naccho.org/blog/articles/pandemic-and-all-hazards-preparedness-and-advancing-innovation-act-signed-into-law
10. US Government Accountability Office (GAO), 'COVID-19–Pandemic Lessons Highlight Need for Public Health Situational Awareness Network', Report to Congressional Addressees, June 2022. https://www.gao.gov/assets/730/721426.pdf
11. Office of Nuclear Reactor Regulation, 'NPR Pandemic Response Plan', US Nuclear Regulatory Commission, 12 March 2009. https://www.nrc.gov/docs/ML0812/ML081210310.pdf
12. Nuclear Energy Institute, 'Pandemic Licensing Plan', NEI White Paper, Revision 1, December 2007, 4. https://www.nrc.gov/docs/ML0800/ML080020173.pdf
13. Findings from author interviews with various UK nuclear industry stakeholders.
14. Findings from author interview with UK nuclear operator, March 2021.
15. Charles Digges, 'Russia's Nuclear Workers Isolated Onsite as Coronavirus Spreads', *Bellona*, 3 April 2020. https://bellona.org/news/nuclear-issues/2020-04-russias-nuclear-workers-isolated-onsite-as-coronavirus-spreads
16. BBC, 'Covid Outbreak at Former Trawsfynydd Nuclear Power Plant', *BBC News*, 11 November 2020. https://www.bbc.co.uk/news/uk-wales-54904337; Sarah Tzinieris, George Foster, and Christopher Hobbs, 'Lessons for Nuclear Security from the UK's Response to Covid-19', CSSS Occasional Paper (London: Centre for Science and Security Studies/King's College London, 2021). https://www.kcl.ac.uk/csss/assets/lessons-for-nuclear-security-from-the-uks-reponse-to-covid19.pdf
17. Benjamin Mallet, 'French Nuclear Plants Tighten Hygiene Procedures over Coronavirus Worries', *Reuters* [online], 20 March 2020. https://www.reuters.com/article/us-health-coronavirus-france-nuclear-idCAKBN2172J1
18. Elodie Broussard, 'How Safety and Security Regulators Addressed Challenges during the COVID-19 Pandemic', *IAEA News* [online], 24 September 2020. https://www.iaea.org/newscenter/news/how-safety-and-security-regulators-addressed-challenges-during-the-covid-19-pandemic
19. Mallet, 'French Nuclear Plants Tighten Hygiene Procedures'.
20. Eszter Zalan, 'Vast Queues, as EU Struggles with Closed Borders', *EU Observer* [online], 19 March 2020. https://euobserver.com/health-and-society/147805
21. BBC, 'Covid: BA Suspends Hong Kong Flights amid Crew Quarantine', *BBC News* [online], 28 November 2021. https://www.bbc.co.uk/news/business-59450416; Helen Coffey, 'Coronavirus: All the Flights That Have Been Cancelled So Far', *Independent* [online], 13 March 2020. https://www.independent.co.uk/travel/news-and-advice/coronavirus-flights-cancelled-ba-ryanair-easyjet-virgin-american-airlines-a9371416.html
22. World Nuclear News (WNN), 'Medical isotope supply chain faces challenges from COVID-19', WNN [online], 21 April 2020. https://world-nuclear-news.org/Articles/Medical-isotope-supply-chain-faces-challenges-from
23. World Nuclear News (WNN), 'Air Transport Improves for Medical Radioisotopes', WNN [online], 28 April 2020. https://www.world-nuclear-news.org/Articles/Air-transport-bottleneck-easing-for-medical-radioi

24. IAEA, 'The Operation, Safety and Security of Nuclear and Radiation Facilities and Activities during the COVID-19 Pandemic', GOV/INF/2021/35-GC(65)/INF/9, in *The IAEA and the COVID-19 Pandemic: Reports by the Director General* (Vienna: IAEA, 2021), 13. https://www.iaea.org/sites/default/files/gc/gc65-inf7-8-9_0.pdf
25. Matt Fisher, 'Nuclear Power Plant Operations: Challenges Faced During COVID-19 Pandemic', *IAEA News* [online], 11 June 2020. https://www.iaea.org/newscenter/news/nuclear-power-plant-operations-challenges-faced-during-Covid-19-pandemic; World Nuclear Association, 'World Nuclear Performance Report 2022', WNA [online], July 2022. https://world-nuclear.org/getmedia/9dafaf70-20c2-4c3f-ab80-f5024883d9da/World-Nuclear-Performance-Report-2022.pdf.aspx
26. World Nuclear Association, 'World Nuclear Performance Report 2022', 4.
27. Hallie Golden, 'A Nuclear Waste Site Where the Biggest Fear Isn't Radiation, But Coronavirus', *The Guardian*, 4 May 2020.
28. Ibid.
29. Sten Spinella, 'Millstone Security Employees Criticize COVID-19 Safety Efforts', *The Day*, 3 May 2020. https://www.theday.com/article/20200503/NWS01/200509848
30. IAEA, 'The Operation, Safety and Security of Nuclear and Radiation Facilities and Activities during the COVID-19 Pandemic', 6.
31. Alex Lawson, 'Boss of Hinkley Point C Blames Pandemic Disruption for £3bn Delay', *The Guardian*, 20 May 2002. https://www.theguardian.com/uk-news/2022/may/20/hinkley-point-c-nuclear-power-station-edf-delayed-covid-costs-rise
32. PwC, 'COVID-19: Making Remote Work Productive and Secure', n.d. https://www.pwc.com/us/en/library/covid-19/making-remote-work-productive-secure.html
33. Jim Harter, 'How Leaders Are Responding to COVID-19 Workplace Disruption', *Gallup Workplace*, 7 April 2020. https://www.gallup.com/workplace/307622/leaders-responding-covid-workplace-disruption.aspx
34. Interpol, 'Global Landscape on Covid-19 Cyberthreat', April 2020, 2. https://www.interpol.int/en/content/download/15217/file/Global%20landscape%20on%20COVID-19%20cyberthreat.pdf
35. Interpol, 'Cybercrime: Covid-19 Impact', August 2020, 4. https://www.interpol.int/en/News-and-Events/News/2020/INTERPOL-report-shows-alarming-rate-of-cyberattacks-during-COVID-19
36. Adil Radoini and Muznah Siddiqui, 'The Cyber-Threat against Chemical, Biological, Radiological and Nuclear (CBRN) Facilities', *Freedom from Fear Magazine* [online], 2021. https://f3magazine.unicri.it/?p=2280
37. See, for example, cyber-attack on US government agencies (including the Department of Energy) in 2020 (William Turton, Michael Riley, and Jennifer Jacobs, 'U.S. Nuclear Weapons Agency Hacked as Part of Massive Cyber-Attack', *Time*, 17 December 2020. https://time.com/5922897/us-nuclear-weapons-energy-hacked); Israel's attack on Natanz in 2021 (Martin Chulov, 'Israel Appears To Confirm It Carried Out Cyberattack on Iran Nuclear Facility', *The Guardian*, 11 April 2021. https://www.theguardian.com/world/2021/apr/11/israel-appears-confirm-cyberattack-iran-nuclear-facility); Russia's attack on Ukrainian nuclear operator's website in 2022 (France24, 'Ukraine Nuclear Operator Reports Russian Cyber-Attack on Website', 16 August 2022. https://www.france24.com/en/europe/20220816-live-first-ship-with-food-aid-bound-for-africa-leaves-ukraine-port).
38. Findings from author interviews with various UK nuclear operators, January–March 2021.

39. Cira Mancuso, 'Nuclear Security is Only as Strong as the Weakest Link: 2020 NTI Index Highlights Cybersecurity and Insider Threat Prevention', Nuclear Threat Initiative [online], 4 August 2020. https://www.nti.org/atomic-pulse/nuclear-security-only-strong-weakest-link-2020-nti-index-highlights-cybersecurity-and-insider-threat-prevention
40. For example, see Matthew Bunn and Scott D. Sagan, *A Worst Practices Guide to Insider Threats: Lessons from Past Mistakes* (Cambridge, MA: American Academy of Arts and Sciences, 2014). https://www.amacad.org/sites/default/files/publication/downloads/insider Threats.pdf
41. See, for example, Spinella, 'Millstone Security Employees Criticize COVID-19 Safety Efforts'.
42. John B. Stewart Jr, John J. Davidson, Harvoy B. Jones, Cynthia H. Fulwiller, and Sarah A. Mullen, 'Generic Adversary Characteristics Summary Report', US Nuclear Regulatory Commission, March 1979. https://www.nrc.gov/docs/ML1927/ML19276F868.pdf
43. Maddle Rosenthal, 'Insider Threats Examples: 17 Real Examples of Insider Threats', *Tessian* [blog], 21 October 2021. https://www.tessian.com/blog/insider-threats-types-and-real-world-examples
44. World Nuclear News (WNN), 'Viewpoint: The Impact of COVID-19 on Nuclear Jobs', WNN [online], 27 April 2020. https://www.world-nuclear-news.org/Articles/Viewpoint-The-impact-of-COVID-19-on-nuclear-jobs
45. Mary Kinsella, 'Nuclear Bursary Open', National Skills Academy Nuclear (NSAN), 8 June 2020. https://www.nsan.co.uk/news/511484/Nuclear-Bursary-Open.htm
46. Christopher Hobbs, Nickolas Roth, and Daniel Salisbury, 'Security Under Strain? Protecting Nuclear Materials During the Coronavirus Pandemic', *The RUSI Journal*, 166/2 (2021), 40–50.
47. LogRhythm, 'NRC RG 5.71 Compliance for Nuclear Facilities', n.d. https://logrhythm.com/solutions/compliance/nrc-rg-571
48. For a time series of Covid-19 restrictions, see the Oxford COVID-19 Government Response Tracker. https://covidtracker.bsg.ox.ac.uk
49. Jan Malec and Luka Snoj, 'Teaching of Reactor Physics Using a Real-Time Research Reactor Simulator During COVID-19 Induced Lockdown', paper presented at the 29th International Conference Nuclear Energy for New Europe, 7–10 September 2020, Portoroz, Slovenia, 1. https://arhiv.djs.si/proc/nene2020/pdf/NENE2020_1603.pdf
50. US Nuclear Regulatory Commission, 'COVID-19 Regulatory Activities for Nuclear Reactors', 2022. https://www.nrc.gov/about-nrc/covid-19/reactors/licensing-actions.html
51. US Nuclear Regulatory Commission (NRC), 'Enforcement Guidance Memorandum 20-002-Attachment 1, Dispositioning Violations of NRC Requirements for Completion Periodicities Associated with Security Training and Requalification Requirements during the COVID-19 Public Health Emergency', 2020, 2–3. https://www.nrc.gov/docs/ML2009/ML20091L385.pdf
52. Nuclear Regulatory Commission, 'Issuance of Multiple Exemptions in Response to COVID-19 Public Health Emergency: A Notice by the Nuclear Regulatory Commission on 06/30/2021', Federal Register National Archives,30 June 2021. https://www.federalregister.gov/documents/2021/06/30/2021-13887/issuance-of-multiple-exemptions-in-response-to-covid-19-public-health-emergency
53. For a complete overview, see US Nuclear Regulatory Commission, 'COVID-19 Regulatory Activities for Nuclear Reactors'.
54. Our World in Data, 'Restrictions on Internal Movement during the COVID-19 Pandemic', 25 September 2020. https://ourworldindata.org/grapher/internal-movement-covid?time=2020-09-25

55. F. A. Ollite, 'Nuclear and Radiation Safety and Security Related Regulatory Challenges Due to the COVID-19 Outbreak: Sharing the Mauritius Experience', Senior Safety and Security Regulators' Meeting, 64th Regular Session of the IAEA General Conference, 24 September 2020, Vienna.
56. Samantha Neakrase, 'Strengthening Nuclear Security with a Sustainable CPPNM Regime', Arms Control Association [online], June 2019. https://www.armscontrol.org/act/2019-06/features/strengthening-nuclear-security-sustainable-cppnm-regime
57. IAEA, *Nuclear Security Report 2021*, Board of Governors General Conference. GOV/2021/35-GC(65)/10 (Vienna: IAEA, 2021). https://www.iaea.org/sites/default/files/gc/gc65-10.pdf
58. IAEA, '*2022 Conference of the Parties to the Amendment to the Convention on the Physical Protection of Nuclear Material: Outcome Document*' ACPPNM/RC/2022/4 (Vienna: IAEA, 2022). https://www.iaea.org/sites/default/files/22/04/english_acppnm_rc_2022_4_outcome_document_approved.pdf
59. IAEA, 'Peer Review and Advisory Services Calendar', 2022. https://www.iaea.org/services/review-missions/calendar?type=3170&year%5Bvalue%5D%5Byear%5D=&location=All&status=All
60. IAEA, *Nuclear Security Report 2021*. 16.
61. Ibid.
62. Broussard, 'How Safety and Security Regulators Addressed Challenges during the COVID-19 Pandemic'.
63. IAEA, *Nuclear Security Report 2021*. 27.
64. Matt Fisher and Aabha Dixit, *IAEA Steps up Support for Nuclear Facility Operators during COVID-19 Crisis* (Vienna: IAEA, 2020). https://www.iaea.org/newscenter/news/iaea-steps-up-support-for-nuclear-facility-operators-during-covid-19-crisis
65. IAEA, 'Impact of COVID-19 Pandemic on the Regulatory Activities for the Safety of Radiation Sources', Version 2, September 2020. https://www.iaea.org/sites/default/files/20/09/covid19_nsrw_report_123_countries.pdf
66. World Institute for Nuclear Security (WINS), 'We Shall Find a Way or Make One: Annual Report 2021'. https://www.wins.org/wp-content/uploads/2021/03/Annual-Report-2021_11.pdf; WINS, 'Hope Despite All Odds: Annual Report 2022'. https://www.wins.org/document/annual-report-2022
67. Anhar R. Antariksawan, 'Delivering Nuclear Security During Times of Crisis', Remarks made at the Virtual Side Event at the 64th IAEA General Conference [41:10], 23 September 2020. https://www.youtube.com/watch?v=PRSDJSQLl20&feature=emb_logo
68. Hobbs et al., 'Security under strain?'.
69. Oleksandr Popov, Anna Iatsyshyn, Norbert Molitor, Andrii Iatsyshyn, Yevhen Romanenko, Iryna Deinega, Petr Sharov, and Gabriel Mnayarji, 'Human Factor in Emergency Occurrence at NPP during the Pandemic COVID-19: New Potential Risks and Recommendations to Minimize Them', *E3S Web of Conferences*, 280/09013 (2021), 9–10. e3s-conferences.org/articles/e3sconf/pdf/2021/56/e3sconf_icsf2021_09013.pdf
70. Carley Willis, 'International Experts Discuss: What Can Radiation Protection Learn from the COVID-19 Pandemic', IAEA [online], 23 November 2020. https://www.iaea.org/newscenter/news/international-experts-discuss-what-can-radiation-protection-learn-from-the-covid-19-pandemic
71. Lewis M. Csedrik, 'Up & Atom: NRC Receives Passing Grade from OIG on COVID-19 Adaptations to Inspections', *Morgan Lewis* [blog], 16 August 2021. https://www.morganlewis.com/blogs/upandatom/2021/08/nrc-receives-passing-grade-from-oig-on-covid-19-adaptations-to-inspections

72. UK Office for Nuclear Regulation (ONR), 'Corporate Plan 2022/23', ONR [online], June 2022. https://www.onr.org.uk/documents/2022/onr-corporate-plan-2022-23.pdf
73. World Nuclear News (WNN), 'ESA Reviews COVID Impacts on Supply and Demand', WNN [online], 8 July 2021. https://www.world-nuclear-news.org/Articles/ESA-reviews-COVID-impacts-on-supply-and-demand
74. IAEA, 'The Operation, Safety and Security of Nuclear and Radiation Facilities and Activities during the COVID-19 Pandemic', 14; Michael Amdi Madsen, 'Pandemic's Impact and Enhancing Regulations Discussed at Nuclear Safety & Security Gathering'. *IAEA News* [online], 22 September 2021. https://www.iaea.org/newscenter/news/pandemics-impact-and-enhancing-regulations-discussed-at-nuclear-safety-security-gathering; UK Office for Nuclear Regulation (ONR), 'Corporate Report: Office for Nuclear Regulation Annual Report and Accounts 2019 to 2020', ONR [online], 29 September 2020. https://www.gov.uk/government/publications/office-for-nuclear-regulation-annual-report-and-accounts-2019-to-2020/office-for-nuclear-regulation-annual-report-and-accounts-2019-to-2020
75. Tristan Palmer, 'Cyberproofing Nuclear Power Plants in the Era of a Pandemic', *Elemendar*, 23 April 2020. https://elemendar.com/cyberproofing-nuclear-power-plants-in-the-era-of-a-pandemic
76. World Nuclear News (WNN), 'Supply Chain Challenges Remain as Kazakh Emergency Ends', WNN [online], 14 January 2022. https://www.world-nuclear-news.org/Articles/Supply-chain-challenges-remain-as-Kazakh-emergency

Suggested Reading

Hobbs, Christopher, Roth, Nickolas, and Salisbury, Daniel, (2021), 'Security Under Strain? Protecting Nuclear Materials During the Coronavirus Pandemic', *The RUSI Journal*, 166/2, 40–50.

IAEA, (2021), 'The Operation, Safety and Security of Nuclear and Radiation Facilities and Activities during the COVID-19 Pandemic', GOV/INF/2021/35-GC(65)/INF/9, in *The IAEA and the COVID-19 Pandemic: Reports by the Director General* (Vienna: IAEA). https://www.iaea.org/sites/default/files/gc/gc65-inf7-8-9_0.pdf

Popov, Oleksandr, Iatsyshyn, Anna, Molitor, Norbert, Iatsyshyn, Andrii, Romanenko, Yevhen, Deinega, Iryna, Sharov, Petr, and Mnayarji, Gabriel, (2021), 'Human Factor in Emergency Occurrence at NPP during the Pandemic COVID-19: New Potential Risks and Recommendations to Minimize Them', *E3S Web of Conferences*, 280/09013, 1–15. e3s-conferences.org/articles/e3sconf/pdf/2021/56/e3sconf_icsf2021_09013.pdf

Schneider, Mycle and Froggatt, Antony, with Hazemann, Julie, et al., 'The World Nuclear Industry Status Report 2020', WNR [online], September 2020. https://www.worldnuclearreport.org/IMG/pdf/wnisr2020-v2_lr.pdf

Tzinieris, Sarah, Foster, George, and Hobbs, Christopher, (2021), 'Lessons for Nuclear Security from the UK's Response to Covid-19', CSSS Occasional Paper (London: Centre for Science and Security Studies/King's College London). https://www.kcl.ac.uk/csss/assets/lessons-for-nuclear-security-from-the-uks-reponse-to-covid19.pdf

INDEX

Tables and figures are indicated by *t* and *f* following the page number.